T0138661

ADVANCES IN UNDERSTANDING ENGINEERED CLAY BARRIERS

PROCEEDINGS OF THE INTERNATIONAL SYMPOSIUM ON LARGE SCALE FIELD TESTS IN GRANITE, SITGES, BARCELONA, SPAIN, 12–14TH NOVEMBER 2003

Advances in Understanding Engineered Clay Barriers

Edited by

Eduardo E. Alonso & Alberto Ledesma
Technical University of Catalonia (UPC), Barcelona, Spain

A.A. BALKEMA PUBLISHERS   LEIDEN / LONDON / NEW YORK / PHILADELPHIA / SINGAPORE

Cover: Picture based on the Febex Experiment. Courtesy of ENRESA and Transedit, Spain

Svensk Kärnbränslehantering AB

Published by: A.A. Balkema Publishers, Leiden, The Netherlands,
a member of Taylor & Francis Group plc
www.balkema.nl, www.tandf.co.uk

ISBN 04 1536 544 9

Printed in Great Britain

Advances in Understanding Engineered Clay Barriers – Alonso & Ledesma (eds)
© 2005 Taylor & Francis Group, London, ISBN 04 1536 544 9

Table of Contents

Fundamental research material behaviour and laboratory testing

Barrier behaviour and THM modelling

Chemical effects. HC and THMC modelling

Closing conference

Advances in Understanding Engineered Clay Barriers – Alonso & Ledesma (eds)
© 2005 Taylor & Francis Group, London, ISBN 04 1536 544 9

Preface

Over the past few years the use of clay barriers for waste isolating purposes has deserved much attention in the geotechnical engineering community. This interest has been linked to a fundamental research activity including the understanding of the behaviour of compacted materials and, particularly, compacted expansive clays. The interaction between the barrier, the waste and the surrounding ground may involve several thermo-hydro-mechanical and chemical coupled processes that have been analyzed by means of "in situ" tests, laboratory experiments and numerical modelling by many research groups. Within this context, some large scale field tests have been developed in recent years by some European Agencies dealing with the management of radioactive waste. These experiments have provided an opportunity to calibrate and to validate computational models and to gain experience on instrumentation and installation techniques. This book constitutes a state-of-the-art of a number of relevant issues concerning engineered clay barriers. It includes about 50 papers presented in a symposium held in Sitges (Barcelona, Spain) in November 12th–14th, 2003. Most of the contributions refer to the following large scale tests being performed in Europe:

- Backfill and Plug Test & Prototype Repository, managed by SKB in Äspö (Sweden).
- Febex, managed by ENRESA (Spain) and carried out at Grimsel underground Laboratory (Switzerland).
- Temperature Buffer Test managed by ANDRA (France) and operating in Äspö Hard Rock Laboratory (Sweden).

All of them include bentonite barriers for radioactive waste isolation in granite environments. However, despite of the specific characteristics of those tests, the majority of the papers focus on the understanding of the processes involved.

Sessions of the symposium were organised according to four main topics:

- Field emplacement and instrumentation techniques (i.e. lessons learned in test installation and dismantling, reliability of instruments, "in situ" determination of barrier and rock properties, advances in rock characterization, site hydrogeology and its effect in test performance, impact of test installation of hydro-geological conditions of host rock, EDZ in granite).
- Fundamental research. Material behaviour and laboratory testing (i.e. advanced constitutive modelling of expansive compacted bentonite, permeability of compacted bentonite, evolution of bentonite structure during saturation and heating, effect of temperature on constitutive behaviour, properties and behaviour of bentonite-based mixtures, role of small-medium scale cells in parameter/properties determination, advanced laboratory techniques, geotechnical parameters and experimental determination, chemical and physical properties of interstitial water, illitization of bentonite, properties of bentonite-based materials).
- Barrier behaviour and Thermo-Hydro-Mechanical modelling (i.e. predicting long term saturation, interaction between barrier and host rock, effect of rock heterogeneity and EDZ on buffer response, canister corrosion effects, effects of environmental conditions, effects of overheating on barrier performance, long-term stress state of the barrier).
- Chemical effects, Hydro-Chemical and Thermo-Hydro-Mechanical and Chemical modelling (i.e. advances in hydro-chemical modelling of bentonite barriers, geochemical models of water-bentonite interaction, transport mechanisms through the barrier, advances in THMC modelling, geochemical modelling of dismantling data).

The symposium was an opportunity to share experiences from those experiments and to identify areas of uncertainties in order to guide future research activities. This book reflects this effort. Its publication has been made possible thanks to the sponsorship of ENRESA and SKB, the Spanish and the Swedish National Agencies for Radioactive Waste Disposal. Their support is gratefully acknowledged. The symposium provided also an opportunity of meeting people from different countries with a common interest, which is nowadays essential in an international and complex issue such as the safe disposal of nuclear waste. We hope that this publication will contribute to a better understanding of clay barriers and to an efficient and safe design of present day concepts and engineering solutions for waste disposal. It is believed also that a large proportion of the topics covered by the different papers have a wide interest in the general field of Geotechnical and Geo-environmental Engineering.

Barcelona, February 7th, 2005
Eduardo Alonso, Alberto Ledesma

Opening

Advances in Understanding Engineered Clay Barriers – Alonso & Ledesma (eds)
© 2005 Taylor & Francis Group, London, ISBN 04 1536 544 9

The FEBEX test as a benchmark case for THM modelling. Historical perspective and lessons learned

Eduardo E. Alonso and J. Alcoverro
Departamento de Ingeniería del Terreno, Universidad Politécnica de Cataluña, Barcelona, Spain

ABSTRACT: The FEBEX (Full-scale Engineered Barriers Experiment in Crystalline Host Rock) "in situ" test was installed at the Grimsel Test Site (Switzerland) to demonstrate the feasibility of the underground storage of high level nuclear waste in granite. The simulated canister is surrounded by compacted bentonite whose thermo-hydro-mechanical properties are described in the paper. The measured performance of the test during three years constitutes the basis for a benchmark exercise involving various research teams and computational models. The following aspects of test performance are discussed in the paper: water pressure changes induced in the host rock during the excavation of the tunnel, the distribution and evolution of relative humidity and swelling-induced stresses within the bentonite barrier, and the stresses and water pressures recorded in the rock as a result of the heating and swelling pressure development inside the excavated tunnel. The paper discusses the merits of different formulations and the difficulties encountered to match some aspects of the observed behaviour.

1 INTRODUCTION

FEBEX was divided into three main parts, namely the "in situ" test, a mock up test and an extensive laboratory program on bentonite properties. The FEBEX in situ test was designed as a demonstration experiment of the Spanish reference concept for disposal of high level radioactive waste in granitic rock. Nuclear canisters are deposited in deep horizontal drifts. A bentonite buffer is provided around the cylindrical canisters to isolate the waste (Fig. 1 shows the general layout and the dimensions of the experiment). In the FEBEX experiment, compacted bentonite "bricks", initially unsaturated (Sr $\cong$ 50%) were manually placed around cylindrical steel heaters, which simulate the nuclear canisters. These heaters maintained the temperature of the bentonite-steel contact at a constant value of 100°C. The bentonite buffer is subjected to two basic phenomena: a progressive hydration induced by the natural water provided by the granitic rock and a desiccation action induced by the heaters. More than 600 sensors were emplaced in the rock, as well as a bentonite buffer, during the installation. They have provided time records of several variables (temperature, rock pore water pressures, bentonite relative humidity (RH), stresses in rock and bentonite, displacements), which provide an accurate image of the performance of the near field rock, and, specially, of the bentonite reaction.

The FEBEX drift was excavated in October 1995. During excavation (using a TBM machine) and immediately afterwards, pore water pressures were recorded in boreholes located in the vicinity of the gallery. Additional small diameter holes were radially perforated from the FEBEX tunnel in order to locate monitoring instruments. Water inflow into the excavated tunnel was measured in February–April 1996. Once the two heaters and the bentonite buffer were in place, a concrete plug was built to seal the test area (Fig. 1). Heating started on February 27th, 1997 (a date identified as Day 0 in the plots given in this paper). The present paper discusses the data recorded during the first 1000 days of test operation.

The FEBEX is described in detail in Enresa (2000). This report provides data on the objectives of the experiment, on the geometry and construction sequence, on the properties of the granitic rock and the bentonite, on the interpretation of monitoring results and on the modelling work performed by the research team participating in the FEBEX experiment.

One of the main concerns raised during the design stage of the experiment was the expected hydration time of the bentonite buffer. In performance evaluation studies it is assumed that a fully saturated bentonite buffer is isolating the canisters. However, no precise information on the time required for this saturation was available. It was clear that the test would provide data on the initial transient regime of the buffer hydration and on the progressive

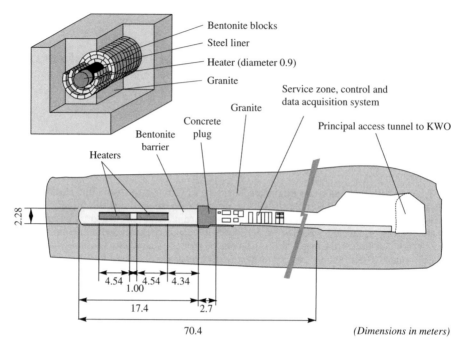

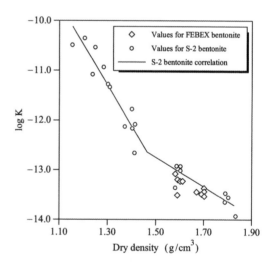

Figure 1. General layout of the FEBEX "in situ" test.

development of swelling pressures but probably not on the time for full saturation.

The preliminary analyses carried out during the design of the test identified the key properties that were expected to control the saturation of the bentonite barrier:

1 Water retention curves of granite and backfill
2 Saturated permeability of granite and backfill
3 Variation of permeability with suction for granite and backfill (relative permeability)
4 Water pressure boundary conditions of the granite in the vicinity of the excavation.

Not all the properties or variables mentioned were known in 1994, when the preliminary analysis was performed. But some tests on rock and compacted bentonite specimens were already available. For instance, Figures 2 and 3 show the hydraulic conductivity and the water retention properties of compacted bentonite at various dry densities. It was expected that the dry density of the blocks would reach $\gamma_d = 1.65 \, \text{g/cm}^3$ in order to guarantee a swelling pressure in excess of 5 MPa. Figure 2 indicates that the saturated hydraulic conductivity of the bentonite was close to $2-6 \times 10^{-13}$ m/s for the expected dry densities prevailing in the buffer. Data on relative permeability was not available and the hydraulic conditions of the granitic rock were roughly known.

Figure 2. Hydraulic conductivity of compacted specimens of S-2 montmorillonite for two different dry densities and varying temperature.

However, some thermo-hydro-mechanical analyses were performed with the purpose of establishing the relative effects and importance of the parameters mentioned. The hydration time of the buffer was defined as the time required for saturating a point

4

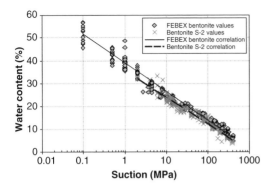

Figure 3. Suction/water content relationship in tests on unconfined samples, for FEBEX and S-2 bentonite (Enresa, 2000).

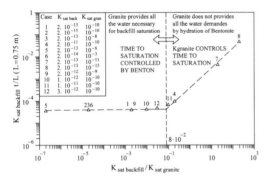

Figure 4. Time to reach saturation in the reference point.

located at a radial distance of 0.15 m from the heater. The results of calculations performed were represented in a dimensionless plot relating the dimensionless time to saturation ($T = K_{bentonite}t/L$, where L is the thickness of the buffer) against the ratio of A finite element program for hydro-mechanical analysis of expansive soils was used to perform the calculations. Several cases were run, characterised by different combinations of granite and backfill permeability. A bentonite buffer thickness $L = 0.75$ m was expected at this time. The results are shown in Figure 4. The plot shows a clear qualitative change in behaviour for a $K_{bentonite}/K_{granite}$ ratio of 8×10^{-2}. Below this ratio, the time for saturation,

$$t = \frac{T \ L}{K_{bentonite}} \tag{1}$$

only depends on the saturated bentonite permeability. Above it, the time for saturation increases and it is also controlled by the granite permeability. It is interpreted that, for $K_{bentonite}/K_{granite} < 8 \times 10^{-2}$ the

granite provides all the necessary water to saturate the buffer. Then, only the bentonite permeability controls hydration times. When granite permeability decreases and $K_{bentonite}/K_{granite} > 8 \times 10^{-2}$, the granite is not able to supply all the water demanded by the bentonite and both, the bentonite and the granite permeabilities, control hydration times.

Grimsel granite has a matrix permeability sufficiently high to provide all the saturation water required by the bentonite. Time for saturation can be easily calculated on the basis of Figure 4. More precise laboratory determinations of the saturated FEBEX bentonite lead to a figure $K_{bentonite}/K_{granite} = 3 \times 10^{-14}$ m/s. Then, for $K_{granite} = 10^{-10}$ m/s and $L = 0.75$ m, a time for saturation $t = 31.7$ yr is obtained. Saturated permeabilities ($K_{bentonite}/K_{granite}$).

It became clear that the expected time length of the experiment (3 to 6 yr) was very short to allow the saturation of the buffer. The question of saturation time has remained as a relevant discussion item as new data coming from the laboratory, the "in situ" and the "mock-up" tests, and the refined analyses performed were considered.

Therefore, the FEBEX in situ test data corresponding to the first 1000 days covers the initial transient response of the bentonite buffer, where thermohydro-mechanical phenomena experience the fastest rates of change.

The main objective of the FEBEX in situ test was to demonstrate the feasibility of the envisaged emplacement procedure. This objective was largely accomplished. The second objective was the study of the THM processes in the near field and it was reasonably well covered. The available data was judged also of interest to conduct a benchmark exercise by interested research groups. This exercise was developed within the context of Decovalex III Project. For the purposes of conducting the benchmark case, in an ordered way, three parts or stages were defined: the first one examines the granite response to the excavation of the rock. The second part refers to the buffer response and it is based on the comprehensive data base obtained from sensors located within the buffer. In the third part, attention is turned again to the rock, affected now by the temperature field generated by the heater and by the swelling pressure of the bentonite against the tunnel walls.

This paper provides some relevant data and conclusions established during the performance of this exercise. Some comparison plots between predictions and actual measurements are also given.

A full report on the benchmark, which includes a comprehensive evaluation of the modelling exercise is also available (Alonso & Alcoverro, 2004). Some insight into some of the main phenomena observed is given this paper.

Table 1. Modelling teams.

Funding organisation	Modelling team/coordinator	Code	Symbol	Color
National Radioactive Waste Management Agency (ANDRA)	• Laboratory "Sols, Solides, Structures" Grenoble. École Polytechnique	ANG	■	red
National Radioactive Waste Management Agency (ANDRA)	• Laboratoire Environnement, Géomécanique et Ouvrages. Ecole des Mines de Nancy	ANN	□	red
Federal Institute for Geosciences and Natural Resources (BGR)	• Federal Institute for Geosciences and Natural Resources	BGR	◆	green
Canadian Nuclear Safety Commission (CNSC)	• Canadian Nuclear Safety Commission • McGill University	CNS	◇	green
U.S. Department of Energy (DOE)	• Sandia National Laboratories	DOE	▲	blue
Institute for Radiological Protection and Nuclear Safety (IRSN)	• Institute for Radiological Protection and Nuclear Safety • Ecole des Mines de Paris	IPS	△	blue
Japan Nuclear Cycle Development Institute (JNC)	• Hazama Corporation • Kyoto University • Japan Nuclear Cycle Development Institute	JNC	✖	brown
Swedish Nuclear Fuel and Waste Management Co (SKB)	• Clay Technology AB • FEM-Tech AB	SKB	▰	black
Swedish Nuclear Power Inspectorate (SKI)	• Lawrence Berkeley National Laboratory	SKI	✚	black
Radiation and Nuclear Safety Authority of Finland (STUK)	• Helsinki University of Technology	STU	●	orange
Empresa Nacional de Residuos Radiactivos SA (ENRESA)	• Technical University of Catalonia (Coordinator)	UPC	○	orange

Profit is taken from discussions and results provided by some of the participants in the benchmark exercise to elaborate the discussion. A maximum number of ten modelling teams have participated in the benchmarks outlined. Their names, codes and symbols used in the paper are given in Table 1.

2 RELEVANT PROPERTIES OF THE COMPACTED FEBEX BENTONITE

The FEBEX bentonite is a calcium-magnesium bentonite extracted from a quarry in Almería (Spain) and subsequently subjected to a process of mechanical conditioning (homogenization, rock fragments removal, drying, crumbling of clods and sieving). The granulated material obtained was compacted in moulds until a nominal specific weight of 1.67 was reached. The average water content in equilibrium with the laboratories which performed most of the tests (CIEMAT and UPC) was 13.3%. The bentonite

has a moderate plasticity ($w_L = 93\%$; $w_p = 47\%$; IP = 46%) and a total specific surface of $725\,m^2/g$.

Experimental properties, which characterise the THM behaviour of this material, are given below. The properties given are purely thermal (T), hydraulic (H), mechanical (M) or involve specific couplings (TH, TM, HM or THM).

2.1 Thermal properties (T)

Thermal conductivities were determined at various dry densities and water contents. Results are given in Figure 5 (the FEBEX bentonite was known as "S-2" in early studies. The S-2 bentonite identifies an initial stock of material from the same quarry area). A good correlation between thermal conductivity, λ, and degree of saturation was found by means of sigmoidal equations of the type:

$$\lambda = A_2 + \frac{A_1 - A_2}{1 + e^{(S_r - x_0)/d_x}} \qquad (2)$$

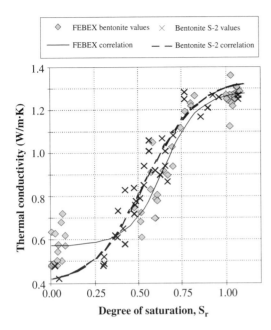

Figure 5. Thermal conductivity of FEBEX and S-2 bentonite.

where λ is the thermal conductivity in W/m · K, S_r is the degree of saturation, A_1 is the value of λ for $S_r = 0$, A_2 is the value of 1 for $S_r = 1$, x_0 is the degree of saturation for which the thermal conductivity is the average value between the extreme values, and d_x is a parameter.

The specific heat (C_s) was found to depend linearly on temperature

$$C_s \left(\frac{J}{kg^{e}C} \right) = 1.38T + 732.5 \quad \left(T \text{ in } °C \right) \qquad (3)$$

2.2 Water retention (II)

Early results were given in Figure 3. However, the associated tests were performed in dessicators or pressure plate apparatus, which do not control the soil volume. The apparent result is a unique relationship between the final water content at equilibrium and the imposed suction, irrespective of the initial dry density. When tests are performed at constant volume, a different result is obtained (Fig. 6). The effect of temperature is small as shown in Figure 7. Temperatures in the range 22–60°C have a very moderate effect on the water retention characteristics of the bentonite.

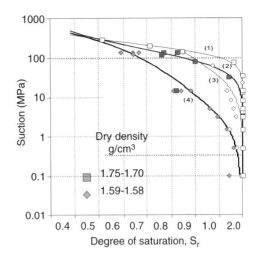

Figure 6. Water retention curves of FEBEX bentonite obtained at different constant dry densities.

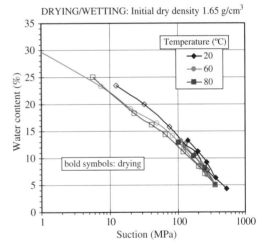

Figure 7. Effect of temperature on water retention properties of FEBEX bentonite (Villar et al., 2004).

2.3 Saturated permeability (H)

Early results are given in Figure 2. Additional tests were performed to investigate the effect of other factors (type of water, anisotropy of compaction and temperature). It turned out that the principal variable controlling permeability is the dry density of the compacted bentonite (Figs. 8 and 9). Temperature, in particular, seems to have a reduced influence in the range 20–80°C.

Relative permeability is not easily determined by direct tests. It usually requires the backanalysis of

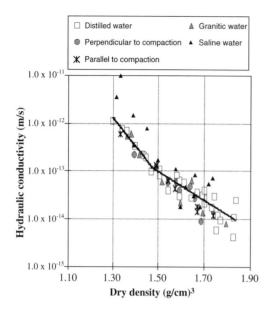

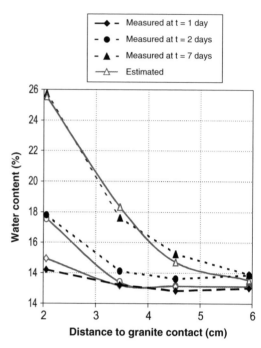

Figure 8. Effect of type of water and direction of flow on saturated hydraulic conductivity.

Figure 10. Measures water content at different times in an infiltration column. The model results ("estimated") correspond to a relative permeability law $k_r = (S_r)^{3.50}$.

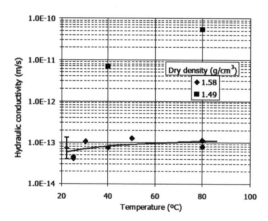

Figure 9. Hydraulic conductivity as a function of temperature for saturated FEBEX clay compacted to different nominal dry densities (the error bars correspond to samples of dry *density 1.58 g/cm³ and 1.49 g/cm³ tested at laboratory temperature*).

classical formulation of the coefficient of permeability as the product of an intrinsic term, and a relative permeability value, which depends on S_r is not appropriate for materials which experience significant changes in microstructure as they are wetted or dried.

2.4 Thermo-hydraulic test in instrumented cells (TH)

Medium-scale tests, which are interpreted as boundary value problems, through an appropriate computer model have the potential of providing good estimates of the material parameters governing the physical phenomena. This technique has been successfully applied in a variety of configurations. One of them is shown in Figure 11. A constant temperature gradient is maintained along the two cylindrical samples located on opposite sides of the central heating unit. A confining latex membrane allows unrestricted lateral deformation and maintains the overall water content. Temperatures at several points are measured during the test. At the end of the test, when steady state conditions are reached, the specimen is cut into six small cylinders and water content is determined. A backanalysis procedure, described in Pintado & Lloret (2002) was used to determine the tortuosity

infiltration tests or the interpretation of transient stages of wetting by means of a suitable model. Figure 10 shows the results and the interpretation of three infiltration tests taken at three different times. This information has been used to derive the exponent "n" of the classical relationship $k_r = (S_r)^n$ for the relative permeability. Unfortunately, a significant variability of n values has been obtained. Probably the

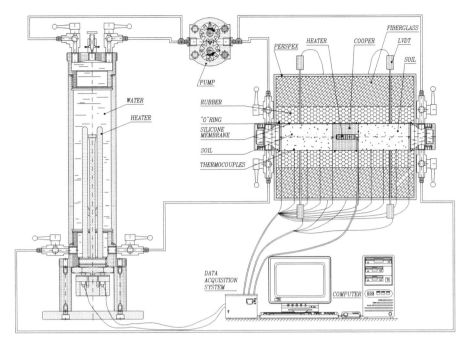

Figure 11. Test arrangement for water transfer in two bentonita columns induced by temperature gradients.

factor, τ, the relative permeability exponent, n (in $k_r = S_r^n$) and the thermal conductivity for saturated conditions, λ_{sat}. The joint consideration of this test and similar tests on cells led to the selection of an optimum set of parameters as follows:

λ (W/m °C) $= 0.47^{(1-Sr)}\ 1.15^{(Sr)}$
$\tau = 0.8$
$K_r = Sr^3$
$K_{sat} = $ (for n $= 0.42) = 2 \times 10^{-21}\ m^2$

This methodology is thought to provide accurate results on water and heat transfer properties of the bentonite.

2.5 Thermal dilation of bentonite (TM)

This is a clear example of thermomechanical coupling, which is usually formulated through the concept of coefficient of linear thermal expansion $\Delta\varepsilon_l/\Delta T$. There was an interest in knowing the influence of temperature and degree of saturation on this coefficient. Cylindrical specimens of varying S_r were isolated in latex membranes and they were subjected to heating and cooling cycles in thermal baths. Changes in longitudinal strains were recorded. The results are collected in Figure 12. It shows the effect of temperature, but does not provide a clear picture on the effect of initial water content. However, it was

noticed that heating/cooling cycles induce significant changes on $\Delta\varepsilon_l/\Delta T$, which are observed in Figure 12.

2.6 Hydromechanical behaviour (HM)

Mechanical tests under varying water content, with or without suction control, provide information of the hydro-mechanical couplings. The swelling pressure determination or, more generally, the effect of confining stress on swelling strains upon wetting are classical examples of this interaction. Figure 13 shows the results of swelling under load tests for specimens of varying dry density. They were performed in oedometer cells. The plot shows also the effect of dry density on swelling pressure (vertical stress for zero deformation).

A more comprehensive evaluation of volumetric behaviour of the compacted bentonite is obtained in suction-controlled tests. Figure 14 shows the stress-suction paths applied in a series of tests designed to characterise the elastoplastic behaviour of the compacted bentonite. A comparison of the measured void ratio for two alternative paths S-1 and S-5, starting and ending in the same stress-suction point (Fig. 15) shows the stress-path dependency of the bentonite. A double structure model (Lloret et al., 2003), which has been proposed to characterise the behaviour of expansive soils provides a satisfactory mechanical representation of the compacted bentonite.

9

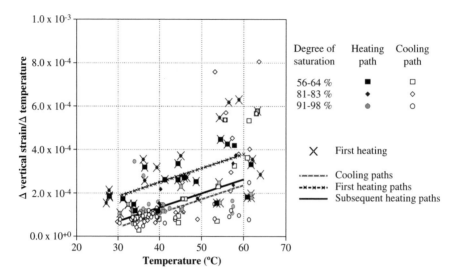

Figure 12. Effect of temperature and number of thermal cycles on the coefficient of linear thermal expansion.

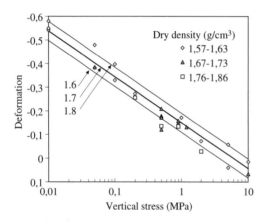

Figure 13. Swelling strains when specimens are wetted at varying vertical stress.

2.7 Coupled thermo-hydro-mechanical tests (THM)

Fully coupled thermo-hydro-mechanical tests were also conducted at a laboratory scale. The interpretation of these tests require the same computational techniques used in the analysis of the large "in situ" test. Figure 16 shows one of the cells built by CIEMAT, in which heating was applied in one edge of the specimen and, at the same time, hydration was provided through an opposite face. Temperature was measured at some locations shown in Figure 16. At the end of the test, the specimen is sliced and physical (and chemical) characterisation tests were performed. The interpretation of these tests is not discussed here.

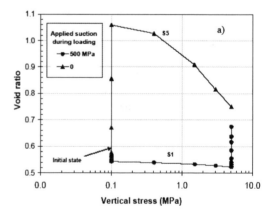

Figure 14. Stress paths in suction controlled oedometer tests.

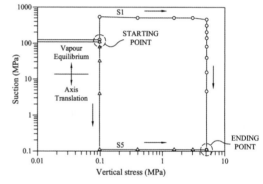

Figure 15. Volumetric behaviour measured in paths S1 and S5.

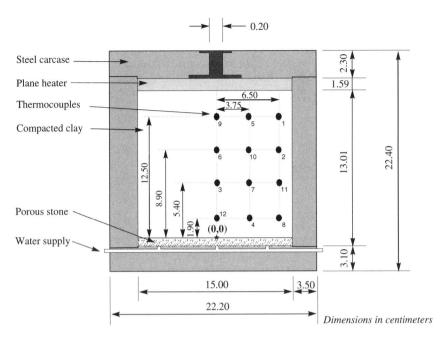

Figure 16. Cell designed by CIEMAT to perform heating and hydration tests.

A description of these tests may be found in Villar et al., 2004.

It is clear that the type of tests outlined above provide large amounts of data and redundant information on bentonite properties. It is believed, however, that a backanalysis of coupled tests of increased complexity provides complementary information to the direct tests. The tests mentioned (and other results not included here) were provided to the teams involved in the benchmark. Data on rock properties was also available. The selection of the material constants, actually used in the prediction exercises, was left to the initiative of the participating teams. Some of them interpreted to their advantage some of the more complex tests mentioned above.

3 WATER PRESSURE CHANGES IN GRANITE DURING TUNNNEL EXCAVATION

Water pressure in the granite was measured in sealed intervals of borings drilled before the FEBEX tunnel was excavated. The measured records were characterised by an increase in water pressure, when the tunnel was being excavated, followed by transient dissipations when the excavation was interrupted (nightshifts, weekends). This behaviour has been related to the volumetric deformations of the saturated granitic rock. Volumetric compressions would result in water pressure increments. The transient dissipations are the result of water flow, which is also affected by the newly created boundary (the open tunnel). The measured water pressure in the measurement interval P-4 is shown in Figures 17a and b (they correspond to the UPC acronym). The scale on the right of the figures shows the tunnel metering. The position and length of the measuring interval P-4 is also shown. Once the tunnel face has gone beyond the P-4 interval, a progressive decay of pore pressure is recorded. The figure also shows the predictions made by different teams. The differences among teams may be explained by the differences in the computer models (which are summarised in Table 2). However, the fundamental contributing factor in this case is the intensity and orientation of the initial stress state of the rock. This issue will be discussed in more detail using results reported by some participating teams.

Most of the "predictions" had difficulties to capture the measured behaviour. Some models did not show any response to the excavation. The accurate prediction of one of the models (SKI) corresponds to an initial state of stress, which is not coherent with the information made available to all the teams. In situ stresses have been measured in the past in the Grimsel test area (Braüer et al., 1989; Pahl et al., 1989; Keusen et al., 1989). It may be concluded from

11

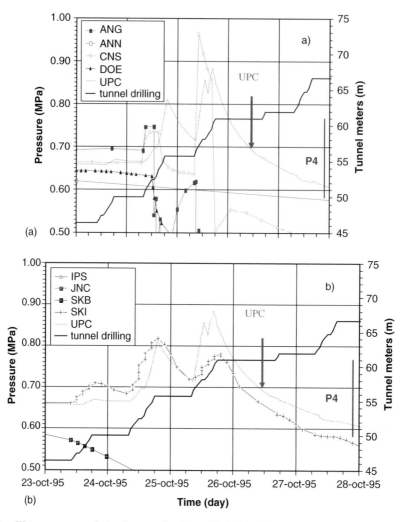

Figure 17a,b. Water pressure evolution in measuring interval P-4 of borehole FEBEX 95.002.

Table 2. Modelling approaches for water pressure response of granite against tunnel excavation.

Team	Coupling	Rock	Observations
ANG	HM	eq.cont + fract	poroelasticity (Biot)
ANN	M→H	fract	steady state
CNS	HM	eq.cont + fract	poroelasticity (Biot)
DOE	HM	eq.cont	compliant joints
IPS	HM	eq.cont	theoretical analysis
JNC	HM	eq.cont	Oda tensor + Barton Bandis
SKB	HM	eq.cont + fract	poroelasticity (Bishop)
SKI	HM	eq.cont	unsat poroelasticity (Biot)

these investigations that the maximum principal stress was horizontal, in the direction NW-SE and it had an intensity of 25–40 MPa. The intermediate principal stress was also horizontal and could reach 15–30 MPa. Both are larger than the lithostatic vertical stress, which is estimated to be around 10 MPa in the FEBEX tunnel, given the overburden height of granite. This interpretation suggests that a horizontal maximum principal stress of around 30 MPa is directed at a 45° inclination with respect to the tunnel axis. This orientation makes the problem of tunnel excavation truly 3-D.

Rutqvist et al. (2003) discuss the effects of initial stress on the development of pore water pressures in the vicinity of tunnel excavations. They considered

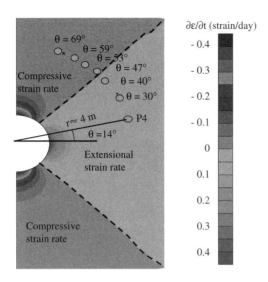

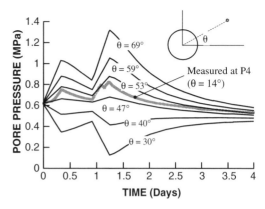

Figure 19. Calculated pore pressure response of the points shown in Figure 18 (Rutqvist et al., 2003).

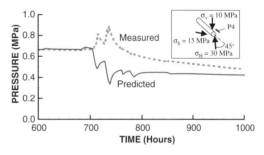

Figure 20. Calculated and measured water pressure evolution at interval P-4 (Rutqvist et al., 2003).

Figure 18. Calculated strain rates, indicating zones of extension and contraction. Also shown are the locations of points for the calculation of pore water pressure histories (Rutqvist et al., 2003).

first a 2-D coupled HM model (Figure 18). A Biot model with E = 24.676 Pa, $\nu = 0.37$, b = 1 (Terzaghi effective stress) and Biot modulus M = ∞ was considered. Saturated intrinsic permeability was set to $7 \times 10^{-18}\,\mathrm{m}^2$. A major horizontal stress of 29.4 MPa was considered. The vertical stress was estimated to be 7.14 MPa. An initial pore water pressure of 0.6 MPa was assumed. The real excavation process was simulated by the progressive decrease of the effective stress and pore pressure on the wall of the tunnel. The computed changes in water pressure are shown in Figure 19, for a series of points identified by the angle θ in Figure 18. This figure shows the areas of compression (positive development of pore water pressure) and extension around the tunnel. The relative position of measuring interval P-4 lies in an extensional zone and, therefore, transient decreases of pore water pressure are predicted. Positive increments are predicted when $\theta > 45°$, as shown in Figures 18 and 19. It appears that the calculated values for $\theta = 1.3°$ are reasonably close to actual measurements. Therefore, any agreement with actual measurements and calculations require a change in the orientation of the initial stress field.

The authors mentioned report also a 3-D analysis using the FE code ROCMAS (Rutqvist et al., 2001). The initial state of stress was described by two horizontal principal stresses $\sigma_H = 30$ MPa, $\sigma_h = 15$ MPa and a third principal vertical stress $\sigma_v = 10$ MPa. σ_H is orientated 45° from the axis of the

tunnel according to observations previously reported. The full history of tunnel excavation was simulated in this case. Predicted pore water response in P-4 was opposite to the actual measurements as shown in Figure 20. This is in agreement with the results of the simplified 2-D simulation and may be interpreted as a strong indication that the assumed initial stress field is not appropriate in the vicinity of the FEBEX tunnel (the presence of a lamprophyre dyke crossing the tunnel may imply strong local variations of the initial state of stress). In order to get a good agreement the assumed stress field had to be rotated around the vertical axis so that the intermediate principal state becomes normal to the tunnel axis and the vertical stress was increased to 225 MPa. Unfortunately no independent information on the state of the stress in the vicinity of the tunnel is available.

4 BENTONITE RESPONSE

Two aspects of the bentonite response are specifically relevant to understand the physical phenomena taking place during the transient heating and wetting stages: the evolution of RH and the development of stresses in the buffer. The temperature field is, in general, well predicted since the heat transfer is dominated by conduction and the coefficient of heat conduction (and its variation with water content) are reliably determined in laboratory experiments. Temperature inside the buffer, after 90 days of heating, changes radially from a value of 100°C at the contact with the heater, to a value of 35°C at the bentonite-rock contact.

Figure 21 shows the evolution of RH measured at three points also indicated in the plot. The measuring points correspond to a section (E1) directly affected by one of the heaters. The point close to the rock (G) was essentially saturated at the beginning of heating (t = 0). The point close to the heater (H) experienced a fast drying followed by a slow recovery of water content. The centered point (C) showed an interesting behaviour: it became initially wetter, to be partially desiccated again and, finally, a progressive increase in RH is recorded. This behaviour is attributed to the change of phase of the vapour generated in the inner parts of the buffer. The hot vapour migrates radially outward, due to its high concentration near the heater. During this migration, it becomes eventually cooler and a condensation phenomenon ensues. This explanation accounts for the initial increase in RH. However,

this phenomenon evolves radially outwards as the temperature increases in the barrier, leading to a transient drying period. At increasing times, the liquid water inflow from the outer rock leads to a progressive saturation.

The models competing in this exercise have been summarised in Table 3. The table indicates the fundamental couplings introduced in the models, the approach used to model the bentonite and some

Table 3. Modelling approaches for the analysis of the bentonite barrier.

Team	Coupling	Rock	Observations
ANG			
BGR	TH→HM	rigid	2D no phase change
CNS	THM	T-poroelast (Biot+s)	3D
DOE	TH→TM	T-poroplast (s)	2D 3D
IPS	T→HM	poroplast (unsat Biot)	2D suct contact
JNC	THM	poroelast (s)	3D swelling press
SKB	T↔HM	T-poroplast (Bishop)	2D swell, no ph ch
SKI	THM	T-poroelast ($\sigma'' + s$)	3D
STU	TH	rigid	1D

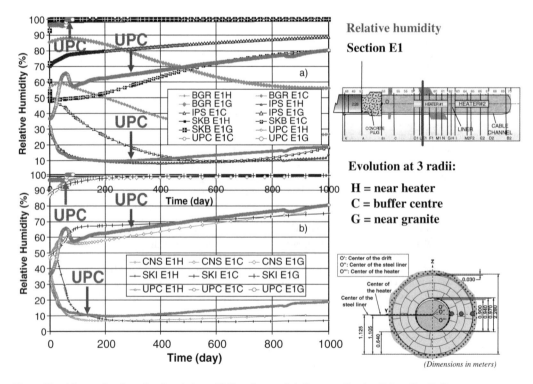

Figure 21. Measured and predicted variation of RH at three radial distances (Section E-1, radius R-4).

14

additional observations concerning the dimensionality of the analysis and other features. Note that two of the models considered a rigid buffer. No phase changes were introduced in two other models. Figure 21 includes also model predictions. Some of the models (those in Fig. 21b) reproduce adequately the fundamental trends of behaviour. Models represented in Figure 21a have more difficulties. Difficulties are, in general, associated with an incomplete formulation of water change of phase and vapour migration.

The distribution of RH in section E-1, 1000 days after the initiation of heating is shown in Figure 22.

Measurements recorded in four perpendicular radii are lumped for the same radial distance. The data shows minor effects of the direction of hydration. A number of modelling teams reproduce reasonably well the radial distribution of RH even if the history of hydration is not captured in detail. Mechanical influence on RH changes is of minor importance in this case. This is shown (see Fig. 23) by the analysis developed by Jussila (2003) who developed, taking into account basic conservation laws and thermodynamic principles, the equations for the evolution of temperature and relative humidity of a rigid porous media. His calculations (continuous line) are compared with the measured changes of RH in Section E-1, Radius R-4. Jussila's model reproduces the main

trends observed in data, although it does not show the marked transient response of the intermediate point ("center": E1C).

Radial stresses were also measured at some points by means of total pressure cells, normally oriented to radial distance. Results recorded in Section E-2, directly affected by Heater 1, are shown in Figure 24.

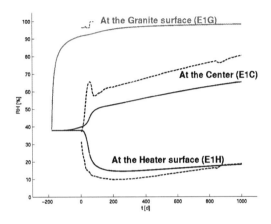

Figure 23. Evolution of RH in section E-1, RD-4. Predictions by Jussila (2003) and measured data (dotted lines).

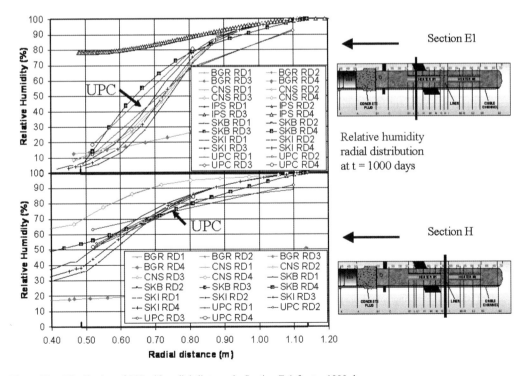

Figure 22. Distribution of RH with radial distance in Section E-1 for t = 1000 days.

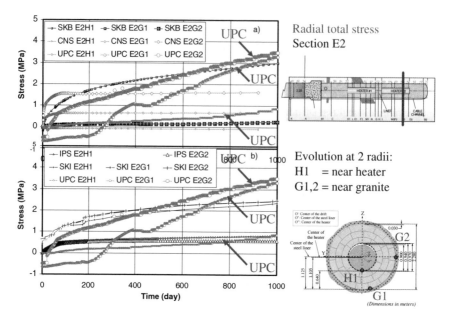

Figure 24. Calculated and measured radial stresses in section E-2 for three points.

One of the cells represented (H-1) is close to the heater. The other two (G-1 and G-2) are close to the granite boundary. Stresses develop very slowly at the hot end (H-1) and, at a faster rate, at the bentonite-granite boundary. This general trend is well-captured by some of the model predictions plotted also in Figure 24. Note, however, that models tend to predict a faster rate of increase of stress (for the wet end) than actually shown by data. This is probably a consequence of the measurement technique, which introduces an apparent delay in the cell response.

Radial stresses in the wet end reach values close to 35 MPa, during the first 1000 days of operation and a lower value (0.88 MPa), near the heater. Their rate of increase, at this time, indicates that the full effect of bentonite swelling is still far from being reached.

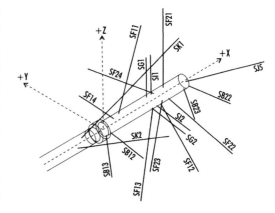

Figure 25. Instrumentation boreholes.

5 ROCK RESPONSE TO HEATING AND BUFFER HYDRATION

Sensors were located in radial (or close to radial) boreholes perforated from the FEBEX tunnel. The layout of these measuring boreholes is shown in Figure 25. Readings are, in general, available for three positions in the borehole (close to the tunnel, intermediate and a distant position). The length of boreholes does not exceed 15 m. Recorded data include temperature, water pressure, stress state and radial displacements. Data was also available on the

pressure of packers used to isolate hydraulically the intervals intended for rock-water pressure measurement. Again, temperature distribution is dominated by conduction and predictions tend to be accurate. On the other hand, the rock stiffness leads to very small values of measured displacements. Therefore, only pore water pressure and (total) stress data will be given here.

The modelling approaches for the rock are summarised in Table 4. One- two- and three-dimensional analyses were performed. The rock is characterised as a poro-elastic continua. The nature of couplings is also shown in the table.

Recorded water pressure in borehole SF-21 (location shown in Fig. 25) for two radial distances (3.03 m and 13.6 m) is given in Figure 26. Recorded values increase very slowly from initial values of 0.4-0.8 MPa to final values of 0.9-1 MPa, recorded at the end of the measuring period. No apparent effect of the increase of temperature is noticed. Some of the participating teams (SKI, IPS) were particularly successful in predicting "flat" response of the rock.

In contrast, Figure 27 shows the recorded pressure in the first packer, located in Borehole SF-21. It shows a significant initial increase in water pressure, from a starting value close to 2 MPa. This increase, which takes place within the first 100 days, carries the

Table 4. Modelling approaches for the THM analysis of the rock.

Team	Coupling	Rock	Observations
ANG	THM	T-poroelast (Biot)	1D 2D
BGR	part THM	POROELAST (Terzaghi)	2D no fluid dilation
CNS	THM	T-poroelast (Biot)	
DOE	TH↔TM	T-poroelast (s)	2D 3D
IPS	THM	T-poroelast	saturated
JNC			
SKB	T↔HM	T-poroelast (Bishop)	3D large mesh
SKI	THM	T-poroelast (Biot)	3D unsaturated

packer pressure to values in excess of 4 MPa. The transient increase in packer pressure is parallel to the increase in rock temperature. In fact, the temperature in a point located in SF-21, at a radial distance of 1.20 m, increases from 12°C to 34°C, within the first 100 days. Later the increase in temperature is slow and it takes 900 additional days to reach 40°C. The packer response reflects, therefore, the water dilation due to temperature increase. The slow pressure decaying after reaching a peak is interpreted as an indication of a leak or delayed deformation of the

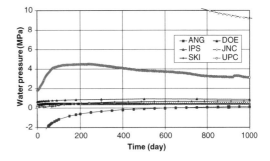

Figure 27. Measured packer pressure response of the first packer located in borehole SF-21 at a radial distance of 1.86 m. Also shown for reference are the calculated water pressures in rock for r = 3.03 m.

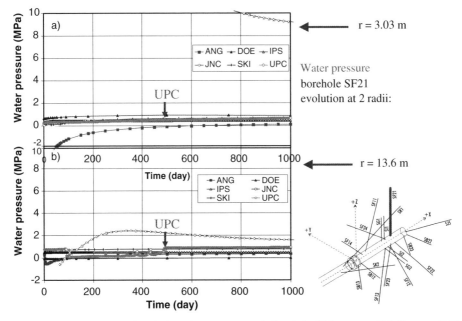

Figure 26. Measured and calculated evolution of water pressure in borehole SF-2 at two radial distances (3.03 m and 13.6 m).

17

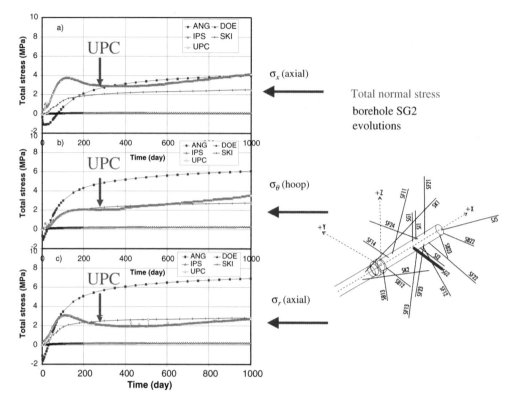

Figure 28.   Evolution of normal stresses in borehole SG-2 and predictions of four research teams.

measuring system. The important point is that the granite is "fast" dissipating any heat-induced pore water pressure and no increase of pore pressure is allowed. The relatively high granite permeability detected is also consistent with the "high" granite matrix permeability, derived from the uniform wetting behaviour of the buffer discussed before. Certainly, a faster rate of temperature change or a different boundary condition may change this result.

Axial circumferential (loop) and radial stresses are compared in Figure 28. They correspond to Borehole SG-2 and they were recorded at three radial distances (2.59 m, 2.97 m and 3.25 m). The measured initial fast increase in stresses is associated with the pattern of increase in temperature at the beginning of heating. The recorded peak and subsequent decrease in stress is not easy to explain as pore water pressures do not react to the temperature changes (total pressure cells measure also the water pressure). In fact, none of the teams participating in the exercise predicted the recorded peak. The measurements records suggest that pore pressure dissipation took place after the initial fast increase of temperature. This may be due to the existence of a local area of low granite permeability around the sensor position or to a

measurement artifact. In all cases, however, stresses show a progressive increase in value in the second half of the considered testing time. Final values for the three normal stresses are different (4 MPa, 3.7 MPa and 2.7 MPa) but not so different as to rise a concern on the yielding of the rock (Alonso & Alcoverro, 2004).

6   CONCLUSIONS

The FEBEX "in situ" test is a well-documented demonstration test for the storage, in a horizontal drift configuration, of nuclear waste. It is backed by an extensive experimental data base on Almeria compacted bentonite. It offers a good opportunity to conduct prediction exercises and to check the capabilities of THM models for barrier performance. The test also provided data on the granite rock response during the excavation of the FEBEX tunnel and, later, once the test started and the rock was subjected to the combined action of temperature increase and bentonite swelling.

The first 1000 days of operation of the FEBEX test have been the subject of a recent benchmark exercise, which involved a maximum of 10 teams from North America, Europe and Japan. A full report of the benchmark exercise performed is available (Alonso & Alcoverro, 2004). Some significant results have been selected for this paper. The following conclusions were reached:

1 The water pressure response due to tunnel excavation requires a fully coupled hydro-mechanical model. The rock response is fundamentally controlled by the orientation and intensity of the initial state of stress. Stresses derived from the back-analysis of pressure data do not match the existing information on initial stresses available before the FEBEX tunnel was opened.

2 A successful prediction of the hydration and stress development of bentonite buffer requires fully coupled THM formulations. Phase changes (liquid water-water vapour) must be included in the formulations.

3 Buffer hydration has progressed in a rather homogeneous and radial manner, independently of the presence of highly conductive rock features (lamprophyre dykes, sheared zones). This is a consequence of the relatively high granite permeability if compared with the buffer saturated permeability. A simple criterion to predict saturation times, in terms of saturated (matrix) permeability of rock and buffer, has been given. The high matrix permeability of the rock prevents also the development of transient increases in rock pore water pressure during the initial heating period.

4 The temperature field introduced by the canister heating results in a progressive increase in granite stresses, which reaches a maximum near the bentonite rock interface. The increase is associated with the thermal dilation of the rock. Most of the models have some difficulties to correctly predict stresses (a conclusion reached also for stresses in the buffer). It is also known that stress recording is not as reliable as other measured variables, such as temperature and relative humidity.

ACKNOWLEDGEMENTS

The support provided by the research project Decovalex III is greatly acknowledged. The authors are also thankful for the help and support received from ENRESA, the Spanish National Agency for Nuclear Waste Disposal, owner and manager of the FEBEX test. The European Community has partially funded the research activities within FEBEX, through projects awarded during the 4th and 5th Research Framework Programs. The authors wish to thank this wide support.

REFERENCES

Alonso, E.E and J. Alcoverro (2004) DECOVALEX III Project. Final Report of Task 1: Modelling of FEBEX in-situ test. SKI Report. Swedish Nuclear Power Inspectorate, Stockholm.

Bräuer, V., B. Kilger and A. Pahl (1989) Grimsel Test Site. Engineering geological investigations for the interpretation of rock stress measurements and fracture flow tests. NAGRA, NTB 88-37E, Apr. 1989.

Enresa (2000) FEBEX Project. Full scale engineered barriers experiment for a deep geological repository for high level radioactive waste in crystalline host rock. Final Report. Enresa. Madrid.

Jussila, P. (2003) Thermomechanical model for compacted bentonite. In: O. Stephansson, J.A. Hudson and L. Jing (editors) GeoProc 2003, International Conference on Coupled T-H-M-C Processes in Geo-systems: Fundamentals, Modelling, Experiments & Applications, Part 1, 120–125. Royal Institute of Technology (KTH), Stockholm, Sweden, 13–15 Oct. 2003.

Keusen, H.R., J. Ganguin, P. Schuler and M. Buletti (1989) Grimsel Test Site. Geology. NAGRA, NTB 87-14E, Feb. 1989.

Lloret, A., M.V. Villar, M. Sánchez, A. Gens and E.E. Alonso (2003) Mechanical behaviour of heavily compacted bentonite under high suction changes. Géotechnique, 53(1): 27–40.

Pahl, A., St. Heusermann, V. Bräuer and W. Glöggler (1989) Grimsel Test Site. Rock stress investigations. NAGRA, NTB 88-39E, Apr. 1989.

Pintado, X. and A. Lloret (2002) Backanalysis of thermo-hydraulic bentonite properties from laboratory tests. Engineering Geology, 64, 91–115.

Rutqvist, J., L. Börgesson, M. Chijimatsu, A. Kobayashi, L. Jing, T.S. Nguyen, J. Noorishad and C.-F. Tsang (2001) Thermohydromechanics of partially saturated geological media: governing equations and formulation of four finite element models. International Journal of Rock Mechanics and Mining Sciences, 38, 105–127.

Rutqvist, J., A. Rejeb, M. Tijani and C.-F. Tsang (2003) Analyses of coupled hydrological-mechanical effects during drilling of the FEBEX tunnel at Grimsel. In: O. Stephansson, J.A. Hudson and L. Jing (editors) GeoProc 2003, International Conference on Coupled T-H-M-C Processes in Geo-systems: Fundamentals, Modelling, Experiments & Applications, Part 1, 114–119. Royal Institute of Technology (KTH), Stockholm, Sweden, 13–15 Oct. 2003.

Villar, M.V., A. Lloret and E. Romero (2004) Final report on Thermo-Hydro-Mechanical Laboratory tests. FEBEX Project. Enresa. Madrid.

Field emplacement and instrumentation techniques

Advances in Understanding Engineered Clay Barriers – Alonso & Ledesma (eds)
© 2005 Taylor & Francis Group, London, ISBN 04 1536 544 9

Transducers and cable connections for measuring THM-processes in engineering barriers – design and experiences

Torbjörn Sandén, Reza Goudarzi & Lennart Börgesson
Clay Technology AB, Lund (Sweden)

ABSTRACT: When designing a repository for spent nuclear fuel, it is important to have knowledge about the properties and behavior of the buffer and the backfill material. At ÄHRL in Sweden, a number of large field tests have been installed. In order to understand the THM-processes in these tests and compare with modeling results, sensors of different types have been installed. The instruments are mainly measuring:

- Temperature.
- Total pressure.
- Pore water pressure.
- Relative Humidity (Water content).

The demands on the instruments are high since the test periods are very long (up to twenty years) and the instruments are exposed to high temperatures and a ground water with a salt content of up to 1.8%. In order to ensure durability of the sensors, it has been necessary to carefully choose materials for the sensors and to encapsulate some of the sensors in protecting houses. It has also been necessary to protect the cables and to make a secure completely water tight sealing when leading them out from the test volume.

Experiences regarding both installation of the sensors and reliability of measured values are accounted for although some of the experiments have been running for only a short while and no experiment is yet terminated. The reliability has been evaluated as % still working sensors of the different types as function of the time from start of the test. The conclusions are that the temperature and relative humidity sensors made by thermocouples are extremely reliable with less than 1% out of order in all tests. Also the sensors for measuring pressure with the vibrating wire technique are very reliable.

1 INTRODUCTION

Four full-scale field experiments, containing bentonite and/or backfill material are installed and still running at Äspö HRL. The experiments have somewhat different purposes but they all contain instruments measuring the THM-processes in the buffer and/or in the backfill material. Besides the instruments measuring THM-processes in buffer and backfill, additional instruments are installed in these experiments e.g. sampling of pore water in buffer, rock and backfill, in the rock there are pore pressure and stress measurements and there are also instruments installed for measuring of canister movements. These sensors are not treated in this article.

There are four principal demands on the sensors:

- They should deliver reliable data with high accuracy.
- The long term stability of the sensors is very important since some of the experiments will run for up to twenty years.

- The sensors and their cables should not influence the experiments and act as leakage ways and prevent the possibility to build up high pore pressures in the system. This means that some of the sensors bodies must be build in protecting houses and the sensor cables must be led in protecting tubes.
- The construction must be robust due to the tough environment with high temperature, salty groundwater and high pressures.

1.1 Prototype repository

The test consists of 6 full-scale KBS-3 deposition holes (diameter = 1.75 m and height = 8 m), copper/steel canisters (diameter = 1.05 m and height = 4.975 m) equipped with electrical heaters for simulating the heating caused by the radioactive decay, a backfilled deposition tunnel (total length of about 60 m) and 2 tunnel plugs. The canisters are embedded in dense buffer clay of compacted bentonite powder. Instruments are placed in the buffer, the backfill and in the rock. A maximum temperature of 100°C is expected in the buffer material. The salt content in the groundwater varies between 0.6% and 1.8%. The maximum total pressure in the buffer can be as high as 15 MPa and the maximum pore pressure 5 MPa. In the backfill the temperature is expected to be 20–40°C and the maximum pore pressure 5 MPa.

1.2 *Backfill and Plug Test*

This test consists of a blasted tunnel with a length of about 50 m, which has been filled with backfill material. The material has been transported into the tunnel where a special designed compaction device has compact it in layers. Instruments are placed in the the backfill and in the rock. The salt content in the groundwater is about 1.2%. The maximum total pressure and pore pressure in the backfill is 5 MPa. No enchanced temperature is produced in this test.

1.3 *Canister Retrieval Test*

The test consists of a full-scale KBS3-type deposition hole (diameter = 1.75 m and height = 9 m), a copper/ steel canister (diameter = 1.05 m and height = 4.975 m) equipped with electrical heaters for simulating the heating caused by the radioactive decay. The canister is embedded in dense buffer clay consisting of compacted blocks. A retaining plug has been built in order to restrict upward movement caused by buffer swelling. Instruments are placed in the buffer and in the rock. An artificial water pressure can be applied in ten filter mats placed on the rock surface. A maximum temperature of 100°C is expected in the buffer material. The salt content in the groundwater is 1.2–1.5%. The maximum total pressure in the buffer can be as high as 15 MPa and the maximum pore pressure 5 MPa.

1.4 *Temperature Buffer Test*

The test consists of a full-scale KBS3-type deposition hole, 2 steel canisters (diameter = 0.6 m and height = 3.0 m) equipped with electrical heaters for simulating the heating caused by the radioactive decay and a mechanical plug at the top. The canisters have been embedded in dense buffer clay consisting of compacted blocks. The upper canister is surrounded by sand in order to reduce the temperature in the clay. A retaining plug has been built in order to restrict upward movement caused by buffer swelling. An artificial water pressure can be applied in the outer slot, which is filled with sand that will function as a filter. Instruments are placed in the buffer and in the rock. A maximum temperature of 170°C is expected

in the buffer material. The salt content in the groundwater is 1.2–1.5%. The maximum total pressure in the buffer can be as high as 15 MPa and the maximum pore pressure 5 MPa.

2 SELECTION OF MATERIAL FOR SENSORS AND TUBES

The corrosivity of different materials offered by manufacturers and used for the parts of the instruments as sensor housing, protection tubes etc, which will be placed in the buffer and backfill, has been investigated and reported by The Swedish Corrosion Institute. According to the report, titanium is recommended as material in buffer conditions. Other materials that have acceptable resistance against corrosion are nickel-based alloys as Inconel 625. There is less demand on the material when the sensors are protected by a housing of titanium as is the case with e.g. the capacitive sensors for relative humidity measurements.

2.1 *Instruments in buffer*

The high temperature and pressure in combination with ground water salinity of 1.2–1.5% have made it necessary to use titanium for all sensor housings except the thermocouples. The cables have been protected by leading them in titanium tubes through the bentonite. The tubes are welded to the sensor houses. The thermocouples are of sheath type where the sheath is made of alloy (70–30 copper nickel).

2.2 *Instruments in backfill*

Considering the lower temperature in the backfill, it has been possible to manufacture or encapsulate all instruments in stainless steel (SS 2343/AISI 316). The cables have been led in polyamide tubes (PA11). The thermocouples placed in the backfill are of the same type as those in the buffer.

3 MEASURING PRINCIPLES AND SUPPLIERS OF EQUIPMENT

A desire when choosing instruments for these full-scale field experiments has been to have at least two different measuring principles for each measured variable. This principle has changed gradually by the increase of experience from the different instruments and suppliers. Some instruments have not fulfilled the demands while others have performed outstanding. Table 1 show a compilation of the measuring equipment, which is installed in the different projects at Äspö HRL for measuring the THM-processes in the buffer and backfill.

Table 1. Compilation of installed measuring equipment.

Measuring quantity	Measuring principle	Supplier	Country	Used in project	Remark
Temperature	Thermocouple	Pentronic	Sweden	Prototype Repository Backfill and Plug Test Temperature Buffer Test	
		BICC	England	Canister Retrieval Test	
	Fibre optic	BICC	England	Prototype Repository Canister Retrieval Test	Placed as a loop on the canisters surface
	Resistive temperature sensor			Prototype Repository Canister Retrieval Test Temperature Buffer Test	Built-in to capacitive humidity sensors and to Geokon pressure cells
Total pressure and pore pressure cells	Hydraulic pressure cells	Glötzl	Germany	Backfill and Plug	
	Vibrating wire	Geokon	USA	Prototype Repository Canister Retrieval Test Temperature Buffer Test	
		Roctest	Canada	Backfill and Plug Test	
	Piezo-resistive	Kulite	Holland	Prototype Repository Canister Retrieval Test	
		Druck	England	Backfill and Plug Test Prototype Repository Canister Retrieval Ttest Temperature Buffer Test	Placed outside the test volume, measuring water pressure
	Fiber optic sensors	DBE	Germany	Prototype Repository Temperature Buffer Test	Only few sensors installed for testing of technology
Water content	Capacative sensors	Vaisala	Finland	Prototype Repository Canister Retrieval Test Temperature Buffer Test	
		Rotronic	Switzerland	Prototype Repository Temperature Buffer Test	
	Psychrometer	Wescor	USA	Prototype Repository Backfill and Plug Test Canister Retrieval Test Temperature Buffer Test	

3.1 Temperature

Three different measuring principles have been used for the temperature measurements:

- **Thermocouple.** The thermocouple consists of two leaders of different metals or alloys which are joined together at one of the ends. A potential is produced at the contact surface between the two metals and is the measure of temperature. When the leaders are connected to a reading device a potential is also produced at its contacts. The result is that the reading device will measure the difference between the two potentials. Thermocouples used in the experiments are delivered by BICC and Pentronic.
- **Resistive temperature detectors.** Resistive temperature detectors are normally made of platinum (Pt 100 sensors) or semiconductor material. A temperature sensor of semiconductor type is often called thermistor but also PTC or NTC resistance.

In the relative humidity sensors from Vaisala and Rotronic, Pt 100 sensors are built in. Each Geokon pressure sensor is also equipped with a thermistor.
- **Fibre optic.** Optical fibre temperature measurements are often called Distributed Temperature Sensing (DTS) or Fibre Temperature Laser Radar (FTR). In the two projects containing copper canisters of KBS3 type i.e. Canister Retrieval Test and the Prototype Repository, temperature measurements are done with this method. On each canister two loops of fibre cables with a protection tube of Inconel 625, are laid in tracks machined out on the canister surface. With this method a temperature profile is achieved along the cable length. The equipment is delivered from BICC.

3.2 Relative humidity (water content)

These equipment measure the relative humidity (RH) in the pore system. RH can be converted into water ratio or total suction (negative water pressure). Suction can then be converted to saturation estimates.

Relative humidity is measured with the following methods:

- **Capacitive sensors.** The sensor consists of a pair of electrodes which are separated by a polymer film. The quantity of moisture which is absorbed by the polymer film is affected by the relative humidity in the surrounding air. The capicitance is affected by the absorbed water quantity and is equivalent to the relative humidity. Sensors using this method is supplied by Vaisala and Rotronic.
- **Psychrometers.** The soil psychrometer is used for measurements of dry and wet temperature in the pore volume of a material. The sensor consists of two thermocouples of which one is used for cooling by the Peltier effect and the other for temperature measurement. The sensor is cooled down below the dew point after which the cooling is interrupted. The knowledge of the wet temperature, which can be read when the condensated water evaporates, and the dry temperature gives the relative humidity and thus the water content in the surrounding material. The measurement range of these sensors is 95–100% RH. The psychrometers used in the experiments are delivered by Wescor.

3.3 *Total pressure and pore pressure*

Total pressure is the sum of the swelling pressure of the buffer or backfill material and the pore water pressure. The pressures have been measured with the following principles:

- **Vibrating wire.** A vibrating wire sensor uses a wire, which in one end is attached to the backside of a pressure diaphragm and held under tension. Pressure acting on the other side of the diaphragm causes wire tension and thus the natural frequency of the wire to change. The principle is that a voltage pulse is applied to an electromagnetic coil to pluck the wire i.e. draw the wire to one side and release it. The sensors also use a permanent magnet, which together with the vibrating wire induces a sinusoidal voltage in the coil. The frequency of the voltage is measured and is a measure of the pressure. Sensors using this technique are delivered by Roctest and Geokon.
- **Piezo-resistive sensors.** The piezo resistive sensor uses a strain gauge of semiconductor material. The strain gauge is bonded to a diaphragm, which is in contact with the surrounding material. The resistance of the strain gauge is measured and represents the pressure of the surrounding material. Sensors of this type is used in all experiments, but mainly at the Backfill and Plug Test, in order to measure water pressure in tubes leading out water

from different positions in the test volume. These sensors are delivered by Druck. The measuring principle is also used in the total pressure cells and the pore pressure cells delivered by Kulite.

- **Hydraulic pressure cells.** The hydraulic pressure measurement is done with an oil-filled pressure cell, which is embedded in the material of which the pressure shall be measured, an oil filled circulation pump system, a control valve and a pressure gauge. The control valve, which regulates the oil flow in the system, is equipped with a diaphragm connected to the outlet pipe from the pump. The pressure increases on the side of the diaphragm connected to the outlet pipe from the pump. The valve opens when the pressure is equal on both sides of the diaphragm. Thus, the pressure change measured in the outlet pipe is equal to the pressure in the surrounding material of the cell. Hydraulic pressure cells are installed in the Backfill and Plug Test and are delivered by Glötzl.
- **Fibre optic sensors.** A few sensors using the fibre optic technique are installed in the Temperature Buffer Test and in the Prototype Repository. The sensors are installed merely for testing the technique. The optical technology is based on the Fibre Bragg grating principle. A Bragg grating sensor consists of an optical fibre, which has been processed by UV radiation to obtain a repeated change from one refractive index to another along the fibre. At each change from the one refractive index to the other a smaller part of the incoming light is reflected back. A pressure sensor is designed in such a way that the pressure to be measured affects the characteristic of the grating and thus the wave length of the reflected light.

4 ENCAPSULATION OF SENSORS

The instruments measuring relative humidity need an additional encapsulation both for mechanically and corrosion reasons. The encapsulation also gives the possibility to cut and plug the tube on the outside if any leakage should occur through the sensor cables after saturation of the system.

4.1 *Vaisala and Rotronic*

The sensors delivered by Vaisala and Rotronic have housings made of stainless steel. The sensor part is covered with filters also made of stainless steel. In order to with stand the total pressure and the corrosive environment each sensor have been built into titanium cases, consisting of housing for the sensor body and a titanium tube for the cable, see Figure 1. The tube is welded to the housing. The sensor cables were thread through the housing until the sensors were in position. The sensors are sealed against the titanium housing with O-rings. The titanium housings are equipped with titanium filters.

4.2 *Wescor*

The Wescor psychrometers are handled in another way. Before positioning the sensor in the housing, the sensor cable was split, exposing the leaders. Then epoxy was injected, sealing and separating the leaders and the psychrometers from the open tube, see Figure 1. This operation was necessary because

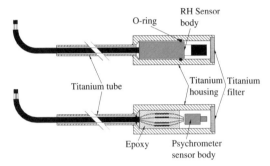

Figure 1. Schematic view showing the methods used when encapsulating the relative humidity sensors. The upper device shows the principle when encapsulating Vaisala and Rotronic sensors and the lower shows the handling of Wescor psychrometers.

there is no firm sensor body to seal against and tests made on the cable showed that it was very permeable.

5 SEALING OF CABLES AND TUBES

In the Temperature Buffer Test and in the Canister Retrieval Test, the lengths of all tubes coming from sensors in the test are adjusted in order to go through the plug located on the top of the deposition hole. The tubes are led in dense bentonite and no additional sealing is needed.

However, in the Prototype Repository and in the Backfill and Plug test, the tubes are also led through backfill material. Depending on the lower total pressure and the lower temperature in the backfill material, the protecting titanium tubes are converted into polyamide tubes at the entrance to the backfill material. This also facilitates the handling of cables during installation. Since these tests include a high pore water pressure in the backfill material during the test period, an additional sealing is needed at the periphery of the test site. In these projects, special bore holes were drilled through the rock to adjacent tunnels. These bore holes are lined with steel tubes, which are sealed against the rock with injected cement and blocks of bentonite (Figure 2). When the cables and tubes from the sensors placed in buffer and backfill are led into the lead-through tubes, they are led through steel flanges equipped with tube fittings of Swagelok type as shown in Figures 2 and 3.

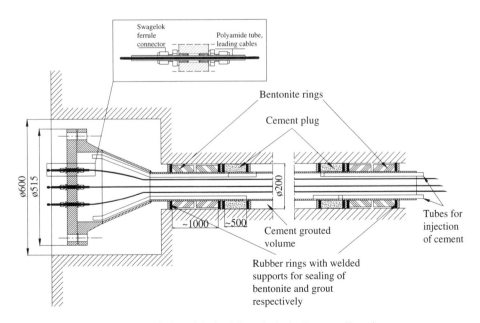

Figure 2. Schematic view showing the design of the lead-throughs in the Prototype Repository.

Figure 3.   Photo showing a lead-through with installed cables in the Prototype Repository.

6   INSTALLATION OF SENSORS

The technique for installing sensors has been different depending on if they are placed in the buffer or in the backfill.

6.1   *Instruments in buffer*

The main principle used for the instrumentation of the buffer has been to place the sensor houses in predrilled holes and cutouts in the blocks (Figure 4). The tubes have then been led out to the block periphery in tracks made on the block surface. At the periphery the tubes have been bent upwards and led along the outer surface and into the backfill material. In the TBT test and the CRT test, slots have been made on the rock wall in order to let the tubes pass the supporting plugs.

6.2   *Instruments in backfill*

The backfill material has been compacted in layers in the tunnel with an inclination of about 35°. On the surface of the layers, recesses have been made by use of handhold tools. The sensor houses have been placed in the recesses and the connecting tubes then led in a track out to the rock wall were the lead-through holes are placed (see Figure 5). During the further compaction the cables are protected from the compaction device and the rock, by leading them in a mixture of bentonite powder and pellets.

7   RELIABILITY OF SENSORS

The reliability of the sensors may be evaluated by plotting the percentage of sensors that are still working as function of the time from start of the tests. Such an evaluation does not tell anything about the accuracy of the sensors. It only tells if they are working and read reasonable values. All sensors and all sensor results are frequently checked and the results reported. Those that are not working or yield obviously erroneous results are noted as being "out of order" and included in the statistics.

It is not always easy to decide whether a sensor is working properly or not. Errors may be caused by data logger problems or similar things that are not depending on malfunction of the sensors themselves. Therefore all malfunctioning sensors or sensors that yield strange results are checked and efforts made to see if the problems can be taken care of.

The results of the reliability evaluation are summarized in 5 diagrams of total pressure, pore pressure, relative humidity and temperature. The diagrams represent the percentage of instruments still working as function of time from start of the respective test. Five large-scale tests are included in the evaluation, namely the Backfill and Plug Test (BaPT) starting 1/6 1999, the Canister Retrieval Test (CRT) starting 26/10 2000, the Prototype Repository sections 1 and 2 (Prototype S1 and S2) starting 17/9 2001 and 8/5 2003 and the Temperature Buffer Test (TBT) starting 26/3 2003. No instruments

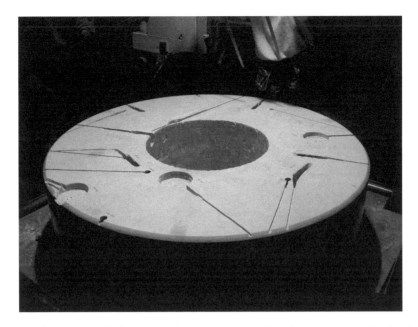

Figure 4. Photo of a bentonite block prepared for instrumentation. The photo is taken during installation of the Temperature Buffer Test.

Figure 5. Photo of the instruments and cables installed in an inclined backfill layer. The photo is taken during installation of the Backfill and Plug Test.

in the rock or on the canister surface are included but only instruments in the buffer and in the backfill.

The diagrams show one curve for each type of sensor for each test. This means that each of the four measuring variables yield many curves. The number of curves in the diagrams has been limited to 8 so one of the variables (relative humidity), which was measured with many instrument types, has been split into two diagrams.

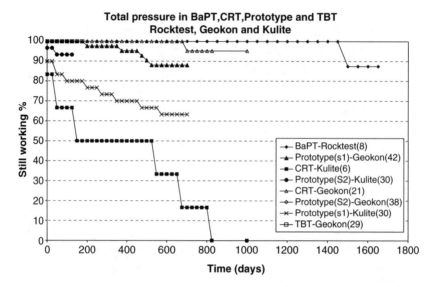

Figure 6.   Reliability of total pressure sensors installed in the buffer and backfill in Äspö HRL. The number of sensors is shown in brackets.

7.1   Total pressure

Figure 6 shows the working percent of the total pressure gauges. Total pressure is measured in all 5 tests.

BaPT. Total pressure in the Backfill and Plug Test is measured with Rocktest and Glötzl transducers but only Rocktest is included in the diagram. The reason is that the Glötzl total pressure cells are of hydraulic type and the automatic measurement and data acquisition system has not worked well but been under reparation during two long periods. The measurements have during these periods been done manually. However, all 13 pressure cells that were installed are still functioning.

CRT. Total pressure in the Canister Retrieval Test is measured with Geokon and Kulite transducers. 21 vibrating wire Geokon sensors were installed. All of them but one are still working. 6 Kulite piezo-resistive sensors were installed. All of them have failed. The reason for the malfunction is interpreted to be caused by brakeage of the transducer connection between the sensor body and the encapsulating titanium tube at high pressure.

Prototype S1. Total pressure in section 1 of the Prototype Repository is measured with Geokon and Kulite transducers. 42 vibrating wire Geokon sensors were installed. 5 of them have failed today. 30 Kulite piezo-resistive sensors were installed. 11 of them have failed mainly in the buffer for the same reason as in CRT.

Prototype S2. Total pressure in section 2 of the Prototype Repository is measured with Geokon and

Kulite transducers. 38 vibrating wire Geokon sensors were installed. All are still working after 125 days. 30 Kulite piezo-resistive sensors were installed. The transducer connections were improved and strengthened. Only 2 of them have failed so far.

TBT. 29 vibrating wire total pressure sensors were installed in the Temperature Buffer Test. After 175 days all are still working.

The conclusion is that the vibrating wire transducers work well and that the problems of other types have occurred either due to bad design of the connection between the tube and the sensor body (Kulite) or failure of the automatic reading system (Glötzl).

7.2   Pore pressure

Figure 6 shows the working percent of the pore water pressure gauges. Pore pressure is also measured in all 5 tests.

BaPT. Pore pressure in the Backfill and Plug Test is with a few exceptions only measured with Glötzl transducers (18). These results are not included in the diagram on the same reason as for the total pressure. The measurements have during these periods been done manually and all 18 cells that were installed are still functioning.

CRT. Pore pressure in the Canister Retrieval Test is measured with Geokon and Kulite transducers. 11 vibrating wire Geokon sensors were installed. All of them are still working. 2 Kulite piezo-resistive sensors were installed. Both 2 have failed probably for the same reason as the total pressure cells.

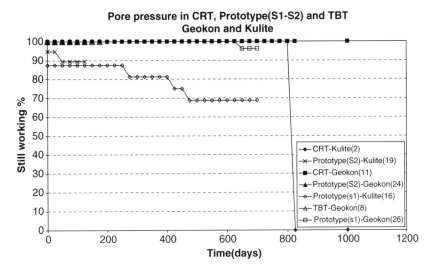

Pore pressure in CRT, Prototype(S1-S2) and TBT
Geokon and Kulite

Legend:
- CRT-Kulite(2)
- Prototype(S2)-Kulite(19)
- CRT-Geokon(11)
- Prototype(S2)-Geokon(24)
- Prototype(s1)-Kulite(16)
- TBT-Geokon(8)
- Prototype(s1)-Geokon(26)

Figure 7.   Reliability of pore pressure sensors installed in the buffer and backfill in Äspö HRL. The number of sensors is shown in brackets.

Prototype S1. Pore pressure in section 1 of the Prototype Repository is measured with Geokon and Kulite transducers. 26 vibrating wire Geokon sensors were installed. Only one of them have failed today. 16 Kulite piezo-resistive sensors were installed. 5 of them have failed.

Prototype S2. Pore pressure in section 2 of the Prototype Repository is measured with Geokon and Kulite transducers. 24 vibrating wire Geokon sensors were installed. All are still working after 125 days. 19 Kulite piezo-resistive sensors were installed. Two of them have failed so far.

TBT. 8 vibrating wire pore pressure sensors were installed in the Temperature Buffer Test. After 175 days all are still working.

Since pore pressure is measured with identical technique as the total pressure the conclusions regarding their function are the same.

7.3   Water content

The working percent of the water content transducers are shown in Figures 8 and 9. Three types of instruments are used and most of them are installed in all tests. All instruments are actually measuring relative humidity (RH). None of the transducers work at full water saturation but are judged to be functioning when they have reached full saturation although they don't yield useful readings afterwards.

BaPT. RH in the Backfill and Plug Test is measured only with Wescor psychrometers since RH in the backfill is higher than 95% from start. 22 transducers were installed in the backfill with the mixture of 30% bentonite and 70% crushed rock. All of them show that the backfill is water saturated today and they have apparently worked the entire time until full saturation.

CRT. RH in the Canister Retrieval Test is measured with Vaisala capacitive sensors and Wescor psychrometers. 29 Vaisala sensors were installed. 8 of them appear to have stopped functioning before reaching full saturation. 26 Wescor psychrometers were installed. Since they can only measure high RH they don't yield results until the buffer is close to saturation. 4 of them have not yet reached high enough saturation and have thus not yielded any results yet. Out of the remaining 22 all seem to have worked properly until full saturation with one possible exception.

Prototype S1. RH in section 1 of the Prototype Repository is in the buffer in the deposition holes measured with Vaisala and Rotronic capacitive sensors and in the backfill in the tunnel with Wescor psychrometers. 8 out of 39 Vaisala sensors and 6 out of 34 Rotronic sensors have failed before water saturation. 45 Wescor psychrometers sensors were installed in the backfill and they all seem to still be working properly.

Prototype S2. RH in section 2 of the Prototype Repository is measured with the same instruments as in section 1. 18 out of 40 Vaisala sensors and 2 out of 34 Rotronic sensors have failed before water saturation so far (125 days). 32 Wescor psychrometers sensors were installed in the backfill and they all seem to work. In addition 35 psychrometers were installed in the buffer in the two deposition holes. So

31

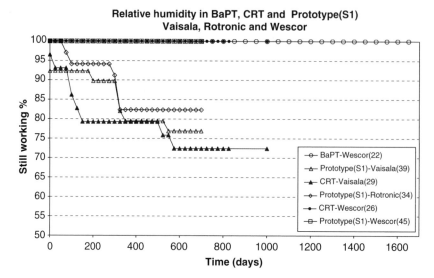

Figure 8. Reliability of relative humidity sensors installed in the buffer and backfill in the three oldest tests in Äspö HRL. The number of sensors is shown in brackets.

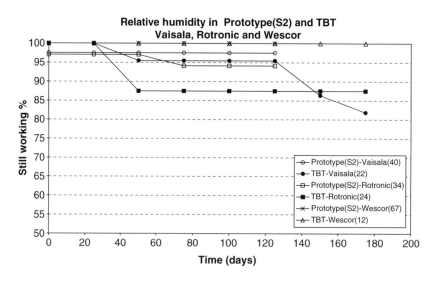

Figure 9. Reliability of relative humidity sensors installed in the buffer and backfill in the two youngest tests in Äspö HRL. The number of sensors is shown in brackets.

far none of them have yielded any readable results due to the slow wetting and short time since start.

TBT. RH in the Temperature Buffer Test is measured with Vaisala and Rotronic capacitive sensors as well as with Wescor psychrometers. 4 out of 22 Vaisala sensors and 3 out of 24 Rotronic sensors have failed before water saturation, the major part being located at high temperature spots. All 12 Wescor psychrometers sensors that were installed

still seem to work although 4 have not yet yielded readable RH.

The conclusion of this compilation of the reliability of the RH sensors is that there is a slow but steady drop-out of the capacitive sensors mainly at high temperature locations and that the Wescor psychrometers seem to work remarkably well both in the backfill and buffer. It must though be mentioned that only a few psychrometers at high temperatures have

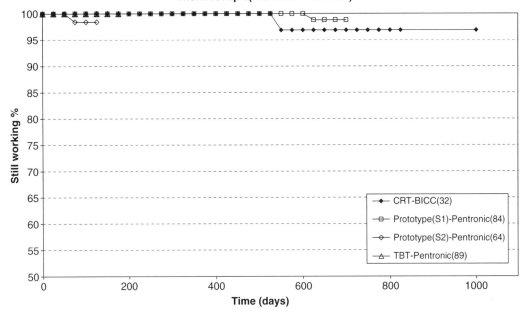

Figure 10. Reliability of temperature sensors installed in the buffer and backfill in Äspö HRL. The number of sensors is shown in brackets.

so far been readable due to the slow wetting and limited measuring range (<95% RH).

7.4 Temperature

Temperature has been measured both by transducers installed with the single purpose to measure temperature (BICC and Pentronic thermocouples) and by transducers built and installed for other measurements, where the temperature measurement is done mostly for temperature compensation purpose. Since the main purpose of this study is to understand the reliability of sensors with the purpose of measuring specified variables only the first-mentioned types are included. Other reasons to include the last-mentioned types are that the transducers are so large that the location of the measured temperature is unclear and that they are very expensive in comparison with thermocouples.

BaPT. No thermocouples were installed in the Backfill and Plug Test.

CRT. Temperature in the Canister Retrieval Test is measured with BICC thermocouples. Only one out of 32 sensors have failed during the test that so far has lasted for more than 1000 days.

Prototype S1. Temperature in section 1 of the Prototype Repository is measured with Pentronic thermocouples. Only one out of 84 sensors have

failed during the test that so far has lasted for about 700 days.

Prototype S2. Temperature in section 2 of the Prototype Repository is measured with Pentronic thermocouples. One out of 64 sensors have failed so far.

TBT. Temperature in the Temperature Buffer Test is measured with Pentronic thermocouples. So far all 89 sensors are still working.

The reliability of thermocouples is thus very high. Only 3 out of 269 thermocouples have so far failed.

8 CONCLUSIONS AND COMMENTS

The article describes the instruments, cable protections and lead through connections made for 5 large-scale tests in Äspö HRL and a compilation of the reliability of the installed instruments. Although no test has yet been terminated and excavated and the time period since start has varied from only 125 days (Prototype section 2) up to more than 1600 days (BaPT) some conclusions can be drawn about the function of the lead through system and the function of the sensors.

The cable protection and lead through systems seem to work well so far but a high water pressure has so far only been reached in the Backfill and Plug Test where 500 kPa have been applied for about 700 days. The sensors and cables have in all tests been protected with covering hoses and tubes with the purpose to withstand high water and total pressure and high temperature. Such protection and watertight lead through systems for carrying the cables through the rock were judged necessary for a proper simulation of water pressure built up and to keep the sensors working for long time.

The reliability investigation, plotting the percentage still working sensors as function of time from start of the different tests, have shown that the sensors generally work well with a few exceptions.

The thermocouples for measuring temperature work extremely well and only 3 out of 269 have so far failed (1%). The same conclusion can be drawn about the psychrometers from Wescor for measuring RH. It is remarkable that out of 172 installed psychrometer no one has so far failed (with one possible exception). Although 43 of the psychrometers have not yet yielded measurable values it is obvious that less than 1% can be said to have failed. The reason for the reliability of both the thermocouples and the psychrometers is that they are very simple and either water intrusion or breakage of the cable is required for making them fail. The psychrometers are made of two thermocouples, which explains the similarity.

The other types of RH sensors (capacitive) seem to fail slowly but steadily. After 1000 days about 25% have failed before being water saturated. It seems that high temperature is one variable that contributes to failure but the statistics is somewhat distorted due to the difference in temperature and difference in time until they are water saturated.

The total and pore pressure transducers show similar behaviour. The vibrating wire sensors (Geokon and Rocktest) seem very reliable with less than 5% failed after 1000 days. The big problems that affected the sensors from Kulite are judged to mainly be caused by the bad design of the connection between the tube and the sensor body. This connection was improved for the sensors installed in section 2 of the Prototype Repository and so far 4 out of 49 have failed after 125 days.

REFERENCES

Börgesson L and Sandén T, (2001). Prototype Repository. Report on instrument positions in buffer/backfill and preparation of bentonite blocks for instruments and cables in section I. IPR-01-20.
Börgesson L and Sandén T, (2003). Prototype Repository. Report on instrument positions in buffer/backfill and preparation of bentonite blocks for instruments and cables in section 2. IPR-03-21.
Gunnarsson D, Collin M and Börgesson L, (2001). Prototype Repository. Instrumentation of buffer and backfill for measuring THM processes. IPR-02-03.
Gunnarsson D, Börgesson L, Hökmark H, Johannesson L and Sandén T, (2003). Report on the installation of the Backfill and Plug Test. IPR-01-17.
Sandén T and Börgesson L, (2000). Canister Retrieval Test. Report on instrument positions and preparation of bentonite blocks for instruments and cables. TD-00-013.
Sandén T and Börgesson L, (2002). Temperature Buffer Test. Report on instruments and their positions for THM measurements in buffer and rock and preparation of bentonite blocks for instruments and cables. R5.
Goudarzi R, Gunnarsson D, Johannesson L-E and Börgesson L, (2003). Backfill and Plug Test. Sensor data report (Period 990601-030101). Report No: 6. IPR-03-14.
Goudarzi R, Börgesson L, Röshoff K and Bono N, (2003). Canister Retrieval Test. Sensor data report (Period 001026-030501). Report No: 6. IPR-03-30.
Goudarzi R and Börgesson L, (2003). Prototype Repository. Sensor data report (Period 010917-030601). Report No: 6. IPR-03-31.

Advances in Understanding Engineered Clay Barriers – Alonso & Ledesma (eds)
© *2005 Taylor & Francis Group, London, ISBN 04 1536 544 9*

Components and processes affecting the recorded performance of clay buffers

Roland Pusch
Geodevelopment AB, Lund, Sweden

Gunnar Ramqvist
SKB Äspö Hard Rock Laboratory, Figeholm, Sweden

ABSTRACT: Instrumentation is necessary for getting information on the important processes in buffer clay in field experiments but the question is whether it can affect these processes or generate artefacts. The gauges are placed in holes drilled in the bentonite buffer blocks and put close to heaters and rock, and cables and tubings connect them with the recording units. This may cause problems by water leakage along the connections and quick water transport to the gauges, giving the impression of much earlier wetting than in buffer without instruments.

A phenomenon that has not yet attracted much interest is the electrical potential that is formed in the buffer in the course of the wetting and that may have an impact on the performance of instruments placed in the buffer. Such a potential exists between the copper canisters and the earth connection in the Äspö experiments and may have three effects: (1) The flow of water in the buffer is counteracted meaning that the recorded wetting rate is lower than predicted by use of the Darcy model, (2) Metal gauges placed in the buffer act as electrodes and may undergo corrosion by which the recordings are affected, and (3) Canister corrosion may accelerate.

1 INTRODUCTION

A number of phenomena appear to have an impact on the recording of important data in field experiments with buffer clay. Two of them have been identified as particularly important: (1) impact of cables and tubings as water conductors in evaluation of swelling pressure and hydration rate, (2) set-up of electrical potentials in field experiments.

2 WATER FLOW ALONG INSTRUMENT CABLES

2.1 Experimental

In certain field and mock-up experiments water has been found to migrate much quicker than expected as predicted by current hydration models. This has brought up the question whether cables and tubings and the joints between buffer blocks can serve as water conductors. The diagrams in Figure 1 represent 165, 270 and 370 days after starting the so-called "Retrieval Test", which represents a true KBS-3 case with the buffer/clay contact being a constantly water-filled filter. Theoretically, the rate of hydration implies initiation of the wetting of the clay close to

the canister after a few months according to models developed and used in the Prototype Repository Project (Pusch, Svemar, 2001), an example being given in Figure 2. The field observation that the wetting was quicker at the canister periphery than 5 cm away from it is a definite proof that the recordings do not accurately depict the true wetting rate. The water pressure was 100 kPa, i.e. a very small fraction of the measured total pressure.

The same phenomenon occurred in the ongoing Czech Mock-up experiment at the Center of Experimental Geotechnics CTU, Technical University in Prague (Pacovsky 2002). This test is similar to the Retrieval Test but with different dimensions. Figures 3 and 4 illustrate the design, which is characterized by the following typical features:

Rock: 8 mm steel cylinder with an inner diameter of 800 mm and inner height of 2230 mm. Two peripheral concentric filters for hydration.

Canister: Heater with 320 mm outer diameter contains two electrical elements powered to yield 90–100°C constant surface temperature.

Buffer: Sector-shaped buffer blocks with 70 mm height and 160 mm radial thickness. RMN bentonite, 10% silica sand, 5% graphite and 7% water. The dry density of the compacted blocks was initially 1700 kg/m^3.

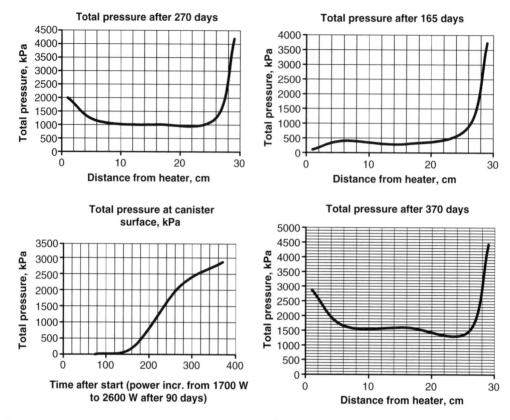

Figure 1. Recorded total pressures in SKB's Retrieval Test at Äspö.

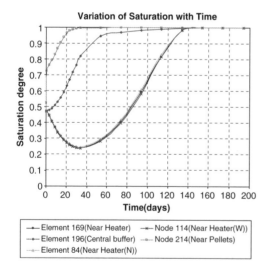

Figure 2. Predicted evolution of the water saturation (Ledesma et al (2002). Other models show complete saturation at the canister surface after up to 2–3 years but they all show the lowest degree of saturation at the canister surface at any time.

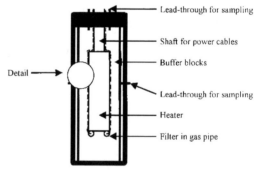

Figure 3. Schematic picture of the Czech Mock-up. (Detail of arrangement for sampling shown in Figure 4).

Pellet fill: The 50 mm gap between the filters and the blocks is filled with manually compacted soil with the same composition as the blocks. The dry density of this fill can be estimated at 1000 kg/m^3.

This experiment implied closed conditions for about 4 months, which led to the water content and temperature distributions shown in Figure 5. The outermost part of the pellet filling (21–23 cm) had been largely water saturated, while the blocks near the heater had dried.

From October 2002 the filters were filled with water under a maximum pressure of about 60 kPa.

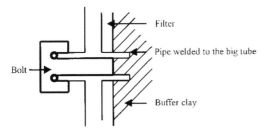

Figure 4. Schematic view of arrangement for taking samples at desired times. The circles in the bolt-head represent an O-ring sealing. After sampling a new suitably prepared clay core is pressed in.

This activated a few moisture sensors like in the Retrieval Test . As shown in Figure 6 the recorded moistening subsequently started to decrease due to diffusion of the intruded water into the surroundings but even after several months the recorded values are higher than in the noninstrumented clay. About one year after the start of wetting all the gauges gave the same value indicating at least 93% RH. In this experiment the true hydration rate is being determined by sampling through specially prepared openings in the steel cylinder with replacement of the samples by new, suitably cured core samples (Figure 4).

2.2 Major findings

Both the Retrieval and Czech Mock-up tests indicate that rock water under very moderate pressure penetrates the buffer along cables and joints between compacted blocks in the first wetting phase. Self-sealing is hence not quick enough to stop it. For the

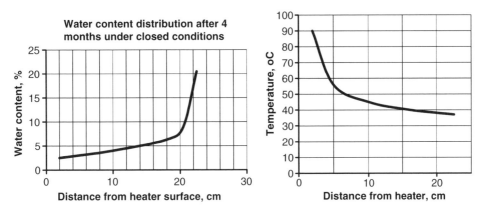

Figure 5. Left: Water distribution in sample extracted at the end of the "closed" phase (4 months). Right: Temperature distribution at the end of the "closed" phase (4 months).

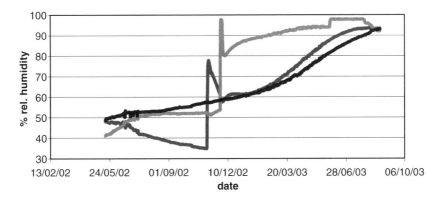

Figure 6. Recorded moistening in the Czech Mock-up experiment. Lowest curve at the start (green) = uppermost part of the clay column; Central curve at the start (violet) = level just below the heater; Uppermost curve at the start (blue) = lowermost part of the clay column (After Cechova et al).

Mock-up test it was assumed that the redistribution of porewater in the first 4 months under closed conditions would make the pellet fill and the most shallow part of the blocks so tight that piping would not take place through the joints between the blocks when water was added. It is therefore believed that the water penetration took place along cables and tubings. Careful attempts of sealing the space around the cables by use of clay powder or paste were not successful in these experiments.

3 ELECTRICAL PHENOMENA IN HYDRATING BUFFER CLAY

3.1 *General*

A DC potential of 0.1 to 1 V/cm is known to drive water through clay soil so effectively that dehydration at the anodes makes the soil desiccate and fissure while wetting at the cathodes causes porewater overpressure and a substantial increase in water content (Casagrande 1953). The opposite to this electro-osmotic effect is the generation of a DC potential by imposed or natural water flow through clay soils. Such effects can be used in practice, for instance for stabilizing deep trenches, and they are known to have an impact on the corrosion of metal objects in the electrical field.

Realization of the electrokinetic/hydrodynamic coupling is believed by the authors to be significant also in recording changes in temperature, water content and ion distribution in buffer and backfill

tests because electrical potentials are generated that should have an impact on microflow in smectitic clay. Their importance with respect to accuracy and relevance of recorded signals emitted by gauges in the buffer and to corrosion phenomena is not yet known but needs to be investigated.

3.2 *Theory*

Following Olsen (1961) one can derive the impact of electrokinetic coupling to hydrodynamics from Onsager's general phenomenological relationships for irreversible phenomena [2]. The following equations apply:

$$q_w = L_{11}(P/L) + L_{12}(E/L) \qquad (1)$$
$$q_e = L_{21}(P/L) + L_{22}(E/L) \qquad (2)$$

where:
L = element length
q_w = water flow rate
q_e = electrical current flow rate
P/L = hydraulic gradient
E/L = electrical potential gradient
$L_{12} = L_{21}$
$L_{11}, L_{12}, L_{21}, L_{22}$ = phenomenological coefficients

Eq. (1) implicitly expresses the hydraulic conductivity and it is obvious that if the influence of electrokinetic coupling is negligible or absent, the electrical flow potential term vanishes and the

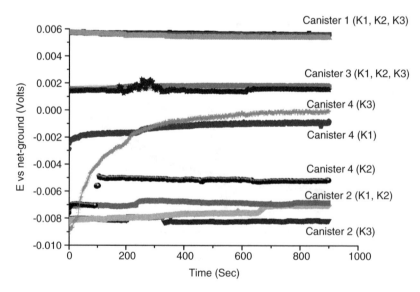

Figure 7. The potential measured versus net-ground of the Prototype canister system. K1, K2 and K3 represent cables 1, 2 and 3 in each canister.

expression gets the form of Darcy's law, i.e. Eq. (3):

$$q_w = L_{11}(P/L) \qquad (3)$$

The influence of electrical coupling on hydraulic flow may hence be expressed by the difference between Eqs.(1) and (2), which yields Eq. (4):

$$(q_2 - q)/q = L_{12}/L_{11}(E/P) \qquad (4)$$

where:
$q_2 =$ flow rate according to Eq. (1)
$q =$ flow rate according to Eq. (4)

The maximum retardation in flow rate due to electrokinetic coupling occurs when the percolate is electrolyte-free. According to Olsen (1961) it is less than 3.3% for dense illite clay in Na form. Assuming that the effect is proportional to the cation exchange capacity one would assume that the corresponding effect on dense smectite clay can be about 10% and hence of some but not very great concern. However, while most of the porewater in illite is mobile under commonly acting hydraulic gradients only some 10–20% is mobile in dense buffer clays and most of it takes place along the surfaces of the stacks of lamellae. This suggests that the coupling of electro-kinetics and microhydraulics is very strong and that the retardation of flow can be even higher than 10%.

3.3 Field data

At Äspö, DC potentials to ground were measured on four of the canisters in the Prototype Repository Project. The measurements show that two of them have a potential that is higher than net-ground, while the others have a potential that is lower than the net-ground. EIS measurements show that the measured system is complicated from electrochemical points of view and that the evaluation is not straightforward. An obvious fact is that the potential is different in holes with no instrumentation in the buffer and in holes where the buffer is instrumented, which means that the instrumentation definitely plays a role. This is exemplified in Figure 7.

4 CONCLUSIONS

Instrumentation of buffer and backfill in experiments can affect major processes or generate artefacts. One problem is that water leakage can take place along cables and tube connections giving the impression of much earlier wetting than in the buffer without instruments. This may have taken place in two of the test series at Äspö, the "Retrieval Test" and the "Proto-type Repository Project". Here, leakage along joints between buffer blocks may have taken place as well.

Another phenomenon is the generation of electrical potentials in the buffer in the course of the wetting with a probable impact on the performance of instruments placed in the buffer. Such a potential exists between the copper canisters and the earth connection in the Äspö experiments implying three effects: (1) The flow of water in the buffer is counteracted meaning that the recorded wetting rate is lower than predicted by use of Darcy models, (2) Metal gauges placed in the buffer act as electrodes and may undergo corrosion by which the recordings are affected, and (3) Canister corrosion may accelerate.

REFERENCES

Casagrande L, 1953. Review of past and current work on electro-osmotic stabilization of soils. Harvard Soil Mechanics Series No 45, USA.

Ledesma A, 2002. Contribution to the Prototype Repository Project STR-3 report. SKB, Äspö.

Olsen H. W, 1961. Hydraulic flow through saturated clays. Dr Thesis, Civ. Eng., MIT, USA.

Pacovsky J, 2002. Mock-Up-CZ: General Report. Centre of Experimental Geotechnics, Czech Technical University in Prague, Faculty of Civil Engineering.

Pusch R, Svemar C, 2001. Selection of THMCB models. Äspö Hard Rock Laboratory. Int. Progr. Report IPR-01-66.

Advances in Understanding Engineered Clay Barriers – Alonso & Ledesma (eds)
© *2005 Taylor & Francis Group, London, ISBN 04 1536 544 9*

Examples of the performance of tunnel plugs with and without recesses – theory and practice

Roland Pusch
Geodevelopment AB, Lund, Sweden

Lennart Börgesson
Clay Technology AB, Lund Sweden

ABSTRACT: Plugs are required for temporary or permanent confinement of backfills in repository tunnels and shafts. Several types have been constructed and tested and two of them are described in this paper, which starts by considering principles for locating and designing plugs. The selection of a suitable location depends on the required plug performance. The design must lead to sufficient tightness and strength to support contacting backfills. So far identification and modelling of degrading processes have been neglected.

One of the concrete plugs described in the paper consisted of two connected units in a blasted tunnel without recesses in the rock but with bentonite "O-ring" seals at the contact with the rock. This contact became tight even at 3 MPa water pressure and the plugs became tight in the course of the almost two year long testing time, while the outflow through the adjacent rock was considerable. The other plug, which also had a bentonite seal component, extended into a peripheral recess. It leaked from start and continued to do so at a successively reduced rate. The flow paths through the plug have not been identified but may be represented by fissures in the concrete.

In the first case a rather few discrete fractures in the excavation-disturbed zone (EDZ) around the plugs were responsible for the larger part of the outflow from the pressurized space. In the second case it is not known where the outflow, which is considerably smaller, actually takes place.

The main conclusion from the tests is that extending plugs into sufficiently deep recesses effectively cuts off the EDZ, and that bentonite seals prevent flow along the plug/rock contact.

1   INTRODUCTION

Plugs are required for confining backfills in repository tunnels and shafts in the waste application phase and in the final phase of closing the repository. Temporary plugs that only need to provide support and tightness for weeks and months can be of simple type, like shotcrete. However, in practice it may be required that many temporary plugs, like permanent ones, must be very strong and tight even for rather short periods of time.

A basic plug case is represented by a bulkhead that must withstand the swelling and water pressures exerted by the contacting backfill. The pressures combine to give a force of more than 10000 t in a KBS-3 repository, which makes it necessary to design the plug so that concrete and reinforcement stresses are acceptable, and to prevent slip of the plug along the plug/rock contact. Concrete plugs can be cast against the tunnel walls over a sufficiently long distance to transfer the force to the rock and be equipped with a grout curtain to seal off the EDZ. Alternatively, the plugs can extend into recesses for transferring the force to the rock and for cutting off the EDZ. A plug test of the firstmentioned type was made as part of the Stripa Project while plugs of the latter type have been constructed in the Äspö URL (HRL). The paper describes construction, theoretical modelling and actual performance of such plugs.

2   LOCATION OF PLUGS

Plugs for isolating backfills in KBS-3 tunnels should be located in fracture-poor rock for causing minimum groundwater flow along the plug and this principle was followed in one of the cases, i.e. the Backfill and Plug Test at Äspö, described here. In the other case, the Stripa test, the plug was placed where a diabase dike with poor contact with the granitic mass intersected the site for testing the sealing potential of the bentonite "O-ring" that was part of the plug construction.

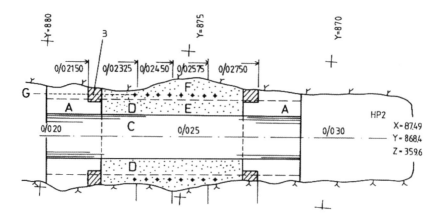

Figure 1. Longitudinal section of the Stripa twin plug. "A" is concrete. "B" represents "O-ring" seals with a cross section area of $0.5 \times 0.5\,m^2$. "C" is a 1.5 m diameter steel tube, which, together with 7 steel anchors kept the plugs together. "E" is a 6 m long sand-filled chamber and "F" a perforated helical pipe for injecting water.

3 DESIGN PRINCIPLES

An important question is the time during which the plug has to serve as an effective isolation. It requires investigation and assessment of the long-term physical and chemical interaction of the plug, backfill and rock. So far only concrete plugs have been used in SKB's underground research laboratories with focus on stability and tightness under high axial pressure, without considering chemical degradation. While the design of the concrete body and reinforcement is a relatively simple issue, the problem of anchoring the plug in the rock without causing stress conditions that favour flow along or close to the plug is more difficult. Thus, the very strong force acting on the plug must be transferred to the rock without causing significant fracturing or displacement. The rock structure and mechanical properties of the rock as well as the initial stress conditions must be known for making the design.

4 THE STRIPA PLUG

4.1 General

The Stripa plug test was made to investigate how a plug can be made for hydraulic separation of a richly water-bearing fracture zone from a backfilled tunnel while making it possible to pass through it (Pusch et al, 1987). It was constructed in normally blasted granitic rock at about 360 m depth and consisted of two concrete plug units held together by steel rods and a 1.5 m diameter steel tube. The plugs were equipped with "O-ring" seals of highly compacted MX-80 bentonite blocks (dry density $1750\,kg/m^3$) placed in recesses in the plugs for eliminating flow

along the rock/concrete interface. The space outside the tube between the concrete units was filled with water-saturated sand, thus simulating a pervious fracture zone. Testing of the tightness was made by pressurizing water in this space and measuring the outflow form it (Figure 1).

4.2 Design

The concrete plugs were cast directly against the cleaned, irregular rock surface of the 4 m wide and high drift with about $16\,m^2$ cross section area using wooden forms prepared to closely fit the rock surface. Seven large bolts served to take a tension force of 1000 t each and they were prestressed stepwise parallel to the successive increase in water injection pressure. Concrete prepared by use of commercial Portland cement, and ordinary reinforcement steel, were used for preparing the plugs. The design implied three-fold safety with respect to failure of concrete and steel components.

4.3 Instrumentation

The most important instrumentation was the pressure-controlled water injection system and associated flow meters with different gauges for different flow intervals. Gloetzl pressure cells were installed for measuring the total pressure in the sand chamber (0–3 MPa) and at the rock/bentonite contact, and extensometers were used for recording axial movements of the concrete plugs relative to the rock and the radial strain of the steel tube. The radial load on the rather thin steel tube implied a risk of buckling at the highest injection pressure 3 MPa.

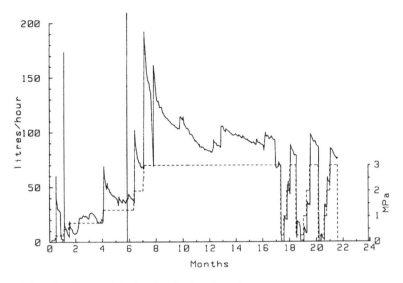

Figure 2. Recorded outflow from the injection chamber into the rock.

4.4 Results

4.4.1 Flow measurements

Recording of the amount of water pumped into the sand chamber per time unit and collected in sumps adjacent to the outer ends of the plug units gave the outflow from the sand-filled chamber and a basis for evaluating the tightness of the plugs (Pusch et al, 1987). The pressure was increased in steps, 0.25, 0.50, 1.0, 2.0 and 3 MPa. Each pressure was maintained constant for about 2 months except in the earliest part of the experiment. When the first pressure, 0.25 MPa, was applied water leaked out at the plug/rock contact but stopped in a few weeks and did not occur in the rest of the experiment. The maximum pressure 3 MPa was maintained for about 10 months, yielding a successively dropping flow. After 17 months the injection pressure was reduced to a very low value and then increased again to 3 MPa and this cyclic change was repeated twice. The diagram in Figure 2 shows the outflow in liters per hour, starting from about 190 l/h and dropping to about 75 l/h, or about 38 l/h per plug, at the end of the test.

4.4.2 Evaluation of the hydraulic conductivity of the near-field rock

FEM calculations were made for estimating the average hydraulic conductivity of the near-field rock. Very good agreement between recorded and actual outflow was obtained by taking the rock to be a porous medium with the hydraulic conductivities according to Figure 3. The conductivities agree well with those obtained from separate large-scale hydraulic tests in Stripa granite.

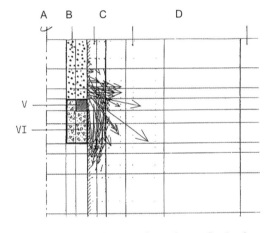

Figure 3. Outflow in vector form of water flowing from the injection chamber at 3 MPa pressure according to a FEM calculation. A (chamber sand) has $K = E-4$ m/s, B (primary EDZ) has $K = E-8$ m/s, C_{axial} (secondary EDZ) $K = E-9$ m/s, C_{radial} $K = E-11$ m/s. D (undisturbed rock) has $K = E-10$ m/s.

The good agreement should not be taken as a proof of the validity of the porous-media model. Thus, clay-grouting of 8 identified water-bearing fractures from 5 boreholes that were strategically positioned to intersect the fractures (Figure 4) gave a reduction to about 1/3 of the value of ungrouted rock (Pusch 1994). The fractures are believed to have been hydraulically activated by the blasting.

In conclusion, the major part of the outflow from the pressurized chamber took place through the disturbed zone with $K = E\text{-}8$ m/s, hence showing that for blasted tunnels this zone has to be cut-off in order to obtain a tight plug.

5 THE ÄSPÖ HRL PLUG

5.1 General

The Äspö plug was of arch type cast in an about 30 m long normally blasted drift with 25 m^2 cross section

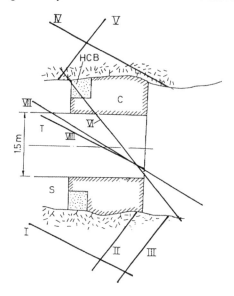

Figure 4. Schematic horizontal section through the plug and the near-field rock with discrete fractures that were grouted successfully.

in granitic rock. The plug was located in fracture-poor rock and extended into a 1.5 m deep recess in the rock. The plug was equipped with a peripheral bentonite seal, similar to that of the Stripa plugs (Gunnarsson et al, 2001). The large size of the plug required cooling pipes in the concrete for keeping the temperature down and hence minimizing cracking. Water pressure on the plug has been applied at different stages and in the last years the pressure has been 0.5 MPa and its tightness evaluated by recording the flow into a sump at the outside of the plug, the flow through the rock taken to be negligible.

5.2 Design

5.2.1 Requirements
The plug was required to (1) provide mechanical support to the compacted backfill material, (2) sustain a high hydraulic pressure, (3) provide a tight seal between the backfilled part and the open part of the tunnel, and (4) cut off the EDZ.

5.2.2 Flow calculations as design basis
FEM flow calculations showed that the hydrostatic pressure acting on the plug may amount to 1.8–2.5 MPa. It depends on the conductivity of the rock and the plug materials (concrete and bentonite), as well as on the hydraulic properties of the EDZ and the plug/rock interface. An example is shown in Figure 5 (Gunnarsson et al, 2001).

5.2.3 Mechanical calculations as design basis
A 3.2 MPa hydrostatic pressure was assumed in the FEM stress analysis. The calculations showed that the plug should have a parabolic shape towards the open tunnel in order to minimize the volume and reinforcement. The thinnest, central part of the plug

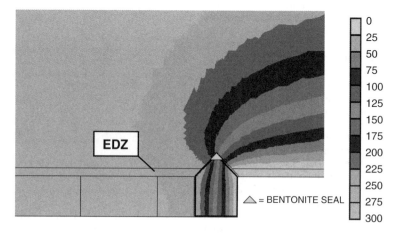

Figure 5. Hydraulic head contours in meters around the axisymmetric plug keyed into the rock. The pressure on the plug is 2.44 MPa.

needs to be 1.2 m thick. The area of loaded rock, i.e. the surface of the recess on the open side, should have 39° inclination relative to the tunnel axis for optimal rock stability as concluded from numerical analyses.

5.3 *Selected design and construction sequence*

Figure 6 shows the selected plug design, implying casting in two steps for finding a practical solution of the problem of applying the bentonite blocks. The aim to minimize the amount of concrete by giving the plug a concave shape led to high stresses and a high ratio of the reinforcement and concrete volumes. This led, in turn, to practical difficulties in placing the reinforcement bars and to a risk of incomplete filling of the form with concrete.

The recess has triangular cross section. The diameter of the tunnel is about 5 m and the diameter of the outer edge 8 m, which gives an average depth of the recess of 1.5 m. The "O-ring" bentonite seal, which has a 0.5 m × 0.5 m square cross section, consists of highly compacted blocks. The target swelling pressure was set at 5 MPa for effective

closing of the various joints. For this purpose the dry density of the blocks was 1600 kg/m^3.

Evaluation of the tightness of the plug made it necessary to avoid excavation damage of the rock at the construction of the recess. It was therefore excavated by use of core drilling of closely located holes for some subsequent smoothing of the wavy surface. A special rig was used for accurate direction of the drilling aggregate and for making a series of equally oriented holes at a time. The excavation of the recess was made by drilling of 340 holes with 0.8 m length., 27 mm rock remaining between adjacent holes after drilling. Release of the rock had to be made by charging the holes with Dynamex and Gurit (70 g/hole).

Figure 7 shows the recess surface in the edge region after final reworking. Figure 8 shows the geometry and appearance of the bentonite blocks, the total number being 1300.

An amount of 65 m^3 concrete BTG 1 K50 was used for the plug body. The concrete was pumped through

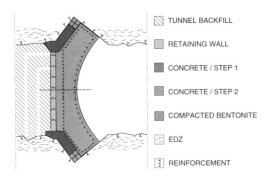

TUNNEL BACKFILL

RETAINING WALL

CONCRETE / STEP 1

CONCRETE / STEP 2

COMPACTED BENTONITE

EDZ

REINFORCEMENT

Figure 6. Longitudinal cross section of rotationally symmetric plug.

Figure 7. Recess surfaces after reworking.

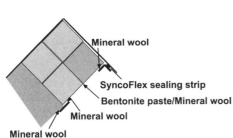

Mineral wool

SyncoFlex sealing strip

Bentonite paste/Mineral wool

Mineral wool

Mineral wool

Figure 8. Geometry and appearance of the bentonite blocks. The mineral wool was used for preventing cement slurry to flow into the narrow space between the blocks and the rock.

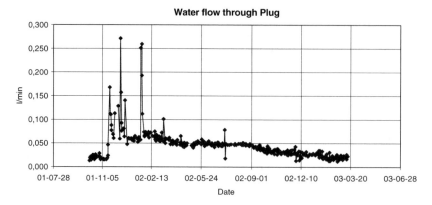

Figure 9. Recorded outflow from the pressurized drift.

holes at different levels in the form. In the lower parts the concrete was vibrated by common tools while a special vibrator attached to the rock was used for vibrating the uppermost part.

5.4 Instrumentation

Gloetzl cells were installed on the surface of the bentonite "O-ring" in the upper and lower parts of the form for measuring the total pressure, i.e. the sum of the swelling (effective) pressure and the water pressure. The most important recording is the frequent measurement of water collected in the sump at the front of the plug.

5.5 Results

5.5.1 Observations
Visual observation of water appearing to have penetrated the concrete and the adjacent rock has been made frequently. After reaching 0.5 MPa pressure water penetrated the plug front in a few spots, the general tendency being a successive reduction in outflow. Leakage along the plug/rock contact has taken place continuously from this event and minor outflow from fine fractures in the rock close to the plug. It is not possible to draw safe conclusions concerning the flow paths from the observations.

5.5.2 Flow measurement
The recorded outflow from the plug is shown in Figure 9 (Goudarzi et al, 2003). Large variations took place in the first three months but the flow then stabilized and successively dropped from about 0.05 to 0.02 liters per minute, i.e. from 3 to 1.2 liters per hour.

6 DISSCUSSION AND CONCLUSIONS

A first issue of interest is to compare the outflow from the pressurized space of the two plug constructions,

i.e. about 38 l/h at 3 MPa pressure for each of the Stripa plugs and 1.2 l/h at 0.5 MPa for the Äspö plug. Assuming proportionality, the outflow from the latter plug would be about 7 l/h at 3 MPa pressure, i.e. about 20% of that of the Stripa plugs. Keeping in mind also the larger cross section area of the Äspö plug, which makes the difference even greater, one concludes that extending a plug into the rock by 1.5 m effectively cuts off the EDZ in normally blasted rock for these drift sizes.

A second matter is the efficiency of bentonite seals. For the Stripa plug it is concluded that it totally stopped leakage along the rock/plug contact. If this was the case also for the Äspö plug the outflow from this plug must have taken place through the concrete.

REFERENCES

Goudarzi R, Gunnarsson D, Johannesson L-E, Börgesson L, 2003. *Backfill and Plug Test. Äspö Hard Rock Laboratory*. Internat. Progress Report IPR-03.14.

Gunnarsson D, Börgesson L, Hökmark H, Johannesson L-E, Sandén T, 2001. *Report on the installation of the Backfill and Plug Test, 2001. Äspö Hard Rock Laboratory*. Internat. Progress Report IPR-01.17.

Pusch R, Börgesson L, Ramqvist G, 1987. *Final report of the borehole, shaft and tunnel sealing test – Volume III: Tunnel plugging. Stripa Project* TR 87-03. SKB, Stockholm.

Pusch R, 1994. Waste Disposal in Rock. *Developments in Geotechnical Engineering*, 76. Elsevier Publ. Co (ISBN0-444-89449-7).

Advances in Understanding Engineered Clay Barriers – Alonso & Ledesma (eds)
© 2005 Taylor & Francis Group, London, ISBN 04 1536 544 9

Emplacement of bentonite blocks and canisters in a KBS-3V fashion – how feasible is the proposed technology?

Christer Svemar

Svensk Kärnbränslehantering AB (Figeholm, Sweden)

ABSTRACT: Performance assessment of the KBS-3 concept starts from an ideal conceptual geometry of a deposition hole with one canister centrally located in a cylinder of homogeneously swollen bentonite buffer. The buffer is deemed to have a saturated density of 1.9–$2.15 \, \mathrm{Mg/m^3}$ in order to provide needed mechanical and chemical properties from start and onwards. The methodology for emplacement to achieve this has been considered to comprise lining of the deposition hole with bentonite blocks and emplacement of the canister in a hole in the centre. The required slot between the bentonite blocks and the rock for installation of the blocks was assumed to be so large that it has to be filled with pellets of bentonite. High initial degree of saturation would be beneficial for a high thermal conductivity.

This conceptual methodology was tested in the installation of the Prototype Repository, with a successful result, but also with question marks raised. The simple idea was to load bentonite blocks with full diameter but only 0.5 m in height on top of each other in the hole, for which a slot of 50 mm against the rock wall is deemed needed. The inner hole would be absolutely vertical and should because of that only need a slot of 10 mm between the canister and the bentonite blocks. Considering 10 cylindrical blocks – on top of a full block – strict requirements would be placed on the deviation from the vertical of the deposition hole, the deviation of the bottom pad from the horizontal, the mm precision in placing each bentonite block, and the accuracy in positioning the canister – weighing 25 tonnes – over the centre hole.

The key to the success was the accuracy with which the boring of the deposition holes was made. Instead of an allowed deviation from the bull target of 25 mm over a depth of 8.5 m was the result less than 12 mm. If this accuracy can be obtained steadily can the bentonite blocks be made larger and the pellets filling in the outer slot excluded. Another major result is the indication that means can be provided for protecting the installed bentonite blocks from water and humid air in the deposition holes in such an efficient way that the installation process will not be the factor determining the maximum acceptable inflow of water into the deposition hole, but the demand on the safety of the repository in the long run. Raised question marks are which construction material to use and how to drain the deposition holes from inflowing water during emplacement.

The overall conclusion is that the earlier conceptual methodology can be developed into a practically feasible method, and that a major potential exists for refinement of details.

1 INTRODUCTION

The most important issue in the evaluation of the KBS-3V (V stands for "vertical") performance is the long term safety of the repository. Such performance assessment focuses on the "steady state" conditions which start at the time when the repository has been saturated and the natural groundwater table has been recovered. The bentonite buffer around the canisters is saturated and homogeneous, and the canister is located exactly in the centre of the buffer. The backfill in the tunnel has been saturated as well and fills the earlier open spaces in the tunnel completely, and has hydraulic properties which are equal to those of the surrounding rock.

The consequent objectives for the different activities taking places before the "steady state" conditions are achieved, like excavation, deposition, backfilling and sealing, are to arrive at these "ideal" conditions, as close as possible.

Many of the practical issues have been studied in the Prototype Repository project in the Äspö Hard Rock Laboratory /Pusch et al, 2000/.

The Prototype Repository (Fig.1) consists of two sections with four respectively two deposition holes with bentonite buffer and canister, the latter having electrical heaters. The sections are separated by a concrete plug, and the whole test is separated from the rest of the laboratory by an outer plug. The project has two objectives.

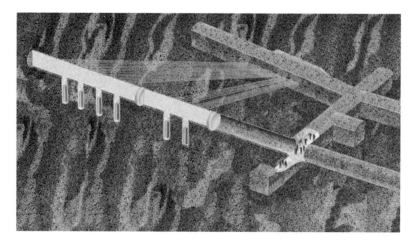

Figure 1. Prototype Repository. All cables are placed in lead-throughs with water-tight connections to the adjacent drift.

1. To demonstrate the integrated function of the deep repository components under realistic conditions and to compare results with models and assumptions.
2. To develop, test, and demonstrate engineering standards and quality assurance methods.

The latter objective is addressed in this paper.

The development, testing and demonstration of engineering issues concern methods and processes which start with boring of deposition holes and end with backfilled and plugged test tunnel. Two parameters are of special concern: temperature on the surface of the canister and saturated density of the bentonite buffer around the canister.

2 TARGET WITH RESPECT TO PERFORMANCE ASSESSMENT

The "ideal" geometry and property of the deposition hole at "steady state", i.e. saturated and homogenised buffer and backfill is shown in the schematic figure (Fig. 2). The bentonite buffer is exactly 350 mm thick around the canister and 500 mm below and 1500 mm above. The target bulk density in saturated stage is $2.0 \, Mg/m^3$, which equals a dry density of $1.57 \, Mg/m^3$. The backfill is designed to provide a swelling capacity which in the salt groundwater environment in the Prototype Repository tunnel (0.7% TDS) requires a bentonite-crushed rock mixture with 30% (by weight) bentonite. The expected average hydraulic conductivity in the rock around the future repository is 10^{-10} m/s or lower, having 10^{-10} m/s as a satisfactory property for a safety case.

3 PRACTICAL ASPECTS ON INSTALLATION

The emplacement technique has always been planned to start with the lining the hole with the bentonite buffer, where after the canister is installed and the top blocks are emplaced, the reason being to provide radiation shielding for man during the entire operation. This is not judged possible if the canister is emplaced first and the bentonite buffer thereafter, around it.

Installation tolerances during deposition have for a long time been assumed to require an outer slot between the bentonite block and the rock of 50 mm and an inner slot between the bentonite block and the canister of 10 mm /SKB, 1983/. One drawback with these slots is that the amount of bentonite in the deposition hole would be too little to reach the target density, and bentonite in the form of pellets or powder would be needed, powder having theoretical difficulties in being filled to the required density.

4 ACCURATE BORING OF DEPOSITION HOLES

The boring of the deposition holes is the key to the whole process. Proven technology was adapted to the specific circumstances and demands of KBS-3V. The technical feasibility of boring holes was, in a first step, tested in the Research Tunnel at Olkiluoto, Finland. For that test a raise boring machine was used. A 5 feet in diameter bore head was pushed downward enlarging the pilot hole to full size in one step. The drawback with the actual raise borer used

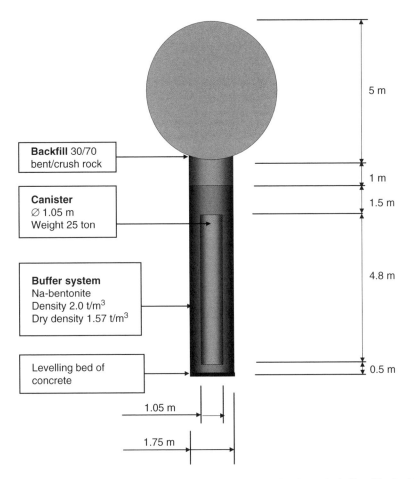

Figure 2. Idealised view of a deposition hole with canister absolutely centred in a bentonite buffer with a backfilled tunnel having a hydraulic conductivity that equals the conductivity of the surrounding rock.

was that the force acting on the wall of the hole was much lower than the expected one in case the maximum force the cutters can take would be applied. The result was, however, very successful and when the set of deposition holes should be excavated in the Äspö HRL the similar technique was considered but with the requirement that full force could be applied (for creating a repository-like EDZ). The least expensive equipment was a converted TBM machine and not a converted raise borer. The very precise requirement of machine performance also made the choice straightforward. It was namely requested that the machine should be able to steer the head in the hole so that the deviation of the hole centre would be no more than 25 mm from the tunnel floor to a depth of 8.5 m. The straightness of the hole was specified to be a maximum deviation of the centre point of 16 mm calculated as the deviation from a theoretical centre line from start and bottom centre points. This value

included a 10 mm sudden change of the centre point in the hole, a change that could happen every time a new casing was put in work.

How was the outcome? The answer is: Better than requested. Of all 13 holes bored no one had a larger deviation between the start and bottom centre points than 12 mm with 5 mm as an average value. Also the other criteria were met. /Andersson et al, 2002/. In some cases became the need of accuracy more demanding than expected from the beginning due to the installation of ultrasonic sound devices, which were installed before the boring of the deposition holes with the task to measure the change in velocity in the EDZ during boring. As this zone is just 20–30 mm wide the deviation had to be less than 25 mm, even if the starting point of the boring machine hit the bull target precisely. Otherwise the ultrasonic ray path would pass outside the EDZ. However, the Robbins machine manufactured in Solon, Ohio, USA

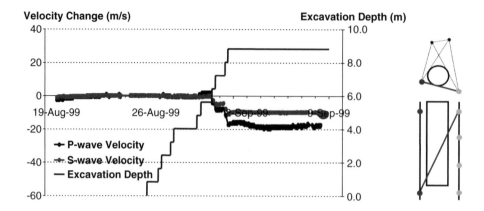

Figure 3.   Ultrasonic ray path (red in the figure to the right) through the EDZ before, under and after boring past the ray path level. Black in the left figure denotes P-wave and red S-wave velocity. Velocity in solid Äspö granite is 5900 m/s and 3400 m/s for P-wave and S-wave respectively.

yielded the wanted accurate performance. How do we know? Figure 3 shows the result of one of the ultrasonic measurements, which changes the velocity when the bore head passes the level of the ray path.

If the experienced low value can be achieved the bentonite blocks can be made a bit wider so that the final density after saturation is achieved without using bentonite pellets in the outer slot. Besides the advantage of excluding the handling of pellets is the exclusion contributing to a more homogeneous buffer around the canister. With only bentonite blocks is the amount of bentonite very homogeneously distributed around the canister. With pellets is even a very small inclination of the hole resulting in an inhomogeneous distribution of bentonite around the canister; more pellets on the side where the slot is the widest.

when backfilling of the tunnel is made. Even small inflows create a water table in the bottom of the hole and have to be directed to a water sump. By casting the bottom pad so that a ditch is made all around it a small sump can be made in this ditch. Drainage can be made by means of vacuum suction as the distance to the tunnel floor is approx 8.5 m. There must, however, be room for the hose with a couple of tenth of mm in diameter in the slot.

The installation in the Prototype Repository took extra long time because of all instruments in the bentonite blocks which were installed once the block was put in place, but the drainage system performed well in all six holes.

For references see /Pusch et al, 2001, Boergesson et al, 2002/.

5   PREPARATION OF DEPOSITION HOLE PRIOR TO EMPLACEMENT

The deposition hole is somewhat concave in the bottom because of the shape of the bore head, and the bottom block needs to be placed on an even surface in order not to break. This surface has to be absolutely horizontal, so that the lining of bentonite blocks can be built up vertically. The material to use in the levelling of the bottom has in earlier discussions focused on either bentonite powder or cast concrete. In pre-tests before installation of the Prototype Repository it was concluded that only concrete can be used and that the surface of the bottom slab need to be cast with a liquefied type that by gravity provides a horizontal surface.

Another complication is the inflowing water which has to be drained under installation and up to the time when the bentonite is allowed to start to swell, i.e.

6   PRACTICAL EMPLACEMENT OF BUFFER AND CANISTER

The inner slot of 10 mm created concern before installation because of the large number of blocks; 10 blocks along the canister having to be placed with mm precision, if not the whole tolerance should be consumed. Pre-tests with concrete blocks verified that the required precision was possible to obtain but that manual adjustment down-the-hole should be needed. The installation process was planned in accordance to these findings and the outcome fulfilled the expectations /Boergesson et al, 2002/.

The 10 mm slot for the canister and the vertical alignment of the inner hole in the bentonite block column was enough in all deposition holes. The deposition machine – handling the 25 tonnes canister – deposited each canister without any problem. The process is robust, which the two retrievals of already deposited canisters bear witness of. In those cases – one in the Canister Retrieval Test, and one in the Prototype Repository – were the electrical contacts with some of the heaters lost when the canister was in place and all connections made. In the Canister Retrieval Test was this discovered before the top blocks had been emplaced but the deposition machine removed. The canister was taken up and sent back to the manufacturing plant, mended and redeposited. In the Prototype Repository were also the three top blocks with instruments emplaced. Nevertheless was also this canister taken up (after removal of the top blocks) and re-deposited. Consequently has the reversibility of the disposal process been tested, but, of course, only in a non-radioactive environment.

7 PRACTICAL EMPLACEMENT OF BACKFILL

The backfilling aims at filling the tunnel above the deposition holes with a material that has a swelling capacity toward the roof during a long time after closure. The rationale is that soil or sand materials settle with time, as the pore volumes between the particles decrease with time. In an underground tunnel in granite this results in the development of channels in the roof region (rock creep is not always fast enough to balance the volume change of the backfill). The solution chosen is to use bentonite, and to mix it with sand or crushed rock for economical reasons. The content of bentonite depends on the salt content of the ground water. In fresh water 10% bentonite is satisfactory for swelling, while swelling needs approx 30% bentonite in water with 0.7% salt (TDS) /Boergesson et al, 2001/.

The technique for emplacement of backfill was developed in the Stripa project resulting in two steps, the first being to fill the bottom half of the tunnel with horizontal layers, which are compacted by a vibrating device, and the second to compact, also with a vibrator, an inclined layer up to the roof. The Stripa test with blowing-in the material in the top part did not provide a high enough density for decreasing the hydraulic conductivity to the level of the surrounding rock /Pusch et al, 1985/. The Strip Mine was, however, very dry.

In Äspö HRL the process had to be changed to backfilling of inclined layers only, because of water dripping from the roof, which created such a slippery surface on a horizontal backfill that it could not work as a base for backfilling of the upper part of the tunnel. The experienced gained in the Backfill and Plug Test /Boergesson et al, 2002/ was utilised and the backfilling of the whole tunnel area in one step worked to satisfaction. The slope angle was set to

35 degrees to the horizontal, which by rule-of-thumb is the angle of repose for crushed rock.

8 PROBLEMS WITH WATER INFLOW

The bentonite for the tests was made with a high water ratio – 17% compared to 10–12% in natural condition – because of the ambition to provide as much water as possible in the deposition hole from start. (More water could also be added by filling all slots with water artificially). These blocks suck water from a humid air, if the relative humidity is above approx. 75% and alternatively shrink if the relative humidity is lower. The experience with bentonite blocks in "non-dry" holes is that the relative humidity quickly becomes higher than the 75% and that conditioning of the air is difficult to do once the blocks have been emplaced. (During emplacement air conditioning was, however, done in the Prototype Repository.)

The rate of swelling was observed in the Canister Retrieval Test, where a floating plug of steel on the top block was anchored to the rock by cable bolts. The swelling was then measured as both the pressure on the plug and as the strain and extension of the cables. The result was that swelling developed quicker than was expected, which creates a time squeeze in a repository.

In order to prevent any swelling during the fairly long time it took from installation to backfilling a "containment" was created around the bentonite blocks by installing a plastic bag, which was attached in the bottom pad and covered the bentonite package. Inside the bentonite it itself conditioned the air to the correct relative humidity. The plastic bag was removed just before the outer slot was to be backfilled with pellets. This worked well also in the heavily instrumented holes, with one exception where horizontal displacement sensors were attached to the rock wall through the plastic sheet. The installed knives did not rip-off the plastic as planned and a lot of extra efforts was necessary in this particular case. This solution indicates the possibility to develop a cover around a bentonite package with the canister inside, that would allow for very long time spans between deposition and backfilling of the tunnel.

From the Backfill and Plug Test /Gunnarsson et al, 2001/ it was concluded that the quality of the backfill, when instruments take time to install, can tolerate no more than an inflow of approx. 5 l/min over a tunnel length of 15 m. Dripping from the roof has to be even lower than that. Sections exceeding these values had to be pre-treated by drainage and diversion of the water inflow. Postgrouting is not a feasible alternative and the question in a repository will be were the limit is and how much engineering that can be tolerated. Conceptual solutions exist but they have to be tried out. Such activities are scheduled to take place in Äspö HRL in the near future.

9 CONCLUSIONS

The Prototype Repository project has verified

1. that boring of deposition holes can be made with the requested quality of holes both with respect to geometry and with respect to working environment. The conclusion is that the geometry of the hole can be made even better than requested in the Prototype Repository project, which would improve the quality of the bentonite buffer
2. that most water needed for saturation of the bentonite buffer can be available in the deposition hole from the beginning
3. the assumption that bentonite blocks should be made as large as possible in order to shorten the emplacement time and to facilitate the achieving of needed geometry of the installed bentonite
4. that betonite pellets can be manufactured and installed in the slot between betonite blocks and the rock in a way that builds up a homogeneous filling
5. that backfill can be in-situ compacted with a mixture of bentonite and crushed rock to the quality needed for yielding the expected average hydraulic properties of repository rock in Sweden,

but the project has also indicated that

1. water inflow causes major problems for the emplacement and backfilling
2. the bentonite blocks swells quicker than assumed in the high humidity environment that is developed in the deposition hole due to water inflow,

and provides routes for improvements such as

1. system for drainage of water inflow during installation
2. system for protecting the bentonite from too high relative humidity in the deposition hole
3 means of avoiding the need of bentonite pellets without jeopardizing the quality of the bentonite buffer at "steady state".

ACKNOWLEDGEMENTS

The work reported has been conducted by SKB and the partners of Äspö International Cooperation. The Prototype Repository Project is co-funded by the European Commission and performed as part of the fifth EURATOM framework programme, key action Nuclear Fission (1998–2002).

REFERENCES

Andersson C, Johansson AE, 2002. Boring of full scale deposition holes at the Äspö Hard Rock Laboratory – Operational experience including boring performance and a work time analysis. Technical Report TR-02-26, Swedish Nuclear Fuel and Waste Management Company, Sweden.

Boergesson L, Johannesson L-E, Gunnarsson D, 2001. Backfilling of the tunnel in the Prototype Repository. Results of pre-tests. Design of material, production technique and compaction technique. Äspö Hard Rock Laboratory International Progress Report IPR-01-11. Swedish Nuclear Fuel and Waste Management Company, Sweden.

Boergesson L, Gunnarsson D, Johannesson L-E, Sandén T, 2002. Prototype Repository Installation of buffer, canisters, backfill and instruments in Section I. Äspö Hard Rock Laboratory International Progress Report IPR-02-23. Swedish Nuclear Fuel and Waste Management Company, Sweden.

Gunnarsson D, Boergesson L, Hoekmark H, Johannesson L-E, Sandén T, 2001. Report on the installation of the Backfill and Plug Test. Äspö Hard Rock Laboratory International Progress Report IPR-01-17. Swedish Nuclear Fuel and Waste Management Company, Sweden.

Pusch R, Nilsson J, Ramqvist G, 1985. Final report of the Buffer Mass test – Volume 1: Scope, preparative field work, and test arrangements. Stripe Project Technical Report TR 85–11.

Pusch R, Svemar C, 2000. Prototype Repository, Project description. Äspö Hard Rock Laboratory International Progress Report IPR-00-30. Swedish Nuclear Fuel and Waste Management Company, Sweden.

Pusch R, Andersson C, 2001. Preparation of deposition holes prior to emplacement of buffer and canister in Section I. Äspö Hard Rock Laboratory International Progress Report IPR-01-64. Swedish Nuclear Fuel and Waste Management Company, Sweden.

SKB, 1983. Final Storage of Spent Nuclear Fuel – SKB 3. Volume I General. Swedish Nuclear Fuel and Waste Management Company, Sweden.

Advances in Understanding Engineered Clay Barriers – Alonso & Ledesma (eds)
© 2005 Taylor & Francis Group, London, ISBN 04 1536 544 9

The Mock-Up-CZ project: a demonstration of engineered barriers

Jaroslav Pacovský
Czech Technical University, Prague (Czech Republic)

ABSTRACT: On 7th of May 2002, the Mock-Up-CZ experiment was launched at the Centre of Experimental Geotechnics, CTU in Prague. It represents a vertical model of a buffer mass test of a Czech Ca-Mg bentonite mixture in the KBS-3 modification (Swedish system). The experiment has been run in several phases with a presumed time duration of 5 years. During its individual phases, the temperature pattern, swelling pressure and saturation degree are continuously monitored. The running experiment is freely accessible on the web-site http://ceg.fsv.cvut.cz. The results of the experiment should verify certain hypotheses or assumptions formulated on the basis of observations made during the test, characterize the Mock-Up components and help to better understand the phenomena occuring during successive operational phases.

1 INTRODUCTION

Research on a bentonite-based engineered barrier requires a non-traditional engineering approach. With respect to the extremely long-term time requirements for the rheological stability and safety of the system, all the available experimental tools and procedures must be used within the research. Here, physical modelling is an indispensable help. The most relevant model types applied have been found to be those made at a scale of 1:1, referred to as Mock-Up models.

Within the research of engineered barriers, a number of URL experiments and Mock-Up experiments have been implemented worldwide. The Mock-Up-CZ belongs to one of them.

The Mock-Up-CZ experiment was constructed at the Centre of Experimental Geotechnics (CEG), CTU in Prague. It is a vertical model of a buffer mass test of a Czech Ca-Mg bentonite mixture in the KBS-3 modification (Swedish system).

2 BASIC DESCRIPTION OF THE MODEL

The model is placed in an underground test silo with dimensions of $3000 \times 3000 \times 3000$ mm. The Mock-Up-CZ model itself consists of a steel tank (an oven) with a cylinder diameter $d = 800$ mm and a height of 2230 mm. The bottom and top covers are made of steel 50 mm in thickness. These individual parts are connected using 16 bolts. The system was designed to withstand an internal pressure of up to 5 MPa.

The canister containing highly radioactive waste is simulated using a heater. The heater consists of a steel cylinder with an outside diameter of 320 mm and a height of 1300 mm. Two thermal radiators have been installed inside the cylinder, each of a 1000 W maximum output. The model container content, the heating medium, is oil.

The potential inflow of granitic water from the natural barrier (it is expected that the Czech underground repository will be constructed in granitic rock) is simulated by using a flooding system installed on the inner side of the bin featuring two concentric filters, the one closer to the bin being coarse-meshed and the other having voids less than 100 µm. Water is let in through four vertical perforated tubes. Hydration system which allows an increase in hydration pressure of up to 1 MPa is placed outside the experimental silo. Synthetic granitic water is used for saturation (CEG/SKB Report, 2002 a).

3 BUFFER MATERIAL

Based on the evaluation of Czech bentonite research achievements, the most appropriate material for the buffer was found to be a mixture of treated bentonite (from the Rokle locality), silica sand and graphite. The basic mixture contains 85% of bentonite, 10% of silica sand and 5% of graphite. This mixture can be found in the Mock-Up-CZ in two different forms – highly compacted blocks (with a dry density of $\rho_d = 1700$ kg/m^3 and a swelling pressure of 3–5 MPa) and a hand-compacted mixture of the same composition as the blocks (dry density of $\rho_d = 1040$ kg/m^3 and swelling pressure of 370 kPa).

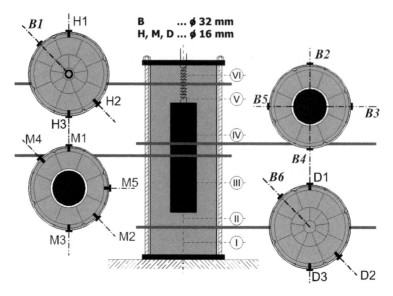

Figure 1. Points for taking core samples.

4 MODEL INSTRUMENTATION AND RECORDING OF PARAMETERS

The Mock-Up model is equipped with 52 thermometers, 50 hydraulic pressure cells, 37 humidity sensors of varying construction and 20 resistive tensiometers (see http://ceg.fsv.cvut.cz).

The construction of the model allows taking core samples of the buffer barrier from 6 positions with a diameter of 32 mm, and 11 positions with a diameter of 16 mm (Fig. 1).

Special filters installed in the Mock-Up serve for testing the buffer's gas permeability.

Temperature, swelling pressure and hydration measurements are taken continuously inside the bentonite barrier throughout the entire duration of the experiment, at 10 minute intervals. These checks are carried out at six horizontal measurement profiles located from the top of the vertical model to its bottom. In addition, there are checks for the heater position changes as well as those for the tightness along the outside surface of the bin and the construction rods. The measurements of energy consumption and the amount of saturation water are also taken. Samples of metals are placed inside the barrier for corrosion research.

The whole system of registration, evaluation and transfer of data is based on the use of 3 small portable AD-SYS data loggers connected to a CEG server. All the data recorded is available on the web-site. The experiment is monitored with the use of 4 webcameras operating 24 hours a day.

5 EXPERIMENTAL PROCEDURES

It is assumed that the duration of the Mock-Up-CZ experiment will not be shorter than 4 years (phase 1, 2 and 3). Then, after a cooling period, the experiment will be dismantled and all the available results will be collected and evaluated.

Phase 1 – the filter is kept dry and the power is switched on to reach a maximum temperature of 90°C in the bentonite (CEG/SKB Report, 2002b).

Phase 2 – the power is maintained at a constant level and the filter is filled with synthetic granitic water (CEG/SKB Report, 2002 c, 2003a,b).

Phase 3 – saturation and temperature loading of the buffer by the heater will be stopped during this phase. When the heater is switched off, the cooling phase will commence lasting approximately 2 weeks.

Then, the dismantling process will be conducted according to a detailed project which will include a scientific program, a sampling plan and a scenario.

6 RESULTS FROM RUNNING THE EXPERIMENT UP TO THE PRESENT (OCTOBER 03)

6.1 *Phase 1*

The experiment began on 7th May 2002 when the heater was switched on. The temperature was increased step by step to reach a maximum temperature in the bentonite of about 90°C (Fig. 2).

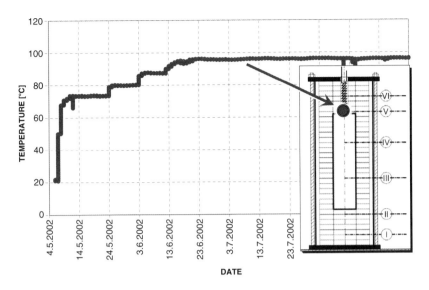

Figure 2. Maximum temperature measured in bentonite.

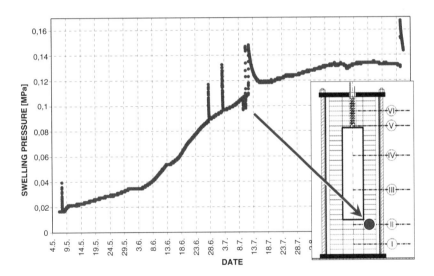

Figure 3. Swelling pressure changes due to water content redistribution – Phase 1.

The required equilibrium temperature was reached after 2.5 months. A maximum temperature of 93°C in the bentonite was measured close to the upper surface of the heater.

The process of moisture redistribution was much slower. The water content transported to the places located further away from the heater caused local increases in the swelling pressure, mainly in the area of the 50 mm filling (Fig. 3).

The results obtained from humidity sensors were of a highly problematic nature from the very beginning. Due to the system's non-homogenity (joints between blocks) and a small difference between the dimensions of the humidity sensors (8 cm) and the thickness of the bentonite barrier (ca 20 cm), the measured values were practically of no use. That is why a decision was made to take the barrier samples at predetermined intervals by means of small diameter

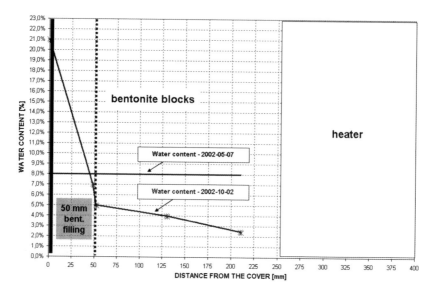

Figure 4.   Moisture redistribution.

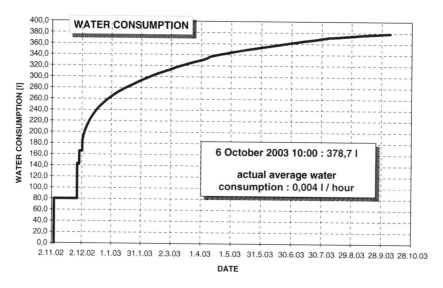

Figure 5.   Water consumption since saturation commencement.

(16 mm) horizontal core drilling. The first samples (diameter of 32 mm) were taken before the end of Phase 1 (2nd of October 02), when a core sample was taken from half-way up the bin. Moisture redistribution was determined (Fig. 4).

At the same time, the sample was used to determine permeability and swelling pressure. The results of these tests (performed by Geodevelopment a.s. Sweden) did not exibit significant differences as compared to the initial parameters. Before saturation commenced, a gas sample was taken from the bin

and exposed to analysis with the following results: 87.5% N_2, 15% O_2, 5% CO_2.

6.2   Phase 2

The saturation process commenced on 4th November by filling the hydration system with synthetic granitic water. Up to now (15th of October 03) 380 l of water have been consumed (Fig. 5).

The water caused a shock drop in the buffer temperature. A new equilibrium temperature was

56

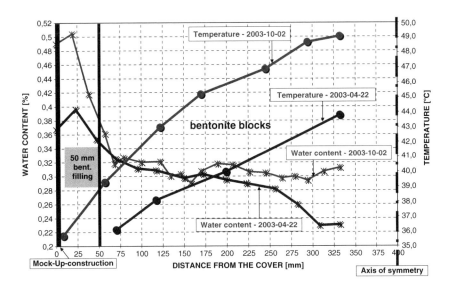

Figure 6a. Temperature and water content changes in the upper part of the barrier (April–October 2003).

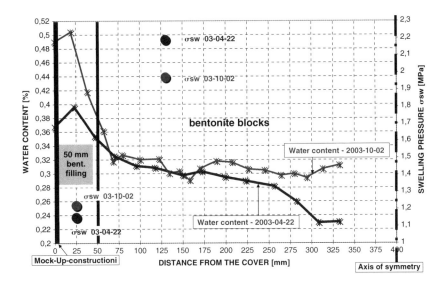

Figure 6b. Swelling pressure and water content changes in the upper part of the barrier (April–October 2003).

established approximately 15 days later. At the same time, the swelling pressure and saturation degree started to increase significantly. The swelling pressure and temperature changes in time in relation to the water content measured at three height levels (upper, medium and lower level) are shown in (Fig. 6–88b).

In January 2004, following an evaluation of the barrier saturation process, a decision will be made, with the participation of SKB and Geodevelopment Sweden, whether to increase the hydration pressure of granitic water in order to accelerate the saturation

process. Our hydration system allows an increase in hydration pressures up to 1 MPa. The current pressure is only due to the height difference between the water tank and the experimental bin (40 kPa).

7 CONCLUSION

The Mock-Up-CZ experiment has been successfully operated for 18 months. Due to incomparably lower costs spent on its construction and operation as

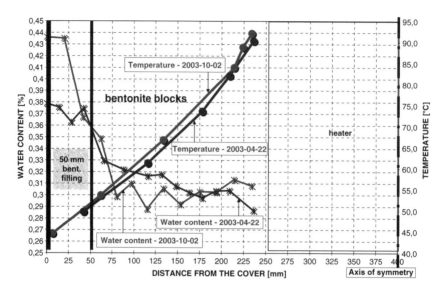

Figure 7a. Temperature and water content changes in the medium part of the barrier (April–October 2003).

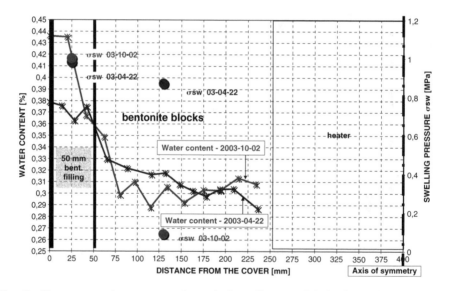

Figure 7b. Swelling pressure and water content changes in the medium part of the barrier (April–October 2003).

compared to analogical experiments running abroad, the results obtained are considered to be of great success for the CEG. From 7th May 2002 when the experiment was launched to 15th October 2003, the total data measured (due to a continuous data measurement technique) amounts to 8.5 million. We presume that the technical condition of the experimental apparatus will allow us to operate the experiment for such a long time which will produce usable outputs for a safe design of the bentonite-based engineered barrier.

Another positive fact is the decision to make all the data and information related to our research freely accessible to everyone interested, not only to a narrow circle of professionals. We hope that this approach will contribute to better dissemination of information among the general public on the safety in handling high-level radioactive waste.

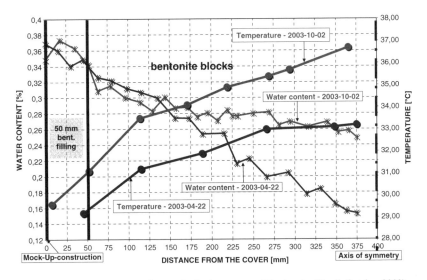

Figure 8a. Temperature and water content changes in the lower part of the barrier (April–October 2003).

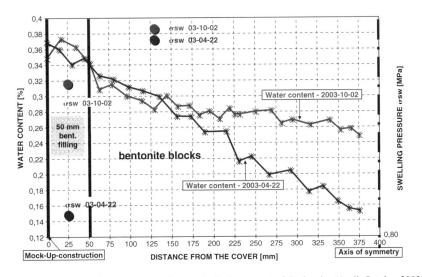

Figure 8b. Swelling pressure and water content changes in the lower part of the barrier (April–October 2003).

ACKNOWLEDGEMENT

The experiment would not have taken place without the support of the CTU, Faculty of Civil Engineering, Prague (VZ J04/210000004), RAWRA (Radioactive Waste Repository Authority in the Czech Republic), the Czech Grant Agency (G103/02/0143) and without co-operation with SKB (the Swedish Nuclear Fuel and Waste Management Co.) and Geodevelopment AB Sweden. We would like to thank all these bodies for their support.

REFERENCES

CEG/SKB Report (June, 2002) Mock-Up-CZ General Report. CTU CEG in Prague, 31 pp.

CEG/SKB Report (September, 2002) Mock-Up-CZ Quarterly Report 1. CTU CEG in Prague, 44 pp.

CEG/SKB Report (December, 2002) Mock-Up-CZ Quarterly Report 2. CTU CEG in Prague, 36 pp.

CEG/SKB Report (March, 2003) Mock-Up-CZ Quarterly Report 3. CTU CEG in Prague, 78 pp.

CEG/SKB Report (June, 2003) Mock-Up-CZ Quarterly Report 4. CTU CEG in Prague, 46 pp.

Advances in Understanding Engineered Clay Barriers – Alonso & Ledesma (eds)
© *2005 Taylor & Francis Group, London, ISBN 04 1536 544 9*

The FEBEX "In Situ" Test: lessons learned on the engineering aspects of horizontal buffer construction and canister emplacement

José-Luis Fuentes-Cantillana
AITEMIN, Madrid, Spain

ABSTRACT: One of the objectives of the FEBEX project, and in particular of the "In Situ" Test being carried out at the Grimsel underground laboratory, was the demonstration of the feasibility of the Spanish concept AGP for horizontal layout HLRW repositories in granite formations. Being the first experience of this kind, a number of engineering aspects were simplified in this Test in relation with a real repository, as for instance radiation shield was not considered, and the buffer was constructed manually. In any case, the test has provided, since its construction in 1996, a series of valuables experiences about many technological aspects associated to the construction of a full scale horizontal buffer and the handling and emplacement of canisters in these conditions.

The paper describes the main lessons learned along the different phases of the project, including the Test installation, the dismantling of part of the Test, and the continuation with the heating operation in the remaining part. Problems associated to the emplacement and dismantling of the buffer are discussed, as well as those associated with the handling, emplacement and retrieval of canisters. The experiences had with the different types of concrete plugs tested in the project are also described. The paper makes also a qualitative assessment of the behaviour of the buffer along the duration of the project, based on data obtained by the installed instruments and also in direct observations.

1 INTRODUCTION: PHASES OF THE EXPERIMENT

The FEBEX "In Situ" Test, installed at the Grimsel underground laboratory in Switzerland, simulates at a real scale the Spanish concept AGP for a high level radioactive waste repository in crystalline rock formations.

The experiment has had different phases. Figure 1 shows the initial configuration, corresponding to the installation completed in October 1996 Two electrical heaters, with the same dimension and similar weight than the reference canisters, were emplaced in a horizontal drift, within a buffer made of compacted bentonite blocks having a total length of 17 m. This zone was closed with a plain concrete plug cast in a recess excavated into the rock.

The heaters were switched on in February 1997. Since then, the power at each heater has been adjusted at all times by the automatic control system so to keep a constant temperature of 100°C at the contact with the bentonite. At the same time, the information provided by more than 600 sensors installed in the buffer and the rock has been recorded by the monitoring system, and has generated a large data base that is being used to validate the models applied for predicting the evolution of THM processes in the near field.

After exactly 5 years of continuous operation, heater n° 1 was switched off on February 2002, in order to proceed with the partial dismantling of the Test, which was carried out during the summer of 2002. This operation included the demolition of the plug, the excavation of the first 4.3 m of buffer, the extraction of heater n° 1, and the removal of the buffer around it, up to 2 m before the second heater. A dummy steel cylinder was placed in the gap left by the heater extracted, and a new concrete plug was then built to close the remaining part of the experiment, which still continues in operation. The new plug was made by shotcreting, without excavating any recess in the rock, and was constructed in two phases to enable the installation of new instruments in the buffer. The final configuration of the test after all these operations is shown in Figure 2.

The main objective of the partial dismantling operation was to confirm the information provided by the instruments and to get direct data about the evolution of the THM and THG processes in the buffer and the surrounding rock mass. For this reason, almost 900 samples from the different materials and components used in the test were taken for analysis during dismantling.

On the other hand, the partial dismantling operation was carried out taking all precautions for minimising the impact on the remaining part of the experiment. In fact, heater n° 2 was kept in operation at all times. Data obtained from the remaining sensors has confirmed that the impact has been very low and that the rest of the experiment continues in normal operation.

Along all these phases, and even considering the simplifications introduced in the test in relation with the real case, a number of lessons have been learned on many engineering aspects associated to the construction of a repository and to the behaviour of the different materials that compose the EBS system.

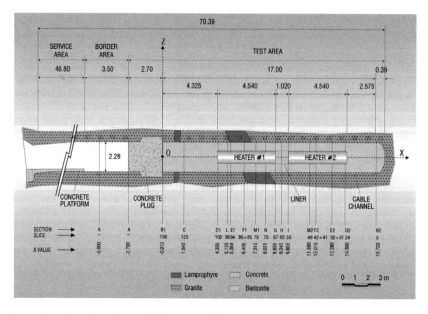

Figure 1.   Test configuration from October 1996 to April 2002.

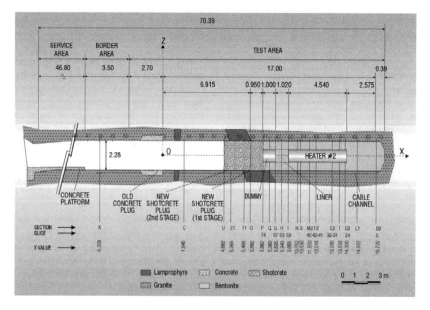

Figure 2.   Current test configuration.

The following sections include a discussion about the experience gained and the main conclusions obtained.

2 BUFFER CONSTRUCTION

2.1 Buffer characteristics

The buffer was made of blocks of the reference material, a calcium-magnesium bentonite from the Cabo de Gata region in Southeast Spain, compacted to a dry density of $1.70\,g/cm^3$. The construction was done manually, and therefore the weight of the bentonite blocks was limited to less than 25 kg each. A total of 5,331 blocks were used in the barrier construction, with a total mass of 115.7 t of bentonite.

2.2 Alignment of the deposition hole

The AGP concept considers the use of a steel liner to enable a safe an easy insertion of the canister in the deposition hole. The difference of diameter between the canister and the liner is only 40 mm (see Figure 3), so the alignment of the base line of the deposition hole with the one of the canister during the insertion is a critical parameter. Taking into account the length of the canister, the maximum allowable misalignment would be 0.8% in any direction. In a horizontal emplacement, the bentonite blocks resting on the drift floor determine the position of the deposition hole, and this may lead to alignment problems if the floor surface has, as normally happens, irregularities.

In the case of the FEBEX "In Situ" Test this problem was solved installing first the liner in the correct position and then constructing the buffer around it (Figure 4), but in principle this procedure would not be feasible in the real case. Some of the options in a real repository could be:

- Prepare an artificial, smooth, and alignment controlled surface at the drift floor, before installing the bentonite blocks
- Make bigger the gap between the canister and the deposition hole so to admit a larger tolerance in the alignment

2.3 Construction gaps and voids

Another characteristic of a horizontal emplacement is that all construction joints and gaps are accumulated at the upper part of the buffer, creating an air gap the top of the drift cross-section. In the case of FEBEX, this gap ranged from 0.5 to 4 cm, depending of the rock irregularities (see Figure 5). In principle it could be thought that this gap could play an important role in the water conduction process around the buffer, at least during the initial saturation phase, but the experience in FEBEX is that this gap closed rather

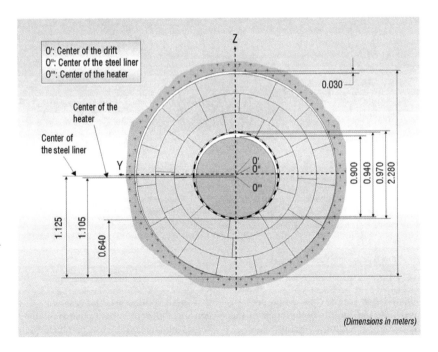

Figure 3. Cross section of the test geometry.

Figure 4.   Buffer construction process.

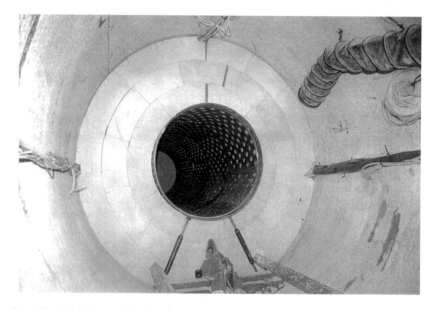

Figure 5.   Deposition hole before canister insertion.

quickly, and that there are no big differences of humidity between the top and the bottom part of the buffer, after five years of natural hydration.

During the installation, the actual mass of the emplaced bentonite was carefully recorded in order to perform a continuous quality control of the final dry density and gaps ratio obtained in the buffer. The values obtained for these parameters along the full buffer length are shown in Figure 6. It may be seen from this figure that lower densities are concentrated

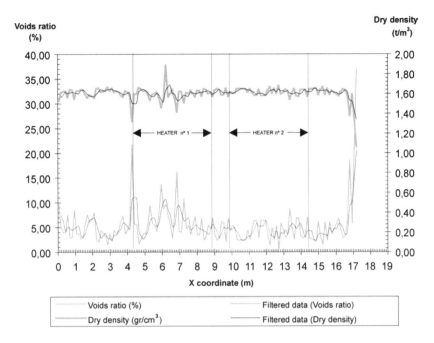

Figure 6. Profile of dry density and voids along the test length.

at the drift dead-end (concave space) and around Heater n° 1, where rock surface was more irregular because of the presence of a lamprophyre dike, but in general terms both dry density and voids ratio are rather uniform, and fall within the planned limits. The overall values finally obtained for the entire buffer are 1.60 t/m3 for the dry density and 5.53% for the construction voids ratio.

2.4 *Effect of humidity and water*

The bentonite blocks proved to be very sensitive to the environmental humidity. Laboratory tests carried out before the construction showed that the mechanical integrity of the blocks deteriorates quickly when exposed to the humid environment normally found underground. For this reason, the work area at the buffer face was kept dry during the construction process using forced ventilation and warm air heaters.

However, some free water was present during the last part of the buffer construction. It was a limited amount, flowing out from an instrumentation borehole and channelled along one of the cable bundles, and could be managed without problems. In any case, the impression got is that a certain amount of water is compatible with the buffer construction works, provided that the rhythm of advance is fast enough.

3 CANISTER EMPLACEMENT

A special piece of equipment was designed and constructed to transport the heaters along the drift and to insert them in the deposition holes (see Figure 7). This was a simple and robust machine, as the objective of this equipment was not to demonstrate the procedure to follow in the real case, but only to perform the functions required for the test. Therefore, some aspects such as radiation shield were not taken into account in the design. Also some other functional requirements such as speed, mobility, etc., were very much simplified.

Probably the insertion equipment to be used in a real repository would look quite different to the one used in FEBEX, but in any case, important lessons were learned in many aspects, as for instance the previously mentioned criticality of the alignment with the deposition hole, and, as a consequence of this, the need of including position regulating devices embedded into the insertion equipment. The efforts required for pushing a canister into the steel liner, and the reaction forces transmitted to the machine anchoring system were also appraised.

4 BUFFER DISMANTLING

A partial dismantling of the buffer was carried out after 5 years of natural hydration and continuous

Figure 7.   Heater transport and insertion machine.

Figure 8.   Aspect of the buffer after dismantling.

heating. At this stage, the buffer was only saturated at the outer part, whereas the inner part, near the heater, was still dry. However, all block joints and construction gaps were found closed (see Figure 8). The buffer saturation pattern was quite uniform, and no significant differences were found between for instance the upper and bottom part of the buffer, or between areas of the drift that supposed to have a greater water inflow than others. All this suggests that at least up to this moment the hydration process was controlled by the bentonite itself, and did not depend so much on the amount of water that the rock formation can provide.

The dismantling was carried out manually, to enable the simultaneous sampling of the different materials and components in the test. The experience served to test different types of tools for cutting and boring the swelled bentonite, and to experience the problems of handling wet clay.

An important conclusion of the partial dismantling operation is that it did not affect much to the remaining part of the experiment. The general conditions of humidity and temperature in that part did not show any variation during the dismantling, and only the stress field showed small changes, which in any case recovered very rapidly when the new plug was constructed (see Figure 9). This not only confirms the validity of the remaining part of the test, but may also be relevant when considering

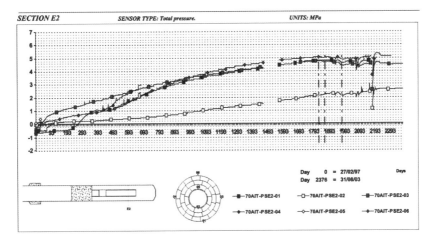

Figure 9. Evolution of stresses at Section E2. Vertical dotted lines indicate the partial dismantling period.

Figure 10. Heater extraction.

potential selective retrieval operations in a future repository.

5 CANISTER EXTRACTION

As mentioned before, the inner part of the buffer was still rather dry when the partial dismantling was carried out. The liner was also dry, with little corrosion, and only slightly deformed (the vertical diameter was about 1 cm larger than the original one). The bentonite had penetrated in the liner perforations, but the space between the heater and liner appeared quite clean.

In these conditions, the retrieval of the heater was carried out without major problems, and the pulling force required for extraction was quite low (about 2 t). The extraction was carried out with the same equipment that was used for insertion, which was modified for this function (Figure 10).

6 PLUGS

Two different types of concrete plugs have been tested in the project. The first operational phase, with two heaters, had a plain concrete plug, without any kind of steel reinforcement, cast in a recess excavated in the rock (see Figure 1). The recess was excavated with rotating saw, in order to minimise the alteration of the rock, and the concrete was cast in three phases to control the correct filling of all gaps, especially at the top part. An especial concrete formulation was used, to reduce setting temperature and minimise shrinkage.

The experience with this plug has been very good. It had an excellent mechanical behaviour, and no cracks or faults have appeared during its service life (the pressure at the inner side of the plug reached a peak value of 4.6 MPa). As checked during dismantling, no water leaks have occurred through the concrete, or at the concrete rock/interface. All recorded leaks have been produced along the cable passes, which were not well sealed, and through the inside of some cables.

The second phase of the project required a different type of solution, as it was necessary to build the new plug quickly after completing the partial dismantling, to minimise the impact in the rest of the experiment. For this reason, complex excavation or drilling operations near the bentonite front were considered not convenient, and also they were difficult to carry out due the space limitations. It was therefore decided to build a parallel plug, without any recess excavated in the rock, nor anchors or bolts. The plug would rely basically upon the friction of the concrete with the rock surface (see Figure 2).

The plug was constructed using shotcreting techniques, which provide a better filling of gaps and voids than conventional concreting, and also a high adherence of the concrete to the rock surface.

Another advantage of this method is the speed of construction and its simplicity, as it does not require the installation of forms.

Given the lack of previous experiences in the use of shotcreting to this type of application, a number of preliminary trials were carried out both at the FEBEX drift and in other sites, to check the feasibility of the method and the quality of the obtained material and of the contact surface between the gunite and the rock surface. Also different gunite formulations and spraying equipment were tested.

The plug was constructed in two phases, as it was necessary to install new instruments in the buffer before completing the plug. A first section with a thickness of 1 m was built immediately after completing the partial dismantling operation, in July 2002. The second section, with a thickness of about 2 m, was built one year later, in June 2003, once the installation of additional instrumentation was completed (see Figure 11).

The experience gained with this type of plug has been very positive, however some practical difficulties had to be solved along the different trials and construction phases. In particular, the reduced space available at the drift makes necessary to reduce to a

Figure 11. Final aspect of the shotcrete plug in the second phase of the project.

minimum the rebound material, otherwise the operation becomes extremely complicated, and the quality of the obtained material becomes poor in the bottom part. This requires an optimisation of the gunite formulation and the adequate selection of the spraying equipment, that must be adapted to the work requirements (long distance pumping, low setting heat, short time between layers, ...).

The results obtained so far have been very interesting:

– The method proved to be feasible, and may be of high interest in a real repository, especially if speed of construction is critical. In addition, this construction technique can be easily teleoperated or robotized.
– The quality of the hardened material is reasonably good. Tests carried out show a permeability value in the same range than the rock matrix.
– The mechanical behaviour of the plug has been so far excellent. The first section of the plug, with 1 m of thickness, supported without problems an internal pressure of up to 3.5 MPa, for almost one year.

Nevertheless, this technique must still be further developed for its use in a real repository. New formulations must be developed, to improve the quality of the hardened gunite, or to include low pH cements. The mechanical limits of this type of construction, in relation with the quality of the rock surface, must also be determined.

7 EQUIPMENT BEHAVIOUR

The behaviour of the different types of equipment installed in the FEBEX "In Situ" Test (heaters,

instruments, monitoring and control system, ...) has been excellent, and this was one of the reasons supporting the decision of extending the project operational phase beyond the initially planned period of 3 years.

The heaters have been working in a continuous mode during 6 years, and in fact unit n° 2 still does. The power required by the system to maintain the temperature of 100°C at the bentonite contact has been growing slowly during this period, in parallel with the increase of the thermal conductivity of the buffer caused by the progressive saturation process (see Figure 12). The few anomalies recorded in the applied power have been always very short and had no influence in the tests, and were normally caused by faults in the mains supply or in the external equipment (instruments, computers, ...) No internal leaks or electrical faults have been detected in the heaters, and only one of the three electrical resistors that were installed on each unit for redundancy purposes has been used so far. External corrosion has been very small, given the dry conditions in the deposition hole during this period.

The instruments have also performed quite well, and most of them have passed by far their guaranteed operational life. After 6 years of operation, about 87% of the originally installed instruments still work and provide coherent signals. From the other 13%, only the 9.5% are really faulty, as the rest are either saturated (in the case of humidity sensors), or were damaged during the installation phase.

The behaviour of the instruments has been however different depending on the type of sensors and their location. Most failures have been produced in the buffer, where conditions are much more

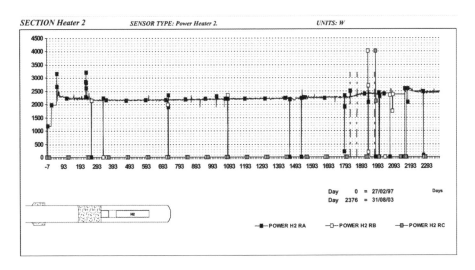

Figure 12.   Power delivered by heater n° 2.

difficult than in the rock. Corrosion has been more important in the sensors located in the intermediate part of the buffer, were both temperature and humidity are high, than those located near the heater (high temperature but low humidity) or in the outer part of the buffer (high humidity but low temperature).

The monitoring and control system has also worked quite satisfactory. The experiment has been running (and continues) in an automatic, unattended mode, being remotely controlled and monitored from the AITEMIN main office in Madrid. Several upgrades have been introduced in the system along the life of the experiment, especially on computers and power supply systems, but one of the most important factors for this good operational record has been the redundancy that was introduced in the system design for the most critical components.

8 CONCLUSIONS

The FEBEX "In Situ" Test has been the first experience of a HLRW repository with horizontal emplacement at a realistic scale in granitic formations. The experiment has proven the technical feasibility of this type of repository concept, even many engineering and construction aspects have been simplified for the Test. The experience gained during the Test installation and the partial dismantling carried out after six years of continuous operation, has provided a very important knowledge about the technological problems that can be expected in the construction phase of a repository of this type. Many lessons have been learned in aspects such as the buffer construction, handling and emplacement of canisters, construction of concrete plugs, etc.

The experiment will still continue in operation for a not yet defined period of time, in order to achieve a higher degree of buffer saturation. This extension will enable to validate the acquired knowledge for a longer time interval, and will provide an excellent opportunity for demonstrating other technological aspects, such as canister retrieval operations.

REFERENCES

AITEMIN (2003) Data Sensor Report n° 30. 10/03. FEBEX Report No 70-AIT-L-6-07.
Bárcena, I., Fuentes-Cantillana, J.L., and García-Siñeriz, J.L. (2003) "Dismantling of the heater 1 at the FEBEX "In Situ" Test. Description of operations". FEBEX Report 70-AIT-L-6-03r1.
Fuentes-Cantillana, J.L., and García-Siñeriz, J.L. (1998) FEBEX: Final Design and Installation of the "In Situ" Test at Grimsel. ENRESA Technical Publication 12/98.

Development of backfilling techniques

David Gunnarsson & Lennart Börgesson
Clay Technology AB, Lund (Sweden)

Lennart Hallstedt
Halltek AB, Karlskrona (Sweden)

Matti Nord
Oskarshamns Maskinteknik AB, Oskarshamn (Sweden)

ABSTRACT: The article describes the development of backfilling techniques performed in Sweden for the sealing of the deep repository for nuclear waste.

In the concept for disposal of the Swedish nuclear waste (KBS-3) it is vital that the drifts can be backfilled with sufficiently good material at high density to fulfil the following requirements:

– to obstruct upwards swelling of bentonite from the deposition holes for the repository life-time
– to prevent or restrict water flow in the tunnel and close to the canisters for the repository lifetime
– not to cause any significant chemical conversion of the buffer surrounding the canister.

A backfilling concept can be divided in the following three parts: (1) layout (2) backfill material and (3) backfilling technique. This article focuses on the development of the backfilling technique.

In the project Field Test of Tunnel Backfilling accomplished in 1996 in the Äspö HRL compaction of 30 cm thick horizontal layers to 1.5 m from the tunnel floor and compaction of inclined layers in the rest of the tunnel was tested. Different mixtures of crushed rock and 0–30% bentonite were tested as backfill material. Conclusions from these tests were that the horizontal layers were very sensitive to water inflow, that compaction of inclined layers over the entire cross-section of the tunnel seemed to be a promising technique and that the backfilling equipment needed to be further developed.

As preparation for the subsequent Backfill and Plug Test a number of tests were made in order to develop compaction equipment and choose carrier. The development resulted in a slope and a roof compactor that were used for the Backfill and Plug Test and, with some minor modifications, in the Prototype Repository. Conclusions from the installation of these tests were that the equipment worked well and that the backfill material containing 70% crushed rock and 30% bentonite can be compacted over the entire cross section of the tunnel to a high enough density (although with low mariginal) to fullfil SKB's requirements for the ground water conditions in Äspö HRL (TDS of about 1% about equal parts NaCl and CaCl$_3$).

In SKB's recent design bases for the backfill a ground water TDS of 3.5% is used as a guideline. This results in a need for a higher density of the clay fraction in the backfill. Two main alternatives for doing this have been suggested:

1. Further development of backfilling equipment in combination with further optimisation of the backfill material or choice of new material.
2. Emplacement of prefabricated blocks, either for the entire tunnel volume or in combination with in situ compacted parts.

Both methods are described in this article. They seem to be feasible and are currently being investigated.

1 INTRODUCTION

In the concept for disposal of the Swedish nuclear waste (KBS-3) it is vital that the drifts can be backfilled with sufficiently good material at high density to fulfil the following requirements:

– to obstruct upwards swelling of bentonite from the deposition holes for the repository life-time
– to prevent or restrict water flow in the tunnel and close to the canisters for the repository lifetime

– not to cause any significant chemical conversion of the buffer surrounding the canister.

Backfilling of tunnels were tested by SKB already in the Stripa projects in the early nineteeneighties. Backfill material (10% bentonite and 90% sand) was compacted in horizontal layers with a vibrating roller until 1.5 m. was left at the roof. The remaining volume was filled with a mixture of 30% bentonite and 80% sand that was blown or shotcreted in place. The density of the latter material was not sufficient.

In the field experiments performed in the Äspö HRL the backfilling technique has been further developed. In this article this development of technique and equipment is described in chronological order starting with the first prototype equipment in the project *Field Test of Tunnel Backfilling* and ending with the equipment as it was designed for the backfilling of the *Prototype Repository*. The plans for continued work are also described briefly.

The text in chapters 3 "Field Test of Tunnel Backfilling" and 4 "Backfill and Plug Test and Prototype Repository" are based on the article "Development of a backfilling concept for nuclear waste disposals" by Gunnarsson et al. (2003).

The text in chapter 7, "Further development of backfilling technique" is based on work performed with in the SKB-POSIVA project *Backfill and closure of the deep repository*.

2 FIELD TEST OF TUNNEL BACKFILLING

The project *field test of tunnel backfilling* was conducted in a TBM-tunnel in the Äspö HRL (The same tunnel that was later used for the *Prototype Repository*). In these tests horizontal layers were compacted to a distance of about 1.55 m from the floor. The rest of the tunnel was compacted with inclined layers with about 35 degrees angle to the roof.

2.1 *Backfill procedure*

The compaction of the horizontal layers was made in the following manner (Fig. 1).

A telescopic truck brought the of backfill material into the tunnel. The material was evened out and then compacted with a vibrating roller to a thickness of 30 cm (layers 1 to 4), 30 cm (layer 5) and 15 cm (layer 6). The surface of the horizontal layers was intended to be slightly convex in order to prevent water from dripping from the roof and form puddles. However, the roller had a tendency to sink into the backfill, especially in the first layers. The material underneath the roller was pushed sideways up the walls of the tunnel and had to be moved back to the center.

The sequence for the inclined layers is shown in Figs 2 and 3 and photos of the backfilling are shown

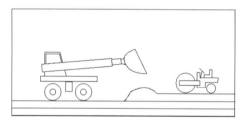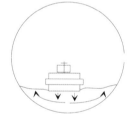

Figure 1.   Backfill procedure for the horizontal layers.

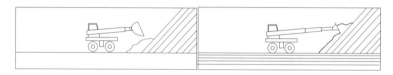

Figure 2.   Bringing material to the tunnel and pushing it in place.

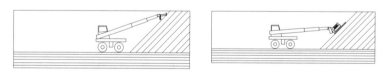

Figure 3.   Pushing the material in place and compacting the layer.

in Fig. 4. The telescopic truck weighed the material and brought it into the tunnel. Three or four buckets of material were required for each layer. The telescopic truck distributed the material evenly over the sloping surface. A small pushing tool was used for bringing the material all the way up to the roof. When the material for a complete layer had been brought in position, the inclined surface was compacted with the vibrating plate.

The compaction of horizontal layers showed to be sensitive to water inflow. Water accumulated on the surface of the horizontal layers and the truck tires turned the backfill into mud. The compaction with inclined layers was more successful even though the density close to the walls and roof was low. The original idea was to make the inclined layers flat. However, the problems close to the rock surface showed that better results would probably be obtained if the layers were bowl-shaped (Fig. 5). The carrier could reach the edges of the inclined layers much better in this way.

In order to investigate if inclined layers could be compacted in the wet parts a compaction test with backfill material containing 10% bentonite was made. The material was applied in the wet zone where the most severe problems with water had appeared. The water inflow was about 4 l/min in a 10 m long section of the tunnel. Three layers of 10/90 were compacted with the inclined compaction technique. When the material for the first layer had

been brought into the tunnel and pushed in place there were some electrical problems with the compactor and it took 40 minutes before the layer could be compacted. In spite of the extra water that the first layer had adsorbed the compaction went quite well. The second and the third layer looked almost as good as the layers compacted in the dry parts of the tunnel. The exception was one area on the left side of the surface where the water caused a bulge of material with high water content in the new material during compaction. The size of the bulge area was approximately 0.5 m^3. This effect became smaller with every layer but was still visible in the last (third) layer. The density measured in this area was lower than in the rest of the cross-section.

The conclusion was that it is possible to apply the material in the tunnel with the actual water inflow if the backfilling rate is high and constant. The test did not give information on how sensitive the backfill is to piping.

2.2 Equipment

The compaction device intended for the inclined layers was originally a Dynapac LG700 vibrating plate. Dynapac modified the vibrating plate before the start of the test (see Fig. 6).

It was early realized that this design of the vibrating plate did not work very well in the tunnel (see Fig. 6), primarily because it was impossible to

Figure 4. Photos of the emplacement and compaction of the backfill material.

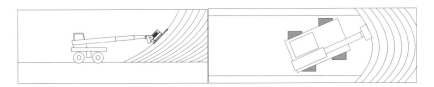

Figure 5. Shape of the compacted layers (slightly exaggerated).

Figure 6.   The first version of the vibrating plate.

Figure 7.   The modified vibrating plate.

compact the material closer than half a meter from the tunnel roof. There were also electrical and various mechanical problems that had to be solved. The compactor was modified on site. The improved version (Fig. 7) worked properly and made it possible to compact backfill material rather close to the roof and walls of the tunnel. The carrier for the vibrating plate was a Matbro Tristar telescopic truck commonly used for construction work (Fig. 8). For compacting the horizontal layers a Dynapac CA 131 D vibrating roller with a weight of 4.5 t was used. A hydraulic hammer was also tested for compacting the inclined layers close to the walls of the tunnel. The hammer was mounted on the telescopic truck the same way as the vibrating plate and connected to the hydraulic system. Two compaction plates were tested. One of the plates served to increase the density close to the wall but the effect was not significant and the compaction procedure took much time (about 40 minutes per layer). The other plate penetrated the material and did not work properly. The backfilling equipment is described in detail by Gunnarsson et al. (1996).

2.3 Experiences regarding backfill technique and equipment from the project Field Test of Tunnel Backfilling

The following experiences were made:

– The technique of backfilling with horizontal layers was very sensitive to water inflow.
– The technique with inclined compaction worked well although it was not possible to compact material close to the walls and roof of the tunnel very efficiently since the design of the vibrating plate was not suitable. The compaction equipment needed further development. It was also necessary to improve reliability and safety.
– The carrier should also be more flexible and be able to move the plate sideways and at the same time adjust for the change in angle between the arm and the inclined surface.

3   BACKFILL AND PLUG TEST AND PROTOTYPE REPOSITORY

For the *Backfill and Plug Test* two new compactors, a roof compactor and a slope compactor were developed in order to make it possible to efficiently compact the backfill material close to the roof and walls of the tunnel. A new carrier for the compactors was also chosen and adapted. The development of the equipment and technique for the *Backfill and Plug test* is described in this chapter. The same equipment (slightly modified) and procedure were also used in the *Prototype Repository*.

The *Backfill and Plug Test* was installed in a drill and blast tunnel in the Äspö HRL. In this test a 38 m long tunnel section was backfilled and instrumented. Half of the length was backfilled with a mixture of 30% bentonite and 70% crushed rock and the other half was backfilled with crushed rock and bentonite

Figure 8.   The carrier and the hydraulic hammer.

blocks placed at the roof. The installation of the test is described by Gunnarsson et al. (2001[1] and 2003).

The *Prototype Repository* comprises six full-scale deposition holes, copper canisters equipped with electrical heaters, bentonite blocks , and a TBM-drilled deposition tunnel backfilled with a mixture of 30% bentonite and 70% crushed rock. It consists of two sections. The first section is 40 m long, comprises 4 deposition holes with canisters and is planned to run for 30 years. The second section is 33 m long, comprises two deposition holes with canisters and is planned to run for 5 years. The buffer, backfill and rock have been instrumented and each section ends with a concrete plug. The installation of section 1 is described by Börgesson et al. 2003.

3.1   Equipment

3.1.1   Carrier
A Volvo BM 6300 backhoe loader (Fig. 9) was selected for the test since it has superior stability, strength and since the boom is designed in a way that facilitates compaction at the roof. The booms of other tested carriers tended to come in contact with the roof during compaction. The chosen carrier is small enough to manoeuvre in the tunnel and is at the same time powerful enough to reach over the entire slope while exerting normal force on the vibrating plates. Also, with the aid of a hydraulically controlled unit (referred to as a rotor / tilting unit) it can position the plate flat against all parts of the backfill layer. The hydraulic system of the carrier is powerful and able to simultaneously move and power the compactors. The hydraulic system of the carrier was adjusted to fit the requirements of the two compactors. The rotor/tilting unit was mounted on the boom and

Figure 9.   The carrier used in the Backfill and Plug Test and the Prototype Repository.

the compactors were in turn connected to the rotor/tilting unit.

3.1.2   Slope compactor
The slope compactor was used for compacting the entire surface of the backfill layers except for very close to the roof. A schematic drawing of the slope compactor and a photo are shown in Fig. 10. The slope compactor was designed to be low to make it possible to compact close to the roof. It is powered with a hydraulic motor to make the low design possible and to optimise the working safety. The total weight of the plate was 900 kg and the vibrating weight 414 kg. The vibrating frequency was 43 Hz and the amplitude 3.7 mm (peak to peak 5.4 mm). The slope compactor is described further by Gunnarsson et al. (2001[1] and 2003).

In the schematic drawing of the slope compactor shown in Fig. 10 the upper frame is red, the vibrating

75

Figure 10.   The slope compactor.

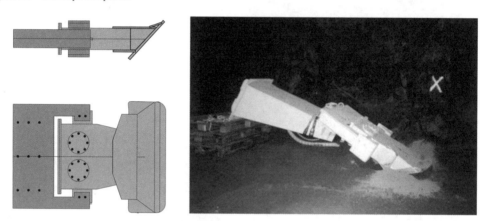

Figure 11.   The roof compactor.

part is green, the safety arrangements (catching hooks) are blue and the eccentric element, which is hidden the picture, is represented by a dotted line.

3.1.3 *Roof compactor*

The backfill adjacent to the roof could not be compacted satisfactorily with the slope compactor. In order to solve this, another vibrating compaction machine, the roof compactor, was designed and manufactured. A principal drawing and a photo are shown in Fig. 11.

The roof compactor was designed to move and compact small amounts of backfill material up to the tunnel roof. The total weight of the roof compactor is 400 kg and the vibrating weight is 340 kg. The frequency is 43 Hz and the amplitude 3 mm (peak to peak 6 mm).

In the schematic drawing in Fig. 11 the vibrating part is green-coloured while the isolated support is marked red.

The tool for pushing backfill material in place was developed as part of the project "*Field Test of Tunnel Backfilling*" (Gunnarsson et al. 1996). Based on experience from this and later tests a new prolonged version was manufactured (see Fig. 12).

Figure 12.   The tool for pushing the backfill material in place.

3.2 *Backfilling procedure*

A schematic drawing of the backfilling procedure is shown in Fig. 13 and comprises the following steps:

1. Removing temporary road and cleaning the floor.
2. Transporting the backfill material for the layer into the tunnel (Fig. 13 a).

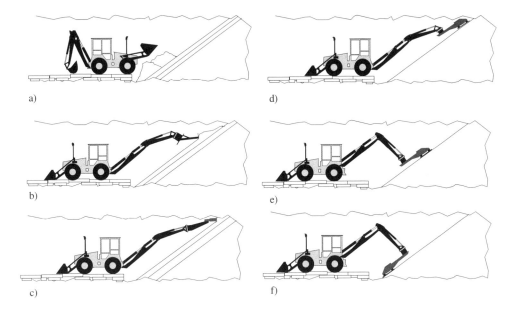

a)

b)

c)

d)

e)

f)

Figure 13. The backfilling sequence.

3. Pushing the material into position with the pusher (Fig. 13 b). This operation gave the desired shape of the layer, which was made slightly concave to simplify compaction close to the walls.
4. Backfill material is pushed towards the roof and compacted with the roof compactor in order to achieve a high density at the top of the backfill and to ensure good contact with the roof (Fig. 13 c).
5. The rest of the layer is compacted with the slope compactor (Fig. 13 d–f). Starting about 1.5 m above the floor it is moved sideways across the layer from one wall to the other. After one sweep the compactor was moved up the slope about half meter for the next sideways sweep. This was repeated to the top of the layer so that the areas treated with the slope compactor and the roof compactor overlapped. The compactor was then moved in the same way, side to side, down the layer to the starting position. This procedure was repeated three times. To further improve the compaction close to the walls the compactor was moved from top to bottom along the rock wall. As a last step, the compactor was turned 180° and the material towards the floor was compacted (Figs. 13 f).

The intention was to give the compacted layer a concave shape to facilitate the compaction close to the walls and roof of the tunnel. As the compaction with the new equipment was tested it was found that it was impractical to have a highly concave shape of the layers and that a slightly concave shape was optimal. This allowed the carrier to move the vibrating plate in

a simple way optimising the compaction work. The slight concave shape still seemed to improve the density of the backfill close to the walls and decrease the negative effect of irregularities of the rock walls. The concave shape also resulted in a steeper inclination of the layer near the roof, which made it easier for the vibrating plate and roof compactor to reach and compact the material here.

3.3 *Practical experiences from the Backfill and Plug Test and modifications for the Prototype Repository*

Below examples experiences from the backfilling of the *Backfill and Plug Test* and the modifications that they brought for the *Prototype Repository* are listed. The backfilling technique requires a lot of skill from the operator of the Backfilling equipment.

6. In the Backfill and Plug Test the total effective operating time for the slope compactor was about 75 hours and about 15 hours for the roof compactor.
7. The isolated U-frame and the vibrating part of the roof compactor had hit each other. This resulted in quite heavy deformations of the U-frame and in considerable vibrations transferred from the roof compactor to the carrier. This caused high ware on the carrier. A new design to remedy this was made and implemented.
8. The temperature of the carrier's hydraulic system rose to a high level after compaction.

77

An extra oil cooler was installed in the oil tank of the carrier to eliminate the problem.

9. All of the catching hooks on the slope compactor were deformed and some of them had to be repaired during the backfilling. The catching hooks were replaced with a more robust system.

10. The rubber isolators broke quite frequently. A new design for increasing the life length of the catching hooks were made and implemented on both the slope compactor and the roof compactor.

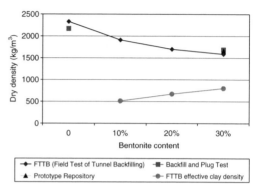

Figure 14. Comparison of densities achieved in the tests of backfilling technique performed at Äspö HRL by Gunnarsson et al. (1996).

4 DENSITY ACHIEVED IN THE FIELD

In this chapter the field densities of the backfill materials consisting of 70% crushed rock and 30% bentonite backfilled in the projects Field Test of Tunnel Backfilling, Backfill and Plug Test and Prototype Repository are compared.

Based on the density measurements made in the different tests it can be concluded that the variation in density is similar for all tests; the density is highest in the middle of the tunnel cross section and then decreases towards the rock floor, walls and roof.

In Fig. 14 the average densities achieved in the different tests are shown. The total dry density decreases and the density of the clay fraction increases with increasing bentonite content. The average density achieved in the backfill material containing 30% bentonite was higher in the Backfill and Plug Test and in the Prototype Repository than in the inclined layers in the project Field Test of Tunnel Backfilling. This is due to the improved compaction technique and equipment. The densities presented in Fig. 14 for both the Backfill and Plug Test and Prototype Repository are valid for parts were the compaction work proceeded without disturbances.

One important experience from the tests is that the higher the fraction of clay is the more difficult it is to compact the material in the field in relation to the results from the proctor compaction tests. The reason for this is that the Proctor technique is quite optimal for compacting clayey material. While this type of compaction (high impact, high surface pressure) is hard to achieve in the field. In Fig. 15 the percent proctor achieved in the project Field Test of Tunnel Backfilling are shown.

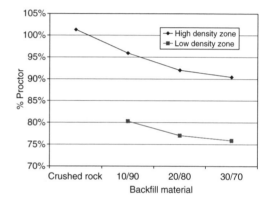

Figure 15. The percent modified Proctor achieved for different backfill materials in the Field Test of Tunnel Backfilling.

5 PRESENT EXPERIENCES

The main conclusion from the performed development is that inclined compaction over the entire tunnel cross section is the most suitable backfilling technique. It is possible to achieve as high density in inclined as in horizontal layers and it is an optimal method of applying and compacting backfill material close to the tunnel roof. Inclined layers extending to the floor are also optimal for handling water inflow. Once the layer has been compacted the backfill material is less sensitive to water, which can drain off the surface down to the floor and out of the tunnel. Even more importantly the backfill is exposed to running water for a much shorter time period than for the case with the horizontal layers.

No results or discussion concerning the properties of the backfill have been given in this article. The most important properties are the hydraulic conductivity, the compressibility and the swelling properties (the latter for a good backfill/rock contact). The dry density of 30/70 reached in the field tests ($1700 \, \text{kg/m}^3$) yields the following properties of the backfill (Johannesson et al. 1999 and Börgesson, 2001).

Hydraulic conductivity with 1.3% salt content of the added water:

$K = 4.10–10 \, \text{m/s}$
Compressibility: $M \sim 30\,000 \, \text{kPa}$
Swelling pressure: $\sigma s = 150–300 \, \text{kPa}$

Since these values are acceptable it is concluded that the tested technique and 30/70 backfill can be used for backfilling tunnels in a repository if the salt content in the ground water is 1.0% or less, while 30/80 with the same compaction result (87% Proctor) yields a hydraulic conductivity that probably is too high.

Since the full scale tests have not yet been finished and analysed it is not possible to conclude if the requirements set on the backfilled tunnels have finally been fulfilled. However, the results from the measurements made during installation in combination with results from laboratory tests indicate that the requirements can be met for ground water salinities up to about 1% as discussed earlier. Improved compaction at the roof may still be required at this salt content. If the salt content of the groundwater is higher it is necessary to use another backfill material and to increase the compaction efficiency of the equipment.

At present other backfilling concepts and materials suitable for locations with high salt content in the ground water are also being investigated. The main challenge is to design a backfill concept that can meet the requirements of hydraulic conductivity, stiffness and swelling properties at a ground water salt content equal to the oceans.

6 FURTHER DEVELOPMENT OF BACKFILLING TECHNIQUE

With in the SKB-POSIVA project "Backfilling and closure of the deep repository". possible techniques for backfilling tunnels are being further investigated and developed. In phase 1 of the project that runs during 2003 a desk study on the backfilling technique has been made.

Two main techniques of applying the backfill material in the tunnels have been investigated: In situ compaction and placement of pre-compacted blocks. In the backfilling concepts these techniques are used in different combinations and with different backfill materials.

6.1 *Further development of compaction equipment*

6.1.1 *Introduction*
To be able to compact natural clays or mixtures of crushed rock and bentonite in inclined layers to densities that are high enough to fulfil the requirements further development of the compaction equipment is necessary. The new equipment should also combine all the functions of the slope compactor, the roof compactor and the pusher in order to avoid changing tools during the backfilling and thereby increasing the compaction efficiency.

While non-cohesive soil can be efficiently compacted by vibration fine-grained material such as clay or clayey soil require more powerful shock waves through the material. This can be accomplished by high surface pressure and a great momentum of the compacting mass.

A model for calculating theoretical surface pressure and momentum for different compaction devises and there by compare the compaction effect have been developed.

35 different concept ideas for finding new solutions for compaction devises have been listed. 8 concepts out of 35 were chosen and further developed. This development resulted in that the three following compaction concepts were recommended for further development.

1. Improved roof compactor.
2. Spring accelerated compaction weight.
3. Mechanic/hydraulic rammer.

6.1.2 *Improved roof compactor*
The compaction principle is vibration but the amplitude has been increased to be more than twice as high as for the present roof compactor. This results in high enough surface pressures to use the equipment for compacting clayey material. The design of the vibrating part is made so that compaction of the slope and the area towards the roof is facilitated. The compactor could also be used for pushing material in place and thus the roof compactor could perform all operations and it would not be necessary to change tools during the backfilling operation.

The compaction time for one backfill layer was calculated to about 30 min.

Relatively high static forces, from the carrier to the compactor, are necessary to get the necessary surface pressure, to be able to push material in place and to maneuver the compactor. The forces will be conveyed by the vibration isolators, which will have a low spring constant to be able to isolate the vibrating part while a high spring constant is necessary to convey the static forces. This will have to be solved and optimized. The problem of endurance strength of the construction will have to be given special attention since the amplitude is very high.

In Fig. 16 the improved roof compactor is shown together with a suggested carrier.

6.1.3 *Spring accelerated compaction plate*
In this compaction concept a compaction plate is accelerated towards the backfill surface with springs. After the impact the plate is lifted back in position by

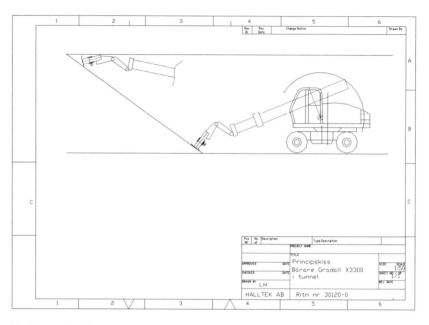

Figure 16.   The improved roof compactor and a suggested carrier (Gradall X3300).

two hydraulic cylinders. During this operation the carrier moves the compactor to a new position. If it is possible to push material in place and to compact material towards the roof will have to be further investigated.

It is relatively easy to vary the impact speed of the compaction weight by varying the tensioning of the springs or by exchanging the springs. The design can be made rather simple but some problems, such as the gripping device for the hydraulic cylinders, the parallel operation of the cylinders and to keep the slide surfaces clean, will have to be solved.

The compaction time for one layer is estimated to approximately 70 min. A drawing of the compactor is presented in Fig. 17.

6.1.4   Mechanic/hydraulic rammer

A rammer has the advantage that it has a high frequency, many impacts per time unit at the same time as the surface pressure is quite high. The surface of the compaction plate is however normally quite small. The rammer is suitable for compacting the slope. It should be possible to adapt it so that the material at the roof can be compacted. It should also, to some extent, be possible to use it for pushing material in place. To achieve the intended compaction effect the compaction plate should hit the backfill surface with the maximum velocity. To ensure this a number of factors have to be considered: The spring packages in the compactor, the pressure from the carrier, the spring constant of the compacted backfill material and

the impact velocity. The endurance strain in the compaction plate will be great calling for extensive development and testing.

The compaction time has been estimated to approximately 60 min. A drawing of a hydraulic rammer is shown in Fig. 18.

6.2   Emplacement of pre-compacted blocks

6.2.1   Introduction

An alternative to insitu compaction is to fill the entire cross-section of the backfilled tunnel is filled with pre-compacted blocks. The remaining slots are if needed filled with bentonite pellets or granules. The floor has to be levelled to provide a stable base for the blocks and to facilitate transport. This can be solved by compacting suitable clay that fulfils the requirements of low hydraulic conductivity and is relatively insensitive to the water that will pass from the inner part. Another solution would be to use a perforated steel plate at the floor. The steel plate would provide a stable foundation for the blocks. It would also act as drainage during installation. The water from the inner part of the drift would run under the plate out of the tunnel. Once the temporary plug is in place in the end of the tunnel the water will fill all voids and the clay will swell through the holes in the plate and completely seal the tunnel cross section. This is also the principle used for the KBS-3-(horizontal deposition holes) and experience from tests made for this could be utilized.

80

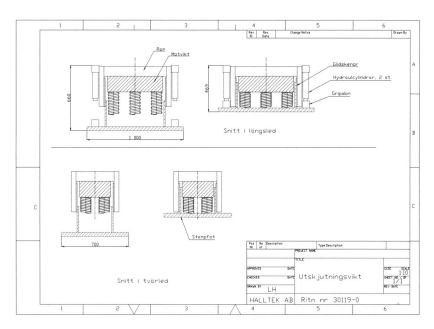

Figure 17.   Spring accelerated compaction plate.

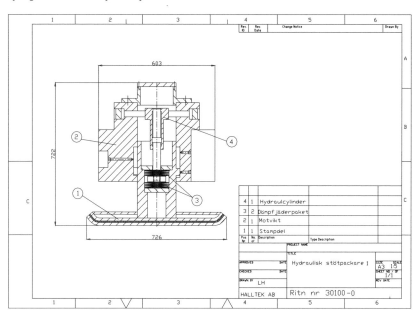

Figure 18.   Drawing of the hydraulic rammer.

A number of different types of materials are considered:

– MX-80 or similar bentonite.
– Friedton or other swelling clay of lower quality.
– Different types of mixtures of ballast and bentonite.

6.2.2 Emplacement technique

Relatively small (about brick size) blocks can be used but also larger blocks may be considered. In order to increase the backfilling rate pallets of blocks will be handled and emplaced. The blocks are placed on the pallet and the pallets are transported to the deposition

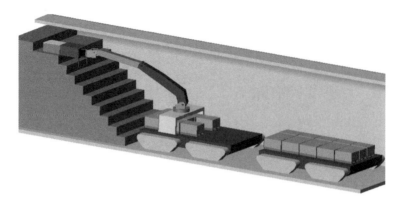

Figure 19.   The two units for emplacement.

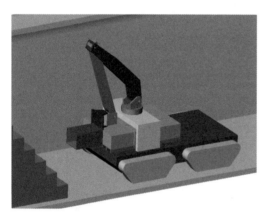

Figure 20.   Loading the placement tool.

tunnel with a truck. From the transport truck the pallets would be shifted onto a specially designed transport device that moves the pallets to the backfilling front. One alternative for the block placement is presented in Fig. 19 to 21.

The emplacement equipment is made up by two units, a transport unit and the emplacement unit itself (Fig. 19). Both units are equipped with conveyor belts so that pallets can continuously be brought forward. The transport unit moves into the tunnel and conveys the pallets to the placing unit and then moves out of the tunnel to be reloaded. The emplacement unit places the pallet in the placement tool (Fig.20) and moves it to the desired location (7-6) where the blocks are placed.

The vehicles will be propelled with caterpillar tracks or by rubber wheels and will be powered by electricity. The cables will be led along the roof or walls. A ditch in the floor facilitates the outflow of water. The solution of having steel plates would facilitate the drainage even more. Once the tunnel is cut off with the

temporary plug in the end of the deposition tunnel the clay will swell through the holes in the plate and the result would be a competent backfill in the entire tunnel cross-section. The plates would also provide a very flat and stable basis for the block stapling.

The volume filling rate is higher if smaller pallets are used. If half pallets can be used this also increases the volume filling rate. The backfilling speed can be increased if larger pallets are used. These factors will have to be investigated and the pallet size optimised. The remaining space can be filled by pellets if the average density in the tunnel has to be raised. If MX-80 blocks with the bulk density of $2050\,\text{kg/m}^3$ and a water ratio of 10% are used a filling degree of 80% would be neccesary to achieve an average density at saturation of $1950\,\text{kg/m}^3$ in the tunnel cross section.

The general impression is that the possibility of using this system is good. A number of areas for further work and questions that have to be answered have been identified and will be addressed in the continued work within the project backfilling and closure of the deep repository.

7   SUMMARY AND CONCLUSIONS

The main conclusion from the performed development for the Backfill and Plug Test and the Prototype Repository is that inclined compaction over the entire tunnel cross section is the most suitable backfilling technique. It is possible to achieve as high density in inclined as in horizontal layers and it is an optimal method of applying and compacting backfill material close to the tunnel roof. Inclined layers extending to the floor are also optimal for handling water inflow. Once the layer has been compacted the backfill material is less sensitive to water, which can drain off the surface down to the floor and out of the tunnel. Even more importantly the backfill is exposed to

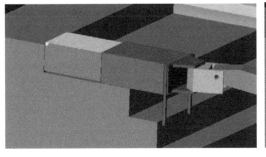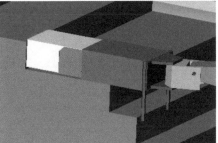

Figure 21.   Retracting the bottom part of the placing tool.

running water for a much shorter time period than for the case with the horizontal layers.

Since the tests (Backfill and Plug Test and Prototype Repository) have not yet been finished and analysed it is not possible to conclude if the requirements set on the backfilled tunnels have finally been fulfilled. However, the results from the measurements made during installation in combination with results from laboratory tests indicate that the requirements can be met for ground water salinities up to about 1% as discussed in chapter 4. Improved compaction at the roof may still be required at this salt content. If the salt content of the groundwater is higher it is necessary to use another backfill material and to increase the compaction efficiency of the equipment.

The conclusion from the work with further developing the compaction equipment is that there are a few promising concepts for improving the compaction efficiency but that these probably crave quite extensive development and testing.

The general impression concerning the block emplacement technique is that the possibility that the system will work in practice is high. A number of areas for further work and questions that have to be answered have been identified but the judgement is that the problems can be solved with existing technology.

REFERENCES

Börgesson, L (2001) Compilation of laboratory data for buffer and backfill materials in the Prototype Repository.

Börgesson, L, Gunnarsson, D, Johannesson, L-E, Sandén, T; (2003) Prototype Repository. Installation of buffer, canisters, backfill and instruments in Section 1. SKB report IPR-03-33.

Gunnarsson, D, Johannesson, L-E, Sandén, T, Börgesson, L; (1996) Field Test of Tunnel Backfilling. SKB report HRL-96-38.

Gunnarsson, D, Börgesson, L, Hökmark, H, Johannesson, L.E, Sandén, T; (2001). Report on the installation of the Backfill and Plug Test. SKB report IPR-01-17.

Gunnarsson, D, Börgesson, L; (2003) Installation of the Backfill and Plug Test. Article for International meeting in REIMS December 9–13, 2003: Clays in natural and engineered barriers for radioactive waste confinement.

Gunnarsson, D, Börgesson, L; (2003) Development of equipment, material and technique for backfilling tunnels in a nuclear waste disposal. Article for International meeting in REIMS December 9–13, 2003: Clays in natural and engineered barriers for radioactive waste confinement.

Johannesson, L.E, Börgesson, L, Sandén, T; (1999) Backfill materials based on crushed rock (Part 3). Geotechnical properties determined in the laboratory. SKB IPR-99-33.

Advances in Understanding Engineered Clay Barriers – Alonso & Ledesma (eds)
© 2005 Taylor & Francis Group, London, ISBN 04 1536 544 9

Reliability of THM instrumentation for underground research laboratories

José-Luis García-Siñeriz & Ignacio Bárcena
AITEMIN, Madrid (Spain)

ABSTRACT: Safety and long-term behaviour of underground permanent repositories depend on a combination of several engineered and geological barriers. The properties of the geological barriers are the natural conditions of the formation, granite in this case, while the performance of the engineered barriers is a result of their design and construction. The properties of the engineered barriers are deeply influenced by the interactions between both geological and engineered barriers in response to the conditions expected in a high level waste repository. These interactions need to be identified and fully understood to allow their input in models describing the behaviour of the near field to predict reliably the long-term performance and safety of a repository.

Besides the necessary testing of the engineered barriers components in the laboratory and, also in realistic conditions but at a reduced scale, long-term and full-scale experiment are required for a correct assessment of these processes and especially for the determination of their evolution with time. One of the main objectives of the experiments carried out in URLs is to obtain data on the evolution of the different processes that take place in the engineered barrier system and in the host rock, in conditions similar to those found in a real repository.

Therefore, THM instrumentation is usually one of the most important aspects of these experiments. The information provided by the instruments is used to improve the understanding of processes and to provide a set of data against which the models can be validated and, if required, further refined. For this reason, the instruments installed in the experiments must provide as much as possible, realistic and accurate data.

Given the difficult working conditions for testing instrumentation: high mechanical and pore pressures, high temperatures and corrosive media; and the expected tests duration (several years), it is not possible to guarantee the operation of instruments and the accuracy for the full lifetime of tests. For that reason, the information provided by the partial dismantling of the FEBEX "in-situ" test is relevant to evaluate the real performance of the installed instrumentation. It will help to answer several questions related to obtained data as for instance: how accurate is the gathered information?, how to correct or re-interpret these data and future data if required?, why some of the instruments failed and how to improve their performance for future tests?

The paper describes the experience gained so far at the FEBEX "in-situ" test carried out in granite, with especial emphasis in the results from the partial dismantling operation. The confidence on recorded and future gathered data increased due to the partial dismantling operation, given that results of calibrated sensors showed negligible or very low drift for most of them. The causes for the sensor failures and possible improvements were also determined. I was stated that the sensor failure rate was very low for non-experimental sensors. Failure rate could be minimised for future experiences, in similar conditions, by improving mechanical and corrosion protection of some components.

1 INTRODUCTION

Permanent deep disposal in underground permanent repositories is generally accepted by the international scientific community as the most realistic option for the long-term management of high level radioactive waste.

Safety and long-term behaviour of underground permanent repositories will depend on a combination of several engineered and geological barriers. The properties of the geological barriers are the natural conditions of the formation, granite in this case, while the performance of the engineered barriers is a result of their design and construction. The properties of the engineered barriers are deeply influenced by the interactions between both geological and engineered barriers in response to the conditions expected in a high level waste repository. These interactions or THM (Thermo-Hydro-Mechanical) and THG (Thermo-Hydro-Geochemical) mechanisms need to be identified and fully understood to allow their input in models describing the behaviour of the near field in order to reliably predict the long-term performance and safety of a repository.

Besides the necessary testing of the engineered barriers components in the laboratory and, also in realistic conditions but at a reduced scale, long-term and full-scale experiment are required for a correct assessment of these processes and especially for the determination of their evolution with time. Therefore, the new test program launched by ENRESA in 1994 included also field tests in the considered formations (granite, clay), the objectives being mainly the analysis of the technical feasibility of the concept, and the validation of predictive codes.

From the various tests planned in this program, the FEBEX "in-situ" experiment is the oldest one in granite of the running experiments, as well as one of the most instrumented. Furthermore, FEBEX is the first one, among such experiments, that has undergone a partial dismantling. This operation provided first hand information, never obtained before, on the real status of the sensors after several years of "in situ" operation, as well as on the reliability of the obtained data by means of "post mortem" re-calibration.

The FEBEX test is a full-scale demonstration of the Spanish repository concept in granite, leaded by ENRESA. It includes two large thermal tests: one "in situ" test at the Grimsel Test Site in Switzerland, and one "Mock up" test in the CIEMAT facilities in Madrid, Spain. The main objectives of FEBEX are the demonstration of the technical feasibility of the concept and the refining and validation of THM and THG codes, especially in the buffer material.

2 URL INSTRUMENTATION, THE FEBEX "IN-SITU" EXPERIENCE

2.1 Test configuration during FEBEX I

According to the Spanish concept, two electrical heaters, of the same size and of a similar weight as the reference canisters, were placed in the axis of a horizontal drift excavated in the Grimsel granodiorite. The drift was constructed especially for this experiment using a TBM and has a diameter of 2.28 m. The gap between the heaters and the rock was backfilled with compacted bentonite blocks, up to a length of 17.40 m, this requiring a total 115,716 kg of bentonite. The backfilled area was sealed with a plain concrete plug keyed into a recess excavated in the rock and having a length of 2.70 m and a volume of 17.8 m^3 (Fig. 1).

A total of 632 instruments were placed into the system along a number of instrumented sections, both in the bentonite buffer and in the host rock, to monitor relevant parameters such as temperature, humidity, total and pore pressure, displacements, … etc. The instruments were of many different kinds, and their characteristics and positions are fully described in the FEBEX Final Design and Installation report (Fuentes-Cantillana & García-Siñeriz, 1998).

A Data Acquisition and Control System (DACS) located in the service area of the FEBEX drift collects

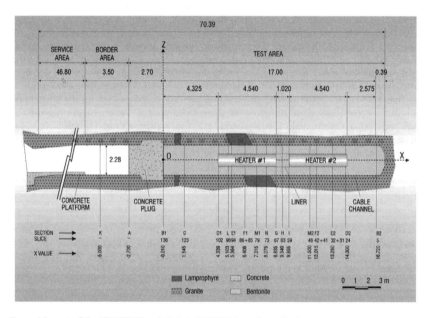

Figure 1.   General layout of the FEBEX "in situ" test (FEBEX I configuration).

data provided by the instruments. This system records and stores information from the sensors and also controls the power applied to the electrical heaters, in order to maintain keep a constant temperature at the heaters/bentonite interface. The DACS allows the experiment to be run in an automated mode, with remote supervision from Madrid via modem. Data stored at the local DACS are also remotely dumped from Madrid by modem and are used to build up the experimental Master Data Base.

The construction of the concrete plug was completed in October 1996, and the heating operation started on 28 January 1997. The system has been in continuous operation since that date, without major disturbances, for a period of more than 5 years, two years more than initially planned. A constant temperature of 100°C has been maintained at the heaters/bentonite interface throughout this period, while the bentonite buffer has been slowly hydrating with the water naturally issuing from the rock. The information provided by the instruments has made it possible to create an experimental database having more than 2,000,000 records (Huertas et al, 1998).

2.2 FEBEX "in-situ" instrumentation

One of the main objectives of the experiments carried out in URLs (Underground Research Laboratories), is to obtain data on the evolution of the different processes that take place in the engineered barrier system and in the host rock, in conditions similar to those found in a real repository. Therefore, instrumentation is usually one of the most important aspects of these experiments.

The information provided by the instruments is used to improve the understanding of processes and to provide a set of data against which the models can be validated and, if required, further refined. For this reason, the instruments installed in the experiments must provide to the extent possible, realistic and accurate data. Consequently, the instrumentation for the FEBEX test was carefully selected and installed.

2.2.1 Rock instrumentation

Nineteen boreholes were drilled in the "in situ" test zone from the interior of the drift, for the placement of 299 sensors in the surrounding rock mass. This instrumentation measured thermal, hydraulic and mechanical parameters. The orientation and length of these boreholes were established as a function of the specified position of each instrument, and especially of those relative to the hydraulic parameters. The length of the boreholes ranged from 7 m to 22 m, resulting in a total of 233 m drilled, with a diameter of 66 mm for all of them but for two, which housed total pressure cells inside and had a diameter of 146 mm.

Four additional long boreholes drilled in the surrounds of the drift were used specifically for measuring hydraulic and mechanical parameters. Finally, other sensors, as psychrometers and TDR probes were installed in smaller boreholes drilled from the drift in areas closer to the rock wall (up to 2.5 m). A breakdown of rock instrumentation in FEBEX "in situ" test is shown in Table 1.

2.2.2 Buffer instrumentation

A total of 310 sensors, out of the whole 632 sensors installed in the entire FEBEX "in situ" test, were located within the buffer, grouped into 16 measuring sections perpendicular to the drift axis, for monitoring different thermal, hydraulic and mechanic parameters. A breakdown of these buffer instrumentation sensors can be found in Table 2.

For measuring the gas pressure and the permeability of the barrier to this gas, and for taking gas samples, 10 ceramic tube type sensors were installed at different points into the buffer (not included in table 2). Four 0.2 m long, 60 mm diameter gas pressure sensors, and six 3 m long 60 mm diameter gas flow sensors, all connected to measuring and sampling equipment located at the service area outside the test zone.

For rock and bentonite instruments, metallic bodies and thermocouple sheaths were made in AISI 304L or AISI 316L, when possible. Cable

Table 1. Rock instrumentation in FEBEX "in situ" test.

Parameter	Measuring principle	Qtty
Temperature (boreholes)	"T" type Thermocouple	62
Total pressure (boreholes & drift wall)	Vibrating wire	34
Hydraulic pressure (boreholes)	Piezoresistive	81
Packers pressure (boreholes)	Piezoresistive	81
Relative humidity + temperature or	Psychrometer	28
Water content	TDR	4
Rock deformations	Vibrating wire extensometers	2×3
Crack evolution	LVDT	1×3
Total		299

Table 2. Buffer instrumentation in FEBEX "in situ" test.

Parameter	Measuring principle	Qtty
Temperature	"T" type Thermocouple	91
Total pressure (heaters surface)	Vibrating wire	6
Pore pressure	Vibrating wire	52
	Capacitive	58
Relative humidity + temperature or	Psychrometer	48
Water content	TDR	20
Heater displacement	Vibrating wire	9
Expansion of bentonite block	Vibrating wire	8
Displacement between bentonite barrier	Potentiometer	2 × 3
Clinometer	LVDT	6 × 2
Total		310

insulations were made in Halar for vibrating wire sensors and polyurethane for capacitive type or TDR humidity sensors and PVC for remaining ones, but protected with an external Teflon insulation at the bentonite buffer area.

2.2.3 FEBEX "in-situ", instrumentation performance

The data collected from instrumentation during all this period has allowed the creation of a large database that is being used to validate the modelling codes. Results have been very positive so far, with a good correlation between model predictions and real data. Hydration has progressed but saturation was reached only in the outer part of the buffer.

Most of the sensors have passed well beyond its expected operational lifetime. By March 2002, before the partial dismantling of the tests, more than 85% of the sensors remained operative, after more than 5 years of operation. A qualitative sensors performance can be extracted at that moment (AITEMIN, 2002):

(a) Temperature sensors (thermocouples): data obtained seems to be accurate when compared with measures from temperature sensors incorporated in other instruments and with the expected evolution provided by models. The temperature trend has been quite uniform during the operational phase, this being the parameter with the highest symmetry along the experiment.

(b) Humidity (capacitive type, psychrometers and TDRs): in general, the readings provided by each sensor type match quite well with the others and also with the prediction from the models. The data obtained from capacitive sensors, given their wide range, is clearly more useful for modellers when the hydration process starts from low water contents and it is sufficiently slow, as in this case.

Psychrometers seemed to be very sensitive to contamination, which may come from the salts in the bentonite, and their signals became difficult to interpret in this case. Only some of them at the outer buffer ring reached the measuring range (above 95% RH). The psychrometers in the granite are giving reasonable readings (very low suction), and seem to be less affected by salt contamination.

The readings from TDRs installed in the buffer are good, although these sensors provide an integrated measurement. The performance of TDRs installed in granite is not so accurate, as they are operating in the limit of their resolution.

The evolution of humidity was more three-dimensional than that of the temperature, probably due to the effect of the air gap at the top of the buffer, and the channelling effect of the cable bundles.

(c) Vibrating wire sensors: the rest of instruments are mostly vibrating wire type. These sensors are used to measure pore pressure in bentonite, total pressure in bentonite, in rock interface, or in boreholes (also called 3D cells), displacements in rock (extensometers), heater displacements, and displacement in bentonite blocks.

In general, their behaviour has been as good as expected, that is: high accuracy, long term stability, reduced temperature drift and no hysteresis. The only difficulty came from the weakness of the embedded thermistor, which has failed in some cases. That thermistor is used for temperature compensation of the vibrating wire signal. Fortunately in all events it has been possible to substitute that temperature by the reading of another temperature sensor (thermocouple) located close enough.

(d) Hydraulic pressure in rock boreholes: internal packer pressures and borehole interval pressures are measured by means of conventional pressure transducers installed outside the sealed area, at the end of nylon type plastic tubes. These sensors are working under standard conditions and can be checked or replaced if necessary, as it was the case with some of them.

Table 3. List of sensors dismantled.

Parameter	Measuring principle	Qtty
Temperature	"T" type Thermocouple	37
Total pressure (buffer area)	Vibrating wire	8
Pore pressure	Vibrating wire	24
	Capacitive	27
Relative humidity + temperature or	Psychrometer	24
Water content	TDR	10
Heater displacement	Vibrating wire	2
Expansion of bentonite block	Vibrating wire	4
Displacement between bentonite barrier	Potentiometer	2×3
Clinometer	LVDT	6×2
Total		154

2.2.4 *FEBEX "in-situ", instrumentation long-term accuracy*

Given the difficult working conditions for testing instrumentation: high mechanical and pore pressures, high temperatures and corrosive media; and the expected tests duration (several years), it is not possible to guarantee the operation of instruments and the accuracy for the full lifetime of tests.

For that reason, the information provided by the partial dismantling of the FEBEX "in-situ" test, is relevant to evaluate the real performance of the installed instrumentation. It will help to answer several questions related to the obtained data, as for instance: how accurate is the gathered information?, how to correct or re-interpret these data and future data if required?, why some instruments failed and how to improve their performance for future tests?

2.3 *FEBEX "in-situ" partial dismantling*

2.3.1 *Partial dismantling basis*

The partial dismantling of the FEBEX "in-situ" test was carried out as planned during the summer of 2002, after 5 years of continuous heating. The operation included the demolition of the concrete plug, and the removal of the section of the test corresponding to the first heater. A large number of samples from all types of materials have been taken for analysis (Fuentes-Cantillana, García-Siñeriz & Bárcena, 2003).

All dismantling operations: bentonite removal, drilling operations, etc, were designed with the aim of preserving the planned samples in the best possible conditions. The bentonite buffer was removed by blocks layers and when a bentonite layer containing sensors was reached, the bentonite layer was sampled according to the corresponding procedures (Fuentes-Cantillana, García-Siñeriz, Bárcena & Tuñón, 2002). Samples were finally packed and documented.

All the instruments contained in the dismantled bentonite buffer section were sampled for analysis.

The ones installed in the rock boreholes (hydrotesting, some psychrometers and TDRs) were left in place. The total pressure cells installed in the granite also remained in place, in order to carry out an "in situ" testing. Therefore, basically only those installed at the bentonite buffer (around 150) were retrieved. Some of them already failed during test operation and others were damaged during dismantling, but the number of operative ones is sufficiently high as to perform the planned analysis and to extract the required conclusions.

Additionally, the 18 internal thermocouples of heater 1 were removed along with it, and some samples of ceramic tube type sensors installed for measuring the gas pressure and the permeability of the barrier to this gas, as well as for taking gas samples, were collected (not included in Table 3).

Out of the dismantled sensors, 76 were out of order before dismantling (around 49%). By sensor type, apart from the 100% of LVDT type clinometers and potentiometers for displacement between bentonite barrier, that failed due to the complete flooding of some of them up to the corresponding electronics (18 sensors), the major percentage (almost 54%) corresponds to the humidity sensors (41 in total).

The utmost care was taken during the dismantling, but this was a difficult operation due to the status of the buffer (humid and highly pressurised) and some functioning instruments were damaged during the dismantling process. Two total pressure cells were damaged during the dismantling of the concrete plug and one pore pressure sensors was damaged during the removal of the buffer. Besides, two thermocouples failed during the dismantling due to the corrosion effects and one psychrometer resulted damaged as well, because of the weakness of these sensors and the high pressure from the bentonite swelling.

Most part of the operation was developed without major disturbances. A number of difficulties had to be solved on site, especially in relation with the methods used for sampling in the bentonite. The only

Table 4. List of sensors failed[1].

Parameter	Measuring principle	Qtty
Temperature	"T" type Thermocouple	7
Total pressure (buffer area)	Vibrating wire	4
Pore pressure	Vibrating wire	4
	Capacitive	16
Relative humidity ǀ temperature or	Psychrometer	17
Water content	TDR	8
Heater displacement	Vibrating wire	2
Expansion of bentonite block	Vibrating wire	0
Displacement between bentonite barrier	Potentiometer	2×3
Clinometer	LVDT	6×2
Total		76

[1] Some relative humidity sensors did not strictly fail, they became out of order after reaching the full saturation.

Figure 2. View of a capacitive type humidity sensor showing mechanical deformation.

significant delay was caused by the mercury spills occurred when the total pressure cells in the inner part of the concrete plug were damaged during the plug demolition.

2.3.2 Status of dismantled sensors
The appearance and consistency of the buffer were very good, and the degree of humidity observed is consistent with the readings provided by the instruments during the operational phase. The joints between blocks were still clearly visible, but all construction gaps were sealed. Sensors and cables were found in good contact with the bentonite and those close to the rock walls were highly pressed towards the rock. No noticeable changes in sensors positioning were found due to the bentonite swelling nor movements of the heater were registered.

The mechanical effects of the bentonite swelling on some sensors were clearly visible. Some capacitive type humidity sensors (Fig. 2), psychrometers and TDRs showed squashed cables (but still operative)

and bent, broken or loose filters. The ceramic filter of the psychrometers measuring head was found broken and/or separated from the cable for several sensors (Fig. 3).

Several TDRs were also found with the cable cut at the connection with the probe (Fig. 4). The gas flow pipes in the contact bentonite/rock, as well as the gas pressure sensors (made in ceramics) were found totally destroyed by the bentonite pressure. The ceramic material seemed to be degraded into a humid mass. Only the gas flow pipes in the inner bentonite ring, which were installed in perforated steel tubes, were extracted intact.

Several sensors at one instrumented section were found severely corroded. For instance, the LVDT type clinometers, the potentiometers for displacement between bentonite barrier and the vibrating wire type heater displacement sensors (Fig. 5 and Fig. 6).

They showed squashed cables and the external isolation was lost for some of them (clinometers), furthermore, they showed clear corrosion effects at

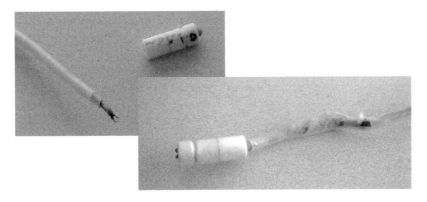

Figure 3.   View of psychrometers showing mechanical effects.

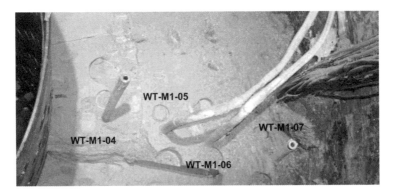

Figure 4.   View of some TDRs sensor showing mechanical deformation a cut cables (source: NAGRA).

Figure 5.   View of an instrumented section with corroded sensors.

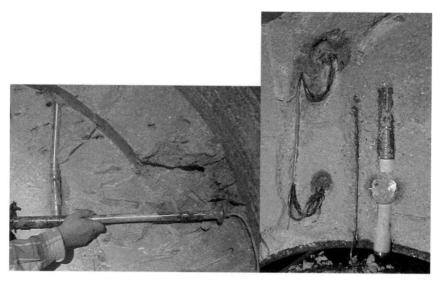

Figure 6. Detail of corroded sensors.

metallic parts and some pieces were found bent and squashed.

Finally, many of the sampled thermocouples at this section showed corrosion effects too (Fig. 7), some were broken in the middle of the buffer section when dismantled.

As a conclusion, only fragile sensor probes (ceramics, metal filters, plastic pieces) or cables (especially PVC ones) were affected by the swelling pressure of bentonite.

High corrosion damage and general rust was observed on the devices for anchoring the sensors where not made of stainless steel. The highest corrosion effects were found at instrumented section D1, located at the front side of heater 1. According to the performed corrosion studies, the significant corrosion damage suffered by these components is due to the following causes: the presence of sulphate reducing bacteria (from the bentonite blocks), and the high humidity of the bentonite blocks housing them. However, the simultaneous action of stresses either residual or applied, and a corrosive agent, normally chlorides, collaborated to the process. Indeed, this section was a border between the hot and cold bentonite buffer areas and it had a lot of sensors (extensometers and thermocouples) connecting the rock walls with the liner and heater body. These factors could increase the water-vapour circulation and the bacteria growing.

2.3.3 Calibration and failure analysis of dismantled sensors

Operative retrieved sensors were calibrated again in order to know their accuracy after more than 5 years of operation. These works were not totally finished yet but some preliminary conclusions can be extracted.

Thermocouples: except for those damaged by corrosion (most of them in contact with liner or heater), the behaviour was excellent and the accuracy of the dismantled operative ones was similar to that of the installation date (below $\pm 1°C$).

Vibrating wire sensors (total pressure, pore pressure and displacements sensors): in general these probes have proved to be very robust, both the sensor head and the cable. A zero point measurement was carried out after the retrieval, before packing the sensors, and the readings obtained in general were good. The only difficulty with these sensors came from the weakness of the embedded thermistor, which has failed in some cases. Apart from this, the heater displacement sensors were found severely corroded, as mentioned above. By the time this paper was written the calibration of these sensors was still in progress.

Humidity (capacitive type): these sensors widely overpassed their expected operative lifetime (one-year maximum in such environment). A 67% of the failed ones were due to sensor saturation after reaching the full measuring range (100% RH), two of them failed due to mechanical problems and one due to electronics malfunction (Fig. 8). The calibration carried out to the still operative sensors showed that in general they were giving readings below the real value (3–4% RH) in the range from 35% to 100% RH. This was also appreciated when comparing the readings from the sensors with the values obtained from the in-situ analysis of bentonite samples (Lloret, A., Daucousse, D., & Alonso, E., 2003). The values

Figure 7. Corroded thermocouples.

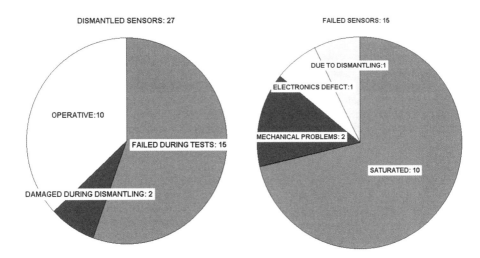

Figure 8. Capacitive humidity sensors analysis.

match fairly well, but those from the sensors were a little bit lower as later stated by the calibration.

Humidity (psychrometers): they proved to be very fragile for such environment (only four of them were operative after dismantling) and also seemed to be very sensitive to contamination, which may come from the salts in the bentonite. More than 50% of sensors failed due to mechanical effects and major percentage was at outer ring location (Fig. 9).

Analysed (calibrated) sensors showed typically contaminated signals, difficult to correlate with real values. Also the only available method for automated continuous measurement with psychometers (psychrometric method) may induce the sensor failure

due to the heating-cooling cycles applied to the measuring head, which could accelerate the chamber contamination by salts deposition. The availability of an automated measuring system using the dew point method will increase the measurement quality and reliability as well as the sensors operative lifetime.

Humidity (TDRs): Also they demonstrate to be very fragile for such environment (only two of them were operative after dismantling). After being calibrated, new calculations have been applied to already gathered data and for future data coming from remaining ones. It seems that the performance of installed TDRs was improved (Albert, Weber, Meier and Dubois, 2003).

93

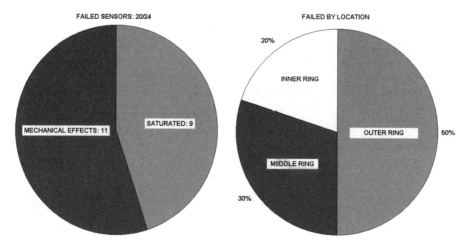

Figure 9. Psychrometers analysis.

3 CONCLUSIONS

Some learning lessons can be extracted from the operation of the FEBEX experiment and from the partial dismantling carried out.

The behaviour of the heating system, the data acquisition and control system, and the instrumentation has been excellent. In fact most of the instruments have exceeded their expected lifetime. The careful design and components selection resulted in a valuable and very robust managing structure that helped the FEBEX experiment to become a reference in this field.

No significant failures were registered for the heating system and the data acquisition and control system, due to the quality of the applied components and their adequate integration. The use of the redundancy basis was also critical for the achieved success.

The confidence on recorded and future gathered data has increased due to the partial dismantling operation, given that the results of calibrated sensors showed negligible or very low drift for most of them. Hence, as a conclusion and giving response to previously raised up questions, the present gathered information is very accurate, and despite some sensors data which are still being processed, it seems that no significant data correction or re-interpretation will be needed.

Regarding the causes for the sensor failures causes and possible improvements, it was stated that the sensor failure rate was very low for non-experimental sensors. The failures for experimental sensors did not cause data loss problems due to the redundancy of signals obtained by using well proven sensors in parallel. Failure rate for the experimental ones could be minimised for future experiences in similar conditions, by avoiding the use of exposed ceramics and plastics (or if not avoided they should be at least well protected mechanically). The use of more corrosion resistant metals, or non-corrosive external protections, will help to extend the lifetime of several components. The installation of passive probe type sensors demonstrates to be very appropriate (for instance the vibrating wire sensors).

In principle, considering the state-of-the-art in sensor technology the operation and accuracy of instruments cannot be guaranteed beyond 10 or 20 years. The use of cables for transmitting data is also a difficulty when safe water and gas proof closures are required.

The probable requisite, from Nuclear Regulatory Commissions, of a monitoring period of 50 to 100 years for future repositories, but keeping the safety of the repository seals, will require further work to find the required monitoring solutions.

In this sense, the further development and testing of innovative non-cable linked sensors (wireless technology), as well as the use of multiple fiber optics sensors linked with a unique fiber optic cable, iare absolutely essential if those requisites should be complied.

Furthermore, the results from on going experiments, as Backfill and Plug Test, Prototype Repository and the Temperature Buffer Tests, will help to increase the knowledge about the reliability of THM instrumentation for URL testing.

REFERENCES

AITEMIN (2002) Data Sensor Report n° 26. *FEBEX Report No 70-AIT-L-5-34.*

Albert, W, Weber, HP, Meier, E, & Dubois, D (2003) Grimsel Test Site: FEBEX II. Excavation of TDR probes, section M1; Laboratory analysis of bentonite samples; New calibration plus calculation of water content from TDR data. *NAGRA internal report 03-03.*

Fuentes-Cantillana, JL, & García-Siñeriz, JL (1998) FEBEX: Final Design and Installation of the "In Situ" Test at Grimsel. *ENRESA, Technical Publication 12/98.*

Fuentes-Cantillana, JL, García-Siñeriz, JL, Bárcena, I. & Tuñón, S (2002) Sampling Plan for Dismantling of Section 1 of "in situ" Test". *AITEMIN, Internal Project Technical Report 02/02.*

Fuentes-Cantillana, JL, García-Siñeriz, JL, & Bárcena, I (2003) Dismantling of the heater 1 at the FEBEX "In Situ" Test. *AITEMIN, Internal Project Technical Report 09/03.*

Huertas, F et al (1998) FEBEX Pre-operational Stage. Summary report. *ENRESA, Technical Publication 01/98.*

Lloret, A, Daucousse, D, & Alonso, E (2003) Results of "in situ" measurements of water content and dry density. *CIMNE, Internal Project Technical Report No 70-UPC-L-5-012.*

Advances in Understanding Engineered Clay Barriers – Alonso & Ledesma (eds)
© 2005 Taylor & Francis Group, London, ISBN 04 1536 544 9

In situ saturated hydraulic conductivity of a backfill by means of pulse tests

Clemente Mata
INITEC NUCLEAR SA, Madrid (Spain)

Alberto Ledesma
Technical University of Cataluña, Barcelona (Spain)

ABSTRACT: Saturated permeability is a key parameter when designing an engineered barrier for nuclear waste in a deep vault. Nonetheless, it is still difficult to estimate the permeability in situ in compacted and low permeability media. In situ measurement of permeability is commonly performed by analysing constant and variable head tests. Variable head tests are preferred in low permeability media because of the long duration of the constant head tests in such soils. Within the framework of the Backfill and Plug Test, held at the Äspö Hard Rock Laboratory (ÄHRL), the in situ hydraulic conductivity of a bentonite-based mixture is being determined with a new layout designed to carry out pulse tests in situ. The results obtained in situ are comparable with those obtained from the laboratory scale. The numerical analyses of the pulse tests performed in situ also show that the equipment is suitable to estimate the permeability of compacted clayey soils.

1 INTRODUCTION

In the context of nuclear waste storage, saturated hydraulic conductivity becomes a fundamental parameter when analysing the long-term behaviour of a deep repository. Despite that, determination of hydraulic conductivity is still far from being a routine for very low permeability materials and its measurement is not straightforward in such soils. Standard techniques become sometimes not suitable because of the long time required for the experiment or the difficulties when assessing the accuracy of the measurements. In addition, the validity of Darcy's law has been a controversial issue for such low permeability soils (Mitchell, 1993).

As flow tests are difficult to handle for very low permeability soils, it is better to perform flow tests under transient conditions (typically falling head tests). In this way, the pulse test concept may be used. The use of pulse tests in geotechnical laboratories or practice is not common, and therefore, the experimental procedure had to be defined. In this case, the decay of a pulse of water pressure applied to a boundary of the soil sample is measured, which is related to the soil permeability and compressibility. That relation has been extensively used for the "in situ" characterisation of geological media.

A new mini-piezometer, called DPPS, has been developed by UPC and built by AITEMIN (AITEMIN, 1999) to perform pulse tests in highly compacted clayey soils. Fourteen minipiezometers were built; one of them was used to perform pulse tests in laboratory. A previous calibration of the new device was performed in the laboratory scale (Mata & Ledesma, 2003a). The other thirteen devices were placed at the ZEDEX gallery at the ÄHRL. Similar techniques and procedures were used to study and analyse the pulse tests performed at the ZEDEX gallery where ENRESA placed them (AITEMIN, 1999).

2 BACKFILL AND PLUG TEST PROJECT

Within the Backfill and Plug Test Project, carried out at the ZEDEX gallery in the ÄHRL, an MX-80 sodium bentonite and crushed granite rock mixture is being investigated as a sealing material. The material considered is a mixture of 30% MX-80 sodium bentonite (Wyoming bentonite) and 70% (by mass) of crushed granite rock from the excavation of the gallery. MX-80 has 75% of montmorillonite content and 85% of its grains are smaller than $2\,\mu m$ (Madsen, 1998). The maximum grain size of the crushed granite rock is $20\,mm$. For particle size slower than sieve 200 ASTM (equivalent to $0.074\,mm$) the grain size distribution was obtained by laser diffraction technique.

The main objective of the project is to test the backfill hydro-mechanical behaviour in situ. Figure 1 shows the vertical section of the ZEDEX gallery where the project is being developed. The test region

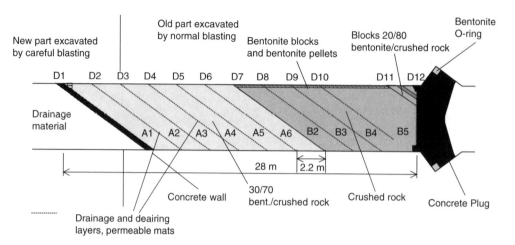

Figure 1. Vertical section of ZEDEX gallery showing the final layout and numbering of sections and mats (Gunnarsson et al, 2001).

can be divided in four different parts: the inner zone with drainage material only, then a zone filled with compacted backfill followed by a third area with crushed granite rock, and finally, the concrete plug closing the gallery. The part of the tunnel filled with the mixture is the aim of the current study. After mixing the crushed granite and the bentonite, using a paddle mixer, initial water content of the backfill was between 12% and 13%. Six different backfill sections (A1 to A6) were compacted in six different layers. Pore water pressure transducers and psychrometers mainly make up the instrumentation placed in situ. The compaction was carried out with a vibrating plate developed and built for this purpose and the final slope of the section was 35°. Among the sections, permeable geotextiles were placed. The purpose of the permeable mats is to inject and/or to collect water when performing flow tests. The Backfill and Plug Test Project is mainly a full-scale flow test and backfill hydraulic conductivity is the key parameter to be investigated.

3 PULSE TESTS IN LABORATORY

The DPPS works in a similar way as the "piezocone" testing method: a controlled positive pressure pulse is applied to the sensors, and the evolution of the water pressure drop, which is controlled by the local permeability, is monitored and analysed. The equipment necessary to perform pulse tests in laboratory is composed of:

– A new designed "mini-piezometer" or DPPS with a pressure transducer inside.
– Two pressure systems (GDS type), including volume change measurement.

– A Personal Computer (PC), which controls the test and acquires the data.
– A stainless steel cell, where soil is compacted within and the DPPS is placed.
– A high-speed valve to get the shut-in effect on the system when closed suddenly.

The scheme of the mini-piezometer and the cell where the pulse tests were performed in laboratory are shown in Figure 2. The piezometer is a cylinder (180 mm height, 50 mm diameter) connected to a pressure transducer and to two small tubes for ingoing and outgoing water. The metallic parts and tubes were made of stainless steel to avoid corrosion. The measuring pressure range of the piezoresistive transducer (Kulite Semiconductor) varies from 0 to 5.88 MPa, the pore diameter of the ceramic filter (outer porous stone) was 60 μm and its hydraulic conductivity was 3.10^{-4} m/s. The DPPS was built by AITEMIN (AITEMIN, 1999).

Backfill was dynamically compacted into the stainless steel cell. The estimated soil volume was 14846 cm^3 and the mass of backfill 27220 grams. The height of the test specimen was 304 mm and the radius was 127 mm. A permeable mat was placed in the outer contour of the cell. Backfill was compacted in twelve layers and the energy applied was 357 kJ/m^3 ($\approx$63% normal Proctor). The initial water content was 12.0%, close to the optimum value on the dry side (Mata, 2003). After the compaction process the dry specific weight was 15.9 kN/m^3 and void ratio 0.643. During compaction, a metallic cast with the same shape and size as the DPPS was used. The cast allowed the compaction of the backfill without ramming the DPPS. After ten layers of compacted

98

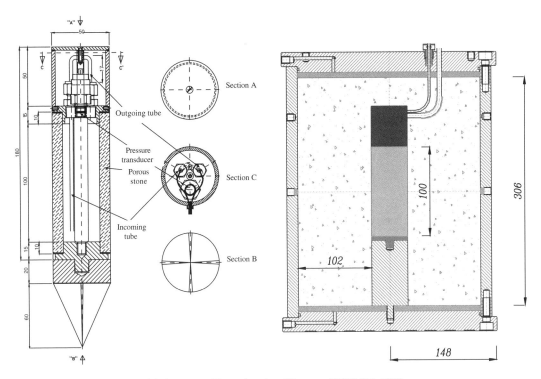

Figure 2. Side view of the mini-piezometer. Dimensions in millimetres (AITEMIN, 1999).

soil, the cast was replaced by the DPPS very carefully in order to avoid disturbing the soil. The compaction process continued for the last two layers. The extraction process of the metallic cast and the introduction of the DPPS did not create a significant remoulded zone in the surroundings (checked in a dismantled test). The soil specimen was saturated with de-ionised water ($<20\,\mu$S/cm).

Once the soil specimen was compacted, the saturation phase of the backfill started. Tests were performed after full saturation. The saturation phase took more than 3500 hours, due to the hydraulic gradient applied, less than 30 during the first month and less than 60 during the second month, in order to prevent hydraulic fracture or crack opening. Initial water pressure inside the backfill was increased up to $p_0 = 700$ kPa. The steps to perform an injection pulse test are the following:

– Water pressure is prescribed at the outer boundary of the cell ($p_0 = 700$ kPa). A pressure increment is applied inside the pressure system with the high-speed valve closed. Thus, water pressure in the pressure system is equal to $p_1 = p_0 + \Delta p$.
– At that moment the high-speed valve opens (controlled by the PC) until water pressure is equal to p_1 in the pressure transducer inside the DPPS.

Table 1. General data of the pulse tests carried out.

Pulse	p_0 (kPa)	p_1 (kPa)	ΔV (cm^3)	f_{obs} (m^3/MPa)
1	703	803.9	0.102	$1.010 \cdot 10^{-6}$
2	709	802.4	0.090	$0.963 \cdot 10^{-6}$

– When water pressure, p_1, in the DPPS is reached, the high-speed valve closes. The sudden close of the high-speed valve introduces the influence of water compressibility and sensor deformability on system flexibility.
– The evolution of water pressure in the sensor is monitored. The necessary amount of water to increase the pressure in the minipiezometer is also estimated from the registered data at the GDS pressure system.

4 SIMULATION OF THE PULSE TESTS IN LABORATORY

Data from the pulse tests carried out in laboratory is shown in Table 1. In both pulse tests the water pressure increment was around 100 kPa. The observed system flexibility, f_{obs}, was calculated by

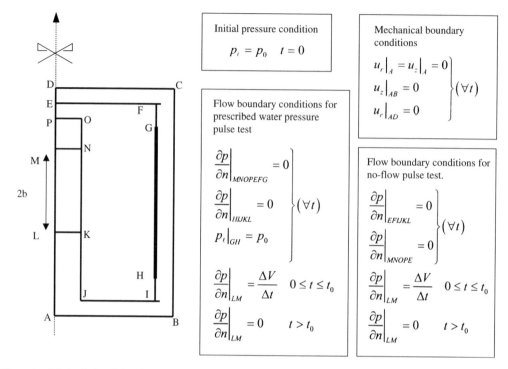

Figure 3. Mechanical and flow boundary conditions applied when simulating the pulse tests.

means of the expression proposed by Neuzil (1982) as the volume of water injected to reach the water pressure increment:

$$f_{obs} = \frac{\Delta V}{\Delta p} \qquad (1)$$

The interpretation of pulse tests is not straightforward and the procedure used is described here in detail. The two pulse tests were simulated with CODE_BRIGHT (Olivella et al, 1996) a finite element code to solve THM problems in geological media in a coupled way.

4.1 Initial and boundary conditions

A standard flow boundary condition between piezometer and soil in two dimensions (cylindrical coordinates, r and z) may be written as (Brand & Premchitt, 1982):

$$2\pi a \frac{k}{\gamma_w} \int_{-b}^{b} \frac{\partial p}{\partial r} dz = f_{obs} \frac{\partial p}{\partial t}\bigg|_{r=a} \qquad (2)$$

where $2b$ is the total length of the ceramic filter of the mini-piezometer, a its radius, and p is the water pressure. A fictitious rigid material was used to simulate this flow boundary condition. The properties of this material were always the same: a very high hydraulic conductivity and Young's modulus relative to backfill. The aim of this material is to store water during the injection as the water cavity of the piezometer does. Figure 3 summarises the initial, flow and mechanical boundary conditions used in the calculations.

4.2 Material properties

An isotropic, homogeneous and linear elastic model was used to simulate the behaviour of the materials considered in the calculations (steel, fictitious material and backfill). To assue a backfill isotropic and homogeneous behaviour for the analyses is supported by different results found in the literature. Even strongly compacted MX-80 bentonites (up to 50 or 100 MPa of compacting pressure) did not show any marked tendency of particles to be oriented after the compacting pressure was applied (Push, 1982). All the performed tests were injection tests so the expected behaviour should be elastic because injecting water is an unloading-reloading process.

Initial backfill void ratio was in all cases 0.643. The fictitious material void ratio was variable

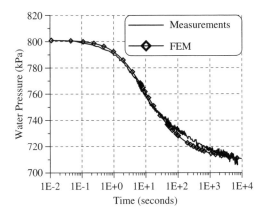

Figure 4. Comparison of the numerical simulation of the pulse number 1 and the pulse measured.

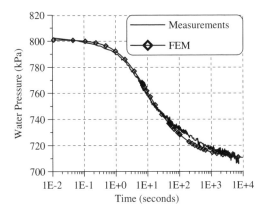

Figure 5. Comparison of the numerical simulation of the pulse number 2 and the pulse measured.

depending on the water volume injected and backfill properties. The Poisson's ratio used was always 0.3. Table 3 shows the backfill hydro-mechanical properties for every test simulation. These parameters were obtained by means of a trial and error procedure. The steel equivalent bulk and shear modulus were $K = 7596 \, \text{MPa}$ and $G = 3506 \, \text{MPa}$ after estimating the cell compressibility. Water compressibility was assumed to be $4.5 \cdot 10^{-4} \, \text{MPa}^{-1}$ in all four pulse tests.

4.3 Numerical results

The finite element simulations of the pulse tests are depicted in figures 4 and 5. A good agreement is observed for both cases, which indicates that the parameters used in the simulation are adequate. Table 2 summarises the results of the numerical analyses performed where k is the soil permeabiltiy, K is the

Table 2. Backfill constitutive parameters used in the pulse test simulations by means of CODE_BRIGHT.

Pulse	k (m/s)	K (MPa)	G (MPa)	m* (MPa^{-1})
1	$9 \cdot 10^{-12}$	2.08	2.27	0.196
2	$8 \cdot 10^{-12}$	1.67	1.82	0.245

bulk modulus and G is the shear modulus and m* is the backfill compressibility including the water compressibility, C_w, (equation 3).

$$m^{\cdot} = \frac{1}{\left(K + \frac{4}{3} \cdot G\right)} + nC_w \quad (3)$$

5 BRIEF DESCRIPTION OF THE IN SITU LAYOUT

The acquisition and control system is divided into three parts. Figure 6 shows a simplified scheme of the whole layout built and placed by AITEMIN for the Backfill and Plug Test Project. The three main components are the acquisition and control system (made up by a N_2 tank, a PC and a complex valve panel where pressure transducers and a flow meter are placed); the second part of the system is made up by the interphase between water and N_2, the vacuum pump and some deposits, which allow changing the system flexibility depending on the expected hydraulic conductivity. Finally, the third part is the switching panel where the electric valves of the different mini-piezometers are. There are three tanks to modify the flexibility of the system (auxiliary tanks) and a big one to refill the interphase. Stainless steel AISI 316L was used for building the metallic parts of the control and the acquisition system.

From the interphase a metallic tube goes to the switching panel placed at the DEMO tunnel, and from the switching panel, thirteen inflow tubes go to each one of the mini-piezometers, and thirteen outflow tubes from the mini-piezometers reach a glass tank within the control room through the switching panel. The system is closed, and, in this way, it is possible to circulate water and remove trapped air from the mini-piezometers (AITEMIN, 1999). Pulse tests have to be performed in series because of the acquisition software and hardware. There are around 100 meters of metallic tube from the acquisition and control system to the mini-piezometers. Nevertheless, the first 50 meters are not directly involved in the system flexibility. System flexibility is made up of the tube from the switching panel to the DPPS and the DPPS itself.

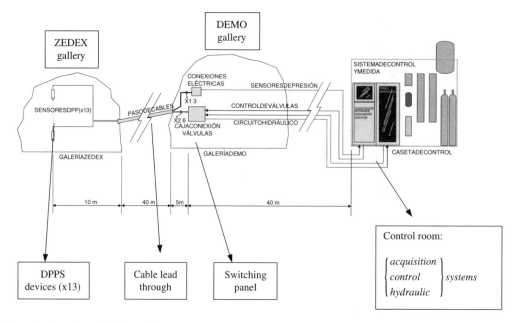

Figure 6. Overview of the layout installed at the ZEDEX gallery (AITEMIN, 1999).

6 POSITIONING OF THE MINI-PIEZOMETERS IN SITU

Three mini-piezometers were placed at layer 2, seven at layer 3 and three at layer 4. Three different orientations were conceived to install them: perpendicular to the layer, parallel to the layer and horizontal (parallel to the axis of the tunnel). Figure 7 shows, for instance, DPP1 after its collocation in the ZEDEX gallery close to the roof. Piezometers were differently oriented in order to study the backfill local hydraulic conductivity in different directions. During the installation of the DPPS, a dummy was used to avoid damaging the DPPS .

Disturbing effects as skin effects, preferential flow paths, etc., around sensors and tubes, could appear, but backfill swelling capacity should be enough to erase these effects in those areas where backfill was well compacted as demonstrated in the laboratory scale (Mata & Ledesma, 2003a). Areas close to the roof and ground of the tunnel were less compacted (Gunnarsson et al, 2001) and as a result, backfill swelling capacity could become almost negligible. Thus, it is expected higher backfill hydraulic conductivity and higher backfill compressibility in the areas where smaller backfill dry specific weight was obtained after the backfill compaction.

The saturation process of the gallery started in mid 1999 and water used to saturate the backfill in situ contains 16 g/L in order to increase backfill hydraulic conductivity and therefore, to speed up the saturation process. Water used in the pulse tests comes from the surrounding aquifer and it contains 12 g/L in average. As the amount of water involved in the test is small, it is not assumed any variation of hydraulic conductivity due to different salt concentration. Full backfill saturation was expected by late 2002 or beginning of 2003 (Mata, 2003).

Since the installation of the sensors, six devices seem to be broken down (late March 2003). Devices DPP1, DPP2, DPP4, DPP6, DPP12 are not working and no response was obtained from them. DPP13 is not working properly as it presented erratic behaviour. Sensors placed at inner parts of the layers of section A4 were subjected to higher compaction efforts as, for instance, DPP2, DPP4, DPP12 or DPP13. The higher compacting effort appears to be the most likely reason to justify why they broke down.

It was decided to perform some pulse tests even before full backfill saturation might have been reached in order to set a procedure and to check the usefulness of the layout. Nevertheless, it is believed that full backfill saturation has been reached as positive water pressure was measured in all the devices. Eighteen injecting pulse tests were performed. Some of the pulse tests were numerically analysed in order to determine the local backfill hydraulic conductivity in different areas of section A4. A complete in situ pulse test campaign will be completed throughout 2003 and 2004. After the

Figure 7. Detail of a DPPS placed in parallel with the layer (AITEMIN, 1999).

analysis of the pulse tests, a map of local backfill hydraulic conductivity will be obtained and compared with global backfill hydraulic conductivity estimated from global flow tests to be carried out throughout 2003 and 2004 in the gallery.

7 IN SITU TESTS IN THE ZEDEX GALLERY

Prior to the tests, some water was flowed into all the piezometers and air might have been trapped was removed. Moreover, an external measurement of the water pressure within the minipiezometers was done in all of them. The results showed that, apparently, backfill saturation had been already reached. The maps of relative water pressure show clearly that areas close to the wall of the gallery are fully saturated and measured water pressure is similar to water pressure applied in the mats (around 510 kPa, Goudarzi et al, 2002).

Pulse tests were performed in devices DPP3, DPP5, DPP7, DPP8, DPP9 and DPP10. An increase of water pressure (750 kPa) was applied in the interphase and transmitted to the switching panel. Once the mini-piezometer has been selected, the valve connecting the pressurised system to the DPPS opens during some seconds (t_0) and then closes. Depending on the permeability of the surrounding backfill, longer time is necessary to increase water pressure within the DPPS. It is difficult to prescribe a priori the maximum water pressure reached in the pulse test

in situ, p_1. Some water is injected during t_0 and thus, water pressure within the DPPS increases. The increase of water pressure is measured a posteriori and depends on system flexibility (tube and DPPS deformation), water head losses within the system, and backfill compressibility and hydraulic conductivity. The flow meter installed in the system measures the mass of water involved in the tests and the maximum measurable flow rate is 65 kg/h (18 g/s). The total amount of water used in each test is measured with a maximum precision of 1 ml.

The water pressure evolution at two sensors where pulse tests were performed is shown in figures 8 and 9 (from 03-25-2003 to 04-02-2003). It is important to note that dissipation of excess of water pressure after the tests in areas close to the roof or the ground are faster than those performed in inner parts of the section. The minimum variation of water pressure detected by the piezorresistive transducers is 5 kPa due to its accuracy and above all, to its high range of measurement (0–5.88 MPa). Therefore, in those devices, where the increase of water pressure was small (DPP8 or DPP10), the dissipation process is "stairs" shaped, which makes difficult its analysis.

It is important to find out the amount of water that does not flow into the backfill but it is included in the measurement. The "extra" amount of water would produce an overestimation of hydraulic conductivity. Flexibility of the tube used in the system was experimentally determined. Flexibility of 5 metres of tube was measured and flexibility of 50 metres is estimated as $1.416 \cdot 10^{-3}$ ml/kPa if linear elastic behaviour of the tube is assumed. As it is shown in figure 4, the distance between the switching panel and the pressure system in the control office is around 50 meters. Therefore, it is possible to estimate the amount of water used to pressurise this part of the system, which is not directly involved in the test as ΔV_r (ml) $=$ ΔV_m (ml) $- \Delta p$ (kPa) $\times 1.416 \cdot 10^{-3}$ ml/kPa. Table 3 summarises the data of the pulse tests analysed.

8 NUMERICAL ANALYSIS OF THE PULSE TESTS

In order to analyse the pulse tests, a finite element code to solve THM problems in a coupled way was used (CODE_BRIGHT, Olivella et al, 1996). Parameters were modified in a trial error procedure to fit the dissipation processes. Two different geometries were considered at this preliminary stage. Both geometries were 2D with axisymmetric geometry. Gravity was not taken into account and consequently, a quasi-3D analysis could be done. However, the real geometry of each sensor is 3D as most of them are placed very close to the host rock. Figure 10 shows the geometry considered to analyse pulse tests where boundary effects of host rock are thought to be negligible (DPP7 and DPP9). The geometry used for pulse tests in sensors DPP3, DPP5, DPP8 and DPP10 is similar but smaller in size. Boundary conditions were the same in both cases. Nevertheless, it is difficult to know the real boundary conditions in those sensors close to the host rock. Figures 11 and 12 show the different positions

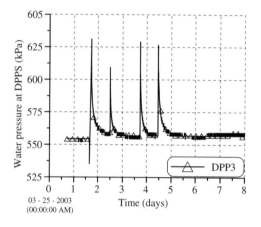

Figure 8. Evolution of water pressure at DPP3. Three pulse tests were carried out, and the simulated one was the fourth one.

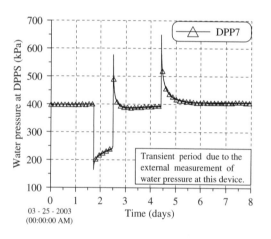

Figure 9. Evolution of water pressure at DPP7. Two pulse tests were carried out in this sensor and both were simulated.

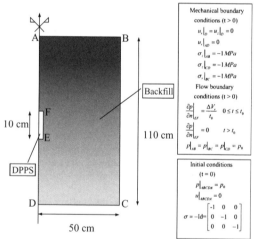

Figure 10. Geometry, boundary and initial conditions (both mechanical and flow) used in the numerical simulation of the in situ pulse tests far away from the rock.

Table 3. General data of the pulse tests carried out in section A4.

Sensor	p_0 (kPa)	p_1 (kPa)	Δp (kPa)	ΔV_m (ml)	ΔV_r (ml)	f_{obs} (m³/MPa)	t_0 (s)
DPP3	552.0	635.7	83.7	4	3.88	$4.633 \cdot 10^{-5}$	20.2
DPP7	403.0	650.0	257.0	4	3.47	$1.355 \cdot 10^{-5}$	20.1

and arrangements of the sensors. Another important difficulty arises when estimating backfill porosity around each sensor. In the inner part of section A4 it is assumed that dry specific weight is 16.6 kN/m³ (porosity, n = 0.368, or void ratio, e = 0.582). Nevertheless, close to the roof or ground, porosity is much lower (Gunnarsson et al, 2001). Backfill dry specific weight in this area was chosen as 13.7 kN/m³ (n = 0.480, e = 0.921). The specific weight of water used in the calculations was 0.0098 MPa/m and water compressibility was $4.5 \cdot 10^{-4}$ MPa^{-1}. Again an isotropic, homogeneous and linear elastic model was used to simulate the behaviour of the backfill.

– **DPP3**: It is placed at layer 2 close to the rock and measured initial water pressure is 552 kPa. Initial water pressure is a little bit higher than the expected one, probably due to the calibration factor relating the electric signal and water pressure. The parameters used in the finite element simulation were $k = 1 \cdot 10^{-10}$ m/s and m* = 0.150 MPa^{-1}. Comparison among the measured values and both calculations is shown in figure 14. The backfill porosity used in the calculation was 0.48.

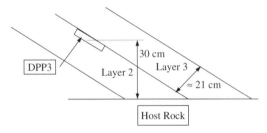

Figure 11. Location of DPP3 in layer 2 of section A4. It is relatively close to the host rock and its position is parallel to the face of the compacted layer.

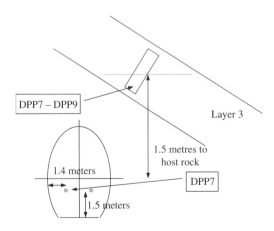

Figure 12. Location and position of sensors DPP7 and DPP9 close to the center of layer 3 in section A4.

– **DPP7**: It is placed at layer 3 far away from the rock and measured initial water pressure is 405 kPa. The backfill porosity used in the calculation was 0.36. Comparison among the measured values and the finite element calculation is shown in figure 14. The parameters obtained were $k = 1.5 \cdot 10^{-11}$ m/s and m* = 0.052 MPa^{-1}. The agreement between the model and the measurement is very good. Moreover, the estimated parameters are in very good agreement as well with estimated backfill hydraulic conductivity and compressibility from an oedometer test performed in laboratory in similar conditions of compaction and salt concentration in water used to saturate the specimen (Mata & Ledesma, 2003b).

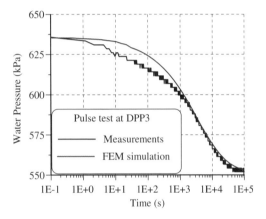

Figure 13. Comparison between the measured pulse test performed in DPP3 and the finite element simulation.

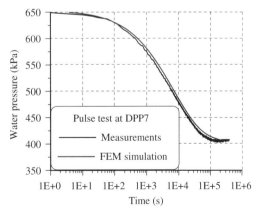

Figure 14. Comparison between the measured pulse test performed in DPP7 and the finite element simulation.

9 CONCLUSIONS

A new mini-piezometer with a pressure transducer inside has been designed and built to perform pulse tests and constant head tests in compacted clayey soils. Moreover, the whole system necessary to perform such tests in situ was also developed and built. The purpose of the system is to investigate the hydraulic conductivity of the backfill compacted at the ZEDEX gallery by means of pulse tests. Those tests have been numerically analysed with a THM coupled finite element code.

Eighteen pulse tests were performed in the ZEDEX gallery in late March 2003. Prior to the pulse tests, trapped air was removed from the mini-piezometers and water pressure was measured in an external pressure transducer. Measured water pressure within the mini-piezometers shows that the saturation process is, probably, finished. Six out of the thirteen mini-piezometers are actually out of order after almost 4 years since the installation. Six pulse tests out of eighteen were analysed by means of CODE_BRIGHT. From the analyses, it was possible to plot a map of backfill compressibility and hydraulic conductivity at section A4.

The calculations confirmed that lower hydraulic conductivity is measured in areas where the compaction effort was higher than the estimated permeability in areas where the compaction effort was not so high (close to the host rock). Estimated backfill hydraulic conductivity in sensor DPP7 is in agreement with estimated backfill hydraulic conductivity after oedometer tests performed in specimens compacted at a dry specific weight of 16.6 kN/m^3 and saturated with water containing 16 g/L of salts (Mata & Ledesma, 2003b). The range of estimated local backfill hydraulic conductivity is large (up to two orders of magnitude) and that will complicate the analysis of the global flow tests to be performed throughout 2003 and 2004. Influence of 3D effects on the problem is important and real boundary conditions (flow and mechanical ones) are not known in those sensors close to the host rock.

The most important consequences are the validity of the pulse test in the laboratory and in situ to estimate local hydraulic conductivity of low permeability media in such conditions and, finally, the reliability of the layout designed and developed by AITEMIN and the Department of Geotechnical Engineering and Geosciences of UPC.

ACKNOWLEDGEMENTS

First author thanks the economical support provided by Catalonian Government through a FI grant. The financial support from ENRESA throughout the project is also acknowledged. The authors want to thank AITEMIN; SKB and Clay Technology for their cooperation. The authors also thank the cooperation of Tomás Pérez.

REFERENCES

AITEMIN (1999). Äspö, Backfill and plug test. Sistema de medida local de permeabilidad mediante ensayos de pulso. Desarrollo, puesta a punto e instalación. Internal Report. Cod. Adj. 703267, Madrid, Spain (In Spanish).

Brand, E.W. & Premchitt, J. (1982). Response characteristics of cylindrical piezometers. Géotechnique, 32 (3): 203–216.

Goudarzi, R., Gunnarsson, D., Johannesson, L-E. & Börgesson, L. (2002). Äspö Hard Rock Laboratory. Backfill and Plug Test. Sensors data report (period 990601-020101). Report N° 4. International Progress Report IPR-02-10. SKB, Sweden.

Gunnarsson, D., Börgesson, L., Hökmark, H., Johannesson, L.E. & Sandén, T. (2001). Äspö Hard Rock Laboratory. Report on the installation of the Backfill and Plug Test. International Progress Report IPR-01-07.

Madsen, F.T. (1998). Clay mineralogical investigations related to nuclear waste disposal. Clay Minerals, 33: 109–129.

Mata, C. (2003). Hydraulic behaviour of bentonite based mixtures in engineered barriers: The Backfill and Plug Test at the Äspö HRL (Sweden). PhD thesis. Dept. of Geotechnical Engineering and Geosciences. UPC, Barcelona, Spain.

Mata, C. & Ledesma, A. (2003a). Permeability of a bentonite-crushed granite rock mixture using different experimental techniques. Géotechnique, 53 (8): 747–758.

Mata, C. & Ledesma, A. (2003b). Hydro-chemical behaviour of bentonite – based mixtures in engineered barriers. This issue.

Mitchell, J.K. (1993). Fundamentals of Soil Behaviour. 2nd edition, John Wiley, NY.

Neuzil, C.E. (1982). On conducting the modified 'Slug' test in tight formations. Water Res. Res., 18 (2): 439–441.

Olivella, S., Gens, A., Carrera, J. & Alonso, E. E. (1996). Numerical formulation for a simulator (CODE_BRIGHT) for the coupled analysis of saline media. Engineering Computations, 13: 87–112.

Push, R. (1982). Mineral-water interactions and their influence on the physical behaviour of highly compacted Na bentonite. Canadian Geotechnical Journal, 19: 381–387.

Advances in Understanding Engineered Clay Barriers – Alonso & Ledesma (eds)
© 2005 Taylor & Francis Group, London, ISBN 04 1536 544 9

Measurements of water uptake in the buffer and the backfill of the Äspö Prototype Repository using the geoelectric method

K. Wieczorek & T. Rothfuchs

Gesellschaft für Anlagen- und Reaktorsicherheit (GRS) mbH (Germany)

ABSTRACT: Water uptake of the bentonite buffer in the deposition holes and of the bentonite/crushed rock backfill in the disposal gallery is one of the central issues to be investigated in the frame of the Prototype Repository Project performed at the Äspö URL. GRS uses the geoelectric method to monitor changes in water content, as the electric resistivity of the materials is determined by the solution content. In order to measure the electric resistivity distribution, two cross sections of the gallery and the buffer at the top of one of the deposition boreholes were instrumented with electrodes connected to an automatic multi-channel geoelectric system. From several hundreds of individual dipole-dipole measurements, the resisitivity distribution is derived by inverse modelling. For relating the obtained resistivities to water content, calibration measurements using the real buffer and backfill materials and Äspö solution were performed in the laboratory. The results obtained up to now illustrate the continuous water uptake of the backfill, while saturation of the buffer is not detectable during the first four months after installation. Measurements in the rock indicate that there has been a slight desaturation during the time the deposition holes were open and pumped out regularly which is now recovering.

1 INTRODUCTION

Within the framework of the Swedish nuclear programme, SKB has constructed a full-scale replica of the deep repository planned for the disposal of spent nuclear fuel. This Prototype Repository has been constructed at the Äspö Hard Rock Laboratory at a depth of 450 m below ground. The Prototype includes six deposition holes in which full-size canisters with electrical heaters have been placed and surrounded by a bentonite buffer. The deposition tunnel has been backfilled with a mixture of bentonite and crushed granitic rock.

Water uptake of the bentonite buffer in the deposition holes and of the bentonite/crushed rock backfill in the disposal gallery is one of the central issues to be investigated in the frame of the Prototype Repository Project performed at the Äspö URL. GRS uses the geoelectric method to monitor changes in water content, as the electric resistivity of the materials is determined by the solution content. Several electrode arrays in the buffer, the backfill, and the rock serve for determining the resistivity distributions in cross sections of the different materials. From the resistivity the solution content can be derived using the results of calibration measurements on samples at defined conditions.

2 OBJECTIVES

The objectives of the measurements are to determine the water uptake with time in the backfill and in the buffer and the potential desaturation of the rock around the deposition holes as a consequence of buffer saturation.

3 GEOELECTRIC INSTALLATION IN THE PROTOTYPE REPOSITORY

An overview of the electrode locations in the prototype repository is given in Figure 1. In Section I on the left hand side of the figure, there is only an electrode array in the backfill above deposition hole #3. The electrodes were arranged on an inclined plane during backfilling of the section in October 2001. The electrode arrangement in the plane is a double cross, as shown in Figure 2.

Three electrode arrays were installed in Section II (Fig. 1):

– Three chains of 30 electrodes each with a spacing of 10 cm were grouted in boreholes between the deposition holes #5 and #6 at depths between 3.45 and 6.35 m in January 2002.

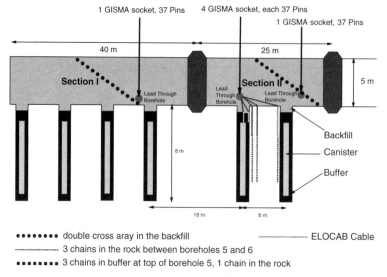

double cross aray in the backfill ⟶ ELOCAB Cable

3 chains in the rock between boreholes 5 and 6

3 chains in buffer at top of borehole 5, 1 chain in the rock

Figure 1. Overview of electrode arrangements in the Prototype Repository.

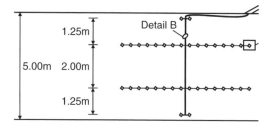

Figure 2. Arrangement of backfill electrodes in the inclined plane (view from above).

– A total of 33 electrodes were installed in the buffer at the top of deposition hole #5 and in an adjacent borehole in the rock, as shown in Figure 3. The installation was performed in April 2003.

– A double cross array identical to that installed in Section I was installed above deposition hole #6 in June 2003.

The electrodes consist of stainless steel. All electrodes placed in boreholes in the rock were grouted with concrete. The backfill electrodes were pressed directly into the backfill material and afterwards covered with backfill. The electrodes in the buffer were fixed to a plastic rod placed in boreholes which were afterwards filled with pounded bentonite powder, except for those on the buffer surface (Fig. 3) which were pressed into the buffer material.

All electrodes were connected via watertight cables to the automatic multichannel measuring system RESECS (Geoserve, Kiel/Germany).

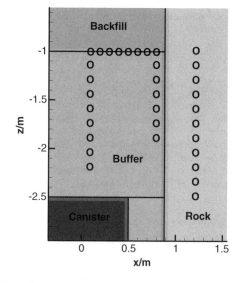

Figure 3. Arrangement of electrodes in the buffer at the top of deposition hole #5 and in the adjacent rock.

4 MEASUREMENT AND EVALUATION TECHNIQUES

The technique most frequently applied for geoelectric measurements in the field is the four-point method (Fig. 4). An electric current is supplied to the formation at two input electrodes (C1, C2). The magnitude and direction of the resulting electric field are dependent on the resistivity conditions in the

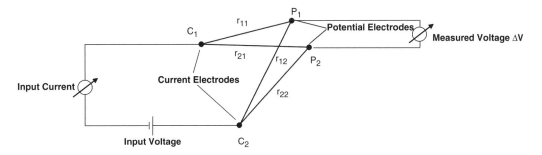

Figure 4. Principle configuration of a dipole-dipole measurement.

rock. The measurement is performed using a second pair of electrodes (P1, P2) at which the potential difference is determined. The input electrodes and the output electrodes are arranged as dipoles.

The result of a single dipole-dipole measurement is the apparent resistivity of the formation, which equals the true resistivity if the formation has an overall constant resistivity. It is calculated by

$$\rho_{app} = \frac{4\pi}{1/r_{11} - 1/r_{12} - 1/r_{21} + 1/r_{22}} \cdot \frac{\Delta V}{I} \quad (1)$$

where

ρ_{app} = apparent resistivity
ΔV = measured potential difference
I = injected current

In the normal case of a spatial varying resistivity, a large number of single measurements with different injection and measurement dipoles is needed. The resistivity distribution can then be derived from the vector of apparent resistivities as best fit between measured data and the response calculated by inverse finite element modelling.

Dipole-dipole measurements are being performed daily with the electrode arrays in the backfill and in the buffer. The evaluation is done using the two-dimensional inversion code SensInv2D (Fechner, 2001), yielding the resistivity distribution in the plane of the electrodes. The optimization method applied is the MSIRT (multiplicative simultaneous iterative reconstruction technique, Kemna, 1995). In the rock boreholes, so-called Wenner measurements are performed, which are a special case of dipole-dipole measurements in that all electrodes are on a line, and the measurement dipole lies between the two injection electrodes with same spacings, which simplifies the geometry factor in equation (1). The measurements in the rock are evaluated in the same way as the measurements in the backfill and the buffer.

The resolution of the inversion method is limited by half the electrode spacing, which therefore defines the finite element size. The accuracy of the model

resistivity distribution as a result of inversion varies with the location. It is controlled by the cumulative sensitivity distribution. Each single measurement configuration gives a so-called sensitivity distribution, which is the matrix of partial derivatives of measured impedance against resistivity. Thus, the sensitivity at a special finite element describes how sensitive the measurement result is to changes in resistivity. The cumulative sensitivity is the sum of the sensitivity matrices of all single measurements, thus describing where a resistivity change has a high influence on the overall results, and where not. Cumulative sensitivity is usually highest near the electrodes and lowest farther away.

In order to relate resistivity distributions to solution content, laboratory testing was performed on backfill, buffer, and rock samples with defined contents of Äspö solution at different temperatures. The measurement procedures are described in a report (Rothfuchs et al., 2001). As an example, Figure 5 shows the resistivity decrease with water content and temperature for the backfill material.

5 RESULTS AND INTERPRETATION

5.1 Backfill Section I

The array in the backfill in Section I was the first one to be installed. Measurements started in October 2001. Figure 6 shows on the left side the resulting resistivity distribution of the first measurement. The initial resistivity of the backfill ranges around 10 to 14 Ωm, corresponding to a water content of 13 to 14%. In the lower part of the backfill the resistivity is somewhat higher, which is due to a lower density of the backfill near the floor (a consequence of the installation and compaction procedure).

Since the first measurement the resistivity of the backfill has been steadily decreasing, starting near the walls of the tunnel and continuing to the center. The right side of Figure 6 shows the resistivity distribution at the end of August 2003. The resistivity has decreased to 3 to 5 Ωm near the tunnel walls and to

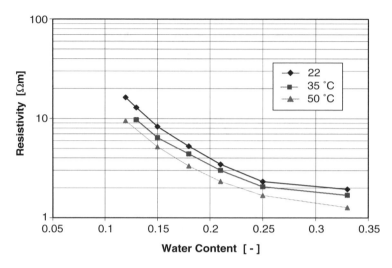

Figure 5. Resistivity of Compacted Backfill versus Water Content at different temperatures.

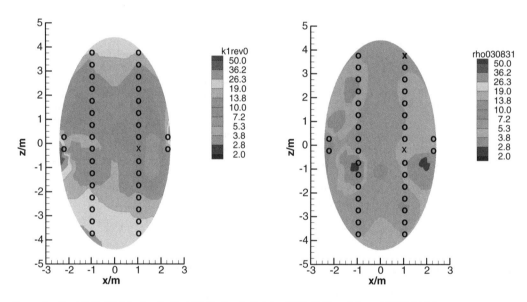

Figure 6. Resistivity Distribution in Backfill Section I, October 2001 (left) and August 2003 (right).

about 6 Ωm in the center. In terms of water content, this corresponds to 16 to 17% of water. There has, however, also been a slight temperature rise in the tunnel (to maximal 32°C) which may cause an additional resistivity decrease. This is limited, however, to less than 1 Ωm (Fig. 5), which means water content may be lower by 0.5 to maximal 1%.

The water uptake from the tunnel walls and the lower saturation in the center is also confirmed by suction measurements performed by Clay Technology (Goudarzi & Börgesson, 2003).

5.2 Backfill Section II

The array in the backfill in Section II has only been installed in June 2003. The results of the first measurement (Fig. 7, left side) show a much lower resistivity than the early measurements in Section I. Obviously, the backfill had a considerably higher water content already during installation. This observation was also made during instrumentation. Resistivity is decreasing further from the drift walls (Fig. 7, right side). Close to the walls it ranges around

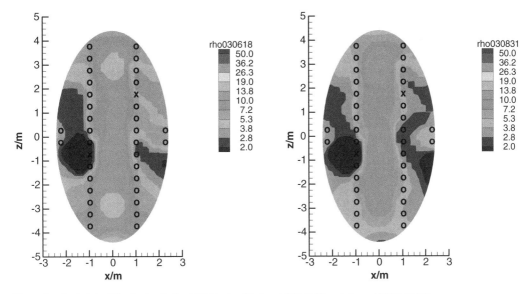

Figure 7.   Resistivity Distribution in Backfill Section II, June 2003 (left) and August 2003 (right).

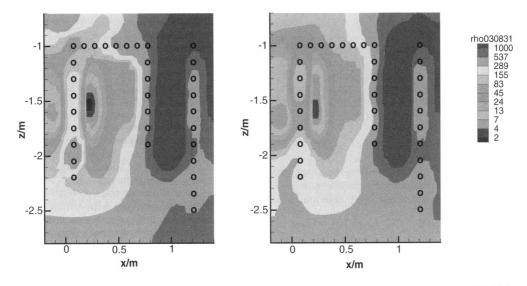

Figure 8.   Resistivity Distribution in the buffer at the top of deposition hole #5, May 2003 (left) and August 2003 (right).

3 Ωm; the backfill is therefore not far from full saturation. In the center resistivity is between 7 and 10 Ωm corresponding to a water content of about 14 to 16%.

5.3   *Buffer*

The results of the first measurement taken in May 2003 (Fig. 8, left side) show the high resistivity (above 1000 Ωm) of the rock on the right side and the low resisitivity of the buffer (below 80 Ωm) on the left. The picture is somewhat distorted by the fact that along the electrode chains the resistivity is increased compared to the undisturbed buffer. The increased resistivity along the electrode chains can be attributed to the refilling of the electrode boreholes with bentonite powder produced during borehole drilling. It is, however, expected that the difference will diminish with time, especially if the buffer takes up water. Until August 2003, however, such an effect

111

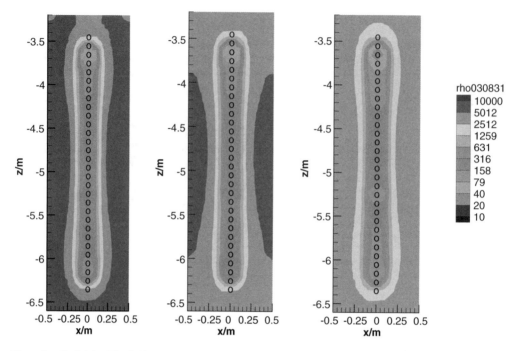

Figure 9. Resistivity Distribution along electrode chain KA 3550 G04 in the rock near deposition hole #5, August 2002 (left), April 2003 (center), and August 2003 (right).

cannot be seen clearly, meaning that the buffer has not taken up significant amounts of water yet.

5.4 Rock

The resistivity distributions along the three electrode chains installed in the rock are quite similar to each other and show no significant variation in time until April 2003. Close to the electrodes, the resistivity ranges around 200 Ωm. This value characterizes the water-saturated concrete used for backfilling the electrode boreholes. Further away from the boreholes, the resistivity rises to values of 2000 to 7000 Ωm which is characteristic for granite. Figure 9 (left side) shows the resisitivity distribution around the chain KA 3550 G04 in the rock near deposition hole #5 as an example. From April 2003 on, there is a slight decrease in resistivity in the rock near deposition hole #5 (Fig. 9, center and right side). This coincides with installation of the buffer which also stopped the pumping of water from the open deposition hole. Apparently, this had caused a slight desaturation of the rock which is now recovering. This interpretation is supported by the fact that the other electrode chains in the rock show the same behaviour with a temporal delay.

6 CONCLUSIONS

The results of the geoelectric measurements in the Prototype Repository obtained up to now show that the method is capable of monitoring water content changes in the backfill, the buffer, and the rock. The results are plausible and coherent.

The backfill continuously takes up water from the rock. In Section II, the water release from the rock takes place at a much higher rate than in Section I. Saturation of the buffer has not been detectable during the first four months after installation and geoelectric measurement. Measurements in the rock indicate that there has been a slight desaturation during the time the deposition holes were open and pumped out regularly. Saturation of the rock is now recovering.

ACKNOWLEDGEMENTS

The work was co-funded by the German Bundesministerium für Wirtschaft und Arbeit (BMWA) under contract No. 02E9279 and by the Commission of the European Communities (CEC) under contract No. FIKW-CT2000-00055. The authors gratefully thank for this support.

REFERENCES

Fechner, T. (2001) SensInv2D-Manual, Geotomographie, Neuwied.

Kemna, A. (1995) Tomographische Inversion des spezifischen Widerstandes in der Geoelektrik, Master Thesis, Universität Köln.

Rothfuchs, T. Komischke, M., Miehe, R., Moog, H., Wieczorek, K. (2001) Äspö Hard Rock Laboratory – Prototype Repository – Geoelectric Monitoring in Buffer, Backfill and Rock, IPR-01-63, Svensk Kärnbränslehantering AB (SKB), Stockholm.

Goudarzi, R., Börgesson, L. (2003) Äspö Hard Rock Laboratory – Prototype Repository – Sensors data report (Period: 010917-030301), IPR-03-23, Svensk Kärnbränslehantering AB (SKB), Stockholm.

Hydro geological and hydro mechanical monitoring of the rock mass in boreholes during the operation phase of the Prototype Repository, Äspö HRL, Sweden

Ingvar Rhén, Torbjörn Forsmark & Patrik Alm
SWECO VIAK, Gothenburg (Sweden)

ABSTRACT: The Prototype Repository is an international, EC-supported activity with the objective to investigate, on a full-scale, the integrated performance of engineered barriers and near-field rock of a simulated deep repository. This is done in crystalline rock regarding heat evolution, rock mechanics, water flow, water chemistry, gas evolution and microbial processes under natural and realistic conditions at approximately 450 m depth below the ground surface. The test site is a 65 m long TBM-bored drift from which six 1.75 m diameter deposition holes extended downwards to about 8 m depth in accordance with the KBS-3 concept. The test site is divided in two parts; an inner 40 m long section (Section I) with 4 deposition holes and an outer section (Section II) with two deposition holes. Stiff and tight plugs will separate the sections and Section II from the rest of the Äspö Hard Rock Laboratory.

A large number of boreholes have been drilled to characterize the rock mass. These boreholes will be used for the long-time monitoring of the Prototype Repository. Packers, 1–5 in each borehole, are installed to facilitate monitoring of the water pressure and water chemistry in borehole sections. Temperature and deformation sensors are installed in some of the boreholes sections. Tubes and cables from the borehole sections are lead to the nearby drift called the G-tunnel, where the pressure, deformation and temperature are measured and the water is sampled. Some sections are also equipped to facilitate dilution measurements (circulation sections) to get approximate flow rates through rock during undisturbed conditions or during hydraulic tests.

Continuous monitoring is made of the water pressure in 144 borehole sections (12 of these are currently planned to be made). For hydromechanical studies, the rock deformation and temperature is measured in 11 borehole sections, of which one is a reference section at some distance from the Prototype tunnel. There are two sensors in each hydromechanical borehole section, one sensor measuring over the intact rock (unfractured) and one sensor measuring over one or more fractures. 17 circulation sections and 8 hydrochemical borehole sections are available for test or sampling campaigns. PEEK tubes are connected to the hydrochemistry sections to get a diffusion tight system allowing studying the redox conditions.

Several hydraulic test campaigns will be performed from the G-tunnel by flowing of borehole sections (one by one) and measuring the pressure responses in the rock mass. Most tests will be made during the first years of operation when the largest rock deformations are expected to be measured.

In order to investigate the hydro mechanical response of the fractures as a result of the increased thermal load, two different approaches are considered. The first approach is to measure the change of the fracture width as function of temperature and time. The second approach implies that the mechanical response is evaluated indirect by using the results from hydraulic tests. Hydro tests will be performed in the same sections as the mechanical measurements are made. Dilution measurements can also be performed in the same sections.

1 BACKGROUND

The Äspö Hard Rock Laboratory is an essential part of the research, development, and demonstration work performed by SKB in preparation for construction and operation of the deep repository for spent fuel. Within the scope of the SKB program for RD&D 1995, SKB has decided to carry out a project with the designation "Prototype Repository Test". The aim of the project is to test important components in the SKB deep repository system on a full scale and in a realistic environment.

The Prototype Repository Test is focused on testing and demonstrating the function of the SKB deep repository system. Activities aimed at contributing to development and testing of the practical, engineering measures required to rationally perform the steps of a deposition sequence are also included. However, efforts in this direction are limited, since these matters are addressed in the Demonstration of Repository Technology project and to some extent in the Backfill and Plug Test.

This article covers parts of the hydrogeological investigations made during the operation phase of the Prototype Repository and the monitoring in tunnel section II is described in more detail.

For establishing models of larger hydraulic features defined in space and the conductive fracture system, the mapping of cores, tunnels and deposition holes is of course essential information needed besides the hydraulic tests. Hydraulic tests and geological investigations made for the hydrogeological descriptive model of the Prototype repository is not included in the article but is summarized in Rhén et al (2001).

1.1 Objectives

The Prototype Repository should simulate as many aspects as possible of a real repository, for example regarding geometry, materials, and rock environment. The Prototype Repository is a demonstration of the integrated function of the repository components. Results will be compared with models and assumptions to their validity.

The major objectives for the Prototype Repository are:

– To simulate part of future KBS-3 deep repository to the extent possible regarding geometry, design, materials, construction and rock environment except that electric heaters simulate radioactive waste.
– To test and demonstrate the integrated function of the repository components under realistic conditions on a full scale.
– To develop, test and demonstrate appropriate engineering standards and quality assurance methods.
– To accomplish confidence building as the capability of modelling EBS performance.

The objectives for the characterisation program are:

– To provide a basis for determination of location of the deposition holes.
– To provide data on boundary and rock conditions for enabling the interpretation of the experimental data.

The objectives with the instrumentation presented in this article are mainly linked to the last point above. The objectives with hydromechanical measurements and hydraulic tests during the operation phase is to get quantitative measurements of the deformation of fractures and transmissivity changes of those fractures

due to the stress changes caused by the heat from the canisters and compare these results to theoretical models for this deformation and transmissivity changes. Hydraulic tests and the long-time monitoring are also considered valuable for the analysis and groundwater flow modelling of the highly heterogeneous fractured rock that surrounds the deposition tunnels with backfill and the canister holes.

2 HYDROGEOLOGICAL INSTRUMENTATION, OVERVIEW

During the characterisation of the rock around the Prototype Repository a large number of core boreholes have been drilled (Fig. 1). Most of these boreholes have been equipped with packer systems to allow for:

– Pressure measurements.
– Water sampling.
– Dilution measurements.
– Hydraulic single-hole and interference tests.
– Hydro mechanical (HM) measurements.

In Table 1 the type of measurement sections and packer systems are summarized.

The bentonite packers were made up by compacted bentonite with rubber coverage. For chemical reasons the bentonite is not allowed to be in contact with the surrounding water in the rock mass and the packers have a cover made of polyurethane (PUR-rubber). The expected long duration of monitoring of tunnel section No I is the reason for chosing this type of packers, which were developed by SKB. PEEK tubes are connected to the "hydrochemistry sections" to get a diffusion tight system allowing studying the redox conditions.

The inflow to the tunnels is measured is at several points. Measurement dams have been constructed in A, G and F-tunnels. Measurement dams are also planned to be constructed near the outer plug in the Prototype repository (close to I-tunnel). These data are needed for the groundwater flow modelling of the Prototype Repository volume.

3 EQUIPMENT AND INSTRUMENTATION OF BOREHOLES IN TUNNEL SECTION II

Boreholes in tunnel section II were equipped with hydraulically expanded packers of one meters length to seal off at most five sections in one borehole. In ten of these boreholes one section also were instrumented with hydro-mechanical equipment adapted to measure small deformations in the solid rock and over selected fractures. Mechanical packers were installed in short boreholes. Another borehole in the G-tunnel

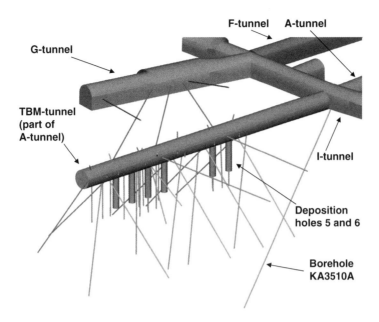

Figure 1. View of the drilled core holes in the Prototype Repository. The length from the I-tunnel to the end of the TBM-tunnel is 90 m. The diameter of the TBM-tunnel is 5 m and the diameter of the deposition holes is 1.75 m. The depth of the deposition holes is 8.37 m in the centre and 8.15 m along the deposition hole wall. The diameter of the core holes is 76 mm except for the short core holes in the roof of the TBM-tunnel that have a diameter of 56 mm.

Table 1. Hydrogeological instrumentation in the Prototype Repository for the operation phase.

Type of instrumentation*	Prototype tunnel		G-tunnel + KA 3510A	Outer plug (planned)	Sum
	Section I	Section II			
Bentonite packers (1–2 m long, 2–5 packers in each 8–50 m borehole)	49	–	–	–	49
Inflatable packers (1 m long, 1–5 in each 3–30 m borehole)	–	46	15	12	73
Mechanical packers (one in each 2 m borehole, stainless)	16	6	–	–	22
Pressure measurement sections	65	52	15	12	144
Circulation sections (two tubes)	5	10	2	0	17
Flow sections (one-two tubes)	7	3	1	0	11
Hydrochemical sections	6	0	2	0	8
Hydromechanical sections	0	10	1	0	11

* The tube types between and from packers:
- Pressure: Polyamide
- Circulation (Dilution measurements):
  - Circulation (two pipes):   Polyamide
  - Pressure: Polyamide
- Flow (one pipe):   Polyamide
- Hydrochemical
  - Flow (for sampling):   PEEK
  - Pressure: Polyamide

(Fig. 1) was instrumented with HM equipment as a reference. The borehole was drilled in the north tunnel wall and is not expected to be influenced by the stress changes around the Prototype tunnel.

3.1.1 *Inflatable packers*

The packers are of the type PU53 or PU72 designed by SKB. All packers have an inflation length of one meter and the minimum and maximum packer

117

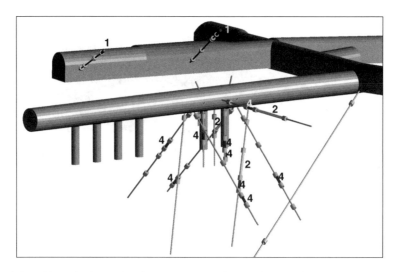

Figure 2. Boreholes with monitoring sections in Prototype Repository Section II. Wide cylinder: Packer, NO NUMBER: Pressure section (P), 1: Hydrochemical section (HC) + P, 2: Flow section (F) + P, 3: Circulation section (C) + P, 4: Hydro Mechanical section (HM) + C + P.

expansion pressure is 6.5 bar and 65 bar respectively. They are expanded by means of water, pressurised by nitrogen gas in a pressure vessel. A regulator controls the magnitude of the inflation pressure. The stainless steel pressure vessel is connected to the packers by a high-pressure 6/4 mm polyamide tube, type Tecalan. A check valve unit with a manometer is mounted on the packer inflation line. In order to avoid accidental deflation the check valve unit also includes a stop valve.

All tubes and pipes were temporarily plugged at the borehole collar until the backfill phase when they were connected to tubes leading to the G-tunnel via the lead-trough boreholes. The electrical cables from the deformation transmitters end up in a special "box" prepared for casting after connection during the backfill phase.

All tubes, pipes and cables coming out of the boreholes were labelled with a plastic tab with a unique identification number. This was needed later on, during the backfill phase, when connection to the cable/tube bundles leading to the G-tunnel via the lead-trough boreholes takes place. All tubes and pipes inside the boreholes were attached to the stainless steel rods by plastic tape.

In the G-tunnel, section tubes from the cable/tube bundles were connected to pressure transducers (one per section at a transducer panel), circulation tubes to valves and the cables from the deformation sensors to data loggers. When the pressure lines were complete they were de-aired before connection to the transducers. If the flow from a section was very small it was necessary to fill the system with water from the outside. In such a case formation water from a nearby borehole/section was used.

3.1.2 Hydro-mechanical equipment

To examine the hydro-mechanical effects connected to the warming, emanating from the canisters with heating elements that are placed in the deposition holes no 5 and 6, deformation measurements will be performed in ten borehole sections around the Prototype tunnel near these deposition holes. In these borehole sections also pressure will be measured and circulation of water will be possible. In one borehole section in the G-tunnel a reference for deformation measurements is also installed (Figs 1 and 2). The hydro-mechanical equipment is built up of three anchors with two deformation sensors. The anchors are made of stainless steel. In pairs the anchors will measure the movement both in intact rock and over a fracture (or a few fractures) + intact rock (Figs 3 and 4).

Geosigma AB modified the anchors for SKB from a commercial available set of anchors and sensors from Geokon. Special designed packers have lead-through to room the signal cables from the deformation sensors out from the borehole. To be able to position and fasten the anchors to the borehole wall another two pipes delivering pressure from outside is connected to positioning-and lock-cylinders in these sections.

The installation was as follows:

– HM equipment was in laboratory mounted on a centre rod between two packers. The locking screw

118

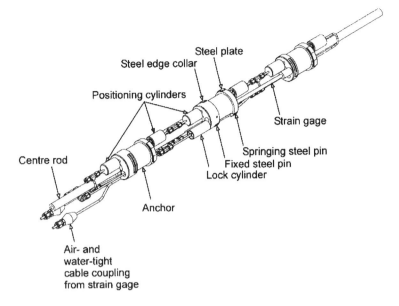

Figure 3. Hydro-mechanical equipment.

Figure 4. Hydro-mechanical equipment mounted on a centre rod between two packers.

for the strain gage was fastened and the position of the anchor was adjusted to give the reading required. Positioning cylinders were attached to the centre rod for locking position of the anchors when pressurerised.

– During installation of the equipment the positioning and lock cylinders were pressurerised with about 85 bar. The springing sheet metal on the anchors holds back the steel pins towards the centre of the anchor. The pressurised position cylinders keep the anchors firmly to the centre rod while installing the equipment.

– The inflatable packers are expanded and tested to assure proper function. Tubes from the measurement section are open.

– The pressure is released to the lock cylinders connected to springing rod. The springing rod presses the solid and springing steel pins against the rock wall and locks the anchor against the rock wall. (The springing steel pin and solid steel plate is expected to glide along the rock wall while the solid steel pin and the steel edge collar is the position of the anchor that is firmly connected to the rock wall (Figs 5 and 6). According to tests in a steel tube, the tensile force is 550–750 N for the section with the solid steel pin compared 150–200 N for the section with the springing steel pin.)

– The pressure is released to the positioning cylinders releasing the anchors from the centre rod.

– The inflatable packers are deflated. (The packers are inflated during the backfilling of the tunnel.)

119

3.1.3 *Connection to lead-through holes and instrumentation in the G-tunnel*

All instrumentation tubes were lead to lead-through holes (Fig. 7) connecting sensors cables and tubes in Tunnel A to Tunnel G. Cables and tubes were embedded in the backfill during the backfilling operation.

In the G-Tunnel the instrumentation, panels with transducers etc. are situated (Fig 8). Some borehole sections can be flowed through separate tubes (circulation and flow tubes) and in a few sections it is possible to flow a pressure tube without disconnecting the pressure transducer. The equipment for pressurizing the inflatable packers is in the G-tunnel. A pressure vessel filled with water and connected to several packers. The packers are expanded by pressurizing the waterfilled pressure vessel with nitrogen gas.

3.1.4 *Measurement accuracy of deformation measurements*

The measurement technique is vibrating wire. The deformation sensors resolution is $<0.26\,\mu m$, the accuracy $<2.1\,\mu m$. (Dependent on the length of the sensors.)

4 HYDRAULIC TESTS DURING THE OPERATION PHASE

Hydraulic tests are performed by flowing a borehole section (mostly from G-tunnel) measuring the drawdown followed by closing the test valve from the test section and measuring the pressure build-up. Single-hole tests will be made a number of times for the HM-sections, see Chapter 5. Interference tests for approximately 7 different flowing sections will be made a few times during the operation phase using all monitoring sections as observation sections.

Dilution tests (the measurement of the groundwater flow through the borehole section during natural gradient or during a hydraulic test. With some assumptions the flow in an intersecting fracture can be made.) will also be made a number of times during the operation phase as well as water sampling for analysis of the water composition. In all HM-sections dilution tests can be performed.

Data is expected to show hydraulic changes (if any significant) during the operation phase and that the data can both be used to evaluate those changes and

Figure 5. The springing sheet metal on the anchors holds back the steel pins towards the centre of the anchor when the springing rod is pressurized.

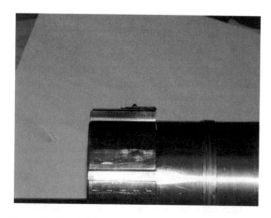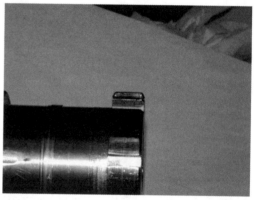

Figure 6. To the left: Steel edge collar. To the right: solid steel plate. The springing steel pin and solid plate is expected to glide along the rock wall while the solid steel pin and the steel edge collar is the position of the anchor that is firmly connected to the rock wall.

serves as data set for groundwater flow simulations in the rock mass.

5 EXPECTED RESULTS FROM THE HM MEASUREMENTS

5.1 Background

During storage of nuclear waste in the rock mass the temperature will increase due to the heat loss from the canisters with spent fuel. This will increase the rock stresses and the fractures will close.

It is of great interest to investigate the magnitude of this effect on the fracture transmissivity since the fracture transmissivity is essential of two reasons. First, enough transmissivity is needed to provide the bentonite buffer with water if no artificial moistening of the buffer is arranged. Secondly, the transmissivity should be as low as possible in order to minimise the hydraulic contact with the canisters. The increased temperature will decrease the transmissivity, which in principal is positive in perspective of Safety Assessment. The last effect is however limited in time and may not be of any greater importance in Safety Assessment.

5.2 Experimental set-up and expected results

In order to investigate the hydro mechanical response of the fractures as a result of the increased thermal load, two different approaches are considered.

Figure 8. In the G-tunnel, tubes from the borehole sections are connected to pressure transducers (one per bh-section at the panel for transducers), circulation and flow tubes to separate valves and the cables from the deformation sensors to electronic devices.

Figure 7. Left: Tubes and cables being installed in a lead-through hole. Right: Lead-through holes with lid fixed.

121

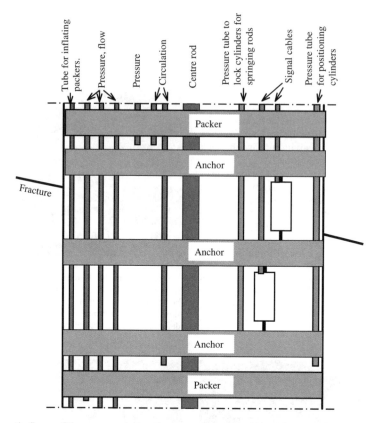

Figure 9. Schematic figure of the measurement equipment and measurement sections.

The first approach is to measure the change of the fracture width as function of temperature and time. The deformation is both measured for the intact rock as for a section with one or more fractures (Fig. 9).

The second approach implies that the mechanical response is evaluated indirect by using the results from hydraulic tests. Hydro tests will be performed in the same sections as the mechanical measurements are made.

Deformation measurements will be made continuously. Hydraulic tests will be made a number of times during the operation period for the ten measurement sections. Most tests will be made during the first years of operation when the largest deformations are expected to be measured.

The stresses expected from the increased temperature have been modelled (Claeson et al, 2001) and these results have been used to estimate the expected decrease in transmissivity of the fractures in Alm and Rhén (2003). The conclusions were:

– The hydraulic response will in general be decreased transmissivity. The decrease will be larger the closer the fractures are to the canisters.

– The transmissivity will be reduced to something between 10 and 80 percent of the in-situ values.
– The fracture closure will be in the range of 5–70 micrometer.
– The hydraulic and mechanical responses on the fracture depend on the orientation of the fracture relative the stress field.
– Major part (~80%) of the increase in stress and temperature is reached within two years.
– The temperature load will be uniformly distributed in the rock mass. This implies that the ratio between $\sigma 1$ and $\sigma 3$ will be more or less constant and the risk of shearing will be low in the rock mass.

The measurements are intended to make it possible to compare the results with the theoretical models used in Alm and Rhén (2003).

The designed HM-equipment has only been tested under conditions in laboratory and not in long-term measurements in the rock, which of course give some uncertainty of its function during the operation phase. However, one part of the HM equipment, the deformation sensor (strain gage in Fig 3) is a commercial product, which is expected to work. If the deformation measurements turn out to be difficult to evaluate of some reason, the hydraulic tests should give indications of any significant hydraulic changes during the operation phase.

REFERENCES

Alm, P, Rhén, I, (2003) (In prep) Hydro-mechanical behaviour of fractures due to excavation and thermal load, Prototype Repository, SKB IPR XX-XX

Rhén, I, Forsmark, T, (2001) Äspö HRL – Prototype Repository, Hydrology, Summary report of investigations before the operation phase, SKB IPR-01-65.

Rhén, I, Forsmark, T, Torin, L, Puigdomenech, I, (2001) Hydrogeological, hydrochemical and temperature measurements in boreholes during the operation phase of the Prototype Repository, Tunnel section I, Prototype Repository, SKB IPR 01-32

Rhén, I, Forsmark, T, Magnusson, J, Alm, P, (2003) Hydrogeological, hydromechanical and temperature measurements in boreholes during the operation phase of the Prototype Repository, Tunnel section II, Prototype Repository, SKB IPR 03-22

Claeson, C, Dahlström, L-O, Sandén, M, (2001) (In prep) Finite element analyses of mechanical consequences due to the rock excavation and thermal load. SKB International Progress Report, IPR-XX-XX, Stockholm.

Advances in Understanding Engineered Clay Barriers – Alonso & Ledesma (eds)
© 2005 Taylor & Francis Group, London, ISBN 04 1536 544 9

Identification of "in situ" stresses from pore water pressure measurements

A. Martínez-Parra, A. Ledesma & E.E. Alonso
Technical University of Catalonia (UPC), Barcelona, Spain

ABSTRACT: The determination of "in situ" stresses in rock massifs is one of the fundamental issues in Rock Mechanics. In many cases, field experiments based on the stress relief of rock samples are used to estimate the orientation and magnitude of principal stresses. In this paper, a method based on the backanalysis of field measurements performed during the excavation of a tunnel is presented. The procedure has been already applied in some cases, but using measured displacements as input data. Here, a similar procedure but based on pore water pressure measurements is described. This is particularly useful when dealing with very hard rocks, for which measured displacements are too small in order to have a reliable interpretation in terms of initial stresses. In those cases, pore water pressure changes due to the stress relief produced by the excavation are important enough to allow performing a backanalysis of the coupled hydromechanical problem. An application to the measurements performed during the excavation of the FEBEX tunnel in Grimsel (Swiss Alps) is also described. The preliminary results shown in the paper are promising and the procedure may eventually become a new method for determining natural stresses in hard rocks.

1 INTRODUCTION

The determination of "in situ" stresses in Rock Mechanics problems is one of the fundamental works that are still far from being a routine in practical applications. Indeed in many underground excavations they constitute one of the bottlenecks in the design process, due to the difficulties encountered when measuring stresses in the field. Recently, a special issue of the Journal of Rock Mechanics and Mining Sciences has been devoted to that topic (IJRMMS, 2003).

Most of the procedures described in the literature for determining initial stresses are based on the stress relief of rock samples. The work by Ljunggren et al (2003) presents a comprehensive review of the methods available for that purpose. They are summarized in table 1, including the volume of rock involved in each method. This aspect is particularly important due to the scale effects that play an

Table 1. Methods for measuring/estimating "in situ" stresses (after Ljunggren et al, 2003).

Category	Method	Rock volume involved (m^3)
Methods performed in boreholes	Hydraulic fracturing	0.5–50
	Overcoring	10^{-3}–10^{-2}
	Hydr. tests on pre-existing fractures	1–10
	Borehole breakouts	10^{-2}–100
Methods performed using drill cores	Strain recovery methods	10^{-3}
	Core-discing	10^{-3}
	Acoustic methods (Kaiser effect)	10^{-3}
Methods performed on rock surfaces	Jacking methods	0.5–2
	Surface relief methods	1–2
Analysis of large scale geological structures	Earthquake focal mechanism	10^9
	Fault slip analysis	10^8
Other	Relief of large rock volumes (back analysis)	10^2–10^3

important role when characterizing rock massifs. Note that back-analysis of the relief of large rock volumes is included in the list in the category "Others". That is, backanalysis of full scale problems do not constitute a typical procedure to determine the natural stress state of the rock. Usually, the rest of categories are more common in practical applications. However, backanalysis of large scale problems involves a volume of rock which is large enough to have a good representation of the stress state at an "engineering scale". Other methods either apply to very small samples or to geological structures, and the change of scale may become a drawback in some cases (for instance in intense faulted massifs).

The use of backanalysis or "inverse problem" in Geomechanics has been usually linked to the determination of strength parameters after a particular failure. This procedure has been always applied in an "ad hoc" manner. However, the development of optimization techniques and numerical tools (Fletcher, 1981) in last decades has allowed performing backanalysis in a more systematic manner. Some of the initial works are due to Cividini et al (1983), Gioda (1985) and Gioda & Sakurai (1987). In general they use field displacements to estimate stiffness parameters (i.e. elastic modulus).

The mathematical problem is defined as a minimization one, where the best parameters are those that minimize a function which depends on the differences between measured and computed variables. In many cases, due to the nature of field data, a probabilistic framework is more convenient in order to deal with the measurement errors. Some contributions based on this approach are due to Ledesma et al (1996a,b) and Gens et al (1996). They include an example of identification of initial stresses from measured displacements in a tunnel excavation problem. A comprehensive description of several application cases may be found in Gens and Ledesma (2000).

It should be pointed out that almost all the works use field displacement as measured variables from which to estimate rock parameters or the initial stresses. An example of application of these techniques considering displacements and pore water pressure as input data may be found in Ledesma and Gens (1997). However, that work refers to an excavation of a tunnel in soft clay. In fact we have not found any reference regarding the use of pore water pressure changes as input data in the identification of rock parameters or "in situ" stresses. This paper investigates the feasibility of that approach considering the probabilistic framework above mentioned. It is considered that for very hard rocks, for which displacements are too small, pore water pressure changes may constitute an appropriate variable to relate with the stress relief, and therefore to the initial stress state of the massif.

The 2D case is considered first, because for elastic conditions some analytical solutions may be used. That allows simplifying some of the mathematics involved, and it is used to understand some aspects of the procedure. Then, a simple 3D case is considered, although more analyses will be done in the future in order to check the capabilities of the method. A final application to the interpretation of the pore water pressure changes produced during the excavation of the FEBEX tunnel is described. This practical case has been in fact the motivation of this work. The piezometers installed in the rock were not designed to measure those changes, but after analyzing the field data it was found out that there was a close relationship between the measured values and the stress relief during excavation of the tunnel. This paper shows the preliminary results obtained in this ongoing research.

2 BASIC FORMULATION OF THE IDENTIFICATION PROBLEM

A deterministic model, M, relating variables that can be measured (i.e. rock displacements or rock pore water pressures), $\mathbf{x}$, and a set of parameters, $\mathbf{p}$, is assumed: $\mathbf{x} = M(\mathbf{p})$. Usually the model includes the geometry of the problem and the constitutive laws of the materials involved. Measurements are represented by $\mathbf{x}^*$, and then differences between measurements and predictions of the model $(\mathbf{x}^* - \mathbf{x})$ are considered as an error, that can be defined in a probabilistic manner.

The best estimation of the parameters is then found by maximizing the likelihood, L, of a hypothesis, $\mathbf{p}$, given a set of error measurements, defined as (Edwards, 1972):

$$L = kf(\mathbf{x}/\mathbf{p}) \tag{1}$$

where $f(\mathbf{x}/\mathbf{p})$ is the conditional probability of $\mathbf{x}^*$ given a set of parameters, $\mathbf{p}$, and k is a proportionality constant. This probability can be assumed as multivariate Gaussian. Also, maximizing (1) is equivalent to minimize the function $S = -2\ln L$, and the "objective function" to be minimized becomes:

$$S = (\mathbf{x}^* - \mathbf{x})^t \mathbf{C}_\mathbf{x}^{-1} (\mathbf{x}^* - \mathbf{x}) + \ln|\mathbf{C}_\mathbf{x}| + m\ln(2\pi) - 2\ln k \tag{2}$$

where $\mathbf{C}_\mathbf{x}$ is the covariance matrix which contains the error structure of the measurements. Assuming that the error structure is fixed, only the first term in (2) must be minimized. The structure of the covariance matrix will depend on the instrument used for measurement purposes. Expressions for that matrix for different instrumentation equipments have been derived elsewhere (Ledesma et al, 1996a).

If measurements are independent and their errors have the same variance, σ^2, then $\mathbf{C}_\mathbf{x} = \sigma^2 \mathbf{I}$, where $\mathbf{I}$ is the identity matrix. In this case, the function to be minimized takes the form:

$$J = (\mathbf{x}^* - \mathbf{x})^t \_ (\mathbf{x}^* - \mathbf{x}) \tag{3}$$

which is equivalent to a "least – squares" criterion. Note that J in (3) is a function on the parameters. Thus the parameters that minimize J give the best estimation within the likelihood framework, and therefore, they are assumed as the solution of the backanalysis problem. This formulation can be generalized to include also prior information on the parameters in the identification procedure (Ledesma et al, 1996b).

Minimization of (2) and (3) can be performed by means of any suitable algorithm. As function J is nonlinear on the parameters, the minimum is reached by means of an iterative procedure. For this type of functions, the use of the Gauss-Newton algorithm has been proved to be efficient. It is based on an expansion of the objective function into a Taylor series, and gives the value of the increment of parameters, $\Delta \mathbf{p}$, in each iteration:

$$\Delta \mathbf{p} = (\mathbf{A}' \mathbf{C}_\mathbf{x}^{-1} \mathbf{A})^{-1} \mathbf{A}' \mathbf{C}_\mathbf{x}^{-1} \Delta \mathbf{x} \tag{4}$$

where $\Delta x = (\mathbf{x}^* - \mathbf{x})$ and the matrix

$$\mathbf{A} = \frac{\partial \mathbf{x}}{\partial \mathbf{p}} \tag{5}$$

is called sensitivity matrix. If the number of measurements is m and the number of parameters is n, the size of this matrix is $m \times n$.

In general Gauss-Newton algorithm tends to exhibit a rapid convergence in this type of problems, although it can be unstable sometimes. In those cases, an improvement of the algorithm proposed by Levenberg and Marquardt (Marquardt, 1963), can be used, in which (4) is modified to:

$$\Delta \mathbf{p} = (\mathbf{A}' \mathbf{C}_\mathbf{x}^{-1} \mathbf{A} + \mu \mathbf{I})^{-1} \mathbf{A}' \mathbf{C}_\mathbf{x}^{-1} \Delta \mathbf{x} \tag{6}$$

where μ is an arbitrary real number. If μ approaches zero, the Gauss-Newton procedure is recovered. As μ increases, the correction of parameters provided by (6) becomes smaller and tends towards the gradient direction of the objective function.

The algorithm described involves the calculation of the sensitivity matrix, $\mathbf{A}$, which becomes a fundamental issue in these procedures. It can be seen that for linear elastic constitutive laws, it is possible to compute that matrix using the finite element formulation. Thus, the same subroutines employed for the analysis of the direct problem can be used to compute $\mathbf{A}$. Details of this coupling between (6) and the finite element method can be found in Ledesma et al (1996a).

The estimation of matrix $\mathbf{A}$ can also be performed by means of a simple difference, as:

$$\frac{\partial \mathbf{x}}{\partial p} \approx \frac{\mathbf{x}(p + \Delta p) - \mathbf{x}(p)}{\Delta p} \tag{7}$$

The "a posteriori" covariance matrix of the parameters is defined also from the sensitivity matrix as:

$$\mathbf{C}_\mathbf{p} = (\mathbf{A}' \mathbf{C}_\mathbf{x}^{-1} \mathbf{A})^{-1} \tag{8}$$

The analysis of that matrix can give some insight about the relation between input data and output results. If a measurement is not sensitive to a change of a particular parameter, the corresponding column of $\mathbf{A}$ will be very small, and therefore the variance of the parameter computed from (8) will be very large. Therefore, a good design of the instrumentation should provide with large values of that matrix, that is, field measurements should be performed at locations where measured values are very sensitive to changes in the corresponding parameters.

3 THE 2D DIRECT PROBLEM

First a brief review of the direct problem is described in this section. Linear elastic constitutive laws and 2D plane strain conditions are considered. Assuming an initial stress state defined by a vertical stress and a horizontal stress with a ratio: $K_0 = \sigma_H / \sigma_V$, the excavation of a circular hole produces a change of the stresses that can be computed analytically (Pender, 1980):

$$\Delta \sigma_r = -\frac{1}{2}(\sigma_V + \sigma_H) * \left(\frac{a^2}{r^2}\right) - \frac{1}{2}(\sigma_V - \sigma_H) * \left(\frac{3a^4}{r^4} - \frac{4a^2}{r^2}\right) * \cos 2\theta$$

$$\Delta \sigma_\vartheta = \frac{1}{2}(\sigma_V + \sigma_H) * \left(\frac{a^2}{r^2}\right) + \frac{1}{2}(\sigma_V - \sigma_H) * \left(\frac{3a^4}{r^4}\right) * \cos 2\theta$$

$$\Delta \tau_{r\vartheta} = -\frac{1}{2}(\sigma_V - \sigma_H) * \left(\frac{3a^4}{r^4} - \frac{2a^2}{r^2}\right) * \sin 2\theta \tag{9}$$

where the symbols are explained in figure 1. The radial displacements (u) and the circumferential displacements (v) are:

$$u = \frac{(1 + \upsilon)}{E} * \left\{ \frac{(\sigma_V + \sigma_H)}{2} * \frac{a^2}{r} - \frac{(\sigma_V - \sigma_H)}{2} * \left[(1 - \upsilon) * \frac{4a^2}{r} - \frac{a^4}{r^3} \right] * \cos(2\theta) \right\}$$

$$v = \frac{(1 + \upsilon)}{E} * \frac{(\sigma_V - \sigma_H)}{2} * \left[(1 - 2\upsilon) * \frac{2a^2}{r} + \frac{a^4}{r^3} \right] * \text{sen}(2\theta) \tag{10}$$

where E is the elastic modulus and ν the Poisson's ratio. This result is a classical analytical solution in elasticity. It should be noticed that the effect of water is not included yet. Due to its simplicity, this result may

127

be used to investigate the relation between initial stresses and stress changes originated by the excavation, and the corresponding pore water pressure changes.

When the tunnel is excavated in a saturated hard rock, an undrained response could be expected. Then the change in pore water pressure is related directly to the change in stresses. In classical Soil Mechanics there are different formulae for predicting the increment of pore water pressure, Δp_w, under undrained conditions. A general one is the Henkel expression:

$$\Delta p_w = B(\Delta\sigma_{oct} + D\Delta\tau_{oct}) \qquad (11)$$

where B is a coefficient close to 1 for saturated conditions, D is the dilatancy coefficient, $\Delta\sigma_{oct}$ is the mean stress and $\Delta\tau_{oct}$ the deviatoric stress. For pure linear elastic models, $D = 0$, and the increment of

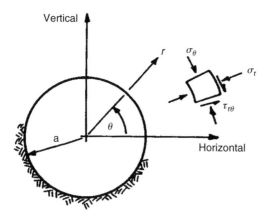

Figure 1. Geometry and main variables used in the analytical solution of the 2D problem.

water pressure is directly related to the change in mean stress at each point.

From equations (9) it is possible to compute the change of mean stress in this problem:

$$\Delta\sigma_{oct} = -\frac{(1+\upsilon)}{3} * \sigma_v * (1 - K_0) * \frac{2a^2}{r^2} * \cos(2\theta) \qquad (12)$$

Note that for $K_0 = 1$ (isotropic initial stress state), there is not change in mean stress, and therefore, pore water pressure remains constant in this case and it is impossible to use pore water pressure measurements to estimate stresses. Additionally, there are some points for which expression (12) becomes zero, as for instance when $\theta = 45°$ or $135°$, and also it is not useful (in this context) to measure water pressure at those locations.

To illustrate this effect, a simple analysis concerning the excavation of a tunnel in a saturated elastic medium has been solved using the code CODE_ BRIGHT (Olivella et al, 1996; Gens and Olivella, 2000). That finite element program solves thermo-hydro-mechanical geotechnical problems (Gens et al, 1998) and for this case only the H-M coupling has been considered. Figure 2 shows the pore water pressure developed just after the excavation of the tunnel, assuming a value of $K_0 = 0.4$. Note that in the zone close to $\theta = 45°$ there is no change in pore pressure, as expected.

4 THE 2D AND 3D INVERSE PROBLEM

In the 2D problem and assuming elasticity, equation (12) can be used directly to estimate the change in pore water pressure, because for saturated conditions

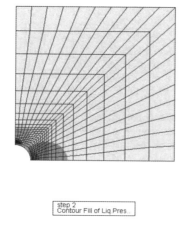
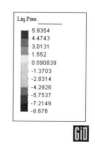

Figure 2. Contours of pore water pressure around a tunnel (case $K_0 = 0.4$).

B = 1 and within elasticity D = 0 in expression (11). That is,

$$\Delta p_w = -\frac{(1+\upsilon)}{3} * \sigma_V * (1-K_0) * \frac{2a^2}{r^2} * \cos(2\theta) \qquad (13)$$

The objective function when the least squares criterion is used becomes:

$$F = \sum_m (\Delta p_w^* - \Delta p_w) \qquad (14)$$

where Δp_w^* refers to the measurements. Thanks to (13), expression (14) has an analytical form, and the

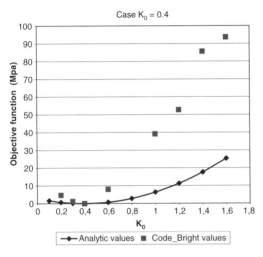

Figure 3. Objective function versus K_0 obtained using the analytical expression (cont. line) or the Code_Bright values of pore water pressures (points). Minimum for $K_0 = 0.4$.

minimum of F with respect K_0 is straightforward. It should be pointed out that measurements located on the line $\theta = 45°$ will not provide any information to (14), as those points are not sensitive to K_0 changes. Figure 3 presents the curve objective function – K_0 for a synthetic case in which the real K_0 value is 0.4. Both the analytical curve and the points obtained using Code_Bright program have been depicted. Differences between both curves are due to the procedure used in Code_Bright to evaluate the undrained change in pore pressure (a particular time step is assumed to be the "undrained time" in those analyses). 4 measurement points (2 vertical and 2 horizontal) have been considered, at distances of $r/a = 2$ and 3.

A simple 3D case has been also analysed, involving the excavation of a tunnel in different steps while measuring pore water pressures in a control plane as in the previous case. Measurements correspond to water pressures computed when excavating the tunnel face at 1 diameter before and after the control plane. The initial stresses are assumed to be defined by $K_{0x} = 0.8$ and $K_{0z} = 1.2$, where K_{0x} refers to the 2D plane considered before, and "z" refers to the tunnel axis. The procedure should be able to identify both parameters from the 4 pore water pressure measurements corresponding to 3 excavation steps (total of 12 measurements).

Figure 4 presents the objective function obtained for this case. Although the minimum corresponds in fact to the values mentioned, it is difficult to identify K_{0z} from that surface. This is due to the fact that for this particular case, measurements on the control plane when the tunnel face coincides with that plane are much more important and have an important relative weight in the objective function. That is, although the problem has been formulated in 3D, the

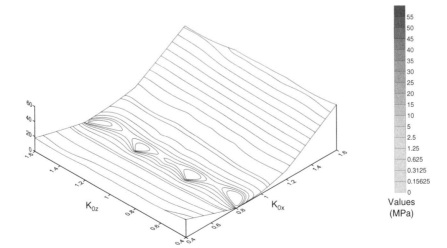

Figure 4. View of the objective function for the 3D case considered.

effect 2D dominates in this particular example. For other situations the result could be different, and that is currently being investigated.

The analysis of the sensitivity matrix for this case gives a consistent result with the shape of the objective function. Indeed, the variance linked to the parameter K_{0z} is very large, whereas the variance corresponding to K_{0x} has reasonable values. Note that the shape of the function as well as the variance behaviour depends not only on the model, but also on the location and distribution of the measurement points. Additionally, the values of the parameters may have some influence on that shape as well.

5 APPLICATION TO THE FEBEX TUNNEL

During the excavation of the Febex tunnel in Grimsel (Swiss Alps), some changes in pore water pressure were measured in two piezometers located in boreholes close to the gallery. Details of the Febex project including some information on the location, geometry and geology of the emplacement may be found in FEBEX (2000). Some additional details of the pore pressure measurements are described in UPC (2000).

For the application presented in this paper, the geometry considered is the one depicted in figure 5. Borehole 95.002 included two packers (P3 and P4) for which some measurements where available. The tunnel was excavated using a TBM machine, and it was checked that the stresses generated by that process could not explain the change in pore water pressure measured in those sections. Therefore it was assumed that they were generated by the stress relief produced after the excavation.

The central point of packers P3 and P4 have been used when assuming the location of the measurement, and 3D conditions have been considered for the whole geometry. Both packer zones are about 40 m away from the tunnel axis. For section with packer

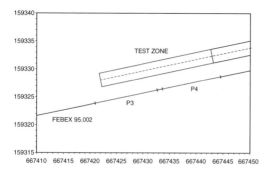

Figure 5. Geometry of the FEBEX tunnel and the borehole where pore water pressure measurements were performed.

P3, the increment of pore water pressure after the excavation of the tunnel was around 0.05 MPa; whereas for section P4 that change was 0.35 MPa.

Different analyses considering only one or several excavation steps have been performed. In these cases only one parameter, K_0, was considered in the identification process, and therefore, the minimization problem was simple. The results in terms of K_0 do not depend substantially on the steps considered in the excavation.

For the P3 zone, the identified value of K_0 was close to 3. This is a large value, but it is consistent with the values provided by other authors based on the local experience of the site and on field experiments. For section P4, the identified result has been $K_0 = 1.9$.

The difference between both parameters identified could be explained in many terms. Apart from the simplifications assumed, there was a dyke separating both zones in the tunnel. Indeed any joint, fracture or discontinuity may create a different stress field on each zone while maintaining equilibrium conditions. That could be the case in the Febex tunnel. However, it should be pointed out that the amount of data available was scarce for this analysis, due to the fact that it was not the purpose of those piezometers to measure water pressure for obtaining initial stresses. Obviously, the quality of the identification is directly related to the quality and amount of input data considered in the analysis, and for future cases, before starting the excavation, a field campaign of measuring pore pressure should be designed.

6 CONCLUSIONS

The paper describes a procedure to estimate the "in situ" stresses of a rock mass from pore water pressure measurements produced during the excavation of a tunnel. This backanalysis method has been applied in some cases in Rock Mechanics problems, but always considering field displacements as input data. However, for hard rocks, displacements tend to be small, whereas pore water pressures may exhibit important changes due to the stress relief.

A standard mathematical procedure to perform backanalysis in a systematic manner has been used for this purpose. In 2D plane strain conditions and using linear elastic constitutive laws, an analytical expression for the identification problem has been presented. For other geometries (3D) or other constitutive laws, the finite element program Code_Bright has been used to perform the coupled hydro-mechanical analyses.

The results show that the procedure described may be an option when investigating the initial stress state of a rock massif. The advantage of this method if compared with other standard procedures is that it involves a large volume of rock (if compared with laboratory or borehole measurements), and may have some advantages with respect to the backanalysis from field movements. This paper presents some preliminary results and future work will be devoted to the analysis of 3D problems considering different geometries and measuring strategies.

ACKNOWLEDGEMENTS

This work has been performed with the support of the European Union through FEBEX Research Project. Moreover, the support from ENRESA (Spanish Agency for radioactive waste disposal) is also gratefully acknowledged.

REFERENCES

Cividini, A., Jurina, L., Gioda, G. (1983). Some aspects of 'characterization' problems in Geomechanics. *Int. J. Rock Mech. Min. Sci. & Geomech. Abst.*, 20, 215–226.

Edwards, A.W.F. (1972). *Likelihood.* Cambridge University Press, Cambridge, UK.

FEBEX. (2000). FEBEX Project. Full-scale engineered barriers experiment for a deep geological repository for high level radioactive waste in crystalline host rock. Final Report. ENRESA Technical Publication 1/2000. Madrid, Spain.

Fletcher, R. (1981). Practical Methods of Optimization. 1- Unconstrained optimization. 2- Constrained optimization. Wiley, Chichester.

Gens, A., García-Molina, A.J., Olivella, S., Alonso, E.E. & Huertas, F. (1998). *Analysis of a full scale "in situ" test simulating repository conditions.* Int. Journal Numerical and Analytical methods in Geomechanics, 22, p. 515–548.

Gens, A., Ledesma, A., Alonso, E.E. (1996). Estimation of parameters in geotechnical backanalysis – II Application to a túnel excavation problem. *Computers and Geotechnics*, 18, 29–46.

Gens, A., Ledesma, A. (2000). Análisis inverso e identificación de parámetros en Geotecnia. *Libro Homenaje a J.A.*

Jiménez Salas, "Geotecnia en el año 2000". CEDEX, Ministerio de Fomento, Madrid.

Gens, A. & Olivella, S. (2000). *Non-isothermal multiphase flow in deformable porous media. Coupled formulation and application to nuclear waste disposal.* In Developments in Theoretical Geomechanics. Smith & Carter eds., Balkema, Rotterdam, p. 619–640.

Gioda, G. (1985). Some remarks on back analysis and characterization problems in geomechanics. *5th Int. Conf. Num. Methods in Geomechanics*, Nagoya, 1301–1307.

Gioda, G. & Sakurai, S. (1987). *Back analysis procedures for the interpretation of field measurements in Geomechanics.* Int. J. Num. Analyt. Meth. In Geomechanics, 11, p. 555–583.

IJRMMS (2003). International Journal of Rock Mechanics and Mining Sciences. Special Issue on rock stress measurement. 40, p. 957–1088.

Ledesma, A., Gens, A. (1997). Inverse analysis of a tunnel excavation problem from displacements and pore water pressure measurements. In *Material Identification using mixed numerical experimental methods.* Sol & Oomens eds., Kluwer, 163–172.

Ledesma, A., Gens, A., Alonso, E.E. (1996a) Parameter and variance estimation in Geotechnical backanalysis using prior information. *Int. J: Num. Anal. Meth. Geomechanics*, 20, 119–141.

Ledesma, A., Gens, A., Alonso, E.E. (1996b). Estimation of parameters in Geotechnical backanalysis – I. Maximum likelihood approach. *Computers and Geotechnics*, 18, 1–27.

Ljunggren, C., Chang, Y., Janson, T., Christianson, R. (2003). *An overview of rock stress measurement methods.* Int. J. Rock Mech. & Mining Sci., 40, p. 975–989.

Olivella, S., Gens, A., Carrera, J., Alonso, E.E. (1996). *Numerical formulation for a simulator (CODE_BRIGHT) for the coupled analysis of saline media.* Engineering Computations, 13, n. 7, p. 87–112.

Pender, M.J. (1980). Elastic solution for a deep circular tunnel. *Géotechnique*, 30, 216–221.

UPC (2000). Decovalex III, Task 1: Modelling of FEBEX in-situ test. Part A: hydromechanical modelling of the rock. Internal Project Report.

Hydraulic characterisation of the FEBEX granite: test performance and field interpretation

F. Ortuño & G. Carretero

AITEMIN, Hydrogeology Area, Toledo (Spain)

L. Martínez-Landa & J. Carrera

Department of Geotechnics and Applied GeoScience, School of Civil Engineering,
Technical University of Catalonia (UPC), Barcelona (Spain)

ABSTRACT: Several hydraulic testing campaigns have been performed in the context of the FEBEX Project. Their main objectives are to obtain the flow model around the tunnel, get the parameters of the main features and the matrix granite, evaluate the distribution of the water inflows to the tunnel, and provide boundary conditions for hydrothermal modeling. Hydraulic campaigns were done in 1995 and 1996, before and after the tunnel construction, and in 2001 before extraction of one of the heaters. A fourth campaign is currently (2003) underway. Twenty-three boreholes were initially available around the FEBEX tunnel, and most of them were equipped in 1996 with packers to get isolated sections. Strategy of the hydraulic characterisation always includes pulse tests in 42 isolated intervals in the radial boreholes to the FEBEX tunnel, and 5 long crosshole tests at low flow constant rates in the most permeable features. Most of them are repeated in each campaign to test if the heating experiment leads to changes in hydraulic properties of the rock.

Hydraulic testing in tunnel boreholes has some advantages over conventional, as the pressure sensors, electric cables and valves are outside the boreholes. But a project like FEBEX, with a large number of sensors and a complex data acquisition system (DAS), implies several difficulties, mainly related to the process of getting the data itself. Pulse tests need to record all the data initially at a high rate, and some 116 of signals need to be recorded (62 of pressure and 54 of temperature). A new software connected to the established AITEMIN DAS can record all signals every 2 seconds, allows real time graphics, prevents loss of data, and recovers records directly in ASCII format from anywhere by modem. In addition, volumes of injection or extraction water in pulse tests are very small (some ml) due to the low transmissivities, and they need to be measured accurately with a small tube or a high precision balance. Regarding the long pumping tests, extraction is preferred to injection; so as to avoid water contamination in granite and temperature induced changes. The use of a HPLC pump has been found to be an efficient way of maintaining low constant rates (some ml/min).

Pulse tests interpretation yields local transmissivity around the boreholes, and is performed in the field with inverse method assuming homogeneous conditions. A preliminary interpretation of the long term pumping tests is done in the field by using the Theis model one-by-one separated on the observation wells. The method is fast and simple and, although field conditions are far from the Theis conditions, it gives a good estimation of the effective transmissivities at a large-scale. On the other hand, the storativity reflects the degree of connection between pumping and the observation well, so that this approach can identify the most permeable features and even flow paths, which is essential for identifying dominant fractures.

1 INTRODUCTION

1.1 *Motivation*

Hydraulic characterisation of the FEBEX (Full-scale Engineered Barrier Experiment) granite is aimed at general objectives: (1) defining the flow model around the tunnel, (2) obtaining hydraulic parameters of main fractures and matrix, (3) evaluating the distribution of water inflows to the tunnel, and (4) providing boundary conditions for hydrothermal modeling. Available data for these purposes include geological mapping of the tunnels, geophysics, measurements of water inflow into the tunnels, hydraulic measurements at boreholes, and pumping tests. Pulse and crosshole tests are the most important set of data for hydraulic characterization.

Up to now, four hydraulic campaigns have been carried out in the FEBEX context coinciding with some of the most important milestones of the project: in 1995 and 1996, before and just after the tunnel construction, in 2001 before the extraction of one of the heaters, and in 2003 (currently underway) after its extraction. Several problems regarding the data acquisition process and the field interpretation of the tests were identified during the three first campaigns. They all have been overcome for the 2003 campaign.

1.2 Objectives and scope

The main objective of this paper is to outline the state-of-the-art on this issue, summarize experiences and explain the most recent advances. Test performance is described in section 2. It provides a brief description of the general methodology and instrumentation, as they are similar to those of previous campaigns in FEBEX, but is foccused on the new software and methods for data acquisition and handling implemented in the last campaign. A preliminary interpretation of pulse and crosshole tests is performed in the field using well-known methods. They are aimed at a very short time interpretation. These are described in section 3. This paper concludes with a summarizing section.

2 TEST PERFORMANCE

2.1 General methodology

Strategy of the hydraulic characterisation is the same in each campaign, and includes pulse tests in 42 isolated intervals in the radial boreholes to the FEBEX tunnel (Figure 1), and 5 long crosshole tests at low flow constant rates in the most permeable features. Most of them are repeated in every campaign to test if the heating experiment leads to changes in hydraulic properties of the rock.

A pulse test is performed in an isolated borehole interval by subjecting it to an instantaneous change in pressure, and then observing the transient pressure response as the system recovers towards its static pressure (Bredehoeft & Papadopulos, 1980). The recovery response is related to the diffusivity of the formation, the geometry of the borehole and the compressibility of the equipment, and yields the local hydraulic transmisivity around the test interval. Crosshole tests are conducted by withdrawing water from the test interval at a constant flow rate, and then observing the pressure responses in all the isolated intervals. These tests can last between 12 hours to 10 days. Flow conditions, effective transmisivities and hydraulic connections could be frequently identified during the tests.

2.2 Borehole instrumentation

Borehole instrumentation of FEBEX was defined by UPC and installed by Solexperts in 1996 (Fierz, 1996). This instrumentation has not had significant modifications and is outlined here. Boreholes were divided in some isolated sections with inflatable packers, each one with three connecting lines, one for the injection or extraction of fluid, another for pressure measurements, and the third for packer inflation (Figure 2). Temperature thermocouples were also installed. Radial boreholes, as well as FBX 1 and 2, and BOUS1 and 2 boreholes (sub-parallel to the tunnel) were equipped with this multipacker system. In all, 19 boreholes are available, and 116 signals (62 interval pressures and 54 temperatures) need to be recorded.

2.3 Hydrotesting instrumentation

The so-called "hydrotesting instrumentation" is placed outside the borehole (Figure 2). It comprises the pressure sensors, electrical connections, valves and pumps needed to perform the tests. It has been the same with practically no modifications in the four testing campaigns from 1996.

The requirements of pulse tests related to the instrumentation are basically two: a high data scan frequency at the very beginning of the test, and a great accuracy in the measurement of extraction or injection volume (some ml). It should be noted that changes in this parameter have influence in the wellbore storage coefficient, and fall into error in the estimation of the hydraulic transmissivity of the borehole interval. A small polyamide tube (3 mm inn. diam.) connected to the interval line was used for the measurement of the volume of water in the extraction pulse tests. Volume was deduced from the length of water inside the tubbing after closing the valve. Injection pulse tests were done with an HPLC pump, measuring the injected water with a high precision balance.

Crosshole pumping tests are prefered to injection tests; so as to avoid temperature induced changes and water contamination in granite. They require a constant flow rate, which is specially complicated when they are very small (some ml/min) as is the case. This was solved in an efficient way with the use of an HPLC pump. One must pressurize the extraction line with a hand pump at exactly the same pressure that the borehole interval before starting the test.

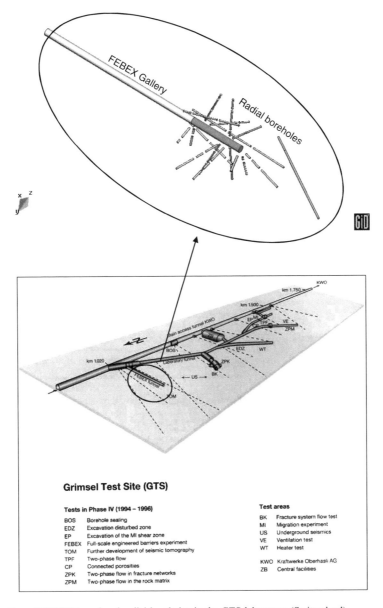

Figure 1. Location of FEBEX tunnel and radial boreholes in the GTS laboratory (Switzerland).

2.4 *Data acquisition*

Data acquisition is perhaps the main problem in the hydraulic characterisation of FEBEX. Pulse tests need to record early time data at a high rate (some seconds), and some 116 of signals must be recorded (62 of pressure and 54 of temperature). But FEBEX has hundreds of more sensors, and all of these signals are connected to the global AITEMIN

Data Adquisition System (DAS) of the project. During normal operation, all sensors are scanned every 30 minutes.

In practice, it would be impractical to disconnect the 116 pressure and temperature sensors and connect them to another system, due to the complexity of the cables in the global AITEMIN DAS. This system has the A/D cards duplicated, so we decided to use one

135

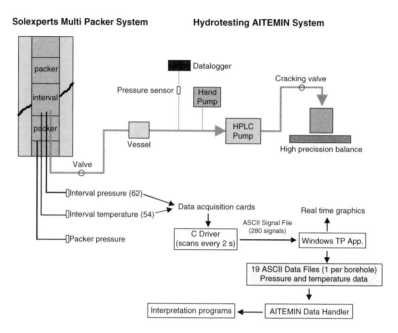

Solexperts Multi Packer System

Hydrotesting AITEMIN System

Figure 2. Borehole, hydrotesting instrumentation and flow chart for data acquisition.

set with an independent PC for the hydraulic tests. The other set of A/D cards can continue recording, in such a way that we could prevent loss of data and avoid disturbing the normal data acquisition process.

Once we had the pressure and temperature signals of the boreholes in a separated system, the problem was restricted to use a software that were able to record all data in a useful format at a high rate using these A/D cards. A commercial program was used in the 2001 campaign, but some problems were identified with the software: (1) data were stored temporarily in the computer memory, so that a PC fault could cause the loss of some data, as it actually occurred; (2) records were written in a binary file, and it took a lot of time for us to convert the binary file in an ASCII file; (3) we got a sheet of data with 116 columns and thousands of files (one for every time scanned), and it was very difficult to manage it; (4) real time graphics were very difficult to obtain, so it was also quite complicated to follow the tests; and (5) the time-lag between obtaining the field data and having the files ready for their interpretation was too long, and in practice it was impossible to have a field interpretation of the tests.

For the 2003 hydraulic campaign, we decided to develop a new software to overcome these problems. AITEMIN developed a C driver that reads all the selected cards connected to the pressure and temperature sensors of the boreholes, and overwrites a single ASCII file every 2 seconds with all the signals. A Windows application (TestPoint, 1994) reads this

file, converts the signals to engineering units, puts them in order, allows real time graphics, and writes 19 files, one per borehole, in a single ASCII format that can be read by a Data Handler program. The user can choose the scan rate. The software can be managed by modem. This allows us to transfer data, see the hydraulic crosshole tests and change the scan rate from Spain once the test has started.

2.5 Data handling

Data handling is an important step in field interpretation. Every data acquisition system has its own format, and also every software for interpretation, and a bridge between them is needed. Additionally, we always get a large number of data, but programs for interpretation only accept a limited number of records. A selection is necessary. This process may be time consuming. It made the field interpretation of hydraulic tests a virtually impossible task in previous campaigns. Note that we can get about 30000 records of every sensor in each complete hydraulic campaign, and we have 116 signals.

We also developed an own Data Handler to overcome these difficulties. The new software allows us to quickly read the data generated by the data acquisition system, cut and paste files, make graphics, select records (normal or logarithmically), correct trends, and export the data in formats ready to the programs or methods of interpretation. Now, once a pulse test has finished, data is ready to the

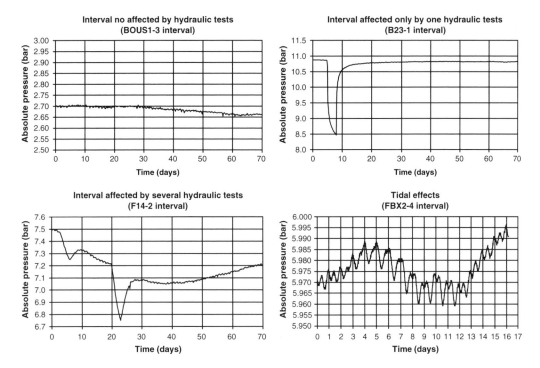

Figure 3. Examples of responses to pumping tests. During 70 days (horizontal axis) some extractions tests were performed. The interval in the upper left displays no response. The one on the upper right responds clearly to only one extraction. The lower left interval responds to three tests and involves a complication to establish the real drawdown in every test. Tidal effects (down right) are always superimposed.

interpretation in a few minutes. For the long term tests, all pressure data (62 sensors) can be reviewed and ready for interpretation in less than one hour.

Field interpretation of crosshole pumping tests only includes those intervals with pressure responses, and the data handling is necessary for their identification in the conducted tests. Figure 3 shows three different types of responses obtained during the last hydraulic campaign. Once the response sections have been identified (which is not always easy because one interval could be affected for more than one test), the data handler also makes the conversion of the pressures into drawdowns.

3 FIELD INTERPRETATION

We present here some examples of field hydraulic interpretation in the FEBEX Project, as part of the hydraulic characterization of the granite block around the tunnel. Field interpretation must be taken as a preliminary interpretation, but it is an easy, simple and fast way to provide local transmissivities, in the case of pulse tests, or to identify hydraulic active features and fracture connections in crosshole pumping tests.

3.1 Pulse tests

Pulse tests interpretation yields local transmissivity around the boreholes, and was performed in the field with inverse method with the code MariaJ_IV (Carbonell, Pérez-Paricio & Carrera, 1997) and the Ephebo PC interface, software developed by the Department of Geotechnics and Applied GeoScience of the UPC. The code uses an automatic optimisation procedure (maximum likelihood theory) for hydraulic parameter estimation based on analytical solutions. The software and the interface are fast and easy. Models used for pulse interpretation assume radial homogeneous conditions, using the general equation of Barker (1988). Figure 4 displays the graphical adjustment and the results of one of this tests at the B22-1 interval.

In a few cases, mainly in the intervals with high transmissivities, interpretations with homogeneous conditions were not good, and the adjustments were not in agreement. Nevertheless, a fracture model can reproduce the test in most of these cases (Figure 4, K2-2 interval), and it is more consistent with the conceptual hidrogeological model around the borehole.

Local transmissivity values obtained at each borehole interval are shown in Figure 5. The mode value obtained is 10^{-11} m^2/s, two orders of magnitued lower than the transmissivities estimated in crosshole tests. This could be considered as the value of the granite rock mass. Higher T values in pulse tests corresponds to the main features identified in the hydrogeological model, as shown in Figure 5: the lamprophyre dyke, a normal fracture (FR-1), an "en echelon" fracture (FR-2), and a shear-breccia zone (FR-3). The heterogenity of the main lamprophyre dyke is another characteristic that may be noticed.

3.2 Crosshole pumping tests

Preliminary field interpretation of crosshole tests is based on Theis (1935) model. Each observation borehole is interpreted separately. Drawdowns are matched using the same inverse method code as in the pulse tests interpretation. One may argue that field conditions are far from meeting the Theis assumptions, but the method is fast and easy and provides the order of magnitude in the effective transmissivity, and helps to identify the hydraulic active features and the fracture connections at the local site (Carrera & Martinez-Landa, 2000).

Several pumping tests were performed in each hydraulic campaign, but we only present here the interpretation of the F14-3 crosshole test of the last

B22-1. Extraction pulse test.

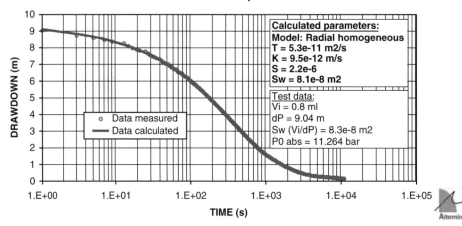

K2-2. Extraction pulse test.

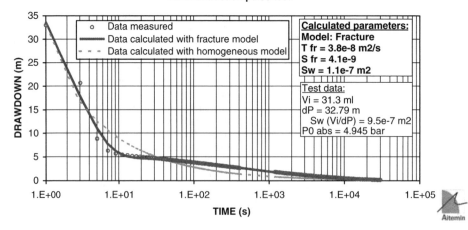

Figure 4.   Inverse method interpretation of pulse tests. A radial homogeneous model has been used above. In cases with high transmissivities, a fracture model may be much better (below).

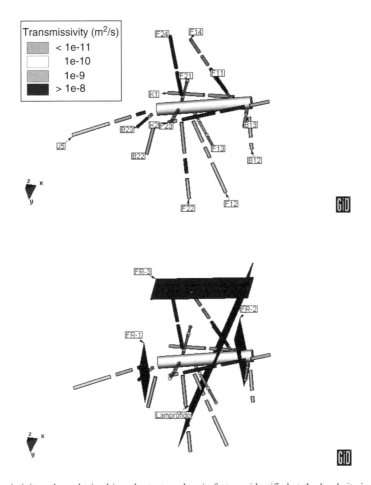

Figure 5.   Transmissivity values obtained in pulse tests and main features identified at the local site in FEBEX.

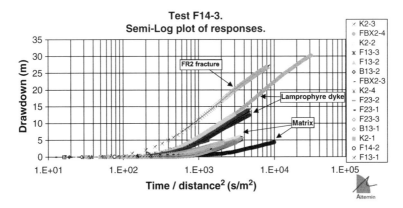

Figure 6.   Drawdowns observed in response to pumping interval F14-3. Notice that the horizontal axis is t/r^2 to filter out distance effects.

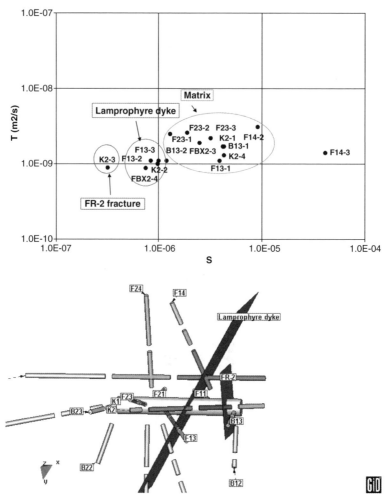

Figure 7. Log-log T versus S plot showing the results obtained with the one-by-one (separate) Theis model interpretation in the F14-3 test (above), and main features related with the observed connections (below). Intervals related with the FR-2 fracture are blue, those with the lamprophyre are red, and intervals within the matrix granite are green. Pumping interval is purple.

campaign. Pumping interval was F14-3, and the extraction flow rate was 4.5 ml/min during 3 days, starting the 18/07/2003. Flow was constant during all the extraction period.

Drawdowns of the intervals with response to this test are shown in Figure 6 (notice that the horizontal axis is t/r^2 to filter the distance effects). All curves display similar slopes, but intercept the X axis at different times. Slope is related with the transmissivity T, and the intercept time with the storativity S. In fact, if we represent the results obtained with the interpretation one-by-one in a log-log graphic T versus S (Figure 7), we obtain similar values of T, between $8.8e-10$ and $3.1e-9\,m^2/s$, while S range between $3.2e-7$ and $9.1e-6$.

Transmissivities derived this way can be taken as the large scale representative value (Meier et al, 1998), independent of the local effects around the pumping and observation intervals. Storativity reflects the degree of hydraulic connection between the pumping and observation points. Fast response is represented by low S values, and indicates a good connection. Note that the pumping interval always yields high S values due to compliance and equipment effects.

In this test, interval K2-3 displays the smallest S value, indicating the fastest response. This is consistent with the fact that it is intersected by fracture FR-2, more transmissive than the lamprophyre dyke (Figure 7). A second group of fast

140

connections is represented by the FBX2-4, K2-2 and F13-3 intervals, which are intersected by the main lamprophyre dyke that supports such connections. F13-2 and B13-2 intervals are also well connected with the pumping borehole F14-3, but connections are arguable and could be related to the two main features mentioned above. Some other intervals with response are worse connected to the F14-3 interval, and they could be considered as part of the granite rock mass.

4 SUMMARY AND CONCLUSIONS

The objective of this paper is to show recent advances in the performance and field interpretation of the hydraulic characterization of the FEBEX granite. Several hydraulic testing campaigns have been performed in the radial boreholes of the tunnel, and the experience has resolved the main problems identified. General methodology, borehole and hidro-testing instrumentation are practically the same used in previous hydraulic campaigns, but new software and methods for data acquisition and data handling has been implemented in the last campaign due to the difficulties of getting and handling the data itself.

The large number of sensors to be recorded (62 of pressure and 54 of temperature) and the existence of a well-established global data acquisition system, prompted us to develop new software for data acquisition in testing campaigns. This software can record all signals at a high rate for pulse tests, produce real time graphics, prevents loss of data, and write files directly in ASCII format. Moreover, it can be managed from anywhere by modem. Additionally, a new data handler allows us to read quickly the data generated, cut and paste files, see graphics, select data and export them to the interpretation programs. All pressure data of a crosshole pumping test (62 signals) can be easily reviewed and are ready to the interpretation in less than one hour. A pulse test can be interpreted in a few minutes.

Interpretation of the pulse tests is performed in the field with inverse method assuming homogeneous conditions. A new PC interface for the interpretation codes allows us to obtain results in an easy and fast way, yielding local transmissivity around the boreholes in the field. The preliminary interpretation of the long term pumping tests is done using Theis model for each observation interval separately. The

method produces good estimates of effective transmissivities at a large-scale, and also the degree of connection between pumping and the observation wells. This approach can identify the most permeable features and dominant fractures at the local site.

ACKNOWLEDGEMENTS

This work is produced under the framework of the FEBEX Project, funded by ENRESA (Spanish Nuclear Waste Management Co.), NAGRA (Swiss National Cooperative for the Disposal of Radioactive Waste) and the CEC (Economic European Community). Authors express their gratitude to Javier Sanz and Jose Luis García Siñeriz (AITEMIN, Spain) for the collaboration in the software programs and the installation of the data acquisition system to perform the 2003 hydraulic campaign. Allende Alcahud and Jordi Jordan (UPC, Spain) have collaborated in the performance of the pulse tests.

REFERENCES

Barker, J (1988) *A generalized radial-flow model for pumping tests in fractured rocks*. British Geological Survey. Wallingford, Oxfordshire, ox10 8BB, U.K. 55 pp.

Bredehoeft, JD & Papadopulos, SS (1980) A method for determining the hydraulic properties of tight formations. *Water Resources Research* 16(1): 233–238.

Carbonell, JA, Pérez-Paricio, A & Carrera, J (1997) *MARIAJ_IV: Programa de calibración automática de ensayos de bombeo. Modelos analíticos y numéricos para medios 2D y 3D*. E.T.S.E.C.C.P.B., UPC, Barcelona.

Carrera, J & Martinez Landa, L (2000) Mixed discrete-continuum models: a summary of experiences in test interpretation and model prediction. *Dynamics of fluids in fractured rocks, Geophysical monograph* 122, 251–265.

Fierz, T (1996) Instrumentation of BOUS 85001 and 85002, FBX 95001 and 95002, and radial boreholes. Solexperts AG. FEBEX internal report.

Meier, PM, Carrera, J & Sánchez-Vila, X (1998) An evaluation of Jacob's method for the interpretation of pumping tests in heterogeneous formations. *Water Resources Research* (34)5: 1011–1025.

TestPoint for Windows (1994) Capital Equipment Corporation.

Theis, CV (1935) The relation between lowering of the piezometric surface and the rate and duration of discharge of a well using groundwater storage. *EOS Trans.* AGU. (16): 519–524.

Advances in Understanding Engineered Clay Barriers – Alonso & Ledesma (eds)
© *2005 Taylor & Francis Group, London, ISBN 04 1536 544 9*

Hydraulic surface packer tests on the EDZ in granite – comparison of results from PR-drift (Äspö) and FEBEX-drift (Grimsel)

L. Liedtke & H. Shao

Federal Institute for Geosciences and Natural Resources (BGR), Hanover (Germany)

ABSTRACT: Drifting in fractured rock in contrast to unfractured rock can give rise to changes in hydraulic and mechanical rock characteristics in the proximal tunnel zone. A disturbed zone, the EDZ, develops in the proximal tunnel zone around the cavity. Local hydraulic properties in the immediate vicinity of the tunnel may be changed by drift excavation and under THM conditions during heating-hydration processes. To characterise the hydraulic connectivity of the fracture networks in the tunnel near-field and to determine the permeability distribution along artificial/natural fractures which intersect the test drift, the BGR surface packer system was developed and used within the projects Prototype Repository (Äspö −450 m about see level) and FEBEX II (Grimsel + 1730 m about see level). Both test sites are overlain by more than 400 m granite. The tunnel sections were drifted with a full face headers. Both projects are co-funded by the European Commission. Water and gas were injected into the surface packer to investigate the two-phase flow phenomena in the EDZ in access tunnels leading to the Prototype Repository test site in Äspö. In the TBM-drilled FEBEX tunnel, seventeen locations, including granite matrix and lamprophyre dykes in the heated area and undisturbed zones, were selected and tested by gas injection.

1 THE PROTOTYPE REPOSITORY TEST SITE IN ÄSPÖ

The SKB (Svensk Kärnbränslehantering AB) Prototype Repository project focuses on testing and demonstrating repository system function in full scale. The experiment comprises six full size canisters

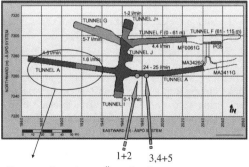

Prototype Repositoty - Äspö Sweden
Surface Packer location

Figure 1. Prototype Repository tunnel.

with electrical heaters surrounded by bentonite (Äspö Hard Rock Laboratory-Annual Report 2002; Technical Report TR-03-10). The inner section with 4 canisters was completed during 2001. The outer section will be installed during 2003. A large number of sensors are installed in the buffer, backfill and surrounding rock (Pusch, R., 2001).

The test site is in the access section of the Prototype Repository Laboratory Tunnel (Fig. 1). 5 tests were carried out at different locations with separations of a few metres. This involved positioning a surface packer on an identifiable fracture. The drilled tunnel has a constant diameter of d = 5.0 m, just like the tunnel in the Prototype Repository Tunnel. Figure 1 indicates the test sites and the inflowing water volumes from the rock in the tunnel sections as marked.

2 THE TBM-DRILLED FEBEX TUNNEL

The FEBEX experiment is a full scale simulation of a HLW[1] disposal facility after the Spanish

---

[1] High level radioactive waste.

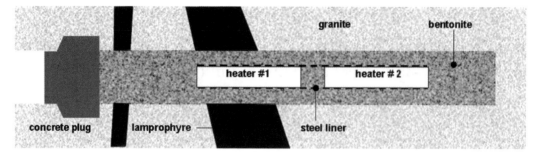

Figure 2. FEBEX experiment concept.

ENRESA[2] concept and is co-funded by the European Commission (Huertas, F., et al. (2000)). Performance of this test involved the placing of two electrical heaters, of dimension and weight equivalent to those of the canisters in the concept, in a 2.28 m diameter drift excavated in granite, the entire space surrounding the heaters being filled with blocks of compacted bentonite to complete the 17.4 m barrier for the test section. The test zone was closed with a concrete plug (Fig. 2).

A drift with a length of 70 m was specifically excavated for this test (FEBEX drift) in an area previously selected in accordance with the existing Grimsel laboratory database (Fig. 2). Two exploratory boreholes drilled from the main tunnel and 19 boreholes drilled from the test drift were the basis for a detailed hydrogeological investigation.

In the first phase of the project, 632 sensors of very diverse type were installed to measure total pressure, displacement, temperature, pore pressure, humidity in the granite rock, the heaters and in particular the clay barrier, made up of 5331 bentonite blocks with a total mass of 115.7 t. The heaters supplied a constant temperature of 100°C to simulate the canister and the various thermo-hydro-mechanical processes occurring in both the clay barrier and the surrounding rock for more than four years were monitored throughout the entire test period to gather data for analysis.

3 EQUIPMENT AND PROCEDURE

In situ Surface Packer Tests arrangement comprised a hollow metal cylinder fixed by a metal ring to the tunnel wall (Liedtke, L. & Shao, H., 1998). The horizontal holes were oriented perpendicularly to the tunnel wall. The equipment (Fig. 3) was basically a distance ring glued to the granite and the surface packer pressed onto the ring by a traverse. The amount of water to be injected is measured by weighting a storage vessel V = 1 litre made of

[2] Empresa Nacional de Residuos Radioactios, S.A.

stainless steel. The initial pressure is built up using an air cushion in the pressurised storage vessel. The pressure in the compressed air bottle is controlled by a pressure regulator. Before the tests started, tunnel wall surface was smoothed using special drilling bit for the installation of the surface packer.

A pulsed pressure test was conducted in the matrix area with lower permeability to detect possible water saturated sectors creating step-wise pressure increases of 0.2, 0.3, 0.4, and 0.5 MPa for 2 hours respectively. Pressurised air was used as injection medium. A gas flow meter with measurable range from 0 to 2 l/min and a pressure transducer with 0.5 MPa were installed. Also a temperature sensor was mounted to control the test conditions.

In the fractured zone the attempt was made to establish a constant gas flow rate test. To cope with the higher permeability, a gas flow meter was installed with measurable range of 0–20 l/min under standard conditions as was a pressure transducer with 0.2 MPa. However the inflow was so high that the measured flow rate was over the measurable range.

4 EXPERIMENTAL RESULTS

4.1 Äspö

Water and gas were used at all of the five test sites. Two different methods were applied.

4.1.1 Water flow
After checking the equipment for tightness, the injection pressure was held constant in the reservoir tank using a pressure control valve. Because of the low water take-up of the rock, the volume flow was determined by continuous weighing of the reservoir tank. The results are summarised in Table 1.

The difference in temperature between the tunnel and the inner side of the packer surface was approx. 2 [°C]. The average temperatures fluctuated between 18 and 23°C, and the air pressure fluctuated between 1074 and 1081 [hPa].

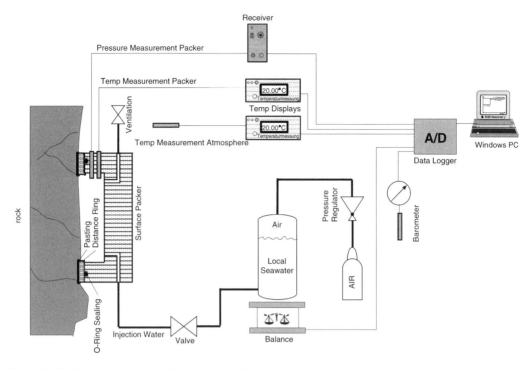

Figure 3. Surface Packer system and test equipment for water tests.

Table 1. Results of the water flow tests.

Location	$p_{i(t=0)}$ [MPa]	$p_{i(t=end)}$ [MPa]	∂_{packer} [°C]	$\Delta\partial_{packer}$ [°C]	∂_{drift} [°C]	$\Delta\partial_{drift}$ [°C]	Δp_{drift} [hPa]	ΔT [s]	ΔV [cm³]
1	0.7976	0.7919	18.7	1.4	20.5	1.2	1	68520	+62.9
2	0.7993	0.7791	18.7	1.3	20.7	2.2	5.6	266042	–
3	0.6039	0.6038	18.4	0.8	20.1	0.9	4	72720	+44.2
4	0.5014	0.5006	18.7	0.5	21.3	1.1	3.2	65520	–
5	0.5470	0.5380	18.6	1.8	23.0	2.5	9.7	94080	–

p_i = injection pressure.

During the tests at locations 2, 4 and 5, the direction of volume flow reversed. Since the pressure of the rock water was not measured, this value was also not available for evaluation.

4.1.2 Gas flow

After detailed control of the gas tightness of the equipment, air was pumped into the cavity of the surface packer. The injection overpressure in the 5 tests was roughly 5 bar. The volume of air flowing through the granite was determined on the basis of the pressure drop and the control volume of the surface packer.

A control valve was used to regulate the injection pressure. After the target pressure was reached, and the closure of all packer lines, the pressure fell to

almost its original level within a day depending upon the permeability of the rock. The volume flow in the rock was calculated from the packer, the line volumes and the pressure drop (Fig. 4).

In contrast to the water tests, the values measured in the gas tests illustrated clear changes during the prescribed test period of about one day. The pressure drop in the surface packer was relatively uniform in all tests. Only a minor permeability change is assumed. Test location 3 with a hydraulically non-effective fracture zone also shows no significant deviations.

4.2 FEBEX

The temperatures during the experiment are generally constant. The temperature in the test area was 14°C in

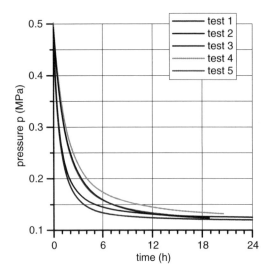

Figure 4.   Pressure drop curves derived from measurement results.

the undisturbed zone, fluctuating during injection phases, while the temperature in the test drift was somewhat higher (16°C) and was influenced by the presence of staff. However, the fluctuation of the temperature has no significant effect on the gas pressure evolution. In the heated area the packer and drift temperature are 2°C higher than those in the undisturbed zone.

The surface of granite in the heated area is generally dry. In the undisturbed granite matrix, there was a visibly moist area. The gas pressure increases during shut-in at the pressure levels 0.2, 0.3, and 0.4 MPa. This phenomenon indicates a relatively high natural pressure and high water saturation.

In test location OP03, there is a minor fracture on the wall. The test result shows that it is a closed fracture of lower permeability. The entry pressure in the area is also relative low and estimated at 0.4 MPa. The higher the permeability, the lower the gas entry pressure will be. The gas entry pressure amounted to 0.3 MPa and the water content 18%, Excavation of a drift redistributes the primary stress of the rock and may induce a change of permeability in the immediate tunnel wall. For this purpose two sections, one in the 54.5 m (lamprophyre dike: tests OP14, OP15, and OP16) and the other in the 56.5 m (granite: tests OP6, OP7, and OP8), were chosen to investigate the stress dependent permeability. In each section, three locations (+45° up, horizontal, −45° down) were tested using surface packers. The stress state through excavation release and swelling load did not induce any change of permeability distribution in the granite. The difference of permeability values from

different directions in the same section is the result of tectonics rather than excavation.

Compacted lamprophyre dike self has low permeability. If a fracture exists, the permeability will be high. For example, in location OP11 there is clearly a fracture in a pre-drilled borehole in the lamprophyre. This fracture belongs to a fracture network which was detected by leakage searching method using spray can and has an extension of 60 cm. A step-wise gas injection test has been performed. The pressure and gas injection flow rate are fairly concurrent and the pressure disappeared quickly after the injection stopped.

Special attention should be paid to the interface between lamprophyre and granite. One test was chosen in such a position. In OP15, the surface packer was directly installed on the fracture line, which intersects the fracture plane and the drift wall. A high injection rate was used and a low pressure build-up measured.

4.3   Numerical interpretation of test using rockflow

4.3.1   Gas flow

The development over time of the pressure drop in the control space together with the finite element method enabled the determination of the permeability. The computer program ROCKFLOW (Zielke, W., et al. 1984–1994) calculated the following differential equation:

$$\frac{\partial (n\,\rho)}{\partial t} + div\left(\left(\rho\,\frac{K}{\mu}(grad\,(p)) - \rho\,g)v \right) \right) - \rho\,q = 0$$

n = total porosity of the porous media
K = permeability
ρ = gas density = $\rho_0 = \frac{p}{p_0} \times \frac{T_0}{T}$
p = pressure
T = temperature
t = time
q = volumetric flow
g = gravity
v = velocity (vector)
μ = dynamic viscosity

Inhomogenic, anisotropic gas flows can be calculated using the generalised DARCY 3D flow law. The calculation results indicate the pressure, the density and the source strength concentrated at the nodes.

4.3.2   Two phase flow (gas and water)

$$S_\alpha = \frac{volume\,of\,Fluid\,phase\,\alpha}{Volume\,of\,pore\,space}$$

The saturation S_α of a phase α denotes the volumetric amount of the pore space that is occupied by regarded

phase in relation to the total pore space (Thorenz, C. 2001):
The water pressure was used as reference pressure.

$$\sum_{\alpha=0}^{phases-1} S_\alpha = 1$$

Thus, the saturation of a phase is always between zero and one. Furthermore, all phase saturation sum up to unity:

In the simulation of multiphase flow systems several formulations are feasible. The most common ones are:

– Pressure formulations
– Pressure-saturation formulations

The pressure formulation requires a capillary pressure function that is strictly monotonic. Only in this case it is possible to replace the saturation in the equations with the reverse capillary pressure function. This is a massive restriction, as the gradient of the capillary pressures function can change rapidly or be very small for some materials.

As the superset pressure equation still contains multiple pressures p_α, it is necessary to replace these with relative pressures versus a reference pressure p_{ref} in order to obtain a single primary variable. This reference pressure can either be one of the phase pressures or a weighted intermediate pressure. Here, one of the phase pressures is used as reference pressure:

$$p_\alpha = p_{ref} - \left(p_{c_{0,\alpha}} - p_{c_{0,ref}}\right)$$

Thus, now the primary variable is p_{ref} [Thorenz, C. Uni Hannover 2001]:

$$\sum_{\alpha=0}^{phases-1}\left[\left(\frac{nS_\alpha}{\rho_\alpha}\frac{\partial\rho_\alpha}{\partial p_\alpha}+S_0 S_\alpha + n\frac{\partial S_\alpha}{\partial p_\alpha}\right)\frac{\partial p_{ref}}{\partial t}\right.$$
$$\left.-\frac{1}{\rho_\alpha}div\left(\rho_\alpha\frac{k_{r_\alpha}\mathbf{k}}{\mu_\alpha}\left(grad\ p_{ref} - \rho_\alpha\mathbf{g}\right)\right)\right]=$$
$$\sum_{\alpha=0}^{phases-1}\left[\left(\frac{nS_\alpha}{\rho_\alpha}\frac{\partial\rho_\alpha}{\partial p_\alpha}+S_0 S_\alpha + n\frac{\partial S_\alpha}{\partial p_\alpha}\right)\frac{\partial\left(p_{c_{0,\alpha}} - p_{c_{0,ref}}\right)}{\partial t}\right.$$
$$\left.-\frac{1}{\rho_\alpha}div\left(\rho_\alpha\frac{k_{r_\alpha}\mathbf{k}}{\mu_\alpha}grad\left(p_{c_{0,\alpha}} - p_{c_{0,ref}}\right)\right)+Q_\alpha\right]$$

The ROCKFLOW software allows 1D, 2D and 3D finite elements to be spatially combined.

4.3.3 *Äspö*

As shown in Fig. 5, the test locations are located at a high level in a surface packer arranged vertically in the rock wall. The network described in the following has been rotated through 90° in the graphic, such that

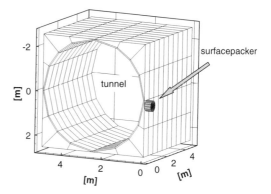

Figure 5. Tunnel with location of Surface Packers.

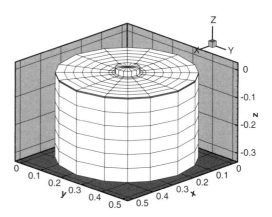

Figure 6. Network of three element groups.

the packer is actually situated as an upside-down pot on the flat rock (Fig. 6).

The network used comprises 8 nodes and 1168 spatial elements degenerated in the centre to 6 space element nodes. The 1322 nodes have rotational symmetry and describe a rock volume of 50/50/30 cm, the 1 cm thick EDZ and the packer. The packer is modelled with actual dimensions (diameter 10 cm; height 2 cm) (Fig. 6) and is located on the elements of the EDZ. The network comprises of these three element groups, each with a different flow law. The elements are spatially internetworked.

4.3.4 *FEBEX*

To evaluate the experimental results and to determine the rock permeability, numerical singlephase flow code *RockFlow-GM2* was applied (Shao, H., 2002). For the simulation of gas flow in the matrix area, a mesh, coupled with three dimensional elements for the rock block and two dimensional elements for the packer volume, was used (Fig. 7). As boundary

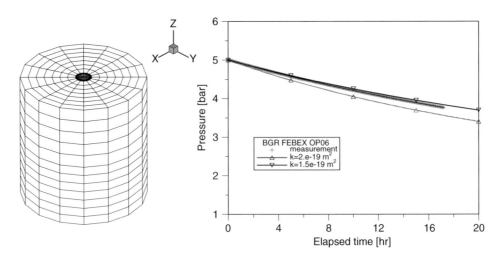

Figure 7. Finite element mesh for simulation of experiment in the matrix area (left) and the modelled and measured results e.g. OP06 (right).

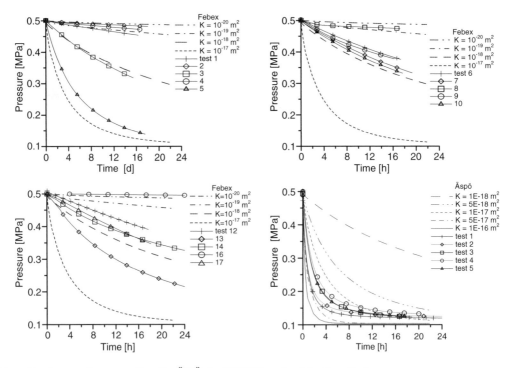

Figure 8. Pressure drop curves from the ÄSPÖ, the FEBEX tunnel and the theoretical curves.

condition, atmospheric pressure was used on the tunnel surface and Neumann's conditions on the remaining boundaries. Rock permeability was varied in the model and can be less than $2 \cdot 10^{-19}\,\mathrm{m}^2$ (Fig. 7 right). Rock porosity was varied from 0.001 to 0.01

and can be amount to 0.002, but it has less influence on the pressure development.

A single 2D fracture model was used in the case of OP11. The same boundary conditions were assigned like in the matrix case. Fracture aperture was

Table 2.   Results of the test.

	Designation	Permeability [m^2]	Transmissibility [m^3]	Geology	Property
Äspö	1	$2.5 \cdot 10^{-17}$		Matrix	
	2	$2 \cdot 10^{-17}$		Matrix	
	3	$2 \cdot 10^{-17}$		Matrix/hydraulic closed fracture	
	4	$1 \cdot 10^{-17}$		Matrix	
	5	$2 \cdot 10^{-17}$		Matrix	
FEBEX	1	$1 \cdot 10^{-19}$		Matrix	Dry
	2	$<1 \cdot 10^{-19}$		Matrix	Humid
	3	$<5 \cdot 10^{-19}$		Matrix/minor fracture	Dry
	4	$<1 \cdot 10^{-19}$		Lamprophyr	Dry
	5	$8 \cdot 10^{-18}$		Lamprophyr	Bentonite rest
	6	$<5 \cdot 10^{-19}$		Matrix	
	7	$<5 \cdot 10^{-19}$		Matrix	
	8	$<1 \cdot 10^{-20}$		Matrix	
	9	$<5 \cdot 10^{-19}$		Matrix	
	10	$8 \cdot 10^{-19}$		Lamprophyr	
	11		$2.5 \cdot 10^{-15}$	Lamprophyr	Humid
	12	$<5 \cdot 10^{-19}$		Matrix	
	13	$2 \cdot 10^{-18}$		Matrix	Humid
	14	$<5 \cdot 10^{-19}$		Lamprophyr	
	15		$2.0 \cdot 10^{-14}$	Matrix/Lamprophyr	
	16	$<1 \cdot 10^{-20}$		Lamprophyr	Humid
	17	$<5 \cdot 10^{-19}$		Matrix	Humid

estimated to 1.0 mm and varied from 0.1–1.0 mm. Fracture permeability was varied in the model and the best fitting can be obtained by a fracture transmissibility (T) of $3–5 \cdot 10^{-15} \, m^3$.

In the model the recovery phase with an initial pressure of 0.5 MPa is simulated by variation of permeability. The maximum measured pressures before the recovery phase were scaled to 0.5 MPa. All measured data was corrected by extraction of system error value, which was determined by testing on an aluminium plate. The most evaluated permeability was lower than $5 \cdot 10^{-19} \, m^2$, which is the reference value of the granite at the Grimsel Test Site. Only the permeability of OP5 was greater than the reference value, where the bentonite residue adhered to the tunnel wall after the dismantling of the bentonite blocks.

The constant rate injection tests in the two highly permeable zones were simulated using a fracture model. The relationships between pressure and transmissibility under the defined gas injection rate were calculated and compared with the measured pressure values. Therefore these zones were determined to have a high permeability, e.g. permeability of $1 \cdot 10^{-10} \, m^2$ in the interface of the lamprophyre and the granite according to the 'cubic-law'.

4.3.5 Permeabilities

Figure 8 shows the pressure drop curves for the tests in the 3 FEBEX diagrams and one Äspö diagram. The plots also show the pressure drop curves as calculated. Table 2 summarises the permeabilities in digital form. They fluctuate between $K = 10^{-15} \, m^2$ and $10^{-20} \, m^2$. The large scatter range of the FEBEX values is assumed to be due to the deliberate selection of test sites with large permeability differences. One of the Äspö test locations and one in the FEBEX tunnel, in which a fracture was deliberately integrated into the test, demonstrate no increased permeability. It is assumed that fractured rock does not necessarily have to be hydraulically effective.

5   CONCLUSIONS

Investigating the excavation disturbed zone (EDZ) is of primary importance for the concept of a nuclear waste repository in deep geological formations. Such damage may occur at first during the opening of the cavities and then perhaps evolve depending on the thermo-hydro-mechanical loading in the post closure phase. Unlike salt or clay, which has some self-healing capacity, the structural damage in crystalline rock is irreversible. Therefore, the EDZ around the underground opening in granite, which may provide enhanced permeability pathways for radionuclide migration, is a long-term safety issue. Within the framework of the FEBEX II and Äspö projects a multidisciplinary team studied the mechanisms creating a potentially damaged zone around the test drift and its evolution after a period of 6 years of heating. The research programmes include laboratory and *in-situ* investigations as well as the numerical modelling of the observed phenomena.

One of the BGR contributions within the Äspö and FEBEX-II projects are the performance of hydraulic *in-situ* investigations in the tunnel near-field for the determination of the hydraulic parameter values of the EDZ and boundary conditions for the numerical THM modelling.

After the dismantling of the bentonite blocks and the heater 1, seventeen test positions in FEBEX II project, located in the granite rock mass and the lamprophyre dike, were selected for the gas injection tests using the surface packer. Six of them are located in the front area of the tunnel, where is not affected by the heating-saturation processes of bentonite, in order to have reference values. The experimental data is evaluated using the Rock Flow numerical single and two-phase flow program. From the *in-situ* investigations it can be summarised that:

– The permeability of the granite matrix in the EDZ of the tunnel drilled by TBM FEBEX is comparable to that of the undisturbed rock mass ($<5 \cdot 10^{-19} \, m^2$).
– The permeabilities of the granite matrix in the EDZ of the Äspö Tunnel are noticeably higher than in the undisturbed granite.
– The probability of an axial flow pathway existing along the tunnel is lower, because the observed fractures are closed or filled with secondary materials.
– The thermal (temperature of the tunnel wall of about 40°C) and mechanical (swelling stress of about 4 MPa) conditions during heating and hydration have no significant influence on the hydraulic parameters (permeability and porosity), but may affect in the residual water saturation of the granite in the EDZ and long-term re-saturation.

– The distribution of the hydraulic parameters around the tunnel cannot be correlated with the redistribution of initial stress.
– Further attention should be focused on the natural fracture system (with relative high permeability in the test zone) which intersects the disposal tunnel. This is even more important for radionuclide migration. This applies to both the Äspö and the FEBEX tunnel.

REFERENCES

ÄSPÖ Hard Rock Laboratory-Annual Report 2002; Technical Report TR-03-10, Stockholm, Sweden

Huertas, F., et al. (2000): Full-scale engineered barriers experiment for a deep geological repository for high-level radioactive waste in crystalline host rock (FEBEX project), Final report, European Commission, Brussels, Belgium

Liedtke, L. & Shao, H. (1998): Bestimmung der Permeabilität des geklüfteten Gebirges mit Gas, Zeitschrift Geotechnik, 1998/3, DGGT, Essen, Germany

Liedtke, L. (2002): The saturation and resulting swelling pressure of a deposition hole – tunnel system filled with bentonite under consideration of the excavation disturbed zone (EDZ) with the help of the two phase flow theory; International Meeting, Reims, December 9–12. 2002; ANDRA; Paris, France

Pusch, R. (2001): Preparation of deposition holes prior to emplacement of buffer and canister in section I; IPR-01-64, SKB, Stockholm, Sweden

Shao, H. (2002): Numerical Calculation of Two-Phase Flow (GMT-Gas Migration Test), Project Report, B2.1-11935/02, BGR, Hannover, Germany

Thorenz, C. (2001): Model Adaptive Simulation of Multiphase and Density Driven Flow in Fractured and Porous Media, UNI Hannover Bericht Nr. 62/2001; ISSN 0177-9028

Zielke, W., et al. 1984–1994: Theorie und Benutzeranleitung zum Programmsystem Rock Flow, University of Hannover, Germany

EDZ in granite

Behrooz Bazargan Sabet
G.3S (France)

Hua Shao
BGR (Germany)

J. Autio
Saanio & Riekkola OY (Finland)

Fco. Javier Elorza, Israel Cañamon & J. Carlos Perez
UPM (Spain)

Roland Pusch
Geodevelopment AB (Sweden)

Christer Svemar
Svensk Kärnbränslehantering AB (Sweden)

1 INTRODUCTION

Various field experiments have been planned and performed to find out whether EDZ is really of importance for transport of water and contaminants in repositories. The lessons learned by conducting such tests, both in EC funded projects and in others, are described and commented below.

2 EXPERIENCE FROM THE FEBEX PROJECT

FEBEX is the acronym of an European Research Project driven by ENRESA and held at Grimsel site (Switzerland). FEBEX test drift was excavated in 1995 using a tunnel boring machine (TBM). The gallery has a circular profile, 70.40 m long and 2.28 m in diameter. It is intersected by two bended lamprophyres at 54 m and 60 m from the entrance. The experiment aims at simulating a radioactive waste storage in real scale. For this purpose two electrical heaters (0.90 m in diameter and 4.54 m long) providing the total power of 4.2 kW and surrounding by three layers of bentonite blocs were installed into the drift. The experiment area was isolated from the rest of the drift by a concrete plug of 2.70 m thick placed at 17 m from the drift dead-end.

After the partial dismantling of the experiment, the mechanisms of creation of the potential damaged zone around the test drift and its evolution after a period of 6 years (1997–2003) of heating were studied within the framework of the FEBEX II project. The on going research program includes laboratory and in-situ investigations as well as the numerical modelling of the observed phenomena.

2.1 Laboratory investigations

Several tests were performed on the core samples taken from the test area (heated zone), and from the service area.

2.1.1 Microfracturing and porosity tests

The objective of the work performed by SKB and Saanio&Riekkola OY was to analyse the porosity and microfracturing both quantitatively and qualitatively. The ^{14}C-PMMA technique was applied to study the spatial distribution of porosity. In addition, complementary microscopy and scanning electron microscopy (SEM) studies were performed to make qualitative investigations of the pore apertures and minerals in porous regions.

There was not found any clear increased porosity zone adjacent to the tunnel wall with the PMMA method in the granite samples. The total porosities of the samples varied between 0.6–1.2%. Grain boundary pores and trans-granular porosity were detected. The samples of unplugged region did not differ from the samples of plugged region.

Figure 1. Photo image of Aare granite section 4.1B (left) and corresponding autoradiograph (right) from the unplugged region in Febex tunnel. The section is perpendicular to the axis of the tunnel wall the excavated surface being on top. Sample width is 10.6 cm.

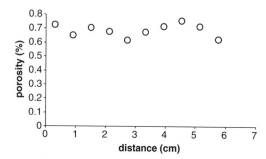

Figure 2. The porosity is calculated with respect to distance from the tunnel wall and presented as porosity histogram.

A clear increase in porosity to depths of 10–15 mm from the tunnel wall was detected in lamprophyre samples. The EDZ consisted of micro fractures perpendicular to the tunnel wall. A strong fracturing was found in some lamprophyre samples at depths of several cm from the tunnel wall. Some samples contained quartz inclusions, where plenty of micro fractures transacted the mineral grains parallel to the tunnel wall and numerous intra-granular fissures transacted the quartz grains. Quartz grains were found porous having 0.5% porosity. The granite matrix, that was surrounded by the quartz inclusion being adjacent to the tunnel wall had clearly higher total porosity (i.e. 1.3–1.5%) than the granite matrix at depths of 10 cm from the tunnel wall, which was 0.9–1.1%.

According to the SEM/EDX analyses the excavation disturbed zone in the granite matrix extended to depths of 1–3 mm from the wall surface. A few quartz grains were crushed and some micro fractures were found. Some trans-granular fissures were found parallel to the wall surface having apertures of

about 5–30 μm. The micro fracture apertures varied between 20–50 μm in EDZ of lamprophyre samples trending to depths of 5–25 mm from the tunnel wall.

2.1.2 Gas permeability test

The permeability measurement is a good indicator for assessing the existence and the evolution of the EDZ. Gas permeability tests were carried out by G.3S on two core samples of about 1 m long each taken from the granite wall perpendicular to the drift. The first sample was cored in the service area far from the heated zone and the second one at the level of the heater. Core samples were first machined in order to obtain hollow cylinders with an internal diameter of 24 mm. The tests were performed at constant gas pressure by setting a steady state radial flow through a section of 1 cm wide isolated by means of four mini-packers. By moving mini-packers along the sample, one can establish the profile of the gas permeability according to the core length.

The magnitude of the permeability measured ranges between $6.5 \cdot 10^{-18}$ and $8.4 \cdot 10^{-19} m^2$, pointing out the absence of a marked damage. The highest value is obtained on the sample taken in the service area and corresponds to a natural sealed joint.

Even if the flow rate profile related to the sample taken in the heated area shows an increase of the permeability over the first 20 centimetres, this evolution cannot be linked with the existence of a damaged zone. Indeed, whereas natural rock heterogeneities provide more variation, the maximum discrepancy in terms of permeability remains less than one order of magnitude.

2.1.3 Acoustic investigations

Measurements of the travel times of ultrasonic waves through a cylindrical granite block have been performed by CIEMAT and UPM. The aim of this experiment has been to quantify the variations in

Figure 3. Photo image of the granite section 3.1A (left) and corresponding auto-radiograph (right) from the unplugged region in Febex tunnel. The section is perpendicular to the axis of the tunnel wall the excavated surface being on left. Sample width is 14 cm.

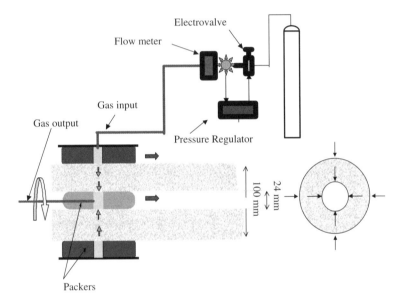

Figure 4. Gas permeability measurement set-up.

propagation velocities of acoustic waves through the rock. The granite cylindrical block has been taken from the service area, and is 38.8 cm. in diameter and 40 cm. long (from the gallery wall towards the rock mass). For this study, the block has been analysed along 2D transversal sections in six radial directions (section 1 (R1-R7) coincides with the FEBEX gallery direction). More than 900 measurements have been taken by combining different positions in the transmitter-receiver pairs.

Different inverse tomographic strategies have been used to analyse the measured data. Studying signal propagation velocities through horizontal trajectories

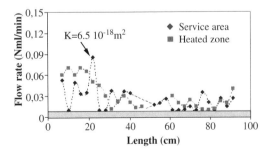

Figure 5. Results of gas permeability tests.

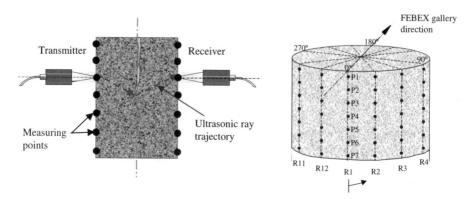

Figure 6. Measuring system set up with a static transmitter and a portable receiver (left), and measuring points grid in the granite sample (right).

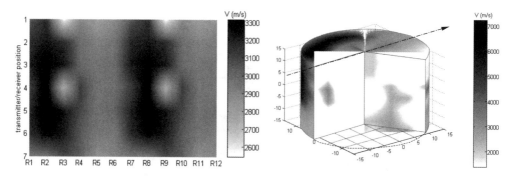

Figure 7. 2D velocities graph, where a preferential direction (R2-R8) with higher velocities can be localized (right); and 3D interpolation of the tomographic inverse problem 2D results in sections 1–6.

(Figure 6), a preferential direction R2-R8 with higher velocities can be observed, although the velocity differences are not significant. The 2D tomographic inversion (Figure 7) shows some heterogeneity in the propagation velocity through the granite. However, no special decrease of velocity towards the gallery wall (top part of the sections) is appreciated. Therefore, no evidence of the existence of EDZ in the analysed sample has been found. Anomalous highest velocities are associated with measurements artefacts, since there were broken parts in the granite sample where measurements were not taken.

Results have been contrasted with those of the Febex site geologic characterisation report (70-IMA-M-2-01). The sampled block is not intersected by any of the fracture traces of the cartography of the Febex Gallery (Figure 8 left), and a visual analysis of this granite block does not show any fracture or fissure. Moreover, pole distributions of the fracture families (Figure 8 right) are not correlated with the pole of the preferential direction R2-R8. Therefore, this preferential direction could be due to any heterogeneity on the fabric of the granite.

2.2 In situ investigations

To characterise the hydraulic connectivity of the potential fracture networks in the tunnel near-field and to determine the permeability distribution along artificial/natural fractures which intersect the tunnel, the BGR surface packer system was developed and used in the FEBEX drift.

On the basis of the geological circumstances and from the viewpoint of the possible mechanical influences of the excavation processes, seventeen locations, including granite matrix and lamprophyre dykes in the heated area (11) and in the service zone (6) have been selected and tested. In the matrix area with lower permeability, pulse test using pressurised air with stepwise pressure increasing was conducted to determine gas entry pressure. In the fractured area, a gas constant flow rate injection test was conducted.

Only two locations with higher permeability were detected; one is a natural fracture in the lamprophyre dyke and the other is the interface between lamprophyre and granite. To interpret the experimental results, the numerical code RockFlow for one

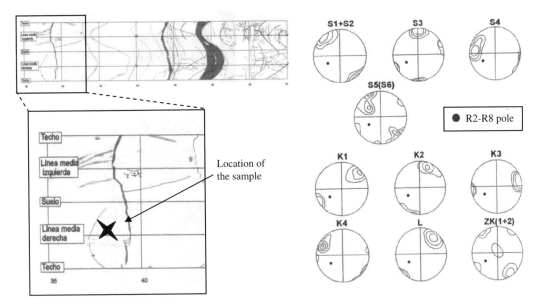

Figure 8. Location of the granite sample in the Febex drift (left) and pole distributions of the fracture families and preferential direction R2-R8. (Based on 70-IMA-M-2-01).

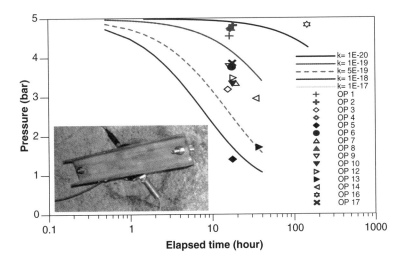

Figure 9. In situ gas permeability measurement.

and two phase flow was applied. The permeability measurements obtained on the granite wall are comparable to that of the undisturbed rock mass ($<5E\text{-}19\,m^2$).

2.3 *In situ investigations*

To characterise the hydraulic connectivity of the potential fracture networks in the tunnel near-field

and to determine the permeability distribution along artificial/natural fractures which intersect the tunnel, the BGR surface packer system was developed and used in the FEBEX drift.

On the basis of the geological circumstances and from the viewpoint of the possible mechanical influences of the excavation processes, seventeen locations, including granite matrix and lamprophyre dykes in the heated area (11) and in the service zone

155

(6) have been selected and tested. In the matrix area with lower permeability, pulse test using pressurised air with stepwise pressure increasing was conducted to determine gas entry pressure. In the fractured area, a gas constant flow rate injection test was conducted.

Only two locations with higher permeability were detected; one is a natural fracture in the lamprophyre dyke and the other is the interface between lamprophyre and granite. To interpret the experimental results, the numerical code RockFlow for one and two phase flow was applied. The permeability measurements obtained on the granite wall are comparable to that of the undisturbed rock mass ($<$5E-19 m^2).

The thermal (temperature on the tunnel wall of about 40°C) and mechanical (swelling pressure of about 4 MPa) loadings during the heating period and hydration have no significant influence on hydraulic parameters (permeability and porosity), but may affect in the residual water saturation of the granite in the EDZ and again long-term re-saturation. However the distribution of the hydraulic parameters around the tunnel cannot be correlated with the re-distribution of stress state. The investigations show that the main

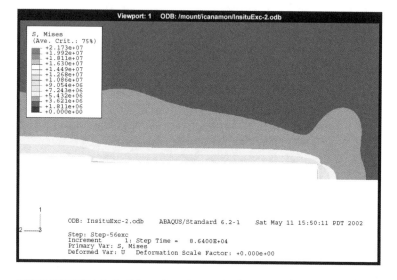

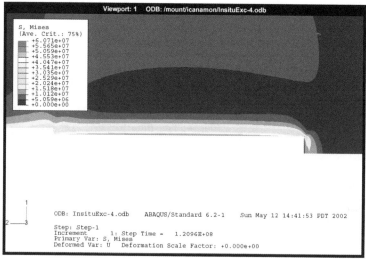

Figure 10.   Results of 2D thermo-mechanical calculation: Von Mises stresses after excavation (upper) and after the six years heating period (lower).

attention should further be focused on the natural fracture system with relative higher permeability, which intersects the disposal tunnel.

2.4 Numerical investigations

2D and 3D thermo-hydro-mechanical and thermo-mechanical analysis have been performed both by UPM and G.3S in order to assess the effects of the excavation and the heating phases on the possible development of an EDZ around the drift.

In the 2D analysis, the main hypotheses adopted were:

– Homogeneous porous medium
– Isotropic mechanical state conditions
– Axisymmetric conditions

It could be observed (Figure 10) that the maximum developed stresses are around 22 MPa. after the excavation phase and of 62 MPa. at the end of the hypothetical worst case heating phase, leading to the conclusion that no mechanical reason exists for the development of an EDZ on the Febex drift.

In order to take into account the strong anisotropy of the initial state of stress, a 3D modelling was performed. According to the in situ measurements, the compressive tectonic regime leads to a horizontal stress 4 or 5 times larger than the vertical lithostatic pressure (9–12 Mpa). In addition, there is a difference of more than 10 Mpa between the minimum and the maximum horizontal stress that makes an angle of 60° with the drift axis The rock mass, concrete plug, heaters and bentonite buffer were modelled. The excavation phase, followed by the thermo-mechanical loading of the test drift were simulated.

The maximum stress developed around the gallery wall compared to the yielding criterion (Drucker Prager) show (Figure 11) that the state of stresses induced by the excavation and the heating phases remains far below the critical curve.

2.5 Conclusions related to the FEBEX experiment

All the experimental results show that the granite matrix has not been damaged neither by the excavation nor by the THM loading undergone by the rock mass. Except a zone of about 3 mm deep put to the fore by the SEM/EDX analyses, the granite seems to preserve its mechanical and hydraulic properties. The evidence of a potential EDZ concerns only the lamprophyre area and the interface between lamprophyre and granite.

The absence of EDZ in the Grimsel granite matrix is also confirmed by the numeric simulation which shows that the maximum stress developed is much lower than the minimum resistance of the rock. Even if at the local scale some uncertainty may exist on the behaviour of the natural joints, the creation of a continuous pathway for radionuclide migration along the drift seems to be highly improbable.

3 SKB EXPERIMENTS IN BLASTED TUNNELS AND DRIFTS

3.1 General

SKB's studies of the hydraulical and mechanical behaviour of rock have consistently included the rock structure and this made detailed examination of the walls of blasted tunnels natural for getting a basis for developing conceptual models of the EDZ. An early finding was that the degree of mechanical degradation varies along the length of a blasted tunnel because of the distribution of charge, a typical schematic illustration being that in Figure 12.

It means that spot-wise determination of the hydraulic conductivity in short boreholes drilled normal to the rock walls is expected to give strongly varying data and that it is impossible to get definite information on the water transport capacity of the most shallow, disturbed rock over longer axial distances by performing such tests.

3.2 Stripa experiments

The first experimental evidence that the EDZ of blasted tunnels has a significant water-bearing capacity was offered by the Buffer Mass Test in the first phase of the international Stripa Project. It was conducted at 360 m depth in granite and represented a half-scale version of the Swedish KBS-3 concept, consisting of a blasted drift with 25 m^2 cross section and a length of 35 m, the inner 12 m part being

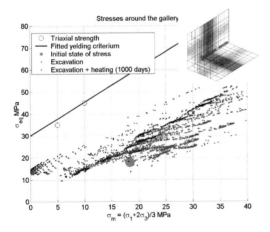

Figure 11. Stresses around the test drift compared to Drucker Prager yielding criterion.

isolated from the outer part by a concrete bulwark (Figure 13). Two 760 mm diameter holes with 3 m depth and 6 m spacing, simulating canister deposition holes, were core-drilled in the inner part that was backfilled with clayey soil, and four ones in the outer part. The proof of the existence of a significantly water-bearing EDZ was the strong inflow into the temporarily empty hole (No 3) located just outside the bulkhead when the backfill behind it started to be

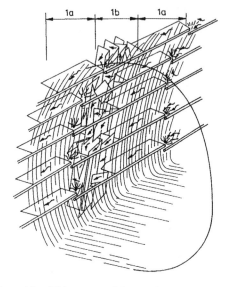

Figure 12. Major types of damage by tunnel blasting. 1a-zones are characterized by regular sets of plane fractures extending radially from the central parts of the blast-holes. 1b-zones represent strongly fractures parts at the tips of the blast-holes [Pusch, 1994].

saturated. This indicated that water flowing from the rock to the backfilled drift was redirected by the EDZ and discharged into this hole.

Experiments performed in the Canadian Shield at about the same time suggested that the EDZs of blasted tunnels are not very conductive. Thus, large-scale EDZ experiments made in AECL's underground rock laboratory by constructing dams in blasted tunnels for measuring axial flow along them indicated that the hydraulic conductivity of the near-field rock below the floor was not significantly higher than that of the virgin rock. The contrast to the Stripa data were explained by the very high rock stresses in the AECL rock, yielding a tight skin zone, and by difficulties in defining the pressure and flow distributions. Still, uncertainty remained with respect to the actual hydraulic performance of the EDZ in blasted tunnels in crystalline rock.

The contradictory results from the field experiments called for large-scale field tests tailored for determination of the hydraulic performance of the EDZ of tunnels that had been excavated by use of common drill-and-blasting, and such an experiment was made in the second phase of the Stripa project (1986–1992). The blast-holes had 3 m length and 0.5 m spacing, the charge being 0.5 kg Gurite per meter of the contour holes and 0.5 kg Dynamex as bottom charge. The test arrangement in the so-called BMT drift is shown in Figure 14. It comprised drilling of 76 radially oriented 2″ boreholes with 7 m length at the inner end of the drift and a corresponding set at the bulkhead. They overlapped to about 0.75 m depth thus forming a slot of this depth, which was separated from the holes and the drift by packers. At their outer ends the boreholes had a spacing of 0.9 m. The holes extending from the slot were

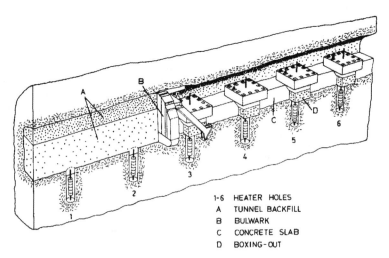

1-6	HEATER HOLES
A	TUNNEL BACKFILL
B	BULWARK
C	CONCRETE SLAB
D	BOXING-OUT

Figure 13. Main features of the Stripa Buffer Mass Test [Pusch, 1989].

equipped with packers at 3 m depth and mutually connected in 4 separated sets to represent the roof, the two walls, and the floor so that in- and outflow of the nearest 0.75 m rock annulus, the 0.75–3 m annulus and the outer 3–7 m annulus could be measured sector-wise. Each sector at the inner end of the drift was equally and simultaneously pressurized while measuring the inflow sector-wise into the outer slot and holes with respect to the amount and distribution. For eliminating inflow of water into and along the open part of the drift it was coated with epoxy and rubber liners and filled with a 100 m³ rubber bladder embedded by Na bentonite slurry. The bladder was pressurized by water to a level corresponding to the highest rock water pressure along the tested part. The rock in the tested drift was equipped with a large number of piezometers extending to different depths and recording of the pressures in the course of the test gave the hydraulic gradient along it.

A finite element flow model was worked out and applied for evaluating the hydraulic conductivity using the continuously recorded pressure and flow data. The prerequisite for this evaluation was a linear drop in pressure from the inner to the outer galleries, which was verified by the piezometer readings at steady state.

Assuming straight horizontal flow paths from the inner to the outer borehole galleries, and applying Darcy flow, it was concluded that the shallow 0.75 m zone, where the number of fractures with water-bearing capacity in drillcores was 3 to 7, had an

average hydraulic conductivity of 1.2×10^{-8} m/s, i.e. 2–3 orders of magnitude higher than that of the virgin rock. The average conductivity of the virgin rock was evaluated as 3×10^{-11} to 10^{-10} m/s. The evaluation showed that the rock from 0.75 to 3 m depth, which was taken to represent the stress-induced EDZ, had an axial average conductivity that was 10 times higher than that of the virgin rock, and a radial average conductivity of about one fifth of that of the virgin rock, hence indicating a "skin" zone. The Stripa BMT flow test is still the only experiment that has been performed on a sufficiently large scale to verify that tunnel excavation by blasting has a significant effect on the conductivity of the near-field rock.

3.3 Äspö experiments – ZEDEX, a blasted tunnel

Further information of the hydraulic properties of the EDZ over a longer distance of blasted drifts at 300–400 m depth can be drawn from ongoing tests in the so-called Zedex drift in the Äspö underground laboratory. Figure 15 is a schematic illustration of the field test with the prime purpose to investigate the conductivity and rate of saturation of tunnel backfills. The innermost part of the Zedex drift is filled with very permeable crushed rock that can be pressurized to several hundred kPa, while the rest of the drift contains inclined compacted layers of mixed bentonite/crushed rock (30/70). Both parts are about 14 m long. The outermost part consists of crushed rock over which blocks of highly compacted bentonite and pellets are placed for compensating the expected settlement of the underlying crushed rock fill that would otherwise lead to an open gap at the roof.

The Zedex drift was excavated even more carefully than the Stripa test drift, using smooth blasting technique designed by ANDRA [Bauer et al, 1996a] and the EDZ would hence be less conductive but measurements indicate a hydraulic behaviour of the EDZ that is similar to that at Stripa. Both rock masses behave similarly as indicated by the figure 6×10^{-11} m/s for the bulk conductivity of undisturbed AEspoe rock used in ongoing flow modelling. The Zedex experiment hence supports the conclusion from the Stripa test that the EDZ in blasted tunnels in crystalline rock is a very important water conductor over longer distances.

Spot-wise determination of the conductivity of shallow rock in the Zedex drift has been made in a comprehensive and careful study performed by ANDRA through Ecole de Géologie de Nancy, using the Seppi Tool [Bauer et al, 1996b, Bauer et al, 1996c]. The evaluated conductivity has a maximum close to the rock – around E-10 to E-9 m/s – but the most shallow 10 cm part, which has the highest conductivity, could not be investigated as in the Stripa experiments. Since the rock around the Zedex drift is

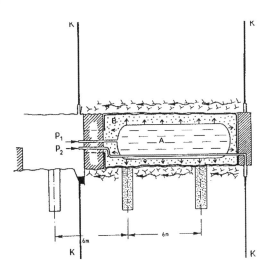

Figure 14. Test set-up for determination of the arrangement for evaluating the hydraulic conductivity of the EDZ of the BMT test drift [Börgesson et al, 1992]. A is the water-filled bladder and B the bentonite slurry that prevented water in the rock to flow into the drift.

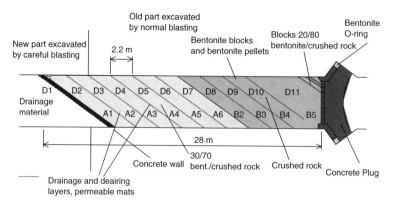

Figure 15. Longitudinal section of the blasted Zedex drift [Gunnarsson et al, 2001].

estimated to have an average hydraulic conductivity of at least E-8 m/s to a depth of a few decimetres, ANDRA's study verifies that spot-wise determination cannot give a reliable value of the average conductivity over a long distance.

3.4 *AEspoe experiments – TBM-excavated drifts and tunnels*

While full-scale testing of the entire near-field is required for adequate interpretation of the extension and nature of the EDZ in blasted crystalline rock, spot-wise determination of the hydraulic conductivity can give good information on the conductivity of the TBM-induced EDZ because of its relative homogeneity. Techniques for determining the hydraulic conductivity by spot measurements have been developed by BGR in Germany, and SKB.

The field experiments were made at about 450 m depth in crystalline rock in SKB's underground research laboratory at Äspö, Sweden. In situ Surface Packer Tests were carried out with an equipment consisting of a hollow metal cylinder fixed by a metal ring to the tunnel wall. The horizontal holes were oriented perpendicularly to the tunnel wall. At each test spot one experiment with pressurizing the packer with water and one with gas were made. The recorded pressure drop over time in the control space was interpreted in terms of conductivity by applying "Two Phase Flow Theory" using a finite element method. This gave values in the interval 10^{-10} to 2.5×10^{-10} m/s for the most shallow 10 mm part of the EDZ. No significant differences were found between water and gas. For rock extending from the rock wall to 100 mm depth the average hydraulic conductivity is estimated at 10^{-11} to 10^{-10} m/s.

The rock samples used for determining the hydraulic conductivity of the shallow rock in the same tunnel where the field tests took place were prepared from 100 mm cores with a length of 250 to 500 mm taken perpendicularly to the rock wall. Several series of 10 mm diameter cores were extracted by diamond drilling perpendicular to the large cores, i.e. parallel to the tunnel at different distances from the tunnel wall, and several series of 3 mm discs were sawed from the large cores to allow determination of hydraulic conductivity perpendicular to the tunnel wall. Hydraulic testing of the samples used a triaxial apparatus and a hydraulic gradient of about 100. The measurements showed that the isotropic hydraulic conductivity was in the range of 2×10^{-9} to 5×10^{-12} m/s from the tunnel wall to 4.5 mm depth, and 5×10^{-12} to 5×10^{-13} m/s from 4.5 to 10 mm depth. The undisturbed crystal matrix has a hydraulic conductivity of about 5×10^{-13} m/s. Fluorescence microscopy of epoxy-impregnated samples gave the porosity and showed the pattern of fissuring caused by the bits. The majority of the fissures formed an angle of ± 25–$45°$ to the rock wall and were responsible for the increase in porosity from less than 0.5% of the undisturbed crystal matrix to 2–5.6% within 10 mm distance from the wall. A number of macroscopic fractures were also identified, their depth and spacing being 10–50 mm, and they are assumed to cause an additional increase in hydraulic conductivity of larger volumes of the matrix than represented by the small samples and hence yield values on the same order of magnitude as the ones obtained from the small-scale packer tests.

Comparison of the field measurements and the lab testing of small samples shows that the lab tests at depths of less than 4.5 mm from the tunnel wall have a higher conductivity than in the field tests but from a depth of 4.5 mm onwards the relationship is inversed. This verifies the belief that stress-generated fissures in the larger rock volume played a major role in the in situ tests. The high uniform hydraulic conductivity of the most shallow part of the EDZ serves to distribute water flowing in from discrete water-bearing

Table 1. The approximate water flux across the assumed $80 \, m^2$ near-field for the hydraulic gradient $i =$ unity distributed over the various EDZ components. The backfilled tunnel is assumed to be impermeable.

Case	Permeated cross section, m^2	Hydraulic conductivity, m/s	Water flux, m^3/s
Virgin rock, no tunnel	80	10^{-11}	8×10^{-10}
Blasted tunnel	20		2.6×10^{-7}
• Blast-EDZ	20	10^{-8}	2×10^{-7}
• Stress-EDZ	60	10^{-9}	6×10^{-8}
TB tunnel	80		2.6×10^{-9}
• Stress-EDZ	18	10^{-10}	2×10^{-9}
• Virgin rock	62	10^{-11}	6.2×10^{-10}

fractures over the entire periphery of deposition holes and tunnels in the early period of water saturation of buffer clay embedding canisters with HLW.

3.5 Experience from attempts to seal the EDZ of blasted tunnels

The high axial hydraulic conductivity of the blast-induced EDZ means that rather much water may flow into the tunnels before, during and after backfilling despite the "skin" effect in the stress-induced EDZ because inflow to the EDZ can take place through discrete water-bearing fractures and fracture zones that intersect the tunnels more or less perpendicularly. Where this takes place the backfilling operation may be difficult or even impossible and attempts have therefore been made to investigate if the blast-induced zone can be sealed by grouting. The work, which was conducted in the Stripa BMT drift after completion of the large-scale conductivity test, comprised drilling of 345 radially oriented holes with about 1 m length and 0.7–0.9 m spacing and grouting using an earlier developed "dynamic injection" technique [Börgesson & Pusch, 1989]. Very fine-grained cement slurry with plasticizer was used for the purpose. The sealing effect was poor, which was primarily explained by the clogging effect of debris that prevented cement penetration to more than a few centimeters depth and by too large spacing of the holes as well as by the fineness of the fractures. A possible additional reason is that the lack of counter-forces at the injection resulted in loss in injection energy, which is the reason for avoiding after-injection in practical rock construction and instead injecting grout prior to excavation.

3.6 Conclusive remarks – major lessons learned

The typical property of undisturbed crystalline rock to transport water through relatively few discrete fractures with a considerable spacing, often several meters, is changed in the vicinity of drifts and tunnels. Here, the dynamic impact and very high gas pressure caused by blasting produces rich fracturing around the blast-holes and can activate discrete fractures and make them propagate and become effective flow paths at several meters distance from the excavated room. In the near-field the frequency of fractures with a potential to transport water hence increases and the blast-induced change in hydraulic performance can be very obvious, especially where one fracture set is parallel or close to parallel to the axis of the tunnel. Within a distance of 1–1.5 m from a tunnel wall blasting even by using careful technique can increase the net bulk hydraulic conductivity by orders of magnitude.

Stress-induced excavation disturbance has much less influence on the hydraulic conductivity but it affects the rock to a larger distance. It is estimated that the axial conductivity of this zone can increase by 10 times while the radial conductivity may drop to one fifth of the conductivity of the virgin rock. The effect most probably depends on the rock structure.

The overall hydraulic effect of the EDZ is illustrated by the Table 1 above showing the estimated change in axial flux across the entire nearfield, $80 \, m^2$, around a blasted tunnel with $25 \, m^2$ cross section area. It is composed of the blast-affected EDZ equalling $20 \, m^2$ and of the surrounding stress-induced EDZ of $60 \, m^2$. For TBMs with the same tunnel size, the EDZ extending to 0.1 m depth represents $18 \, m^2$. For comparison of the two cases the net flux across the same section area, $80 \, m^2$, is given in the table, implying that the larger part of the section is made up of virgin rock for the TBM case.

Assuming the conductivity of the virgin rock to be 10^{-11} m/s and that of the blasted zone to be 10^{-8} m/s and taking the axial conductivity of the stress-affected EDZ around the blasted tunnel to be 10^{-9} m/s, one finds the total flux across the $80 \, m^2$ near-field section area to be about 100 times higher than for a TBM tunnel, for which the conductivity of the EDZ is taken as 10^{-10} m/s.

Comparing these data one finds the expected water transport capacity of a cross section corresponding to the total near-field of TBM tunnels to be no more than about 1% of that of blasted tunnels. The importance from the point of safety analysis of this fact is obvious.

The following important additional conclusions can be drawn from the various attempts to characterize the EDZ from a hydraulic point of view:

– Careful blasting of the type applied to the Stripa BMT experiment causes a blast-induced EDZ that extends to about 1 m from the periphery and is at least 100 times more conductive than the virgin rock. The surrounding stress-induced EDZ extends to about 3 m from the periphery and has an axial conductivity that is about 10 times higher than that of the virgin rock, while the radial conductivity is about 5 times lower than this conductivity. The consequence of this is that a plug intended to totally cut off the EDZ of a blasted tunnel must extend to about 3 m from the periphery. For cutting off the blast-induced EDZ it must extend to 1.5 m depth in the floor and 1 m in the walls while 0.5 m may be sufficient in the roof.
– The blasting techniques employed in the discussed cases were careful but improved methods and even greater care may reduce the disturbance caused by this excavation method.
– Unlike the EDZ in salt and certain clay layers no self-sealing is expected in the EDZ of crystalline rock. However, the hydrothermal conditions that will prevail for a couple of thousand years in certain HLW repository concepts may lead to formation of clay minerals in fractures by converting feldspars and certain heavy minerals. This matter is worth studying.
– Sealing of the EDZ in blasted tunnels by grouting of short holes has a poor effect. For minimizing water inflow into tunnels during backfilling grouting must be made prior to the excavation focusing on significantly water-bearing fracture zones that can be identified before the excavation starts.

REFERENCES

Bauer C, Homand F, Ben Slimane K, 1996a. Proceeding of the EDZ Workshop "Designing the Excavation Disturbed Zone for Nuclear Repository in Hard Rock". Winnipeg, September 1996. Canada. CNS pp. 87–96.

Bauer C, Homand F, Ben Slimane K, 1996b. Disturbed zone assessment with permeability measurements in the ZEDEX tunnel.
Bauer C, Homand F, Ben Slimane K, Hinzen K.G, Reamer S.K, 1996c. Damage zone in the near field in the Swedish ZEDEX tunnel using in situ and laboratory measurements. EUROCK Congress in Turin, September 1996.
Berryman J.G. Lectures Notes on nonlinear inversion and tomography: Borehole Seismic Tomography. University of California. October, 1991.
Börgesson L, Pusch R, 1989. Rock sealing by dynamic injection; Stripa Project Phase III. Proc. NEA/CEC Workshop on Sealing of Radioactive Waste Repositories at Braunschweig, OCDE/OEDC, Paris.
Börgesson L, Pusch R, Fredriksson A, Hökmark H, Karnland O, Sandén T, 1992. Final Report of the Rock Sealing Project – sealing of Zones Disturbed by Blasting and Stress Release. Stripa Project Technical Report 92–21. SKB, Stockholm.
Gunnarsson D, Börgesson L, Hökmark H, Johannesson L-E, Sandén T, 2001. Äspö Hard Rock Laboratory, Report on the installation of the Backfill and Plug Test. SKB IPR-01-17, SKB Stockholm.
Jolly R.J.H, WEI L, Pine R.J, Stress sensitive fracture flow modelling, IPC Villa hermosa, Mexico, 2000.
Pardillo J, Campos R, Guimerá J, Caracterización geológica de la zona del ensayo Febex (Grimsel-Suiza). CIEMAT Tech. Report: 70-IMA-M-2-01. May 1997. (Draft).
Pine R.J, Nicol D.A.C, Conceptual and numerical models of high pressure fluid-rock interaction, Proc. 29th US Symposium on Rock Mechanics, Minneapolis, 1988.
Pusch R, 1989. Alteration of the hydraulic conductivity of rock by tunnel excavation. Rock Mech. & Mining Sciences, Vol. 26, No.1 (pp.71–83).
Pusch R, 1994. Waste Disposal in Rock, Developments in Geotechnical Engineering, 76. Elsevier Publ. Co. ISBN: 0-444-89449-7.
Young R.P, Martin C.D, Potential role of acoustic/microseismicity investigations in the site characterization of nuclear waste repositories. Int. J. Rock Mech. Min Sci. & Geomech. 1993.

Advances in Understanding Engineered Clay Barriers – Alonso & Ledesma (eds)
© 2005 Taylor & Francis Group, London, ISBN 04 1536 544 9

Temperatute Buffer Test instrumentation

Torbjörn Sandén & Harald Hökmark
Clay Technology AB, Lund, Sweden

Michel de Combarieu
ANDRA, Paris, France

ABSTRACT: The Temperature Buffer Test, TBT, is a heated full scale field experiment which ANDRA and SKB are presently carrying out at the SKB Äspö Hard Rock Laboratory. The experiment is intended to run under thermal load conditions that are relevant for vitrified waste deposition concepts, i.e. under high temperatures. Two buffer design options are tried: with and without a sand shield around the heat generating canister. The test was installed in the beginning of 2003 and will run for several years. The deposition hole used is located beside the CRT-experiment in the TASD-tunnel on the 420 meter level. In order to monitor the THM processes in the system, sensors of various types are placed in the buffer, the sand, the heaters and in the rock. This paper describes the strategy for the instrumentation, the sensor types and the instrumentation layout. Examples of results and a preliminary attempt of interpretation are provided.

1 INTRODUCTION

When storing vitrified waste in KBS3-type deposition holes, the buffer material surrounding the reprocessed waste canisters could be subjected to temperatures exceeding 100°C unless long cooling time is allowed before storage. No full scale tests have been carried out with temperature of the buffer exceeding 100°C so far. This is the main reason for the initiation of TBT.

An existing full scale KBS3 deposition hole at the HRL in Äspö has been selected for the test. An important part of the work is to measure the thermal, hydraulic and mechanical processes in the bentonite during saturation and afterwards. The objectives may be summarized as follows:

Observation of the THM behavior of the clay during the initial thermal shock, the following water saturation phase and the final pressurization under elevated temperature.

The test will check the models concerning the THM processes in the engineered barriers.

When dismantling the buffer, samples will be taken of the bentonite in order to determine the temperature limit for the buffer degrading processes.

One method that could be used in order to reduce the temperature in the bentonite buffer is to place a sand screen between canister and bentonite. The test will evaluate the feasibility and effectiveness of this type of composite barrier, comparing its behavior during saturation with that of a homogenous barrier under identical experimental conditions.

2 TEMPERATURE BUFFER TEST – GENERAL DESCRIPTION

The test consists of a full scale KBS3-type deposition hole, 2 carbon steel canisters equipped with electrical heaters for simulating the heating caused by the radioactive decay and a mechanical plug at the top (Figure 1). The canisters have been embedded in dense buffer clay consisting of blocks (cylindrical and ring-shaped) of compacted bentonite powder. An artificial water pressure has been applied in the outer slot, which is filled with sand to provide controlled and uniform hydraulic boundary conditions.

The upper canister is surrounded by sand in order to reduce the temperature in the bentonite and to test the general prospects of a composite buffer.

A retaining plug has been built in order to restrict upward movement caused by buffer swelling.

The main part of the instrumentation for the TBT consists in measuring devices for:

– Temperature
– Total pressure
– Pore pressure
– Moisture content

In the deposition hole, temperature is measured in 89 points, total pressure in 31 points, pore water

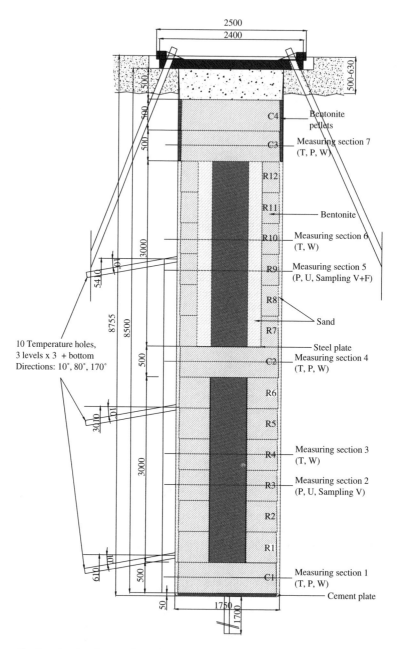

Figure 1. Schematic view showing the experiment layout.

pressure in 10 points and relative humidity in 35 points. In the surrounding rock, temperature is measured in 40 points. Additional temperature measurements are done inside and on the heaters surface, in the pressure sensors and in the relative humidity ones. Beside these measurements, a number of measuring devices are installed:

– equipment for sampling of gas/fluid in the buffer and in the sand
– hydraulic load cells measure the load on the retaining plug anchors

164

Table 1. List of measuring equipment.

Measuring quantity	Measuring principle	Supplier	Number of sensors	Remark
Buffer and rock				
Temperature	Thermocouple	Pentronic	89	
Total pressure	Vibrating wire	Geokon	29	Containing resistive temperature sensor
Total pressure	Fiber optic	DBE	2	Containing resistive temperature sensor
Pore pressure	Vibrating wire	Geokon	8	Containing resistive temperature sensor
Pore pressure	Fiber optic	DBE	2	Containing resistive temperature sensor
Water content	Relative humidity (capacitive method)	Rotronic	12	Containing resistive temperature sensor
	Relative humidity (capacitive method)	Vaisala	11	Containing resistive temperature sensor
	Soil psychrometer	Wescor	12	Temperature measurements with thermocouple
Gas/water pressure	Piezo electric	Druck	6	
Additional measurements				
Temperature	Thermocouple	Pentronic	40	Placed in boreholes in the rock
Gas/water pressure	Piezo electric	Druck	2	
Load cells for anchors	Piezo electric	Glötzl	3	
Displacement	Inductive	Solartron	3	

– displacement of the plug is measured with three sensors
– an artificial water saturation system is installed measuring applied water pressure and injected water volume.

3 INSTRUMENTATION

The different instruments used in the experiment are described in this section. The instruments were chosen based on experience from instrumentation of the Prototype Repository and other projects going on at the Swedish Hard Rock Laboratory. The higher temperature of this experiment has been considered. Table 1 shows the list of measuring equipment used in TBT.

3.1 *Buffer and sand*

3.1.1 *Conditions and materials*
Temperature is expected to be up to 170°C. Salt content in the water is 1.2–1.5%. Total pressure could be as high as 15 MPa and pore water pressure as high as 5 MPa. With these conditions and with the experience from the above mentioned experiments it was decided that all instruments, except the thermocouples should be manufactured in or encapsulated in titanium (grade 2) housing and the cables should be led in titanium tubes the whole way up through the bentonite. The tubes are welded to the sensor housing. The thermocouples are of sheath type where the sheath is made of, 70–30 copper nickel alloy.

3.1.2 *Temperature*
Only one type of instruments has been used for temperature measurements, since experience of this type of sensors and the supplier are outstanding. In addition to the thermocouples, resistive temperature sensors are built-in the capacitive relative humidity sensors as well as in the pressure gauges of vibrating wire type. Temperature is also measured with the psychrometers.

3.1.2.1 Pentronic
Thermocouples type K, class 1. Sensors of sheath type (made of CuNi) protected all the way from the measuring point to the data collecting equipment. Outer diameter is 4.0 mm. The thermocouples are calibrated at 0°C and at 100°C.

89 thermocouples have been placed in the buffer.

3.1.3 *Moisture content*
Moisture content is measured with the following methods:

– Capacitive sensors
– Psychrometers

These devices measure the relative humidity (RH) in the pore system. RH can be converted into water ratio or total suction (negative water pressure). The psychrometers are complementing the capacitive sensors since the measuring range is different. The relative humidity is measured in 35 points in the buffer.

3.1.3.1 Rotronic
Measuring with the capacitive method. The measurement range of the sensors is 0–100% RH and −50°C–200°C. Sensor body is made of stainless steel. The

bodies are filled with resin in order to prevent water leakage. Sensor bodies and cables have been encapsulated in titanium. The sensors after encapsulation have a 22 mm outer diameter and a 135 mm length. The sensors were delivered calibrated. No additional calibration has been done but a function control was done before installation.

12 Rotronic relative humidity sensors have been positioned in the buffer.

3.1.3.2 Vaisala
Measuring with the capacitive method. The measurement range of the sensors is 0–100% RH and −40°C–180°C. The sensor body is made of stainless steel. The bodies are filled with. Sensor bodies and cables have been encapsulated in titanium. The sensors after encapsulation have a 22 mm outer diameter and a 63 mm length. The sensors were delivered calibrated. An additional calibration has been done due to the fact that a minor soldering work was done when leading the sensor cables in the protection tubes.

11 Vaisala relative humidity sensors have been positioned in the buffer.

3.1.3.3 Wescor psychrometer
The measurement range of the sensors is 95–100% RH. The temperature range is unknown but the sensors have been used in the CRT-test up to 70°C. These sensors have not been placed in the hottest part of the buffer. Sensors and cables were encapsulated in titanium and sealed with resin. The sensors after encapsulation have a 22 mm outer diameter and a 70 mm length. The sensors were delivered calibrated. A function control has been performed before installation.

12 Wescor psychrometers sensors have been positioned in the buffer.

3.1.4 *Total pressure*
Total pressure is the sum of the effective stress and the pore water pressure.

The total pressure is mainly measured by Geokon sensors which are using the vibrating wire technique. Geokon sensors are used in both the Prototype Repository and in the Retrieval test and the return on experience is good. In addition, two fiber optic sensors from DBE are installed as a demonstration and as a test of the technique.

Total pressure is measured in a total of 31 points in the test hole.

3.1.4.1 Geokon
The measuring principle is vibrating wire. The pressure cells are specially designed and are made of titanium. The cells have an outer diameter of 76 mm and a thickness of 16 mm. The sensors are dimensioned for a pressure range of 0–15 MPa and a working temperature of 70–170°C. The cables from the sensors are led out in titanium tubes.

3.1.4.2 DBE
The measuring principle is fiber optic based on fiber Bragg grating technology. The pressure cells are specially designed and are made of titanium. The cells are rectangular with a size of 140×70 mm. The sensors are dimensioned for a pressure range of 0–15 MPa and a working temperature of 70–170°C. The cables from the sensors are led out in titanium tubes.

3.1.5 *Pore pressure*
When pore water pressure is measured, the pressure cell is placed behind a ceramic or metallic filter. The pore water pressure is mainly measured with Geokon sensors. In addition, two optic fibersensors from DBE are installed as a demonstration and as a test of the technique.

Pore pressure is measured in a total of 10 points in the test hole.

3.1.5.1 Geokon
The measuring principle is vibrating wire. The pressure cells are specially designed and are made of titanium. The sensors are shaped as a cylindrical tube with a 25 mm outer diameter and a 127 mm length. The sensors are dimensioned for a pressure range of 0–5 MPa and a working temperature of 70–170°C. The cables from the sensors are led out in titanium tubes.

3.1.5.2 DBE
The measuring principle is fiber optic based on fiber Bragg grating technology. The sensors are specially designed and are made of titanium. The sensors are cylindrical shaped and have an outer diameter of 26 mm and a length of 150 mm. The sensors are dimensioned for a pressure range of 0–5 MPa and a working temperature of 70–170°C. The cables from the sensors are led out in titanium tubes.

3.1.6 *Gas pressure/sampling in sand*
In the sand surrounding the upper canister, two titanium tubes with filter tips have been positioned. This gives a possibility to measure the gas pressure in the sand and also to take in situ samples of the gas/fluid in the sand. The tubes have three way valves connected in their ends, where one way is connected to a pressure transducer and the other will be used when taking samples. The pressure is measured by Druck pressure transducers.

3.1.7 *Gas/fluid sampling in bentonite*
Two sample collectors have been installed. They are of the same type as those installed in the Prototype Repository.

A sample collector consists of a titanium cup with a titanium filter placed on the top Pore gas/fluid from the bentonite will flow through the filter and into the cup. A tube made of PEEK is connected to the bottom of the cup. This makes it possible to take in situ samples during the test period. The tubes have three way valves connected in their ends, where one way is connected to a pressure transducer and the other will be used when taking samples. The pressure is measured by Druck pressure transducers.

3.1.8 Gas tests in the bentonite

In the bentonite surrounding the upper canister, two titanium tubes with filter tips are positioned. At full water saturation one of the filters could be used as gas inlet and the other as gas outlet. The exact test procedure will be decided later. The tubes have three way valves connected in their ends, where one way is connected to a pressure transducer and the other will be used when taking samples. The pressure is measured by Druck pressure transducers.

3.2 Additional measurements

3.2.1 Temperature

Temperature is measured in 40 points in the rock. The thermocouples used are of the same type as those in the buffer and sand.

3.2.2 Temperature measurements inside heaters and on the surface

6 thermocouples have been installed inside each heater. They are of type "J" class 1. Two thermocouples are installed 200 mm from each end, at opposite sides and the third pair is placed in the middle of the heater.

Additional 11 thermocouples are placed on each heaters surface. They are of the same type as the inner. Three groups of three thermocouples are installed. One group is placed at 100 mm from each end of the heater and one at the middle of the heater. The thermocouples are placed with a disposition of 120°. Two additional thermocouples are installed at the bottom and at the top of the heater.

3.2.3 Artificial water saturation

An artificial water pressure will be applied in the outer slot witch is filled with sand. Titanium tubes equipped with filter tips have been placed in the sand on two levels, 250 mm and 6750 mm from bottom, four at each level. The applied pressure is measured with a Druck pressure transducer and the injected volume is measured with a sensitive pressure transducer measuring the difference in pressure in the water container. The pressure is applied with help of gas.

3.2.4 Hydraulic load cells for the anchor system

Totally nine anchors are installed in order to keep the retaining plug in position. Three of the anchors are equipped with hydraulic load cells. The cells are manufactured by Glötzl.

3.2.5 Displacement of plug

Three displacement sensors are positioned on the top of the deposition hole in order to measure the displacement of the plug. The sensors are manufactured by Solartron.

3.3 Sensor performance

The experiment was installed during February and March 2003. The data acquisition started on the 26 of March. At the test start all instruments were working. At present (October 2003), we can observe that 3 RH sensors from Rotronic and 4 RH sensors from Vaisala are not working. The reason for this is not yet confirmed. For sensors close to the sand filter, the reason could be that the sensors are measuring a RH of about 100%, which means that the sensors are out of range. For other sensors the reason could be connection problems with the Datascan units which are collecting the data or actual sensor failure. This will be investigated in the near future.

4 STRATEGY WHEN POSITIONING THE INSTRUMENTS

The main objective of the experiment is to observe the THM behavior of the buffer under high temperature conditions. The following had been taken in account when positioning the different sensors:

- The initial thermal shock were the evolution of the hydration will be very fast
- The long re-hydration where evolution will be slow and trends can be defined accurately by long time observations
- Due to the conditions and the artificial water saturation, the system is considered having a cylindrical symmetry. This means that the temperature gradients in the buffer could tell us a lot about the hydration status.
- The observations are concentrated to the crucial points which are the parts at the middle of each heater and to the extremities where we need information in order to define the heat flux.

The measurements in the buffer and sand are concentrated to 7 measuring sections placed on different levels (Figure 1). On each level, sensors have been placed in eight main directions A, AB, B, BC, C, CD, D and DA according to Figures 2 to 6.

4.1 Positions of instruments in the bentonite

The instruments are located in three main levels in each instrumented block, the surface of the block (only total pressure cells measuring the horizontal pressure) and 50 mm and 250 mm from the upper block surface.

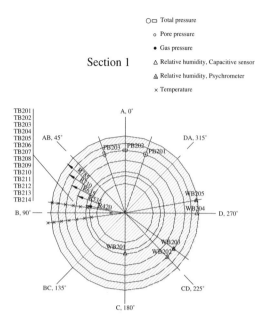

Section 1

○□ Total pressure
○ Pore pressure
● Gas pressure
△ Relative humidity, Capacitive sensor
▲ Relative humidity, Psychrometer
× Temperature

TB201
TB202
TB203
TB204
TB205
TB206
TB207
TB208
TB209
TB210
TB211
TB212
TB213
TB214

Figure 2. Schematic view, showing the instrumentation in measuring Section 1. The instruments in Section 4 and 7 are placed in the same way as in this section.

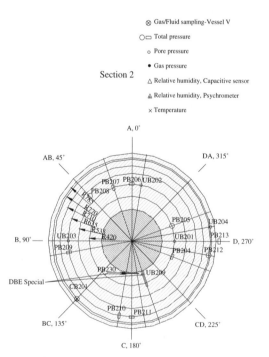

⊗ Gas/Fluid sampling-Vessel V
○□ Total pressure
○ Pore pressure
● Gas pressure
△ Relative humidity, Capacitive sensor
▲ Relative humidity, Psychrometer
× Temperature

Section 2

Figure 3. Schematic view, showing the instrumentation in measuring Section 2.

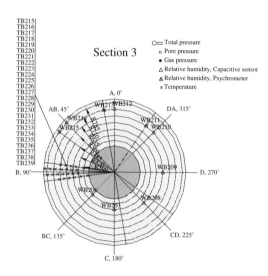

Section 3

○□ Total pressure
○ Pore pressure
● Gas pressure
△ Relative humidity, Capacitive sensor
▲ Relative humidity, Psychrometer
× Temperature

TB215
TB216
TB217
TB218
TB219
TB220
TB221
TB222
TB223
TB224
TB225
TB226
TB227
TB228
TB229
TB230
TB231
TB232
TB233
TB234
TB235
TB236
TB237
TB238
TB239

Figure 4. Schematic view, showing the instrumentation in measuring Section 3.

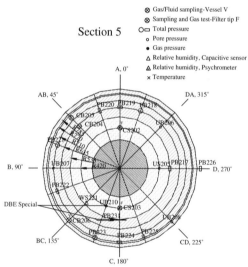

Section 5

⊗ Gas/Fluid sampling-Vessel V
⊗ Sampling and Gas test-Filter tip F
○□ Total pressure
○ Pore pressure
● Gas pressure
△ Relative humidity, Capacitive sensor
▲ Relative humidity, Psychrometer
× Temperature

Figure 5. Schematic view, showing the instrumentation in measuring Section 5.

The thermocouples and the total pressure cells are placed in the 50 mm level for practical reasons and the other sensors in the 250 mm level.

4.2 Positions of thermocouples in the rock

The thermocouples are located in ten bore holes in the rock, Figure 2–1. The depth of each borehole is 1.5 m. In each borehole 4 thermocouples are placed at different distances from the rock surface: 0, 0.375, 0.75 and 1.5 m.

168

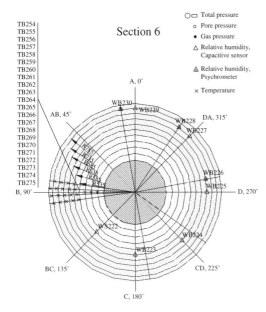

TB254
TB255
TB256
TB257
TB258
TB259
TB260
TB261
TB262
TB263
TB264
TB265
TB266
TB267
TB268
TB269
TB270
TB271
TB272
TB273
TB274
TB275

Section 6

○□ Total pressure
○ Pore pressure
● Gas pressure
△ Relative humidity, Capacitive sensor
▲ Relative humidity, Psychrometer
× Temperature

Figure 6. Schematic view, showing the instrumentation in measuring Section 6.

4.3 *Leading sensor cables out of the deposition hole*

All cables and tubes, coming out from the instruments in the bentonite blocks plus the power cables and the thermocouples from the heaters, were led up along the bentonite block periphery surface. In the same levels as the top of each heater, they were led into slots made in the rock wall. The cables were divided into different slots depending on whether they come from the lower heater and its surrounding bentonite or from the upper part. This distribution was done to allow for separate excavation of the upper part of the test leaving the lower part for a longer test period.

Since a lot of cables and tubes were led in the gap between rock and bentonite in the deposition hole (about 260 units), it was important to distribute them on the block periphery in a prescribed order. Every cable or tube has a coordinate which is the angle from direction A (Figures 2–6). Each cable was led out to this position from the sensors position in pre-manufactured tracks on the blocks surface, up along the bentonite block periphery and into the prescribed slot in the rock.

5 TBT – EXAMPLES OF RESULTS AND PRELIMINARY EVALUATION

Section 3 through Ring 4 and Section 6 through Ring 10 are particularly densely instrumented. Being

located at mid-height of the two heaters with almost purely radial conditions, these sections provide convenient test grounds for various interpretation techniques.

5.1 *Measured and interpreted results*

Figure 7 shows temperatures recorded after 20, 60, 115 and 170 days in Ring 4. Recordings obtained by use of the dense array of thermocouples as well as recordings obtained from the RH sensors are shown. Also readings from extra thermocouples built in to the heater surface are included. The thermocouple readings are used to find the thermal conductivity λ as function of distance r from the heater axis using the steady-state expression:

$$\lambda = \frac{Q}{2 \cdot \pi \cdot (T_r - T_{r+dr})} \cdot \ln\left(\frac{r + dr}{r}\right) \tag{1}$$

Here, Q is the heater power per unit heater length. It is difficult to determine the conductivity quantitatively with high accuracy and high resolution, because the steady-state expression is sensitive to errors in instrument position and temperature readings. This is in particular the case here where it is applied to the small radial steps between neighboring instruments in the thermocouple array. In addition, there are some uncertainties in the value of the power Q, which is not a fixed and constant quantity. Qualitatively, however, there is a clear indication of a reduced conductivity around the heater.

The conductivity profile shows without doubt that there is a dehydrated zone with an annular width of about 0.15 m in the hottest buffer part.

Figure 8 shows results of RH measurements in Ring 4. The RH readings were translated into suction values using the psychrometric law and then converted to saturation estimates using two different suction-saturation models. The initial saturation in this section was about 80%.

The saturation in the vicinity of the heater seems to have dropped to about 60% some 5 cm from the heater surface and probably even lower, perhaps to 50%, at smaller distances.

Figure 9 shows measured temperatures and calculated conductivities in Ring 10. Note that there is temperature offset between bentonite and rock after 20 days. This is probably due to incomplete saturation of the sand filter in the beginning of the test. The different slopes of the sand-shield parts and the bentonite parts of the temperature curves reflect the difference in heat transport conditions between these two materials in a lucid way.

Figure 10 shows results of RH measurements in Ring 10. Similar to the results in Ring 4 (Fig.8) the RH readings have been converted to saturation.

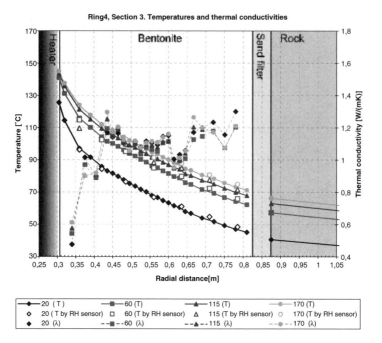

Figure 7. Temperatures measured by thermocouples (filled symbols, solid lines) and by use of RH sensors (unfilled symbols, no lines). Thermal conductivities calculated using Eq. 1(filled symbols, dotted lines). The legend numbers give elapsed time in days.

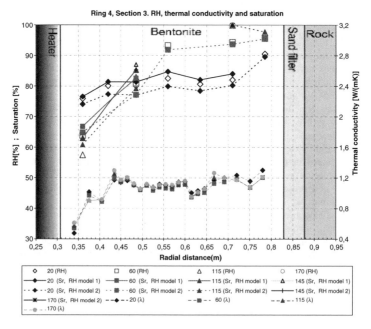

Figure 8. RH readings (unfilled symbols). Estimated saturation (filled symbols, with solid and dotted lines for model 1 and model 2, respectively). Thermal conductivities are shown here only for comparison (c.f. Fig. 7). The legend numbers give elapsed time in days.

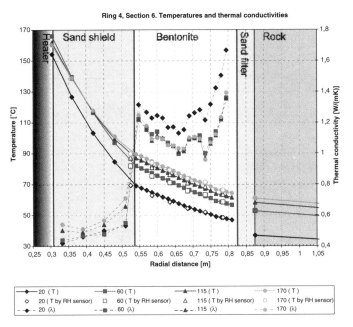

Figure 9. Temperatures measured by thermocouples (filled symbols, solid lines) and by use of RH sensors (unfilled symbols, no lines). Thermal conductivities calculated using Eq. 1(filled symbols, dotted lines). The legend numbers give elapsed time in days.

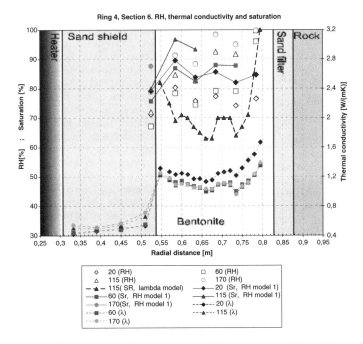

Figure 10. RH readings (unfilled symbols). Estimated saturation (filled symbols, solid lines). Saturation estimated from conductivities (filled symbols, dashed line). Thermal conductivities are shown here only for comparison (c.f. Fig. 9). The legend numbers give elapsed time in days.

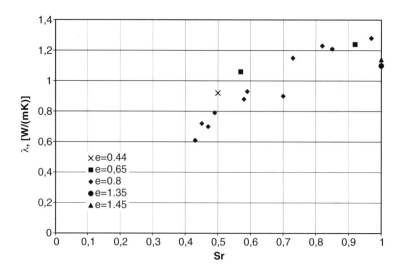

Figure 11. Thermal conductivity as function of saturation for MX80 bentonite. The legend gives the void ratio.

However, in this section we only use one of the two models tried in Section 3 for suction-saturation translation. Instead, an attempt is made here to convert the thermal conductivity estimates to saturation using an experimentally established conductivity-saturation relation.

In Ring 10, there does not seem to be any indications of thermally driven water redistribution.

5.2 *Discussion*

The following observations can be done:

– There is a zone of significant desaturation around the lower heater. The drop in saturation may be from 80% down to 50%, and the zone seems to be about 15 cm in radial extension.
– In the bentonite surrounding the sand shield around the upper heater, we see no signs of desaturation.
– In both sections we see no, or only very slow, changes in thermal conductivity after 2 months, which may indicate that there are only minor changes in the hydrations status. This is not entirely correlated with the results from the RH readings which show continuously increasing values, at least in the upper section. (In the lower section most RH-signals have failed after 115 days, some of them even earlier).

There are the following aspects on the validity and relevance of the interpreted results:

– The power is 1500 W/heater. If that power is distributed uniformly over the heater surface, the radial output at mid-height of the canister

corresponds to about 440 W per meter of heater length. This value was used here for Q in Eq. 1, but is in reality valid only close to the heater, where the heat flux is purely radial. This means that the conductivity values in general must be taken as qualitative indications of how conditions vary with radial distance rather than as absolute values.
– Saturation estimates based on bentonite thermal conductivity calculations (Fig. 10) are uncertain. There is not only the uncertainty in the conductivity determination described in the previous point, but also an uncertainty in the conductivity/saturation relation (c.f. Figure 11). It is obvious that saturation estimates based on that relation are sensitive to errors in conductivity, in particular in the high saturation range.
– The two models used here to translate RH readings to saturation estimates seem to give reasonable results. The initial saturation is just above 80%. The two models appear to bracket this value nicely (Fig. 8, 20 days). This supports the general confidence in the conclusion that there is in fact a dehydrated zone around the lower heater, as shown by the RH-based results for 60 days and 115 days (Fig. 8).

Finally it should be noted that the suction-saturation models used here are based on new work and look different from the one suggested in the modeling program (Hökmark and Fälth, 2002). The models are shown in Figure 12. Model 1 is based on a direct relation between water content and suction. This model is not valid in high saturation ranges. Model 2 is a vanGenuchten fit to that relation. For saturation ranges below 50%, both models are uncertain.

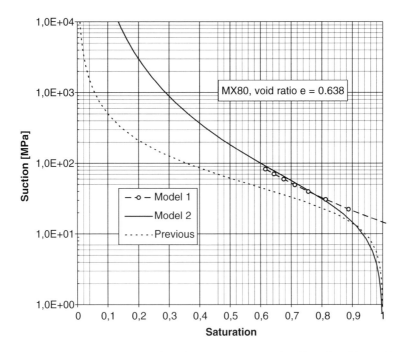

Figure 12.   Models for the saturation-suction relation. The estimated ranges of validity of models 1 and 2 are uncertain.

REFERENCES

Hökmark H., Fälth B., (2003). TBT predictive modeling program, rev3.

Sandén T., Börgesson L., (2002). Temperature Buffer Test, Report on instruments and their positions for THM measurements in buffer and rock and preparation of bentonite blocks for instruments and cables. R5.

Fundamental research material behaviour and laboratory testing

Advances in Understanding Engineered Clay Barriers – Alonso & Ledesma (eds)
© 2005 Taylor & Francis Group, London, ISBN 04 1536 544 9

Thermo-mechanical and geochemical effects on the permeability of high-density clays

M.V. Villar

Centro de Investigaciones Energéticas, Medioambientales y Tecnológicas, Madrid (Spain)

E. Romero & A. Lloret

Universitat Politènica de Catalunya, Barcelona (Spain)

ABSTRACT: Several factors having a significant influence on the permeability of compacted expansive materials have been experimentally investigated in the context of projects regarding the near field of a high level radioactive waste repository. Highly expansive materials – MX-80 and FEBEX bentonites – and moderately swelling materials – such as Boom clay and a granite/bentonite mixture – have been analysed. The effect of void ratio and degree of saturation of the bentonite, salinity of the permeant, hydraulic gradient and temperature has been analysed. The main results obtained and conclusions drawn are summarised in this paper.

1 INTRODUCTION

The clay blocks of an engineered barrier are initially unsaturated, and an important percentage of their pores (40–50%) will be filled with air. Thermal and hydraulic gradients will be generated in the repository, due to the heat emitted by the wastes and to the resaturation of the near field. Consequently, a combined movement of gas and water will take place in the clay barrier, that must be considered as a confined system in which no volume change is allowed.

Despite the complications involved in the different physicochemical interactions between the three phases present in unsaturated soils, the laws governing the flow of the different constituents are not very different from the classical laws applied to saturated soils. In the case of water, the validity of Darcy's law has been experimentally verified, the coefficient of permeability depending on the degree of saturation. In any soil, permeability to water increases with the degree of saturation. Taking this into account, relative permeability (k_{rw}) is defined for a given porosity and degree of saturation as the ratio between the unsaturated permeability (k) for this degree of saturation and the saturated permeability (or hydraulic conductivity, k_w) for the same porosity. It is generally accepted that relative permeability is related to the degree of saturation (S_r) in accordance with a potential law ($k_{rw} = S_r^\lambda$), in which λ is an empirical constant usually related to pore size distribution.

On the other hand, air flow is governed by Darcy's law, with a coefficient of permeability to gas (k_g). Given that there are no important interactions between the air and the other constituents of the soil, air flow depends only on the air pressure gradient and on the volume of air in the soil, which in turn depends on the degree of saturation. The flow of air requires the continuity of the gas phase. Water-filled pores block gas flow, consequently, gas permeability decreases as soil water content increases.

Hence, water and gas advective fluxes ($\mathbf{q}_w$, $\mathbf{q}_g$) follow Darcy's law with relative permeabilities (k_{rw}, k_{rg}) dependent on liquid degree of saturation:

$$\mathbf{q}_w = -\frac{\mathbf{k}_i k_{rw}}{\mu_w}\left(\nabla P_w - \rho_w \mathbf{g}\right) = -\mathbf{k}_w k_{rw}\left(\frac{\nabla P_w}{\rho_w \mathbf{g}} + \mathbf{i_z}\right) \quad (1)$$

$$\mathbf{q}_g = -\frac{\mathbf{k}_i k_{rg}}{\mu_g}\left(\nabla P_g - \rho_g \mathbf{g}\right) = -\mathbf{k}_{g0} k_{rg}\left(\frac{\nabla P_g}{\rho_g \mathbf{g}} + \mathbf{i_z}\right) \quad (2)$$

where $\mathbf{k_i}$ is the intrinsic permeability tensor, P_w, P_g, ρ_w, ρ_g, μ_w and μ_g are the pressure, density and dynamic viscosity of water and gas phases, $\mathbf{g}$ the gravity vector, z the vertical co-ordinate and $\mathbf{i_z}$ the unit vector in the z direction, $\mathbf{k}_w$ is the hydraulic conductivity in saturated conditions and $\mathbf{k}_{g0}$ the gas permeability in dry conditions.

The value of the intrinsic permeability is generally associated to pore diameter and pore size distribution. Qualitatively, the effect of pore size and total porosity on intrinsic permeability may be evaluated through Poiseuille's equation.

Ideally, intrinsic permeability (k_i) depends only on soil structure and has the same value for gas and liquid flow. This value can be obtained either from gas flow tests in the totally dry soil (k_{ig}), or from water flow tests in the saturated soil (k_{iw}), and should be the same regardless the fluid employed in its determination. However, if fluid-media interactions alter the medium structure, the intrinsic permeability can be greatly altered. This is the case of expansive soils, in which water reacts with clay minerals causing the swelling of the clay lattice, thereby reducing the pore space available for flow (Tindall & Kunkel 1999).

From the values of the saturated hydraulic conductivity (k_w), the intrinsic permeability evaluated from water flux (k_{iw}), can be calculated by the following expression:

$$k_{iw} = \frac{k_w \times \mu_w}{\rho_w \times g} \qquad (3)$$

Under saturated conditions, *i.e.* when all the pores are occupied by water, hydraulic conductivity decreases exponentially with dry density. Temperature and the type of permeating fluid may also modify the values of hydraulic conductivity. Usually, changes in permeability with temperature are attributed to the decrease of fluid viscosity with temperature increase.

This paper summarises the work carried out by CIEMAT and UPC relating to the hydraulic behaviour of bentonites and other expansive materials in the framework of the experiments FEBEX, Bacchus2, Prototype and Backfill and Plug Test. The main results and conclusions presented in reports and papers issued during the last years have been pulled together.

2 MATERIALS

The tests referred to in this paper have been performed with different materials: two highly expansive bentonites, the FEBEX and the MX-80, and two moderately swelling materials: the Boom clay and a mixture of 70% granite and 30% MX-80 bentonite.

The FEBEX bentonite, from the Cortijo de Archidona deposit (Almería, Spain), has a content of dioctahedric smectite of the montmorillonite type higher than 90 percent. Besides, it contains variable quantities of quartz, plagioclase, cristobalite, potassium feldspar, calcite and trydimite. The cation exchange capacity is of 111 ± 9 meq/100 g, and the exchangeable cations are Ca (38%), Mg (28%), Na (23%) and K (2%). The liquid limit of the bentonite is $102 \pm 4\%$ and 64–70% of particles are smaller than $2 \, \mu m$. The hygroscopic water content of the clay at laboratory conditions (relative humidity $50 \pm 10\%$, temperature $21 \pm 3°C$) is about $13.7 \pm 1.3\%$. The value obtained for the external specific surface using BET technique is 29–35 m^2/g and the total specific surface obtained using the Keeling hygroscopicity method is about 725 m^2/g (ENRESA 1998, 2000).

Boom clay is a sedimentary material from Mol (Belgium) with a mineralogical composition of 20–30% illite, 20–30% kaolinite, 10–20% smectite and 15–25% quartz. This clay has a liquid limit of 56%, and 50% of particles less than $2 \, \mu m$. The hygroscopic water content in equilibrium with the laboratory atmosphere (relative humidity about 47%, total suction about 100 MPa) is 2.5–3.0%. External specific surface, determined using BET technique, is about 40 m^2/g (Volckaert et al. 1996).

The MX-80 bentonite is a commercial clay coming from Wyoming (USA). It has a content of dioctahedric smectite of the montmorillonite type between 65 and 82%. Besides, it contains 4–12% quartz, 1–3% calcite, feldspars, pyrite and cristobalite. The cation exchange capacity is around 74 meq/100 g, and the exchangeable cations are Na (61 meq/100 g), Ca (10 meq/100 g) and Mg (3 meq/100 g). The liquid limit of the bentonite is 350–570% and 80–90% of particles are smaller than $2 \, \mu m$. The hygroscopic water content of the clay at laboratory conditions (relative humidity $50 \pm 10\%$, temperature $21 \pm 3°C$) is 7–9%. The value obtained for the total specific surface is about 512 m^2/g.

3 METHODOLOGY

3.1 *Measurement of hydraulic conductivity at laboratory temperature*

The permeability of expansive materials cannot be determined in standard permeameters, mainly due to the important changes in volume undergone by the clay during saturation. As a result, a method based in the constant-head permeameter principle was developed at CIEMAT (Villar & Rivas 1994). It consists on the measurement, as a function of time, of the water volume that traverses the saturated specimen, while a constant hydraulic gradient is imposed between top and bottom. The specimen is confined in a cylindrical rigid cell that prevents any change of the clay volume. The swelling of the saturated clay against the cell wall guarantees a perfect contact between clay and cell, avoiding a preferential pathway.

A sketch of the test arrangement is shown in Figure 1 (Villar 2002). The cell is made of stainless steel and has an inner section of 19.63 cm^2 and a height of 2.5 cm. The granulated clay, with its hygroscopic water content, is uniaxially compacted to the desired dry density directly inside the permeability cell. Porous stones are placed on its top and bottom. The sample is saturated by injecting distilled water through the porous stones at a pressure of 0.6 MPa.

Once the sample is saturated, the lower inlet of the cell is switched to a piston pump. The injection pressure in the lower part of the cell is then increased, while

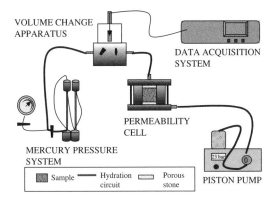

Figure 1. Schematic representation of the set-up to measure hydraulic conductivity.

the back-pressure on top is kept constant and equal to 0.6 MPa. The automatic volume change apparatus, connected to a data acquisition system, measures the water outflow through the upper outlet of the cell. When the volume of water that crosses the specimen for a given time is constant, the hydraulic conductivity (k_w) is calculated applying the Darcy's law.

3.2 Measurement of hydraulic conductivity at high temperature

The determination of hydraulic conductivity as a function of temperature was performed in high-pressure oedometer equipments (Figure 2, Villar & Lloret, in press). Granulated clay was compacted uniaxially and statically at room temperature in the oedometer ring, which has an inner diameter of 5.0 cm, the length of the resulting specimen being 1.2 cm.

The oedometer assemblage is placed inside a silicone oil thermostatic bath that keeps target temperature. Once the temperature stabilises, the sample, confined between porous stainless steel sinters, is hydrated at constant volume through the bottom face with deionised water injected at a pressure of 0.6 MPa, while the upper outlet remains open to atmosphere. Once the sample is completely saturated, hydraulic conductivity is determined. For that, the water pressure at the bottom of the samples is increased, while a backpressure of 0.6 MPa is applied on top, resulting in hydraulic gradients between 11700 and 20000. The water outflow is measured and the hydraulic conductivity is calculated applying Darcy's law.

3.3 Measurement of gas permeability

The method used for gas permeability measurement is a non-steady-state method whose basic principle was developed by Kirkham (1946). It consists on the pressurisation of a gas tank, and the release of this

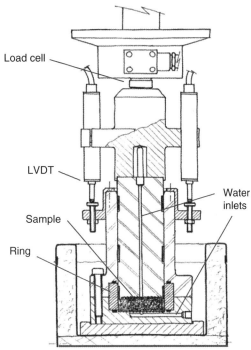

Figure 2. Schematic layout of the oedometer cell inside the thermostatic bath.

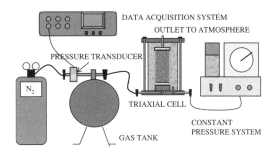

Figure 3. Schematic representation of the set-up to measure gas permeability.

pressure to the atmosphere through the soil column of unknown permeability. The gas permeability is calculated from the rate of decrease of the tank pressure.

A schematic representation of the experimental set-up is shown in Figure 3. The cylindrical clay sample is placed in a triaxial cell, covered by two latex membranes. A confining pressure of 1.6 MPa is applied to the cell, in order to assure a perfect adherence of the membranes to the sample surface. The bottom of the sample is connected to a hermetic tank of known volume, in which nitrogen gas has been previously injected at a pressure slightly higher than the atmospheric one. A pressure transducer connected to a data acquisition system measures the gas pressure in the tank. The upper outlet of the cell, connected to the sample top, is open to the atmosphere. During the test, the gas in the tank is allowed to flow through the sample, while the pressure decrease in the tank is monitored as a function of time.

The gas permeability is calculated according to the following expression (Yoshimi & Osterberg 1963):

$$k_g = 2.3 \times \frac{V \times L \times \rho_g \times g}{A \times \left(P_{atm} + \dfrac{P_0}{4}\right)} \times \frac{-\text{Log}_{10}\left(\dfrac{P(t)}{P_0}\right)}{t - t_0} \qquad (4)$$

where k_g is the gas permeability (m/s), V is the volume of the deposit (m^3), L is the height of the sample (m), A is the section of the sample (m^2), ρ_g is the gas density (kg/m^3), P_{atm} is the atmospheric pressure (N/m^2), P_0 is the excess pressure in the tank over the atmospheric one for time t_0 (s), and $P(t)$ is the excess over the atmospheric pressure for time t (Villar & Lloret 2001).

Gas permeability (k_g) can be expressed as the product of the gas permeability in dry conditions (k_{g0}) by the relative permeability to gas (k_{rg}). Taking into account the dynamic viscosity of nitrogen, the following relationship between gas permeability (k_g, m/s) and the product of the intrinsic permeability measured with nitrogen gas (k_{ig}, m^2) by the relative permeability to gas (k_{rg}), can be established:

$$k_g = \frac{\rho_g \times g}{\mu_g} \times k_{ig} \times k_{rg} = 6.2 \cdot 10^5 \times k_{ig} \times k_{rg} \qquad (5)$$

The specimens have been prepared by uniaxial compaction of the granulated clay with different pressures in order to obtain different dry densities. Both the clay with its hygroscopic water content and with added water have been used. In the latter case, the clay is mixed with the appropriate quantity of distilled water, according to the target water content, and the mixture is allowed to stabilise for several days in closed plastic bags, in order to facilitate a homogeneous distribution of humidity. To manufacture specimens with water contents lower than the hygroscopic one, the granulated clay has been oven-desiccated for short periods of time before compaction.

The tests have been performed at room temperature.

3.4 Measurement of water relative permeability

The methods used to determine permeability to water in the unsaturated state are based on the measurement or control of suction, this being necessary in order to determine the value of the gradient. To measure the permeability of moderately expansive materials – such as the Boom clay – a transient method based on

the principle of axis translation has been used. The sample inside an oedometric cell with matric suction control is subjected to wetting/drying by stepwise variation of suction while the water inflow/outflow is measured versus time (Romero 1999). Figure 4 shows the layout of the temperature and controlled suction system and a number of auxiliary devices necessary to perform the tests. A controlled suction oedometer cell surrounded by a thermostatically controlled heater (silicone oil bath and heater) was employed. Measurements of the water permeability were performed under transient conditions in wetting (or inflow procedure) and drying (or outflow procedure) paths. Isothermal wetting and drying cycles were performed at constant net vertical stresses or constant volume conditions. The same matric suction steps were followed in the wetting and drying paths: 0.45, 0.20, 0.06 and 0.01 MPa. The increases in suction must be sufficiently small for the hydraulic conductivity to be considered constant, and the ratio between water content and suction linear. The results obtained are interpreted through application of the Richards equation, with a number of simplifications and assumptions.

Neither the steady-state nor the transient methods are easily applicable to measurement of unsaturated permeability of highly expansive materials. An alternative method for calculation of the permeability of these materials is the back-analysis of infiltration tests performed under transient conditions, through the application of a theoretical model for solution of the inverse problem. This technique is similar to those used in groundwater engineering and geophysics (Pintado et al. 2002). Tests backanalysis allows the exponent of the unsaturated permeability law (λ) to be identified and, if thermal flux is applied, the vapour tortuosity factor (τ). Infiltration tests performed in compacted FEBEX bentonite with and without thermal gradient have been used for this purpose (Lloret et al. 2002). Distilled water was employed in all the tests.

Isothermal infiltration tests have been performed using both small cylindrical cells and samples surrounded by latex membranes and confined in the triaxial cell. In all cases distilled water is injected through the lower end of the compacted bentonite samples at different pressures and water contents along the specimen were measured at some times after the initiation of the infiltration process.

Figure 5 shows the device used to identify by means of a backanalysis technique, the vapour tortuosity factor and the exponent of the unsaturated permeability law using water content measurements obtained at the end of a thermal flux test. A controlled flux of heat was applied to one end of a cylindrical specimen (38 mm diameter, 76 mm height). In the tests, a constant power of 2.17 W was injected, reaching

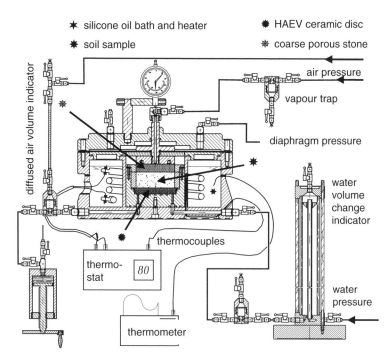

Figure 4. Experimental layout of the temperature and controlled suction system (Romero *et al.* 1995).

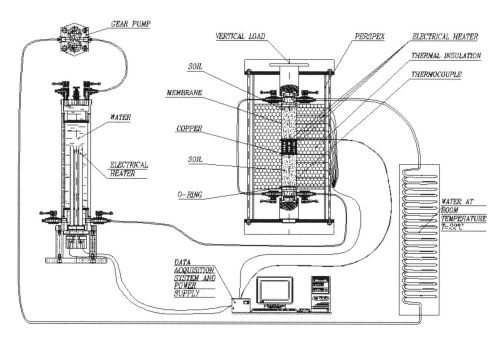

Figure 5. Scheme of the experimental device used to obtain vapour tortuosity factor and unsaturated water relative permeability exponent by backanalysis of a thermal flux test (Pintado *et al.* 2002).

steady temperatures in the range of 70–80°C in the hotter end of the specimen. In the cold end, a constant temperature of 30°C was maintained. A latex membrane that allows the deformation and keeps constant the overall water content surrounds the specimen. During the tests, temperatures in both ends and in three internal points of the specimen are monitored by means of a data acquisition system. At the end of the tests, diameter change was measured in some points of the specimen. Finally, the soil samples were cut in six small cylinders to determine water content.

4 FACTORS AFFECTING PERMEABILITY

It is generally accepted that the main factors having a significant influence in the value of permeability are the liquid degree of saturation, the void ratio, the temperature and the salinity of the permeant in the case of liquid fluid. There is also the discussion about the possible influence of the hydraulic gradient applied to determine permeability on the result obtained. Tests have been done to check all these issues and the main results obtained are shown below.

4.1 Variation of hydraulic conductivity with void ratio

Hydraulic conductivity was determined using cylindrical specimens of FEBEX clay compacted to different density from the clay with its hygroscopic water content (Villar 2000, 2002). The range of dry densities studied was from 1.30 to 1.84 g/cm³. Distilled water was used as permeating fluid.

Figure 6 shows all the values obtained versus dry density. There are significant variations in permeability depending on the dry density of the sample,

the former decreasing clearly as the latter increases. The greater dispersion observed among the values corresponding to high densities is probably due to the difficulty involved in determination, since in view of the low permeability of bentonite, highly accentuated by the increase in density, water flows are very small and their measurement difficult. On the other hand, it is necessary to apply high hydraulic to make the measurement possible with the available techniques.

The values of hydraulic conductivity (k_w, m/s) are exponentially related to dry density (ρ_d, g/cm³) and a distinction may be made between two different fittings – also shown in the figure – depending on the density interval:

for dry densities of less than 1.47 g/cm³:

$$\log k_w = -6.00\, \rho_d - 4.09 \tag{6}$$

for dry densities in excess of 1.47 g/cm³:

$$\log k_w = -2.96\, \rho_d - 8.57 \tag{7}$$

The values of intrinsic permeability (k_{iw}, m²) obtained through Equation 3 from the measurements shown above are plotted in Figure 7 as a function of soil porosity (n). The prediction of Kozeny's model for this porosity interval is also plotted in the same figure. It can be observed how the variation of permeability with pore volume is higher than that predicted with Kozeny's model (Lloret et al. 2002). The equations of the lines plotted in the figure are:

Exponential law for porosities between 0.46 and 0.52:

$$k_{iw} = 5.08 \cdot 10^{-28}\, e^{37.33\, n} \tag{8}$$

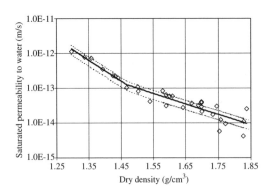

Figure 6. Permeability of FEBEX bentonite to distilled water versus dry density of the clay and fittings obtained, along with the average range of deviation of experimental values with respect to theoretical values (Villar 2002).

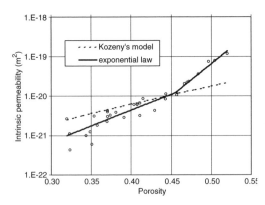

Figure 7. Intrinsic permeability of saturated FEBEX bentonite for different porosities (Lloret et al. 2002).

Exponential law for porosities between 0.46 and 0.32:

$$k_{iw} = 2.79 \cdot 10^{-24}\, e^{\,18.39\,n} \tag{9}$$

Kozeny's model for porosities between 0.52 and 0.32:

$$k_{iw} = 7.72 \cdot 10^{-21} \frac{(1-0.420)^2}{0.420^3} \frac{n^3}{(1-n)^2} \tag{10}$$

Pusch (1979) explains the extremely low permeability of the bentonite compacted at high density in terms of the low water contents required to saturate the highly compacted clay, since under these conditions, the thickness of the film of interlaminar water is very small (3–5 Å) due to its high specific surface, this meaning that the molecules of water are strongly adsorbed to the surfaces of the clay minerals, leaving only very tortuous interparticle channels for the transport of water. Besides, in the calcium montmorillonite compacted at dry densities in excess of 1.60 g/cm³, more than 90 percent of the total volume of the pores is occupied by this interlaminar water of restricted mobility (Pusch et al. 1990).

The existence of two intervals of density for which the increase in hydraulic conductivity with decreasing density is different was also found by Villar & Rivas (1994) for the S-2 bentonite (from the same deposit than the FEBEX clay) and interpreted as being caused by the sharp decrease in swelling pressure for densities below this value, this leading to a more significant increase in the size of the flow paths. Given that intrinsic permeability, and consequently hydraulic conductivity, is related to pore diameter by the Poiseuille equation, the increase in

the average diameter of the pores due to lower swelling, from a given density threshold value down, leads to a sharper increase in permeability.

4.2 Variation of permeability with the degree of saturation

4.2.1 Results obtained from gas permeability tests

The gas permeability has been measured in samples of FEBEX bentonite with nominal dry density between 1.50 and 1.90 g/cm³, and degrees of saturation between 25 and 92 percent (Villar 2000, 2002, Villar & Lloret 2001).

The gas permeability values obtained in the tests can be plotted as a function of the dry density of the clay for different water contents, as shown in Figure 8. For each water content, the gas permeability decreases as dry density increases, following an exponential law.

The variation of the permeability to gas (k_g, m/s) is shown in Figure 9 as a function of the accessible void ratio, which indicates the ratio between gas accessible pore volume and particle volume and can be expressed as the product of the void ratio (e) by the unit minus the degree of saturation ($1 - S_r$). This factor, $e(1 - S_r)$, shows in fact a higher correlation with the permeability values than the dry density or the water content do. The correlation found, plotted in Figure 9, can be expressed as:

$$k_g = 2.29 \cdot 10^{-6}\, (e\,(1-S_r))^{4.17} \tag{11}$$

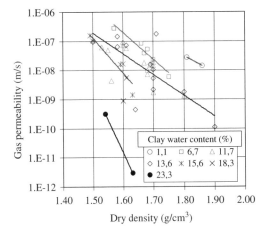

Figure 8. Permeability to gas of FEBEX bentonite compacted with different water contents as a function of dry density (Villar & Lloret 2001).

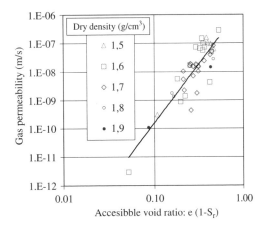

Figure 9. Relationship between the gas permeability of FEBEX bentonite and the volume of pores accessible to gas (Villar & Lloret 2001).

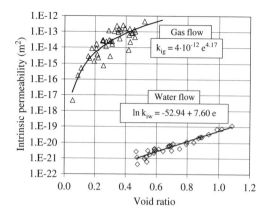

Figure 10. Intrinsic permeability of compacted bentonite obtained from saturated water flow and from unsaturated gas flow tests (Villar & Lloret 2001).

An estimation of the intrinsic permeability derived from gas flow tests can be performed assuming a value of degree of saturation equal to zero in Equation 11, in order to obtain the gas permeability in dry bentonite (k_{g0}). Afterwards, from Equation 5, the intrinsic permeability can be evaluated, resulting the values plotted in Figure 10. This figure shows also the variation of the values of intrinsic permeability obtained from water flow with the void ratio (a joint correlation for the two density intervals pointed out in Figure 6 is shown). For the same volume of pores, differences of about eight orders of magnitude in the intrinsic permeability values, for dry or saturated clay, can be observed.

These observations suggest that a fundamental difference in the microstructural arrangement of the saturated and the unsaturated sample exists, due to the swelling of the clay as it saturates, which would not happen in non-expansive materials. The hydration of clay particles under constant volume reduces the size of pores between aggregates of clay. In dry conditions, the diameter of macropores accessible to gas flow may be up to more than $10 \, \mu m$. Hydrated clay in confined conditions presents the same global volume of pores, but the spaces between aggregates have been reduced or eliminated due to the swelling of the clay particles. That is to say, during the hydration of the clay at constant volume, the volume occupied by small mesopores and micropores increases, while the volume occupied by macropores decreases. In saturated conditions, accessible mean pore diameter is in the range of meso and micropores, i.e. more than three orders of magnitude smaller than the big interaggregate pores. Most of the water in a bentonite saturated under constant volume will be in the interlamellar space (Pusch et al. 1990). The water molecules constituting this film will be strongly adsorbed to the clay mineral surfaces, what leaves only narrow, tortous interparticle passages for water and ion transportation. The change in the mean pore diameter available to flow explains the big differences between the values of measured intrinsic permeability for dry and saturated clay (Villar & Lloret 2001).

Changes in pore size can be observed in the study of the compacted clay by environmental scanning electron microscopy (ESEM). This technique allows the observation of the microstructure of the material under different relative humidity conditions and consequently, different degrees of saturation. Figure 11a shows the aspect of the clay aggregates in a sample compacted to dry density $1.70 \, gcm^3$ with the hygroscopic water content and observed under a relative humidity of 50 percent (similar to that of the laboratory). This sample was progressively hydrated during 5 hours in the microscope chamber, by increasing the relative humidity while keeping the temperature constant. The volume of the sample tested is very small, ensuring its quick saturation. Figure 11b shows the final aspect of the same sample under a relative humidity of 100 percent. Despite the absence of confinement, a diminution of the size of some voids can be observed.

4.2.2 *Results obtained from infiltration tests*
The water infiltration tests carried out by CIEMAT and UPC (ENRESA 2000, Lloret et al. 2002) provide data allowing unsaturated hydraulic conductivity and its dependence on the degree of saturation to be obtained. By means of parameter identification techniques, intrinsic permeability (k_{iw}) and the exponent of the law used to obtain the relative permeability (λ) may be estimated. In addition, both the unsaturated water permeability (k) and the tortuosity factor (τ) may be estimated from thermo-hydraulic experiments with prescribed heat and water flows.

In the numerical modelling, the effect of temperature on water viscosity and the effect of pore volume on intrinsic permeability allow to explain the different infiltration velocities observed in the infiltration tests performed under different temperatures and with different dry densities. As a result, the unsaturated permeability can be expressed by the following expression:

$$k \, (S_r, \, T, \, \rho_d) = k_{iw} \, (S_r = 1, \, \rho_d = \rho_{d0}) \, 10^{-3.6 \, (\rho_d - \rho_{d0})}$$
$$k_{rw} (S_r) \, (\gamma_w / \mu_w(T)) \; ; \quad k_{rw} (S_r) = S_r^{\lambda} \qquad (12)$$

where k_{iw} is the intrinsic permeability obtained from water flow, ρ_d is the dry density, ρ_{d}, is the reference dry density, μ_w and γ_w are the viscosity and specific unit weight of water and $k_{rw}(S_r)$ is the relative permeability.

Figure 11. Environmental scanning microscope image, taken under relative humidity 50% (a) and 100% (b), of the bentonite compacted to dry density 1.70 g/cm³ (Villar & Lloret 2001).

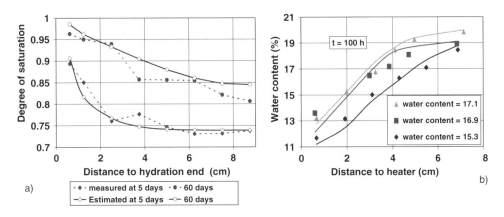

Figure 12. Comparison between computed and measured water contents in (a) isothermal infiltration test and (b) prescibed heat flow in samples with constant average water content.

In heat flow tests, the backanalysis procedure showed that there are a few combinations of parameters that give similar differences with the test measurements. This is reasonable, as measured water content is a global quantity, and it is difficult to distinguish between water transported by liquid flow (controlled by "λ") and by vapour diffusion (controlled by τ).

After the backanalysis of several tests (ENRESA 2000, Lloret *et al.* 2002), the values of the best hydraulic parameters obtained for a porosity of 0.4 are: intrinsic permeability of $2 \cdot 10^{-21}\,\text{m}^2$, λ exponent of 3 and tortuosity factor of 0.8. Figure 12 shows the comparison between the values computed using a numerical model with the set of parameters mentioned and the measurements performed in the isothermal infiltration tests and heat flow tests described in section 3.4.

4.2.3 Results obtained from wetting/drying tests with suction control

The unsaturated water permeability was determined for compacted Boom clay by evaluating the transient inflow/outflow of water in a controlled suction oedometer cell, as described in section 3.4 (Romero *et al.* 2001). The values obtained at a temperature of

185

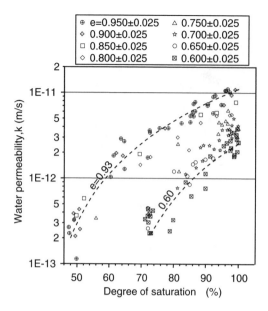

⊕ e=0.950±0.025	△ 0.750±0.025
◇ 0.900±0.025	☆ 0.700±0.025
□ 0.850±0.025	○ 0.650±0.025
◇ 0.800±0.025	⊠ 0.600±0.025

Figure 13. Water permeability of Boom clay vs. degree of saturation at 22°C (Romero *et al.* 2001).

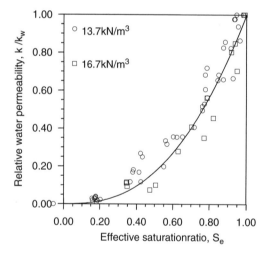

Figure 14. Relative water permeability values of Boom clay vs. effective saturation ratios for different void ratios (Romero *et al.* 2001).

22°C are represented in Figure 13 for different degrees of saturation and void ratios. Data have been fitted with curves representing two different void ratios: 0.60 and 0.93. An important dependence of water permeability on void ratio and degree of saturation is observed. At $S_r \approx (75 \pm 5)\%$, the coefficient of permeability presents a variation of one order of magnitude when changing the void ratio from $e = 0.90$ to 0.60. This variation is just as important as the change of the degree of saturation from $S_r = 72\%$ to 50% for a fixed $e = 0.93$.

Relative water permeability values k_{rw} (S_r) are shown in Figure 14 as a function of the effective saturation ratio (S_e) for different constant void ratios (e). Relative permeability values have been fitted to the expression $S_e^{\lambda e}$. A parameter $\lambda_e = 2.6$ adequately fits experimental data corresponding to the different void ratios tested for Boom clay. Effective saturation ratio is defined as

$$S_e = \frac{w - w_{ref}}{w_{sat} - w_{ref}} \tag{13}$$

where w, w_{sat} and w_{ref} are the clay water content, the water content in saturated conditions and the water contained inside the intra-aggregate porosity, 13% (Romero *et al.* 1999).

4.3 *Influence of hydraulic head on permeability – threshold gradient*

The application of Darcy's law for the calculation of the coefficient of permeability requires that the velocity of the flow be proportional to the hydraulic gradient, that is to say, that the value of the coefficient obtained be independent of the hydraulic gradient applied during determination. This means that the relation between flow and hydraulic gradient is linear, and that this linear relation passes through the origin. For different reasons, this condition may not be fulfilled, thus invalidating the use of the Darcy's expression.

In the tests presented in section 4.1, the values of hydraulic head applied were between 70 m, for dry densities of 1.30 g/cm³, and 660 m, for dry densities of 1.84 g/cm³. Taking into account that the length of the specimen is 2.5 cm, the average hydraulic gradient was 15200. All the samples were tested with at least two different hydraulic gradients suitable for their dry density, in other words, sufficiently high so as to provide a measurable flow but below swelling pressure, in order not to cause compression of the material (Pusch 1994). The values of hydraulic conductivity obtained for the two hydraulic gradients applied in each test are represented in Figure 15. The points joined by lines correspond to the same test; *i.e.* to the measurements performed on a specific sample of a given dry density. It may be observed that, although there may be a certain difference between the value of conductivity obtained with the different gradients for the same sample, such variations are probably the result of the experimental method, since there is no trend for one variable with respect to the

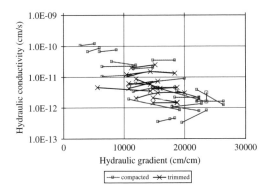

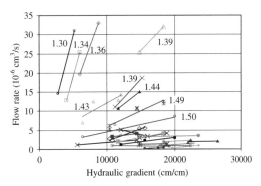

Figure 15. Variation in hydraulic conductivity obtained for different samples of FEBEX bentonite versus the hydraulic gradient as a function of the kind of sample preparation (points joined by lines correspond to the same sample) (Villar 2002).

Figure 16. Variation in flow rate obtained for different FEBEX samples versus the hydraulic gradient (points joined by lines correspond to the same sample) (Villar et al. 2002).

other. This would confirm the independence of the flow observed from the hydraulic gradient applied, and therefore the validity of Darcy's law for calculation of the coefficient of permeability.

It may also be appreciated in the figure that, as the permeability of the sample decreases, in other words, as its dry density increases, the value of the hydraulic gradients applied for performance of the measurement increases, this being necessary in order to be able to measure very low flows. However, it is not possible to determine whether no flow occurs in the case of lower hydraulic gradients or whether it is simply not possible to measure it with the available technique.

In order to deepen in the investigation on the existence of a threshold gradient, the values of flow rate for the two hydraulic gradients applied in each of the tests mentioned above have been plotted in Figure 16 and Figure 17, this last being a detail of the previous one. Again, the points joined by lines correspond to the same test; i.e. to the measurements performed on a specific sample of a given dry density. If we assume that the relation between gradient and flow is linear and we extrapolate these lines towards flow 0, despite the variation found among the different tests, we would observe a tendency to find gradients higher than 0 to have a measurable flow, around 5000, and even higher in the case of the highest densities (Figure 17). However, this aspect should be confirmed with more refined and specific tests.

4.4 Influence of temperature on saturated and unsaturated permeability

The hydraulic conductivity at different temperatures was determined in the equipment shown in Figure 2 on FEBEX samples with an average dry density of

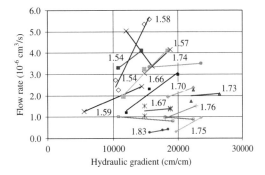

Figure 17. Variation in flow rate obtained for different FEBEX samples of high density versus the hydraulic gradient (points joined by lines correspond to the same sample) (Villar et al. 2002).

$1.58 \, \mathrm{g/cm^3}$ (Villar & Lloret, in press). The results obtained are shown in Figure 18. The error bar shown was obtained from values measured in tests performed at laboratory temperature to obtain Equation 7. The hydraulic conductivity slightly increases in a logarithmic way with temperature, as expected from the decrease in water kinematic viscosity.

The increase of hydraulic conductivity with temperature resulting from the decrease in water kinematic viscosity, was calculated for FEBEX bentonite, taking as a starting point the permeability value determined at laboratory temperature (22°C) from Equation 7. The results are also plotted in Figure 18. Despite the scarce experimental data, the increase in hydraulic conductivity with temperature observed could be considered as slightly lower than that predicted solely on the basis of water viscosity change.

On the other hand, the unsaturated water permeability was determined for compacted Boom clay at

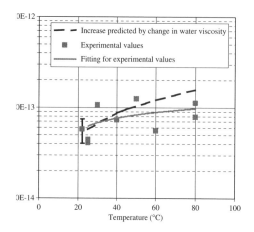

Figure 18. Hydraulic conductivity values as a function of temperature for the saturated FEBEX bentonite compacted at an average dry density of $1.58\,g/cm^3$ and comparison with those predicted for the same temperatures taking into account the change in water kinematic viscosity (Villar & Lloret, in press).

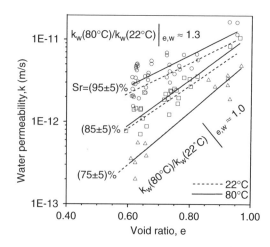

Figure 19. Water permeability vs. void ratio for constant degree of saturation at 22 and 80°C (Romero et al. 2001).

different temperatures (22 and 80°C) by evaluating the transient inflow/outflow of water in a controlled suction oedometer cell, as described in section 3.4 (Romero et al. 2001). In Figure 19, the water permeability values obtained have been grouped into different degree of saturation ranges and plotted versus void ratios for both temperatures. Temperature effect is more important under near-saturated conditions with preponderance of bulk water. The slight increase of permeability with temperature at $S_r = (95 \pm 5)\%$ of $k_w(80°C)/k_w(22°C) = 1.3$ cannot be explained in terms of a reduction of free water viscosity in the same interval of temperature (solid lines at 80°C and dashed lines at 22°C). Furthermore, relative water permeability values at constant void ratio showed no temperature dependence.

The temperature effect on hydraulic conductivity of compacted bentonites or additive/bentonite mixtures has been studied by several authors, who generally observed an increase in permeability with temperature, mostly attributed to the thermal decrease of kinematic viscosity (Mingarro et al. 1991, Cho et al. 1999). However, the changes of hydraulic conductivity with temperature can be above or below those expected on the basis of the change in water viscosity, depending on the kind of material (Romero et al. 2001, Villar & Lloret, in press). To better analyse this aspect, the effect of temperature on hydraulic conductivity in the interval of 20–80°C can be approximated through a factor β_T, as suggested by

Romero et al. (2001):

$$k_w(T) / k_w(T_0) \approx 1 + \beta_T (T\text{-}T_0) \tag{14}$$

where k_w is the saturated hydraulic conductivity and T and T_0 are the temperature of the measurement and the temperature of the reference measurement (usually room temperature).

The value of β_T evaluated taking into account the change in water density and viscosity is of $0.030\,K^{-1}$. For FEBEX bentonite the value of β_T is $0.0097\,K^{-1}$, i.e. much smaller than the theoretical one. The experimental data for Boom clay are associated with $\beta_T = 0.005\,K^{-1}$ at nearly saturated conditions $S_r = (95 \pm 5)\%$, while at lower bulk water contents, in the proximity of the intra-aggregate zone at $S_r = (75 \pm 5)\%$, experimental results show $\beta_T \rightarrow 0$ (Figure 19).

This increase of permeability smaller than it could be expected from the thermal change in water viscosity could be explained by thermo-chemical effects altering clay fabric and porosity redistribution (Romero et al. 2001).

4.5 Influence of the salinity of the permeant on permeability

It is an accepted fact that the type of water used as a permeating agent, and especially its salinity, may have an impact on the coefficient of permeability of a soil. With a view to verifying this issue, the saturated permeability of specimens of different density was determined, using different kinds of water as permeating agent.

Figure 20 shows the values for FEBEX bentonite permeated with granitic and saline water along with

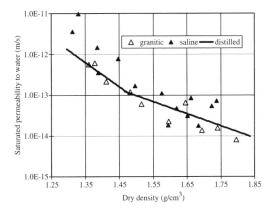

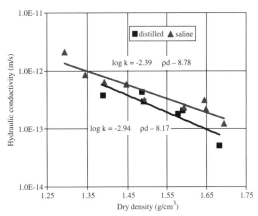

Figure 20. Hydraulic conductivity for granitic or saline water versus the dry density of the FEBEX clay and fitting obtained for distilled water (Villar 2002).

Figure 21. Results of the hydraulic conductivity tests with compacted MX-80.

the fittings obtained for distilled water (section 4.1). The granitic water used has a salinity of 0.02 percent, while the saline water – which simulates the composition of the FEBEX pore water – has a salinity of 0.8 percent (see ENRESA 2000 for detailed chemical composition). No clear trend is observed concerning the variation of the values obtained with granitic water with respect to those obtained with distilled water. The values obtained with saline water are, however, 184 percent higher on average than those expected for a sample of the same density tested with distilled water, and in addition, they show greater dispersion. This higher permeability to saline water with respect to that expected for distilled water is more significant for low densities.

In the context of the Prototype project, the same kind of tests have been performed with the MX-80 bentonite permeated with distilled and saline water. The saline water used in this case was obtained in the laboratory by mixing $CaCl_2$ and NaCl at a relationship 32/68 percent to give rise to a solution of 0.5 percent salinity. The results obtained are plotted in Figure 21, in which the exponential correlations fitted between dry density and hydraulic conductivity are also shown. The hydraulic conductivity is a 135 percent higher for saline water than for distilled water.

On the other hand, in the context of the Backfill & Plug Test project a series of infiltration tests on a 70/30 weight percent Äspö granite/MX-80 bentonite mixture compacted at a dry density of $1.70\,g/cm^3$ were performed. The tests were carried out for different periods of time, between 10 and 100 days, and with two kinds of infiltration water, to check the influence of the salinity in the hydration behaviour.

The infiltration tests were performed in stainless steel cells of internal diameter of 15.0 cm and internal height of 12.9 cm. The mixture was uniaxially

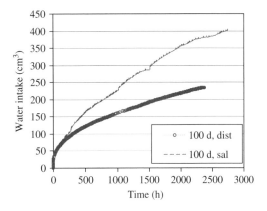

Figure 22. Water intake of a granite/bentonite mixture in two tests performed with distilled (dist) or saline (sal) water of duration 100 days.

compacted directly in the cell in one layer. The two hydration ports at the base of the cell were connected to a water deposit placed 1.5 m above. Between the deposit and the cell an automatic volume change apparatus measured the water intake. The hydration water was either distilled or saline. In the last case, the salinity of the water was 1.2 percent, and it was obtained by mixing $CaCl_2$ and NaCl in a weight ratio 32/68.

Figure 22 shows the evolution of water intake in the tests performed with a duration of 100 days. It becomes clear the higher water intake capacity in the tests performed with saline water, what reveals a higher permeability. This is confirmed by the greater final water content of the tests performed with saline water (Figure 23).

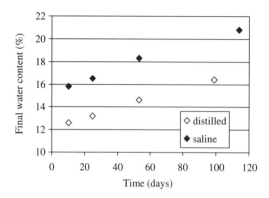

Figure 23. Water content of a granite/bentonite mixture after infiltration tests with different kind of water of different duration.

The increase in permeability with the salinity of the permeating fluid has been underlined by various authors (Rolfe & Aylmore 1977) and attributed to: (1) alterations in pore dimension distribution as a result of variations in swelling pressure in the clay matrix, (2) variations in the mobility of the molecules of water associated with the exchangeable cations adsorbed on the surfaces or forming diffuse double layers (DDL), (3) alterations in the viscous behaviour of the structure of the water. As a result of these mechanisms, as the concentration of the electrolyte increases there is a reduction in the swelling pressure of the clay particles, the size of the flow channels increasing to the detriment of the number of small channels, this causing flow – and therefore, permeability – to increase enormously. On the contrary, the increasing development of diffuse double layers on reduction of the concentration of the electrolyte causes a decrease in permeability. The reduction of effective porosity with the decreasing salinity of the permeating agent would be due to the fact that the pore space is occupied by the bound water (DDL), the viscosity of which is higher than that of free water. According to the diffuse double layer theories, the thickness of it decreases as the concentration of water in the pores increases, as a result of which, for a given porosity, the effective porosity of the clay would increase with increasing concentration of the solution, with the corresponding increase in permeability.

5 CONCLUSIONS

Several factors having a significant influence on the value of permeability of compacted expansive materials have been experimentally analysed by determination of the saturated and unsaturated permeability. Highly expansive materials – MX-80 and FEBEX bentonites – and moderately swelling

materials – such as Boom clay and a granite/bentonite mixture – have been used. From the results obtained the following conclusions can be drawn:

– The increase of the dry density of the sample gives rise to an exponential decrease in saturated water permeability (and of intrinsic permeability). In the FEBEX bentonite the increase in hydraulic conductivity with decreasing density is sharper for low dry densities, this being due to the fact that, as the swelling capacity of the clay decreases, there is a growth in the size of the flow channels and, therefore, in effective porosity and permeability. On the other hand, the permeability of artificially prepared fabrics of the less expansive Boom clay also clearly depends on void ratio.

– The parameter that presents the best correlation with the value of permeability to gas in the FEBEX bentonite is the accessible pore volume ($e(1 - S_r)$). A potential relation may be established between the two.

– The intrinsic permeability of FEBEX samples having the same porosity varies with the degree of saturation, with differences of up to eight orders of magnitude between dry and saturated samples. This is due to the fact that the water causes the clay laminae to expand, thus reducing the size of the pores between aggregates. During the hydration of expansive clays at constant volume, the pore size distribution changes, due to the swelling of the clay particles: clay hydrated under confined conditions shows the same overall volume of pores as a dry sample of the same dry density, although the spaces between aggregates are reduced or disappear due to the swelling of the clay particles. This variation in the average diameter of the pores available to flow explains the major difference in the values of intrinsic permeability in dry and saturated samples. On the other hand, the intrinsic permeability for a given degree of saturation depends on the dry density of the sample.

– The back-analysis of infiltration tests performed in compacted FEBEX bentonite with distilled water, has allowed to obtain the following values for a porosity of 0.4: intrinsic permeability (k_{iw}) of $2 \cdot 10^{-21} \, m^2$, exponent of the relative law (λ) of 3 and tortuosity factor (τ) of 0.8. An exponent, λ_ε, of 2.6 can describe the influence of effective saturation ratio, S_e, on relative permeability of Boom clay for different dry densities.

– An analysis of the permeability tests performed with FEBEX bentonite has shown the independence between hydraulic gradient applied and hydraulic conductivity obtained. However, there seems to be a tendency to need gradients higher than 0 to have a measurable flow, what would point to the existence of a threshold gradient. This aspect should be confirmed with more refined and specific tests.

– It has been measured how the temperature increases the water saturated permeability of the FEBEX bentonite and the unsaturated and saturated permeability of Boom clay, although this measured increase is slightly lower than what would be expected on the basis of the thermal changes in water kinematic viscosity. For the Boom clay, the temperature influence on water permeability is more relevant at low suctions corresponding to bulk water preponderance (inter-aggregate zone). Below a degree of saturation of 75% (in the proximity of the intraaggregate zone) no clear effect of temperature on permeability is detected. Thermo-chemical effects altering clay fabric (flocculation or dispersion), porosity redistribution (creating preferential pathways or blocking macropores) and pore fluid chemistry (affecting viscosity) could be relevant issues regarding the effect of temperature on permeability.

– Hydraulic conductivity of bentonite (FEBEX and MX-80) increases slightly when saline water is used instead of distilled water as the permeating agent, especially in the case of low densities. The salinity of the permeant clearly increases the unsaturated permeability of less expansive materials, such as granite/bentonite mixtures.

REFERENCES

Cho, WJ, Lee, JO & Chun, KS (1999) The temperature effects on hydraulic conductivity of compacted bentonite. *Applied Clay Science* 14: 47–58.

ENRESA (1998) FEBEX. Bentonite: origin, properties and fabrication of blocks. *Publicación Técnica ENRESA* 4/98, Madrid. 146 pp.

ENRESA (2000) Full-scale engineered barriers experiment for a deep geological repository for high-level radioactive waste in crystalline host rock (FEBEX project). *Nuclear Science and Technology Series. EUR* 19147. Luxembourg. 362 pp.

Kirkham, D (1946) Field method for determination of air permeability of soil in its undisturbed state. *Soil Sci. Soc. Am. Proc.* 11: 93–99.

LLoret, A, Villar, MV & Pintado, X (2002) Ensayos THM: Informe de síntesis. *Internal Report* CIEMAT/DIAE/54520/1/02. *FEBEX Report* 70-UPC-M-0-04. 160 pp.

Mingarro, E, Rivas, P, del Villar, LP, de la Cruz, B, Gómez, P, Hernández, AI, Turrero, MJ, Villar, MV, Campos, R, Cézar, JS (1991) Characterization of clay (bentonite)/crushed granite mixtures to build barriers against the migration of radionuclides: diffusion studies and physical properties. Task 3-Characterization of radioactive waste forms. A series of final reports (1985–1989)-No.35. *Nuclear Science and Technology Series.* Commission of the European Communities, Luxembourg. 136 pp.

Pintado, X, Ledesma, A & Lloret, A, (2002) Backanalysis of thermohydraulic bentonite properties from laboratory tests. *Eng. Geol.* 64: 91–115.

Pusch, R (1979) Highly compacted sodium bentonite for isolating rock-deposited radioactive waste products. *Nuclear Technology* 45: 153–157.

Pusch, R (1994) Waste disposal in rock. *Developments in Geotechnical Engineering* 76. *Elsevier.* Amsterdam. 490 pp.

Pusch, R, Karnland, O & Hökmark, H (1990) GMM – A general microstructural model for qualitative and quantitative studies on smectite clays. *SKB Technical Report* 90–43. Stockholm. 94 pp.

Rolfe, PF & Aylmore, LAG (1977) Water and salt flow through compacted clays: I. Permeability of compacted illite and montmorillonite. *Soil Sci. Soc. Am. J.* 41: 489–495.

Romero, E (1999) Characterisation and thermo-hydro-mechanical behaviour of unsaturated Boomclay: An experimental study. *Ph. D. Thesis,* Universidad Politécnica de Cataluña, Barcelona. 405 pp.

Romero, E, Lloret, A & Gens, A (1995) Development of a new suction and temperature controlled oedometer cell. In: Alonso, EE & Delage, P (eds) Proceedings of the 1st International Conference on Unsaturated Soils Vol. 2, Paris. *Balkema/Presses des Ponts et Chaussées.* pp. 553–559.

Romero, E, Gens, A. & Lloret, A, (2001) Temperature effects on the hydraulic behaviour of an unsaturated clay. *Geotechnical and Geological Engineering* 19: 311–332.

Tindall, JA & Kunkel, JR (1999) Unsaturated Zone Hydrology for Scientists and Engineers. *Prentice Hall.* Upper Saddle River. 624 pp.

Villar, MV (2000) Caracterización termo-hidro-mecánica de una bentonita de Cabo de Gata. *Ph. D. Thesis,* Universidad Complutense de Madrid, Madrid. 396 pp. (In Spanish).

Villar, MV (2002) Thermo-hydro-mechanical characterisation of a bentonite from Cabo de Gata. A study applied to the use of bentonite as sealing material in high level radioactive waste repositories. *Publicación Técnica ENRESA* 01/2002. Madrid. 258 pp.

Villar, MV & Rivas, P (1994) Hydraulic properties of montmorillonite-quartz and saponite-quartz mixtures. *Applied Clay Science* 9: 1–9.

Villar, MV & Lloret, A (2001) Variation of the intrinsic permeability of expansive clays upon saturation. In: Adachi, K & Fukue, M (eds) Clay Science for Engineering. *Balkema,* Rotterdam. pp. 259–266.

Villar, MV & Lloret, A (in press) Temperature influence on the hydro-mechanical behaviour of a compacted bentonite. *Applied Clay Science.*

Volckaert, G, Bernier, F, Alonso, E, Gens, A, Samper, J, Villar, MV, Martín, PL, Cuevas, J, Campos, R, Thomas, IIR, Imbert, C & Zingarelli, V (1996) Thermal-hydraulic-mechanical and geochemical behaviour of the clay barrier in radioactive waste repositories (model development and validation). *Nuclear science and technology. EUR* 16744. Commission of the European Communities. Luxembourg. 722 pp.

Yoshimi, Y & Osterberg, JO (1963) Compression of partially saturated cohesive soils. *J. Soil Mechanics and Foundations Division.* ASCE 89, SM 4: 1–24.

Advances in Understanding Engineered Clay Barriers – Alonso & Ledesma (eds)
© 2005 Taylor & Francis Group, London, ISBN 04 1536 544 9

Microstructural changes of compacted bentonite induced by hydro-mechanical actions

E. Romero, C. Hoffmann, E. Castellanos, J. Suriol & A. Lloret
Universitat Politècnica de Catalunya, Barcelona (Spain)

ABSTRACT: This paper presents a quantitative study of the microstructure of a compacted bentonite, carried out using mercury intrusion pore size distribution measurements. Compacted samples of artificially prepared granules (pellets) and natural aggregations (pseudomorphs) were investigated. The paper includes the test results and interpretation of the microstructural changes induced by loading and wetting/drying paths.

1 INTRODUCTION

In the context of the investigation of bentonite-based engineered barriers for the enhancement of radioactive waste repository systems, a series of tests has been conducted to give a better insight into the understanding of the microstructural characterisation of these materials. From a fundamental point of view, measurements and observations at this microstructural level involving clay units and their aggregations are very important, since they should help in further understanding of higher structural levels and their consequences on material properties and behaviour under various hydro-mechanical stress state conditions (Tessier *et al.* 1992, Delage *et al.* 1996).

The quantitative examination of the microstructural level has been usually focused on the distribution of the sizes of the soil voids. The pore size distribution is an essential fabric element, which is related to some soil properties and behavioural characteristics, such as water, air and heat conductivity properties, absorption and desorption properties, and skeleton deformability (rearrangement of fabric units). Predictive equations for the saturated permeability based on the pore size distribution have been presented in Juang & Holtz (1986) and Lapierre *et al.* (1990). Romero (1999), Romero *et al.* (1999) and Aung *et al.* (2001) used porosimetry data to predict the water retention properties of a soil. Delage & Lefebvre (1984), Griffiths & Joshi (1989) and Delage *et al.* (1996) used the porosimetry technique to investigate the fabric changes of soils during consolidation and static compaction. Al-Mukhtar *et al.* (1996) analysed the pore size distribution changes of a clay under controlled mechanical and hydraulic stress states. Recently, Simms &

Yanful (2001) studied the morphology changes of the pore network of a soil during drying.

A very useful technique for the quantitative study at this microscale is the mercury intrusion porosimetry (MIP) (Webb & Orr 1997), which has been used in this research to describe the microstructure of a compacted bentonite and to provide information about factors influencing its microstructural changes. The wide range of pore sizes that can be examined (around 10 nm to 400 μm) allows MIP to be very suitable for the study of the microstructure of clays. In the section that follows, the background, limitations and interpretation of this technique will be briefly described.

Two types of compacted samples were studied in this research using the same bentonite powder: compacted samples of artificially prepared granules (pellets) and samples made of natural aggregations (pseudomorphs). The influence of various mechanical (loading) and hydraulic (wetting/drying) stress paths on the pore size distribution of both types of compacted bentonites were analysed.

2 BACKGROUND OF MERCURY INTRUSION POROSIMETRY

In the MIP technique an absolute pressure p is applied to a non-wetting liquid (mercury) in order to enter the empty pores. The following Washburn equation applies (Juang & Holtz 1986, Webb & Orr 1997) for pores of cylindrical shape and parallel infinite plates (fissure-like microstructure):

$$p = -\frac{n\,\sigma_{Hg}\cos\theta_{nw}}{x} \tag{1}$$

where σ_{Hg} is the surface tension of mercury ($\sigma_{Hg} = 0.484\,\text{N/m}$ at 25°C), θ_{nw} the contact angle between mercury and the pore wall, and x the entrance pore diameter ($n = 4$) or the width between parallel plates ($n = 2$). The contact angle with the clay minerals, usually taken between 139° and 147° (Diamond 1970), was assumed $\theta_{nw} = 140°$. Delage & Lefebvre (1984) chose 141° for natural clay, while Griffiths & Joshi (1989) used 147° for illite and kaolinite clays.

The main limitations of MIP are: isolated pores enclosed by surrounding solids (enclosed porosity) are not measured; pores that are accessible only through smaller ones (constricted porosity) are not detected until the smaller pores are penetrated; the apparatus may not have the capacity to enter the smallest pores of the sample (non-intruded porosity); and the minimum practical pressure of the apparatus limits the maximum pore size to be detected (non-detected porosity). In this way, when the clay sample is intruded by mercury, the intruded void ratio estimated under the maximum applied pressure does not coincide with the estimated void ratio of the sample. Differences mainly arise due to the non-intruded porosity with entrance pore sizes lower than 10 nm and the non-detectable porosity for pore sizes larger than 400 μm. It is assumed that the enclosed porosity is not significant in soils, and that the small pores display a non-constricted porosity (Delage & Lefebvre 1984).

The high pressures applied to the samples may result in temporary or permanent alteration in the pore geometry. However, Lawrence (1978) observing similar first and second intrusion curves (after being removed all the mercury of the first intrusion by heating under vacuum), concluded that this technique did not cause any permanent damage to the soil. This last statement could be accepted due to the incompressibility of mercury and to the fact that, although the semi-hydrostatic pressures are large, the unbalanced forces acting to cause damage are only moderate, as the pore size is very small (Reed et al. 1980, Griffiths & Joshi 1989).

A typical graph of the MIP technique includes a cumulative intrusion of pore volume vs. entrance pore diameter. For convenient mathematical treatment, it is preferable to express the ordinate of the cumulative distribution in a dimensionless term, e.g. the cumulative intruded mercury volume divided by the volume of solids, which is equivalent to the intruded void ratio e_{nw}. Furthermore, the plot of the derivative of the cumulative intrusion curve is not useful for the interpretation of MIP data, since pore sizes x extend over several orders of magnitude and the class width dx reported by the intrusion record is not constant, emphasising the smaller pores. To overcome these distortions, the pore size density (PSD) function at the midpoint of each class $\log x_m$ can be defined as follows:

$$f(\log x_m) = -\frac{de_{nw}}{d(\log x)} = \frac{de_{nw}}{d(\log p)} \qquad (2)$$

MIP tests were performed on a 'Micromeritics-AutoPore IV' equipment, attaining maximum intrusion pressures of 220 MPa.

3 EXPERIMENTAL PROGRAMME

3.1 Material

The research for this paper was carried out on FEBEX bentonite (Almería, Spain). The processing of the powder at the factory has consisted in disaggregation and gently grinding, drying at 60°C and sieving by 5 mm. This material has a content of montmorillonite higher than 90% with small and variable quantities of accessory minerals (quartz, calcite and feldspars). The cation exchange capacity is 102–120 meq/100 g (42% Ca, 33% Mg, 23% Na, 2% K). The liquid limit of the bentonite is 93–106%, the plastic limit 47–56%, the specific gravity 2.70, and 64–70% of particles are smaller than 2 μm. The hygroscopic water content in equilibrium with the laboratory atmosphere (relative humidity about 47%, total suction about 100 MPa) is 14%. The value obtained for the external specific surface using BET technique is 29–35 m²/g and the total specific surface obtained using the Keeling hygroscopic method is about 725 m²/g (ENRESA 2000). The low values of the Atterberg limits and the small quantity of less than 2 μm granulometric fraction can be explained taking into account the large quantity of silt-sized aggregates that are 'pseudomorphs' of the volcanic minerals transformed into smectite (Villar 2000).

3.2 Sample preparation techniques

Tests were performed on two different structures prepared from FEBEX bentonite powder: compacted samples of artificially prepared granules (pellets) and samples made of natural aggregations (pseudomorphs).

The following procedure, which is described in Figure 1, was followed at the factory to fabricate the pellet-based material. Naturally-aggregated bentonite was compacted using a roller press. In order to facilitate this operation, the bentonite powder was pre-heated to 120°C. The particle size distribution of this pre-heated and naturally-aggregated material is presented in Figure 2. As a result of this process, granules with a dry density of around 1.95 Mg/m³ and reduced water content of 3–4% were obtained. Finally, the fraction of granules with sizes larger

194

than 10 mm was used as the basic material for the production of pellets with a jaw crusher. Afterwards, these pellets were sorted, sieved and mixed to arrive to the optimal grain size distribution. Laboratory tests were performed on mixtures with a maximum pellet size $D_{max} = 4$ mm. A minimum pellet size of

I) PREHEATED Ca-BENTONITE
120°C (w=3-4%)

ROLLER PRESS

II) HEAVILY COMPACTED GRANULES

$(\rho_d = 1.95$ Mg/m^3 ; e=0.38; $S_r = 28\%)$

III) GRANULES WERE CRUSHED AND CLASSIFIED (PELLETS)

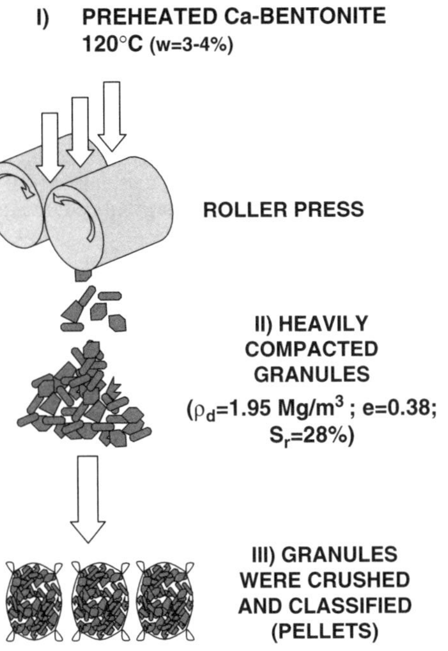

Figure 1. Fabrication process of artificially-prepared granules.

$D_{min} = 0.4$ mm was selected to avoid the tendency towards segregation of the finer particles. Figure 2 shows the pellet size distribution curve that is based on a modified Fuller curve.

The particle size distribution curve of the material made of natural aggregations is also presented in Figure 2, which exactly follows the granulometry of the source material used for the fabrication of the pellet-based material. The original grain size distribution of the naturally-aggregated material has been preserved, without the application of energy (impact or stirring) that could destroy the pseudomorphs. The maximum size of the natural aggregations has been limited to ASTM No.4 (4.75 mm). Despite presenting a finer particle fraction, no appreciable granulometric differences exist between both structures.

Both structures were then statically and one-dimensionally compacted at constant hygroscopic water content to different target dry densities ranging between 1.35 and 1.95 Mg/m^3 for the pellet-based material and between 1.40 and 1.80 Mg/m^3 for the naturally-aggregated bentonite. The water content was maintained around 4% for the pellet-based material and around 14% for the naturally-aggregated material. Figure 3 shows the maximum vertical net stresses reached in these paths at constant water content for both structures. As observed, the pellet-based material displays a somewhat higher compressibility on loading. The artificially-prepared pellets, despite presenting a higher suction, exhibit a lower strength and rigidity, and are prone to display more particle crushing and deformation during compaction.

To avoid excessive shrinkage a freeze drying process was used to prepare the dehydrated samples

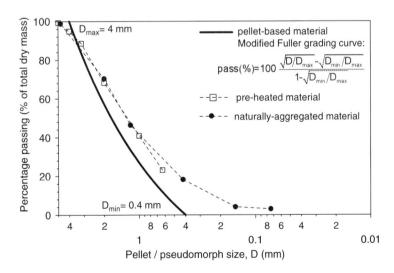

Figure 2. Grain size distribution curves for pellet-based and naturally-aggregated materials.

required by the MIP tests. Details of the procedure can be found in Delage *et al.* (1982) and Delage *et al.* (1996).

3.3 *Stress paths followed*

The evolution of the microstructure of the pellet-based and naturally-aggregated materials was investigated using mechanical and hydraulic stress paths. The mechanical paths involved loading paths at constant water content. In the case of the pellet-based material, the water content was maintained at 4%, which corresponded to a total suction of around 250 MPa. On the other hand, the naturally-aggregated material was loaded at a water content of 14% (total suction of around 100 MPa). Maximum vertical net stresses of the pellet-based material varied between 3 and 50 MPa to reach the desired dry density between 1.35 and 1.95 Mg/m³ (Fig. 3). The corresponding maximum compaction stresses reached on the naturally-aggregated material varied between 8 and 38 MPa to arrive to the target dry densities 1.40 to 1.80 Mg/m³ (Fig. 3).

Figure 4 shows the stress path followed on loading (L-U), which also includes an unloading stage prior to the performance of the MIP test. However, it is assumed that irreversible loading mechanisms (crushing and rearrangement of pellets) predominate over the reversible mechanisms of the unloading stage. The stress path does not include the freeze-drying stage, in which it is assumed that suction is increased without appreciable change of soil volume.

A series of hydraulic stress paths included wetting under isochoric (constant volume) or constant vertical stress conditions. In the case of the pellet-based material, all the wetting paths were carried out under isochoric conditions, starting from the as-compacted state at a dry density of 1.35 Mg/m³ (path WIC-U in Fig. 4). The three wetting paths were performed in a single-stage procedure reaching the following final states: suction of 3 MPa, 400 kPa and saturated condition. The maximum swelling pressure reached in the saturation path of the pellet-based material was 1.2 MPa. On the other hand, the wetting paths on the naturally-aggregated material at an initial dry density of 1.65–1.68 Mg/m³ were performed under constant stress states: under unconfined conditions (path W-L-U) and under a constant vertical net stress of 2 MPa (L-W-U). Under unconfined conditions, the sample was saturated in a single stage tending to slurry, which was then loaded to a maximum vertical net stress of 1.5 MPa. The remainders under constant vertical net stress were wetted in a single-stage procedure reaching the following final states: suction of 39 MPa (1 test) and saturated conditions (3 tests). A series of hydraulic stress paths on the naturally-aggregated material also included drying paths under unconfined conditions (D-W) starting from the as-compacted state at a dry density of 1.65 Mg/m³. The maximum total suction applied was 310 MPa, before returning back to 100 MPa. All the hydraulic paths include an unloading stage prior to the performance of the MIP test. It is assumed that irreversible mechanisms on suction reduction and loading predominate over the reversible mechanisms of the unloading stage. The freeze-drying stages are not plotted in the figure.

4 MIP RESULTS AND ANALYSIS

4.1 *Mechanical paths*

Figure 5 shows the PSD function of artificially-prepared pellets with a grain size lower than $D = 2$ mm that were poured into the penetrometer

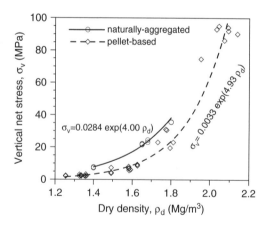

Figure 3. Maximum vertical net stresses reached in the static compaction paths.

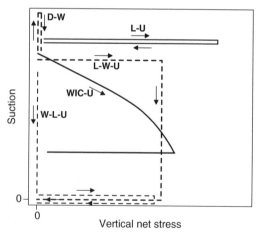

Figure 4. Stress paths followed.

for mercury intrusion at an estimated overall dry density of $1.15\,Mg/m^3$. Three different types of pore spaces were identified within the structure based on size classifications. The largest pore size mode at a value around $250\,\mu m$ corresponds to the inter-pellet porosity (pores between pellets), which is in agreement with an estimated value of the inter-pellet pore size of $d = 0.2\,D$ (Kamiya & Uno 2000). The remainders are associated with the pores inside and between aggregations contained in a pellet. The pore size mode at $3\,\mu m$ corresponds to the inter-aggregate porosity (between aggregations), while the lowest value is associated with the intra-aggregate porosity inside aggregations. This lowest value with pore size

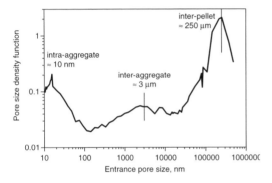

Figure 5. PSD function of poured artificially-prepared pellets.

around $10\,nm$ is not easily determined with MIP. The compressibility corrections for the various components of the high-pressure system could make difficult its determination at mercury intrusion pressures $>100\,MPa$. Pore structure investigation by gas adsorption on solid surfaces is a better technique in this low pore size range (Webb & Orr 1997).

Tessier *et al.* 1992 identified three different types of pore spaces within the microstructure of clays. The pore size ranging between 1 and $2.5\,nm$ was identified as the intra-particle (inter-platelet) pore space. The inter-particle (intra-aggregate) pore space was considered between 2.5 and $150\,nm$. The inter-aggregate pore domain was identified for pore sizes larger than $150{-}200\,nm$.

Figure 6 shows the PSD functions of the pellet-based material after undergoing a loading stage (path L-U in Fig. 4). As observed, the PSD curves of the as-compacted states at dry densities varying between 1.35 and $1.70\,Mg/m^3$ only display two modes at pore sizes $<10\,nm$ corresponding to the intra-aggregate pore space and at pore sizes between 5 and $40\,\mu m$ (inter-pellet porosity). It appears that the dominant inter-pellet pore mode overlaps the inter-aggregate porosity that cannot be distinguished. Following this hypothesis, it is admitted that the as-compacted states still present three pore modes; even so only two are distinguished. On loading, two irreversible mechanisms are assumed: crushing of pellets and macropore reduction due to rearrangement and deformation of pellets. Crushing of pellets is mainly detected by the

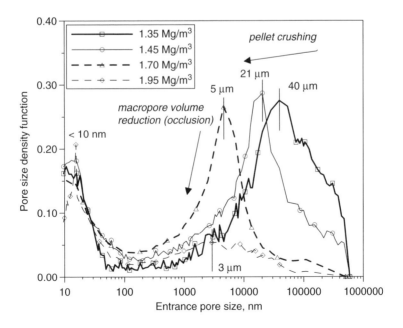

Figure 6. Loading paths (L-U) on pellet-based material.

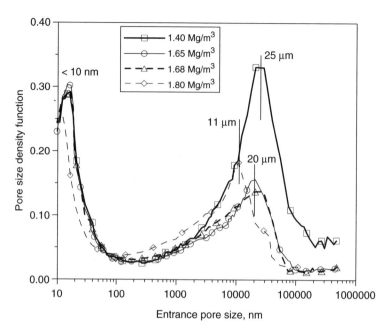

Figure 7. Effect of loading paths (L-U) on naturally-aggregated material.

important shifting towards lower inter-pellet pore size modes (dry densities varying from 1.15 to 1.70 Mg/m^3 in Figs. 5 and 6), as a consequence of the particle size changes of the material. On the other hand, macropore reduction due to the rearrangement and deformation of pellets is linked to the important variation of the height of the pore mode peak with some shifting towards lower values (dry densities varying from 1.70 to 1.95 Mg/m^3). When the inter-pellet porosity is occluded at a dry density > 1.95 Mg/m^3, the two modes of the pellet (inside and between aggregations) are recovered.

Figure 7 shows the PSD functions of the naturally-prepared material after undergoing a loading stage (path L-U in Fig. 4). As observed, only two predominant modes can also be distinguished in the naturally-aggregated material. Again, it appears that the dominant inter-granular pore mode at around 11–25 hinders the inter-aggregate porosity that cannot be distinguished. On loading, the rigid pseudomorphs are less prone to crushing and it is assumed that the main irreversible mechanism of deformation is associated with the rearrangement of the pseudomorphs.

Figure 8 presents a summary of the inter-granular pore sizes as a function of the dry density for the artificially-prepared and naturally-aggregated materials in the as-compacted state. As observed, the pellet-based material undergoes an important change of the inter-pellet probably related to particle crushing. The data of the naturally-aggregated material have

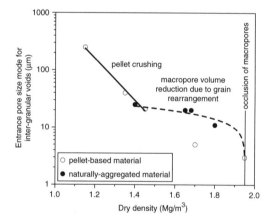

Figure 8. Pore size mode for the inter-granular voids as a function of the dry density.

been fitted to the following geometrically-based expression for inter-granular volume reduction, which takes into account the dry density required for the occlusion of this macroporosity, ρ_{do}:

$$D_M^3 = \chi \frac{\left(\dfrac{\rho_{do}}{\rho_d} - 1 \right)}{\rho_s} \quad \text{for} \quad \frac{\rho_{do}}{\rho_d} \geq 1 \qquad (3)$$

198

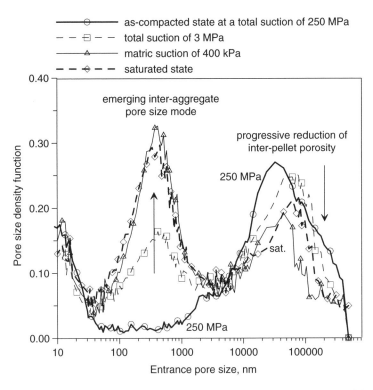

Figure 9.   Evolution of PSD functions on wetting at constant volume (WIC-U). Pellet-based material.

where D_M is the dominant inter-granular pore size, $\rho_s = 2.70\,\mathrm{Mg/m^3}$ the solid density and χ a gravimetric-geometric constant. Fitted parameters were $\rho_{do} = 1.952\,\mathrm{Mg/m^3}$ and $\chi = 8.929 \times 10^{-14}\,\mathrm{Mg}$.

4.2   Hydraulic paths

The evolution of the PSD function of the pellet-based material on wetting under isochoric conditions is presented in Figure 9. It can be seen that the hindered and latent inter-aggregate pore size mode emerges and reaches systematic higher peaks on progressive wetting. Simultaneously, and as a consequence of the constant volume condition, the inter-pellet porosity is drastically reduced, tending to its occlusion. Due to limitations of the equipment, the evolution of the intra-aggregate pore size mode cannot be registered, which is expected to increase on wetting. It is important to indicate that the emerging inter-aggregate pore mode systematically displays the same value at around 400 nm. This fact is difficult to understand, since a slight increase of this pore size mode should be expected on wetting (Romero et al. 2002). It appears that the wetting induces the progressive splitting of the original structure of the pseudomorphs, which are

purely smectitic piles (Villar 2000). The progressive subdivision originates the increase in the number of reactive surfaces and the emerging porosity. The progressive subdivision of clay particles on wetting has been suggested by Saiyouri et al. (1998) to explain suction-induced strain accumulation on successive wetting-drying cycles.

Figure 10 shows the PSD function of the naturally-aggregated material, which was saturated in one step to slurry condition under zero confining stress and then loaded (path W-L-U in Fig. 4). The state after the loading path presented a dry density of around $1.15\,\mathrm{Mg/m^3}$. The PSD function displays a clear dominant pore size mode at 430 nm, which corresponds to the inter-aggregate pore space. Important differences are observed between the PSD functions of the as-compacted states presented in Figure 7 and the saturated material. Again, the hindered inter-aggregate porosity emerges on wetting.

Figure 11 shows the PSD functions of the wetting paths under constant vertical load (path L-W-U in Fig. 4) performed on the naturally-aggregated material. Following an equivalent explanation to that previously described, the hindered inter-aggregate mode at entrance pore sizes between 750 and 1100 nm emerges under saturated conditions. However, at the wetting stage corresponding to a total suction of 39 MPa (vapour transfer), the initial as-compacted microstructure with dominant inter-granular and intra-aggregate pore size modes is nearly preserved and the inter-aggregate mode is still latent. It appears that the predominant changes occur due to the transfer of liquid water. The different dry densities of the materials before the performance of the freeze-drying process are indicated in the figure.

Finally, Figure 12 presents the results of the drying path carried out on the naturally-aggregated material. As observed, the as-compacted microstructure is still preserved after the drying path, in which a dominant inter-granular pore size mode is identified. The inter-aggregate porosity remains hidden.

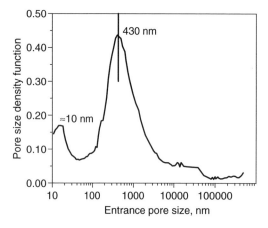

Figure 10. PSD function of the saturated state attained after the stress path (W-L-U). Naturally-aggregated material.

5 SUMMARY AND CONCLUSIONS

A series of mercury intrusion porosimetry tests were performed to characterise the microstructural changes observed on compacted bentonite subjected to different hydro-mechanical paths. Test results were presented for two types of soil structures: compacted samples of artificially prepared granules (pellets) and compacted samples made up of natural aggregations

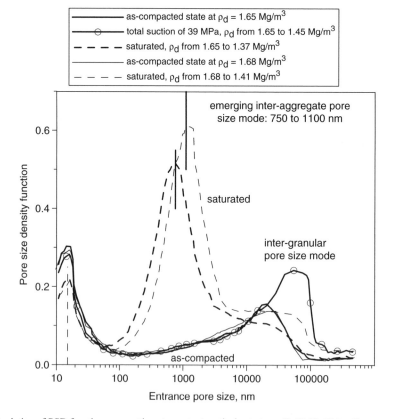

Figure 11. Evolution of PSD functions on wetting at constant vertical net stress (L-W-U). Naturally-aggregated material.

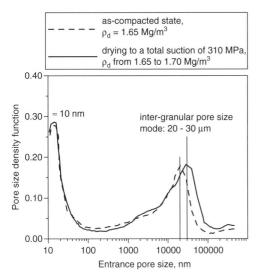

Figure 12. Evolution of PSD functions on drying at zero confining stress (D-W). Naturally-aggregated material.

(pseudomorphs). Hydro-mechanical paths included loading at constant water content, wetting under constant volume or constant vertical stress, and drying at zero confining stress.

The loading paths reduce the larger pores (i.e., inter-granular pores) before tending to the limit condition of the pellet or pseudomorph structure with intra- and inter-aggregate porosity. In the as-compacted state, the dominant inter-granular pore mode hid the inter-aggregate porosity that could not be distinguished. Two irreversible mechanisms were identified on loading: pellet crushing that was predominant in the pellet-based material and pore volume reduction due to the rearrangement of pellets and natural aggregations.

Suction reduction influenced the inter-aggregate pore size mode. On progressive wetting the hidden and latent inter-aggregate pore size mode systematically emerged, probably due to the progressive splitting of the original structure of the pseudomorphs. The emerging inter-aggregate porosity was only detected when liquid water was transferred. As a consequence of the wetting path at constant volume, the inter-pellet porosity was drastically reduced in the pellet-based material. Due to limitations of the equipment, the evolution of the active intra-aggregate pore size mode during wetting could not be followed.

Further studies may be necessary to complement and provide more information about the mechanisms influencing pore size distribution changes on high active clays.

REFERENCES

Al-Mukhtar, M., Belanteur, N., Tessier, D. & Vanapalli, S.K. (1996). *Applied Clay Science*, 11: 99–115.

Aung, K.K., Rahardjo, H., Leong, E.C. & Toll, D.G. (2001). Relationship between porosimetry measurement and soil-water characteristic curve for an unsaturated residual soil. *Geotechnical and Geological Engineering*, 19: 401–416.

Delage, P., Tessier, D. & Audiguier, M.M. (1982). Use of the cryoscan apparatus for observation of freeze-fractured planes of a sensitive Quebec clay in scanning electron microscopy. *Can. Geotech. J.*, 19: 111–114.

Delage, P. & Lefebvre, G. (1984). Study of the structure of a sensitive Champlain clay and its evolution during consolidation. *Can. Geotech. J.*, 21: 21–35.

Delage, P., Audiguier, M., Cui, Y.-J. & Howat, D. (1996). Microstructure of a compacted silt. *Can. Geotech. J.*, 33: 150–158.

Diamond, S. (1970). Pore size distributions in clays. *Clays and Clay Minerals*, 18: 7–23.

ENRESA (2000). FEBEX Project. Full-scale engineered barriers experiment for a deep geological repository for high level radioactive waste in crystalline host rock. Final Report. Publicación Técnica ENRESA 1/2000, Madrid, 354 pp.

Griffiths, F.J. & Joshi, R.C. (1989). Change in pore size distribution due to consolidation of clays. *Géotechnique*, 39(1): 159–167.

Juang, C.H. & Holtz, R.D. (1986). A probabilistic permeability model and the pore size density function. *Int. J. Numer. Anal. Meth. Geomech.*, 10: 543–553.

Kamiya, K. & Uno, T. (2000). Grain size and void diameter of sands. *Proc. Asian Conf. on Unsaturated Soils*. Rahardjo, Toll & Leong (eds). Balkema, Rotterdam: 399–404.

Lapierre, C., Leroueil, S. & Locat, J. (1990). Mercury intrusion and permeability of Louiseville clay. *Can. Geotech. J.*, 27: 761–773.

Lawrence, G.P. (1978). Stability of soil pores during mercury intrusion porosimetry. *J. Soil Science*, 29: 299–304.

Reed, M.A., Lovell, C.W., Altschaeffl, A.G. & Wood, L.E. (1980). Replay: Frost-heaving rate predicted from pore size distribution. *Can. Geotech. J.*: 639–640.

Romero, E. (1999). Characterisation and thermo-hydro-mechanical behaviour of unsaturated Boom-clay: An experimental study. *Ph. D. Thesis*, Universidad Politécnica de Cataluña, Barcelona. http://www.tdcat.cesca.es/TDCat-0930102-092135/

Romero, E., Gens, A. & Lloret, A. (1999). Water permeability, water retention and microstructure of unsaturated Boom clay. *Engineering Geology*, 54: 117–127.

Romero, E., Castellanos, E. & Alonso, E.E. (2002). Lead nitrate tests on compacted sand-bentonite buffer material GMT Emplacement Project. Internal Report, NAGRA, Switzerland.

Saiyouri, N., Hicher, P.Y. & Tessier, D. (1998). Microstructural analysis of highly compacted clay swelling. *Proc. 2nd Int. Conf. on Unsaturated Soils*, Beijing, International Academic Publishers, 1: 119–124.

Simms, P.H. & Yanful, E.K. (2001). Measurement and estimation of pore shrinkage and pore distribution in a clayey till during soil-water characteristic curve tests. *Can. Geotech. J.*, 38: 741–754.

Tessier, D., Lajudi, A. & Petit, J.C. (1992). Relation between the macroscopic behaviour of clays and their microstructural properties. *Appl. Geochem.*, Suppl.1: 151–161.

Villar, M.V. (2000). Caracterización termo-hidro-mecánica de una bentonita de Cabo de Gata. *Ph. D. Thesis*, Universidad Complutense de Madrid.

Webb, P.A. & Orr, C. (1997). Analytical methods in fine particle technology. Micromeritics Instrument Corp, Norcross.

Advances in Understanding Engineered Clay Barriers – Alonso & Ledesma (eds)
© *2005 Taylor & Francis Group, London, ISBN 04 1536 544 9*

Hydro-chemical behaviour of bentonite – based mixtures in engineered barriers

Clemente Mata
INITEC NUCLEAR SA, Madrid, Spain

Alberto Ledesma
Technical University of Cataluña, Barcelona, Spain

ABSTRACT: Among the different concepts of engineered barriers for nuclear waste, bentonite-based mixtures are attractive for economic, technical and environmental reasons. One of the agencies betting on such mixtures is SKB, the Swedish Agency for nuclear waste. In this paper, an overview of the hydro-mechanical behaviour of a sodium bentonite and crushed granite rock mixture that is being used in a project carried out at the Äspö Hard Rock Laboratory, is provided. The influence of salinity in hydrating water on the hydro-mechanical behaviour of such a mixture was also taken into account since the mixture might be used in a saline groundwater environment. After the experimental study carried out, a hydro-mechanical characterisation of the mixture was achieved. Backfill permeability is low enough ($k < 1 \cdot 10^{-11}$ m/s) even for the higher salt concentration in hydrating water and at a dry specific weight of 16.6 kN/m^3. Moreover, swelling pressure is high enough for the purpose of this material within the concept of a nuclear barrier in similar conditions ($p_s > 180$ kPa).

1 INTRODUCTION

Active clayey soils as bentonites are widely used in engineering. This kind of soil is common in environmental engineering, for example, in liners (Alther, 1987), and it is being considered as sealing material in future engineered barriers for nuclear waste. Different mixtures of bentonite and sand, or other granular materials as crystalline rocks or salt rock, have been characterised (e.g. Gray et al, 1984; Radhakrishna et al, 1989; Mingarro et al, 1991; Pfeifle & Brodsky, 1991; Komine & Ogata, 1999; Dixon, 2000; Engelhardt, 2002, among others).

While pure bentonite is being considered as a primary barrier for the canisters containing the radioactive waste, mixtures of bentonite and granular materials are considered for the final backfilling of tunnels. Within the Backfill and Plug Test Project, carried out at the ZEDEX gallery in the Äspö Hard Rock Laboratory, an MX-80 sodium bentonite and crushed granite rock mixture is being investigated as a sealing material. The main objectives of the project are to test the backfill hydro-mechanical behaviour, to study backfill interaction with the host rock in the tunnel, and to develop and test techniques and materials of backfilling of tunnels. Figure 1 shows the vertical section of the ZEDEX gallery where the project is being developed. The test region

can be divided in four different parts: the inner zone with drainage material only, then a zone filled with compacted backfill followed by a third area with crushed granite rock, and finally, the concrete plug closing the gallery. The part of the tunnel filled with the mixture is the aim of the current study. Six different backfill sections (A1 to A6) were compacted in six different layers. Pore water pressure transducers and psychrometers mainly make up the instrumentation placed in situ. The compaction was carried out with a vibrating plate developed and built for this purpose. The final slope of the section was 35° due to compaction technique requirements. Among the sections, permeable geotextiles were placed. The purpose of these permeable mats is to inject and/or to collect water when performing flow tests. The Backfill and Plug Test Project is mainly a full-scale flow test and backfill hydraulic conductivity is the key parameter to be investigated.

In early 1997, different ratios of bentonite and crushed granite had been tested (Börgesson et al, 1996; Johannesson et al, 1999). The main backfill characteristics had to be low permeability and cost. A ratio of 30/70 (30% of sodium bentonite and 70% of crushed granite rock) was finally chosen after this testing campaign. The granite rock came from the blasting of the ZEDEX gallery. Groundwater at ÄHRL comes from different fractures connected to

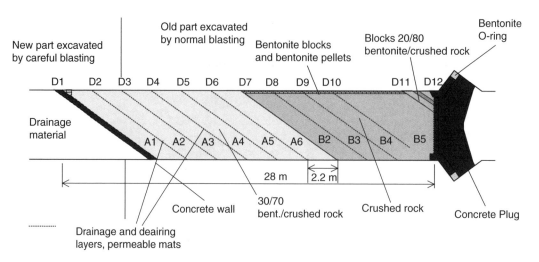

Figure 1. Vertical section of ZEDEX gallery showing the final layout and numbering of sections and mats (Gunnarsson et al, 2001).

the Baltic Sea. Different salt concentrations are measured according to the age of the stored water in different fractures. Mixing water used when preparing the backfill had an average salt concentration of 6 g/L and sodium chloride and calcium chloride were the main chemical species (50/50 by mass approximately). When the mixing process finished, backfill was stored keeping constant the water content until the moment of its compaction in the gallery. By mid 1998, backfill compaction and the installation of whole layout started and the process ended in early 1999. The initial average dry specific weight was 16.6 kN/m³. However, close to the rock walls, dry specific weight was lower because of compaction difficulties (Gunnarsson et al, 2001). Backfill hydration process started in June of 1999. Due to the expected low backfill permeability, salt concentration in injected water during the saturation was increased up to 16 g/L to speed up the hydration as backfill hydraulic conductivity is sensitive to concentration of salt water in hydrating water.

Previous studies show the important changes on hydro-mechanical properties of active clayey soils that may occur when they are permeated with different electrolyte (organic or inorganic), different pH or different concentrations (e.g. Bolt, 1956; Mesri & Olson, 1970; Mesri & Olson, 1971; Sridharan & Venkatappa Rao, 1973; Goldenberg & Magaritz, 1983; Fernandez & Quigley, 1985; Moore, 1991; Miller & Nelson, 1992; Barbour & Yang, 1993; Yanful et al, 1995; Di Maio, 1996; Daupley, 1997; Petrov & Rowe, 1997; Dixon, 2000; Mata et al, 2002; Guido et al, 2002 among others).

In order to experimentally investigate and characterise the mixture, taking into account the influence

of the pore fluid chemistry, different tests were carried out in laboratory (Mata, 2003, Mata & Ledesma, 2003, Börgesson et al, 1996 and Johannesson et al, 1999):

– Oedometer tests in Rowe cells: two different initial packings and specimens saturated with water containing different salt concentrations.
– Proctor tests: two levels of energy were applied and both kinds of water (de-ionised water and salt water) were added to the specimens.
– Saturated flow tests to characterise the backfill saturated hydraulic conductivity.
– Determination of osmotic suction by means of psychrometers.
– Chemical analysis of collected pore water from compacted backfill specimens.
– Water uptake tests to estimate the unsaturated hydraulic conductivity.
– Matric and total suction in different conditions of compaction.

Only the results of some oedometer tests and the chemical analysis of the collected water from some specimens of compacted backfill are shown here.

2 DESCRIPTION AND PROPERTIES OF THE MIXTURE

The backfill is a mixture of MX-80 sodium bentonite and crushed granite rock coming from the excavation of the ZEDEX gallery. The proportion of the mixture is 30% of bentonite and 70% of crushed granite rock (by mass). MX-80 has 75.5% of montmorillonite content and 85% of its grains are smaller than 2 µm

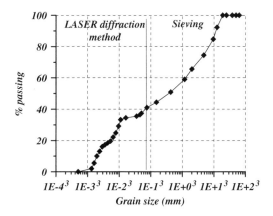

Figure 2. Backfill grain size distribution.

Table 1. Initial conditions of the backfill samples. C is the total concentration of salts used in the water used to saturate the specimens.

Specimen number	γ_d^{ac} (kN/m^3)	e^{ac}	e^{as}	w_0 (%)	C (g/L)
1	16.7	0.577	0.571	10.57	0
2	16.9	0.551	0.543	12.13	6
3	17.1	0.534	0.531	12.63	16

(Madsen, 1998). The specific surface of the MX-80 is 562 m^2/g and its cation exchange capacity (CEC) is 76 meq/100 g. The particle unit weight, γ_s, of MX-80 bentonite is 27.05 kN/m^3 (Madsen, 1998). The backfill particle unit weight is 26.4 kN/m^3 calculated as a weighted average of content by mass of bentonite and crushed granite rock and their particle unit weights. The grain size distribution of the 30/70 mixture is shown in Figure 2. The maximum grain size of the crushed granite rock is 20 mm. For particle size slower than sieve 200 ASTM ($\phi = 0.074$ mm) the size distribution curve was obtained by LASER diffraction technique. Bentonite and crushed granite rock were mixed up with salt water containing 6 g/L of NaCl + CaCl$_2$ (approximately 50/50 by mass) until initial water content ranged from 12% to 13% (w_0). After the mixing process, the backfill was stored and finally compacted in the gallery. The backfill material used in this experimental characterisation came from the initial amount prepared for the project.

3 OEDOMETER TESTS

Three consolidation tests were performed on statically compacted specimens. The equipment used for these tests was a modification of the well-known Rowe's cell (diameter = 152 mm). Two stainless steel cells were designed, built and used to carry out these six tests. Deformation of the two cells was calibrated to take it into account when calculating soil compressibility.

Three different solutions were used to saturate the backfill: de-ionised water, 6 and 16 g/L of salt (50/50 NaCl and CaCl$_2$ by mass). The initial dry specific weight was 16.6 kN/m^3. The specimens were compacted in one layer of 50 mm height and they were saturated from bottom to top with the different permeants. During the saturation phase of the specimens,

null total strain of the specimens was prescribed, thus an external load was applied depending on the observed swelling of the specimen. In this way, a swelling pressure, p_s, was calculated as the difference between the applied external load and the water pressure in the sample (when saturated).

Table 1 summarises the conditions of the three specimens after the compaction process (as superscript) and after the saturation process (as superscript). Initial water content was expected to be around 12%, however, heterogeneous behaviour was observed in some specimens when the initial water content was measured and it ranged from 10.57% to 12.63%. This wide range produced differences on the initial conditions among the three compacted specimens. After saturation, more than 1200 hours in all cases, the consolidation tests started from the swelling pressure as initial effective stress in the specimens. Double drainage of the soil specimens was allowed to speed up the consolidation process. Initially it was assumed that a consolidation step was finished when vertical deformation velocity was less than 5 μm/day (specimens 2 and 3), but later on the velocity was increased up to 20 μm/day (specimen 1) because the previous limit produced very long loading steps ($\approx$300 hours).

The Casagrande's log t method was used to estimate consolidation coefficients. The method does not provide very different consolidation coefficients from those calculated from pore water pressure dissipation measurements and it can be used to estimate the consolidation coefficient in the absence of pore water pressure data (Robinson, 1999). However, for specimens compacted at the small dry specific weight, Terzaghi's approach is only an approximation due to the big strains applied.

3.1.1 Swelling pressure
Table 2 summarises the measured swelling pressure of the specimens. A small decrease of the measured swelling pressure is noticed in specimen number 3 (16 g/L) if compared with specimen number 1 (0 g/L). However, due to the "low" backfill bentonite content (30%) and the low range of salinity considered (up to 16 g/L), swelling pressure is not strongly dependent on the pore fluid chemistry at this dry specific weight.

Table 2. Swelling pressure and pre-consolidation pressure measured for the six specimens.

Specimen	1	2	3
p_s (kPa)	196	202	183

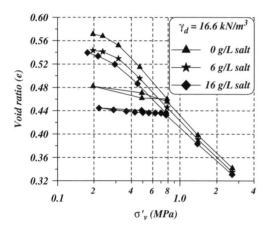

Figure 3. Effective stress versus void ratio relations for the highest dry specific weight (16.6 kN/m³) for different salt concentrations in the water used to saturate the soil specimens.

3.1.2 Void ratio vs. effective stress relationships

In average, each loading step lasted more than 150 hours due to the low backfill permeability and the big size of the specimens. Figure 3 shows the stress-strain relationships measured after the consolidation tests. The pre-consolidation stress was around 220 kPa in the soil specimens compacted at this dry specific weight. Soil compressibility was calculated as (1) and compressibility vs void ratio is depicted in Figure 4. Only the results of the loading process were depicted.

$$m_v = \frac{-(e_2 - e_1)}{(1 + e_1)(\sigma_2 - \sigma_1)} \qquad (1)$$

From the results is it clear that backfill compressibility was not dependent on the pore fluid chemistry at this bentonite content, dry specific weight and salt concentrations. No secondary compression was observed during the consolidation steps.

3.1.3 Consolidation coefficient

The consolidation coefficient was calculated by means of Casagrande's log t method as $C_v = 0.196 H^2 \cdot t_{50}^{-1}$. Figure 5 shows the evolution of the consolidation coefficient. The variation of the consolidation coefficient is lower than one order of

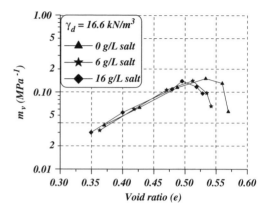

Figure 4. Calculated soil compressibility.

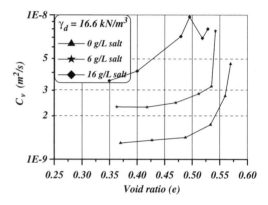

Figure 5. Consolidation coefficient versus void ratio.

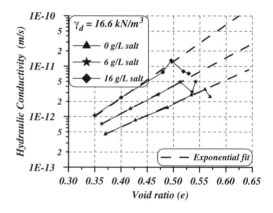

Figure 6. Estimated hydraulic conductivities at the high dry specific weight.

Table 3. Results of the chemical analysis performed in pore water collected from soil specimen numbers 1, 2 and 3. Clay Technology (1999) provided data of 6 g/L mixing water.

Water	Na^+ (mg/L)	Ca^{2+} (mg/L)	Mg^{2+} (mg/L)	K^+ (mg/L)	EC (μS/cm)	pH
6 g/L mixing	1391.8	1077.5	–	–	–	–
De-ionised water	3.0	5.6	2.6	0.0	20	7.9
Specimen number 1	6078.8	460.9	48.6	111.2	22800	7.2
6 g/L incoming	1269.2	1162.3	9.7	1.5	11000	6.5
Specimen number 2	6012.0	497.0	51.1	79.4	22000	7.2
16 g/L incoming	4141.6	2468.9	2.4	2.3	26700	7.0
Specimen number 3	7281.2	958.0	143.5	127.1	33600	7.4

Table 4. Results of the performed chemical analysis on collected pore fluid from soil specimen numbers 1, 2 and 3. Clay Technology (1999) provided data of 6 g/L mixing water. $I = 0.5\Sigma C_i z_i^2$ is the ionic strength in mol/L and z_i is the charge of the ion. $C = \Sigma C_i$ is the total molar concentration of chemical species in the pore fluid.

Water	Cl^- (mg/L)	HCO_3^- (mg/L)	SO_4^{2-} (mg/L)	NO_3^- (mg/L)	NO_2^- (mg/L)	I (mol/L)	C (mol/L)
6 g/L mixing	3104.5	–	–	–	–	–	–
De-ionised water	14.2	30.5	1.7	0.8	0.0	0.0009	0.0013
Specimen number 1	3467.0	115.9	8516.0	594.8	23.5	0.3928	0.4790
6 g/L incoming	4310.7	42.7	10.2	0.8	0.0	0.1472	0.2069
Specimen number 2	4948.5	85.4	4767.0	228.3	12.1	0.3324	0.4727
16 g/L incoming	11032.0	36.6	12.7	1.6	0.0	0.3688	0.5534
Specimen number 3	10628.0	158.6	4621.0	524.4	60.1	0.4712	0.7095

magnitude. Tendency of curves corresponding to specimens 1 and 2 are in agreement with normal response: important changes for applied stresses smaller than the pre-consolidation stress, and smooth variations for stresses larger than the pre-consolidation stress. The curve corresponding to specimen number 3 (16 g/L) had some deviations from the expected response. The oscillations of the consolidation coefficient were due to the evolution of t_{50} for this soil specimen at the different loading steps.

3.1.4 Hydraulic conductivity

Hydraulic conductivity was calculated as $k = C_v m_v \gamma_w$. The estimated variation of hydraulic conductivity is depicted in Figure 6. It is clear that the influence of pore fluid chemistry strongly influences the backfill hydraulic behaviour. The maximum difference of hydraulic conductivity was bigger than 6 times when salinity of water used to saturate the specimens was increased up to 16 g/L.

4 CHEMICAL ANALYSIS OF COLLECTED WATER

During the saturation process some pore water was collected from the specimens compacted at both dry specific weights and chemically analysed. Concentrations of predominant species as sodium, calcium,

potassium, magnesium, chloride, bicarbonates, sulphates, nitrates and nitrites were determined as well as pH and electric conductivity, EC. The results clearly showed an enrichment of sodium concentration and a decrease of the measured calcium concentration in outgoing pore water. This sodium enrichment is due to the cation exchange of sodium by calcium. It means that the sodium bentonite is transforming into a calcium bentonite. Tables 3 and 4 summarise the results of the chemical analysis of mixing, incoming and outgoing water. An important increase in the concentration of sulphates is also noticed. Gypsum, as a secondary mineral, gets dissolved when the bentonite is hydrated. Nitrates and nitrites have an uncertain origin, however, nitrate concentration was higher than, for example, bicarbonate concentration. Possible reasons could be fertilizers used in the excavation area, organic substances or acid treatment (Muurinen, 2001).

5 CONCLUSIONS

Chemical influence on hydro-mechanical properties for a MX-80 sodium bentonite and crushed granite rock (30/70 by mass) was studied by means of different tests performed in laboratory. Three oedometer tests in Rowe cells were carried out at a dry specific weight of 16.6 kN/m^3 and specimens were saturated with three different permeants. Results of these tests clearly show that the backfill compressibility is not dependent of changing the salt concentration in the water used to saturate the sample. However, the hydraulic behaviour is strongly dependent on the salt concentration at water used to saturate the soil sample. The hydraulic conductivity increased up to 6 times for the higher salt concentration (compared with obtained ones with de-ionised water). Moreover, swelling pressure is also sensitive to the hydration with water containing different salt concentrations but to a lesser degree. The chemical analysis of collected water from some of the specimens oedometrically tests confirmed the existence of the cation exchange between Na^+ and Ca^{2+} what means that the sodium bentonite is transforming into a calcium bentonite. This exchange reaction alters the backfill behaviour and its hydraulic properties.

ACKNOWLEDGEMENTS

First author thanks the economical support provided by Catalonian Government through a FI grant. The financial support from ENRESA throughout the project is also acknowledged. The authors want to thank SKB and Clay Technology for their cooperation. The authors are indebted to Tomás Pérez and José Álvarez.

REFERENCES

Alther, G.R. (1987). The qualifications of bentonite as a soil sealant. *Engineering Geology*, 23: 177–191.

Barbour, S.L. & Yang, N. (1993). A review of the influence of clay-brine interactions on the geotechnical properties of Ca-montmorillonite clayey soils from western Canada. *Canadian Geotechical Journal*, 30: 920–934.

Bolt, G. (1956). Physico-chemical analysis of the compressibility of pure clays. *Géotechnique*, 6 (2): 86–93.

Börgesson, L., Johannesson, L.E. & Sandén, T. 1996. Backfill materials based on crushed rock. Geotechnical properties determined in laboratory. SKB Progress Report **HRL96-15**.

Clay Technology, (1999). Personal communication.

Daupley, X. (1997). Etude du potentiel de l'eau interstitielle d'une roche argileuse et des relations entre ses propriétés hydriques et méchaniques, Thèse Doctoral. Ecole Nationale des Mines de Paris.

Di Maio, C. (1996). Exposure of bentonite to salt solution: osmotic and mechanical effects. *Géotechnique*, 46 (4): 695–707.

Dixon, D.A. (2000). Pore water salinity and the development of swelling pressure in bentonite-based buffer and backfill materials. POSIVA report **OY 2000–04**, ISBN 951-652-090-1, ISSN 1239–3096, Helsinki, Finland.

Engelhardt, I. (2002). Thermal-Hidrologic experiments with bentonite/crushed rock mixtures and estimation of effective parameters by inverse modelling. In: Preprints of Contributions to the Workshop on Microstructural Modelling of Natural and Artificially Prepared Clay Soils with Special Emphasis on the Use of Clays for Waste Isolation, Pusch, R. (Ed.), Lund, 15–17 October, Sweden.

Fernandez, F. & Quigley, R.M. (1985). Hydraulic conductivity of natural clays permeated with simple liquid hydrocarbons. *Canandian Geotechnical Journal*, 22: 205–214.

Goldenberg, L.C. & Magaritz, M. (1983). Experimental investigation on irreversible changes of hydraulic conductivity on the seawater-freshwater interface in coastal aquifers. *Water Resources Research*, 19 (1): 77–85.

Gray, M.N., Cheung, S.C.H. & Dixon, D.A. (1984). The influence of sand content on swelling pressures and structure developed in statically compacted Na-bentonite. AECL-7825, Manitoba, Canada.

Guido, M., Romero, E., Gens, A. & Castellanos, E. (2002). The role of structure in the osmotic shrinking-swelling on FEBEX bentonite. In: Preprints of Contributions to the Workshop on Microstructural Modelling of Natural and Artificially Prepared Clay Soils with Special Emphasis on the Use of Clays for Waste Isolation, Pusch, R. (Ed.), Lund, 15–17 October, Sweden.

Gunnarsson, D., Börgesson, L., Hökmark, H., Johannesson, L.E. & Sandén, T. (2001). Äspö Hard Rock Laboratory. Report on the installation of the Backfill and Plug Test. International Progress Report **IPR-01-07**.

Johannesson, L.E., Börgesson, L. & Sandén, T. (1999). Backfill materials based on crushed rock (part 2). Geotechnical properties determined in laboratory. SKB ÄHRL **IPR-99-23**, Sweden.

Komine, H. & Ogata, N. (1999). Experimental study on swelling characteristics of sand-bentonite mixture for nuclear waste disposal. *Soils and Foundations*, 39 (2): 83–97.

Madsen, F.T. (1998). Clay mineralogical investigations related to nuclear waste disposal. *Clay Minerals*, 33: 109–129.

Mata, C. (2003). Hydraulic behaviour of bentonite based mixtures in engineered barriers: The Backfill and Plug Test at the Äspö HRL (Sweden). PhD thesis. Dept. of Geotechnical Engineering and Geosciences. UPC, Barcelona, Spain.

Mata, C., Romero, E. & Ledesma, A. (2002). Hydro-chemical effects on water retention in bentonite-sand mixtures. *Proc. 3rd Int. Conf. on Unsaturated Soils*, UNSAT 2002, Jucá, J.F.T., de Campos, T.M.P. & Marinho, F.A.M. (Eds.), Recife, Brazil, Vol. 1, pp. 283–288.

Mata, C. & Ledesma, A. (2003). Permeability of a bentonite-crushed granite rock mixture using different experimental techniques. *Géotechnique*, 53 (8): 747–758.

Mesri, G. & Olsen, R. (1970). Shear strength of montmorillonite. *Géotechnique*, 20 (3): 261–270.

Mesri, G. & Olsen, R. (1971). Consolidation characteristics of montmorillonite. *Géotechnique*, 21 (4): 341–352.

Miller, D.J. & Nelson, J.D. (1992). Osmotic suction as a valid state in unsaturated soils. *Proc. 7th Int. Conf. Expansive Soils*, Dallas, 179–184.

Mingarro, E., Rivas, P., del Villar, L.P., de la Cruz, B., Gómez, P., Hernández, A., Turrero, M.J., Villar, M.V., Campos, R. & Cozar, J. (1991). Characterization of Clay (Bentonite)/Crushed Granite Mixtures to Build Barriers

Against the Migration of Radionuclides: Diffusion Studies and Physical Properties. Technical Report EUR 13666 EN. Luxembourg: Commission of the European Communities.

Moore, R. (1991). The chemical and mineralogical controls upon the residual strength of pure and natural clays. *Géotechnique*, **41** (1): 35–47.

Muurinen, A. (2001). Personal Communication.

Pfeifle, T.W. & Brodsky, N.S. (1991). Swelling pressure, water uptake and permeability of 70/30 crushed salt/bentonite. *SAND91-7070*. Albuquerque, NM: Sandia National Laboratories.

Petrov, R.J. & Rowe, R.K. (1997). Geosynthetic clay liner (CGL) – chemical compatibility by hydraulic conductivity testing and factors impacting its performance. *Canadian Geotechnical Journal*, **34**: 863–885.

Radhakrishna, H.S., Chan, H.T., Crawford, A.M. & Lau, K.C. (1989). Thermal and physical properties of candidate buffer-backfill materials for a nuclear fuel waste disposal vault. *Canadian Geotechnical Journal*, **26**: 629–639.

Robinson, R.G. (1999). Consolidation analysis with pore pressure measurements. *Géotechnique*, **49** (1): 127–132.

Sridharan, A. & Venkatappa Rao, G. (1973). Mechanisms controlling volume change of saturated clays and the role of the effective stress concept. *Géotechnique*, **23**: 359–382.

Villar M.V. (2002). Thermo-hydro-mechanical characterisation of a bentonite from Cabo de Gata. A study applied to the use of bentonite as sealing material in high radioactive waste repositories. Technical Publication 04/2002, ENRESA, Madrid.

Woodburn, J.A., Hold. J. & Peter, P. (1993). The transistor psychrometer: a new instrument for measuring soil suction. Unsaturated Soils Geotechnical Special Publications N° 39, Dallas, S.L. Houston and W.K. Wray (Eds). ASCE, p. 91–102.

Yanful, E.K., Shikatani, K.S. & Quirt, D.H. (1995). Hydraulic conductivity of natural soils permeated with acid mine drainage. *Canadian Geotechnical Journal*, **32**: 624–646.

Advances in Understanding Engineered Clay Barriers – Alonso & Ledesma (eds)
© 2005 Taylor & Francis Group, London, ISBN 04 1536 544 9

Hydro-mechanical properties of backfill material

L.-E. Johannesson, Lennart Börgesson & David Gunnarsson
Clay Technology AB (Sweden)

ABSTRACT: Laboratory tests on backfill material of crushed rock and mixtures of crushed rock and bentonite have been performed as preparation for the Backfill and Plug Test in ÄHRL. The bentonite content of the backfill varied between 0 and 30% and the ballast material consisted of crushed TBM-muck in all tests. Parameters essential for the function of the backfill material such as hydraulic conductivity, swelling pressure, compressibility of the saturated backfill materials and the suction of unsaturated backfill materials have been measured. The following results are notable:

The measured hydraulic conductivity of the tested backfill materials decreased with increasing clay content and density and decreased substantially when salt was added to the saturating water.

The measured total suction described as a function of the clay water ratio was independent of the clay content but influenced by the salt content of the pore water. The calculated osmotic suction for 30/70 mixtures was about 1.5 MPa when mixed with Äspö water (1.2% salt). The matric suction was found to be rather low at water ratios higher than 15%.

The measured swelling pressure of the backfill material depended very much on the clay content, the density and the salt content of the added water.

The salt content has an insignificant influence on the measured compression properties. The bulk modulus was found to be a function of the percent of maximum Proctor density irrespective of the clay content.

1 INTRODUCTION

Tunnels and shafts in the Swedish KBS-3 concept need to be backfilled before closure of the repository. There are several purposes with the backfill. Following functions of the backfill are considered necessary:

– to obstruct upwards swelling of the bentonite buffer from deposition holes
– to prevent or restrict the water flow
– to stay intact during a long period of time
– not to cause any significant chemical conversion of the buffer surrounding the canister

In Äspö HRL a full scale in situ project (Backfill and Plug Test) is running with the purpose to e.g. study the function of the backfill and its interaction with the near field rock in a tunnel excavated by blasting. Furthermore the tunnel in the *Prototype Repository* is backfilled with a mixture of 30% bentonite and 70% crushed TBM muck. A laboratory program for investigating the hydromechanical properties of possible backfill materials is in progress. The laboratory program contains investigations of mainly the following properties:

– Compaction properties
– Compression properties
– Swelling properties
– Shear properties
– Hydraulic properties of water saturated and unsaturated backfill materials
– Piping resistance.

This paper is mainly focusing on results achieved so far on compression properties swelling and hydraulic properties and the suction of unsaturated backfill material. The tests have been made on mixtures of bentonite and crushed rock. Crushed TBM-muck with a maximum grain size of 20 mm has been used. Different composition of the mixing and saturation water has been used with salt content varying from 0 to 1.8%. This salinity corresponds to water salinity in the Äspö site. This water is called Äspö water.

2 HYDRAULIC TESTS

The hydraulic conductivity of soils is generally measured in oedometers. The maximum grain size of the soil determines the choice of oedometer dimensions. A generally applied rule when determining the hydraulic conductivity is that the diameter of the oedometer should not be smaller than 5 times

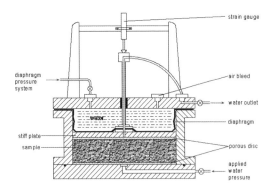

Figure 1.   Rowe oedometer.

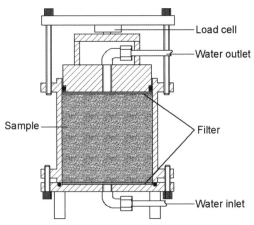

Figure 2.   Proctor cylinder adapted for measuring hydraulic conductivity.

the maximum grain size. The maximum grain size for the tested backfill materials is 20 mm, which means that the oedometer should have a diameter larger than 100 mm. Most of the tests were made using two different kinds of equipment; a so-called Rowe oedometer with 250 mm diameter and a "Proctor cylinder" with a diameter of 101 mm.

A schematic drawing of a Rowe oedometer is shown in Fig. 1. The backfill is manually compacted into the sample holder in 5 layers to the desired density. The 80 mm high sample with a 250 mm diameter is confined by a filter at both ends. The filters in the through connections were removed. After completed preparation a diaphragm is pressurized with water to a pressure equal to the estimated swelling pressure of the backfill sample at water saturation.

After installation the samples were deaired with a vacuum pump and water supplied from both ends. During water saturation, the displacement of the upper filter was measured and recorded. After completed saturation a hydraulic gradient was applied to the sample and the water percolation measured. The hydraulic conductivity was evaluated with Darcy's law when steady flow has been reached.

The same installation procedure was used for proctor cylinder (see Fig. 2) i.e. the backfill was manually compacted into the cylinder in 5 layers to the desired density to a height of about 100 mm. The sample was confined by filters, a fixed bottom plate and a movable piston at the top. An effective stress was applied on the top of the sample by loading the piston. The load was applied by bracing the bolts.

In these tests, the samples were saturated by applying a small water pressure at the bottom and letting air seep out from the top of the sample. After saturation a hydraulic gradient was applied to the sample and the hydraulic conductivity determined.

One test was also made where a sample was saturated and reached equilibrium in a triaxial cell.

Finally the hydraulic conductivity on one sample was determined in an oedometer test, by examining

the time-deformation curve of the sample at a change of the total stress. The hydraulic conductivity was determined with the Eqn 1.

$$k = T_v \frac{h^2}{t} \frac{g \rho_w}{M} \tag{1}$$

where

M = compression module (kPa)
T = time factor (=1 at 95% consolidation)
h = half of the height of the sample (at drainage at both ends of the sample) (m)
t = time for reaching 95% consolidation (s)
ρ_w = the density of water (t/m^3)
g = acceleration due to gravity (m/s^2).

The results from the tests are summarized in Table 1. The tests performed on the 30/70 mixture are also plotted in Fig. 3.

The following comments about the results can be made:

– The scatter in measured hydraulic conductivity is quite large for most of the investigated materials.
– The hydraulic conductivity of the mixtures is strongly affected by the salt content of the pore water. The hydraulic conductivity increases 1 to 2 orders of magnitude when Äspö water is used instead of distilled water.
– The hydraulic conductivity is strongly affected by the dry density. It decreases about 20 times when the dry density increases 10%.
– The hydraulic conductivity is strongly influenced by an increase in bentonite content. It decreases about 10 times at an increase in bentonite content with 10 percentage points.

Table 1. Summary of the hydraulic conductivity tests. The tests marked * are performed in a Rowe oedometer. Tests marked ** are evaluated from oedometer tests while the test marked *** is made in a triaxial cell. The rest of the tests are performed in a Proctor cylinder.

Test no.	Clay cont. (%)	Water	w_{ini} (%)	Final properties Proct. (%)	w (%)	ρ_d (t/m³)	e	Sr (%)	K (m/s)	K with back Pr. (m/s)
1	0	Dist.	5.6	96	8.6	2.21	0.24	97	1.40E-08	1.96E-08
2	0	Dist.	5.6	91	10.4	2.11	0.31	95	6.19E-08	1.33E-07
3	0	Dist.	5.6	101	5.7	2.33	0.19	90	1.66E-09	4.55E-09
4*	0	Dist.	5.6	94	8.3	2.18	0.26	88	8.50E-08	–
5*	0	Dist.	5.6	92	7.8	2.12	0.30	73	1.00E-07	–
6	10	Dist.	7.0	94	12.4	2.02	0.37	96	2.87E-10	3.40E-10
7	10	Äspö*	7.0	94	12.1	2.04	0.36	96	5.12E-09	1.13E-08
8	10	Dist.	7.0	94	13.4	2.02	0.36	102	2.79E-10	5.62E-10
9	10	Äspö	7.0	94	12.4	2.04	0.35	99	2.67E-09	5.59E-09
10	10	Dist.	7.0	88	16.0	1.90	0.45	98	2.02E-10	1.34E-10
11	10	Äspö	7.0	90	14.1	1.95	0.42	95	4.26E-09	2.70E-08
12*	10	Äspö	7.0	98	10.7	2.11	0.30	97	1.77E-09	–
13	20	Dist.	8.0	93	15.5	1.91	0.45	97	4.62E-11	5.22E-11
14*	20	Äspö	8.0	94	15.2	1.93	0.43	98	4.14E-10	9.46E-10
15	20	Dist.	8.0	87	18.1	1.79	0.57	93	5.53E-11	6.25E-11
16*	20	Äspö	8.0	90	17.1	1.85	0.50	96	3.27E-09	5.51E-09
17*	20	Äspö	8.0	87	18.8	1.78	0.54	95	1.46E-09	2.17E-09
18	20	Äspö	8.0	79	23.1	1.62	0.69	91	5.36E-08	–
19	30	Äspö	13.0	91	19.7	1.77	0.56	98	4.71E-11	7.84E-11
20	30	Äspö	13.0	89	21.5	1.72	0.61	99	4.25E-10	4.49E-10
21*	30	Äspö	13.0	89	19.6	1.72	0.60	91	9.73E-11	2.01E-10
22	30	Äspö	13.0	88	21.5	1.71	0.61	96	6.32E-10	9.09E-10
23	30	Äspö	13.0	78	27.3	1.52	0.81	93	7.22E-09	–
24*	30	Äspö	4.6	96	15.9	1.86	0.47	95	6.00E-11	–
25*	30	Dist.	13.7	96	17.3	1.87	0.48	101	4.09E-12	–
26*	30	Dist.	13.7	95	17.1	1.85	0.48	99	1.10E-12	–
27	30	Äspö	6.5	89	20.8	1.73	0.59	97	2.86E-09	2.98E-09
28**	30	Äspö	12.2	89	–	1.72	0.60	–	–	7.93E-12
29**	30	Äspö	12.2	97	16.5	1.88	0.47	97	–	1.89E-12
30***	30	Äspö	10.7	94	17.2	1.83	0.50	94	–	1.51E-12

* Total salinity of the pore water 1.2%.

– The results from the tests performed on 30/70 mixture indicate that the initial water content at the compaction of the material affect the hydraulic conductivity at saturation (see Fig. 5). Also other factors like the mixing and compaction technique may have an influence. Additional tests to study these phenomena are required.
– The tests evaluated according to Equation 3 (marked green in Fig. 5) differ substantially from the rest of the tests performed with Äspö water. A possible explanation is that the consolidation time (and the evaluated hydraulic conductivity) is mostly a function of the hydraulic conductivity of the clay matrix and the influence of heterogeneity and preferential flow paths are unimportant in contrary to constant gradient flow tests where a few paths may lead a lot of water (see e.g. Börgesson et al 2003).

– The reason for the very low hydraulic conductivity measured for the test performed in the triaxial cell is not clear. More tests need to be done.
– Hydraulic conductivity measurements performed by Clemente (2003) on 30/70 mixtures with Äspö water and evaluated from oedometer tests fits well with the oedometer results shown in Fig. 3. However, other tests by Clemente with other techniques indicate less effect of salt water than shown in Fig. 3. The influence of test technique needs to be further investigated.

3 SUCTION TESTS

There are two reasons why it is important to measure the suction of the backfill material as a function of its water ratio. One reason is that it is the suction (matric suction) together with the unsaturated hydraulic conductivity that regulates the rate of the water uptake of the backfill material. The second reason is that the suction can be measured in field and, in order to be able to evaluate the wetting in a large full-scale experiment this relation must be known. The total suction or the water potential can be expressed with the Kelvin equation;

$$S_t = \frac{RT}{V_w} \ln(p / p_0) \qquad (2)$$

where

R = molar gas constant
T = the Kelvin temperature
V_w = the molar volume of water
p/p_0 = the relative humidity of the pore air.

The total suction can according to soil mechanical standards be divided into matric suction and osmotic suction;

$$S_t = S_m + S_o \qquad (3)$$

where

S_t = total suction
S_m = matric suction
S_o = osmotic suction
The matric suction can be calculated as;

$$S_m = u_{air} - u_w \qquad (4)$$

where

u_{air} = air pressure in the porous system
u_w = water pressure in the porous system.

The Kelvin equation (Eqn. 2) shows that it is possible to determine *total suction* by measuring the relative humidity in the pore system and the temperature of the sample. The following two techniques have been used for determination of the total suction of different mixtures of bentonite and crushed TBM muck:

– measurement of equilibrium water content of soil samples for given relative humidity
– to use a calibrated soil psychrometer.

In the first type of tests both compacted and uncompacted samples were placed in a chamber where a constant relative humidity was established. The constant relative humidity (RH) was reached by use of saturated salt solutions. Saturated aqueous solutions of salt produce a stable equilibrium with the relative humidity (RH) in the air above the surface of the solution. RH in the air can thus be controlled by selecting an appropriate salt. The samples reach equilibrium with the constant relative humidity by loosing or taking up water from the surrounding air. The range of RH in these measurements is 81.3–96%, which correspond to the suction 5.000–28.000 kPa according to the Kelvin equation.

The second technique to determine the total suction of the materials is to measure the relative humidity in a small volume of air, which is in equilibrium with the pore system of the mixtures. This type of test has been made with WESCOR soil psychromters and is in accordance with in situ measurements of the wetting process in large-scale field tests.

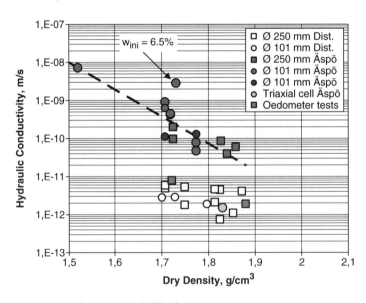

Figure 3. Hydraulic conductivity determined on 30/70 mixtures.

Tests in order to determine the *matric suction* were made by use of the equipment shown in Fig. 4 (Axis translation technique). The sample is put on a high air entry disk and placed in a triaxial cell with a water-filled compartment below the disk. The compartment can be flushed with water during the tests. It is of great importance that the compartment is completely filled with water for the entire duration of the test. A pore pressure transducer measures the water pressure (u_w) in the compartment below the high air entry disk. The measured pressure is assumed to correspond to the pore water pressure in the sample. By increasing the air pressure (u_{air}) in the triaxial cell it is possible to produce a pore water pressure above atmospheric pressure and the matric suction can be calculated with Eqn. 4. Since the samples used in the tests only had a diameter of about 50 mm the TBM-muck with its maximum grain size of 20 mm was replaced with sand.

Another technique for determine the matric suction is the "filter paper technique" where a filter paper is placed in direct contact with compacted samples of the material. The filter paper will take up water from the sample until it reaches equilibrium with the sample. After the test the water ratio of the sample and the filter paper are determined. With the known relation between the suction of the filter paper and its water ratio the matric suction of the sample can be determined.

All results from the measurements on backfill with a mixture of 30% bentonite 70% crushed rock and Äspö water added have been compiled and plotted in one diagram in Fig. 5. The figure shows that the RH and psychrometer methods, which measure the total suction, agree well and that the matric suction is about 1.5 MPa lower than the total suction. If the osmotic suction is evaluated according to Eqn 3 the

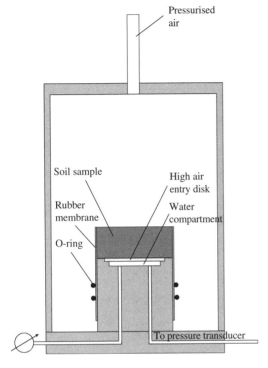

Figure 4.   Equipment used for measuring matric suction.

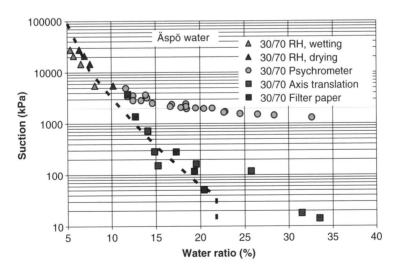

Figure 5.   Compiled suction measurements made on 30/70 mixtures with Äspö water. The square dots imply matric suction while all other dots are results from measurement of total suction. The hatched line is the relation used for modeling where w = 22% correspond to 100% degree of saturation.

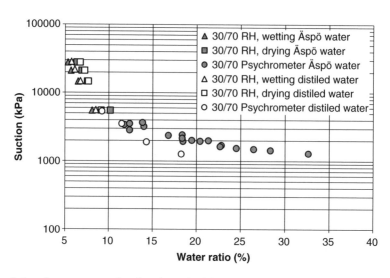

Figure 6.   Compilation of measurements of total suction on backfill material with 30% bentonite and 70% crushed rock and water with different chemical composition.

osmotic suction will be approximately 1.5 MPa and does not change very much with water ratio.

The results of total suction in Fig. 5 are used for evaluating the water ratio from the RH field measurements and the results of matric suction are used for the water flow modelling.

Parameters assumed to be significant for the measured total suction of the different mixtures are

– salt content in the pore water
– whether the material is absorbing or desorbing water (wetting or drying)
– clay content of the mixtures
– initial void ratio
– water ratio.

Typical results from measurements on 30/70 mixtures with both Äspö water and distilled water in the pore system are shown in Fig. 6. The figure indicates that backfill with 1.2% salt in the added water has a suction that is about 1 MPa higher than backfill with distilled water added. The water potential of a 1.2% NaCl solution is in the same rang as this difference in suction between the two backfill materials. The figure also indicates the difference in behavior in water uptake and drying.

Since the suction of the ballast material of crushed rock is very small compared to the suction of the bentonite it is reasonable to believe that the clay alone is responsible for the magnitude of the suction of a mixture. With the use of following equation the water ratio in the clay phase for any mixture of bentonite

and crushed TBM-muck can be calculated:

$$w_{cl} = \frac{[w - (1 - k) \times w_b]}{k} \qquad (5)$$

where

w_{cl} = clay water ratio
k = ratio between the dry weight of the solid mass of bentonite and the dry weight of the total solid mass of the mixture. For a 30/70 mixture k is = 0.3
w_b = water ratio of the ballast material. At lower suction than 6.000 kPa the water ratio in the ballast material is assumed to be 0.6%
w = water ratio of the mixture.

All results from measurements of total suction in different mixtures of bentonite and crushed TBM muck with Äspö water added expressed as function of the water ratio in the clay phase are compiled in Fig. 7. The plot confirms that the suction of any mixture can be expressed as function of the water ratio of the clay phase.

The measurements of suction have yielded data and information valuable for the modeling and understanding of the behavior of unsaturated backfill materials. The results yield some preliminary conclusions and observations:

– total suction of the different soil mixtures is a function of the clay water ratio irrespective of the bentonite content

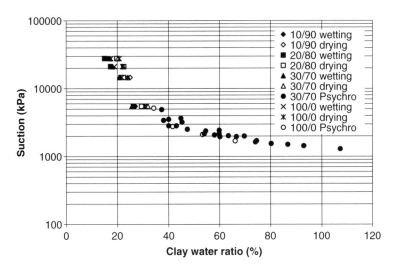

Figure 7. Compilation of results from measurements of total suction on backfill material with different bentonite content expressed as function of the clay water ratio.

Table 2. Summary of the swelling pressure tests performed with Äspö water.

Test no.	Clay cont. (%)	w_{ini} (%)	Final properties					
			Proctor (%)	w (%)	ρ_d (t/m^3)	e	Sr (%)	Swelling pressure (kPa)
1	30	6.3	89	21	1.73	0.59	97	220
2	30	13	88	21	1.71	0.61	96	244
3	30	13	78	27	1.52	0.81	93	68
4	20	8	79	23	1.62	0.69	91	21

– total suction are influenced by the salt content of the pore water
– matric suction is significantly lower than total suction.

4 SWELLING PRESSURE

The tunnel backfill serves several purposes. Except for preventing the buffer material from swelling the backfill also needs to support the roof and minimize the axial water flow in the tunnel. When placing and compacting the backfill in the tunnel it is difficult to obtain a good contact between the backfill and the rock surface and a high density of the backfill close to the roof. A swelling backfill material facilitates a good contact between the rock and the backfill. Previous tests on mixtures of bentonite and ballast material indicate that the swelling pressure of a mixture of bentonite and crushed rock is higher than the swelling pressure of pure bentonite at the same clay density (average density of the clay in the voids between the ballast particles).

The following test technique has been used for measuring the swelling pressure: The backfill material is compacted in the Proctor cylinder. A load cell is applied on the top of the piston. The sample is saturated with Äspö water by applying a low water pressure at the bottom of the sample and allowing air to seep out from the top during the water inflow. The swelling pressure is recorded by continuous readings of the load cell. After completion of the test, the density (ρ) and the water ratio (w) are determined. The measured values are then used to calculate the degree of saturation (S_r), the void ratio (e) and the dry density (ρ_d).

Three tests with a mixture of 30% bentonite and 70% crushed TBM muck and one test with 20/80 mixture have been made. The data and results from the tests are summarized in Table 2. The results show that:

– the swelling pressure is low and increases as expected with increasing density and bentonite content

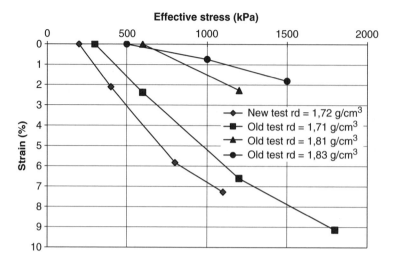

Figure 8. Strain as a function of applied vertical stress for the compression tests on 30/70 bentonite/crushed TBM muck.

- a density corresponding to 80% Proctor is not enough to reach 100 kPa in swelling pressure for any of the tested materials
- the initial water ratio at compaction, which had a significant influence on the hydraulic conductivity, does not seem to affect the swelling pressure very much
- 30/70 mixture with a dry density of about $1.6 \, t/m^3$ is required in order to reach a swelling pressure of 100 kPa.

5 COMPRESSION PROPERTIES

In order to be able to judge the possibility for the backfill material to withstand large upwards swelling of the buffer in the deposition holes the knowledge of the compression properties of the material is required.

The characteristics of a soil during one-dimensional consolidation or swelling can be determined by means of oedometer tests. The test sample is held in a rigid ring and filter stones are mounted on the top and bottom of the sample. The vertical effective stress is stepwise increased while horizontal deformation of the sample is prevented by the stiff ring. From the oedometer test also the *compression modulus* and the *coefficient of consolidation* can be determined by measuring the deformation of the sample with time. A momentary increase of the total load of the sample will result in an increase of the pore pressure in the sample and pore water will seep out from the sample through the filter stones on the top and bottom of the sample. The pore pressure will decrease with time and be accompanied by a compression of the simple.

A so called Rowe oedometer with the diameter $\sim$250 mm and height $\sim$80 mm was used for the tests (see Fig. 1). The samples were compacted in the oedometer ring in five layers. Most of the samples were compacted at their optimum water content according to the modified proctor test. Two samples were compacted at a lower water ratio than the optimum. The samples were after compaction saturated in the oedometer. Some of the tests were performed with a back pressure. After equilibrium the samples were loaded in steps and the vertical displacement was measured continuously.

The compression module was evaluated as a secant module according to Eqn. 6:

$$M = \frac{\Delta \sigma'_v}{\Delta \varepsilon_v} \qquad (6)$$

where

M = compression module
$\Delta \sigma'_v$ = vertical stress increment
$\Delta \varepsilon_v$ = vertical strain increment.

Tests were made on mixtures of crushed TBM muck and bentonite with bentonite content varying between 0 and 30%. Fig. 8 shows the measured strain as a function of applied pressure for the 30/70 mixture. The basic data and the evaluated compression modulus M are shown in Table 3 together with results from tests with other mixtures. The compression modulus has been evaluated for the entire test, that is from $\varepsilon_v = 0$ to the last load step. p_i is the initial vertical stress corresponding to the swelling pressure of the backfill. The densities of the samples are also expressed in percent of the maximum modified Proctor density.

Table 3. Basic data and evaluated modules for the compression tests.

Mtrl	Proctor (%)	Water	ρ_d (g/cm^3)	e	M (MPa)	p_i (kPa)
Crushed TBM	88	Distilled	2.026	0.343	26.5	100
Crushed TBM	92	Distilled	2.109	0.29	25.1	100
Crushed TBM	91	Distilled	2.095	0.329	41.4	100
Crushed TBM	98	Äspö*	2.247	0.219	91.5	90
Crushed TBM	90	Äspö	2.068	0.325	29.2	50
10/90	89	Distilled	1.907	0.423	24.9	100
10/90	93	Distilled	2.006	0.286	46.8	200
10/90	94	Äspö	2.016	0.347	68.5	175
20/80	85	Distilled	1.736	0.579	17.9	150
20/80	93	Distilled	1.909	0.435	40.4	400
20/80	93	Äspö	1.914	0.432	38.9	300
30/70	88	Distilled	1.707	0.619	16.4	300
30/70	93	Distilled	1.813	0.516	26.5	600
30/70	94	Äspö	1.826	0.492	55.3	500
30/70	88	Äspö	1.721	0.604	12.4	200

* Total salinity of the pore water 1.2%.

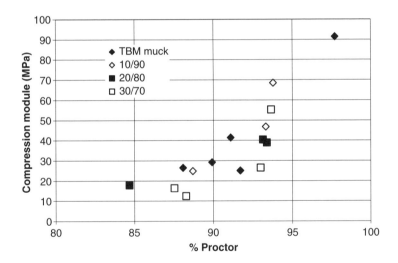

Figure 9. Compression modulus as function of the compaction energy (% modified proctor) for all tests.

The tests have shown that the compression properties are mainly a function of the compaction energy and not very much dependent of the density and the bentonite content (percent of maximum density for proctor tests, see Fig. 9). It should be noted that in field test, where the backfill materials are compacted with a dynamic compaction device, the average density expressed as percentage of maximum Proctor density is decreasing with increasing clay content. The salt content has an insignificant influence.

REFERENCES

Börgesson L, Johannesson L-E and Gunnarsson D, 2003. Influence of soil structure heterogeneities on the behavior of backfill material based on mixtures of bentonite and crushed rock. Applied Clay Science 23 (2003): 121–131.
Börgesson L, Johannesson L-E and Sandén T, 1996. Backfill material based on crushed rock. Geotechnical properties determined in the laboratory. SKB Progress Report HRL-96-15.
Börgesson L, Johannesson L-E, Sandén T and Hernelind J, 1995. Modelling of the physical behaviour of water

saturated clay barriers. Laboratory tests, material models and finite element application. SKB Technical Report 95–20.

Börgesson L, and Stenman U, 1985. Laboratory determined properties of sand/bentonite mixtures for WP-cave. SGAB IRAP 85511.

Clemente M M, 2003. Hydraulic behaviour of bentonite based mixtures in engineered barriers: The Backfill and Plug Test at Äspö HRL (Sweden). PhD Thesis. Technical University of Catalonia. Department of Geotechnical Engineering and Geosciences School of Civil Engineering.

Fredlund D G, and Rahardjo H, 1993. Soil mechanics for unsaturated soils. John Wiley & Sons, Inc.

Gunnarsson D, Johannesson L-E, Sandén T and Börgesson L, 1996. Field tests of tunnel backfilling. SKB ÄHRL Progress Report HRL 96–28.

Johannesson L-E, Börgesson L and Sandén T, 1999. Backfill material based on crushed rock (part 2). SKB ÄHRL IPR-99-23.

Pusch R, 1995. Consequences of using crushed crystalline rock as balast in KBS-3 tunnels instead of rounded quartz particles. SKB Technical Report TR 95–14.

Rowe P, and Barden L, 1966. A new consolidation cell. Geotechnique, 16, 162–70 Wan A., 1996. The use of thermocouple psychrometers to measure in situ suction and water content in compacted clay. Department of Civil and Geological Engineering. University of Manitoba, Winipeg, Mantiba.

Öberg A- L, 1997. Matrix Suction in Silt and Sand Slopes. Department of Geotechnical Engineering. Chalmers University of Technology.

Advances in Understanding Engineered Clay Barriers – Alonso & Ledesma (eds)
© 2005 Taylor & Francis Group, London, ISBN 04 1536 544 9

A constitutive model for compacted expansive clay

M. Sánchez, A. Gens & S. Olivella
Geotechnical Engineering Department, Technical University of Catalunya, Barcelona, Spain

L. do N. Guimarães
Federal University of Pernambuco, Recife, Brazil

ABSTRACT: Compacted expansive clays are usually used in the design of engineered barriers and seals for high level radioactive waste isolation. Expansive clays generally present a clear double pores size distribution. This paper presents an elasto-plastic model for expansive clays based on the general approach proposed by Gens & Alonso (1992). The model considers explicitly the two basic pores levels experimentally detected. Additionally, the model is developed for the general case of non-equilibrium between the water potentials of the two structural levels. The model is formulated in the space of stresses, temperature, macrostructural and microstructural suctions.

1 INTRODUCTION

The clay barrier is a basic component in the disposal concept for high level radioactive waste (Figure 1). The behavior of clay barriers is highly complex, since it involves coupled *THM* (Thermo-Hydro-Mechanical) phenomena that take place due to the simultaneous heating (generated by the waste radioactive decay) and hydration of the barrier (due to the contribution of the surrounding rock). Coupled *THM* analyses are required to achieve a good understanding of the clay barrier behavior. The mechanical constitutive law is a key element in these modeling and it can be viewed as the nucleus of the formulation.

In the last few years, a number of tests have been performed revealing the strong influence of the pore structure on the behavior of expansive clays (i.e. Lloret et al., 2003). Two dominant pore levels have generally been distinguished in the fabric of swelling clays. A general approach that takes explicitly into account the pore structure of expansive clays was proposed by Gens & Alonso (1992). In accordance with experimental observations the framework considers two basic structural levels within the material: the microstructure and the macrostructure. The main interactions between both structural levels are also contemplated in the model.

The double structure approach has been widely used as a conceptual framework to analyze swelling materials and also to perform numerical analyses (i.e. Alonso, 1998; Alonso et al., 1999, Lloret et al., 2003, Sánchez, 2005).

In Sánchez et al. (2005), a general mathematical framework for the double structure model is proposed in the context of elasto-plasticity for strain hardening materials. Concepts of classical and generalized plasticity theories are used to develop the model. A basic hypothesis of this model is the assumption of hydraulic equilibrium between the water potentials of both structural levels. However, in some cases, a more complete description of the problem can be made when this hypothesis is released. This paper presents the extension of the constitutive model, and the mathematical formulation, to handle problems in which non-equilibrium between water potentials exists. This is an important aspect in the transient *THM* modeling of expansive materials.

2 EXPANSIVE CLAY FABRIC

The fabric of expansive clays has been actively studied (Push, 1982; Romero 1999; Cui et al., 2002;

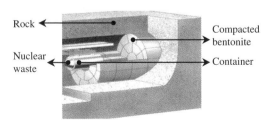

Figure 1. Scheme of an engineered barrier made up of compacted clay for a high level radioactive waste repository.

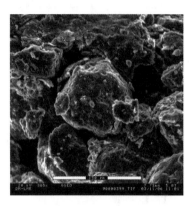

Figure 2. Micrograph of a compacted bentonite sample obtained using SEM (FEBEX II Report, 2004).

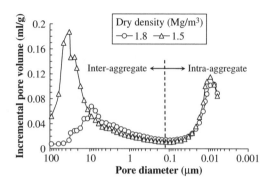

Figure 3. Distributions of incremental pore volume obtained using MIP technique (Lloret et al., 2003).

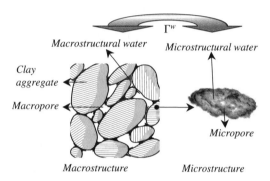

Figure 4. Schematic representation of the two structural levels considers.

Lloret et al., 2003). Expansive clays generally present a double structure, made up from clay aggregates and large macrostructural pores (i.e. Pusch, 1982). The pore space inside the aggregates was constituted by voids of a much smaller size. The same double structure has been identified in several compacted clays intended for engineered barriers in radioactive waste isolation. For example, some results obtained by Lloret et al. (2003) are presented below.

The FEBEX bentonite has been selected by ENRESA (Spanish national company for radioactive waste) for the backfilling and sealing of a high level waste repository. The bentonite has a montmorillonite content higher than 90 percent. It also contains variable quantities of quartz and other minerals. Figure 2 shows a micrograph of FEBEX bentonite, obtained using SEM (Scanning Electron Microscope), where the presence of aggregates, the inter-aggregate voids and the typical fabric of this material are readily apparent (FEBEX II Report, 2004).

Mercury intrusion porosimetry (MIP) tests were performed to examine the pore size distribution of the statically compacted material. Figure 3 shows the measured incremental pore volume for two samples compacted to very different values of dry density (ρ_d), $1.5\,\mathrm{Mg/m^3}$ and $1.8\,\mathrm{Mg/m^3}$. It can be observed that the pore size distribution is clearly bi-modal, very characteristic of this type of materials (Alonso et al., 1987). The dominant values are $10\,\mathrm{nm}$ that would correspond to the pores inside clay aggregates and a larger pore size that depends on the compaction dry density and ranges from $10\,\mu\mathrm{m}$ (for $\rho_d = 1.8\,\mathrm{Mg/m^3}$) and $40\,\mu\mathrm{m}$ (for $\rho_d = 1.5\,\mathrm{Mg/m^3}$). These larger voids would correspond to the inter-aggregate pores. The boundary between the two pore size families can be seen to be around $0.13\,\mu\mathrm{m}$, as pores smaller than this size do not appear to be affected by the magnitude of the compaction load.

These two dominant pores size could be associated with two basic structural levels (Figure 4):

- The macrostructure, associated with the global arrangements of clay aggregates (the skeleton of the material), with macropores between them.
- The microstructure, which corresponds to the active clay minerals and their vicinity.

Evidently, the microstructure organization of expansive clays is very complex and more pore levels could be distinguished (Push & Karland, 1996; Saiyouri et al., 2000; Hueckel et al., 2001). However, in this work for the sake of simplicity, the two basic structural levels identified above are considered. The approach is open enough and it could be extended to include more structural levels in the analysis, if it deemed relevant.

3 MASS WATER TRANSFER

The water retained at each pore level considered in the model has generally been identified as macrostructural water and microstructural water (Figure 4).

The hypothesis of instantaneous hydraulic equilibrium between the two pore levels can be assumed when the characteristic time of the local water transfer between media is so fast, compared with other processes (such as advection). However, the more general case corresponds when this hypothesis is removed, therefore, the potentials of the water stored at each structural level could be different.

Starting from an equilibrium state, if a perturbation induces non-equilibrium between the water potentials of the two pores levels, transfer of water mass between them occurs until the initial equilibrated state is recovered. There are different alternatives to model such kind of process (Huyakorn et al., 1983; Sánchez, 2004). A simple model for the mass transfer of water can be expressed as:

$$\Gamma^i = \gamma \left(\Psi_M - \Psi_m \right) \tag{1}$$

where Ψ_M and Ψ_m represent the thermodynamic forces involved in the mass transfer, γ is the leakage parameter and the subscripts M and m identifies the macro and micro levels respectively. When the water mass transfer is considered, the total water potential is the variable involved in Equation (1). The evaluation of the water potential can be made considering the chemical potential of the water. Expressions based on the chemical potential of an ideal gas are generally used to its evaluation. The chemical potential of an ideal gas is derived from the Gibbs free energy and can be expressed as (Appelo & Postma, 1993):

$$\mu_i = \mu_i^0 + RT \ln(a) \quad where \quad a = P_i / P^0 \tag{2}$$

With Equation (2) the chemical potential (μ_i) of the gas i can be evaluated at the current gas pressure (P_i) and under isothermal conditions (T). μ_{0i} is the reference chemical potential of the gas at the reference pressure P_0 (and Temperature), R is the gas constant (8.314 J/deg.mol) and a is the activity.

Expressions like (2) have been extended to other substances and conditions in which the activity is defined likewise from a concentration measured. Several approaches adopt the liquid chemical potential as the thermodynamic force to model the process of mass transfer with satisfactory results (e.g. Navarro & Alonso, 2001; Loret et al., 2002). However, it is important to have in mind that Equation (2) has been derived for a perfect gas; its extension to other liquids or solids is an assumption (Appelo & Postma, 1993). On the other hand, in compacted expansive clays the microstructural water is highly affected by the psychico-chemical phenomena occurring at clay particle level, especially by the presence of exchangeable cations. These conditions are far from the ideal ones and it seems not simple to find the expression for the chemical potential of the microstructural water. Some attempts to its evaluation have been made recently by Fernandez (2004) for the case of the FEBEX bentonite, but it seems not easy to manipulate numerically these concepts.

In Gens & Olivella (2001) a well-established alternative approach is presented to evaluate the water potential. The total water potential, Ψ, can be defined as the variation of the Gibbs free energy of water per unit change of mass when the rest of the thermodynamical variables remain constant. Therefore, the potential can be envisaged as the energy required (under isothermal conditions) to extract a unit mass of water and take it to a reference state. Total water potential controls the mass transfer of water. When two water volumes are in contact and there is no flow, it necessarily means that the water potentials are the same in the two zones (Gens & Olivella, 2001). The total water potential can be divided into four different components:

$$\Psi = \Psi_{gravitational} + \Psi_{gas\ pressure} + \Psi_{matric} + \Psi_{osmotic} \tag{3}$$

The gravitational potential is related to the difference in elevations. The gas pressure potential is linked to the applied gas pressure. The osmotic potential is associated to the differences in solute concentration. Finally, the matric potential is a measure of the attraction that a soil matrix has for liquid moisture. Only the matric and osmotic potentials affect the mechanical response of unsaturated soils.

The concept of suction is generally used instead of potential (Gens & Olivella, 2001). Suction is equal to potential but with the opposite sign. Therefore matric suction is $s = -\Psi_{matric}$ and osmotic suction is $s_o = -\Psi_{osmotic}$. The sum of these two suctions is the total suction (Equation 3). The matric suction is often equated to the capillary pressure ($s = p_g - p_l$). This definition has only real meaning in the framework of the capillary model of soil and is not directly applicable to many soils, as for example the expansive clays, in which active clay fractions are significant. In these materials, when conditions of high suctions prevail, it is not certain that the water is in a state of true tension. So, the high negative pore pressures reflect the degree of attraction of the water by the soil matrix and are often more related to the physico-chemical effects than to some virtual capillary meniscus existing in the soil (Gens & Olivella, 2001).

Generally, when dealing with soils in unsaturated conditions, two stress fields have been adopted: net stresses and matric suction. However, when non-equilibrium between water potentials is considered, it is necessary its extension to include the non-balanced microstructural suction. Therefore, there are at least two different suctions (one for each pore levels considered). In this work the macrostructural suction (s_M) is related to the matric suction, while a generic microstructural suction (s_m) is associated with the water potential at the micro-level. In Guimarães et al. (2001), the model has been extended to consider explicitly the osmotic suction.

4 DOUBLE STRUCTURE MODEL

The swelling behavior of unsaturated expansive clays has often been reproduced through relatively simple and empirical laws, which relate the material response to suction changes and applied stresses. However, in Gens & Alonso (1992), a general approach for expansive soils is proposed in which particular attention is placed on the clay structure and how it can be integrated in the constitutive modeling of expansive soils. The mechanical model presented herein is based on the general framework proposed by Gens & Alonso (1992) and considers some of the improvements proposed by Alonso et al. (1999). In this paper, the constitutive model introduced in Sánchez et al. (2005), has been extended for the case of non-equilibrium hydraulic between the water potential of both structural levels.

The model has been formulated using concepts of elasto-plasticity for strain hardening materials. A series of modifications and developments have been performed in order to enhance the constitutive law and also to formulate the model in a more suitable form for its implementation in a finite element code. One of the aims is to provide a more general mathematical framework in order to achieve a more general interpretation of the phenomena that take place in expansive clays when they are subjected to complex *THM* paths. With this objective, concepts of generalized plasticity theory have been included in the formulation of the model.

The mathematical framework of the model is presented in detail in Sánchez (2004). The model is formulated in terms of the three stress invariants (p, J, θ), macrostructural suction (s_M), microstructural suction (s_m) and temperature (T). Where p is the mean net stress, J is the square root of the second invariant of deviatoric stress tensor and θ is the lode angle.

The complete model formulation requires the definition of laws for:

- The macrostructural level,
- The microstructural level and
- The interaction between the structural levels.

These laws are briefly introduced in the following sections.

4.1 *Macrostructural model*

The inclusion of this structural level in the analysis allows the consideration of phenomena that affect the skeleton of the material, for instance deformations due to loading and collapse. These phenomena have a strong influence on the macroscopic response of expansive materials. The macrostructural behavior can be described by concepts and models of unsaturated non-expansive soils, such as the elasto-plastic Barcelona Basic Model (*BBM*) (Alonso et al., 1990). The *BBM* considers two independent stress variables to model the unsaturated behavior: the net stress (σ) computed as the excess of the total stresses over the gas pressure ($\sigma_t - \mathbf{I}p_g$), and the matric suction (s_M). The *BBM* extends the concept of critical state to the unsaturated conditions.

The inclusion of the thermal effects has been made according to Gens (1995). In this way it is considered that temperature increases reduce the size of the yield surface and the strength of the material. Therefore, the *BBM* yield surface (Figure 5) depends not only on the stress level and history variables (as in a critical state model) but also on the matric suction and temperature. The *BBM* yield surface (F_{LC}) is given by:

$$F_{LC} = 3J^2 - \left[\frac{g(\theta)}{g(-30°)}\right]^2 M^2 (p + p_s)(p_0 - p) = 0 \quad (4)$$

where M is the slope of the critical state, p_0 is the apparent unsaturated isotropic pre-consolidation pressure, $g(\theta)$ is a function of the lode angle and p_s considers the dependence of shear stress on suction and temperature. A basic point of the model is that the size of the yield surface increases with matric suction. The trace of the yield function on the isotropic $p - s_M$ plane is called *LC* (Loading-Collapse) yield curve, because it represents the locus of activation of irreversible deformations due to loading increments or collapse (when the suction reduces). The position of the *LC* curve is given by the pre-consolidation yield stress of the saturated state, p_0^* (hardening

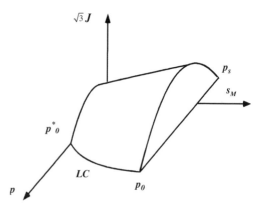

Figure 5. *Barcelona Basic Model*, schematic representation of the yield surface in the space: p, s_M and J.

variable), according to the following expression:

$$\dot{p}_0^* = p_0^* \frac{(1+e)}{(\lambda_{(0)} - \kappa)} \dot{\varepsilon}_v^p \qquad (5)$$

where e is the void index, $\dot{\varepsilon}_v^p$ is the volumetric plastic strain, κ is the elastic compression index for changes in p and $\lambda_{(0)}$ is the stiffness parameter for changes in p for virgin states of the soil in saturated conditions.

4.2 Microstructural model

The microstructure is the seat of the basic physical-chemical phenomena occurring at clay particle level. It is assumed that these phenomena are basically reversible (Gens & Alonso, 1992). So, the strains arising from microstructural phenomena are considered elastic and volumetric. The microstructural effective stress is defined as:

$$\hat{p} = p + \chi s_m \qquad (6)$$

It is assumed that χ is a constant ($\chi > 0$). In the $p - s_m$ plane the line corresponding to constant microstructural effective stresses is referred to as Neutral Line (NL), since no microstructural deformation occurs when the stress path moves on it (Figure 6). The increment of microstructural elastic strains is expressed as:

$$\dot{\varepsilon}_{vm}^e = \frac{\dot{\hat{p}}}{K_m} = \frac{\dot{p}}{K_m} + \chi \frac{\dot{s}_m}{K_m} \qquad (7)$$

where the subscript e refers to the elastic component of the volumetric (subscript v) strains and K_m is the microstructural bulk modulus.

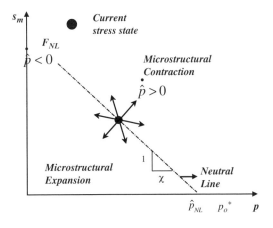

Figure 6. Directions of microstructural swelling and contraction directions in the isotropic plane $s_m - p$.

According to equation (7), the Neutral Line divides the $p - s_m$ plane into two parts (Figure 6), defining two main generalized stress paths. These two paths are identified as: *MC*, microstructural contraction path (increment of $\hat{p}$); and *MS*, microstructural swelling path (decrease of $\hat{p}$).

4.3 Interaction between macro and micro structure

In expansive soils there are other mechanisms in addition to the ones included in the *BBM* which induce plastic strains. This irreversible behavior is ascribed to the interaction between macro and micro structures.

Analyzing the behavior of expansive clays under cycles of suction reversals (e.g. Pousada, 1984), two main aspects can be highlighted: the irreversible behavior appears independently of the applied suction and it is difficult to determine the initiation of the yielding. These facts aim the use of the more general framework of generalized plasticity theory to formulate the model. In a generalized plasticity model the yield function is not defined or it is not defined in an explicit way. This is particularly advantageous in expansive materials, because no clear evidence exists concerning the shape of the internal yield surfaces corresponding to the interaction mechanisms between the two structural levels (Sánchez et al., 2005).

It is assumed that the microstructural behavior is not affected by the macrostructure but the opposite is not true, i.e. macrostructural behavior can be affected by microstructural deformations, generally in an irreversible way. An assumption of the model is that the irreversible deformations of the macrostructure are proportional to the microstructural strains according to interaction functions f. The plastic macrostructural strains are evaluated by the following expression:

$$\dot{\varepsilon}_{vM}^p = f \dot{\varepsilon}_{vm}^e \qquad (8)$$

Two interaction functions f are defined: f_c for microstructural contraction (*MC*) paths and f_S for microstructural swelling paths (*MS*). In the case of isotropic load, the interaction function depends on the ratio p/p_o (p_o is the net mean yield stress at current macrostructural suction and temperature). This ratio is a measure of the degree of openness of the macrostructure. The coupling between both plastic mechanisms is considered mathematically assuming that:

$$\dot{\varepsilon}_v^p = \dot{\varepsilon}_{vLC}^p + f \dot{\varepsilon}_{vm}^e \qquad (9)$$

where ε_{vLC}^p is the plastic strains induced by the yielding of the macrostructure (*BBM*). In fact the coupling is given by p_o^*, hardening variable of the macrostructure (Figure 5), which depends on the total plastic volumetric strain (5). In this way it is considered that microstructural effects can affect the global arrangements of aggregates (macrostructure).

Note that the material response will depend strongly on the direction of the microstructural stress path relative to the *NL*, which delimits two regions of different material behavior. A proper modeling of this behavior requires the definition of specific elasto-plastic laws for each domain. This would allow describing correctly the material behavior according to the microstructural stress path followed (*MC* or *MS*). Generalized plasticity theory can deal with such conditions, allowing the consideration of two directions of different behavior and the formulation of proper elasto-plastic laws for each region (Sánchez 2004, Sánchez et al., 2005).

In summary, the behavior of the macrostructure is modeled in the context of classical plasticity (*BBM*). This is a proper framework because the yield surface associated to this behavior could be generally inferred by the usual methodology of classic plasticity. The microstructural effects have been modeled using a nonlinear elastic model. The interaction between both structural levels has been model using the more general framework of generalized plasticity theory.

The following section presents the elasto-plastic tensors related to the increment of strains, suctions and temperature; which define the stress-strains relations used in the implementation of the model in the finite element program CODE_BRIGHT.

5 ELASTO-PLASTIC STRESS-STRAIN RELATIONS

The behavior of the soil described by the double structure model can be regarded as the consequence of joint action of several mechanisms that can act simultaneously. A procedure similar to the proposed in Sánchez et al. (2005) has been followed hereafter. That is, the first step is the assumption of an additive decomposition of the strains into elastic and plastic components, so, the total strain increment can be expressed as:

$$\dot{\boldsymbol{\varepsilon}} = \dot{\boldsymbol{\varepsilon}}^{e} + \sum_{n=1}^{i=na} \dot{\boldsymbol{\varepsilon}}_{n}^{p} \qquad (10)$$

where *na* is the number of active plastic mechanisms that correspond to one subset of the total plastic possible mechanisms. The model has three inelastic mechanisms: *lc*, due to yield of the *BBM*, and *mc* or *ms* when one of the two interaction mechanisms is active (two is the maximum number of simultaneous active plastic mechanisms, i.e. *lc* plus *mc* or *ms*). In classical plasticity theory, it is assumed that the material behaves either as an elastic or a plastic solid. The yield surface defines the transition from elasticity

to plasticity, stress states inside the yield surface are considered as elastic ($F < 0$). In generalized plasticity theory the state of the material is determined from the control variables: generalized stresses, strains and a finite number of internal variables. A process of loading is defined as elastic if the set of internal variables remains unchanged (Lubliner & Auricchio, 1996).

In the case of an elastic loading, the stress increment is related to the increment of strains, temperature, macrostructural and microstructural suctions by the following relation:

$$\dot{\boldsymbol{\sigma}} = \mathbf{D}_{e} \cdot \dot{\boldsymbol{\varepsilon}} + \boldsymbol{\alpha}_{s_{M}} \dot{s}_{M} + \boldsymbol{\alpha}_{s_{m}} \dot{s}_{m} + \boldsymbol{\alpha}_{T} \dot{T} \qquad (11)$$

where $\mathbf{D}_{e}$ is the global elastic matrix which considers the elastic component of the two structural levels. $\boldsymbol{\alpha}_{T}$, $\boldsymbol{\alpha}_{sM}$ and $\boldsymbol{\alpha}_{sm}$ are the elastic vectors associated to temperature, macro and micro suctions respectively (Sánchez, 2004).

When a loading process is inelastic, the plastic strain rates are assumed to be governed by a flow rule. For the macrostructural model, the strain increment can be expressed as:

$$\dot{\boldsymbol{\varepsilon}}_{LC}^{p} = \dot{\lambda}_{LC} \frac{\partial G}{\partial \boldsymbol{\sigma}} = \dot{\lambda}_{LC} \, \mathbf{m}_{LC} \qquad (12)$$

where $\dot{\lambda}_{LC}$ is the plastic multiplier associated to the *lc* plastic mechanism, G is the plastic potential (see Appendix) and $\mathbf{m}_{LC}$ is the flow rule direction. However, when the plastic mechanism related to the interaction between both structural levels is active, the plastic strain increment related to the stress increment can be expressed as (Sánchez et al., 2005):

$$\dot{\boldsymbol{\varepsilon}}_{\beta}^{p} = \dot{\lambda}_{\beta} \, \mathbf{m}_{\beta} \qquad (13)$$

where $\dot{\lambda}_{\beta}$ is the plastic multiplier associated to the β plastic mechanism, and $\mathbf{m}_{\beta}$ is the flow rule direction.

The material behavior is described by elasto-plastic mechanisms that can be activated during the loading process. The set of active plastic mechanisms is not known in advance. Therefore it is necessary to use an iterative procedure to find them (i.e. Simo & Hughes, 1988; Carol & Prat, 1999). A possibility is to assume that all the plastic mechanisms are initially active. Herein, it is assumed that both plastic mechanisms are initially active, that is: *lc* and β (i.e. *mc* or *ms*). Following a procedure similar to the one proposed in Sánchez et al (2005), the system of equation for two active mechanisms is given by:

$$\begin{cases} \dot{\lambda}_{LC} \overline{H}_{LC} + \dot{\lambda}_{\beta} \overline{b}_{\beta} = \dot{e}_{LC} + \dot{s}_{LC}^{M} + \dot{s}_{LC}^{m} + \dot{t}_{LC} \\ \dot{\lambda}_{LC} \overline{b}_{LC} + \dot{\lambda}_{\beta} \overline{H}_{\beta} = \dot{e}_{\beta} + \dot{s}_{\beta}^{M} + \dot{s}_{\beta}^{m} + \dot{t}_{\beta} \end{cases} \qquad (14)$$

where $\dot{\lambda}_{LC}$ and $\dot{\lambda}_\beta$ are the unknowns. $\overline{H}_{LC}, \overline{h}_\beta, \overline{h}_{LC}$ and are moduli related to the lc and β plastic mechanisms, while $s_{LC}^M, s_{LC}^m, t_{LC}, s_\beta^M, s_\beta^m$ and t_β are variables linked to the increment of strains, suctions and temperature. More details can be found in Sánchez (2004). System (14) can be expressed in a compact form, as:

$$\overline{H}.\dot{\lambda} = \dot{e} + \dot{s}^M + \dot{s}^m + \dot{t} \qquad (15)$$

The solution of the system (15) requires the inversion of the $\overline{H}$ matrix, which would be a P-matrix (Rizzi et al., 1996, Sánchez et al., 2005). In this case, the unknowns are obtained as follows:

$$\dot{\lambda} = \overline{H}^{-1}.\left(\dot{e} + \dot{s} + \dot{t} \right) \qquad (16)$$

The choice of the plastic mechanisms assumed initially active should be verified by checking that they are actually active (Carol & Prat, 1999). If one of them is not active, the problem becomes a single dissipative model. Finally, the net stress increment can be expressed as:

$$\dot{\sigma} = D_e \cdot \left(\dot{\varepsilon} - \dot{\varepsilon}_{SM}^e - \dot{\varepsilon}_{sm}^e - \dot{\varepsilon}_T^e - \sum_{n=1}^{nu} \dot{\varepsilon}_n^p \right) \qquad (17)$$

where $\dot{\varepsilon}_T^e$, $\dot{\varepsilon}_{SM}^e$ and $\dot{\varepsilon}_{Sm}^e$ are the elastic strain increments due to temperature and suctions changes. After some algebra (Sánchez, 2004) the following general form is obtained:

$$\dot{\sigma} = D_e \cdot \dot{\varepsilon} + \gamma_{s_M} \dot{s}_M + \gamma_{s_m} \dot{s}_m + \alpha_T \dot{T} \qquad (18)$$

where D_{ep} is the global elasto-plastic matrix, γ_T, γ_{sM} and γ_{sm} are the elasto-plastic vectors associated to temperature, macro and micro suctions, respectively. In Sánchez (2004) the expressions of these matrixes are presented in detail.

6 CLOSURE

A double structure model, based on the general framework for expansive materials proposed by Gens & Alonso (1990) has been presented. In order to be closer to the typical fabric of expansive materials, the existence of two pores structures has been explicitly included in the formulation. The distinction between the macrostructure and microstructure provides the opportunity to take into account the dominant phenomena that affect the behavior of each structure in a consistent way.

This constitutive model extends the one presented in Sánchez et al. (2005) for the case of non-equilibrium hydraulic between the two pore levels. When non-equilibrium between the water potentials

occurs, a process of mass transfer is induced in order to re-establish the equilibrium. The inclusion of the mass transfer of water allows a more general representation of the behavior of expansive soils. For example, a characteristic response of this phenomenon is the exhibition of more than one swelling stage under hydration (Alonso et al., 1991; Volckaert et al., 2000).

Figure 7 shows the results obtained in swelling pressure tests, carried out over samples made up from a mixture of clay pellets and clay powder (Volckaert et al., 2000). This kind of material presents a clear double structure. The clay used in the experiments is the FEBEX bentonite (Section 2). Despite of the observed differences (due to the different densities and mixtures composition of the samples), the tests exhibit a typical behavior with two main swellings stage in all cases. The initial swelling corresponds to the hydration of the clay powder and sponds to the hydration of the clay powder and also of the more external part of the pellets. The second delayed swelling is due to the hydration of the more inaccessible clay, inside the pellets.

The model presented in this work has been implemented in the finite element program CODE_BRIGHT (Olivella et al., 1996) and has been used to model typical problems involving expansive materials. Figure 7 also presents the qualitative response obtained with the model. It can be observed that the model result is satisfactory, because the main stages of test are qualitatively well reproduce. It seems that the relevant physical phenomena are well captured

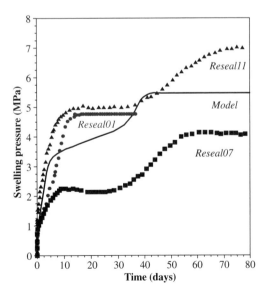

Figure 7. Time evolution of the swelling pressre. Test data from Volckaert et al. (2000).

by the double structure framework. A more detailed analysis of the mixtures of clay pellets and clay powder can be found in Sánchez (2004).

REFERENCES

Alonso, E. 1998. Modeling expansive soil behavior. Second International Conference on Unsatured Soils. Beijing, China. (1): 37–70. Beijing: International Academic Publisher

Alonso, E., Gens, A. & Josa, A. 1990. A constitutive model for partially saturated soils. *Géotechnique*, 40(3): 405–430.

Alonso, E., Gens, A. & Lloret, A. 1991. Double structure model for the prediction of long-term movements in expansive materials. In Beer, Booker, Carter, (eds), *Computer Methods and Advances in Geomechanics*: 1 541–548. Rotterdam: Balkema.

Alonso, E., Vaunat, J. & Gens, A. 1999. Modeling the mechanical behavior of expansive clay. *Eng. Geol* 54: 173–183.

Appelo, C. & Postma, D. 1993. *Geochemistry Groundwater and Pollution*. Balkema: Rotterdam.

Carol, I. & Prat, P. 1999. A multicrack model based on the theory of multisurface plasticity and two fractures energies. *Proceedings IV Conference on Computational Plasticity, Barcelon*a 1583–1594.

Cui, Y.J., Yahia-Aissa, M. & Dalage, P. 2002. A model for the volume change behavior of heavily compacted selling clays. *Engineering Geology*, 64: 233–250.

FEBEX II Report 2004. Final report on thermo-hydro-mechanical laboratory tests. Deliv. D17/3. UPC-L-7-13.

Gens, A. 1995. Constitutive Laws. In Gens, A., Jouanna, P. & Schrefler, B. *Modern issues in non-saturated soils*: 129–158. Wien New York: Springer-Verlag.

Gens, A. & Alonso, E.E. 1992. A framework for the behaviour of unsaturated expansive clays. *Can. Geotech. Jnl.*, 29: 1013–1032.

Gens, A. & Olivella, S. 2001. THM phenomena in saturated and unsaturated porous media. *Fundamentals and formulation. Revue.française de génie civil*, 5(6): 693–717.

Guimarães, L., Gens, A., Sánchez, M., Olivella, S. 2001. Chemo-mechanical modelling of expansive materials. *6th International Workshop on Key Issues in Waste Isolation Research; Proc. Symp.*: 463–465 November 2001. Paris.

Hueckel, T., Loret, B. & Gajo, A. 2001. Swelling materials as reactive, deformable, two-phase continua: basic modelling concepts and options. Clay Behaviour: Chemo-mechanical coupling, Workshop. Maratea, Italy.

Huyakorn, B., Lester & Faust. C, 1983. Finite element techniques for modelling groundwater flow in fractured aquifers. *Water Resources Research*, 19(4): 1019–1035.

Lloret, A., Villar, M.V., Sánchez, M., Gens, A., Pintado, X. & Alonso, E. 2003. Mechanical behaviour of heavily compacted bentonite under high suction changes. *Géotechnique*, 53(1): 27–40.

Lubliner, J. & Auricchio, F. 1996. Generalized plasticity and shape-memory alloys. *Int. J Solids Struct.*, 33(7): 991–1003.

Olivella, S., Carrera, J., Gens, A. & Alonso, E.E. 1994. Non-isothermal multiphase flow of brine and gas through saline media. *Transport in porous media*, 15: 271–293.

Olivella, S., Gens, A., Carrera, J. & Alonso, E.E. 1996. Numerical formulation for a simulator (CODE_BRIGHT) for the coupled analysis of saline media. *Engineering Computations*, 13(7): 87–112.

Pousada, E. 1984. *Deformabilidad de arcillas expansivas bajo succión controlada*. PhD Thesis, Technical University of Madrid, Spain.

Push, R. 1982. Mineral water-interaction and their influence on the physical behaviour of highly compacted Na bentonite. *Can. Geotech. Jnl.*, 19: 381–387.

Rizzi, E., Giulio, M. & Willam, K. 1996. On failure indicators in multi-dissipative materials. *Int. J. Solids Structures*, 33 (20–22): 3187–3124.

Romero, E. 1999. Characterization and thermal-hydro-mechanical behavior of unsaturated Boom clay: an experimental study. PhD Thesis, Technical University of Catalonia, Spain.

Sánchez, M. 2004. *Thermo-hydro-mechanical coupled analysis in low permeability media*. Ph. D. Thesis, Technical University of Catalonia. Barcelona.

Sánchez, M., Gens, A., Guimarães, L. & Olivella, S. 2005. A double structure generalized plasticity model for expansive materials. *Int. Jnl. Numer. Anal. Meth. In Geomech* (in print).

Simo, J. & Hughes, T. 1998. *Computational Inelasticity*. New York: Springer-Verlag.

Volckaert, G., Dereeper, B., Put, M., Ortiz, L., Gens, A., Vaunat, J., Villar, M.V., Martin, P.L., Imbert, C., Lassabatère, T., Mouche, E., Cany, F. 2000. A large-scale in situ demonstration test for repository sealing in an argillaceous host rock Reseal project – Phase I. EUR 19612 EN, Brussels: European Commission.

8 APPENDIX

The *BBM* plastic potential (*G*) is expressed as:

$$G = \alpha 3 J^2 - \left[\frac{g(\theta)}{g(-30°)} \right]^2 M^2 (p + p_s)(p_0 - p) = 0 \qquad (A1)$$

where α is determined according to Alonso et al. (1990). The dependence of the tensile strength on suction and temperature is given by:

$$p_s = ks \; e^{-\rho \Delta T} \qquad (A2)$$

where k and ρ are model parameters. The dependence of p_0 on suction is given by:

$$p_0 = p_c \left(\frac{p_{0T}^*}{p_c} \right)^{\frac{\lambda_{(0)} - \kappa}{\lambda_{(s)} - \kappa}}; \quad p_{0T}^* = p_0^* + 2 \left(\alpha_1 \Delta T + \alpha_3 \Delta T |\Delta T| \right) \qquad (A3)$$

where p_c is a reference stress, α_1 and α_3 are models parameters. $\lambda_{(s)}$ is the compressibility parameter for changes in net mean stress for virgin states of the soil; which depends on suction according to:

$$\lambda_{(s)} = \lambda_{(0)} \left[r + (1 - r) \exp(-\zeta s) \right] \qquad (A4)$$

where r is a parameter which defines the minimum soil compressibility. ζ is a parameter that controls the rate of decrease of soil compressibility with suction. The macrostructural bulk modulus (K_M) for changes in mean stress is evaluated as follows:

$$K_M = \frac{(1+e_M)}{\kappa} p \qquad (A5)$$

The microstructural bulk modulus (K_m) is evaluated as follows:

$$K_m = \frac{e^{-\alpha_m \hat{p}}}{\beta_m} \qquad (A6)$$

where α_m and β_m are model parameters. The shear modulus G_t is obtained from a linear elastic model:

$$G_t = \frac{3(1-2\mu)K}{2(1+\mu)} \qquad (A7)$$

where μ is the Poisson's coefficient. The macrostructural bulk modulus for changes in suction (s) is computed considering the following law:

$$K_s = \frac{(1+e_2)(s + p_{atm})}{\kappa_s} \qquad (A8)$$

where κ_s is the macrostructural elastic stiffness parameter for changes in s. The macrostructural bulk modulus for changes in T is computed as follows:

$$K_T = \frac{1}{(\alpha_0 + \alpha_2 \Delta T)} \qquad (A9)$$

where αo and α_2 are parameters related to the elastic thermal strain. More details are given in Sánchez (2004) and Sánchez et al. (2005).

Advances in Understanding Engineered Clay Barriers – Alonso & Ledesma (eds)
© 2005 Taylor & Francis Group, London, ISBN 04 1536 544 9

Short and long term stability of selected bentonites in high saline solutions

J. Kasbohm
University of Greifswald, Greifswald (Germany)

H.-J. Herbert
Gesellschaft fuer Anlagen- und Reaktorsicherheit (GRS), Braunschweig (Germany)

K.-H. Henning[1]

ABSTRACT: The mineralogical changes of bentonites were studied under the boundary conditions of a repository in salt formations by means of laboratory experiments.

In long term laboratory experiments the fraction $<2\,\mu m$ of the MX-80 bentonite was reacted with two high saline solutions, a NaCl and a Mg rich IP21 brine. The experiments were conducted in batch reactors using a solid-liquid ratio (1:1) without any stirring ("closed" system). The temperatures were fixed at 25°, 90° and 120°C and samples were taken after 2, 10, 50, 100, 122, 214, 307, 402 and 580 days. The mineral composition of the initial material and the run products was studied by XRD and TEM.

In all experiments montmorillonite remained at all temperatures and all pH the predominant mineral over 580 days, with full ethylene-glycol expandability to 17 Å. However significant changes could be detected by such parameter as morphology, crystallinity, particle height, particle surface, interlayer charge, and chemistry of octahedral layers. MX80 bentonite showed two transformation processes already after the first 10 days: an increasing illitisation and an alteration to Al-rich end-member of montmorillonite. The illitisation process transformed temporary about 20% of montmorillonite to randomly interstratified illite-montmorillonite (S_{max}: 20–30%). However, the main process was the substitution of octahedral Mg by Al in montmorillonite. The composition of the montmorillonitic tetrahedral layer remained unchanged (very close to $Al_0\,Si_4\,O_{10}$). This alteration reduces the charge in the interlayers. These processes caused also considerable changes of the swelling capacity. Smectites in salt solutions lost about half of their water uptake capability compared with pure water. The reduction is dependent on the reaction time and temperature.

Dissolution and precipitation according to an "Interlayer-by-Interlayer"-transformation (Altaner & Ylagan 1997) are believed to be the main reaction for the two observed transformation processes.

Short-term experiments (20 days) with other bentonites (1 n NaCl-solution, at ambient operating temperature and arranged in a shaker equipment) show (="dynamic system") alterations in form of illitisation, vermiculitisation or beidellitisation. In all cases the Si:Al ratio in the tetrahedral layers was decreased, too.

Using the stability diagram of clays published by Lippmann (1979) it is possible to explain these differences. Illites are more likely to occur in open "dynamic" geological systems with lower Si and Al concentrations. In the closed repository systems the Si and Al concentrations can get higher because of its more limited mobility and therefore kaolinites and pyrophyllites are expected. The concentration of Si, Al and K in the aqueous solutions determine the reaction's direction of montmorillonite, whereas salinity of the solutions influences only the intensity of alteration. High salinities tend to accelerate the appearance of new phases. The composition of such neoformed phases is also dependent from the ion constitution of applied brines.

1 INTRODUCTION

In the German disposal concept compacted bentonites are candidate materials for sealing elements of technical barriers in repositories for radioactive and chemical wastes in salt formations.

The high content of expandable smectites in bentonites is responsible for the swelling and sealing capacity of these materials. The mostly described process which may reduce the swelling and sealing capacity of smectites in the long run is illitisation. The process of illitisation is accompanied by changes of the layer charge and changes in the lattice due to decreasing contents of Si, Mg, and Fe^{3+} and increasing Al, Fe^{2+} and K contents (Cuadros & Linares 1996, Amouric et al. 1991). From over 140 reviewed papers dealing with experimental studies we have found only few in which the influence of high saline solutions was addressed. The most published experimental studies were conducted at temperatures $>120°C$ or even $>250°C$, which is far above the temperatures expected in a high level waste repository.[1]

Despite the abundance of literature and experimental data on the transformation processes of smectites the reviewed literature revealed that no experiments have been performed under the very specific boundary conditions of a repository in salt formations.

As natural analogues for the transformation of smectites under saline condition may serve bentonites in sedimentary basins with saline porewaters in order to demonstrate that illitisation (direction of montmorillonite transformation) and acceleration of illitisation are diagenetic main processes in salt-bearing bentonites.

Such an analogue was found in the East Slovak Basin which is part of the Transcarpathian Basin between the Western and Eastern Carpathians, NE of Budapest. In this basin bentonites of volcanoclastic origin were studied in several drill holes up to 4000 m depth (Šucha et al., 2000). An interesting feature of the basin is the presence of evaporate complexes. Sodium and chloride are the most abundant ions. The expandability of the investigated clays in this basin decreases with depth and thus with increasing temperature. But this feature is even more dependent on the presence or absence of NaCl in the pore water. The existence of salty bentonites with the conventional expandability from the upper part of the basin implies, that the effect of salty environment probably plays significant role after achieving certain (so far unknown) temperature. By K/Ar age dating these authors published a duration of illitisation of 1.7–2 Mio. years in salty bentonites (for more details see also in Herbert et al., 2003). Either Šrodoń (1984) found the accelerated illitisation in saline environment of the shallow parts of the Miocene Carpathian foredeep.

The objective of this study was the investigation of the long term stability of bentonites in contact with high saline brines which may occur in German salt mines. These conditions were considered: highly saline brines with a very specific chemical composition, wide range of possible pH, high solid/liquid

ratio, temperature, and a "closed" reaction system. On this background the tradeware MX-80 (Wyoming bentonite) was selected for the long-term experiments. Additional short term experiments were caried out in order to compare mineralogical alteration processes between a "closed" and a "dynamic" reaction system.[2]

2 EXPERIMENTS

2.1 Long term experiments

The reactions of the $<2\,\mu m$ fraction of the Wyoming bentonite MX-80 with high saline solutions were investigated over about 580 days in batch experiments. Two brines were used which are typical for the German Zechstein salt formations. The pure brines had an initial pH of 6.5. Beside the experiments with the pure brines two more sets of experiments were conducted using the same brines but with additives which resulted in a low and high initial pH. pH 1 was obtained by adding $FeCl_3$. By adding cement the initial pH was set to 13. All experiments were conducted at $25°C$, $90°C$, and $150°C$. In order to simulate the reactions in compacted bentonites as close as possible the smallest practicable volume of fluid was added to the clay material. By adding 10 ml solution to 10 g of clay it was still possible to recover enough fluid for chemical analyses. All experiments were conducted with this high solid/liquid ratio. The reaction vessels were of teflon for the $90°$ and $150°C$ experiments and of polypropylene for the $25°C$ experiments. No stirring or shaking was done during the reaction. Samples of the resulting solutions and the run products were taken and analysed after 2, 10, 50, 53, 100, 122, 214, 307, 402, and 580 days. Each experiment was carried out seperately in different vessels for each solution and experiment duration.

The employed solutions were a NaCl saturated solution with a density of $1.2\,g/cm^3$ and a Mg and K rich IP21 solution with a density of $1.3\,g/cm^3$. The NaCl solution was saturated at $25°C$ with respect to the two main minerals in the rock salt, halite (>95 wt-%), anhydrite (<5 wt-%).

The mineralogical composition of the bentonite MX-80 used in the experiments (Table 1) was analysed after grain size separation by means of XRD, DTA/TG and TEM (see also Kasbohm et al. 1999). The composition of mineral matter in the presently traded bentonite MX80 is different from

---

[1]A complete discussion of reviewed literature is presented in Herbert et al., 2003.

[2]In this paper only results concerning interaction of MX-80 betonite with NaCl-solution in a "closed" systems are presented here in order to focus to the observed main alteration processes. The full description for the "closed" reaction system is available in Herbert et al., 2003.

	Total bentonite	Fraction $<2\,\mu m$
Minerals wt-%	100	84
Quarz	4	2
Albite	2	n.d.
Calcite	<2	<2
Pyrite	<1	n.d.
Cristobalite	<2	<2
Montmorillonite	90	96

n.d. – not detected.

the mineralogy of MX80 published by Mueller-Vonmoos & Kahr (1982, 1983). These differences could have been caused by changed exploitation conditions in Wyoming (e.g. considering the identification of cristobalite).

The main composition of MX-80 montmorillonite was determined by TEM-EDX (Kasbohm et al., 1998):

$$(Ca_{0,07}Na_{0,22}K_{0,04})(Al_{1,54}Fe_{0,17}Mg_{0,26})(Al_{0,05}Si_{3,95}O_{10}|(OH)_2).$$

The reaction products were analysed by X-ray diffraction and transmission electron microscopy (morphology, element distribution, electron diffraction, HR-TEM).

2.2   Short term experiments

We have carried out also short-term experiments (20 days) with other bentonites to compare their behaviour with MX-80 run products (1 n NaCl-solution, at ambient operating temperature, solid/liquid ratio = 1:5). Simulating a "dynamic" reaction system these experiments were arranged in a overhead shaker equipment with two different velocities (lower energy system: 25 rpm; higher energy system: 60 rpm).

Following bentonites are used in these short term experiments: Pioche (Nevada), Chambers (Arizona), Belle Fourche (South Dakota), Otay (California), Garfield (Washington), Hectorite (California).

3   METHODS

3.1   Sample preparation after the experiments

The salt was removed from the run products after the experiments by dialysis (up to a constant value of the conductivity). For the storage and the further investigations the samples have been freeze-dried before.

3.2   X-ray diffraction (XRD)

For randomly oriented powder samples a SIEMENS D5000-equipment with Cu-K$_{alpha}$ wavelength and secondary curved graphite-monochromator (40 kV, 30 mA; variable slits: 20 mm, soller: 0.5/25, step & time: 0.02° 2Θ for 3 s) was used. Oriented mounts were analysed in a Präzitronic Freiberg HZG 4-diffractometer controlled by SEIFERT-C3000 unit (30 kV, 30 mA; Co-radiation, Fefilter, fixed slits: 1.09 mm/6.0 mm, soller: 0.5/25, detector slit: 0.35 mm; step & time: 0.03 0.02° 2Θ for 2 s).

In order to recognize any asymmetries of individual interferences the WinFit-software (Krumm, 1994) was used for deconvolution procedures. The Fourier-transformation implemented in this software was used for the determination of "coherent scatter domains" (CSD). The trend of CSD-values was verified by the MudMaster-software (Eberl et al., 1996).

3.3   Transmission electron microscopy (TEM)

The clay suspension was prepared after 10 min ultrasonic treatment on carbon-coated Cu-grids by air drying.

A JEOL-microscope with a JEM-1210 the individual particles were charactrized by means of morphology, electron diffraction and element distribution (voltage: 120 kV, LaB6-cathode). An OXFORD-LINK EDX-system was connected with the TEM-equipment. Morphology of particles was described according to Henning & Störr, 1986. The electron diffraction allows an evaluation of the stack order. Ring-like structures of the diffracted beam indicate here a turbostratical order of the layers. Zöller (1993) has demonstrated the possibility to distinguish between a 2M- or 1M-order for 2:1 sheet silicates, based upon the comparison of the intensities of (020)- and (110)-interferences in a convergent beam-system.

2M-polytype: $|110|/|020| > 1$

1M-polytype: $|110|/|020| < 1$

In order to evaluate the mineral formula of the individual particles the photons of each analysed particle were detected for 20 sec. The excited XRF-area of each particle was less than 100 nm in a diameter. The EDX-measurements were carried out standardless and by using of the Cliff-Lorrimer-factors.

The procedure of Köster (1997) for the conversion of the chemical analysis to a mineral formula was adopted also to EDX-analysis (a total charge of 22 for 2:1 sheet silicates was assumed).

A sophisticated system described by Środoń et al. (1992) was used to distinguish between montmorillonite and illite-smectite mixed layer structures (IS-ML) by the mineral formula. Środoń et al. (1992) described a strict correlation of "fixed" cations with the expandability of IS-ML. They found a good correlation also with the amount of Al in the tetrahedral layer. These authors are of the opinion that in IS-ML montmorillonite and illite are characterized by a typical total charge of 0.4 for montmorillonite and 0.89 for illite. This situation allows a verification of the diversity of expandability in illite-smectite mixed layer phases.

The calculation of mineral formula from TEM/EDX-analysis is described in details in Kasbohm et al. (2002).

3.4 BET-measurement

The BET-surface was determined using N_2-adsorption-technique. After heating of samples up to 300°C this method offers only values for the external surface and allowed on this way an impression of the growth situation of the run products. Caused by heating interlayer cations compositions do not affect the results.

For the verification of this BET-technique the certified clay standard Plessa (Lusatia) was involved (measured: $12.8 \, m^2/g \pm 4\%$; certified: $12.4 \, m^2/g \pm 3.5\%$). The untreated MX-80 bentonite ($<2 \, \mu m$) has a BET-surface of $41.4 \, m^2/g$.

3.5 Water uptake

The water uptake of the run products was measured with an Enslin-Neff-like technique. The water adsorbed (in ml) by capillary forces by 1 g samples powder was measured.

4 RESULTS OF SELECTED EXPERIMENTS

4.1 Long term experiments ("closed" reaction system)

4.1.1 Montmorillonite

A strong increasing of Si and Al in the solutions at the beginning of the experiments engaged, that first processes of dissolving of montmorillonite start immediately after the first contact of clays with the solutions. However, The X-ray traces of the randomly oriented samples of run products demonstrate that montmorillonite the main mineral phase over the entire experimental time frame. In air dried mounts the position of the (001)-montmorillonite interference shifts oscillating from 1.5 nm at the beginning of the experiments via 1.2–1.3 nm to 1.5 nm again mostly after 580 days (in room humidity). The peak intensities of (001)- and (003)-montmorillonite interferences as well as the ratio between both peak intensities differ largely during the experiments. It is not possible to recognize any trend of alteration from these intensities. By ethylene-glycole saturation the X-ray diffractograms of oriented mounts visualize the full expandability to 1.7 nm of the run products over

the total time of the experiments. The positions of (002)- and (003)-peaks at 0.856 nm (12.01° 2Θ) and 0.566 nm (18.19° 2Θ) confirm the identification of montmorillonite as main phase in the run products (Fig. 1).

Generally the "coherent scattering domains" (CSD) follow a lognormal distribution with a maximum for 5–6 layers per stack. The run products have only small differences for the CSD-distribution in the areas of the domains-maximum. The particles from experimental series with cement-additives change to an asymmetric distribution of the CSD-values with a maximum at 2–3 nm. That means, only 2–3 layers per stack would represent in the average such a "coherent scatter domain" (Fig. 2). But in this case it is not allowed to equate the CSD-value with the particle thickness, because HR-TEM-images from these samples show not a dominance of mono- and bilayer-stacks.

By XRD-techniques no obvious differences between the starting montmorillonite and the montmorillonite of the run products could be observed. Transmission electron microscopy however reveals that the former film-like matrix of the initial montmorillonite is changed in the run products to discrete, xenomorphous particles without any morphological swelling features as folds and curled edges (Fig. 3). In addition the stacks of the untreated montmorillonite which arranged in a turbostratical order are rearranged in the run products to an 1M-polytype (Fig. 4). Only thick and flaky aggregates have retained the original turbostratic order in the stacks. Furthermore the TEM-EDX-analyses of montmorillonite particles show a higher content of Al as well as a lower amount of Mg in the octahedral layer in comparison to the untreated montmorillonite (Fig. 5; Table 2). The exchange of Mg by Al can increase continuously with the time of shows increased values, but without an unique trend (e.g. NACl-solution system at 25°C)

4.1.2 Illite-smectite mixed layer structures

In few samples the (002)- and (003)-interferences of ethylene-glycole saturated montmorillonite are characterized by an asymmetric peak shape. Peak fitting procedures interpretate these asymmetries as (001)/(002)- or (002)/(003)-interferences of illite-smectite mixed layer structures (IS-ml) or as partly collapsed layers of montmorillonite (Fig. 1). Due to the difference of the positions of both peak-pairs the possible IS-mixed layers have about 30–40% smectitic layers in the run products of non-cement-series. By means of the element distribution, measured by TEM-EDX, is it possible to distinguish between collapsed layers and true IS-mixed layers. The result of this differentiation in samples of the NaCl-series at 25°C is visualized in Fig. 6, where IS-mixed layers

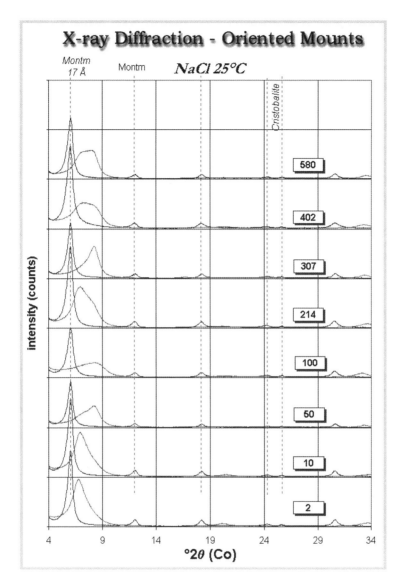

Figure 1. XRD-traces of run products from the NaCl-solution series at 25°C (oriented mounts) after a contact time from 2 d, 10 d, 50 d, 122 d, 214 d, 307 d, 402 d; and 580 d (bright line: air dried, room humidity; dark line: ethylene-glycole saturation) [from Herbert et al., 2003].

appear and disappear several times during the experiments (1M-polytype).

4.1.3 Technical parameter (water uptake, BET-surface)

In order to verify the influence of the detected mineralogical alterations on the technical properties of the MX-80 bentonite the ability for water uptake of the run products (Fig. 7) was analysed. Generally a decreasing water uptake ability is to observe with increasing temperature and reaction time for these run products.

For the untreated MX-80 $<2\,\mu m$ it was analysed a N_2-BET-value with $41.4\,m^2/g$. Otherwise the BET-values of the run products cover a large range of 20–$80\,m^2/g$. The lowest changes were found in the low temperature series (NaCl-series at 25°C: 41–$63\,m^2/g$). It is not possible to describe an unique trend of these changes.

5 DISCUSSION

5.1 What was happened with MX-80 bentonite?

In the run products montmorillonite was the main mineral also after 580 days time of contact between bentonite and high concentrated saline solution. But MX-80 bentonite in NaCl-solutions already showed two transformation processes after the first 10 days (!): (i) an alteration to Al-rich end-member of montmorillonite and (ii) temporary also an illitization process.

The solutions are characterized by increased Si and Al contents.

In all experiments we observed a high variability of the parameters, for example, for particle thickness and BET-surface. The changes however were not in one direction only. The measured values oscillated. Rassineux et al. (2001) and Bauer et al. (2001) reported increased standard deviation for different

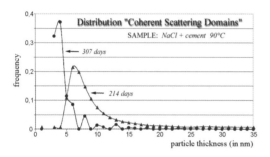

Figure 2. "Coherent Scattering Domains" (CSD) distribution by XRD (oriented mounts, ethylene-glycole solvated; data of the (001)-interference of montmorillonite by means of MudMaster [Eberl et al., 1996]). (triangle – lognormal distribution (example: NaCl+cement 90°C, 214 days); circle – asymptotic distribution (example: NaCl+cement 90°C, 307 days)) [from Herbert et al., 2003].

parameters (CEC, element distribution in the solution etc.) with values higher than 10%.

The above mentioned N_2-BET-measurements of the run products show generally also extremely increased mean deviation (fig. 8: e.g. for NaCl + cement-series at 25°C: $22–58 \, m^2/g$). The untreated MX-80 bentonite ($<2 \, \mu m$) has a BET-surface of $41.4 \, m^2/g$. We conclude, that we have actual variations in the BET-surfaces of the run products and not such a high standard deviation. Especially the CEC- and total surface-data sheets, published in Rassineux et al. (2001), indicate a similar oscillating behaviour of these two parameters. Such variations may be explained with dissolution processes at the mineral surface, which lead to the so-called "activation".

5.1.1 Alteration to Al-rich end-member of montmorillonite (= kaolinization or pyrophyllitization)

The main process was a substitution of octahedral Mg by Al. The composition of tetrahedral layers remained unchanged (very close to $Al_0Si_4O_{10}$). This substitution reduces the charge in the interlayers (from >0.3 downward to 0.2 and smaller). This process runs finally to an Al-rich stadium of the montmorillonite group without any charges in interlayers (=pyrophyllitization). The remarkable reduced charge of montmorillonitic interlayers is confirmed by alkylammonium treatments. These processes influence remarkably technical parameters of montmorillonite and reduce, for example, the water-uptake capacity of the run products up to 50% of the untreated MX-80 bentonite.

5.1.2 Illitisation process

Illitisation process is characterized by a temporarily (!) transformation of about 20% of the original montmorillonite to randomly interstratified illite-montmorillonite mixed layers (%Smax: 70 ... 80%).

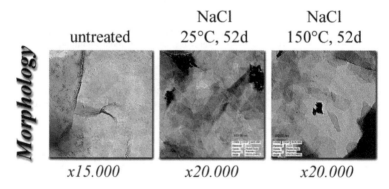

Figure 3. Experimental series with NaCl-solution after 52 days: transmission electron microscopy (TEM) – transformation of the former film-like matrix (left) in xenomorphous, often elongated plates (center and right).

Such IS-mixed layers are not stabil under selected experimental conditions.

5.2 Comparative short-term investigations at compacted MX-80 and other bentonites in NaCl-solutions

In Pusch & Kasbohm (2002) results are presented to alteration of compacted MX-80 bentonite after NaCl-treatment. During the only 30 days of continuous experiments identical mineralogical alterations (Al-enrichment and reduced charges in the interlayers) were observed also in such compacted material. Saline solutions increase the rate of alteration for the mineralogical and also technical parameters (hydraulic conductivity, swelling pressure) compared to attempts with distilled water. However, simultaneous annealing on 110°C leads to no further acceleration.

Short-term experiments (20 days) with other bentonites (1 n NaCl-solution, at ambient operating

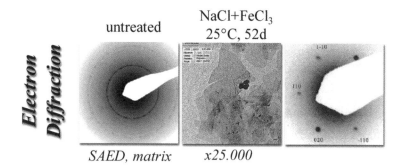

Figure 4. Experimental series with NaCl-solution after 52 days: transmission electron microscopy (TEM) – transformation of the former turbostratical matrix structure (left) to 1M-polytype (right).

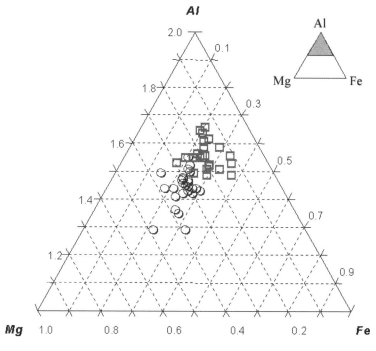

Figure 5. Experimental series with NaCl-solution after 52 days – TEM-EDX-analyses: Ternary presentation of the octahedral occupation by cations for montmorillonite particles of the starting MX-80 material (circle) and the run products (square) [from Herbert et al., 2003].

Table 2. Averaged mineral formulas of all analysed montmorillonite particles from run products of NaCl-series at 25°C (analysed by TEM-EDX, calculated to $(OH)_2$ O10)

time (in days)	Interlayer				Octahedral layer			Tetrahedral layer		
	Ca	Mg	Na	K	Al	Fe^{3+}	Mg	Si	Al	XII
MX80 start material ($<2\,\mu m$)										
	0.07		0.22	0.04	1.54	0.17	0.26	3.95	0.05	0.40
NaCl 25°C										
10 d	0.04	0.01	0.04	0.05	1.60	0.20	0.17	3.99	0.01	0.19
50 d	0.03	0.02	0.03	0.04	1.61	0.21	0.15	3.98	0.02	0.17
122 d	0.03	0.04	0.06	0.05	1.62	0.19	0.19	3.99	0.01	0.25
214 d	0.03	0.04	0.00	0.02	1.69	0.18	0.11	3.98	0.02	0.16
307 d	0.01	0.03	0.02	0.02	1.64	0.22	0.13	3.99	0.01	0.12
402 d	0.02	<0.01	0.02	0.05	1.67	0.20	0.13	3.99	0.01	0.13
580 d	0.03	<0.01	0.16	0.07	1.52	0.19	0.21	3.98	0.02	0.29

XII – charge of the interlayer space.

Diversity of IS

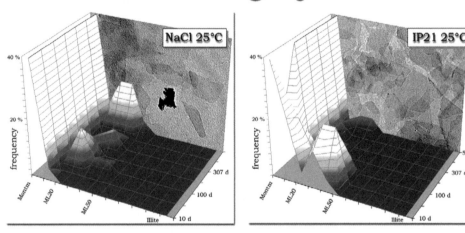

Figure 6. Visualization of the appearing and the disappearing of IS-mixed layers and its diversity depending from the contact time [by TEM-EDX-analyses]. (left: NaCl-series at 25°C between 10–580 days; right: IP21-series at 25°C between 10–580 days) [from Herbert et al., 2003].

temperature and arranged in a shaker equipment) show (= "dynamic system"):

– no detectable alterations in bentonite Belle Fourche
– increasing amount of IS-mixed layers in Pioche bentonite
– Fe^{3+}-substitution by Mg in Otay bentonite with an increasing charge for interlayers
– complete transformation of former montmorillonite of Chambers bentonite to IS-mixed layers (with 40% of illitic layers)
– nontronite of Garfield with Al-substitution by Mg accompanied by an increasing charge for interlayers

– caused by percolation an increasing proportion of illitic layers in the former IS-mixed layer of Friedland (Germany)

The result of these additional short-time experiments is: the alteration of montmorillonite caused by contact with saline solution is not limited only to the MX-80 Wyoming-bentonite. But it is possible to describe different rates of the hydrolytic reactions of montmorillonite.

5.3 Thermodynamical aspects

The behaviour of montmorillonite in contact with saline brines is to distinguish in an "dynamic" system

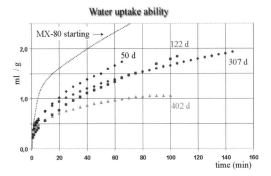

Figure 7. Water uptake ability of run products (NaCl-series at 25°C) by a Enslin-Neff-like technique (powder preparation).

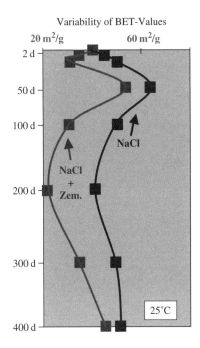

Figure 8. NaCl-solution series: Oscillating behaviour of BET-values.

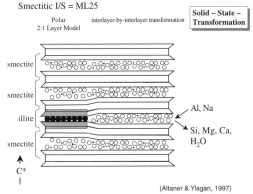

(Altaner & Ylagan, 1997)

Figure 9. "Interlayer-by-Interlayer"-transformation in according to Altaner & Ylagan (1997) as explanation for the alterations in montmorillonite and the only temporary stability of illitesmectite mixed layer phases in "closed" rection systems.

"Dynamic" System

unchanged (in 20 days)	*Belle Fourche bentonite*
beidellitization	*MX-80 bentonite*
vermiculitization	*nontronite of Garfield*
illitization	*Chambers bentonite to IS completely; Pioche bentonite partially; Otay bentonite in first steps*

By this generalization it is obviously, that concentration/mobility of free available Si in the interlayer space will play an important role for the different directions of alteration processes in "closed" compared to the "dynamic" systems.

5.4 Mechanism

The rapid transformation of the particle morphology, the fast rearrangement of the layers and the complete alteration of the octahedral layer of all fine montmorillonitic particles however seems to be in contradiction with the low differences in crystallinity and particle thickness distribution of montmorillonite in the run products. Fast mineralogical alterations due to dissolution/crystallisation processes should result in greater differences. The repeated appearing and disappearing of IS-mixed can also not be explained easily by dissolution/precipitation alone. The combined processes of dissolution/crystallisation and the "interlayer-by-interlayer"-transformation model of Altaner & Ylagan (1997) possibly offer a better explaination for these open questions. Furthermore the appearing and disappearing of IS-mixed layers can be discussed again under this viewpoint.

and in a "closed" system concerning the physico-chemical reactions. In a "closed" system a process is to recognize, which will run in direction of a pyro-phyllitization. Otherwise in "open" systems such transformations are to expect as beidellitization, kaolinitization, vermiculitization or an illitization.

Under these viewpoints the mentioned experiments could be classified as followed:

"Closed" System

pyrophyllitization	*MX-80 bentonite*

The source for the starting Al for its increased incorporation in the octahedral layer is to find in the dissolution of montmorillonitic monolayers.

6 CONCLUSIONS

During our experiments with MX-80 bentonite and high saline solutions no new stable mineralogical composition was reached or came in sight. The TEM-results (morphology, electron diffraction, mineral formula by TEM-EDX) indicate a nearby complete alteration of all montmorillonitic particles in our experiments already after 2 days. During the experiments montmorillonitic particles undergoing continuously again such described processes of alteration.

The experiments indicate a significant dissolution of the original montmorillonite and the precipitation of an altered montmorillonite already during the first few days. The observed mineralogical alterations follow thermodynamical pathways: kaolinitisation/pyrophyllitisation in "closed" systems with high Si concentrations. We believe that under repository conditions the transformation of montmorillonite will result rather in a kaolinitisation/illitisation process than in illitisation. In open "dynamic" geological systems with lower Si concentrations however illitisation and beidellitisation is believed to be the main process.

ACKNOWLEDGEMENTS

This work financed by the German ministry for research and technology under contract Nr. 02 C 06590. The authors highly appreciate the contributions and fruitfull discussions of our colleagues Helge Moog (GRS), Herbert Kull (GRS), Carsten Venz (Greifswald University), Heinz Sprenger (GRS) and Udo Ziesche (GRS) as well as Manfred Zander (Greifswald University) and Monika Schäfer (Greifswald University) for their assistance in the laboratories.

REFERENCES

Altaner, S.P., Ylagan, R.F. (1997) Comparison of structural models of mixed-layer illite/smectite and reaction mechanisms of smectite illitization. *Clays and Clay Minerals* 45(4): 517–533.

Bauer, A., Schäfer, T., Dohrmann, R., Hoffmann, H., Kim, J.I. (2001) Smectite stability in acid salt solutions and the fate of Eu, Th and U in solution. *Clay Minerals* 36(1): 93–103.

Bauer, A., Velde, B. (1999) Smectite transformation in KOH solutions. *Clay Minerals* 34(2): 261–276.

Cuadros, J., Linares, J. (1996) Experimental kinetic study of the smectite-to-illite transformation. *Geochim. Cosmochim. Acta* 60: 439–453.

Eberl, D.D., Drits, V., Środon, J., Nüesch, R. (1996) MudMaster: a program for calculating crystallite size distributions and strain from the shapes of X-ray diffraction peaks. *U.S. Geological Survey Open File Report* 96–171: 55 p.

Henning, K.-H., Störr, M. (1986) Electron micrographs (TEM, SEM) of clays and clay minerals. *Akademie-Verlag Berlin [Schriftenreihe für geologische Wissenschaften, Bd. 25]*: 352 S.

Herbert, H.-J., Kasbohm, J. & Henning, K.-H. (2003) Long term behaviour of the wyoming bentonite MX-80 in high saline solutions. *Applied Clays Science* (in press).

Kasbohm, J., Henning, K.-H., Herbert, H.-J. (1998) Transmissionselektronenmikroskopische Untersuchungen am Bentonit MX80. In: *Berichte der DTTG, Jahrestagung Greifswald, 3.-5.9.1998*: 228–236.

Kasbohm, J., Henning, K.-H., Herbert, H.-J. (1999) Mineral composition of the bentonites MX-80 and Calcigel. *Proceed. EUROCLAY'99, 5.-10.9.1999, Kraków*: 98–99.

Kasbohm, J., Tarrah, J., Henning, K.-H. (2002) Transmissionselektronenmikroskopische Untersuchungen an Feinfraktionen der Ringversuchsprobe "Ton Stoob". in: *Ottner, F., Gier, S. (Hrsg.) Beitröge zur Jahrestagung Wien, 18.-20.9. 2002 [Berichte der Deutschen Ton- und Tonmineralgruppe e.V., Band 9]*: 71–84.

Köster, H. M. (1977) Die Berechnung kristallchemischer Strukturformeln von 2:1-Schichtsilikaten unter Berücksichtigung der gemessenen Zwischenschichtladungen und Kationenaustauschkapazitäten, sowie die Darstellung der Ladungsverteilung in der Struktur mittels Dreieckskoordinaten. In: *Clay Miner.* 12: 45–54.

Lippmann, F. (1979) Stabilitorenbeziehung der Tonminerale. *N. Jb. Min. Abh.* 136: 287–309.

Krumm, S. (1994) WINFIT 1.0 – A Computer Program for X-ray Diffraction Line Profile Analysis: *Acta Universitatis Carolinae Geologica, 38, XIIIth Conference on Clay Mineralogy and Petrology*, Praha: 253–261.

Pusch, R., Kasbohm, J. (2002) Alteration of MX-80 by hydrothermal treatment under high salt content conditions. *Technical Report TR-02-06*, SKB, Stockholm.

Rassineux, F., Griffault, L., Meunier, A., Berger, G., Petit, S., Vieillard, P., Zellgui, R., Munoz, M. (2001) Expandibility-layer stacking relationship during experimental alteration of a Wyoming bentonite in pH 13.5 solutions at 35 and 60°C. *Clay Minerals* 36(2): 197–210.

Šucha, V., Honty, M., Uhlík, P. (2000) Illitisation in the Slovak Basin. *unpublished Report for GRS project 02C06590*

Środoń J. (1984) Mixed-layer illite-smectite in low temperature diagenesis: data from the Miocene of the Carpathian foredeep. *Clay Minerals*, 19: 205–215.

Środon, J., Elsass, F., Mchardy, W.J., Morgan, D.J. (1992) Chemistry of illite-smectite inferred from TEM measurements of fundamental particles. *Clay Minerals* 27(2): 137–158.

Zöller, M.H. (1993) Charakterisierung von Illitkristallen durch konvergente Elektronenbeugung. In: *Berichte der Deutschen Ton- und Tonmineralgruppe e.V.*: 211–220 [Beiträge zur Jahrestagung der DTTG 1992].

Advances in Understanding Engineered Clay Barriers – Alonso & Ledesma (eds)
© 2005 Taylor & Francis Group, London, ISBN 04 1536 544 9

Bentonite swelling pressure in NaCl solutions – Experimentally determined data and model calculations

O. Karnland
Clay Technology AB, Ideon Research Centre (Sweden)

A. Muurinen
VTT Processes (Finland)

F. Karlsson
Swedish Nuclear Fuel and Waste Management Co. (Sweden)

ABSTRACT: Bentonite clays are known to have a high swelling capacity in low electrolyte water solutions, which gives rise to a macroscopic swelling pressure if the volume is restricted. This swelling pressure in combination with the associated low permeability are the main reasons for proposing bentonite clay as buffer material in nuclear waste repositories, e.g. the KBS3-repository for spent nuclear fuel (SKB 1999). However, salt solutions reduce the swelling capacity of bentonite, and no swelling pressure will obviously develop if the capacity is reduced to an extent where the actual volume is not filled by the maximum swollen clay. This paper presents new measured and modeled swelling pressure data for saline conditions. Special respect is paid to the general thermodynamic model:

$$P_s = -\frac{RT}{M \cdot \bar{v}_w} \ln \frac{p}{p_e} \qquad p_e < p_0$$

where P_s is swelling pressure, R is the gas constant, $\bar{v}_w$ is partial specific volume of the water, p, p_e is vapor pressure of clay adsorbed and of external salt solution, and M is the molecular weight of the water. An osmotic component in the clay/water system is introduced in order to adapt the model to a calculable form by use of Donnan equilibrium. The validity of this model approach has been examined by direct measurement of the swelling pressure, water activity and pore-water concentration in the clay at different clay/water/salt conditions. Swelling tests with Na^+-exchanged and purified Wyoming bentonite (MX-80) were made on samples with dry densities ranging from 350 to 2000 kg/m^3 at saturation with pure water and NaCl solutions ranging from 0.1 to 3 M. The measured swelling pressure for pure water conditions may be described by:

$$P_s = A^*(\exp(B/d)-1)$$

where A and B are constants, and d is the mean theoretical spacing between the mineral flakes in the bentonite. A systematic swelling pressure decrease with increasing NaCl concentration in the external solution was observed. The experimental and model data show significant bentonite swelling pressures also in 3 M sodium chloride solution if the density of the clay/water system is high enough. In practice, the results show that there will be a sufficient swelling pressure in the buffer in the KBS3-repository, even if the system is exposed to NaCl brines. However, the sealing effects of bentonite in the tunnel backfill material may be lost if the in situ compaction gives a too low bentonite density.

1 INTRODUCTION

Sodium bentonite clays generally have a high swelling capacity in low electrolyte water solutions, i.e. the bentonite may reversibly take up water into the clay structure to a volume which is several times the original dry clay volume. The swelling gives rise to a macroscopic swelling pressure acting uniformly on the surroundings if the swelling is restricted to a volume smaller than that of maximum swelling. The volume of a KBS3 deposition hole in rock is, in principle, fixed and the bentonite mass may be balanced to give a desired swelling pressure, and the coupled low hydraulic conductivity leads to an efficient seal of the volume. This swelling ability is one of the basic reasons for using bentonite clay as buffer material in a repository for spent nuclear fuel.

Salt solutions reduce the swelling capacity of bentonite, and no swelling pressure will obviously develop if the capacity is reduced to an extent where the actual volume is not filled by the maximum swollen clay. Several studies have consequently been made in order to determine the effects of electrolytes on bentonite swelling in general (Norrish 1954, Low and Anderson 1958, Low 1979/1980, Low 1983, Sposito 1972, Yong and Warkentin 1975, Pusch 1980, Dickson et al. 1996, Karnland et al. 1997).

Effects of very salt solutions (brines) on bentonite swelling may seem an academic problem, since brines are seldom found at proposed repository depth. A fundamental understanding of the system is, however, desirable from at least the following quite different aspects:

– During the saturation phase, saline ground-water is taken up from the surrounding rock by the bentonite and partly evaporated in the warmer parts of the bentonite. A temporary buildup of high salt concentrations in the inner parts of the bentonite buffer may thereby take place (Karnland and Pusch 1995, Karnland et al. 2000).
– The salt content in the ground-water is observed to increase substantially with depth at several possible repository sites. Although there is no obvious scenario, it can not simply be excluded that there will be a draw-up of deep ground-water into a future repository.
– Due to the long life-time of a repository, a glaciation has to be considered in the safety analyses. During a glaciation cycle the coastal areas will undergo large elevation changes and the coastal line will change dramatically. Any coastal localization, in Sweden or Finland, will at some time during a glaciation cycle be under sea-water conditions. The minimum sodium chloride tolerance will thereby be set by sea conditions i.e. 3.5% NaCl by weight, corresponding to a 0.6 molar solution.
– It has been argued that ions from an external solution (ground-water) cannot enter into the inter-lamellar space in the clay, and that no changes in swelling pressure will take place. However, from a thermodynamic point of view, this would rather lead to a maximum effect on swelling pressure, which can be quantified, but not found in laboratory experiments.

A classical view of osmosis, as shown in Figure 1, may be used in order to illustrate the background of the present approach. The two sides of the water system

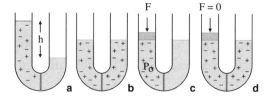

Figure 1. The basic principles of osmosis as an analog to bentonite swelling pressure.

are divided by a semi-permeable membrane only permeable to water molecules. If salt is added to the left side, the water activity is reduced and a transport of water from right to left will take place. The left water table will thereby rise and a hydrostatic pressure develops (a). If salt is now added to the right side, there will be a leveling of the water tables on both sides (b). If the volume change, in the first place, is instead prohibited by force, a pressure will rise in the salt solution (c), corresponding to the previous hydrostatic pressure. Introduction of salt into the right hand side of the membrane will lead to a drop in pressure. A complete drop of pressure will evidently take place when the two concentrations are equaled (d).

The obvious analog is that the left side of the system represents the bentonite clay and the right side an external solution (test solution or ground-water). The origin of the reduced water activity in the clay does not have to be due to osmosis, and from this perspective the cause could be, for example, hydration of counterions or surface hydration. An external solution would reduce the pressure produced by the clay, to an extent equal to the osmotic pressure of the solution, if no ions could pass into the clay. Fortunately, the effect is significantly smaller according to laboratory experiments and one can assume that ions from the external solution do pass into the clay. The size of charge of the anions has been proposed as the cause of hindrance but, for examples, chloride ion radii are 1.81 Å, and including counter-ions the clay is electrically uncharged. On the other hand, if ions could pass freely into the clay, then the final conditions would be equal concentrations on both sides and no effect on swelling pressure would be found, neither of which is in agreement with experimental results. Consequently, we can assume that ions enter the clay-water system to a certain extent, and thereby reduce the water activity in the clay. The introduced ions may be seen as an additional osmotic component acting in the clay-water volume, and the difference in concentration between external and introduced ions leads to a drop in the swelling pressure.

The DLVO double layer model is one of the obvious candidates for modeling the introduction of ions and thereby the effects on swelling pressure in bentonite, since it has proved to be valid for many systems similar to swelling clays (Jönsson and Wennerström 1980). However, results from a straight forward calculation, using the DLVO theory, are qualitatively in accordance with experimental values, but not quantitatively (Komine & Ogata 1996, Karnland 1997).

The hypothesis in the present approach is that the concentration of introduced ions can be calculated by use of basic thermodynamics and the theoretically simple Donnan equilibrium. Here we report the theoretical basis and experimental results including pressure, water activity and ion concentration for a wide range of clay densities and sodium chloride solution concentrations.

2 EXPERIMENTAL

2.1 Bentonite test material

The physical properties of bentonite are determined by the interaction between the water and montmorillonite component with the ideal formula:

$$n(H_2O)\ c_x\ Si_8\ Al_{4-x}Mg_x\ O_{20}\ (OH)_4$$

where c indicates charge compensating cations. The montmorillonite structure is schematically shown in Figure 2. The octahedral Mg (dark circles) substitution for Al results in a net negative charge (x) in the range of 0.4 to 1.2 unit charges per $O_{20}(OH)_4$-unit, by definition. The induced negative layer charge is balanced by inter-lamellar cations (c), and swelling is due to water uptake in the inter-lamellar space, leading to larger spacing.

The actual test material was based on a commercial Wyoming bentonite product from American Colloid Co. The material is a blend of natural bentonite horizons termed "type 6 material", and milled to a grain size distribution which is sold under the product name MX-80. The material was converted into a homo-ionic Na^+ state and coarser grains were removed (Muurinen et al. 2002). The montmorillonite content was thereby increased to above 90% of the total material. The cation exchange capacity (CEC) was measured to be around 0.85 eq/kg clay, and the specific area of the montmorillonite was calculated to be 750 000 m^2/kg. The composition of the montmorillonite was calculated based on ICP/AES element analyses and CEC of the clay fraction according to the technique presented in Newman 1987:

$$(Al_{3.11}\ Fe^{3+}_{0.38}\ Mg_{0.51})\ (Si_{7.86}\ Al_{0.14})\ O_{20}\ (OH)_4\ Na_{0.65}.$$

All charge compensating cations are assumed to be sodium, although small amounts of calcium and magnesium may be present also after the conversion.

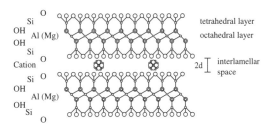

Figure 2. Generalized structure of 2 montmorillonite mineral flakes perpendicular to the c-axis.

2.2 Swelling pressure tests

The main purpose of all the tests was to determine swelling pressures for a wide range of bentonite densities and sodium chloride concentrations. In order to identify possible test artifacts three different types of swelling test series were used;

– **Main swelling test series**, in order to determine swelling pressure, water activity, and to make comprehensive analyses of clay and pore-water,
– **Varying concentration test series**, in order to study the isolated effect of salt concentration by contacting the samples to different NaCl concentration successively,
– **Free swelling test series**, in which the samples were not confined and could swell freely.

In all series, the test solutions were pure water and 0.1, 0.3, 1.0 and 3.0 M NaCl solutions.

2.2.1 Main swelling test series

The series comprised, in total, 23 samples (Table 1). The principle was to have a constant sample volume and constant solution concentration for each sample during the whole test. Relatively large sample holders (40 cm^3) were used in order to get enough material for subsequent analyses of the samples (Figure 3).

The samples were compacted to the intended density and placed in the cylindrical sample holders. The piston, pressure transducer and upper lid were attached and fixed in order to give the predefined density. The water solutions were slowly circulated behind the bottom filters to start with, in order not to trap the original air in the samples. After reaching pressure equilibrium and a minimum test time of 1 month, the test solutions were disconnected and the samples were removed and split in order to make detailed analyses of the water ratio, sample density, pore-water chemistry, water activity and microstructure. Approximately a quarter of the original samples were placed in a test chambers and the water activity was measured by use of a Vaisala HMP237 humidity sensor.

Table 1. Measured water ratio (w), swelling pressure (P_s), pore water concentration (C_p) and water activity a_w. Calculated data are shown for dry density (ρ_d), saturated density (ρ_m), pore ratio (e) and mean basal spacing (d). C_0 indicates external solution concentration.

C_0 M	Sample	w	ρ_d kg/m³	ρ_m kg/m³	e	n	dÅ	P_s kPa	a_w	C_p M
0.0	S2-19	1.91	440	1280	5.25	0.84	38.2	194	1.00	0
	S2-01	0.99	737	1469	2.73	0.73	19.9	555	1.00	0
	S2-03	0.47	1193	1759	1.31	0.57	9.5	2016	0.99	0
	S2-07	0.30	1502	1956	0.83	0.45	6.0	8523	0.93	0
	S2-10	0.23	1672	2064	0.64	0.39	4.7	18 191	0.89	0
	S2-23	0.20	1777	2131	0.55	0.35	4.0	36 514	0.76	0
	S2-20	0.19	1804	2148	0.52	0.34	3.8	39 694		0
0.1	S2-21	0.91	787	1501	2.49	0.71	18.1	337	1.00	0.039
	S2-02	0.90	794	1505	2.47	0.71	17.9	302	0.98	0.072
	S2-04	0.45	1234	1785	1.23	0.55	8.9	1909	0.99	0.025
	S2-17	0.28	1545	1983	0.78	0.44	5.7	8620	0.92	0.024
	S2-18	0.23	1680	2069	0.64	0.39	4.6	18 800	0.85	0.048
0.3	S2-13	0.91	784	1499	2.51	0.71	18.2	133	0.99	0.191
	S2-14	0.44	1248	1794	1.20	0.55	8.8	1664	0.98	0.113
	S2-15	0.28	1559	1992	0.76	0.43	5.6	9313	0.92	0.093
	S2-16	0.23	1680	2069	0.64	0.39	4.6	19 792	0.85	0.134
1.0	S2-22	0.91	786	1500	2.50	0.71	18.2	30		1.000
	S2-05	0.42	1268	1807	1.17	0.54	8.5	1291	0.96	0.570
	S2-08	0.28	1548	1985	0.78	0.44	5.6	7146	0.91	0.482
	S2-11	0.22	1703	2084	0.61	0.38	4.5	15 663	0.86	0.355
3.0	S2-06	0.42	1279	1814	1.15	0.53	8.4	736	0.89	2.16
	S2-09	0.26	1595	2015	0.72	0.42	5.3	4643	0.86	1.64
	S2-12	0.22	1722	2096	0.60	0.37	4.3	14 476	0.80	1.42

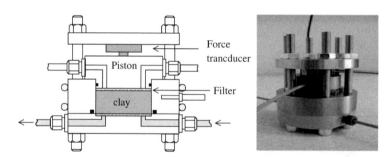

Figure 3. Schematic drawing and photo of the sample holder for the main test series.

2.2.2 Varying concentration test series

The series comprised, in total, 18 samples. Small sample holders (1.6 cm³) were used in order to measure the swelling pressure of a single sample successively exposed to pure water, and the test solutions. The purpose was to ensure that the swelling pressure changes were only dependant on the solution concentration, and not on artifacts in density determination. The small scale was used in order to reach equilibrium reasonably fast and to ensure that there were no obvious scale effects, by comparison with the main swelling test series. Six samples at the time were precompacted to dry densities ranging

from 480 to around 1915 kg/m³. The air-dry samples were placed in the sample holders and the piston, pressure transducer and upper lid were fixed. Deionized water was slowly circulated behind both the upper and lower filters in all samples. At equilibrium, i.e. at constant measured swelling pressure, the water solution was changed to 0.1 M NaCl, and a new equilibrium was awaited. The concentration of the NaCl solution was thereafter changed to 0.3, 1.0, 3.0 M, respectively (Figure 5). In one series the final concentration was changed back to 1.0 M in order to check reversibility. After the tests, the water ratio and the chloride content were determined.

2.2.3 Free swelling test series

The series comprised, in total, 20 samples. Small sample holders (initially 2.5 cm^3) were used in order to determine the maximum swelling of the test material exposed to the standard test solutions, respectively. The sample holders were made of PEEK material and the ratio between the sample diameter (25 mm) and starting height (2.5 mm) was increased to 10, in order to reduce wall friction. Five test samples were pre-compacted at the time, to the same volume, but to different dry densities 790, 1260 (2 samples) and 1570 kg/m^3 (2 samples). The samples were placed in the sample holders, and a piston was placed on top of each sample. At the start of the tests, the actual test solution was circulated behind the common bottom filter, which was in contact with all samples. The swelling was recorded by measuring the piston displacement by use of an electronic vernier caliper (Figure 7). The final equilibrium conditions are expected to represent maximum swelling at the confining pressure of 3 kPa, due to the weight of the piston. After each test the water ratios were determined.

2.3 Water ratio and Chloride analyses

All samples were analyzed with respect to water ratio (w) and Sodium chloride content. The determinations are not straight forward since the samples contain various amount of salt, and the following technique was therefore used. Approximately 2 g of a test sample was used to determine the mass of water (m_w) by the mass loss at oven-drying at 105°C overnight. The remaining dry material was dispersed in 200 ml of deionized water, left overnight, and centrifuged at 2000 rpm for 20 minutes. The supernatant was removed, ultra-filtered and analyzed by use of ion chromatography with respect to Cl$^-$. The porewater concentration (C_{pore}) was calculated from super-natant concentration (C_{disp}) according to:

$$C_{pore} = \frac{C_{disp} \cdot V_{disp}}{m_w \cdot \rho_w} \qquad (1)$$

where V_{disp} is the volume of the dispersion, and ρ_w is water density. The mass of salt (m_{salt}) was calculated according to:

$$m_{salt} = C_{pore} \cdot m_w \cdot \rho_w \cdot M_{NaCl} \qquad (2)$$

where M_{NaCl} is the molar weight of NaCl. The solid clay mass was calculated from:

$$m_s = m_{dry} - m_{salt} \qquad (3)$$

where m_s is the mass of solid clay and m_{dry} is the mass of the total dry material. Finally the water ratio (w)

was calculated according to:

$$w = \frac{m_w}{m_s} \qquad (4)$$

3 EXPERIMENTAL RESULTS

3.1 General

The swelling pressure in bentonite, at equilibrium and pure water conditions, is expected to increase exponentially with increasing sample density. The density determination is therefore critical since a small accuracy problem in density indicates signifi-cant swelling pressure changes. In the main swelling test series, the density was directly determined, by submerging a part of the samples into paraffin oil, and the results were in good agreement with the water ratio data assuming full water saturation. In the other test series only water ratios were systematically determined due to the small sample volumes. In order to be consistent, all the presented dry density values are based on water ratio data according to:

$$\rho_d = \frac{\rho_s \cdot \rho_w}{\rho_s \cdot w + \rho_w} \qquad (5)$$

3.2 Main swelling test series

Figure 4 shows the measured build up of swelling pressure during water uptake. The final equilibrium condition was quite distinct and the final pressure values are shown in Table 1. The subsequent deter-mination of water activity is less accurate for high water activity conditions ($a_w > 0.95$) due to the sensor accuracy of a few percent in this region and to tem-perature sensitivity. The measured lower activities are thought to show actual conditions, since sensors accu-racy is better in this region and the physical changes are larger. The measured pressure, water activity and chloride concentration are summarized in Table 1.

3.3 Varying concentration test series

Measured equilibrium pressure results are shown in Table 2 and swelling pressure changes versus time are exemplified in Figures 5 and 6.

3.4 Free swelling test series

The swelling of the doublet samples generally show only minor variation. The swelling in 0.1 M solution results in a volume increase from 3 to 5 times the original volume (Figure 7, upper left). The densities of the final samples are still quite close, showing that the initial degree of compaction is less important (Figure 8, upper left). To the other extreme, the

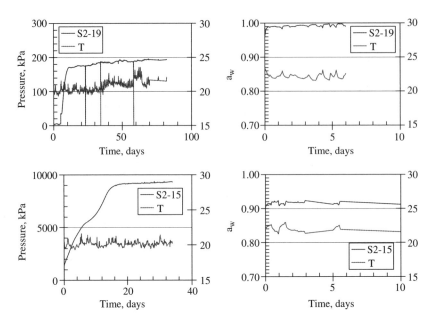

Figure 4. Measured swelling pressure progress (left) and the subsequent water activity determination (right). Legends show sample number, and lower lines show temperature.

Table 2. Measured water ratio (w), calculated dry density (ρ_d), and swelling pressure (P_s), for the equilibrium conditions at gradual exposure for pure water and the four test solutions.

Sample	w	ρ_d kg/m^3	Pressure, kPa				
			0.0 M	0.1 M	0.3 M	1.0 M	3.0 M
S2-41	1.80	463	300	179	61	13	7
S2-42	0.89	795	598	363	160	50	16
S2-43	0.39	1321	2946	2709	2312	1554	473
S2-44	0.29	1529	9321	9170	8775	7302	4906
S2-45	0.25	1634	15 280	15 103	14 753	13 447	11 161
S2-46	0.18	1851	32 825	32 849	32 417	31 122	28 040
S2-51	2.60	338	126				
S2-52	0.91	785	618	400	196	92	42
S2-53	0.42	1272	2000	1735	1335	799	196
S2-54	0.31	1487	5853	5750	5287	4021	1979
S2-55	0.23	1686	13 581	13 463	12 990	11 776	9632
S2-56	0.20	1784	27 380	27 189	26 926	25 069	21 800
S2-61	1.51	535	216	82	41	13	10
S2-62	0.81	849	646	419	194	53	17
S2-63	0.35	1398	3666	3455	3046	2178	669
S2-64	0.27	1584	9455	9275	8904	7614	5269
S2-65	0.21	1730	20 056	19 976	19 800	18 930	16 356
S2-66	0.13	2017	55 588	55 488	54 960	53 626	49 680

swelling in 3.0 M solution results in no swelling for the lowest initial density and a doubling in volume in the highest density samples. Also, in this case, the final densities are close, and we can conclude that there is a typical maximum swelling for all the examined solution concentrations, with only small effects from the initial degree of compaction. The results are plotted in Figure 15 with a fixed pressure of 3 kPa versus measured final clay density. The pressure should be regarded as a minimum value

246

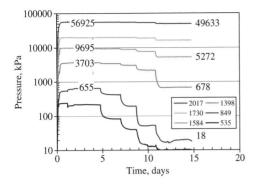

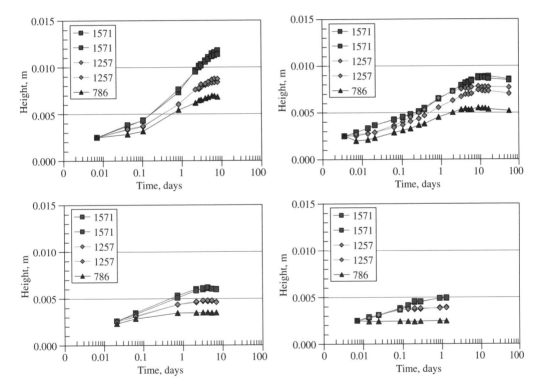

Figure 5. Swelling pressure of samples S2-61 to 66 during the gradual exposure to pure water, 0.1, 0.3, 1.0 and 3.0 M NaCl solutions. The legend shows sample dry densities (kg/m^3).

Figure 6. Detail of Figure 5 for samples S2-61, 62 and 63 showing swelling pressure in linear scale during the gradual exposure to pure water, 0.1, 0.3, 1.0 and 3.0 M NaCl solution. The legend shows sample dry densities (kg/m^3).

Figure 7. Sample height versus time during the swelling course of samples exposed to the test solutions 0.1 M (upper left), 0.3 M (upper right), 1.0 M (lower left), and 3.0 M (lower right). Legends show the initial sample dry density (kg/m^3).

since it is calculated from piston weight only and no friction is taken into account.

The measured final water ratios were used to calculate the saturated density by use of Equation (5). The results were close to the calculated densities based on volume change, showing that the samples were close to saturation.

4 MODELLING AND DISCUSSION

4.1 *Swelling pressure at pure water conditions*

An intra-crystalline and an osmotic region are often referred to in order to describe the swelling behavior of sodium bentonite. The inter-crystalline swelling

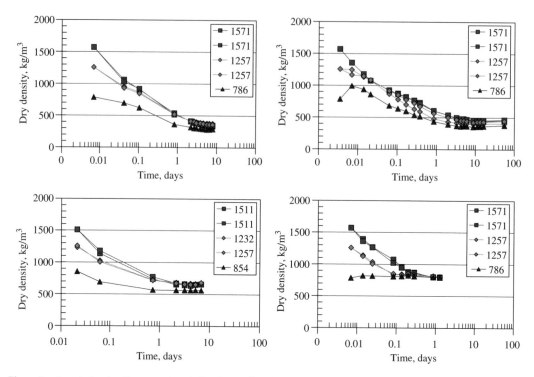

Figure 8. Sample dry density versus time during the swelling course of samples exposed to the test solutions 0.1 M (upper left), 0.3 M (upper right), 1.0 M (lower left), and 3.0 M (lower right). Legends show the initial sample dry density (kg/m³).

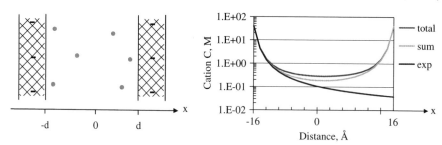

Figure 9. Schematic scaled view of two montmorillonite surfaces with the separating distance $2d = 32$ Å, approximately corresponding to a dry density of $780\,kg/m^3$, and calculated distribution of counter ions in the sample (right).

represents the first uptake of water in a few layers, and osmotic swelling is characterized by larger distances between the mineral flakes in which the colligative properties of the counter-ions dominate. Electrical double-layers are formed and the thermodynamics of such systems is described by the Poisson-Boltzmann (PB) equation, which has an analytical solution for parallel surfaces. For pure water conditions the PB equation may be used in order to calculate the distribution of counter-ions between the mineral flakes according to:

$$\rho_i(x) = \left[2\varepsilon_r\varepsilon_0 k_B T s^2 \left(Zed^2\right)\right] / \left[\cos^2\left(sx/d\right)\right] \quad (6)$$

where $\rho_i(x)$ is the ion density at a distance x from the surface, ε_r and ε_0 are the relative dielectric permittivity of water and permittivity in vacuum, respectively, k_B is Boltzman's constant, T is temperature, Z is the valence of compensating ions, e is the charge of the electron, d is the distance from the surface to mid position, and s is given by the following expression:

$$s\tan(s) = |\sigma| Zed / (2\varepsilon_r\varepsilon_0 k_B T) \quad (7)$$

The latter equation can be solved by simple iteration. Figure 9 (right) shows the calculated

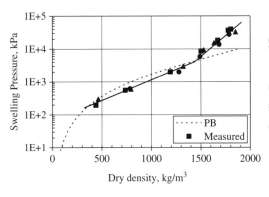 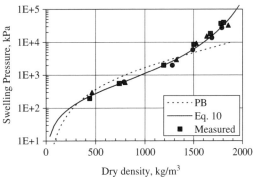

Figure 10. Measured swelling pressure at pure water conditions versus sample dry density. PB indicates results by use of the Poisson-Boltzmann equation, Eq. 10 indicate results by use of Equation 10.

distribution for the present test material at a dry density of $780\,\text{kg/m}^3$. The calculation was made by assuming the distance between the mineral flakes was equal to the theoretical mean distance according to:

$$d = \frac{V_w}{A_m} \tag{8}$$

where V_w is the total water volume and A_m is the total surface of the montmorillonite.

In the osmotic region, the repulsive force between the charged mineral surfaces is expected to be of colligative origin and we can calculate the osmotic pressure P_o according to:

$$P_o = RT \cdot C_{mid} \tag{9}$$

where R is the gas constant, T is temperature, and C_{mid} is the counter-ion concentration at the mid position between the charged surfaces.

Figure 10 shows measured swelling pressure data for pure water conditions versus dry clay density, with the dotted line showing the calculated pressure for the actual test material according to the straight forward use of the Poisson-Boltzmann (PB) theory and Equation 9. Calculations using the full DLVO theory, which includes attraction forces, were made according to the theory presented in Jönsson and Wennerström 1980. This, in principle, gave the same results.

In the present material, the border line between the osmotic and inter-crystalline region should consequently be around a dry density of around $1500\,\text{kg/m}^3$, where the discrepancy between PB results and measured data starts to increase with density. The density dependence in each domain is often generalized by straight lines in a lin-log plot, i.e. there is an exponential increase of swelling pressure with density in both regions (Figure 10, left).

An alternative way to generalize the dependence is to correlate the swelling pressure to the theoretical mean distance between the mineral flakes according to:

$$P_s = A \cdot \left(\exp\left(\frac{B}{d}\right) - 1 \right) \tag{10}$$

where P_s is swelling pressure, A and B are constants and d is the mean distance between the mineral flakes as proposed by Low, 1983. Equation 10 has the same form as the Boltzmann distribution equation:

$$C(x) = C_{mid} \cdot \exp\left(\frac{-ZeN\psi}{RT}\right) \tag{11}$$

where $C(x)$ is the concentration at distance x from the mineral surface, C_{mid} is the concentration at midplane between the mineral surfaces, N is Avogadro's number, ψ is the potential at distance x. The lowest line in Figure 9 (right) may illustrate this distribution. A reduction of the distance between mineral surfaces makes the mid-concentration increase approximately along the distribution line, and thereby also the swelling pressure. The potential ψ is consequently decreasing linearly with distance according to the experimental data, which is in agreement with the extensive experimental data presented by Low for lower pressures. Only at the highest measured density ($\rho_d < 2000\,\text{kg/m}^3$), this generalization is less accurate. The relation between pressure and mean theoretical distance is not unambiguous, since e.g. the inverse mean concentration of charge compensating cations, or sample water ratio are proportional to the distance. The physical meaning may be left aside at the moment, and Equation 10 may still be used in the following calculations, since it well describes the actual swelling pressure for relatively high swelling pressures, including what is proposed for the KBS3 repository. Using the water ratio, A will take the value 350, with the dimension kPa, and B will take the value

0.9, to give the presented line in Figure 10. It should be pointed out that by leaving the physical meaning of distance between the flakes, and the forthcoming use of thermodynamics, we do in principle deprive ourselves of making micro-structural interpretations.

The magnitude of the pressure at equilibrium is governed by the difference in vapor pressure between the clay-water system and the solution, and may also be calculated by means of thermodynamics (Low 1958, Sposito 1972, Kahr et al. 1986).

Chemical potential (μ), or molar Gibbs energy, is described by the individual conditions which affect the energy in a system. In the case of clay swelling in water, the chemical potential of water is of interest, and in liquid states, where volume change by pressure (P) is negligible, we may use:

$$\mu_w = \mu_0 + RT \ln \frac{p_c}{p_0} + RT \ln \frac{p_e}{p_0}$$
$$+ m_w gh + \overline{V}_w \cdot P + zFE + \dots \text{J/K} \quad (12)$$

where μ_0 is the chemical potential of free water. The second and third term represents the energy contribution from the clay and an electrolyte, respectively, where R is the gas constant, T is temperature, p_c, p_0 and p_e represent the vapor pressure set by the clay, pure water, and an electrolyte, respectively. The fourth term ($m_w gh$) represents the contribution from hydrostatic pressure, g is the gravity constant and h is height. The fifth term ($\overline{V}_w \cdot P$) represents water pressure of other origin, where $\overline{V}_w$ is the molar volume of water, the last term represent contribution from electric origin, and dots indicate that contribution from other origins are possible.

At chemical equilibrium in a system, the chemical potential is equal in all positions by definition. Since the chemical potential in dry bentonite clay is lower compared to that of pure water, water will move into the clay until the potentials equal. This leads to water uptake ending in a clay gel if the volume is not restricted and equilibrium is reached by for example increase of the hydrostatic term in the clay volume. If the volume is defined, equilibrium may be reached by an induced pressure in the clay, usually called the swelling pressure.

In the present experimental system with negligible hydrostatic pressure, and the clay saturated with pure water, the expression for the chemical potential is reduced to:

$$\mu_w = \mu_0 + RT \ln \frac{p_c}{p_0} + \overline{V}_w \cdot P \quad (13)$$

The external water potential is μ_0 at equilibrium and we have:

$$0 = RT \ln \frac{p_c}{p_0} + \overline{V}_w \cdot P \quad (14)$$

$$P_s = \frac{RT}{V_w} \ln \frac{p_c}{p_0} \quad (15)$$

We can now introduce the index s for swelling pressure, since the induced pressure is obviously a consequence of the clay.

4.2 Swelling pressure at salt solution conditions

If the clay is in contact with a salt solution, and ions are introduced into the clay (Figure 12), then the chemical potential in the clay is given by:

$$\mu_w = \mu_0 + RT \ln \frac{p_c}{p_0} + RT \ln \frac{p_{ie}}{p_0} + \overline{V}_w \cdot P \quad (16)$$

where p_{ie} is the vapor pressure set by ions introduced into the clay. The chemical potential in the solution is given by:

$$\mu_w = \mu_0 + RT \ln \frac{p_e}{p_0} \qquad p_e < p_0 \quad (17)$$

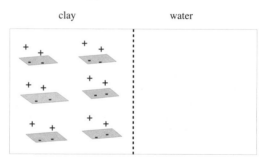

Figure 11. Principle drawing of the clay-water system. The two physical volumes are separated by a semi-permeable membrane permeable to dissolved ions and water, but not to the mineral flakes. The membrane may represent a filter in a laboratory test, the rockbuffer interface or internal membrane functions in the clay volume.

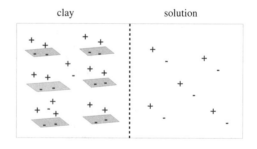

Figure 12. Principle drawing of the clay-water-ion system. The two physical volumes are separated by a semi-permeable membrane (filter) permeable to dissolved ions and water, but not to the mineral flakes. Clay mineral flakes and charge compensating cations on the left side, and sodium and chloride ions on both sides of the filter.

At equilibrium the potentials are the same in the clay and the solution which gives:

$$RT \ln \frac{p_c}{p_0} + RT \ln \frac{p_{ie}}{p_0} + \overline{V}_w \cdot P = RT \ln \frac{p_e}{p_0} \qquad (18)$$

$$P_s = \frac{RT}{V_w} \left(-\ln \frac{p_c}{p_0} - \ln \frac{p_{ie}}{p_0} + \ln \frac{p_e}{p_0} \right) \qquad (19)$$

This obviously leads to a reduction of the swelling pressure (P_s) compared to pure water conditions if p_e is lower than p_{ie}. Since the vapor pressure reduces with increasing concentration, it is obvious from Equation (19) that the more ions entering the clay, the less reduction in swelling pressure there will be.

The above is generally valid and the measured chloride concentrations clearly show that ions from the external solutions enter the clay-water volume (Table 1 and Figure 14). A crucial question is then if these ions contribute to the lowering of water activity and represent the second term in Equation 19, or alternatively, if the introduced ions occupy a volume separated from the original swelling pressure inducing ions, such as larger voids. The ion concentration in such a possible void, which will hardly be higher than in the external solution, induces a corresponding water activity. The pressure inducing volume will have the same water activity as for pure water conditions if no ions can enter. The difference in water activity between the two compartments is possible to maintain, also at equilibrium, if a balancing pressure acts on the void-forming walls in order to keep the chemical potentials equal. Such a two-compartment system with different water activities will give an intermediate water activity after the pressure is released, since water may then be transported from high activity compartments to a lower activity compartments. If water transport by some reason still is prohibited, then the measured water activity will not be below the lower of the two.

However, the measured water activities are systematically lowered with increasing salinity to values below the lowest of what is produced by the salt solution and the clay, respectively. This strongly indicates that the introduced ions enter the clay volume which causes the swelling pressure. The measured water activity of a sample may be used in order to calculate a potential pressure (P_{aw}) (or suction with reversed sign) of the sample according to Equation 19, where P_{aw} represents the pressure effect of the two first terms. In the same way, the external solution in a test corresponds to an osmotic pressure, represented by the third term.

According to Equation 19, the resulting calculated swelling pressure (P_{net}) will be the difference of the potential pressure and the external osmotic pressure

(P_{salt}). The results from the main test series calculated in this way are shown in Figure 13 and compared with the measured swelling pressures (P_m).

The fact that measured P_m and calculated P_{net} results coincide quite well for pure water conditions (Figure 13, upper left) supports Equation (15). The same correspondence for the salt conditions supports Equation (19). Further, the high values of P_{aw} for salt conditions strongly indicate that the introduced ions have entered the swelling pressure inducing volumes. This does, however, not expel the existence of voids in the clay, and an alternative interpretation of the test results compared to the following is given in Muurinen et al. (2002).

4.3 Donnan equilibrium calculation

The montmorillonite part in bentonite clay is composed of the negatively charged mineral flakes and charge compensating cations (Figure 2). The mineral flakes may consequently be seen as macro-ions and the system may be regarded as a soluble polyelectrolyte. Donnan (1911) showed that a system similar to that in Figure 12 may lead to an unequal distribution of ions which are diffusible across a membrane separating the two volumes, i.e. the charge compensating cations and ions from an external solution.

The condition for equilibrium is that the chemical potentials are equal on both sides of the membrane for all components which can permeate. Since the macro-sized mineral flakes cannot pass the membrane, the equilibrium will be established only for ions in the saturating solution and the original montmorillonite cations. Ideally, in this study this will only be Na^+ and Cl^-. The conditions of electrical neutrality in the entire system are:

$$0 = z \cdot C_m + C_{Cl^-} - C_{Na^+} \qquad (20)$$

where z is the valence of the macromolecule, C_m, C_{Na}^+ and C_{Cl}^- is the concentration of the macromolecules, sodium ions and chloride ions, respectively. In terms of ion activity (not water activity) the equilibrium condition for the diffusable ions will be:

$$\{Na^+_c\}\{Cl^-_c\} = \{Na^+_e\}\{Cl^-_e\} \qquad (21)$$

where the indices c and e refer to the clay-water system and to the surrounding electrolyte, respectively. The $\{Na^+_c\}$ factor is the sum of the concentration of original charge compensating ions $\{Na^+_{cc}\}$ and introduced electrolyte ions $\{Na^+_{ie}\}$. Local electrical neutrality has to be established thus $\{Na^+_e\} = \{Cl^-_e\}$ and $\{Na^+_{ie}\} = \{Cl^-_{ie}\}$ and Equation (20) may be written:

$$(\{Na^+_{cc}\} + \{Na^+_{ie}\}) \{Na^+_{ie}\} = \{Na^+_e\} \{Na^+_e\} \qquad (22)$$

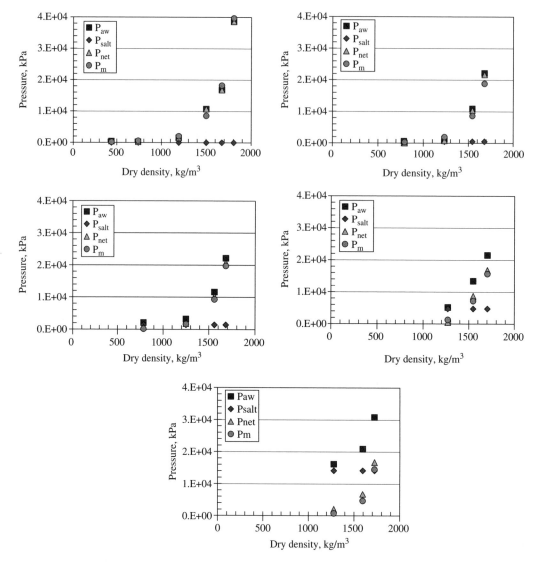

Figure 13. Potential pressure (P_{aw}), osmotic pressure of external solution (P_{salt}), calculated difference of P_{aw} and P_{salt} (P_{net}), and measured swelling pressure (P_m) for samples in the main series. Diagrams show results from pure water conditions (upper left), 0.1 M NaCl solution (upper right), 0.3 M solution (middle left), 1.0 M solution (middle right), and 3.0 M solution (lower left).

which may be rewritten as:

$$\left\{Na_{ie}^{+}\right\} = \frac{-\left[\left\{Na_{cc}^{+}\right\} \pm \sqrt{\left\{Na_{cc}^{+}\right\}^{2} + 4 \cdot \left\{Na_{e}^{+}\right\}^{2}}\right]}{2} \quad (23)$$

Consequently, it is possible to calculate the activity of sodium ions introduced into the clay-water system if the activity of the original cations $\{Na_{cc}^{+}\}$ in the clay and the activity of the ions in the external solution $\{Na_{e}^{+}\}$ are known.

4.4 Activity of original cations in the clay

Consider a symmetrical system with two mineral flakes as in Figure 9. If two double-layers start to overlap, the counter-ion concentrations will increase, which lowers the entropy. A decrease in energy offsets some of the resulting increase in free energy, but at the mid-position between the mineral flakes, where no fields act on the ions, only the entropy effect is significant. At pressure equilibrium, the ions are of course also in equilibrium, although with a concentration distribution. In terms of number of ions they increase towards the surface but are also affected more by the surface, and the activity at different distances are the same. Assuming the repulsive force between the charged mineral surfaces is of colligative origin we can calculate the activity of the counter-ions $\{Na_{cc}^+\}$ from:

$$P_s = kT \cdot \sum_i c_i \,(mid) = RT \cdot \left\{Na_{cc}^+\right\} \qquad (24)$$

where k is Boltzmann's constant, T is temperature, $c_i(mid)$ is the number of ions of types i, at the mid position between the charged surfaces, and R is the gas constant.

We can now use measured values of P_s, or Equation 10, in order to calculate the counter-ion activity. By assuming that the swelling pressure is only of colligative origin, we do not underestimate the counter-ion activity, which then ensures that the Donnan effect is not underestimated. In other words, the calculation of swelling pressure reduction may be overestimated if causes other than counter-ions contribute to the swelling pressure. This approach is consequently conservative with respect to swelling pressure reduction from a safety assessment perspective.

4.5 Activity of ions in the external solution

Sodium chloride dissolves almost completely in water and produces ions. Still there is a discrepancy between concentration and activity of the ions usually described by a mean activity coefficient, which is well below unity for the actual solutions. The discrepancy may be explained by ion interaction, i.e. at a given instant a certain amount of the ions, or alternatively, all ions to a certain degree, interact in such a way that a fraction of the ions may be treated as single particles without ion properties. The conventional way to describe the conditions from a chemical point of view is by use of an activity factor. From a colligative point of view the conditions may be described by use of van't Hoffs factor. According to the latter a sodium chloride solution may be looked upon in the following way:

$$Na^+Cl^- \longleftrightarrow Na^+ + Cl^- \qquad (25)$$

$$C-\alpha C \qquad \qquad \alpha C + \alpha C$$

where C is the concentration, and α is the fraction NaCl which may be described as individual noninteracting ions ($\alpha \leq 1$). The relation to van't

Hoffs factor (i) is:

$$C((1-\alpha) + 2\alpha) = C(1+\alpha) = i \cdot C \qquad (26)$$

The concentration of non-interacting ions is consequently $(2\alpha \cdot C)$ in a NaCl solution with concentration C. The remaining ions in the solution may be looked upon as associated and do not show ion properties but contributes to the i-value by $(C-\alpha C)$. The colligative properties of the solution are given by the i-factor.

In order to calculate the Donnan equilibrium we use:

$$\{Na^+_e\} = \alpha C_e , \qquad (27)$$

Tabulated α-values are available for the whole range of possible concentrations in e.g. Heyrovska (1996). The reason for determining ion activity in this way is that e.g. Debye-Hückel theory and computer codes are not better suited for the present high concentrations. However, PHREEQC v 2.6 gives similar values of the sodium ion activity as those evaluated from Equation 27 and tabulated α-values.

4.6 Donnan equilibrium effects on concentration and pressure

By use of the activity of the original cations in the clay ($\{NA_{cc}^+\}$), as defined by Equation 24, and ions in the external solution ($\{Na_e^+\}$), as defined by Equation 26, we can now calculate the activity of ions introduced into the clay ($\{Na_{ie}^+\}$) according to Equation 23. In order to fully describe the system and to compare with measured values of pressure and concentration, we also have to determine the number of ion association (Na^+Cl^-) as defined by Equation 25. In the external solution we can use the tabulated values according to section 4.5. However, in the clay pore water the activity of ion association is not obvious, since the original charge compensating cations have to be taken into account.

In the external solution, an ion equilibrium constant may be defined as:

$$K = \frac{\left\{Na_e^+\right\}\left\{Cl_e^-\right\}}{\left\{Na_e^+Cl_e^-\right\}} \qquad (28)$$

(Note that this is not the commonly used solubility product equilibrium constant K_{sp}, which describe the relation between solved and not dissolved matter). We assume the same conditions apply to the clay pore-water solution:

$$K = \frac{\left\{Na_c^+\right\}\left\{Cl_c^-\right\}}{\left\{Na_c^+Cl_c^-\right\}} \qquad (29)$$

The prerequisite of system equilibrium, according to Equation 21, is that the ion products in the external

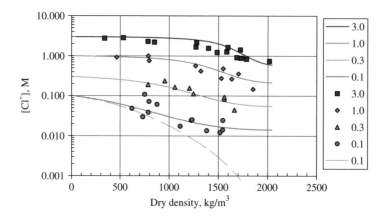

Figure 14. Measured (dots) and calculated (lines) chloride concentrations in the clay versus dry clay density. The lowest line (dashed) shows results from calculations without taking ion association into account for 0.1 M solution. Legends show external solution concentration in mole/L.

solution and in the clay pore-water solution are the same, and we get:

$$\{Na^+_c \, Cl^-_c\} = \{Na^+_e \, Cl^-_e\} \qquad (30)$$

The total concentration of sodium chloride (C_{ie}) introduced into the clay is then given by:

$$C_{ie} = (\{Na^+_{ie}\} + \{Cl^-_{ie}\})/2 + \{Na^+_e \, Cl^-_e\} \qquad (31)$$

where $\{Na^+_{ie}\} = \{Cl^-_{ie}\}$ due to local electrical neutrality request (section 4.3).

Calculations were made by use of Equation (30) for the four test solutions (0.1, 0.3, 1.0 and 3.0 M NaCl) for the whole range of examined clay dry densities, and the results are presented as continuous lines in Figure 14. For comparison, 0.1 M results without adding the contribution from ion association is shown as a dashed line.

A relatively large part of the chloride ions in the clay pore-water appears as associated ions according to this approach. This effect increases with clay density due to the relative increase of original charge compensating sodium ions. For the same reason, the effect also increases with decreasing concentration in the external solution. In the extreme combination of 0.1 M NaCl solution and high clay density (e.g. sample S2-12) the concentration contribution from what can be regarded as associated ions is several times higher than the contribution from what can be regarded as individual ions.

Once again, assuming the repulsive force between the charged mineral surfaces is of colligative origin we can calculate the resulting swelling pressure in the clay-salt-water system by use of Equations (19)

and (24):

$$P_s = RT \, (\{Na^+_{cc}\} + \{Na^+_{ie}\} + \{Cl^-_{ie}\} + \{Na^+_{ie}Cl^-_{ie}\} -$$
$$(\{Na^+_e\} + \{Cl^-_e\} + \{Na^+_e Cl^-_e\})) \qquad (32)$$

The swelling pressure decrease (ΔP) in the clay due to equilibrium with an external NaCl solution is consequently given by:

$$\Delta P_s = RT \, (\{Na^+_{ie}\} + \{Cl^-_{ie}\} + \{Na^+_{ie}Cl^-_{ie}\} -$$
$$(\{Na^+_e\} + \{Cl^-_e\} + \{Na^+_e Cl^-_e\})) \qquad (33)$$

Since the contribution from ion associates is assumed to be equal in the clay and the external solution according to Equation (29) we get:

$$\Delta P_s = RT \, (\{Na^+_{ie}\} + \{Cl^-_{ie}\} - (\{Na^+_e\} + \{Cl^-_e\})) \qquad (34)$$

Calculations of the resulting swelling pressure by use of Equation (31) were made for the four test solutions (0.1, 0.3, 1.0 and 3.0 M NaCl) for the whole range of examined clay dry densities, and the results are presented as continuous lines in Figure 15.

5 CONCLUSIONS

The experimental results cover several orders of magnitude with respect to bentonite swelling pressure and concentration of introduced ions into the clay. The measured data show clear trends and are expected to describe the actual conditions.

For pure water conditions the experimental results may be summarized in the following items:

- Reasonable concordance is found between measured swelling pressure and swelling pressure

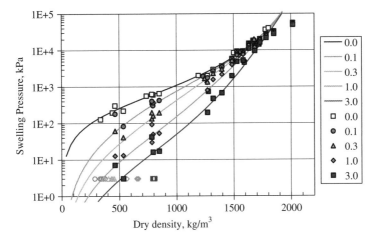

Figure 15. Measured (dots) and calculated (lines) swelling pressure versus dry clay density. Legends show external solution concentration in mole/L.

calculated by use of the Poisson-Boltzmann equation, for clay dry densities up to around 1500 kg/m^3.

- An exponential relationship between measured swelling pressure and calculated mean basal interlamellar spacing was found for clay dry densities up to around 1800 kg/cm^3.

For salt solution conditions the experimental results may be summarized in the following way:

- The degree of compaction has only a minor effect on maximum swelling.
- Ions from an external solution enter the clay volume in a systematic way.
- Water activity in the clay is reduced to a value below that of the original salt solution and original clay, respectively, which indicate that the introduced ions enter the swelling pressure inducing volume.
- Swelling pressure is systematically reduced at all clay densities.
  - The pressure reduction increases with increasing clay density.
  - The relative pressure reduction decreases with clay density.

The measured pressure and concentration effects of sodium chloride solutions show a characteristic pattern for the various combinations of clay density and electrolyte concentration. The proposed analytical model to calculate these effects reproduces measured results quantitatively quite well, on the assumption that the effect of ion association takes the same value in the clay as in the external solution.

The experimental and model data show significant bentonite swelling pressures also in 3 M sodium

chloride solution if the density of the clay/water system is high enough. From a safety assessment perspective, the results show that there will be a sufficient swelling pressure in the buffer in a KBS3 repository, even if the repository is exposed to NaCl brines. However, the sealing effects of bentonite in the tunnel backfill material may be lost if the in situ compaction gives a too low bentonite density.

ACKNOWLEDGMENTS

The authors wish to acknowledge that this paper is a result of work funded by the Swedish Nuclear Fuel and Waste Management Company (SKB), and Posiva OY, Finland.

REFERENCES

Dixon D.A., Gray M.N., Graham J., 1996. Swelling and hydraulic properties of bentonites from Japan, Canada and the USA. Environmental Geotechnics, Kamon (ed), Balkema, Rotterdam.

Heyrovska R., 1996. Physical Electrochemistry of Strong Electrolytes Based on Partial Dissociation and Hydration: Quantitative Interpretation of the Thermodynamic Properties of NaCl(aq) from "Zero to Saturation". Journal of Electrochemical Society, Vol. 143, No. 6, June 1996, pp. 1789–1793.

Jönsson B., Wennerström H., 1980. Ion Condensation in Lamellar Liquid Crystals. Chemica Scripta Vol. 15 (1980) No. 1.

Kahr G., Krähenbühl F., Müller-Vonmoos M., Stoekli H.F., 1986. Wasseraufnahme und Wasserbewegung in hochverdichtetem Bentonit»
NAGRA TR 86-14, Wettingen, Switzerland.

Karnland O., Pusch R., 1995. Cementation phenomena of importance for the performance of smectite clay buffers in HLW repositories. Radioactive Waste Management and Environmental Remediation, ASME, pp. 853–855.

Karnland O. Bentonite swelling pressure in strong NaCl solutions – Correlation between model calculations and experimentally determined data. SKB TR-97-31, Stockholm.

Karnland O., et al. 2000. Long term test of buffer material – Final report on the pilot parcels. SKB TR-00-22, Stockholm.

Komine H., Ogata N., 1996. Prediction of swelling characteristics of compacted bentonite. Canadian Geotechnical Journal Vol. 33:1, pp. 11–22.

Low P.F., Anderson D.M., 1958. Osmotic Pressure Equation for Determining Thermodynamic Properties of Soil Water. Soil Sci. 86, p. 251.

Low P.F. 1979. The swelling of Clay. I: Basic concepts and Empirical Equations. Soil Sci. Soc. Amer. J. 43, p. 473.

Low P.F. 1980. The swelling of Clay. II: Montmorillonites. Soil Sci. Soc. Amer. J. 44, p. 667.

Low P.F., 1983. Direct measurement of the relation between interlayer force and interlayer distance in the swelling of montmorillonite. Journal of Colloid and Interface Science Vol. 96, No. 1, p. 229.

Newman A.C.D., 1987. Chemistry of clays and clay minerals. Mineralogical Society Monograph No. 6, ISBN 0-582-30114-9.

Norrish K., Quirk J.P., 1954. Crystalline swelling of montmorillonite. Nature 173: 255–256.

Muurinen A., Karnland O., Lehikoinen J., Ion concentration caused by an external solution into the porewater of compacted bentonite. Submitted to Applied Clay Science, 2002.

Pusch R., 1980. Swelling pressure of highly compacted bentonite. SKB TR 80-13, Stockholm.

SKB, 1999. Deep repository for spent nuclear fuel. SR 97 Post-closure safety, Main Report Volume I. SKB Technical report TR-99-06, SKB, Stockholm.

Sposito G., 1972. Thermodynamics of Swelling Clay-Water Systems. Soil Sci. 114, p. 243.

Yong R.N., Warketin B.P., 1975. Soil properties and behavior, Elsevier Scientific Publishing Co., Amsterdam, p. 449.

Advances in Understanding Engineered Clay Barriers – Alonso & Ledesma (eds)
© 2005 Taylor & Francis Group, London, ISBN 04 1536 544 9

Analysis and distribution of waters in the compacted FEBEX bentonite: pore water chemistry and adsorbed water properties

Ana María Fernández & Pedro Rivas
CIEMAT, Madrid (Spain)

ABSTRACT: Compacted bentonites are being considered in many countries as a backfill material in high-level radioactive waste disposal (HLW) concepts because of their low permeability, high swelling capacity and high plasticity. For the safety assessment of underground repositories, laboratory experiments and predictive models are needed to describe the process of infiltration and swelling pressure in compacted bentonites.

The pore water chemistry in bentonites is the result of the water-solute-clay interactions in the clay/water system. For this reason, in order to determine the pore water chemistry it is necessary to know the mineralogical and chemical components of the clay system, the physico-chemical characteristics, the hydration mechanisms, as well as, the types of waters, porosity, microstructure or fabric and ion diffusion pathways in compacted bentonites.

Textural and structural features and changes which occur during the intracrystalline swelling of the FEBEX bentonite have been studied using nitrogen adsorption volumetry, water adsorption and desorption gravimetry and X-ray diffraction under controlled humidity conditions. The dry state of the FEBEX bentonite was characterised by nitrogen adsorption-desorption isotherms. Water vapor adsorption/desorption isotherms, together with c-lattice spacing determinations, were used to characterise the saturated state of the FEBEX bentonite and to identify the different states and location of the water.

Different stages of hydration corresponding to the solvation of exchangeable cations were possible to distinguish in the FEBEX bentonite, as well as the accumulation of water molecules adsorbed on the silicate surface in the interlamellar space and on external surfaces. Water is mainly adsorbed in the interlamellar space, but a 3–6% of water is external water adsorbed on the external surfaces of the stacks, edges and OH groups. The uptake of water occurs in three discrete stages, as the montmorillonite structure expands and the internal surface is calculated from these discrete phases of water adsorption. These calculations have allowed to determine the bentonite properties, such as accessible porosity for the chemical reactions and density of the adsorbed water. Furthermore, the concentration of chloride needed to model the pore water chemistry as a function of the dry density has been obtained.

1 INTRODUCTION

Compacted bentonites are being considered in many countries as a backfill material in high-level radioactive waste disposal (HLW) concepts because of their low permeability, high swelling capacity, high plasticity and high sorption. In this context, a knowledge of the pore water chemistry in the clay barrier is essential for the performance assessment since the pore water composition is an important parameter, which affects the release and transport of the radionuclides, the canister corrosion, the dissolution of the waste matrix, the sorption on mineral surfaces, the solubility of radionuclides, etc.

The pore water chemistry in bentonites is the result of the interactions among water, solutes and clay occurring in the clay/water system. For this reason, in order to determine the chemical composition of the pore water, it is necessary to know the mineralogical and chemical components of the clay system, its physico-chemical characteristics, the hydration mechanisms, as well as the types of waters, porosity, microstructure and the ion diffusion pathways of compacted bentonites.

An important issue in calculating the pore water chemistry in high-clay mineral content media is to distinguish between the types of water and their distribution in the clay/water system (absorbed water, interlayer water and pore water), as well as, their respective volumes accessible to cations and anions. Different techniques have been used to estimate the amounts of free water (external water) and available interlayer water: (a) thermogravimetric analysis (TG and DTA); (b) X-ray diffraction (XRD) analysis as a function of the water content and temperature; (c) nitrogen and water vapour adsorption/desorption isotherms and (d) determination of the accessible or geochemical porosity. The results of these techniques are described in this paper. The purpose of this work is to study the quantity, state and location of the water in the FEBEX bentonite.

2 MATERIALS

The Spanish concept for the geological disposal of radioactive wastes considers granite formations as a suitable host rock for underground repositories. The canisters containing the spent fuel are placed horizontally in drifts and surrounded by a clay barrier constructed with highly-compacted bentonite blocks (ENRESA, 1994). The clay material used is a bentonite from Cortijo de Archidona deposit (Almería, SE Spain) which has been selected by ENRESA (Empresa Nacional de Residuos Radiactivos) as a suitable material for the sealing and backfilling of HLRW repositories.

The conditioning of the bentonite in the quarry, and later in the factory, was strictly mechanical (homogeneization, rock fragment removal, crushing, drying at 50–60°C and sieving) to obtain a granulated material with the specified characteristics: grain-size distribution (≤ 5 mm) and 14 wt% water content. Some 300 tons of this bentonite were acquired for characterisation studies and manufacturing of bentonite blocks compacted at 1.65 g/cm^3 of dry density used as an engineered barrier within the FEBEX project during the *in situ* (Grimsel, Switzerland) and *mock-up* (Madrid, Spain) tests (Huertas et al., 2000).

The major mineral phase (92%) of the FEBEX bentonite is a montmorillonite (dioctaedric aluminic smectite with a 2:1 structure, which consists of two silica tetrahedral sheets with a central alumina/magnesium octahedral sheet). This bentonite was formed by alteration of pyroclastic volcanic rocks, such as poligenitic tuffs and porphyritic agglomerates of rhyolitic-rhyodacitic nature (Fernández-Soler, 1992).

The main physico-chemical characteristics are summarised as follows. Other characteristics can be found in Huertas et al. (2000), Villar (2002) and Fernández et al. (in press a,b). The specific weight is 2.70 ± 0.04 g/cm^3, the liquid limit is $102 \pm 4\%$, the plastic limit is $53 \pm 3\%$, the saturated permeability at 1.65 g/cm^3 is $3 \cdot 10^{-14}$ m/s, the gravimetric water content at equilibrium of the clay under laboratory conditions (relative humidity $50 \pm 10\%$) is about $13.7 \pm 1.3\%$.

In spite of the high smectite content, the bentonite contains numerous accessory minerals. Some of them are neoformed minerals (calcite: 0.6 ± 0.1 wt%, cristobalite: 2 ± 1 wt%), and others are rests of the original volcanic rock minerals (quartz: 2 ± 1 wt%, plagioclase: 3 ± 1 wt%, K-felspars: traces, tridimite: traces) which appear nearly unaltered. Other accessory minerals are mica (biotite, sericite, muscovite), chlorite, non-differentiated Al, K, Fe, Mg and Mn silicates, Augite-diopside, hypersthene, hornblende, oxides (ilmenite, rutile, magnetite, Fe-oxides), phosphates (apatite, xenotime, monacite) and non-differentiated titanium minerals and rare earths, which have been determined by weight from dense concentrates and SEM identification, their total content being 0.8 wt%. Some minerals, such as carbonates (calcite, dolomite), chlorides (0.13 ± 0.02 wt%) and sulphates (gypsum (0.14 ± 0.01 wt%), barite, celestite (0.019 ± 0.004 wt%)), have been determined by a normative calculation and SEM identification. Attention is brought to the content of these last minerals present at trace levels in the bentonite because of their influence on the chemistry of the pore water (Fernández et al., 1999, 2001).

Based on XRD and the positions (CuK$_\alpha$) of the 001/002 and 002/003 reflections (Moore & Reynolds, 1989), the smectitic phases of the FEBEX bentonite are actually made up of a smectiteillite mixed layer with $\sim 11\%$ of illite layers ($\Delta 2\theta = 5.502$). In general, the clay mineral in this deposit has an interstratification average value of 7–15% (Cuadros and Linares, 1996). The thickness of the FEBEX smectite quasicrystals is around 102 ± 5 Å, calculated by means of the Scherrer equation. The quasicrystals of the saturated FEBEX smectite consist on 6 lamellae or layers stacked along the crystallographic c-axis. The Biscaye index of the smectite crystallinity is 0.966.

Based on chemical analyses, the structural formula or unit-cell formula of the Ca-conditioned FEBEX smectite is (Fernández et al., in press a,b):

$$(\text{Si}_{7.78}\,\text{Al}_{0.22})^{IV}\,(\text{Al}_{2.77}\,\text{Fe}^{3+}_{0.33}\,\text{Fe}^{2+}_{0.02}\,\text{Mg}_{0.81}\text{Ti}_{0.02})^{VI}$$
$$\text{O}_{20}\,(\text{OH})_4\,(\text{Ca}_{0.50}\text{Na}_{0.08}\text{K}_{0.11})_{1.19}$$

Tetrahedral charge: -0.22, Octahedral charge: -0.97, interlayer charge: $+1.19$.

18% of the charge is due to tetrahedral substitution, and 82% to octahedral replacement. According to the chemical analyses, the theoretical exchange capacity is 1.05 eq/kg and therefore, the total charge per unit-cell is 0.79. The 0.11 K ions per formula unit are not exchangeable and belong to the 11 wt. % of illitic layers corresponding to the smectite-illite mixed layers in the FEBEX bentonite clay. The *surface charge density* (σ), i.e., the excess charge per unit surface area of this smectite is 0.136 C/m^2.

The total specific surface area was determined by two methods. The first method is based on the unit cell parameters ($a = 5.156 \, \text{Å}$ and $b = 9.0 \, \text{Å}$), obtained from the XRD analyses, and the unit weight (752.46 g per 44 atoms of oxygen) (Horseman et al., 1996). The calculated value was $746 \pm 4 \, \text{m}^2/\text{g}$. The second one is based on the Keeling's hygroscopic method from an adsorption isotherm of water in a constant relative humidity atmosphere (75%) of saturated NaCl. In this method, the weight variation of a dried clay material is analysed over a period of several weeks. The total specific surface area of the FEBEX bentonite obtained by this method was $S_{\text{TOTAL}} = 725 \pm 47 \, \text{m}^2/\text{g}$.

3 METHODOLOGY

The **dry density** of the samples was determined by the mercury displacement method.

The percentage water **content** is defined as the ratio between the weight of the water lost after heating the sample at 110 or 150°C for 24 hours and the weight of the dried clay.

The mineralogical analysis was performed by **X-ray diffraction** techniques. Diffractometer patterns from samples dried at 60°C and powdered to a particle size less than 60 μm were analysed. From the <2 μm particle size fraction, obtained by sedimentation, the aggregate oriented diffractometer patterns were obtained on an air dried sample under glass plate (OA), an ethylene glycol solvated (OA + E.G.) and heating at 300° and at 500°C. A Philips PW 1730 diffractometer has been used, with an anticatode of Cu (CuK$_\alpha$), using an acceleration voltage of 35 Kv and a current intensity of 40 mA in the filament. The apertures of divergence and reception were 1 and 0.2°, respectively, using a Ni filter. The exploration velocity was of 2°/min for the disoriented powder sample (total analysis of the sample) and 1°/min for the randomly oriented preparations with the <2 μm fraction. On the powder diffractometer a semiquantitative analysis of the present minerals in the sample was carried out (% in mass). The percentage of the mineral contents of the sample was corrected according to their corresponding reflectance powder (Schultz, 1964).

The X-Ray diffraction data were also used to determine the distribution of the water molecules perpendicular to the (a, b) plane of the layer of clay samples under different states of hydration, immediately after the water vapour adsorption tests.

Thermal Analyses (TG/DTG/ATD) of the bulk sample have been determined using a Perkin-Elmer 1700 DTA (differential thermal analysis) apparatus at a scan rate of 10°C/min and under a dynamic air atmosphere (40 cm^3/min) at temperatures between 25°C and 980°C.

Conventional N$_2$ adsorption/desorption isotherms were obtained at 77 K (nitrogen liquid temperature) in a discontinuous volumetry sorptometer, Micromeritics ASAP 2010. Prior to the measurement, the samples were outgassed by heating at 90°C under a residual vacuum of 0.01 Pa. Depending on the porosity, about 0.05 to 0.25 grams of the sample was used. Surface areas were determined using the standard N$_2$-BET method. The presence of micropores in the samples were assessed using the t-plot method (de Boer et al., 1966).

Mercury intrusion porosimetry has been determined by using a Micromeritics Poresizer 9320 apparatus. The interval of pressure injection was from 7 kPa to 210 MPa, measuring apparent pore diameters from 200 to 0.006 μm. Compacted samples were dehydrated by means of a liophilization process, prior to the tests.

Mercury does not wet most substances and does not penetrate pores by capillary action. It has to be forced to do so. Entry into pore spaces requires applying pressure in inverse proportion to the opening size. When mercury is in contact with a pore opening of circular cross-section, the surface tension of the mercury acts along the circle of contact for a length equal to the perimeter of the circle. Thus the force with which mercury resists entering the pore is equal to $-\pi D \gamma \cdot \cos \theta$, where D is the pore diameter, γ the surface tension and θ the contact angle. The force due to an externally applied pressure acts over the area of the circle of contact and is expressed mathematically by $-\pi D \cdot P/4$, where P is the applied pressure. At equilibrium, the opposing forces are equal; thus,

$$-\pi D \gamma \cos \theta = \frac{\pi D^2 P}{4} \quad \text{or} \quad D = \frac{-4\gamma \cos \theta}{P}$$

This equation is known as the Washburn equation (Webb & Orr, 1997).

A water vapour adsorption/desorption isotherm was measured at 303 K by gravimetry in quasiequilibrium mode using dessicators and an analytical balance. Prior to adsorption, Febex bentonite samples, crushed to <0.1, mm were heated at 150°C for 48 h. Under these conditions, the bentonite was completely dehydrated (the measured interlayer space of the montmorillonite equaled 9.8 Å) without reaching dehydroxylation. Afterwards, 1–2 g of the dried bentonite were placed in crucibles in dessicators at 20°C and 10^{-3} mmHg vaccuum in water vapour atmosphere controlled by means of saturated saline solutions of different chemical compounds (CRC Handbook, 1994). During adsorption, the dessicators were placed in a thermostated room at 293°K. Once the equilibrium was reached (after $\approx$1 month), the samples were weighed ($\pm$0.0001 g) to determine the quantities of adsorbed water (mg H_2O/g of dry clay). Following this, the interlayer distance d_{001} at different vapour pressures or relative humidities was immediately determined using XRD. Surface areas were measured using the BET method. A generalization of the Dubinin-Astahakov isotherm (Dubinin, 1967) and the t-plot method (Hagymassy et al., 1969) were used to describe the water vapour adsorption into micropores.

4 RESULTS AND DISCUSSION

4.1 *Preliminary considerations*

Primary clay particles consist of a coherent stack of silicate layers. These stacks are called quasicrystals which often occurs as aggregates displaying various textures determined by the shape and arrangement of the constituent primary particles. The cluster of the clay aggregates may generate a micro-structure (fabric) often with large pores between the constituent clay aggregates. Thus, a clay may develop three kinds of pores: interlamellar pores within the primary particles, intraaggregate pores (0.01–0.2 μm size) delineated by the boundaries of primary particles within the aggregates, and interaggregate pores (0.2 to several micrometers size) within the microstructure (Göven, 1993; Touret et al., 1990).

Most properties of the clay minerals are due to their small grain size and their sheet-like shape which implies very high surface areas relative to the mass of the material (van Olphen, 1977). Clay particles may present two types of reactive basal surfaces that may be distinguished as "siloxane" (smooth planes of basal oxygens attached to the silica tetrahedra forming pseudohexagonal rings) and "hydroxide" (molecularly smooth planes of hydroxyl ions) surfaces. The surface reactivity of the solid phases derives form the chemical behaviour of these surface functional groups. Smectites and other 2:1 type layer silicates are bound by siloxane surfaces on both sides. Two mechanisms are responsible for the development of electric charges on the surfaces: (1) ionic substitutions in tetrahedral and octahedral sheets, which are assumed to be responsible for the creation of permanent charges on the basal surfaces of the layer silicates and, (2) deprotonation of the two types of edge-surface hydroxyl groups (silanol and aluminol).

Clay hydration involves adsorption of water molecules on the clay surfaces that are exposed in different pore spaces of the clay. As a consequence, the specific surface area will influence the amount of adsorbed water. Three modes of clay hydration can be distinguished and may take place simultaneously with the increasing water activity or relative humidity (Güven, 1993): (a) interlamellar hydration which involves the adsorption of limited amounts of water molecules on the internal surfaces of primary clay particles, named *interlamellar water*; (b) continuous (osmotic) hydration which is related to an unlimited adsorption of the water on the internal and external surfaces of primary particles, named *intraparticle water*; and (c) capillary condensation of free water in micropores within the clay fabric (i.e., in the interaggregate and intraaggregate pores), named *interparticle water*.

The swelling pressure is due to interlamellar hydration of expandable minerals and to osmosis related to interparticle forces. The first mentioned depends on the type of adsorbed interlamellar cation, which determines the maximum number of hydrates and hence the maximum c-dimension of the stacks of lamellae (Push, 1999). In the first stage of hydration of the dry clay particles, water is adsorbed in successive monolayers on the surfaces and pushes the particles or the layers of a montmorillonite clay apart. In this stage, the principal driving power is the adsorption energy of the water layers on the clay surface. The second stage of swelling is due to double-layer repulsion. In this stage, one usually speaks about osmotic swelling and it was found to be inversely proportional to the square-root of the salt concentration in the solution (Norrish, 1954). At plate distances beyond about 10 Å (equivalent to four monomolecular layers of water), the surface hydration energy is no longer important, and the electrical double-layer repulsion becomes the major repulsive force between the plates. Large volume changes accompany this stage of swelling, but there is often a limit to the swell caused by the formation of cross links by edge-face (EF) and face-face (FF) associated particles leading to the formation of gels. The hydration complexes between these distant layers are not organised as in the interlamellar swelling, but consist of a *continuous* diffuse double layer. Such a set of smectite layers that are separated by their overlapping double layers is called tactoid. In a quasycrystal, the silicate layers are held together by attractive electrostatic forces related to the coupling by the hydration complexes of interlayer cations. The tactoid, on the other hand, maintains its integrity by long-range repulsive forces due to overlapping double layers of smectite lamellae. The individual smectite layers are separated by their diffuse double layers in a tactoid but by their interlamellar hydration complexes in a quasicrystal. Diffuse double layer may develop only on the external surfaces of quasicrystals (Göven, 1993).

4.2 Surface physical properties of the FEBEX bentonite powder

4.2.1 Characterisation of the dry state by X-Ray diffraction, DTA/TG and nitrogen adsorption/desorption isotherms and mercury porosimetry

The determination of the amount of water actually available for chemical reactions/solute transport is one of the main issues in bentonite pore water modelling. Different techniques have been used to analyse the different types of waters, e.g. XRD analysis, thermogravimetric (TG) analysis, etc. Textural properties, such as the surface area and the pore size distribution, were undertaken by combining N_2 gas adsorption/desorption and mercury porosimeter (PSD). The association of these techniques has led to a continuous description of porosity at various scales from macroporosity to microporosity. Micro and mesopore size distributions were obtained by using N_2 gas adsorption (range from 0.3 to 300 nm). Meso and macropore size distributions were obtained by using mercury intrusion analysis (range from 360 to 0.003 µm).

4.2.1.1 Nitrogen adsorption/desorption isotherms

The dry state can be further characterised by nitrogen adsorption-desorption isotherms. In the adsorption of non-polar molecules such as N_2, the nitrogen molecules do not penetrate between the layers. The external surface area measured corresponds to the external faces and the edges of the montmorillonite particles. The BET micro, meso and total specific surface areas were determined according to t-plots and Harkins-Jura methods (de Boer, 1966; Cases et al., 1992).

Figure 1 shows the N_2 adsorption-desorption isotherms obtained from the total powder fraction of a FEBEX bentonite, expressed in terms of the volume of gas adsorbed per unit mass, V_a (Fernández, 2003). The adsorption isotherm has the characteristic features of the type-IV isotherm with a very important hysteresis loop due to capillary condensation in mesopores formed between different quasicrystals. The curves exhibit a H3 hysteresis loop in desorption, which is characteristic of the presence of slit-shaped pores. Table 1 shows the relevant parameters deduced from the BET and t-plot treatments. The specific surface area calculated from the BET method is 56.4 m²/g while that obtained from the B point (P/P_o = 0.057) is 57.0 m²/g. The adsorption isotherm was replotted following the procedure of de Boer et al. (1966) in the form V_a versus t (not shown), where t is the mean thickness of the adsorbed layer calculated according to the Harkins & Jura method. This method has been used to analyse the micro-porosity. The original slope of the t plot indicates the total specific surface area, 61.5 m²/g (range t: 0–3.35), whereas a positive intercept (converted to liquid volume) indicates the presence of micropores, $V_o = 8.60 \cdot 10^{-3}$ cm³/g (range t: 3.35–6.9) and mesopores, $V_o = 1.20 \cdot 10^{-2}$ cm³/g (range t: 6.9–12.34) having a total volume i. The FEBEX bentonite is microporous with microporous volumes accounting for ~43% of the BET monolayer capacity.

The average number of clay layers per quasicrystal was calculated from the nitrogen adsorption data assuming that the plates (square parallelipieds = 3000 Å in the lateral direction and a multiple of the d_{001} in height (thickness being 9.6 Å) are perfectly stacked as in a deck of cards. It is possible to

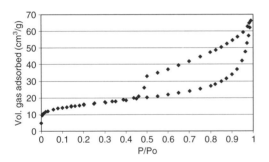

Figure 1. Adsorption-desorption isotherms of nitrogen at 77 K onto FEBEX bentonite.

Table 1. Parameters deduced from BET and t-plot treatment on the adsorption of N_2 at 77 K.

C_{BET}	V_m (cm³/g)	S_{BET} (m²/g)	St_{or} (m²/g)	V_{tot} (cm³/g)	S_{micro} (m²/g)	V_{micr} (cm³/g)	$S_{ext\ micro}$ (m²/g)	$S_{ext\ meso}$ (m²/g)	V_{meso} (cm³/g)
388	$1.99 \cdot 10^{-2}$ (l)	56.38	61.53	$9.6 \cdot 10^{-2}$	20.73	$8.6 \cdot 10^{-3}$ (l)	35.65	29.35	$1.2 \cdot 10^{-2}$ (l)

V_m: Monolayer capacity derived from the BET treatment; expressed as liquid (l); S_{BET}: BET Surface area.
St_{or}: Total surface area derived from the slope of the straight line passing through the origin of the t-plot.
V_{micr}: Liquid microporous volume derived from the ordinate at the origin in the second straight line of the t-plot.
$S_{ext\ micro}$: Surface area of micropores derived from the slope of the second straight line of the t-plot.
V_{tot}: Total pore volume, derived from the amount of nitrogen adsorbed at P/P_o of 0.98.
$S_{ext\ meso}$: Surface area of mesopores derived from the slope of the third straight line of the t-plot.
V_{meso}: Liquid mesoporous volume derived from the ordinate at the origin in the third straight line of the t-plot.
S_{micro}: = $S_{BET} - S_{meso} = S_{BET} - (S_{ext\ micro} - S_{ext\ meso})$: Surface area of the micropores.

determine the mean size of the quasicrystals using the relationship (Cases et al., 1992):

$$S(m^2/g) = 56.38 = 746/n + 5.13 \quad \Rightarrow \quad \text{n} \approx \text{14 layers}$$

When dry, quasicrystals of about 14 elementary condensed montmorillonite layers are formed on average. As was shown in section 4, when the FEBEX clay powder is brought into equilibrium with saturated water vapour, the mean size of the quasicrystals (Scherrer equation) remains constant during the filling of the interlayer spaces, and the size corresponds to particles about six layers thick with two or three monolayers of water in the interlamellar space. It seems that during the first stage of water adsorption, the original 14-layer-thick quasicrystals are split into smaller ones approximately 6 layers thick (Cases et al., 1992).

The analysis of the isotherms by means of the BJH method allows us to determine the mesopore volume and the pore size distribution functions. The pore diameter of the FEBEX bentonite has been estimated using the BJH method (Gregg & Sing, 1982) based on the Kelvin equation and the t values determined by the Frenkel-Halsey-Hill model (FHH) (Webb & Orr, 1997). By application of the Kelvin equation, the minimum radius of pores in which capillary condensation can take place is calculated from the relative pressure at the lower limit of the hysteresis loop:

$$Ln \frac{p}{p_o} = Ln\, a_w = \frac{-2\gamma V_m}{RT} \frac{\cos\theta}{r_m}$$

$$250\,\text{Å} < r_m < 10\,\text{Å} \quad \text{(cylindrical pores)}$$

where, γ is the surface tension of the nitrogen gas at its boiling point (8.85 mN/m at 77 K) and V_m is the molar volume of liquid adsorptive (34.68 mL/mol), R is the gas constant (8.3143 J/mol · K), T represents the boiling point of nitrogen (77 K), P/P_o is the relative pressure of nitrogen where the capillary condensation occurs, and DK is the Kelvin diameter of the pore in which condensation occurs at a relative pressure of P/P_o.

The equation for nitrogen can be reduced to:

$$D_k(\mathring{A}) = \frac{-4\gamma V_m}{RTLn(p/p_o)} = \frac{0.83208}{\log(p_o/p)} \quad \text{and}$$

$$r_k(\mathring{A}) = \frac{-2\gamma V_m}{RTLn(p/p_o)} = \frac{0.41604}{\log(p_o/p)}$$

The distribution of the pore size for the FEBEX bentonite is shown in Figure 2 (desorption branch). The accumulated specific surface area for the adsorption ($69.2\,\text{m}^2/\text{g}$) and desorption branches ($35.5\,\text{m}^2/\text{g}$) are shown in Figure 3.

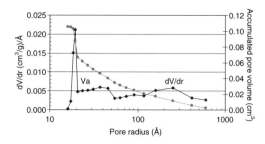

Figure 2. Pore size distribution from the desorption branch (cylindrical pores).

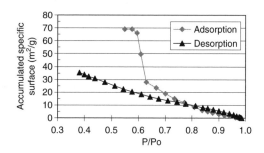

Figure 3. Accumulated specific surface areas from the adsorption and desorption branchs.

The mean radius of a group of mesopores, taking into account a cylindrical pore geometry, can be evaluated from the total pore volume and the specific surface area:

$$r_p = 2\frac{V_{mesopores}}{S_{BET}} = 37.4\,\mathring{A} \quad \text{and} \quad r_p = 2\frac{V_{mesopores}}{S_{total}} = 30.4\,\mathring{A}$$

The porosity (n) in a porous medium is given by:

$$n = \frac{V_{meso} \times \rho_d}{m} \times 100 \quad \text{due to} \quad V_p = \frac{m}{\rho_d} \times n$$

For a dry density of 1.65 g/cm³, the porosity is:

$$n = \frac{V_{mesopores} \times \rho_d}{m} \times 100$$

$$= \frac{1.0523 \cdot 10^{-1}\, cm^3 \times 1.65\, g/cm^3}{1\, g} \times 100 = 17.36\,\%$$

Thus, taking into account the mean pore size of the mesopores, calculated by the BJH method, from a total porosity of ~39%, the effective or interconnected porosity is around 17%.

The main parameters calculated from the mesoporous analysis are summarised in Table 2.

Table 2. Main parameters calculated by the BJH method.

	Desorption branch	Adsorption branch	Total
Accumulated volume	$1.0523 \cdot 10^{-1}\,cm^3/g$	$9.4702 \cdot 10^{-2}\,cm^3/g$	$9.60 \cdot 10^{-2}\,cm^3/g$
Pore surface	$69.209\,m^2/g$	$35.472\,m^2/g$	$56.38/61.53\,m^2/g$
Average pore radius	$30.4\,\text{Å}$	$53.4\,\text{Å}$	$34.0/31.2\,\text{Å}$
Modal pore radius	$19.3\,\text{Å}$	$216\,\text{Å}$	–
Porosity (at $\rho_d = 1.65\,g/cm^3$)	17.4%	15.6%	15.8%

Table 3. Different parameters determined by Hg porosimetry as a function of the dry density.

Parameters	$\rho_d = 1.55\,g/cm^3$	$\rho_d = 1.65\,g/cm^3$	$\rho_d = 1.75\,g/cm^3$
Average pore radius (4V/A) (nm)	20.5	13.1	11.2
Intruded total volume (cm^3/g)	0.0895	0.1402	0.1254
Total pore area (m^2/g)	8.7377	21.3285	22.4691
Average pore radius (Volume) (nm)	822.7	118.3	22.9
Average pore radius (Area) (nm)	4.7	5.0	4.8
Porosity (%)	17.5	28.5	25.7

Table 4. Pore size distribution by Hg porosimetry as a function of the dry density.

	Macropores				Mesopores	
	Pores > 3 µm		Pores 0.05–3 µm		Pores < 0.05 µm	
ρ_d (g/cm³)	%	Mode (µm)	%	Mode (µm)	%	Mode (nm)
1.55	65.81	21	3.35	0.0938	30.84	13.1
1.65	31.24	16.4	20.61	0.858	48.15	8.02
1.75	22.41	12.8	21.20	0.858	56.38	6.27

4.2.1.2 Analysis of the macroporosity by mercury porosimetry

Mercury intrusion porosimetry allows to determine the pore size distribution in compacted material in the dry state. In this study, samples compacted at 1.55, 1.65 and 1.75 g/cm³ have been analysed. As can be seen in Table 3, the accessible porosity to Hg is 17.5%, 28.5% and 25.7% for the dry densities of 1.55, 1.65 and 1.75 g/cm³, respectively. This is equivalent to a 41%, 73% and 66% of the total porosity (42.6% for $\rho_d = 1.55\,g/cm^3$; 35.2% for $\rho_d = 1.65\,g/cm^3$ and; 38.9% for $\rho_d = 1.75\,g/cm^3$, respectively). This doesn't seem to be corresponded to the physical idea that when the material is compacted, the porosity decreases.

There are three pore families for each type of compacted bentonite sample (Table 4, Fig. 4). For a dry density of 1.55 g/cm³, the maximum intrusion is produced to a mean pore size of 21 µm and the percentage of mesopores is ∼31%. For a dry density of 1.65 g/cm³, the maximum Hg intrusion is produced to a mean pore size of 0.86 µm. However, there are still

large pores (∼31%) with a size of 16 µm and the percentage of mesopores increases (∼48%). For a dry density of 1.75 g/cm³, the percentage of mesopores increases and the maximum intrusion is produced to a pore size of 6.27 nm. Nevertheless, there are still a high percentage (∼21%) of macropores of medium size (0.83 µm) and the large pores decrease (∼22%), centered now at a size of ∼13 µm. As it is was expected, when the compaction degree increases, the percentage of large pores decreases and new pore families with smaller pore size are originated. It is worthwhile noting that the macropores and some family of mesopores will disappear when the hydration of bentonite occurs, due to the structural expansion of the smectite as it adsorbs water.

4.2.1.3 Termogravimetry and X-Ray diffraction analyses

Powdered FEBEX bentonite samples in the *as received* state kept under laboratory conditions (relative humidity (RH) $50 \pm 10\%$) were heated to

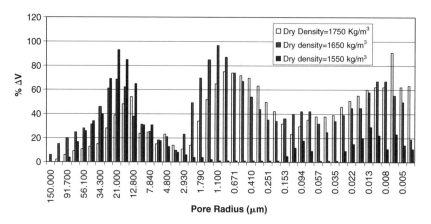

Figure 4.   Distribution of pore size families for different dry densities of compacted FEBEX bentonite.

constant weight in an oven at 105–110°C in order to determine the water content. This value, expressed as a % of the oven dried clay weight, is 13.8 ± 0.8 wt.%.

At ambient conditions (RH = 50–60%), the inter-layers of the FEBEX smectite contain two mono-layers of water, as deduced from the measured basal spacing of around 15.2 Å (FEBEX bentonite is a calcium-magnesium-sodium smectite). The swelling due to interlaminar hydration at different water contents and the shrinkage as a function of the temperature were also analysed. In wetting experiments, the FEBEX montmorillonite expands in the c-dimension to an extent equivalent to the interlayer space containing a maximum of three monolayers of water (see next section). In drying experiments (Fig. 5), the basal spacing corresponded to one monolayer of water at 100–110°C and the structure collapsed at a c-spacing of 9.6 Å at 150–200°C, i.e., essentially no water in the interlayer space. It should be highlighted that the FEBEX bentonite is not fully dehydrated at the standard temperatures at which the water content is determined (105–110°C), since one monolayer of water remains at this temperature. Only at tempera-tures above 150°C (150–220°C) is the total interlayer water lost. The structural water, or dehydroxylation water, is about 4.3 ± 0.6 wt.%, calculated from mass losses between 300 and 950°C as determined from TG analysis (discounting CO_2 and SO_2 contents).

4.2.2   Characterisation of the saturated state by X-Ray diffraction and water vapour adsorption isotherms

4.2.2.1   X-Ray diffraction analysis

The basal spacing of powdered FEBEX bentonite samples was also analysed as a function of the water content, from an initial water content of 14% up to the liquid limit of this bentonite, i.e., 102%. In these wetting experiments, the FEBEX montmorillonite expands or swells in the c-dimension to an extent

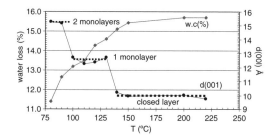

Figure 5.   Evolution of the water loss and the basal spacing as a function of temperature.

equivalent to a maximum of three monolayers of water in the interlayer space (Fig. 6).

As it is known, the swelling capacity of the smectites is due to their property of adsorbing water in the interlayer space (intracrystalline swelling). The basal spacing d(001) expands due to the hydration of smectite and the formation of successive monomo-lecular layers of water. The extent of crystalline swelling is controlled by a balance between relatively strong swelling forces, due to the hydration potential of the interlayer cations and charge sites, and electrostatic forces attraction between the negatively charged 2:1 phyllosilicates layers and the positively charged interlayer cations. Thus, the sorption of water into interlamellar space is governed by the hydration energy of the interlayer cations and their polarization degree (charge/size relationship), the location of the layer charge of the adjacent silicate sheets, the relative vapour pressure, the water content and the salinity of the solutions surrounding the clay particles (Norrish, 1954; van Olphen, 1977; Sposito, 1983; Newman, 1987; Laird, 1999; Cases et al., 1992). Bradley et al. (1937) showed that water molecules could be sorbed in monomolecular sheets

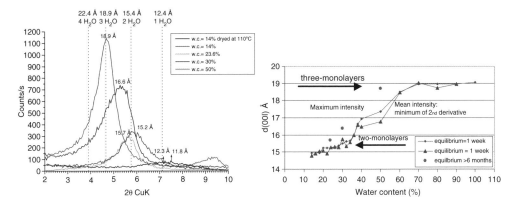

Figure 6.   Variation of the basal spacing, d(001), as a function of the water content.

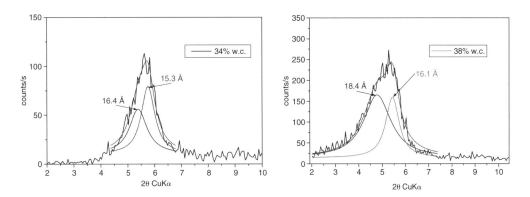

Figure 7.   Decomposition of the d(001) peak for samples with different water contents.

and that apparently definite and discrete hydrates of one, two, three and four layers of water are possible, having cell heights of 12.4, 15.4, 18.4 and 21.4 Å, respectively. However, usually a continuous variation of d(001) with water content have been experimentally found. Hendricks and Jefferson (1938) theoretically showed that the c-axis dimension spacing varies continuously but not uniformly with water content as a result of an averaging effect from a lattice that contains various numbers of water layers in different parts. That is, the apparently continuous change in the c-dimension results from a random alternation of successive discrete hydrates (Grim, 1968; MacEwan and Wilson, 1980). This phenomenon can be observed in the FEBEX bentonite (Fig. 6 and Fig. 7) as a function of the water content. At a water content (w.c.) of 18% the basal spacing is 15.0 Å, which corresponds to two-monolayers of water. The basal spacing changes from 15.3 Å, at 28% of w.c., to 15.7 Å., 16.6 Å and 17.3 Å at w.c. of 34%, 38% and 50%, respectively. At a water content of 60%, the basal spacing is 18.5 Å (three-monolayers

of water), which is maintained in 18–19 Å up to a water content of 100%. This means that, in wetting experiments, the smectite basal spacing increases as a function of the water content from two-monolayers of water to a maximum of three-monolayers, but through a continuous of $2°\theta$ distances. Thus, we observe average values from a lattice that contains a random alternation of successive discrete hydrates in bentonite with water contents from 14% to 100%. For this reason, the d(001) peak can be decomposed in different populations with quasicrystals formed with two and three monolayers of water (Fig. 7). It is worthwhile noting that: (i) according to different tests performed, the water adsorption is also a kinetic process. In samples with the same water content but equilibrated with water during six months, the basal spacing is higher than samples equilibrated during one week (Fig. 6) and, (ii) as we can see later on, the bi-mononalyer state has two domains. One corresponds to the filling of the second solvation shell (with a basal spacing of ∼15.2 Å) and the other to the filling of the free siloxane cavities (∼16 Å).

265

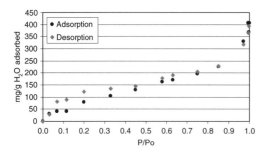

Figure 8. Water vapour adsorption-desorption isotherm for the FEBEX bentonite powder.

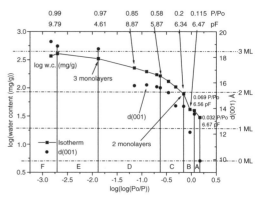

Figure 9. FHH water sorption plot obtained for the FEBEX bentonite describing the different states and locations of the water. Each change of slope in the curve means a different state. Thus, the letters A, B, C, D, E and F correspond to domains where water retention is due to different hydration sites and mechanisms. The basal spacings d(001) are also shown on the right axis corresponding to the filled circles.

4.2.2.2 Water vapour adsorption/desorption isotherms

The saturated state of the FEBEX bentonite was characterised by water-vapour adsorption-desorption isotherms (Fig. 8).

The Frenkel, Halsey and Hill (FHH) formalism was used (Prost et al., 1998) to identify and describe the different states and locations of the water and the mechanism involved in the water retention phenomenon as a function of the activity of the water, a_w (or P/P_o or relative humidity). The FHH curves were obtained by plotting log w as a function of log $(\log(P_o/P))$. This plot is similar to log w versus pF. In Fig. 9, six domains were distinguished from an analysis of the slope changes, which are related to the different states and location of water. Moreover, these domains also corresponded to the d_{001} spacings obtained from XRD measured at each water activity, a_w. The amount of water uptake in each domain was quantified (Table 5) by means of the Dubinin-Radushkevish equations (Kraehenbuehl et al., 1987; Gregg & Sing, 1982; Webb & Orr, 1997):

$$W = W_o \exp\left[-\left(\frac{RT}{\beta E_0}\right)^2 \ln^2\left(\frac{P_0}{P}\right)\right] \quad \text{and}$$

$$\ln W = \ln W_o - D \ln^2\left(\frac{P_0}{P}\right)$$

where, W_o is the total volume of the micropores accessible to the given adsorbate (mmol/g), W represents the volume of the adsorbate condensed within the micropores at temperature T and relative pressure P/P_o (P is the partial pressure of the adsorbate, and P_o is the saturation vapour pressure of the adsorbate.), E_o is the so-called characteristic free energy of adsorption for a standard vapour, which depends on both the vapour and the pore system, i.e., E varies with the pore size; β and n are specific parameters of the adsorbate. β is the similarity coefficient or affinity coefficient and is the ratio of adsorption potentials of the adsorbate to a reference adsorbate, taken as 1 or 0.34 (in this paper a value of $\beta = 1$ has been considered).

Domain A (see Table 5) is observed at the lowest values of a_w (0.032–0.069) and corresponds to adsorption on the external surfaces of the stacks, edges and edge surface sites. This domain corresponds to the original spacing of the montmorillonite in the dry state with virtually no water in the interlayer space (closed state, $d_{001} = 9.8$ Å). Based on the analysis of the data, adsorption on external surfaces is 60.36 mg/g. Thus, taken into account the adsorption isotherm and the calculated value, the external water has a value between 3 to 6%. This quantity, $\sim$5%, is interpreted as the *free water* volume and will be used for modelling of porewater in the *as received* bentonite (5 mL of H_2O in 95 g of dry bentonite, i.e., a solid to liquid ratio of $\sim$19 kg/L dry FEBEX bentonite). *Domain B* ($a_w = 0.069$–0.1) corresponds to the adsorption of water for filling a monolayer of water in the interlayer space. Water is adsorbed around the interlayer cations and the c-spacing increases from 9.6 to 12.1 Å. The amount of interlayer water in the complete one layer hydrate is 102.4 mg/g, which theoretically covers a surface area of 363 m²/g. *Domain C* ($a_w = 0.2$–0.58) corresponds to the solvating of interlayer cations (possibly second solvation shell) in the two-layer hydrate ($d_{001} = 15.2$ Å). The amount of interlayer water in the two-monolayer is 183.4 mg/g, which theoretically covers a surface area of 650 m²/g. *Domain D* ($a_w = 0.58$–0.97) corresponds to the filling of free siloxane cavities, where no interlayer cations exist, and to multilayer adsorption on exterior surfaces. In this domain, the spacings between the unit layers

266

Table 5. Quantification of the amount of water, surface area, energy and number of water molecules per ion in each domain (location of water) by means of Dubinin's equations.

	c-spacing	a_w	mg/g	S_o (m²/g)	E (kJ/mol)	H₂O molecules/ion
Domain A	~9.8 Å	0.032–0.069	60.36	213.89	10.0285	3.3
Domain B	10.3–12.1 Å	0.069–0.1	102.38	362.83	6.3448	5.6
Domain C	15.2 Å	0.2–0.58	183.42	650.01	3.5521	10.8
Domain D	15.8 Å	0.58–0.97	274.16	971.58	1.5400	14.9
Domain E	18.8 Å	0.97–0.99	378.99	1343.8	0.5502	20.6

remain almost constant ($d_{001} = 15.8$ Å) but the system continues adsorbing water. Due to the fact that multilayer adsorption on exterior surfaces could occur simultaneously with interlayer adsorption, it is difficult to determine how much water is present in the completed two-layer complex between the unit layers. The total amount of adsorbed water obtained for this domain is 274.2 mg/g, which theoretically covers a surface area of 972 m²/g. *Domain E* ($a_w = 0.97$–0.99) corresponds to the filling of the three-layer hydrate ($d_{001} = 18.8$ Å) and the total amount of water adsorbed is 379 mg/g. At this point, $a_w > 0.99$, the interlayer space is further filled with water and the capillary condensation in intra-aggregates or inter-aggregates begins to occur (*Domain F*).

In Table 5, the amount of water adsorbed in each domain, as well as the number of water molecules per ion solvating the cations corresponding to the closed state, monolayer, two-monolayer and three-monolayer (3.3, 5.6, 10.0–14.9 and 20.6, respectively), are shown. According to this model, the changes in structure by interlayer swelling can be observed. It is worthwhile noting that the FEBEX bentonite is in Domain C at laboratory or initial conditions (RH = 50 ± 10%) and, for this reason, the amount of external water present in the system corresponds to the adsorbed water in Domain A. Thus, the *as received* bentonite contains 52.6 mL of water per kilogram of dry clay.

Other methods have been used to obtain the amount and types of water in the FEBEX bentonite: Cases's model (Cases et al., 1992), Laird's model (Laird, 1999) and Touret's model (Touret et al., 1990), in order to compare the results obtained.

In Cases's model, if the total external surface determined by nitrogen adsorption/desorption isotherms and the values of the water vapour adsorption isotherms are known, the amount of external, Q_{ext}, and internal water, $Q_{m(i)}$, per gram of montmorillonite can be calculated:

$$Q_{ext} = \theta \frac{S_{ext}}{\sigma N_A} \quad \text{and} \quad Q_{m(i)} = i \frac{S_{int}}{2\sigma_i N_A}$$

where, S_{ext} is the external surface area, θ is the surface coverage, calculated by means of the t,

adsorbed water film thickness (Hagymassy et al. (1969)); N_A the Avogrado number, σ the cross-sectional area of the water molecule strongly adsorbed on the external surface (~14.8 Å), σ_i is the cross-sectional area of the molecule in the hydrated state considered (being, 7.8 Å for $i = 1$ (monolayer state) and 8.7 Å ($i = 2$ and $i = 3$ for bi-monolayer and three-monolayer states, respectively) and, S_{int} is the internal surface area.

The theoretical amount of water in the interlayer space is: 132.5 mg/g, 239.0 mg/g and 358.5 mg/g for the monolayer, bi-monolayer and three-monolayer state, respectively. The amount of water in the external surface varies as a function of the relative humidity, but always, a small amount of external water exists in the system. The results are shown in Figure 10. Most of the water taken up resides in the interlayer space, *interlayer water*, and the rest is *external water*. As can be seen in Fig. 10, there are very low volumes of *free water* in this system.

In Laird's model, the hydration of Ca and Mg-saturated expandable 2:1 phyllosilicates can be calculated by the following relationship:

$$\theta_w = 0.5 \cdot \rho_w \cdot (d - D)(S_h - S_x) + \varepsilon \cdot S_x \sigma$$

where, θ_w is the gravimetric water content (g H₂O/g clay), ρ_w is the density of the interlayer water (~$1 \cdot 10^6$ g/m³), d is the d(001) value (m), D is the unit layer thickness ($9.6 \cdot 10^{-10}$ m), S_h is the hydratable surface area (m²/g), S_x is the external surface area (m²/g), ε is the total amount of water per charge site on external surfaces ($7.18 \cdot 10^{-22}$ gH₂O/e) and σ is the surface charge density (e/m²).

The first term in this equation estimates the quantity of interlayer water as the product of interlayer volume and the density of the interlayer water. The second term calculates the water retained on external surfaces as the product of the number per charge sites located on external surfaces ($S_x \cdot \sigma$) and the water retained by each charge site (ε). In this model, a restricted multilayer adsorption of water on external surfaces is assumed in contrast to Cases's model. By applying this model to the case of the FEBEX bentonite, the amount of interlayer water is

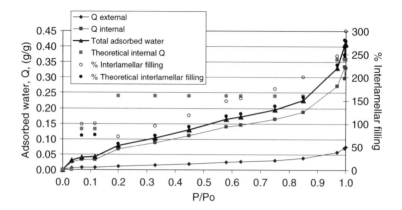

Figure 10. Water adsorption isotherm (V adsorbed) for the FEBEX bentonite, expressed as the amount of water adsorbed on the external surface (Q external) and interlamellar space (Q internal).

Table 6. Internal or interlamellar water as a function of the number of stacked layers per quasicrystal.

	M = 14	M = 7	M = 6	M = 5
Interlayer water (g/g)	0.326	0.301	0.292	0.280

32.2 mg/g and the external water is 38 mg/g. That is, the interlayer water content is ~32.2% and the external water content is ~3.8%.

In the case of Touret's model (1990), the interlayer or internal water is calculated by the relationship:

$$W_i = \frac{S(M-1))}{2M} \times \Delta d(001) \cdot 10^{-4}$$

where, M is the number of stacked layers in a quasicrystal, S is the total specific surface (m²/g), $\Delta(001) = d(001) - d(001)^{dry}$ (Å) and W_i is the internal water of the particle (g/g). If the total adsorbed water and the internal water are known, the interparticle water can be calculated. The results are shown in Table 6 and Table 7. According to this model, there is no water in the external surfaces up to a relative vapour pressure of 0.97. This is different in relation to the other models, in which a proportion of external water always exists for each a_w value.

The main results obtained with the different methods used to distinguish the types and distribution of water in the FEBEX bentonite are shown in Table 8.

4.2.2.3 Types, state and location of waters in FEBEX bentonite powder

The maximum amount of internal or interlayer water in the FEBEX bentonite is ~30–36%, once the results obtained with the different methods have been compared. This amount seems to depend on the number of stacked layers per quasicrystal. There is an amount of external water of about 3–6% at initial conditions (R.H = 50–60%). This external water can increase up to ~8–11% as a function of the relative humidity and the number of stacked layers per quasicrystal. The Dubinin-Radushkevish's equation and Cases's model allow to determine the maximum amount of water in the one-monolayer, two-monolayers and three-monolayers of water states, both results being comparable. According to the literature review, the more stable state in smectites is two-monolayers of water, which corresponds to the one formed by the cation solvation shells (first and second shell) and the filling of the free siloxane cavities. The total amount of internal or interlayer water in this state is around 21–27%. However, there is always a small fraction of external water.

As summary, the different types of waters in the FEBEX bentonite powder, their state, location and quantity are shown in Table 9.

4.2.2.4 Density of the interlayer water

The adsorbed water density (ρ) can be calculated from the measurement of c-axis spacing and the water content plus the assumptions that the surface area is known and the water is spread uniformly over the area (Martin, 1962):

$$\rho = \frac{W}{S\Delta}$$

where, W is the water content (g/g), S is the surface area (m²/g) and Δ is half the interlayer distance (Å).

In the case of the FEBEX bentonite, the X-Ray diffraction values as a function of P/P₀ have been used to calculate the density of the adsorbed water by means of this equation. The results obtained with the Cases's model have been used, where the internal adsorbed water and S_{int} for each P/P₀ values are known. The results are shown in Figure 11 and Table 10. Villar (2002) found values of adsorbed water of FEBEX bentonite samples at different water content

Table 7. Inter-particle (inter-aggregate and intra-aggregatte) water (g/g) as a function of the number of stacked layers per quasicrystal.

P/P$_o$	Total V$_{adsorbed}$ (g/g)	M = 14	M = 7	M = 6	M = 5
1	0.4075	0.082	0.107	0.115	0.13
0.997	0.368	0.042	0.067	0.076	0.09
0.995	0.4067	0.081	0.106	0.115	0.13
0.97	0.3305	0.005	0.030	0.038	0.05
0.85	0.2268	−0.099	−0.0742	−0.065	−0.05

Table 8. External and internal water obtained by means of different methods (in g/g).

Methods	Total external water	Total internal water	One-layer hydrated	Two-layer hydrated	Three-layer hydrated
BET Method	–	–	0.082	–	–
t-plot Method	0.056	0.311	–	–	–
FHH & Dubinin	0.03–0.06	–	0.102	0.183–0.274	0.379
Laird's Model	0.038	0.322			
Cases's Model	0.002–0.075 f(a$_w$, σ_{H_2O}10.6 Å) 0.003–0.105 f(a$_w$, σ_{H_2O}14.8 Å)	–	0.132	0.239	0.358
Touret's Model	0.038–0.115 (f(M = 6))	0.292 (f(M = 6))			

f(a$_w$, σ_{H_2O}) indicates values as a function of the water activity and the cross-sectional area of the water molecule; f(M) indicates values as a function of the number of stacked lamellae to form a quasicrystal.

Table 9. Types, state and location of waters in the FEBEX bentonite powder.

Method	Type of water	Subdivision	Value (%)	Energy (kJ/mol)	d(001) Å	a$_w$
TG	Structural water	–	4.3 ± 0.6	–	–	–
		Monolayer of water	10.2	6.34	10–12	0.06–0.1
		Two-monolayer of water: 2nd solvation shell	18.3	3.55	15.2	0.2–0.58
FHH + Dubinin	Interlayer water	Two-monolayer of water: Filling water of the free siloxane cavities + (external water)	21.4–27.4(*)	1.54	15.8	0.58–0.97
		Three-monolayer of water	31.9–37.9(*)	0.55	18.8	0.97–0.99
FHH + Dubinin	External water	DDL + free	3–6	10.3	∼9.8	0.03–0.06

(*) The values 27.4% and 37.9% can be over-estimated, since accumulated values can be obtained with this method. For this reason, the value in these two states has been calculated by subtracting the amount of external water. Thus, the real values for two-monolayer and three-monolayer would be: bi-monolayer, 21.4% and three-monolayer: 31.9%.

as a function of the dry density which are greater than 1 g/cm^3; changing the water density from 1.094 g/cm^3 to 1.225 g/cm^3 for dry densities between 1.28 to 1.71 g/cm^3.

If the values obtained with the FHH + Dubinin method are considered, the results are somewhat different (Table 11). In any case, the density of the adsorbed water is greater than 1.0 g/cm^3 for activities of water lower than 0.97.

4.3 Distribution of water as a function of the dry density of the compacted bentonite

As it was shown, the compaction of the bentonite originates an intergranular porosity which is a function of the compaction degree (dry density) and the amount and size of the accessory minerals. In a compacted material, the total volumetric water content is equivalent to the total pore volume. The

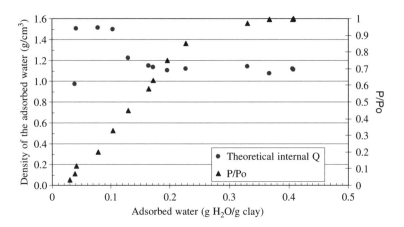

Figure 11. Density of adsorbed water for the FEBEX bentonite as a function of the type of water: external, monolayer, two-monolayers and three-monolayers of water.

Table 10. Location and adsorbed water density for each type of water in the FEBEX bentonite.

Location	State	Basal spacing d(001)	$a_w = P/P_o$	ρ_w (g/cm^3)
Zone A, i = 0	Dry, external water	~9.8 Å	P < 0.069	0.97
Zone B, i = 1	Monolayer	10.3–12.1 Å	$0.115 < P/P_o < 0.33$	1.51
Zone C, i = 2	Bi-monolayer	15.2 Å	$P/P_o = 0.45$	1.23
Zone D, i = 2	Bi-monolayer + external	15.8 Å	$0.58 < P/P_o < 0.85$	1.13
Zone E, i = 3	Three-monolayer	18.8 Å	$P/P_o < 0.97$	1.11

Table 11. Density of the adsorbed water according to the results obtained with the FHH + Dubinin method.

	Monolayers											
	Three				Two							One
$a_w = P/P_o$	1	0.997	0.995	0.97	0.85	0.75	0.63	0.58	0.45	0.329	0.2	0.115
$Q_{int, theoretical}$ (g/g)	0.319	0.319	0.319	0.319	0.214	0.214	0.214	0.214	0.214	0.214	0.214	0.102
ρ_w(g/cm^3)	0.999	0.969	1.006	1.030	1.013	1.003	1.027	1.041	1.110	1.358	1.370	1.170
	Average = 0.991				Average = 1.023					1.110	1.364	1.170

$S_{int} = S_{total} - S_{ext} = 684.47 \, m^2/g$.

distribution and physico-chemical characteristics of the water in the system depend on the types of pores and the pore size distribution.

A bentonite-water system must be defined, among other parameters, through a water/rock ratio, since the concentrations in the aqueous phase strongly depend on this parameter. This water/rock ratio can be conveniently described by a quantity analogous to the porosity. In compacted bentonite, the total porosity (physical porosity) can be divided in: (i) internal porosity (interlayer water) and (ii) external porosity (external water), which corresponds to free

water plus DDL water. The external porosity is called effective porosity (Pusch et al., 1990) or geochemical porosity (Pearson, 1998). The amount of water involved in the geochemical processes in the engineering barrier corresponds to the free water or inter-particle water, i.e., the external water, taking into account that the amount of this water depends on the modifications of the dry density and the salinity concentration in the system. Usually, the geochemical or accessible porosity is supposed to be equivalent to the accessible porosity to chloride in through-diffusion experiments. The chloride anions

270

Table 12. Amount of interlayer water in a quasicrystal formed by six interlayer spaces as a function of the number of hydrated monolayers (g/g).

6T	5T1B	4T2B	3T3B	2T4B	1T5B	6B	5B1b	4B2b	3B3b	2B4b	1B5b
0.319	0.301	0.284	0.266	0.249	0.231	0.214	0.2085	0.2032	0.1979	0.1926	0.1873
6b	5b1M	4b2M	3b3M	2b4M	1b5M	6M					
0.182	0.169	0.155	0.142	0.129	0.116	0.102					

T = Three-monolayer hydrated = 0.37899 − external water g/g; B = Bi-monolayer hydrated = 0.27416 − external water g/g (2nd solvation shell + filling of siloxane cavities; b = 0.182 g/g (bi-monolayer of water, only 2nd solvation shell); M = Monolayer of water = 0.10238 g/g; External water = 0.06036 g/g (data from Table 9).

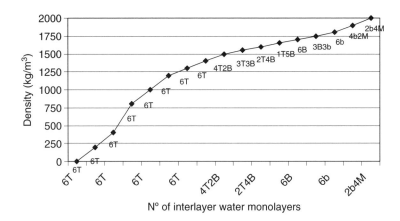

Figure 12. Distribution of monolayers of water in the interlayer space as a function of the dry density of compacted FEBEX bentonite (number of lamellae per quasicrystal = 7).

are excluded from the interlayer space, but they do move relatively easily through compacted bentonite (Pearson, 1998; Bradbury and Baeyens, 2002, 2003).

In this paper, due to the importance of knowing the distribution of waters in compacted bentonite, the amount of water in the internal and external space as a function of the dry density of compacted bentonite has been calculated. For these calculations, the types and amount of water in the FEBEX bentonite powder constituted the principal source of the data used.

For the calculation of the external and internal water in the compacted bentonite at saturated state, a statistical distribution of monolayers of water as a function of the pore volume, fixed by the total porosity, has been performed. To do this, the values obtained from the FHH + Dubinin method have been used. Also, several hypotheses have been considered: (1) in the FEBEX bentonite, a quasicrystal is formed by seven stacked lamellae, which implies six interlayer spaces to be filled with internal water; (2) the volume of water fixed by the total porosity is distributed in internal and external water; (3) a proportion of external water always exists in the system; (4) the tendency of the bentonite-water system is to fill the maximum amount of internal monolayers

of water, which are three. The layers are filled with two-monolayers of water (first complete filling, then only the second cations solvation shell is filled) and, finally, with one-monolayers of water if all the layers of the quasicrystal in the interlayer space cannot be filled with three-monolayers of water (Table 12).

The results of the statistical distribution are shown in Figure 12. As can be observed, the number of maximum water monolayers in the interlayer space depends on the dry density. The total number of layers can be filled with three-monolayers of water up to a dry density of 1400 kg/m^3. At dry densities higher than 1400 kg/m^3, a complete uptake of three-monolayers of water is not geometrically possible. However, most part of water is distributed in the interlayer space.

With this distribution, the internal or interlayer water expressed as a percentage of the internal porosity in the system can be determined (Fig. 13). The internal water distribution is similar to that found by Pusch et al. (1990) for the MX-80 bentonite. Also, the amount of external water can be determined. This quantity can be considered similar to the available water for the chemical reactions that occur in the inter-particle pores. These results have been compared to the effective porosity calculated by means of

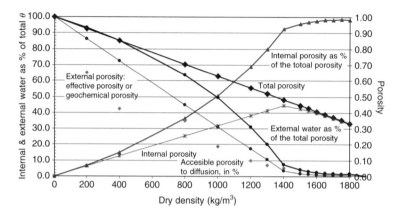

Figure 13. Total porosity, external porosity and internal or interlayer porosity as a function of the dry density.

through-diffusion experiment performed in compacted FEBEX bentonite samples (García-Gutiérrez et al., 2003). Chloride is excluded from interlayer positions and, for this reason, its concentration depends on the external water (Fig. 13). The data differ at low densities, due to the saline concentration and osmosis, since these diffusion tests were performed with distilled water. Muurinen et al. (1989) found that the effective porosity increases with salinity of the injected water in the diffusion tests. At low densities (in this case <1400 kg/m³), where the efficiency of the bentonite functioning as a semipermeable membrane is much lower than 100%, the external fluid composition may have some influence above a certain concentration. However, at high densities and membrane efficiencies approaching 100%, the re-saturating fluid composition plays little or no role (Bradbury & Baeyens, 2002; Dixon, 2000).

If the system resaturates under constant volume with dry densities higher than 1400 kg/m³, the above calculations indicate that the swelling resulting from the uptake of different monolayers of water is sufficient to fill the available porosity. Thus, with an increase of water content, the swelling is sufficiently great to virtually fill all the initially air-filled macroporosity and transform it into predominantly interlayer space. However, a small fraction of external water still exists.

Thus, based on the study of the properties of the bentonite-water system, a key parameter of the medium, called the effective porosity, has been obtained.

4.3.1 Partial specific volume of water and swelling pressure

According to the principles of thermodynamic equilibrium, the swelling pressure, P_s, and relative partial free energy of water in the clay/water system are related by the relationship (Low & Anderson, 1958; Kahr et al., 1990, Karnland, 1997):

$$\bar{g}_w - \bar{g}_w^o = -\bar{v}_w \times P_s \quad \text{(J/mol)}$$

$\bar{v}_w = (\delta V/\delta m_w)_{T,P}$ represents the partial specific volume of water. This quantity can differ substantially from the value for pure water near room temperature ($v_w = 10^{-3}$ m³/kg), in particular in bentonite systems with low contents of water (Oliphant & Low, 1982).

The difference in free energy is also related to P, the partial pressure of water vapour in the system and the vapour pressure, P_o, of pure water at temperature T of the experiment:

$$G_w - G_w^o = M_w\left(g_w - g_w^o\right) = RT \ln \frac{P}{P_o} \quad \text{(J/mol)}$$

where, R is the gas constant (8.31 J/mol · K) and M_w is the molecular weight of water (1.8 · 10⁻² kg/mol). Thus, the changes in free energy of water in bentonites and the swelling pressure can be determined from the water vapour isotherm. The swelling pressure as a function of the water content is:

$$P_s = -\frac{RT}{M_w v_w} \ln \frac{P}{P_o}$$

For low water contents, Oliphant & Low (1982) found an empirical correction for the value of the partial specific volume of water in bentonite-water systems:

$$v_w = 1.002 \exp\left(0.036 \frac{m_w}{m_s}\right)$$

Applying the above relationship and considering the relationship m_w/m_s as a function of the dry

272

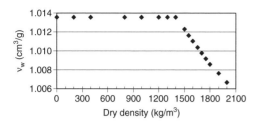

Figure 14. Partial specific volume of water as a function of the dry density.

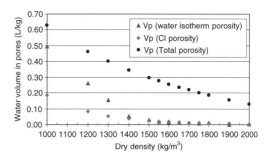

Figure 15. Water volume in interparticle pores as a function of the dry density.

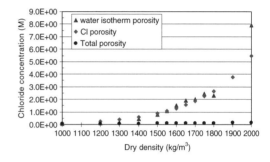

Figure 16. Chloride concentration as a function of the dry density for different porosities.

density (see Fig. 12 and Table 12), the v_w (cm³/g) for the FEBEX bentonite-water system can be easily calculated (Fig. 14). Then, once the water activity at equilibrium in the compacted bentonite-water system is known, the swelling pressure can be determined.

4.4 Pore water chemistry as a function of the dry density of the bentonite

One of the main parameters to know in order to model the chemical composition of the pore water is the water volume available for chemical reactions or the amount of free water. As it was previously mentioned, a bentonite-water system must be defined, among other parameters, by a water/rock ratio, since the concentrations in the aqueous phase strongly depend on this parameter. However, the water/rock ratio would be described in terms of the external porosity, since the conservative anions are not excluded from this space. The interlayer water or internal porosity is only related to the exchangeable cations. The total porosity takes into account the external and internal porosity. Thus, the concentration of chloride to be used in the modelling calculations can be fixed by the external porosity. When the bentonite-water system reaches equilibrium, the chemical composition of the pore water will be controlled by the dissolution of trace minerals in the bentonite, by the influence of the high effective S:L ratio and by the buffering effects of the exchangeable cations and $\equiv SOH$ sites (Bradbury & Baeyens, 2003; Fernández et al., in press-a).

In a previous work (Fernández et al., in press-a), the free water in compacted FEBEX bentonite was taken to be equivalent to the chloride accessible porosity calculated from Cl "through diffusion" experiments. In this work, the free water is taken to be equivalent to the external porosity calculated in section 6.2. In Fig. 15 and Fig. 16, the volume of water in the inter-particle pores (external water) and the chloride concentration as a function of the dry density are shown (called water isotherm porosity).

The values are compared with those obtained from total porosity and chloride accessible porosity. As it can be observed, the chloride accessible porosity and the external porosity calculated in this work are quite similar considering dry densities between 1400 and 1800 kg/m³. At densities lower than 1400 kg/m³, the bentonite water-system is better described by the calculated external porosity, due to the dependance of the through-diffusion experiments on the salinity of the injected water (see Muurinen et al., 1987). At densities higher than 1800 kg/m³, other mechanisms seem to be involved in the compacted FEBEX bentonite-water system. The salinities are greater than 3 M at densities higher than 1800 kg/m³. This must be considered in order to model the pore water system and to understand the equilibrium of the bentonite-water system.

5 CONCLUSIONS

The textural and structural features and changes which occur during the intracrystalline swelling of the FEBEX bentonite have been studied by using nitrogen adsorption volumetry, water adsorption and desorption gravimetry and X-ray diffraction under controlled humidity conditions.

The external surface area, according to the t-plots and Harkins-Jura methods, is $61.5 \, m^2/g$. The nitrogen adsorption isotherm shows the characteristic features of the type-IV isotherm with a very important hysteresis loop due to capillary condensation in mesopores. The curve exhibits a H3 hysteresis loop in desorption, which is indicative of the presence of slit-shaped pores.

The FHH formalism and the Dubinin-Radushkevish equations were used to identify and quantify the different states and locations of water in the FEBEX bentonite powder. Water is mainly adsorbed in the interlamellar space, but a 3–6% of water is external water adsorbed on the external surface of the stacks, edges and OH groups (free water + DDL water).

Uptake of water occurs in three discrete stages as the montmorillonite structure expands. The distribution of water in the interlayer or internal space of the FEBEX bentonite powder can be classified in:

– monolayer hydrated state, which corresponds to a water content of 10.2%
– two-monolayer hydrated state, which can be defined by two substates: (i) formation of the second solvation shell, which corresponds to a water content of 18.3% and, (ii) the filling of the free siloxane cavities and possible adsorption of water at external surfaces, which corresponds to a water content of 21.4%
– three-monolayer hydrated state, which corresponds to a water content of 31.9%.

However, in compacted bentonite, the amount of water depends on the dry density. A statistical model has been used to distinguish between internal porosity (interlayer water) and external porosity (external water) in compacted material. The distribution of the water in the compacted material (external/internal water) as a function of the total porosity of the compacted bentonite-water system has allowed to obtain the external porosities, which were found to be similar to the accessible porosities calculated from through-diffusion tests.

Re-hydration of the interlayer cations is the main mechanism responsible for the generation of swelling pressures in highly compacted bentonite systems. The swelling accompanying the re-saturation of the compacted bentonite tends to fill most of the initially present macroporosity and turns it into predominantly interlayer porosity.

The external water will allow to model the pore water chemistry of the FEBEX compacted bentonite, both as a powder and as a compacted material.

ACKNOWLEDGEMENTS

This work has been supported by CIEMAT-ENRESA, and the European Commission 5th framework programme within the FEBEX project. The contribution of A. Melón, Luis Gutiérrez- Nebot, F. Colmenero and the chemical department of CIEMAT to the experimental work are gratefully acknowledged.

REFERENCES

Bradbury, M.H., Baeyens, B., 2003. Porewater chemistry in compacted resaturated MX-80 bentonite. J. Contaminant Hydrology, Vol. 61, pp. 329–338.
Bradbury, M.H., Baeyens, B., 2002. Porewater chemistry in compacted re-saturated MX-80 bentonite: Physico-chemical characterisation and geochemical modelling. PSI Bericht Nr. 02–10 Paul Scherrer Institut, Villigen, Switzerland and Nagra Technical Report NTB 01–08, Nagra, Wettingen, Switzerland.
Bradley, W.F., Clark, G.F., Grim, R.E., 1937. A study of the behaviour of montmorillonite on wetting, Z. Krist, 97, pp. 216–222.
Cases, J.M., Bérend, I., Besson, G., Francois, M., Uriot, J.P., Thomas, F., Poirier, J.E., 1992. Mechanism of adsorption and desorption of water vapor by homoionic montmorillonite. 1. The sodium-exchanged form. Langmuir, 8, 2730–2739.
CRC Handbook of chemistry and physics. 1994. Ed. Lide David R. CRC Press, Inc.
Cuadros, J., Linares, J., 1996. Experimental kinetic study of the smectite to illite transformation. Geochimica et Cosmochimica Acta, 60, No. 3, pp. 439–453.
de Boer, J.H., Lippens, B.C., Linesen, B.G., Brokhoff, J.C.P., Van der Heuvel, A., Osinga., J., 1966. The t-curve of multimolecular N2 adsorption. J. Colloid Interface Sci., 21, pp. 405–414.
Dixon, D.A., 2000. Porewater salinity and the development of swelling pressure in bentonite based buffer and backfill materials. POSIVA Report 2000–04, Posiva Oy, Helsink, Finland.
Dubinin, M.M., 1967. Adsorption in micropores. J. Colloid Interface Sci., 23, 487–499.
ENRESA, 1994. Almacenamiento geológico profundo de residuos radiactivos de alta actividad (AGP). Conceptos preliminares de referencia. ENRESA. Publicación Técnica 07–94.
Fernández, A.Mª., Cuevas, J., Rivas, P., 1999. Estudio del agua intersticial de la arcilla FEBEX. Febex Project Technical Report 70-IMA-L-0-44. CIEMAT.
Fernández, A.Mª., Cuevas, J., Rivas, P., 2001. Pore water chemistry of the FEBEX bentonite. Mat. Res. Soc. Symp. Pro. 663, pp. 573–588.
Fernández A.M., Rivas P., 2003. Analysis of types and distribution of waters in the FEBEX bentonite used as an engineered barrier. EUROCLAY 2003 congress, Módena (Italy). June 22–26, pp. 96–97.
Fernández, A.Mª., Baeyens, B., Bradbury, M., Rivas, P., in press-a. Analysis of the pore water chemical composition of a Spanish compacted bentonite used in an engineered barrier. Submitted to Applied Clay Science.

Fernández A.Mª., in press-b. Caracterización y modelización del agua intersticial en materiales arcillosos: Estudio de la bentonita de Cortijo de Archidona. Ph. D. Thesis.

Fernández-Soler, J.M., 1992. El volcanismo calco-alcalino de Cabo de Gata (Almería). Ph. D. Thesis. Universidad de Granada. 243 pp.

García-Gutiérrez, M., Missana, T., Cormenzana, J.L., 2002. Diffusion coefficients and accessible porosity for HTO and ^{36}Cl in compacted bentonite. International Meeting at REIMS Clay in natural and engineered barriers for radioactive waste confinement, pp. 331–332.

Grim, R.E., 1968. Clay Mineralogy, 2nd edition. McGraw-Hill. New York. 596 pp.

Gregg, S.J., Sing, K.S.W., 1982. Adsorption, Surface área and Porosity. Academic Press. 303 pp.

Güven, N., Low, P.F., Mitchell, J.K., Sposito, G., van Olphen, H., 1993. Clay-Water interface and its rheological implications. CMS Workshop Lectures. Volume 4. R.E., Ferrel Editor. The Clay Minerals Society. Colorado. USA.

Hagymassy, J., Brunauer, S. Mikhail, R., 1969. Pore structure analysis by water vapor adsorption. I. t-curves for water vapor. J. Colloid Interface Sci., 29, pp. 485–491.

Hendricks, S.B., Jefferson, M.W., 1938. Structures of kaolin and talc-pyrophyllite hydrates and their bearing on water sorption of the clays. Amer. Min., v. 23, pp. 863–875.

Horseman, S.T., Higgo, J.J., Alexander, J., Harrington, J.F., 1996. Water, gas and solute movement through argillaceous media. Report CC-96/1. Nuclear Energy Agency, OECD, Paris, France.

Huertas, F., Fuentes-Cantillana, J.L., Jullien, F., Rivas, P., Linares, J., Fariña, P., Ghoreychi, M., Jockwer, N., Kickmaier, W., Martínez, M.A., Samper, J., Alonso, E., Elorza, F.J., 2000. Fullscale engineered barriers experiment for a deep geological repository for high-level radioactive waste in crystalline host rock (FEBEX project). European Commission Report EUR 19147 EN.

Kahr, G., Kraehenbuehl, F., Stoeckli, H.F., Müller-von Moos, 1990. Study of the water-bentonite system by vapor adsorption, immersion calorimetry and X-ray techniques. II. Heats of immersion, swelling pressures and thermodynamic properties. Clay Minerals 25, pp. 499–506.

Karnland, O., 1997. Bentonite swelling pressure in strong NaCl solutions. Correlation between model calculations and experimentally determined data. SKB technical report 97–31, 30 pp.

Kraehenbuehl, F., Stoeckli, H.F., Brunner, F., Kahr, G., Mueller-Vonmoos, M., 1987. Study of the water-bentonite system by vapour adsorption, immersion calorimetry and X-ray diffraction techniques: I. Micropore volumes and internal surface areas, following Dubinin's theory. Clay Minerals 22, 1–9.

Laird, D.A., 1999. Layer charge influences on the hydration of expandable 2:1 phyllosilicates. Clays and Clay Minerals, vol. 47, 5, pp. 630–636.

Low, P.F., Anderson, D.M., 1958. Osmotic pressure equation for determining thermodynamic properties of soil water. Soil Sci. 86, pp. 251.

MacEwan, D.M.C., Wilson, M., 1980. Interlayer and intercalation complexes of clay minerals. In: Crystal Structures of Clay minerals and their X-ray identification. F.W. Brindley and G. Brown eds. Mineral Soc. Monogr. 5, London, pp. 197–248.

Martin, R.T., 1962. Adsorbed water on clay: a review. Clay and Clay Minerals, v. 9, pp. 28–70.

Moore, D., Reynolds, R., 1989. X-Ray diffraction and the identification and analysis of Clay Minerals. Oxford University Press, N.Y., 332 pp.

Muurinen, A., Penttila-Hiltunen, P., Rantanen, J., Uusheimo, K., 1987. Diffusion of uranium and chloride in compacted Na-bentonite. Report YJT-87-14. Nuclear waste Commission of Finnish Power Companies, Helsinki, Finland.

Newman, A.C.D., 1987. Chemistry of Clays and Clay Minerals. Mineral Society. Monograph Nº 6. Longman Scientific & Technical, 480 pp.

Norrish, K., 1954. The swelling of montmorillonite. Discuss. Faraday Soc. 18, 120–133.

Pearson, F.J., 1998. Geochemical and other porosity types in clay-rich rocks. Water-Rock interaction, Arehart & Hulston (eds.), pp. 259–262. Balkema. Rotterdam.

Oliphant, J.L., Low, P.F., 1982. The relative partial specific enthalpy of water in montmorillonite-water systems. J. Colloid and Interface Sci. 89, pp. 366.

Prost, R., Koutit, T., Benchara, A., Huard, E., 1998. State and location of water adsorbed on clay minerals: Consequences of the hydration and swelling-shrinkage phenomena. Clays and Clay Minerals, 46, 117–131.

Pusch, R., Muurinen, A., Lehikoinen, J., Bors, J., Eriksen, T., 1999. Microstructural and chemical parameters of bentonite as determinants of waste isolation efficiency. European Commission. Nuclear Science and Technology. Project Report EUR 18950 EN.

Schultz, L.G., 1964. Quantitative interpretation of mineralogical composition from X-Ray and chemical data for the Pierre Shale. Geological Survey of U.S.A. Professional Paper 591-C, C1–C31.

Sposito, G., 1983. The surface chemistry of soils. Oxford University Press, N.Y., 234 pp.

Touret, O., Pons, C.H., Tessier, D. et Tardy, Y., 1990. Etude de la reparticion de Léau dans des argiles saturees Mg^{2+} aux fortes teneurs en eau. Clay Minerals 25, 217–233.

van Olphen, 1977. An introduction to Clay Colloid Chemistry for clay technologies, geologists and soil scientists. John Wiley, N.Y., 318 pp.

Villar, M.V., 2002. Thermo-hydro-mechanical characterisation of a bentonite from Cabo de Gata. ENRESA Publicacioen Técnica 04/2002.

Webb, P.A., Orr, C., 1997. Analytical Methods in Fine Particle Technology. Micromeritics Instrument Corporation. Norcross, GA USA. 301 pp.

Advances in Understanding Engineered Clay Barriers – Alonso & Ledesma (eds)
© 2005 Taylor & Francis Group, London, ISBN 04 1536 544 9

Early chemical alteration of buffer smectite in repository environment

Roland Pusch
Geodevelopment AB, Lund, Sweden

Hiroyasu Takase & Toshihide Noguchi
Quintessa Japan, Yokohama, Japan

ABSTRACT: Chemical alteration of smectite in the clay buffer, which serves as a low-permeable and strain-absorbing embedment of HLW canisters in repositories, can lead to shrinkage, fissuring and brittleness. This may happen because of conversion of the smectite minerals to non-expanding ones, primarily illite, and by precipitation of dissolved mineral components or elements contained in groundwater, causing cementation.

Conversion of smectite to non-expanding minerals means that part of the porewater that is originally strongly sorbed in the interlamellar space will be released into widening voids in conjunction with denser stacking of the mineral lamellae. For a constant bulk density this can increase the hydraulic conductivity by orders of magnitude and strongly reduce the expandability and self-sealing potential. Cementation will cause an increase in shear strength but also brittle behaviour with further drop in expandability and self-sealing potential.

These processes are outlined here in the form of conceptual and mathematical models, which are applied to the KBS-3 concept for the first 500 years after closure, *i.e.* the period in which a significant temperature gradient prevails in the clay buffer, and for a period of 10000 years.

1 INTRODUCTION

A commonly considered alteration process is the conversion of the smectite species montmorillonite to non-expanding illite and chlorite, which is not expected to be significant in a short time perspective for a KBS-3 repository if the temperature is below 100°C. However, changes in the form of dissolution and cementation of silicious components can be important as discussed in the present paper.

2 CHEMICAL MODELLING

2.1 *General*

Several attempts have been made to model chemical alteration of montmorillonite under repository-like conditions. Here, we will describe and apply a recent model proposed by Takase and Benbow and assess the practical importance of the derived changes in clay properties. The model implies dissolution of smectite and precipitation of illite and is in better agreement with the actual smectite/illite (S/I) conversion and precipitation of silica of the Kinnekulle bentonite bed than solid-solution models (Pusch and Madsen, 1995). In contrast to the conventional Reynold-type models the one proposed by the present

authors does not preset smectite-to-illite as the basic reaction. The model, which is developed from an earlier version derived by Grindrod and Takase 1994 (Pusch, Takase and Benbow, 1998), takes $O_{10}(OH)_2$ as a basic unit and defines the general formula for smectite and illite as:

$$X_{0.35}\ Mg_{0.33}\ Al_{1.65}\ Si_4O_{10}\ (OH)_2 \text{ and}$$
$$K_{0.5-0.75}\ Al_{2.5-2.75}\ Si_{3.25-3.5}\ O_{10}(OH)_2 \tag{1}$$

where X is the interlamellar absorbed cation (Na) for Na montmorillonite. The reactions in the illitization process are:

Dissolution:

$$Na_{0.33}\ Mg_{0.33}\ Al_{1.67}\ Si_4O_{10}(OH)_2 + 6H^+ =$$
$$0.33Na + 0.33Mg^{2+} + 1.67Al^{3+}\ 4SiO_{2(aq)} + 4H_2O \tag{2}$$

Precipitation of illite and silicious compounds:

$$K\ AlSi_3\ O10\ (OH)_2 = K^+ + 3Al3^+ + 3\ SiO2(aq) + 6H2,$$
$$\text{(logK being a function of temperature)} \tag{3}$$

The rate of the reaction R that will be used in this paper is:

$$R = A\ exp\ (-E_a/RT)(K^+)S^2 \tag{4}$$

where:

A = Coefficient.

E_a = Activation energy for S/I conversion.

R = Universal gas constant.

T = Absolute temperature.

K^+ = Potassium concentration in the porewater.

S = Specific surface area for reaction.

In smectite clay buffer embedding a canister with HLW the thermal gradient will cause more dissolution in the hot clay at the canister surface than at the clay/rock interface, which generates transport of dissolved species towards the cold side.

2.2 The chemical code

The chemical model described in the previous section is now coupled with the transport problem in the one-dimensional cylindrical coordinate representing the buffer material. The kinetic reactions (2) and (3) are linked with diffusion dominated transport of the aqueous species, i.e., silica, aluminum, sodium, magnesium, and potassium to form a set of quasi-nonlinear partial differential equations for the aqueous species and minerals which fall in the following general class:

$$\frac{\partial c_i}{\partial t} = \nabla . D_i \nabla c_i - \sum_j \lambda_{ij} R_{ij},$$

$$\frac{\partial m_k}{\partial t} = \sum_j R_{jk}, \tag{5}$$

c_i = Total concentration of aqueous species including element i.

m_k = Abundance of mineral k.

R_{jk} = Precipitation of mineral k that includes element i.

λ_{ik} = Stoichiometry of element i in mineral k.

We, then, spatially discretize equation (5) by the finite difference scheme to obtain a set of ordinary differential equations for each numerical grid (method of lines). All the aqueous speciation reac""tions are assumed to be in local instantaneous equilibrium. Based on this assumption, we can determine the concentration of each aqueous species by solving the mass action equations with the total concentration of the element that can be specified by solving (5). Thus we have a system of differential and'algebraic equations (DAE), which is potentially stiff, i.e., including multiple processes of hugely variable time scales, and solved numerically by applying the backward difference formulae (Petzold, 1987) to guarantee accuracy and efficiency at the same time.

Temperature dependence of the equilibrium constants and the rate constants for the mineral reactions

and the aqueous speciation is defined in the same manner as in (Pusch, Takase and Benbow, 1998).

3 APPLICATION OF THE MODEL TO THE KBS-3 VERTICAL CASE

3.1 Basic data

We will take a KBS-3 deposition hole as example in the modelling and assume the following conditions:

- Dry density 1700 kg/m³.
- Porosity 0.35.
- Montmorillonite fraction is 75%.
- Initial pore water is seawater.
- Groundwater is of seawater origin and assumed to be in equilibrium with granite at the specified temperature in the surrounding rock, e.g., 60°C during the first 500 years.
- Temperature at canister surface 90°C, temperature at rock wall 60°C.
- Diameter of hole is 1.8 m, diameter of canister is 1.0 m.
- Vertical water flow rate in the 1 cm rock annulus (EDZ) around the hole is 1 mm/day and 1 mm/month, respectively.
- Diffusion of K^+ in the clay is E-9 m²/s.
- Considered time 500 years.

Under the prevailing thermal gradient, which is conservatively taken to be constant during the considered time period, silica is currently released and transported from the hottest to the coldest part of the clay buffer. The possibility of silica being precipitated in the direction of the thermal gradient is in focus of this study.

3.2 Results of calculation

Despite the conservative assumption of constant thermal gradient in the first 500 years, effectively no illite was formed, which is consistent with our expectation based on the previous study at the Kinnekulle. In addition one finds that no precipitation of quartz or amorphous silica takes place in the clay since both minerals are under-saturated (Figure 1). However, there will be a significant loss of quartz due to dissolution/transport in the inner, hot part as illustrated by Figure 2.

One concludes that existence of the temperature gradient alone is not sufficient for precipitation of quartz or amorphous silica. However, cooling will make the situation quite different. To illustrate this, we carried out an additional calculation assuming the same temperature conditions for 500 years and then a linear temperature drop with time to 25°C after 10 000 years. The results are shown in Figures 3 and 4, from

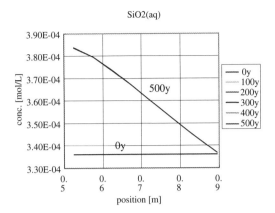

SiO2(aq)

conc. [mol/L]

Legend: 0y, 100y, 200y, 300y, 400y, 500y

position [m]

Figure 1. Evolution of SiO$_2$(aq) concentration profile (~500 y).

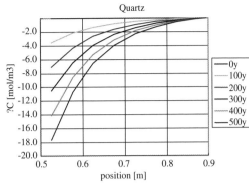

Quartz

?C [mol/m3]

Legend: 0y, 100y, 200y, 300y, 400y, 500y

position [m]

Figure 2. Evolution of quartz abundance profile (~500 y).

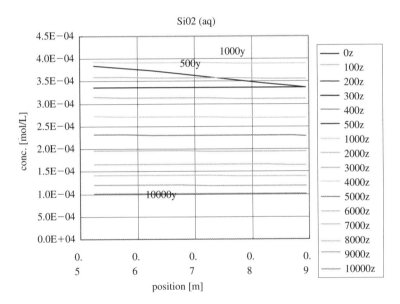

SiO2 (aq)

conc. [mol/L]

Legend: 0z, 100z, 200z, 300z, 400z, 500z, 1000z, 2000z, 3000z, 4000z, 5000z, 6000z, 7000z, 8000z, 9000z, 10000z

position [m]

Figure 3. Evolution of SiO$_2$(aq) concentration profile (~10000 y).

which it is concluded that precipitation of quartz will take place within about 0.1 m from the rock wall. Amorphous silica will not be precipitated. Possibility of the amorphous silica precipitation seems to be very much dependent on the time scale of cooling relative to that of the quartz precipitation, *i.e.*, rapid cooling leaves the system super-saturated with quartz and excess concentration of silica in solution eventually reaches the solubility of amorphous silica at the lower temperature. This might be the case in the laboratory experiment where cooling much more rapid than in the repository environment is expected.

3.3 *Implications of the results*

It is clear that cementation by precipitation of quartz will not take place in a 500 year perspective in KBS-3 buffer clay under the assumed temperature gradient while subsequent cooling to ambient temperature would cause formation of quartz. In a 10 000 year perspective rather comprehensive quartz precipitation is expected in the larger part of the KBS-3 buffer clay. The impact it may have on the physical properties of the buffer clay is illustrated by the experimental investigations described below.

279

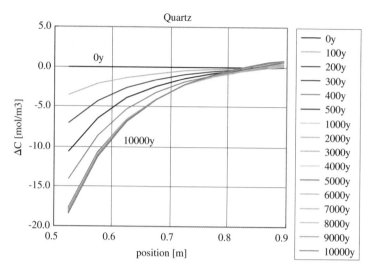

Figure 4. Evolution of quartz abundance profile (~10000 y).

4 EXPERIMENTAL

4.1 Nature of silica precipitations

Comprehensive microstructural studies using scanning and transmission electron microscopy have been performed for identifying and characterizing silica precipitations formed in hydrothermally treated montmorillonite-rich clays (Pusch and Karnland, 1988). These studies were made by confining SWY-1 clay samples that had been saturated with distilled water and homogenized in gold-plated autoclave cells and kept in ovens at temperatures ranging from room conditions to 200°C for a few weeks to 1 year. Examination and analysis of presumable precipitates by EDX indicated that they were cristobalite, quartz of poor crystallinity, and amorphous silica (Figure 5), while in nature, exemplified by the very slowly cooled Kinnekulle bentonite, the precipitates are well crystallized quartz (Figure 6). Figure 5 also illustrates the change in microstructure of dense smectite clays by hydrothermal treatment, i.e. permanent collapse of the stacks of lamellae, yielding dense packets, and formation of dense silicious particles in the heated clay (Pusch and Karnland, 1988).

4.2 Impact on the hydraulic conductivity of the clay by cementation

The impact of silica precipitation on the hydraulic conductivity of montmorillonite-rich clay buffer with the density intended for the KBS-3 concept is not believed to be significant although the loss of mineral mass in the hottest part of the buffer may cause some increase in conductivity and the precipitation of silicious matter in the cold part may reduce the conductivity somewhat. The possible increase in conductivity in the hottest part is probably much less important than that caused by microstructural rearrangement, i.e. widening of voids between the stacks due to collapse of the stacks of lamellae. The belief that the hydraulic conductivity will not change much in KBS-3 buffer clay stems from laboratory tests of samples extracted from the MX-80 clay in the Stripa Buffer Mass Test where the temperature had been about 125°C at the canister (heater) surface and 72°C at the rock wall for 0.9 years. The conductivity and swelling pressure of the samples could not be distinguished from those of unheated MX-80 (Pusch, 1985).

4.3 Impact on the strength of the clay by cementation

Precipitation of silicious matter within particle aggregates is believed to weld them together, which should increase the strength and stiffness, reduce the ductility and enhance brittleness. Such changes are clearly demonstrated in uniaxial compressive testing of hydrothermally treated SWY-1 clay samples (Figure 7). However, since both precipitation and microstructural reorganization combined to yield the changes it is not possible to define which of the processes that is most important.

The axial load was increased stepwise leaving each load on for 5 minutes to let creep strain develop.

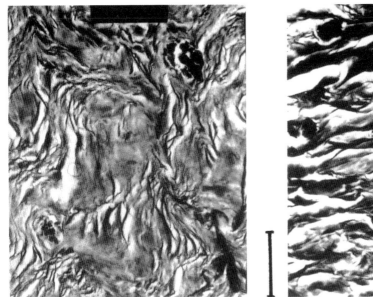

Figure 5. Comparison of montmorillonite-rich clay at room temperature (left) and after hydrothermal treatment at 150°C for 1 year (right). Notice the permanent collapse of the stacks of lamellae, yielding denser stacks, and formation of dense silicious particles in the heated clay. Dry density 1590 kg/m³ (Pusch and Karnland, 1988).

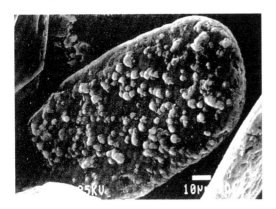

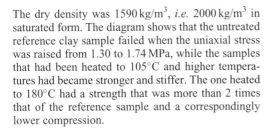

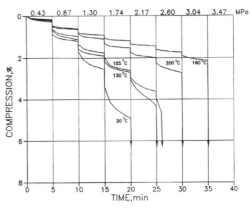

Figure 6. Silt particle coated with small quartz particles believed to have been formed by release and precipitation of silica in Kinnekulle bentonite that was heated to maximum 160°C for some hundred years (After Mueller vonMoos).

Figure 7. Results of uniaxial testing of hydrothermally treated montmorillonite-rich clay (Pusch and Karnland, 1988).

The dry density was 1590 kg/m³, *i.e.* 2000 kg/m³ in saturated form. The diagram shows that the untreated reference clay sample failed when the uniaxial stress was raised from 1.30 to 1.74 MPa, while the samples that had been heated to 105°C and higher temperatures had became stronger and stiffer. The one heated to 180°C had a strength that was more than 2 times that of the reference sample and a correspondingly lower compression.

5 CONCLUSIONS

Application of the Takase/Benbow model for hydrothermal conversion of smectite-rich clay shows that practically no illitization will take place in the first 500 years in KBS-3 buffer under assumed temperature and groundwater chemical conditions, while an increase in solid silicious compounds, primarily quartz, will occur in a large part of the buffer clay in a 10000 year perspective, assuming the temperature to be constantly equal to 90°C at the canister surface and 60°C at the rock wall for the first 500 years, followed by a linear drop of temperature from this state to room temperature in 10000 years.

The expected impact of the physical properties of the KBS-3 clay buffer in the considered 500 year period can be listed as follows:

– Practically no change in hydraulic conductivity since illitization is negligible.
– Increase in shear strength and deformation moduli by around 50% caused by silica precipitation and permanent microstructural reorganization, meaning that the creep rate of the heavy canister embedded in the clay will drop, and that the self-sealing potential drops.

It would be of substantial interest to apply the model using a more realistic temperature history and to quantify the change in physical properties by conducting additional autoclave tests.

REFERENCES

Grindrod P, Takase H, 1994. Reactive chemical transport within engineered barriers. 4th Int. Conf. Chemistry and migration behaviour of actinides and fission products in the geosphere. R. Oldenburg Verlag, Muenchen.
Petzhold L.R, 1983. A Description of DASSL: A differential/algebraic system solver, in Scientific Computing, R.S. Stapleman et al., eds., North-Holland, Amsterdam, 65–68.
Pusch R, 1985. Final Report of the Buffer Mass Test – Volume III: Chemical and physical stability of the buffer materials. Stripa Project Technical Report 85–14. SKB, Stockholm.
Pusch R, Karnland O, 1988. Hydrothermal effects on montmorillonite. SKB Technical Report TR 88–15.
Pusch R, Madsen F.T, 1995. Aspects of the illitization of the Kinnekulle bentonites. Clays and Clay Minerals, Vol.43, No.3 (pp.133–140).
Pusch R, Takase K, Benbow S, 1998. Chemical processes causing cementation in heat-affected smectite – the Kinnekulle bentonite. SKB Technical Report TR-98-25.

Advances in Understanding Engineered Clay Barriers – Alonso & Ledesma (eds)
© *2005 Taylor & Francis Group, London, ISBN 04 1536 544 9*

Chemo-mechanical behaviour of high-density bentonites. Imbibition and diffusion tests

G. Musso
Politecnico di Torino, Torino (Italy)

E. Romero
Universitat Politènica de Catalunya, Barcelona (Spain)

ABSTRACT: The chemo-mechanical behaviour of FEBEX bentonite, compacted at high densities, has been investigated in oedometric conditions. Samples imbibed with distilled water or with NaCl solutions have been exposed to salinisation–desalinisation cycles. The paper presents some preliminary results in terms of reversibility of chemically induced deformations, together with the critical aspects evidenced by a test where diffusion and chemical swelling were jointly monitored.

1 AIM AND FRAME OF THE INVESTIGATION

In virtue of their high surface charge and specific surface, smectitic soils such as bentonites are higly sensitives to environmental changes. In particular the influence of the wetting fluid chemistry on their mechanical behaviour is well known (see e.g. Van Olphen, 1977; Mitchell, 1993). Mechanical changes are therefore expected to occur as far as the wetting fluid chemistry changes. At a macroscopic level the consequences of an increase in the ionic strength of the wetting fluid are a decrease of the liquid limit, chemical consolidation and the increase of the shear strength. Corresponding phenomena at the microscopical level are the decrease of the double layer thickness, with the addition of cation exchange phenomena if also the chemical composition is changing.

Theoretical attempts to relate mechanical effects to chemical composition of the wetting fluid have mostly been based on the extrapolation of theories valid at the microscopic scale to the macroscopic scale (Bolt, 1956; Sridharan & Venkatappa Rao, 1973; Barbour & Fredlund, 1989). As evidenced (Fam & Santamarina, 1996), a common approach has been to modify Terzaghi's effective stresses, adding a term due to forces of chemical interactions acting between clay particles and between clay particles and the pore fluid.

Main drawbacks of this kind of approach are two. On one hand, when upscaling to the macroscale phenomena occurring at the microscale, the strong assumption that clay particles are disposed homogeneously in the porous medium and arranged in a parallel fashion has to be made. As a consequence, it is not possible to handle soils possessing well developed structures. On the other hand, these models are not able to distinguish between the effects exherted by different cationic species. Indeed tests performed by Di Maio (1996) on slurried bentonite samples proved that the persistence of the deformations induced chemically depends on the type of cation, being the effects of Na^+ and Ca^{2+} reversibles and those of K^+ almost irreversibles, as potassium is able to stick more firmly to the mineralogical structure.

A series of tests has therefore been executed in order to investigate the true macroscopic effects of NaCl diffusion on bentonite samples having a well developed structure. In this study, structure was imposed compacting the material at high density in a static fashion. In a first stage the total amount and the reversibility of the chemically induced deformations under oedometric conditions has been checked by imposing different chemical histories.

In a second stage, the role of structure has been investigated comparing the strains developed during chemical cycles by compacted samples with those developed by slurry samples mechanically consolidated. The comparison has been integrated with the characterisation of the two kind of samples at the microscopic scale (ESEM and porosimetry analyses). Chemical swelling development in time resulted being the most critical aspect, as it occurs in a continuous fashion in consolidated samples, while in a discontinuous fashion in compacted samples (see also Musso et al., 2003).

This latter issue has been the object of more detailed analyses during the third stage of the investigation, still ongoing. Here the answers to desalinisation of a compacted sample both in terms of mechanical effects and diffusion are under observation, in order to outline how diffusion at the macroscopic scale is affected by structural arrangements and its relationship with the swelling process.

2 EXPERIMENTAL PROGRAMME

2.1 Preparation of samples

Tests were performed on samples of FEBEX bentonite, constituted mainly of montmorillonite (smectite percentage higher than 90%) and of variable quantities of quartz, plagioclase, K – feldspar, calcite opal CT. The material has a liquid limit $w_l = 102\%$ in distilled water, a plastic limit $w_p = 53\%$, specific gravity $Gs = 2.70$ and a percentage of material smaller than $2\,\mu m$ of 67% (ENRESA, 1998).

Mechanically consolidated samples MC1 and MC2 were obtained starting from slurries of FEBEX powder and distilled water at initial water content $w_i \approx 150\%$. Slurries were placed in traditional oedometers where increasing mechanical stresses were applied to induce mechanical consolidation, up to a vertical effective stress $\sigma'_v = 1500\,kPa$. σ'_v was then reduced down to 200 kPa (sample MC1) and 500 kPa (MC2) so to reach the desired void ratio and the mechanical conditions imposed during the chemical cycling.

Compacted samples were prepared compacting statically the soil powder at the hygroscopic water content of the laboratory environment ($w_c \approx 11\%$), reaching a dry density $\rho_d = 1.40\,Mg\,m^{-3}$ for 'L' samples and a dry density $\rho_d = 1.68\,Mg\,m^{-3}$ for 'H' samples. After compaction samples have been placed in the oedometers, charged with a net vertical stress of 200 kPa and imbibed with distilled or saline water (NaCl solution with a concentration $c = 5.5\,mol\,L^{-1}$). Chemical cycling was then applied.

Direct and reverse chemical cycles were studied. In direct cycles samples were imbibed with distilled water. Once strain equilibrium was reached, distilled water was removed from the bottom porous stone and replaced by a saline NaCl solution ($c = 5.5\,mol\,L^{-1}$), so to induce NaCl diffusion in the sample and therefore chemical consolidation (salinisation path). Afterwards, replacement of the saline solution with distilled water caused diffusion of ions outside the sample and chemical swelling (desalinisation path). In reverse cycles imbibition occurred with the saline solution, followed by the placement of distilled water and chemical swelling. Finally, saline solution was reintroduced, re-consolidating the sample.

Both consolidated and compacted samples had an initial height of 10.5 mm. All tests were run in traditional oedometers, exception made for test H1 (run in a suction controlled oedometer) and for test H4 (run in a traditional oedometer modified to allow control on the diffusion process).

Table 1 provides the initial void ratios and the effective stress conditions applied during the tests.

2.2 Initial pore size distribution, samples structure and main cationic species

Mercury intrusion porosimetry analyses were performed, following a freeze-drying process (see e.g. Delage et al., 1982) to determine the pore size distribution of the samples prepared in different ways. All the samples showed two different peaks in the pore size distribution. Figure 1 gives the results obtained for sample MC1 before the chemical cycles and for sample L1 in the as-compacted state. The peak in the mesoscale is common for both samples and occurs at an entrance pore size of about 12 nm. The peak in the macrostructure occurs at 385 nm for MC1, at 40 000 nm for L1 as compacted.

The existence of more than one significant peak in a given sample could imply the arise of a bimodal behaviour. Indeed, transient phenomena could occur at two different time scales, as the time required for their completion in the micropores could be significantly bigger than the one for its occurrence in the macropores (see e.g. Navarro and Alonso, 2001 for

Table 1. Samples and testing conditions.

Sample	Density $\rho_d\ (Mg\,m^{-3})$	Void ratio before chemical cycling e	Effective stress $\sigma'\ (kPa)$	Type of cycle
L1 (compacted)	1.40	1.22	200	Direct
L2 (compacted)	1.40	1.15	200	Reverse
H1 (compacted)	1.68	0.98	200	Direct
H2 (compacted)	1.68	0.86	200	Reverse
H3 (compacted)	1.68	0.95	500	Direct
H4 (compacted)	1.68	0.72	500	Reverse
MC1 (consolidated)	–	1.42	200	Direct
MC2 (consolidated)	–	1.32	500	Direct

implications on the time evolution of mechanical consolidation). During the tests performed, pores with smaller sizes are expected to be the lasts to reach the final equilibrium concentration during salinisation and desalinisation paths.

Environmental Scanning Electron Microscopy (ESEM) analyses were pursued to highlight the structural differences between the different types of samples. Figure 2 shows photomicrographs of sample MC1 before the chemical cycle and of sample L1 in the as-compacted state, both taken at the same magnification. Sample MC1 possesses a more uniform and homogeneous structure (matrix structure), in which peds are not clearly distinguished. The compacted sample, displaying an aggregated structure, shows denser peds separated by pores of greater dimensions (macropores).

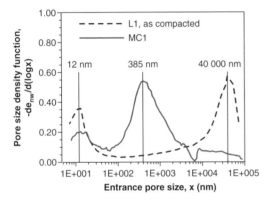

Figure 1. Pore size distribution of samples L1 and MC1.

Squeezing of a slurry prepared mixing FEBEX powder with distilled water at about the liquid limit was performed to check the presence of main cationic species in solution. Data have been referred to higher densities by imposing species mass conservation and neglecting chemical reactions. Table 2 gives the concentrations that have been estimated for $\rho_d = 1.40 \, \mathrm{Mg \, m^{-3}}$.

3 SWELLING DURING IMBIBITION

All the imbibition phases induced swelling, both when executed with distilled water or with the saline solution. Swelling always occurred in a continuous fashion.

Imbibition of samples L1 and L2 (Figure 3) occurred under a vertical effective stress of 200 kPa. Sample L1 was imbed with distilled water and reached a final void ratio e = 1.23, sample L2 was imbed with saline water, its final void ratio being e = 1.15.

Samples H1 and H2 (Figure 4) were imbed as well under a vertical stress of 200 kPa, the former with

Table 2. Relevant cationic species in bulk water after saturation with distilled water.

Cation	Concentration (mol L^{-1})
Na$^+$	5.0×10^{-2}
K$^+$	5.1×10^{-4}
Mg^{2+}	2.1×10^{-3}
Ca^{2+}	1.8×10^{-3}

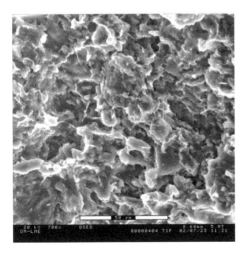
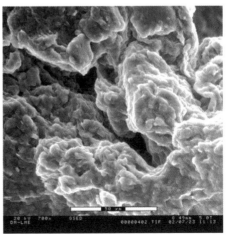

Figure 2. ESEM pictures of sample MC1 (left) and of sample L1 (right). Bar length: 50 μm.

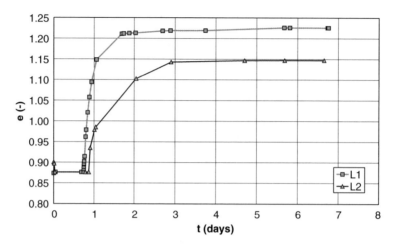

Figure 3. Swelling during imbibition, $\rho_d = 1.40\,\mathrm{Mg\,m^{-3}}$.

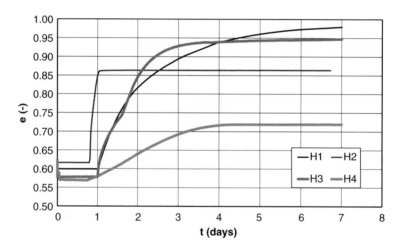

Figure 4. Swelling during imbibition, $\rho_d = 1.68\,\mathrm{Mg\,m^{-3}}$.

distilled water and the latter with the saline solution. Final void ratio was $e = 0.98$ for H1 and $e = 0.86$ for H2. Some considerations can be made comparing these data with the ones of tests H3 and H4, where the vertical effective stress was of 500 kPa. Indeed sample H3, saturated with water, swelled up to $e = 0.95$, that is to say to a void ratio quite similar to the one of H1. On the contrary final void ratio for sample H4 was $e = 0.72$, quite far from the value of H2.

4 CHEMICAL CYCLING

4.1 *Reversibility of chemo mechanical effects in compacted samples*

Given the pore fluid chemistry found in Table 2, the mineralogical composition of FEBEX bentonite and the type of solution used, it is reasonable to expect that chemical cycles will exert reversible effects at microscopic level. Nevertheless, this doesn't ensure reversibility at the macroscopic scale, as evidenced for instance by Guimaraes (2002).

Figure 5 shows the steady state conditions achieved at the end of the imbibition processes and of the various phases of direct and reverse chemical cycling. Void ratio data have been plotted as a function of the molar concentration at equilibrium. It has here been assumed that, when the sample is in equilibrium with boundary conditions of distilled water, NaCl concentration in its pores is the same that has been measured after squeezing.

Tests L1 and L2 are the ones where reversibility is weaker. Indeed, during test L1 the void ratio at the end of a complete cycle is higher than the initial one,

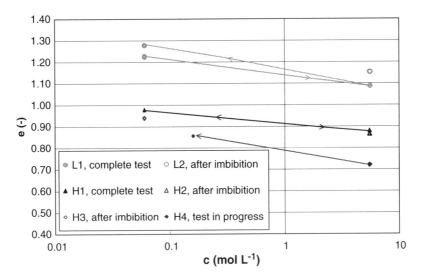

Figure 5.   Void ratios at chemical equilibrium.

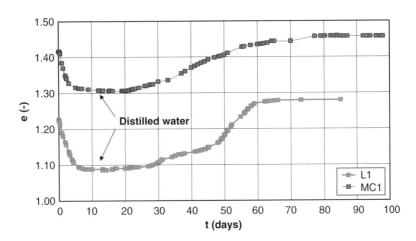

Figure 6.   Direct chemical cycle (tests L1 and MC1).

and also the void ratio achieved in test L2 after imbibition with saline water is higher than the one observed in test L1 after consolidation.

Complete reversibility of deformations was observed during test H1; moreover void ratio after consolidation corresponded to the one of the imbibition test H2. Test H4 is still in progress; nevertheless it has been possible to estimate the present average concentration inside the sample on the basis of its particular set up (see the following paragraph for details). Anyway, the trend seems to confirm that the void ratio after complete swelling will be quite close to the one of test H3 after imbibition.

4.2   Chemical cycles in compacted and consolidated samples

4.2.1   Chemical consolidation (salinisation path)

As far as the total amount of chemically induced deformations is concerned, consolidated and compacted samples showed quite the same behaviour (Figure 6). Chemical consolidation in compacted samples is slower than swelling under saturation, in test L1 the first requiring about 3 days and the second about 14 days. Chemical consolidation times appear on the other hand not to be very much dependent on

287

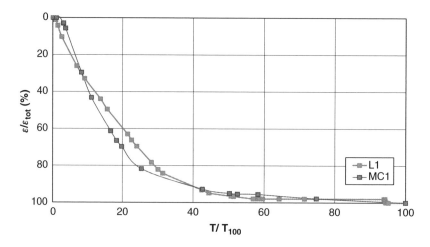

Figure 7. Non dimensional graphs for chemical consolidation (tests L1 and MC1).

the structure, as the time required for the process is as well of 14 days for sample MC1.

Indeed, in order to examine the type of relationship existing between diffusion and mechanical effects in both samples, consolidation curves have been drawn in their non-dimensional form with respect to time and magnitude of the final strain (Figure 7). The evolution in time of the strain ratio is common for all samples and quite continuous, with a quite good agreement between the behaviour of all samples.

Consolidation curves have a shape typical of linear diffusion problems. By assuming a linear constitutive relationship between concentrations and, it is found that the back calculated effective diffusion coefficient ranges between $2 \times 10^{-6}\,cm^2\,s^-$ and $3.6 \times 10^{-6}\,cm^2\,s^{-1}$ (diffusion coefficient for Na^+ in free electrolyte solutions is $13 \times 10^{-6}\,cm^2\,s^{-1}$). These values are consistent with those found for a comparable material in literature (see e.g. Barbour and Fredlund, 1989). It is anyway reasonable to state that a linear relationship between strains and concentration is not accurate enough (see e.g. Alonso et al., 1994 for other relationships). In particular, if referral to the double layer theory is made, volumetric strains should be more sensitive to concentration changes at low concentration values and less sensitive at large values.

4.2.2 Chemical swelling (desalinisation path)
Times for complete chemical swelling are significantly larger than consolidation ones, so that at least from the temporal point of view one phenomenon is not the reversal of the other. Shape of the swelling curves is in both cases quite different from the one of linear diffusion problems and a back calculation such as the one done for the salinisation path cannot be made in this case.

The swelling response in time of compacted samples is very different from the one of the consolidated ones. Indeed, in MC tests swelling occurred continuously without evidences of different times scales, while this doesn't hold for compacted samples. In the non dimensional form of the swelling diagram (Figure 8) no superposition can be made. At least two separate swelling phases can be observed in test L1: during the first one (until day 48 in Figure 6), the sample swells slowly reaching a quasi equilibrium condition; after that swelling starts again and ends around day 72.

Porosimetry analyses were therefore performed on a sample analogous to L1 but saturated with the saline solution (see Musso et al., 2003), outlining a bimodal distribution with macropores peak occurring at 75 000 nm. On basis of these results, a first explication for the bimodal behaviour of compacted samples during the desalinisation paths has been postulated. Pore size distributions would suggest that the change in the swelling rate could be due to the very low percentages of pores whose diameter range between 100 and 10 000 nm, so that micropores are rather inaccessible. Being micropores less accessible after chemical consolidation, their ionic depletion would be more difficult and delayed in time.

4.3 Swelling in time: a test with control on the diffusion process

Test H4 (still ongoing) has been executed lately in order to check the effects of structure not only on the swelling process but also on diffusion, so to verify the relationsip between the two processes. In order to do this a modified oedometer has been realised, connecting each porous stone with a reservoir aimed at hosting small quantities of solution (Figure 9).

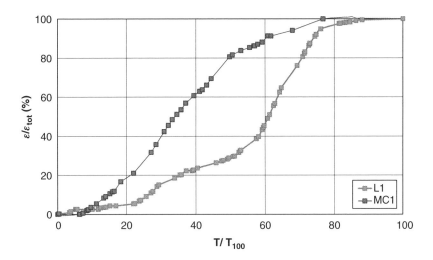

Figure 8. Non dimensional graphs for chemical swelling (tests L1 and MC1).

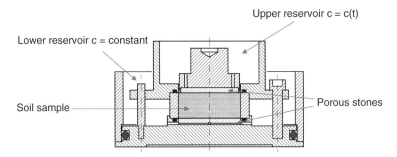

Figure 9. Modified oedometer for diffusion testing.

During salinisation events concentration is kept constant in the bottom reservoir, while evolves freely because of diffusion in the upper reservoir. Measurement of the electrical conductivity values in the upper reservoir can then be used to estimate the NaCl concentration (see e.g. Shackelford et al., 1999).

Electrical measurements have been performed by means of a electrical conductimeter. In Figure 10 the theoretical and experimental calibration curves relating the electrical conductivity to NaCl concentrations are given.

Figure 11 gives the experimental results found. Swelling times are very large, as swelling keeps occurring 240 days after the beginning of the desalinisation path. As detected in previous tests, swelling rate is highly discontinuous, with peaks in the first 20 days and at about 220 days.

Monitoring the concentration appeared to be important not only as far as the intepretation of the diffusion phenomenon is concerned, but also with respect to the correct implementation of the mechanical part of the test. Indeed, between day 150th and

210th no swelling was detected, so that referring only to mechanical aspects the test could have been stopped in this period. On the other hand concentration measurements ranged between 0.8 and 0.4 mol L^{-1}, such to suggest that the swelling process still wasn't complete. Actually a further significant swelling stage began after day 220, developing not negligible deformations.

5 DISCUSSION AND CONCLUSIONS

Well structured smectitic soils, such as compacted bentonites, have been proved to be sensitives to chemical changes, as expected in virtue of their mineralogical nature.

Swelling of compacted bentonites during imbibition depends on the type of fluid, on the dry density and on the applied load. Preliminary results seem to indicate that under high mechanical stresses and for high densities the role of solution molarity increases;

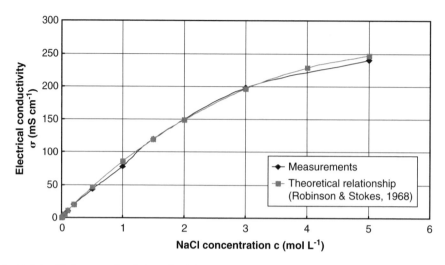

Figure 10. Relationship between electrical conductivity and NaCl concentration.

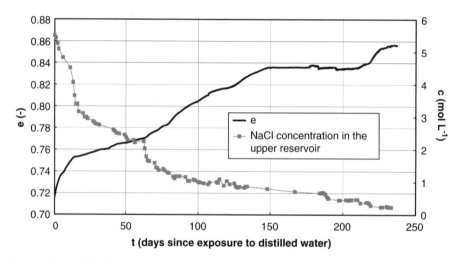

Figure 11. Swelling and NaCl breakthrough during the desalination path (test H4).

in order to better understand this aspect further oedometric tests at different molarities are currently performed. Chemically induced deformations measured during salinisation – desalinisation cycles made with NaCl where largely reversibles.

On the other hand, the fact that during chemical swelling compacted samples have a bimodal behaviour (not evidenced by mechanically consolidated samples) has been confirmed. The non linearity of the swelling process appears to be related to analogous results of the diffusion process, as found out in a proper test.

REFERENCES

Alonso E.E., Gens A. & Gehling W., 1994. Elastoplastic model for unsaturated expansive soils. *Proc. 3rd European Conf. on Numerical Methods in Geotech. Engrg.*, Manchester. I. Smith (ed.). Balkema, Rotterdam: 11–18

Barbour S.L. & Fredlund D.G., 1989. Mechanisms of osmotic flow and volume changes in clay soils. *Canadian Geotechnical Journal* 26, 551–562

Bolt G.H., 1956. Physico-chemical analysis of the compressibility of pure clays. *Geotechnique* 6, 86–93

Delage P., Tessier D. & Marcel-Audiguier M., 1982. Use of the Cryoscan apparatus for observation of freeze-fractured planes of a sensitive Quebec clay in scanning electron microscopy. *Canadian Geotechnical Journal* 19, 111–114

ENRESA, 1998. FEBEX Full scale Engineered Barriers Experiment in crystalline host rock – Pre operational stage summary report. Technical Publication 01/98, Madrid

Fam M. & Santamarina J.C., 1996. Coupled diffusion – fabric – flow phenomena: an effective stress analysis. *Canadian Geotechnical Journal* 33, 515–522

Guimaraes L., 2002. Analisis multi-componente no isotermo en medio poroso deformable no saturado. *Doctoral Thesis*, Technical University of Catalonia

Mitchell J.K., 1993. Fundamentals of soil behavior, 2nd ed., John Wiley, New York

Musso G., Romero E., Gens A., Castellanos E., 2003. The role of structure in the chemically induced deformations of FEBEX bentonite. *Applied Clay Science* 23, 229–237

Navarro V. & Alonso E.E., 2001. Secondary compression of clays as a local dehydration process. *Geotechnique* 51, 10, 859–869

Robinson R.A. & Stokes R.H., 1968. Electrolyte solutions. Buttherworths, London

Shackelford C.D., Malusis M.A., Majeski M.J., Stern R.T., 1999. Electrical conductivity breakthrough curves. *Journal of Geotechnical and Geoenvironmental Engineering* 125, 4, 260–270

Sridharan A., Venkatappa Rao G., 1973. Mechanisms controlling volume change of saturated clays and the role of the effective stress concept. *Geotechnique* 23, 359–382

Van Olphen H., 1977. An introduction to clay colloid chemistry, Wiley Interscience, New York

Advances in Understanding Engineered Clay Barriers – Alonso & Ledesma (eds)
© 2005 Taylor & Francis Group, London, ISBN 04 1536 544 9

Fundamental properties of bentonite pellet for Prototype Repository Project

Y. Sugita
Japan Nuclear Cycle Development Institute (JNC), Tokai (Japan)

M. Chijimatsu
Hazama Corporation, Tokyo (Japan)

H. Suzuki
Inspection Development Corporation, Tokai (Japan)

ABSTRACT: JNC has joined the Prototype Repository Project (PRP). The PRP uses bentonite pellet to fill the clearance between the buffer block and rock mass in the test pits. It is difficult to consider the behavior of a grain of the pellet for analysis of the coupled thermo-hydro-mechanical (THM) on the PRP. We performed the prediction analysis assuming that bentonite pellet region is block or water. Therefore, we examined fundamental properties of the bentonite pellet. Pellet is Japanese bentontie Kunigel V1. Water used for the tests were distilled water and synthetic seawater. Qualitative properties of the pellet were clarified.

1 INTRODUCTION

The PRP is the full-scale and in-situ experiment to validate the engineered barrier system (EBS) on geological disposal of high-level radioactive waste, and is one of the R&D programs of European Commission projects (Svemar and Pusch, 2000). The PRP has studied the vertical emplacement option of the EBS. The PRP has six test holes and the test tunnel over the test holes is backfilled to simulate the expecting repository environment. In the PRP, some numerical codes are applied to assess the barrier performances of the EBS. JNC has joined the PRP and has performed the simulation of the EBS using a fully coupled THM numerical code. However, the numerical code is unable to consider the real experimental conditions in the PRP, especially the clearance, which is filled with bentonite pellet, between buffer block and rock mass. Therefore, a certain assumption on the experimental conditions is required. Such assumptions in the numerical simulation have to be validated. Then, properties of bentonite pellet, which is not considered in the simulation and is filled in the clearance between buffer block and rock mass, were examined by laboratory tests.

2 EXPERIMANTAL CONDITIONS OF THE PRP AND ASSUMPTION IN NUMERICAL ANALYSIS OF THE COUPLED THM BEHAVIOR IN AND AROUND THE EBS

The PRP is in-situ experiment of disposal concept as KBS-3 type. Figure 1 shows the schematic view of the PRP which has two experimental sections (section I and section II). Section I has four test holes and section II has two test holes. Buffer blocks, which are installed in the test holes, are highly compacted bentonite, and are ring shape or cylindrical shape. There is a 50 mm clearance between buffer block and rock mass. The clearance is filled with the high dense compressed bentonite pellets. Test tunnel over test holes is backfilled with backfill material, and the experimental sections are isolated by the reinforced concrete plugs.

The expected behavior in the clearance is as follows as shown in Figure 2. First, water infiltrates into the clearance. The pellet filled in the clearance is swollen by infiltrated water (Fig. 2 II). Then, water infiltrates into the buffer block next to the clearance, the infiltrated buffer block swells out the initial clearance area, low dense swollen pellet is consolidated by the

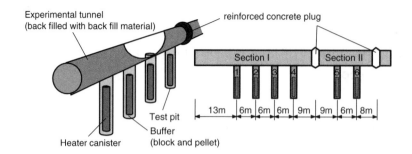

Figure 1. Schematic view of the PRP.

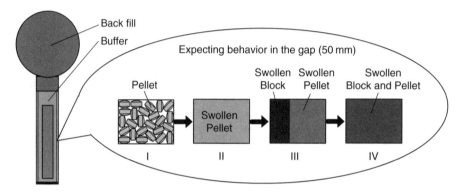

Figure 2. Expected behavior of block buffer and pellet in the clearance.

swollen buffer block. At that time, density of the swollen buffer block is decreasing (Fig. 2 III). Whole density of the buffer including block and pellet becomes uniform gradually (Fig. 2 IV).

Such behavior in the clearance with time is the coupled THM behavior there. However, our numerical code THAMES (Chijimatsu et al., 2000), which is FEM based on the continuous media, is difficult to consider behavior of a grain of the pellet in simulation. Then, we have performed simulations cover such behavior in the clearance, and have assessed the coupled THM behavior in the EBS. In these simulations, two cases of analysis were performed as shown in Figure 3 (Sugita et al., 2002). Case 1 was the condition that buffer block filled up the whole test hole and there was no clearance. In case 1, two kinds of dry density of the buffer block were calculated. One was that initial high dense buffer block fills a whole test hole (high density case), another one was that average dense buffer block considering block and pellet fills a whole test hole (low density case). Case 2 considered the clearance between buffer block and rock mass. In Case 2, buffer block was initial high density. The material in the clearance was not pellet but water. Saturation behaviors of buffer of these cases are considered as shown in Figure 3.

Saturation behavior in Case 2 is expected to act between both cases of Case 1. And, saturation behavior of the experimental condition filled with pellet is expected to act between Case 2 and high density case in Case 1.

When saturation behavior pellet and block in the clearance finishes and becomes stable sooner, the saturation behavior of buffer in the experimental condition is similar to a certain case of Case 1 and Case 2. Eventually, simulations of those cases are validated.

Then, we examined the fundamental properties of bentonite pellet and compared the properties of it with the properties of high density block (Sugita et al., 2003). The examined properties are thermal, hydraulic and swelling properties.

3 FUNDAMENTAL PROPERTIES OF BENTONITE PELLET

Bentonite pellet for the PRP (PRP pellet) is highly compressed sodium bentonite MX-80. In these examines, we used Japanese bentonite Kunigel V1 (Kunigel pellet) which is sodium bentonite. Figure 4 shows the bentonite pellets. Kunigel pellet is next

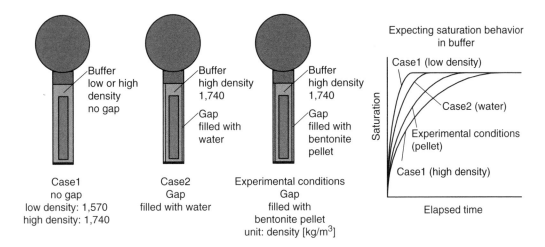

Figure 3. Analysis cases of JNC's prediction analysis.

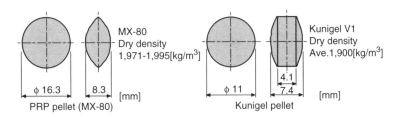

Figure 4. Bentonite pellets.

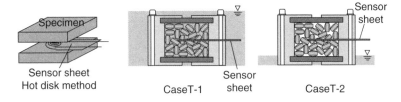

Figure 5. Measurement of thermal properties.

smaller size and lower density than PRP pellet. However, the obtained properties of Kunigel pellet are valid to explain the properties of sodium bentonite pellet qualitatively.

3.1 Thermal properties

Both thermal conductivity and thermal diffusivity were measured by hot disc method (Gustafsson, 1991) at the same time. Sensor sheet was set at the center of the test cell filled with pellet as shown in Figure 5. Water solution was the distilled water. In Case T-1, test cell was set in water bowl. Water infiltrated from both top and bottom ends of the test cell. In Case T-2, water infiltrated from only bottom

end of the test cell. Case T-2 simulates gradual infiltration condition of water at a test hole.

Figure 6 shows measured thermal conductivity and thermal diffusivity variation with time. In Case T-1, both thermal conductivity and thermal diffusivity became stable sooner. In Case T-2, both thermal conductivity and thermal diffusivity became stable gradually. However, both thermal properties became stable in about 400 hours, and measured values were almost the same as measured values in Case T-1.

Figure 7 shows comparison of thermal conductivity between block and pellets. Density of block was almost the same as average density of the cell filled with pellets. Thermal conductivity of pellets, that were initial and fully saturation, were shown in Figure 7.

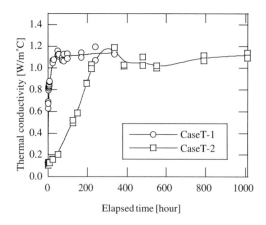

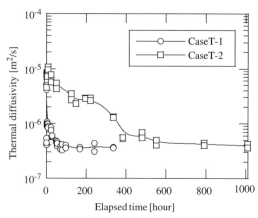

Figure 6.   Thermal conductivity and diffusivity variation with elapsed time.

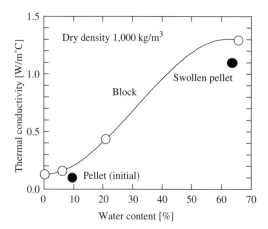

Figure 7.   Comparison of thermal conductivity between block and pellets.

The values of pellets were little smaller than the ones of block, and seemed to be almost the same as the ones of block. Table 1 shows thermal conductivity of swollen pellets and saturated blocks. Thermal conductivity of fully saturated blocks did not depended on dry density, and had almost the same values.

It is concluded that thermal properties in the clearance filled with pellets becomes stable sooner in case with water infiltration, thermal conductivity of saturated and swollen pellets is almost the same as one of the same dry density saturated block. It is also almost the same value as one of high density blocks.

3.2 Hydraulic properties

Hydraulic conductivity was measured by permeability test. Water solutions were distilled water and synthetic seawater. Table 2 shows test conditions of

Table 1.   Thermal conductivity of swollen pellet and block.

Dry density [kg/m^3]	Thermal conductivity [W/m°C]	Remarks
1,000	1.1	Swollen pellet
1,000	1.3	Block
1,600	1.4	Block
1,800	1.4	Block

Fully saturated with distilled water.

Table 2.   Test conditions of permeability test.

Case	Effective clay density [kg/m^3]	Water solutions	Injection periods [day]	Remarks
H-1	1,077	Distilled water		
H-2	1,061	Synthetic seawater		
H-3-1	1,134	Synthetic seawater	20	Swollen
H-3-2	1,130		150	once by
H-3-3	1,137		190	distilled
H-3-4	1,137*		421	water first

*On going (calculation).

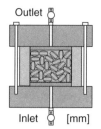

296

permeability test. Case H-1 was test with distilled water, Case H-2 and Case H-3 were tests with synthetic seawater. In Case H-3 series, pellets were swollen once by distilled water for 24 hours first. Then, synthetic seawater was injected.

Figure 8 shows the results of permeability tests with plots of existing value of blocks (distilled water

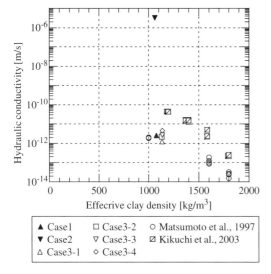

Figure 8. Relationship between hydraulic conductivity and effective clay density.

conditions (Matsumoto et al., 1997) and synthetic seawater conditions (Kikuchi et al., 2003)). In Case H-1 with distilled water, both swollen pellet and block of the same dry density had same hydraulic conductivity. However, hydraulic conductivity differed by six orders of magnitude in Case H-2 with synthetic seawater and in Case H-1 with distilled water. One of reasons was considered that swelling of the pellets was not completed and spaces between grains of pellets were still low dry density as shown in Figure 9. Such low dry density regions caused high hydraulic conductivity. On the other hand, when pellets were swollen once by distilled water (Case H-3), hydraulic conductivity of the swollen pellet is almost the same with the hydraulic conductivity of block at the same dry density. However, hydraulic conductivity of the swollen pellet increased with injection time of synthetic seawater toward the hydraulic conductivity of block injected synthetic seawater. In this case, the relationship of hydraulic conductivity on pellet between distilled water and synthetic seawater is the same as relationship on block (Kikuchi et al., 2003).

It is concluded that hydraulic conductivities of both swollen pellets with distilled water and block of the same dry density were almost the same. However, since hydraulic conductivity depends on dry density, hydraulic conductivity differs by two orders of magnitude in swollen pellets ($1,000 \, \text{kg/m}^3$) and block ($1,800 \, \text{kg/m}^3$) as shown in Table 3.

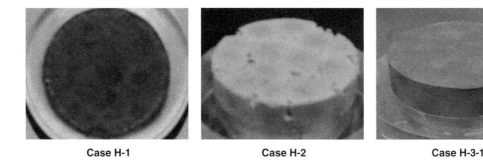

Case H-1   Case H-2   Case H-3-1

Figure 9. Specimen after permeability test.

Table 3. Estimated hydraulic conductivity of pellet and block.

Effective clay density [kg/m³]	Hydraulic conductivity [m/s]		
	Distilled water	Synthetic seawater	Remarks
1,000	1.0E-12	1.0E-12<	Swollen pellet
1,000	1.0E-12	1.0E-11	Block
1,600	1.0E-13	1.0E-12	Block
1,800	1.0E-14	1.0E-13	Block

3.3 Swelling properties

Swelling pressure was measured by swelling test. Water solutions were distilled water (Case S-1) and synthetic seawater (Case S-2).

Figure 10 shows measured swelling pressure variation with time. Two specimens were tested in each case. Average swelling pressure in Case S-1 was 0.23 MPa. Average swelling pressure in Case S-2 was 0.01 MPa.

Figure 11 shows existing swelling pressure of blocks (Suzuki & Fujita, 1999). Swelling pressure at

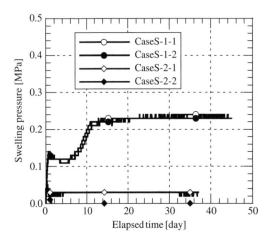

Figure 10. Swelling pressure variation of pellet with elapsed time.

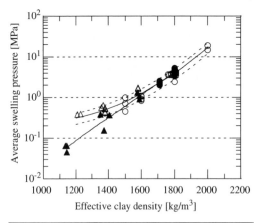

Figure 11. Relationship between swelling pressure and effective clay density of block (Suzuki & Fujita, 1999).

$1,000 \, kg/m^3$ of dry density with distilled water is expected from 0.2 to 0.3 MPa in Figure 11. Such swelling pressure of the block is almost the same as the swelling pressure of the pellet. Swelling pressure at $1,000 \, kg/m^3$ of dry density with synthetic seawater is expected about 0.02 to 0.03 MPa in Figure 11. Such swelling pressure of the block is almost the same as the swelling pressure of the pellet, too. When effective clay density is $1,600 \, kg/m^3$ or higher, swelling pressure depended on not water solution but dry density. However, in case of low dry density such as swollen pellet, swelling pressure depended on also water solution and differed by one order of magnitude in cases of distilled water and synthetic seawater.

It is concluded that swelling pressure of pellet is the same as one of block of the same dry density. However, swelling pressure depends on dry density and also water solution especially in low dry density.

4 CONCLUSIONS

To validate assumption in numerical simulation of the EBS in the PRP, fundamental properties of bentonite pellet, which is filled in the clearance between buffer block and rock mass and is not considered in model in numerical simulation, were examined by laboratory tests. In this study, we examined the fundamental properties of bentonite pellet and compared the properties of the pellet and the high density block. The examined properties are thermal, hydraulic and swelling properties.

Pellet filling region is still low density after pellets were swollen. Regarding thermal conductivity, thermal conductivity of the swollen pellets is almost the same as one of high density block. Therefore, pellet region is possible to be considered to be high density block in simulation. However, since hydraulic conductivity depends on dry density, hydraulic conductivity differs by two orders of magnitude in cases of low density and high density. In this case, it is not adequate that pellet region is considered to be high density block in simulation. Especially, in case that injection water is synthetic seawater, hydraulic conductivity becomes very high. In such case, pellet region should not be considered to be high density block in simulation. However, high hydraulic conductivity in pellet region with synthetic seawater accelerates infiltration of water into block next to pellets. Then, swollen block consolidates swollen pellets, and buffer in a test hole becomes uniform sooner eventually. Therefore, pellet region is possible to be considered to be high density block in simulation.

Fundamental properties of bentonite pellet are concluded that described above qualitatively. However, detail assessment of behavior of pellet in the PRP needs tests that use PRP pellet and water of the PRP.

In experimental conditions of the PRP, each test hole has seepage as some liters per day (Svemar & Pusch, 2000). Therefore, both infiltration of water into pellet region and swelling of pellets begin just after the installation of pellet. So, simulation considers pellet region to be high density block can explain the expecting saturation behavior in experimental conditions of the PRP.

REFERENCES

Chijimatsu, M, Fujita, T, Kobayashi, A & Nakano, M (2000) Experiment and Validation of Numerical Simulation of Coupled Thermal, Hydraulic and Mechanical Behaviour in the Engineered Buffer Materials, *International Journal for Numerical and Analytical Methods in Geomechanics*, 2000, 24, pp. 403–424.

Gustafsson, SE (1991) Transient plane source techniques for thermal conductivity and thermal diffusivity measurements of solid materials, *Rev. Sci. Instrum.* 63 (3), pp. 797–804.

Kikuchi, H, Tanai, K, Matsumoto, K, Satou, H, Ueno, K & Tetsu, T (2003) Hydraulic characteristics of Buffer Material II, JNC TN8430 2003-002 (in Japanese).

Matsumoto, K, Kanno, T, Fujita, T & Suzuki, H (1997) Saturated Hydraulic conductivity of buffer material, PNC TN8410 97-296 (in Japanese).

Sugita, Y, Ito, A, Chijimatsu, M & Kurikami, H (2002) Prediction analysis A for the PRP with the numerical code THAMES, SKB IPR-02-24.

Sugita, Y, Suzuki, H & Chijimatsu, M (2003) Thermal, hydraulic and swelling properties of bentonite pellet – Examine on calculating parameter assessment on PRP-, JNC TN8400 2002–03.

Suzuki, H & Fujita, T (1999) Swelling characteristics of buffer material, JNC TN8400 99-038 (in Japanese).

Svemar, C & Pusch, R (2000) Project description FIKW-CT-2000-00055, SKB International Progress Report IPR-00-30.

Advances in Understanding Engineered Clay Barriers – Alonso & Ledesma (eds)
© 2005 Taylor & Francis Group, London, ISBN 04 1536 544 9

Adsorption site for Cs^+, Ba^{2+} and Sr^{2+} in bentonite by means of EXAFS spectroscopy

Masash Nakano
Res. Inst. of Soil Science and Technology (RISST)

Katsuyuk Kawamura
Tokyo Inst. of Technology, Tokyo (Japan)

ABSTRACT: Extended X-ray absorption fine structure (EXAFS) spectroscopy was applied to estimate adsorption sites for ions Cs^+, Ba^{2+} and Sr^{2+} in bentonite. Samples of EXAFS experiments were prepared using a 0.1 M CsCl, $BaCl_2$ and $SrCl_2$ solution. The K-edge absorptions were measured in transmission mode for the wet pastes and for air-dried compacted samples at pH 10.0 at room temperature. Data analysis was done by a two oxygen-shell fit around adsorbed ions. The used phase shifts and amplitude functions were obtained from the sulfate compounds using ATOMS code and the ab initio program FEFF6.

The derived structural parameters revealed that in the wet pastes adsorbed atoms resided in the vicinity of the edge of a oxygen hexagonal cavities, and they were wrapped up by the water molecules if adsorbed atoms were over the surface of clay minerals. However, adsorbed atoms were seemed to fluctuate intensely, to roam in interlayer water, and to show a large mobility as large values of Deby-Waller factors were obtained. In air-dried compacted samples adsorbed atoms were sandwiched between two clay minerals. The side of adsorbed atoms was surrounded by water.

If adsorbed atoms resided in fractured sections in edge-to-edge contact, they were kept between two fractured sections of aluminol sheets in both the air-dried and the wet past samples. In the case of edge-to-face contact adsorbed atoms were sandwiched between a fractured section of aluminol sheets and an edge of oxygen hexagonal cavities on the mineral surface, regardless of water content.

1 INTRODUCTION

It is essential to elucidate the behavior and adsorption sites for ions in bentonite used as the engineered barrier, in order to understand the mechanisms of preventing the nuclides diffusion to environment. This study focuses to atoms Cs, Ba, and Sr in bentonite because they are key nuclides in a radioactive waste disposal program and show high activities in adsorption.

Extended X-ray absorption fine structure (EXAFS) spectroscopy is considered to be a method estimating adsorption sites for ions in bentonite as it is able to determine the average local molecular environment of atoms in materials. In the past the EXAFS analyses of the adsorbed Cs have been performed in smectite and illite by Kemner et al. (1997), in Zeolites by Doskocil and Davis (1999), and recently, in bentonite paste by Nakano et al. (2003). As regards EXAFS of Ba, Zhang (2001) has studied on adsorption in montmorillonite. Concerning Sr, papers have been published on montmorillonite and illite by Chen and

Hayes (1999), on kaolinite by Parkman et al. (1998) and Sahai et al. (2000), and on Ispra clay (Illite and kaolinite) by Cole et al. (2000). However, very little work has been done to elucidate the active sites for ions in bentonite from EXAFS results.

The objectives of this study were to elucidate the location where adsorbed atoms reside at the solid-water interface in smectite/montmorillonite hydrates, which are a major composition in bentonite, using EXAFS spectroscopy in air-dried samples as well as the wet pastes of bentonite.

2 EXPERIMENTAL

2.1 *Sample preparation and data collection*

Sorption samples were prepared by adding 4 ml of a 0.1 M CsCl, $BaCl_2$ and $SrCl_2$ solution to 2 g dry weight of suspended Kunipia F bentonite in 196 mL of water. The pH of the samples was adjusted to about 10.0 with a 0.1 M NaOH solution. The process was

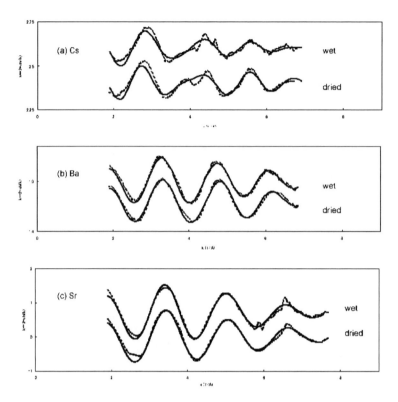

Figure 1.   EXAFS spectra. Dotted line: the observed, and solid line: the best fit.

designed to eliminate CO_2 in the samples by purging with N_2 gas. After centrifuging the samples, the wet pastes were washed once with distilled water. Then, a part of them were sealed to keep them moist in tubes and others were air-dried and compacted for EXAFS measurement.

EXAFS spectra were collected at a synchrotron, the Spring-8, Japan, using a Si(311) and Si(111) monochromator. The K-edge absorptions were measured in transmission mode for the wet pastes of 100 mm in length and a solid/liquid ratio of 0.046, of 100 mm in length and a solid/liquid ratio of 0.041, and of 20 mm in length and a solid/liquid ratio of 0.059, which were sealed in tubes, in Cs-, Ba-, and Sr-bentonite, respectively. The K-edges of air-dried samples were measured for compacted Cs- and Ba-bentonite of about 4 mm thickness and for Sr-samples of about 1.5 mm thickness, with density about 2.1 gcm^{-3} in all cases at room temperature.

2.2   EXAFS analyses and results

Data were analyzed with the program TECHXAS (Technos Co. Ltd. Japan) to examine the local oxygen environment around atoms Cs, Ba, and Sr. Analyses were performed with the steps of background

subtraction, normalization, isolation of backscattering contributions and energy unit conversion and intensity weighting (See Fendorf 1999). The observed EXAFS spectra and the Fourier transform are shown with the dotted lines in Figures 1 and 2, respectively.

EXAFS spectra of Cs were extremely complicated by the presence of a double excitation (Doskocil E.J. and Davis R.J. 1999). Each spectrum was slightly shifted in energy each other. The second peaks in Fourier transformed spectra were remarkably higher in Cs-bentonite than in others. The first and the second peaks were considered to result from oxygen of interlayer water and of clay minerals, respectively (Nakano et al., 2003). It seems that the third peaks in Fourier transform were derived from Si or Al atoms in clay minerals, taking account of the structure of smectite/montmorillonite.

Fitting was done with non-linear, least squares curve-fitting techniques on the filtered spectra back-transformed from Fourier transformed EXAFS function of the R range in which the first and determined by a two oxygen-shell fit. The reference phase shifts and amplitude functions used in the fitting were calculated from structural data for the model compounds Cs_2SO_4, $BaSO_4$ and $SrSO_4$ using ATOMS

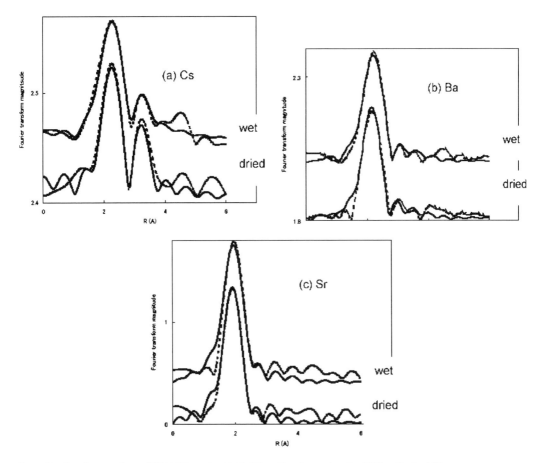

Figure 2.　Fourier transform of EXAFS spectra. Dotted line: the observed and solid line:the best fit.

Table 1.　Local structural information of atoms in bentonite.

Atom	Sample	pH	ΔE_0(eV)	1st shell (Owater)			2nd shell (Oclay)		
				R(Å)	N	σ^2	R(Å)	N	σ^2
	Wet paste		4.72	3.164	6.74	0.04	3.587	5.72	0.05244
Cs	Air-dried	9.8	5	3.195	4.65	0.02074	3.593	6.82	0.0392
	Wet paste		7.4	2.918	9.48	0.0228	3.299	4.24	0.04244
Ba	Air-dried	10.0	7.01	2.872	7.7	0.01769	3.202	4.33	0.03842
	Wet paste		4.62	2.624	7.8	0.01277	2.888	2.39	0.01716
Sr	Air-dried	9.6	4.1	2.591	5.85	0.00723	2.829	2.53	0.01323

ΔE_0 (eV): the difference between threshold energy in reference and sample spectra. R(Å): interatomic distances. N: coordination numbers. σ^2: Debye-waller factors (Å2).

code and the ab initio program FEFF6. The derived structural parameters are summarized in Table 1. The obtained best fits are shown with dotted lines in Figs 1 and 2.

　Bond distances in each shell were little different essentially within the accuracy of the fitting between the wet pastes and air-dried samples. The coordination numbers in the first shell remarkably decrease in air-dried samples. Those in the second shell slightly increase in air-dried samples. Debye-waller factors are extremely large in comparison with those of crystalline compounds, about 0.0036 on average. Then, adsorbed atoms in bentonite were considered to fluctuate intensely and to roam in inter-layer water of smectite/montmorillonite hydrates.

However, there are the multi-excitation effects in the observed EXAFS spectra. This effect decreases amplitude of EXAFS spectra. Moreover, a multiple scattering effects occur from backscattering atoms such as Si, Al, Mg, Fe, S, Cl, and others in clay minerals as well as in inter-layer water. Even if the best background subtraction and the best fit between the observed and calculated curve were performed, the results yield values of R $\pm$ 0.01~0.02, and N $\pm$ 20% (Fendrof 1999), where R is bond distance and N is coordination number.

3 DISCUSSION AND CONCLUSIONS

3.1 Evaluation of EXAFS results

According to a review by Ohtaki and Radnai (1993), X-ray diffraction studies in the past, yielded a Cs$^+$-H$_2$O bond length of 2.95–3.21 Å and a hydration number of either 6 or 8 for Cs$^+$ in an aqueous solution of CsCl. Antonio et al (1997) also reported that Cs-O distances were 3.01 to 3.23 Å in dibenzo-crown ethers. Doskocil and Davis (1999) reported Cs-O distances 2.94 Å in zeolites.

Concerning Ba, an X-ray diffraction study and NMR yielded a Ba^{2+}-H$_2$O bond length of 2.9 hydration number of 9.5 in an aqueous solution of BaCl$_2$ (See. Ohtaki and Radnai 1993). Zhang (2001) decided a Ba-O distance 2.80 Å and coordination number 7.0 at pH8 in montmorillonite.

As regards Sr, the Sr^{2+}-OH$_2$ bond length and coordination number have been determined to be 0–2.64 Å and 7.9–8.0, respectively, in an aqueous solution of SrCl$_2$ by the X-ray diffraction (See Ohtaki and Radnai 1993). In kaolinite, a Sr-O distance 2.58 Å and coordination number 8.2–8.8 were reported by Sahai et al. (2000), while 2.57 Å and 7.2 were reported by Chen and Hayes (1999). A Sr-O distance 2.55–2.58 Å was also estimated by Parkman (1998). However, in Ispra clay, a Sr-O distance 2.61 Å and coordination number 7.3–7.6 were determined by Cole (2000). In montmorillonite, Chen and Hayes (1999) reported 2.58 Å and 5.7, respectively.

The above results have been derived on the nearest neighbors, and the clay the state of a paste. In this paper, bond distance 3.16 Å and coordination number 6.7 for Cs, 2.92 Å and 9.5 for Ba, and 2.62 Å and 7.8 for Sr were obtained in the first shell of the wet paste. These results closely coincide with those of the hydrated ions in an aqueous solution in all cases. It was considered that the results appeared because smectite/montmorillonite had the inter-layer water

that consists of the 3-layer water equivalent to the thickness about 8.0 Å.

The air-dried samples had the 1.5-layer water equivalent to the thickness about 4.0 Å Then, the adsorbed atoms were considered to be sandwiched between two clay minerals, and their sides were surrounded by the water molecules forming the first shells. Therefore, coordination numbers decreased in the first shells.

As regards coordination numbers in the second shells, they appeared due to a difference between the diameters of each atom, and of a slight shift in location between the upper and lower mineral in a stack of minerals.

The fractured sections were considered to form extremely small gaps, which are occupied by the 1.5-layer water in both edge-to-edge and in edge-to-face contact, regardless of water content. Then, it was concluded that the adsorbed atoms were also sand-wiched between two clay minerals, and surrounded by water in the same way as a mode on air-dried samples.

3.2 Estimation of adsorption sites

Therefore, if adsorbed atoms resided over surface of clay minerals, coordinate numbers in the wet pastes revealed that adsorbed atoms were in the vicinity of the edge of a oxygen hexagonal cavities because those in the second shell were less than 6 O of a hexagonal cavity, and that they were wrapped up by the water molecules equivalent to hydration numbers (Fig. 3. (a), (b) and (c)). Then, adsorbed atoms looked forming incomplete inner-sphere-like complexes, while they behaved as a hydrated ion in inter-layer water and behaved like that forming the outer-sphere complexes.

In air-dried compacted samples, adsorbed atoms were sandwiched between oxygen of the upper and of lower clay minerals. The numbers of oxygen in a clay mineral, which contributed to hold adsorbed atoms, were a half of coordination number in the second shells. The sides of adsorbed atoms were also surrounded by the water molecules, which formed the first shells (Fig. 3 (d), (e) and (f)).

If adsorbed atoms resided in fractured sections of clay minerals in the edge-to-edge contact, they were kept by oxygen of aluminol sheets, in which the numbers in a sheet also were a half of coordination number in the second shells in both the air-dried and the wet pasts samples, regardless of a difference in the form of fractured section occurring from a slight shift of minerals in location in a stack (Fig. 3 (g)–(l)).

In a case of edge-to-face contact, adsorbed atoms were sandwiched between oxygen of an aluminol sheet and oxygen of a hexagonal cavitie, regardless of water content. The numbers of oxygen holding adsorbed atoms also were a half of coordination number in the second shells.

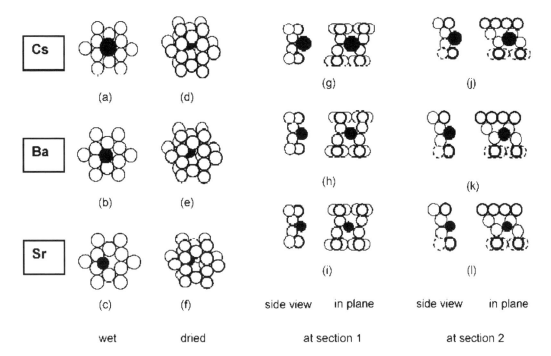

Figure 3. Illustration of adsorption site. Black circle: each atom, white: O of clay minerals.

Their sides of adsorbed atoms were also surrounded by the water molecules, which formed the first shells, in the same way as a mode in the air-dried compacted sample.

ACKNOWLEDGEMENTS

We thank Dr. Shuichi Emura, Osaka University, for valuable advices on EXAFS data collection, analysis, and discussions, Dr. Hajime Tanida and Dr. Tomoya Uruga for his suggestions, discussions and adjustment of the beam line BL01, Japan Synchrotron Radiation Institute, as well as to Dr. Kazunori Uchida, Kobe University, and Dr. Taku Nishimura, Tokyo university of A&T, for data collections.

REFERENCES

Chen C.C. & Hayes K.F. (1999). X-ray absorption spectroscopy investigation of aqueous Co(II) and Sr(II) sorption at clay-water interfaces, Geochim. Cosmochim. Acta, 63(19/29): 3205–3215.

Cole T., Bidoglio G. Soupioni M., O'Gorman & Gibson N. (2000). Diffusion mechanisms of multiple strontium species in clay, Geochim. Cosmochim., 64(3): 385–396.

Doskocil E.J. & Davis R.J. (1999) Spectroscopic characterization and catalytic activity of zeolite X containing occluded alkali species, J of Catalysis, 188: 353–364.

Fendorf S. (1999). Fundamental aspects and applications of X-ray absorption spectroscopy in clay and solid science, cms workshop lectures Vol.9. pp18–67, Synchrotron X-ray methods in clay science, edt. by Schulze D.G., Stucki J.W. & Bertsch P.M., The Clay Minerals Society, Boulder, Co.

Kemner K.M., Hunter D.B., Bertsch P.M., Kirkland J.P. & Elam W.T. (1997). Determination of site specific binding environments of surface sorbed Cesium on clay minerals by Cs-EXAFS, J. Physics ‡W France 7, C2: 777–779.

Nakano M., Kawamura K. & Ichikawa S. (2003). Local structural information of Cs in smectite hydrates by means of an EXAFS study and molecular dynamics simulations. Applied Clay Sci. 23:15–23.

Ohtaki H. & Radnai T. (1993) Structure and dynamics of hydrated ions. Chemical Reviews. 93(3): 1157–1204.

Parkman R.H., Charnock J.M., Livens F.R. & Vaughan D.J. (1998). A study of the interaction of strontium ions in aqueous solution with the surfaces of calcite and kaolinite, Geochimica et Cosmochimica Acta. 62(9): 1481–1492.

Sahai N., Carroll S.A., Roberts S. & O'Day P.A. (2000). X-ray absorption spectroscopy of strontium(II) coordination, II.Sorption and precipitation at kaolinite, amorphous silica, and goethite surface, J. Colloid and Interface Science 222: 198–212.

Zhang Peng-Chu, Brady P.T., Arthur S.E., Zhou Wei-Qing, Sawyer D. & Hesterberg D.A. (2001). Adsorption of barium(II) on montmorillonite: an EXAFS study, Colloids and Surfaces A, 190: 239–249.

Advances in Understanding Engineered Clay Barriers – Alonso & Ledesma (eds)
© 2005 Taylor & Francis Group, London, ISBN 04 1536 544 9

ECOCLAY II. Laboratory experiment concerning compacted bentonite contacted to high pH solutions

Ola Karnland & Ulf Nilsson
Clay Technology AB

Siv Olsson
Geochimica, Lund

1 INTRODUCTION AND OBJECTIVES

1.1 Background

It will be most practical to use concrete in many constructions and cement for sealing rock fractures in a repository for spent nuclear waste. In the Finnish ONKALO underground facility as an example, the estimated quantity of cement is 5366 000 kg (Vienno et al.). However, the high pH in cement pore water may affect the stability of repository components, and are thereby of concern for the performance and safety assessment analyses. In the Swedish KBS3 repository, the high pH may influence fracture minerals, the bentonite buffer, canister components and the spent fuel. The pH evolution in the bentonite is of special interest, since the bentonite buffer completely surrounds the fuel canisters and thereby may protect the canister and the fuel by a chemical buffering of the high pH. This potential pH buffering capacity of bentonite may be divided in three different groups:

– accessory mineral reactions
– montmorillonite surface reactions
– montmorillonite crystal lattice reactions

Depending on the type of cement, type of bentonite and physico-chemical conditions, the following principle conditions may prevail in bentonite contacted to high pH solutions:

– Relatively slow reaction between high pH solution and all components in the bentonite compared to the transport of solutes in the pore-water. The pH propagation front will thereby be governed by diffusion only, leading to a diffuse and fast moving front.
– Relatively fast reaction, between high pH solution and some part of the bentonite (i.e. accessory minerals or montmorillonite surface sites), compared to the transport of solutes in the pore-water. The pH front propagation will thereby be slower compared to pure diffusion and have a steeper concentration profile.
– Fast reaction, between high pH solution and the major part of the bentonite (congruent dissolution of the montmorillonite lattice), compared to the transport of solutes in the pore-water. The pH front propagation will thereby be slow and have a steep concentration profile.

Item one implies a well functioning buffer with respect to physical properties, but with no pH buffering capacity. The third item conditions would give a large buffering capacity but destruction of the bentonite and a reduced effective buffer thickness. Knowing the conditions make it possible to adopt the repository layout to the prevailing conditions, e.g. if the latter case prevails, then the buffer thickness can be adopted to permit a sacrificed part.

Figure 1 shows a simplified stability diagram of montmorillonite with respect to cation/proton activity ratio and silica activity. Alkali hydroxides dissolve in fresh cement which may raise pH to values between 13 and 14, and in matured cement materials pH is governed by Portlandite ($Ca(OH)_2$), which gives a pH of 12.4. Compared to typical bentonite conditions this leads to an increase of the cation/proton activity ratio and into the stability field of Feldspars (Figure 1, left). Further, an increase in pH will lead to a dramatic change in silica solubility. Figure 1 right shows the saturation concentrations for amorphous silica and for quartz versus pH in NaOH solutions as calculated by use of PREEQCI. The alkali hydroxides in fresh cement may consequently lead to four orders of magnitude higher silica activity compared to neutral conditions.

1.2 Objectives

Several studies have been made concerning the interaction between bentonite and cement pore-water. Most of these have been based on batch experiments

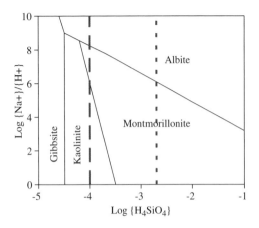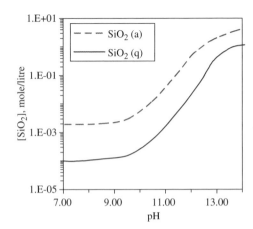

Figure 1. Stability diagrams for Na-montmorillonite (left) simplified from Helgesson et al. 1969. Left dotted line indicate quartz saturation and right dotted line indicate amorphous silica saturation. Solubility of silica versus pH (right). Dotted line indicates amorphous silica (a) and unbroken line indicate quartz (q).

with high water solid ratio. The present study focuses on highly compacted bentonite since e.g. ion-equilibrium and transport restrictions significantly may affect possible reactions and reaction rates. The overarching objectives in this study were to gather information concerning:

- pH transport rate through compacted bentonite
- pH buffering capacity of bentonite
- changes in swelling pressure as a consequence of high pH
- effect of accessory minerals on the above items
- mineralogical changes in bentonite exposed to alkali and calcium hydroxide solutions
- dissolution rates of involved minerals.

2 EXPERIMENTAL

2.1 Test principles

The general test principle was to expose highly compacted bentonite clay to high pH solutions and to measure the response in swelling pressure, transport of the high pH plume through the bentonite, and finally to analyze the mineralogical changes in the bentonite. Both commercial bentonite and almost pure montmorillonite were used, respectively, in order to get information on the effects of the bentonite accessory minerals. After test termination, the test solutions were analyzed by use of ICP/AES and ion chromatography (IC). The clay material was analyzed by use of XRD, ICP/AES and the cation exchange capacity and the type of exchange cations were determined. In total, 24 tests were carried out. All tests

were made at room temperature, and most tests were run in a closed system with limited access to air.

High bentonite dissolution rate and fast transport of a high pH plume are in different ways the most unfavorable conditions from a safety assessment perspective. In order to be conservative from this respect, the concentration gradients over the test samples were kept constant, and the volume of the external solutions were large enough to keep the concentrations of dissolved species far from saturation. These conditions are assumed to minimize precipitation of dissolution products and thereby give a maximum effect on the bentonite and on the transport rates.

2.2 Test material

Material from one batch (MX-80, 2001-01-19) of commercial Wyoming bentonite from American Colloid Co. was used as starting material for all tests. This bentonite grade, termed "type 6 material", is a blend of natural bentonite horizons, which is milled to millimeter sized grains and sold under the product name MX-80.

In order to remove the accessory minerals and produce homo-ionic montmorillonite, ten grams of the original MX-80 material was dispersed in 1 L of de-ionized water. The fraction coarser than 2 micron was settled as calculated by Stoke's law and the supernatant was removed by decantation. The salts KCl, NaCl and CaCl2 were added to a concentration of 1 M in the decanted dispersion, respectively. The treatment was repeated twice and the dispersion was left to settle and the supernatant was removed and new de-ionized water was added. Approximately 0.8 L of the suspension was placed in dialyses membranes (Spectra/pore 3, 3500 MWCO). The membranes were placed in a Plexiglas tube with 5 l of de-ionized water, which was circulated by use of a magnetic stirrer. The electrical conductivity of the external water was measured and changed daily. After approximately one week the conductivity was below 10 μS and the clay material was removed and oven-dried at 60°C.

The four test materials, MX-80 standard, MX-Na, MX-Ca and MX-K, were analyzed by use of XRD, ICP/AES, and the cation exchange capacity (CEC), and the type of exchangeable ions were determined by the technique described in section 4.3.

2.2.1 Original MX-80 material

Five separate XRD scans were made of the powdered MX-80 total material and the mean intensities were used to identify and quantify the main minerals according to the technique described in section 4.4.3. The total MX-80 material contained 83% montmorillonite, 7% albite, 5% quartz, 3% cristobalite, 1% muscovite and gypsum. In addition, grains of pyrite (FeS_2), calcite ($CaCO_3$), siderite ($FeCO_3$), barite ($BaSO_4$) and iron hydroxides in mean quantities less than 1% were occasionally found both by the XRD and SEM analyses.

The CEC was determined to be 0.74 eq/kg for the total MX-80 material, and 0.86 eq/kg for the clay fraction. The montmorillonite fraction of the original MX-80 material, as calculated from the CEC values, is consequently 86% assuming that only montmorillonite contributes to the CEC.

The mole-weight of the montmorillomite based on an $O_{20}(OH)_4$ cell was calculated to 747 g and the layer charge −0.63. The following structural formula of the montmorillonite component was calculated from the cation distribution in the original MX-80 material, and from the clay fraction analyses, by the technique described in section 5.3.3:

$$(Si_{7.86} Al_{0.14}) (Al_{3.11} Fe^{3+}_{0.37} Mg_{0.50} Ti_{0.01}) O_{20} (OH)_4,$$
$$Na_{0.47} Ca_{0.05} Mg_{0.02} K_{0.01}$$

2.2.2 Homo-ionic clay fractions based on MX-80

The XRD analyses of the purified MX-80 material showed minor peaks from cristobalite and quartz, which was quantified to represent less than 2% of each according to the technique described in section 4.4.3. According to the ICP/AES analyses and the calculation technique described in section 5.3.3 the SiO_2 content (cristobalite and quartz) was 3.9% in the MX-K material, 3.4% in the MX-Na material, and 2.3% in the MX-Ca material.

The sodium exchanged material (MX-Na) material had a CEC of 0.85 eq/kg, and based on an $O_{20}(OH)_4$ cell the mole-weight was calculated to be 748 g, the layer charge −0.64, and the structural formula:

$$(Si_{7.82} Al_{0.18}) (Al_{3.13} Fe^{3+}_{0.38} Mg_{0.47} Ti_{0.01}) O_{20} (OH)_4$$
$$Na_{0.59} Mg_{0.02} Ca_{0.01}$$

The potassium exchanged material (MX-K) material had a CEC of 0.86 eq/kg, and based on an $O_{20}(OH)_4$ cell the mole-weight was calculated to be 757 g, the layer charge −0.65, and the structural formula:

$$(Si_{7.82} Al_{0.18}) (Al_{3.11} Fe^{3+}_{0.39} Mg_{0.48} Ti_{0.01}) O_{20} (OH)_4$$
$$K_{0.53} Na_{0.02} Ca_{0.02} Mg_{0.01}$$

The calcium exchanged material (MX-Ca) material had a CEC of 0.78 eq/kg, and based on an $O_{20}(OH)_4$ cell the mole-weight was calculated to be 746 g, the layer charge −0.58, and the structural formula:

$$(Si_{7.88} Al_{0.12}) (Al_{3.13} Fe^{3+}_{0.39} Mg_{0.47} Ti_{0.01}) O_{20} (OH)_4$$
$$Ca_{0.28} Mg_{0.01}$$

2.3 Test solutions

Alkali hydroxides dissolve in fresh cement which may raise pH to values between 13 and 14, and in matured cement materials pH is governed by Portlandite ($Ca(OH)_2$). In order to cover the possible conditions the test solutions were chosen to be 0.1, 0.3 and 1.0 M NaOH, and saturated $Ca(OH)_2$ solution. One test was also made with 1.0 M KOH. The corresponding pH values as calculated by PHREEQC are 12.9, 13.3, 13.7 (13.8, llnl database) for the NaOH solutions, 12.4 for the $Ca(OH)_2$ solution, and 13.7 for the KOH solution.

2.4 Samples and test equipment

A compromise between the need of reasonable fast reactions and enough material for subsequent analyses led to test samples with a diameter of 35 mm, and a height of 5 mm. The proposed KBS3 buffer density of 2000 kg/m³ at full water saturation was aimed at in all tests. This corresponds to a dry density of 1570 kg/m³, water ratio of 27%, a pore ratio of 0.75, and a porosity of 43%. A slightly lower density was generally measured in all samples after test termination due to the sample dimensions and yieldingness in the equipment.

The samples were slightly over-compacted compared to the intended density and placed in the cylindrical sample holders made of titanium and PEEK (Figure 2). The piston, load cells and upper lid were attached and fixed. The volume was initially controlled by distance pieces placed between a flange

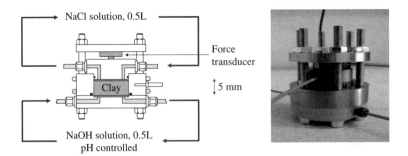

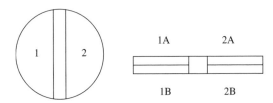

Figure 2.   Schematic drawing and photo of a sample holder.

on the piston and the sample confining ring. Sensotech miniature load cells model 53 were used in order to measure the axial force. The pistons were pre-stressed to a force of around 25% of the expected final force from the sample swelling pressure. The distance pieces were removed when the measured force exceeded the applied pre-stress.

The predefined saturation solutions (0.5 L) were slowly circulated behind the two filters by means of a peristaltic pump. After reaching pressure equilibrium the test solutions were changed to a high pH solution (0.5 L) on the bottom side and a corresponding isotonic chloride solution (0.5 L) on the upper side. The test solutions were placed in polypropylene bottles and the solutions were circulated in closed loops, except for the short time required for the pH measurements. In a few tests the same solution was used on both sides and successively changed on both sides of the sample in order to ensure that the swelling pressure changes were an effect only of the solutions, respectively. At test termination the solutions were disconnected and the samples were quickly removed and split, and the samples and solutions were analysed according to the program in section 5.

3 MEASUREMENTS AND ANALYSES

3.1 Temperature and swelling pressure

The temperature and axial force were automatically measured every half hour, and the swelling pressure (P_s) was calculated from the axial force (F) according to:

$$P_s = \frac{F}{A} \qquad (1)$$

where A is the piston/bentonite contact area. The load cells were calibrated before and after each test. Temperature was measured by use of thermocouples placed outside the samples.

Figure 3.   Principle partition of the test samples for the subsequent analyses.

3.2 pH results

A Metrohm 691 pH meter equipped with and a T-glass electrode was used for all pH measurements. Several hours long recovery time in de-ionized water followed by storage overnight in 3 M KCl solution was needed after measuring in the high pH solutions. The measurements were therefore made from low to higher low pH solutions, and thorough washing of the electrode was made after each measurement in order to minimize the risk for the contamination of the solutions. The pH was measured virtually daily in the low pH solutions, and weekly in the high pH solutions. The electrode was calibrated regularly in reference solutions of pH up to 11.

3.3 Mineralogy

3.3.1 General

The test clay samples were split as shown in Figure 3 immediately after removal from the sample holders. The central part was used for Scanning microscopy. Part number 1 and 2 were split in the axial direction in order to give a low pH (A) and a high-pH side (B). The "1" and "2" parts were split for the subsequent analyses and in some cases used as doublets.

3.3.2 Cation Exchange Capacity, CEC

The type of exchangeable cations was determined by exchange with ammonium ions (0,15 M NH4Cl) dissolved in 80% ethanol (Jackson, 1975). The cation exchange capacity was determined by exchange with copper(II) triethylenetetramine (Meier and Kahr, 1999). The measured CEC values in the calcium saturated clay were not repeatable and generally lower than for the same clay saturated with sodium and potassium. A reference test series were therefore made in order to improvement the technique. The series included also a second montmorillonite (Milos), and Cu concentrations from 8 to 24 mM. Increasing the Cu concentration to 12 mM at a minimum stabilized the values for calcium saturated montmorillonite, but the analyses still gave approximately 10% lower CEC values for divalent ions (Mg and Ca). The values for the calcium saturated clay are still used since possible changes, and not the absolute values, are of main concern. However, the structural formulas for the calcium saturated clays given in section 3.2.2 and 5.3.2 are likely slightly wrong.

3.3.3 X-ray diffraction analyses

All scans were made with a step size of $0.02°$ 2θ by use of a Seifert 3000 TT X-ray diffractometer with a variable slot and CuKa radiation produced at 50 kV/30 mA. The samples were analyzed by use of X-ray diffraction of powdered samples and of oriented samples prepared according to the filter-membrane peel-off technique (Drever). The oriented mounts were scanned after pretreatment with 0.5 M $MgCl_2$ and after exposure to ethylene glycol (EG) vapor for 24 h. Mineralogical quantification was made by use of the Siroquant Software based on the Rietveld technique (Taylor and Matulis, 1994). The results were recalculation to fixed slot conditions in order to facilitate quantification of the clay minerals.

3.3.4 Element analyses

The clay samples were analysed by ACME laboratories, Vancouver, by use of ICP/AES, mainly in order to identify changes in the ratio between the main elements in the clay material, but also in order to verify ion-exchange, and to reveal possible uptake of atmospheric carbon dioxide.

The tests solutions were analysed by use of ICP/AES and ion chromatography (IC) by the Ecology institution at Lund University, mainly in order to quantify dissolved elements from the clay, and secondly in order to detect possible dissolution of the equipment due to the very aggressive conditions.

3.3.5 Scanning Electron Microscopy (SEM) analyses

SEM analyses were made by use of a Phillips 515 SEM microscope, equipped with a LINK ISIS EDX. A small piece, representing the whole section, of the central part of the samples (Figure 3) was freeze dried

for 24 h at $-30°C$ and <0.1 Pa. The specimen was broken to give a fresh surface, glued on to holders and gold-sputtered to a film thickness of 10 Å in order to reduce electrical charging during the analyses. This preparation technique induces large micro-structural changes but prevail the content and distribution of dissolved species. A rather large number of positions, representing the whole section from the low pH to high pH-side were analyzed in each sample. Morphological and the chemical analyses were made in magnifications from 50 to 5000.

4 RESULTS

4.1 General

In total, 24 separate tests were run. Twelve were special tests or doublets to check repeatability, and these tests were not fully analyzed. The measurements and analyses described in the previous chapter were fully carried out for the 12 separate tests shown in Table 1 and the important results are described in this chapter.

4.2 Swelling pressure

Significant reductions in swelling pressure were found for the samples contacted to high concentrations of sodium chloride and hydroxide solutions. A major difference was though that the chloride solutions gave a fast drop to a new stable pressure value, while contact to the hydroxide solution led to a continuous lowering of the swelling pressure. Figure 4 (left) shows the pressure response from one sample successively exposed to pure water, 1.0 M NaCl, 1.0 M NaOH, 1.0 M NaCl, and finally pure water. Figure 4 (right) shows individual samples exposed to the same 1.0 M solutions.

A significant continuous swelling pressure reduction was also observed for all samples exposed to 0.3 M NaOH solutions (Figure 5). However, no swelling pressure change was noticed in the samples exposed to 0.1 M NaOH (Figure 6) or to saturated $Ca(OH)_2$ solution (Figure 4). The minor drop in the lower curve in Figure 4, right, is fully explained by sensor leeway, ensured in the post-calibration, and is consequently not an effect of swelling pressure decrease.

The MX-Na samples (Figures 3 and 4, right) generally gave higher swelling pressure compared to the MX-80 samples (Figures 3 and 4, left). No principle difference was noticed between the MX-80 samples and the MX-Na samples with respect to the response to NaCl or NaOH solutions, except for minor initial ion-exchange responses in the MX-80 bulk samples (Figures 5 and 6, left).

Table 1.   Fully analyzed samples, solutions and calculated pH.

Sample	Clay	Solution A	Solution B	pH solution B
21	MX-80	0.1 M NaCl	0.1 M NaOH	12.9
22	MX-80	0.3 M NaCl	0.3 M NaOH	13.3
23	MX-80	1.0 M NaCl	1.0 M NaOH	13.7
24	MX-Na	0.1 M NaCl	0.1 M NaOH	12.9
25	MX-Na	0.3 M NaCl	0.3 M NaOH	13.3
26	MX-Na	1.0 M NaCl	1.0 M NaOH	13.7
31	MX-80	0.1 M NaCl	0.1 M NaCl	–
32	MX-80	1.0 M NaCl	1.0 M NaCl	–
33	MX-80	0.015 M CaCl$_2$	0.015 M Ca(OH)$_2$	12.4
34	MX-80	0.015 M CaCl$_2$	0.015 M Ca(OH)$_2$	12.4
35	MX-Ca	0.015 M CaCl$_2$	0.015 M Ca(OH)$_2$	12.4
36	MX-Ca	0.015 M CaCl$_2$	0.015 M Ca(OH)$_2$	12.4

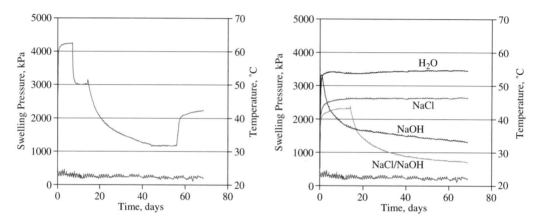

Figure 4.   Pressure response from a MX-80 sample successively exposed to pure water, 1.0 M NaCl, 1.0 M NaOH, 1.0 M NaCl, and finally pure water (left). Pressure response for the same exposures but in four individual MX-80 samples (right). The sample indicated NaCl/NaOH were saturated with 1.0 M NaCl and exposed to 1.0 M NaOH solution after 15 days. Lowest line shows temperature.

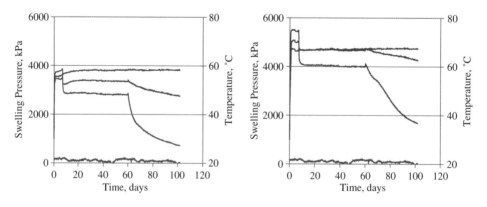

Figure 5.   Swelling pressure evolution in MX-80 bulk samples (left) and Na-montmorillonite (right). All samples were initially saturated with pure water, subsequently exposed to the chloride solutions (5 days), and finally to the hydroxide solutions (60 days). Lowest line at 100 days shows results from 1.0 M solution, middle line from 0.3 M solutions, and uppermost line show 0.1 M solution.

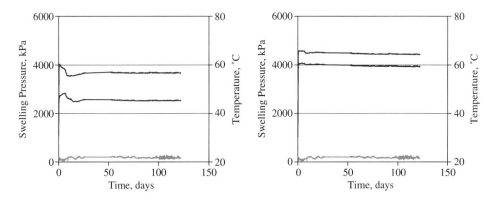

Figure 6. Swelling pressure evolution for two MX-80 bulk samples (left) and two MX-Ca samples (right). All samples were initially saturated with pure water, exposed to the 0.015 M Ca(Cl)$_2$ after 5 days, and finally exposed to saturated Ca(OH)$_2$ solution in the lower filter after 60 days.

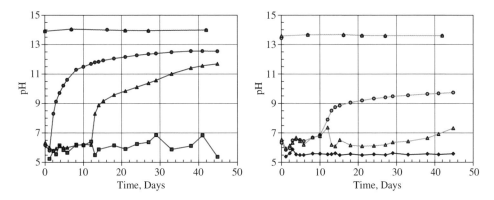

Figure 7. pH evolution in the NaCl solutions of samples exposed to 1.0 M solutions (left), and 0.3 M solution (right). Circles indicate results from MX-80 samples, triangles indicate results from MX-Na samples. Uppermost curves indicate results from the NaOH solutions, and lowest lines results from NaCl reference solutions.

Initial tests with 1.0 M KOH and NaOH solutions gave in principle the same results and no further tests were made with KOH solutions.

4.3 pH results

A pH increase in the NaCl solutions was measured in the four tests exposed to 1.0 M and 0.3 M solutions. Logically, the increase came earlier and was more pronounced in the tests with the highest concentration (1.0 M, sample 23, and 26). For both concentrations, the increase was first measured in the MX-80 samples (sample 22, 23). Figure 7 shows the measured pH in the 1.0 M NaCl solutions (left) and in the 0.3 M NaCl solutions (right). No pH increase was measured after 45 days in the two tests with 0.1 M solutions, nor after an exposure time of 110 days in the four tests with saturated Ca(OH)$_2$ solutions.

4.4 Mineralogy

4.4.1 Cation Exchange Capacity, CEC

A significant increase in CEC was noticed for all samples exposed to the NaOH solutions (Table 2). The increase was significant on both sides of the samples except for the NaCl side of the sample exposed to the lowest concentration (sample 24A).

No significant changes in CEC were measured in the samples exposed to only NaCl solutions (samples 31 and 32), or to the samples exposed to the 0.015 M Ca(OH)$_2$ solutions (Table 3). Dissolution of salt from pore-water leads to much higher content of individual ions in the test samples than in the reference samples. The results are still presented since the distribution of ions show an almost complete ion exchange to calcium in samples 33 and 34, although the calcium concentration is as low as 0.015 M in the test solutions.

4.4.2 *X-ray diffraction analyses*

Significant changes were found in the samples exposed to NaOH solutions with respect to cristobalite and quartz (Figures 8 and 9). No cristobalite were present in the samples exposed to 1.0 M NaOH solution. Loss of cristobalite was found in the sample exposed to the 0.3 M NaOH solution, and a possible decrease is indicated on the high pH side of the MX-80 sample exposed to 0.1 M NaOH solution. Quartz was much less affected and the amount was almost unchanged in the MX-80 samples. A significant decrease of quartz was found in the MX-Na sample

exposed to 1.0 M NaOH solution. No other significant dissolution was noticed.

No neoformation of minerals was found in any of the MX-Na samples. An unidentified peak pattern was found in the MX-80 samples exposed to 1.0 M NaOH solution (23B) (Figure 8), repeated scans did not show this pattern, indicating that the first results were accidental. A small amount of gel material was found outside the low pH side of sample 23A. This material did not show a montmorillonite peak

Table 2. Measured pH and CEC from MX-80 and MX-Na samples exposed to 0.1, 0.3 and 1.0 M NaOH solutions on one side and corresponding NaCl solutions on the other. The CEC values show eq/kg clay.

Sample	NaCl side (A)		NaOH side (B)	
	pH	CEC	pH	CEC
MX-80 reference		0.74		0.74
21	n	0.80	13.0	0.80
22	11.8	0.86	13.6	0.90
23	12.3	0.94	14.0	0.95
MX-Na reference		0.85		0.85
24	n	0.84	13.0	0.90
25	7.5	0.93	13.6	0.99
26	9.5	1.03	14.0	1.07

Table 3. Measured CEC from MX-80 exposed to 0.1 M (31AB) and 1.0 M solutions (32AB), and MX-Ca samples exposed to saturated Ca(OH)$_2$ solutions (A side) and 0.015 M CaCl$_2$ solutions (B side). The CEC and cation values show eq/kg clay.

Sample	Cu-CEC	Ca	K	Mg	Na
MX-80 reference	0.74	0.12	0.01	0.05	0.55
31AB	0.74	0.19	0.01	0.08	0.77
32AB	0.78	0.07	0.01	0.01	1.17
33A	0.77	0.96	0.01	0.01	0.02
33B	0.74	1.67	0.01	0.01	0.02
34A		1.05	0.01	0.02	0.02
34B	0.71	1.51	0.01	0.01	0.02
MX-Ca reference	0.78	0.87	0.00	0.04	0.01
35A	0.78	0.98	0.00	0.01	0.01
35B	0.74	1.41	0.00	0.01	0.00
36A	0.79				
36B	0.81	1.32	0.00	0.01	0.00

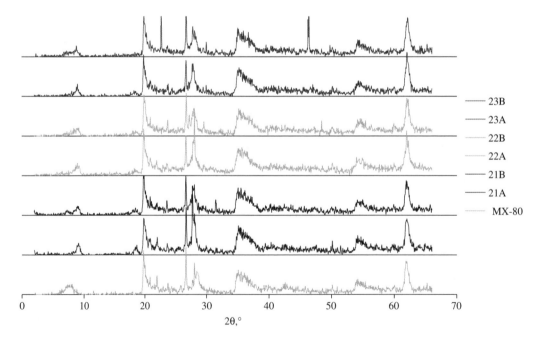

Figure 8. Powder XRD pattern from MX-80 samples exposed to NaOH solutions.

pattern, but a non-distinct pattern indicating an amorphous structure. The oriented and EG treated samples did not show any significant changes in swelling properties compared to the reference clay material (Figure 10).

4.4.3 Element analyses
4.4.3.1 Clay samples
Major results from the ICP/AES element analyses are shown in Appendix 1. The silica/aluminium ratio

decreased significantly in samples exposed to the NaOH solutions, and an Si/Al gradient was found in most samples exposed to the NaOH solutions (Figure 11). No other elements showed a similar trend (Figure 11, right).

The XRD analyses of the purified MX-80 material showed minor peaks from cristobalite and quartz, which was quantified to represent less than 2% of each according to the technique described in section 4.3.3. The structural formulas of the montmorillonite

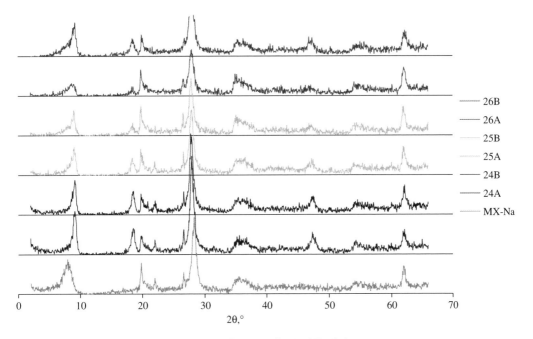

Figure 9.   Powder XRD pattern from MX-Na samples exposed to NaOH solutions.

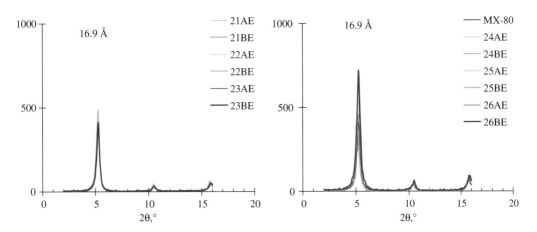

Figure 10.   XRD pattern from all samples exposed to NaOH solutions and MX-80 reference material after pre-treatment with Mg and ethylene glycol.

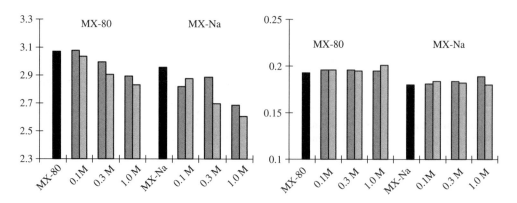

Figure 11. Si/Al ratio (left) and Fe/Al ratio (right) from ICP/AES analyses. Left bars indicate the low pH side of the samples and the right bars indicate the high pH side.

in the purified materials were calculated based on the ICP/AES analyses to match an $O_{20}(OH)_4$ cell according to Newman 1987. The calculated layer charges were lower than what was calculated from the measured CEC as determined by the Cu-method. The silica content was therefore adjusted to give the structural formula which matched the CEC values. The reduction of silica corresponded to an SiO_2 content (cristobalite and quartz) of 3.9% in the MX-K material, 3.4% in the MX-Na material, and 2.3% in the MX-Ca material. The structural formula for the samples exposed to the NaOH solutions were calculated to be:

MX-Na $(Si_{7.82} Al_{0.18}) (Al_{3.13} Fe^{3+}_{0.38} Mg_{0.47}Ti_{0.01}) O_{20} (OH)_4$
        $Na_{0.59} Mg_{0.02} Ca_{0.01}$

24A: $(Si_{7.82} Al_{0.18}) (Al_{3.14} Fe^{3+}_{0.38} Mg_{0.47}Ti_{0.01}) O_{20} (OH)_4$
        $Na_{0.63}$

24B: $(Si_{7.83} Al_{0.17}) (Al_{3.09} Fe^{3+}_{0.38} Mg_{0.51}Ti_{0.01}) O_{20} (OH)_4$
        $Na_{0.67}$

25A: $(Si_{7.78} Al_{0.22}) (Al_{3.10} Fe^{3+}_{0.39} Mg_{0.49}Ti_{0.02}) O_{20} (OH)_4$
        $Na_{0.70}$

25B: $(Si_{7.73} Al_{0.27}) (Al_{3.10} Fe^{3+}_{0.39} Mg_{0.49}Ti_{0.01}) O_{20} (OH)_4$
        $Na_{0.74}$

26A: $(Si_{7.71} Al_{0.29}) (Al_{3.09} Fe^{3+}_{0.41} Mg_{0.49}Ti_{0.02}) O_{20} (OH)_4$
        $Na_{0.77}$

26B: $(Si_{7.65} Al_{0.35}) (Al_{3.10} Fe^{3+}_{0.40} Mg_{0.49}Ti_{0.02}) O_{20} (OH)_4$
        $Na_{0.80}$

and the structural formula for the samples exposed to the $Ca(OH)_2$ solutions were calculated to be:

MX-Ca $(Si_{7.88} Al_{0.12}) (Al_{3.13} Fe^{3+}_{0.39} Mg_{0.47}Ti_{0.01}) O_{20} (OH)_4$
        $Ca_{0.27} Mg_{0.01}$

35/36A: $(Si_{7.87} Al_{0.13}) (Al_{3.14} Fe^{3+}_{0.38} Mg_{0.47}Ti_{0.01}) O_{20} (OH)_4$
        $Ca_{0.29} Mg_{0.004}$

35/36B: $(Si_{7.87} Al_{0.13}) (Al_{3.11} Fe^{3+}_{0.38} Mg_{0.49}Ti_{0.01}) O_{20} (OH)_4$
        $Ca_{0.30} Mg_{0.002}$

4.4.3.2 Test solutions

Test reference solutions and test solutions on both sides of the samples were analysed by use of ICP/AES and ion chromatography (IC) and the main elements are presented in Appendix 2. The IC analyses did not give repeatable results in the solutions with the highest pH, and the chloride concentration results should be used with care.

Silica was released from all samples exposed to NaOH solutions and a clear coupling was found between silica concentration and pH also in the NaCl solutions (Table 4). Small amounts of titanium were noticed in the 1.0 and 0.3 M NaOH solutions, likely emanating from the sample holder filters.

4.4.4 Scanning Electron Microscopy (SEM) analyses

The SEM/EDX analyses showed accessory mineral dissolution and possible minor structural changes in the exposed samples. The typical flaky montmorillonite character was, however, not change significantly in any sample (Figures 12 and 13). Relatively large variation in silica content was found, but the natural variation is normally rather large and no clear conclusions about the silica distribution can be drawn from the present analyses. Table 6 shows two spot analyses from gel material (Figure 13) found outside the filters in sample 23A (MX-80, 1.0 M NaOH).

Table 4. ICP/AES analyses of solutions from tests with MX-80 exposed to NaOH solutions. Figures in mM.

	Sample	pH	Al	Ca	Mg	K	Si	Ti
NaCl side	21	n	0.001	0.185	0.084	0.394	0.119	0.000
	22	11.8	0.000	0.416	0.132	0.421	0.620	0.000
	23	12.3	0.000	0.170	0.000	0.605	1.336	0.000
NaOH side	21	13.0	0.004	0.013	0.000	0.231	0.419	0.000
	22	13.6	0.005	0.000	0.000	0.305	1.890	0.003
	23	14.0	0.013	0.000	0.000	0.463	4.167	0.041

Table 5. ICP/AES analyses of solutions from tests with MX-Na exposed to NaOH solutions. Figures in mM.

	Sample	pH	Al	Ca	Mg	K	Si	Ti
NaCl side	24	n	0.003	0.017	0.065	0.147	0.165	0.000
	25	7.5	0.000	0.048	0.157	0.210	0.310	0.000
	26	9.5	0.000	0.058	0.000	0.322	2.846	0.000
NaOH side	24	13.0	0.007	0.000	0.000	0.105	0.662	0.002
	25	13.6	0.012	0.000	0.000	0.208	1.860	0.005
	26	14.0	0.021	0.000	0.000	0.366	4.243	0.027

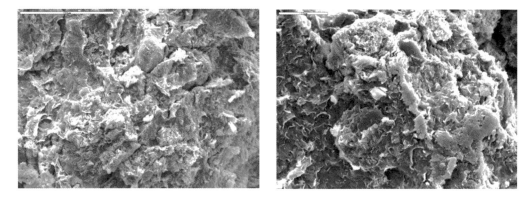

Figure 12. Typical bentonite structure (left) in sample 23A (MX-80, NaCl side, 1.0 M), and a similar structure (right) in sample 23B (NaOH side). Space bars show 50 mm (left) and 20 mm (right).

Figure 13. Pure montmorillonite (left) from sample 26B (MX-Na, NaOH side, 1.0 M). Gel from the precipitation outside the filter in sample 23A (MX-80, NaCl side, 1.0 M).

5 DISCUSSION

The pressure response from the sample successively exposed to pure water, 1.0 M NaCl, 1.0 M NaOH, 1.0 M NaCl, and finally pure water (Figure 4, left) shows several important characteristics. The fast drop to new pressure equilibrium at exposure to NaCl solutions shows the osmotic effects of the salt. The small initial response when the test solutions is changed from NaCl to NaOH, or reverse, indicate that the osmotic response is the same for the NaOH solution. The immediate stop in swelling pressure reduction, as an effect of changing the NaOH solution to NaCl, further indicates that there is a reaction between the NaOH solution and the bentonite, and not just a slow adaptation to equilibrium conditions. The final change to pure water results in an increase in swelling pressure, which is close to the initial decrease at the first contact with the NaCl solution.

Table 6. SEM/EDX analyses of the gel material found outside the filters in sample 23A(MX-80, NaCl side, 1.0 M).

	P23A1 (%)	P23A2 (%)
Na	21.3	38.5
Mg	1.0	0.0
Al	0.4	0.1
Si	21.3	4.7
Cl	15.4	51.8
K	−0.1	−0.1
Ca	39.3	5.3
Ti	0.0	−0.3
Fe	1.4	0.1

This strongly indicates that the swelling pressure reduction as a result of NaOH contact is permanent, and not reversible as the NaCl effect. The individual samples in Figure 4 (right) show that the final difference in pressure between pure water and NaCl solution conditions are approximately the same as the pressure difference between the NaOH solution and the NaOH/NaCl solution conditions, which shows that the pressure effect of the NaCl solution seems to be constant regardless of the impact of the NaOH solution. The pressure reduction in the two samples exposed to NaOH solution is similar, which shows that the effect of NaOH is not affected by the presence of NaCl. The pressure effects of NaOH and $Ca(OH)_2$ solutions are principally the same for MX-80 as for MX-Na (Figures 5 and 6).

The ICP/AES analyses of water and clay, the CEC analyses and the XRD analyses all show that silica dissolves from the bentonite, and the resulting change in mass likely is the cause of the swelling pressure decrease. The best quantitative value is given by the ICP/AES analyses of the solutions. Sample 26, which shows the highest release, has a total loss of 200 mg Si. The total sample mass is 7.56 g of which 7.26 g is montmorillonite corresponding to 9.6 mmole montmorillonite. The structural formulas show a loss of 0.14 mole Si per mole montmorillonite in this sample, which gives 1.35 mmole Si and a corresponding mass of 38 mg. Assuming the additional loss of silica comes from SiO_2 (mainly cristobalite) gives a mass of $162 * 60/28$ mg which gives 347 mg. The total mass loss is then $347 + 38 = 385$ mg. The final sample dry mass is then $7.56 - 0.385 = 7.175$ g. The change in density of the sample was then from a dry

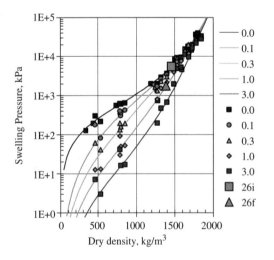

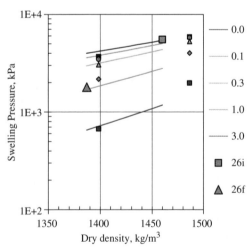

Figure 14. Measured and calculated values for swelling pressure in pure montmorillonite exposed to NaCl solutions. Initial (26i) and final (26f) pressure and density conditions in sample 26. Detail of the diagram to the right. Figures show NaCl concentrations in moles/L.

density of 1460 to 1387 kg/m^3. The swelling pressure drop was from initial 5500 in pure water to the final 1800 kPa in the 1.0 M NaOH/NaCl solution. The pressure conditions are shown in Figure 14, which illustrates that the pressure drop may be fully explained by the mass loss and the osmotic effects of the solutions.

It is obvious that a high pH plum may diffuse through bentonite at a sufficiently high pH gradient. The pH increase on the NaCl solution was first noticed in the MX-80 tests, which indicate that there is no major pH buffering effect from other accessory minerals than cristobalite and quartz.

The diffusion can not be described by a single diffusion coefficient according to the evaluation made by Höglund in work package 5. The evaluated effective diffusivity coefficients for the 1.0 M solutions conditions are two orders of magnitude higher than for 0.3 M solution conditions. It is obvious from the results in this study that dissolution of cristobalite takes place, which is one plausible explanation. Different effective porosity due to the test solutions is discussed by Höglund. The similar osmotic behavior of the hydroxide and chloride solutions in this study indicate that the ion-equilibrium conditions between bentonite and the two solutions are the similar. If this really is the case, Donnan equilibrium play an important role for the actual concentrations in the clay. The chloride concentrations in clay pore-water has been calculated to be 3 times lower than in an external solution for a 1.0 M NaCl external solution, 10 times lower for a 0.3 M solution and 25 times lower for a 0.1 M solution (Karnland et al. 2002).

The release of silica from the montmorillonite was around 40 mg from sample 26 and the exposure time was 40 days. The dissolution rate in this specific geometry is consequently $0.04/7.26/40/86400 = 1.6\text{E-}9\,\text{g}*\text{g}^{-1}\text{clay}*\text{s}^{-1}$ in the 1.0 M NaOH solution, and approximately $5\text{e-}10\,\text{g}*\text{g}^{-1}\text{clay}*\text{s}^{-1}$ in the 0.3 M solution.

6 CONCLUSIONS

The swelling pressure in bentonite is strongly reduced by exposure to 0.3 and 1.0 M NaOH solutions. The decrease can to a large extent be related to dissolution of silica and the subsequent mass loss of the bentonite samples.

The fast initial swelling pressure decrease is reduced after a relatively short time, and a subsequent almost constant decrease rate has been observed. The fast initial effect is likely due to the dissolution of cristobalite.

A significant increase in CEC has been observed in samples exposed to the NaOH solutions. The structural formulas of the exposed montmorillonite have been calculated based on the ICP/AES and CEC data, and show a significant release of silicon from the bentonite. The release rate has been determined to be $1.6\text{E-}9\,\text{g}*\text{g}^{-1}\text{clay}*\text{s}^{-1}$ for the 1.0 M NaOH solutions and $5\text{E-}10\,\text{g}*\text{g}^{-1}\text{clay}*\text{s}^{-1}$. Such change in the tetrahedral layers of montmorillonite may be seen as the first step towards beidellite, which can be seen as a minor problem with respect to the buffer functions. However, the properties of the clay will change dramatically if this reaction goes on until a high charges mineral is formed.

No effects on swelling pressure were found in the samples exposed to 0.1 M NaOH and saturated Ca(OH)$_2$ solutions. No mineralogical/chemical effect except ion exchange from sodium to calcium was found in the samples exposed to the Ca(OH)$_2$ solution.

The similar osmotic behavior of bentonite exposed to NaCl and NaOH solutions indicate that Donnan equilibrium is established also for hydroxide solutions, which may help to explain the lack of chemical/mineralogical reactions in the Ca(OH)$_2$ solution despite the fact that silica saturation concentration is very high also at pH 12.4.

ACKNOWLEDGMENTS

The authors wish to acknowledge that this paper is a result of work funded by the Swedish Nuclear Fuel and Waste Management Company (SKB), and the European Commission. A special thanks to Tomas McMenamin who kindly agreed to the publication of this paper.

REFERENCES

Drever, J.I. American Mineralogist, 62, 553–554.

Helgeson H.C., Brown T.H., Leeper R.H., 1969. Handbook of theoretical activity diagrams depicting chemical equilibria in geological systems involving aqueous phases at one atm. and 0 to 300°C. Freeman, Cooper & Co., San Francisco.

Jackson, M.L. Soil Chemical Analysis – Advanced Cours. 2nd Edition. Published by the author. Madison, Wisc. 991 pp (1975).

Karnland, O., Muurinen A., Karlsson, F. Bentonite swelling pressure in NaCl solutions Experimentally determined data and model calculations. Submitted to Applied clay science 2002.

Meier, L.P., Kahr, G. Clays and Clay Minerals, Vol 47, No. 3, 386–388, 1999.

Slaughter, M. and Early, J.W. Geol. soc. Am. Spec. Pap., 83 (1965).

Vieno T., Lehikoinen J., Löfman J. Nordman H., Mészáros F., 2003. Assessment of disturbances caused by construction and operation of ONKALO. Posiva Report 2003–06.

Taylor, J.C., Matulis C.E. A new method for Rietveld clay analyses. Part I. Powder Diffraction, 9, 119–123. (1994).

	SiO2	Al2O3	Fe2O3	MgO	CaO	Na2O	K2O	TiO2	LOI	Tot C	Tot S	Summa
Mx-80	57.3	18.5	3.6	2.31	1.28	2.04	0.51	0.15	13.7	0.31	0.29	99.7
MX-Na	58.9	20.1	3.61	2.36	0.1	2.47	0.08	0.14	11.8	0.14	0.02	99.6
MX-Ca	53.8	18.2	3.39	2.12	2.35	0.05	0.07	0.13	19.8	0.16	0.02	100.0
MX-K	58.6	19.8	3.66	2.45	0.08	0.11	4.01	0.14	10.8	0.12	0.02	99.7
MX-Mg	53.4	17.9	3.28	3.77	0.03	0.05	0.08	0.12	21.2	0.13	0.02	99.9
C21A	61.4	20.0	3.91	2.35	0.84	3.01	0.48	0.16	7.4	0.36	0.2	99.8
C21B	59.3	19.6	3.83	2.48	1.64	3.27	0.49	0.16	9	0.42	0.21	100.0
C22A	59.6	19.9	3.9	2.38	1.15	3.59	0.47	0.16	8.4	0.4	0.2	99.8
C22B	58.3	20.1	3.91	2.35	1.03	3.79	0.48	0.17	9.2	1.68	0.19	99.5
C23A	58.0	20.0	3.91	2.33	0.88	4.44	0.48	0.17	9.2	0.46	0.19	99.6
C23B	57.7	20.4	4.09	2.37	0.82	4.49	0.49	0.17	8.8	0.4	0.21	99.6
C24A	61.5	21.8	3.94	2.43	0.02	2.77	0.09	0.14	7.4	0.17	0.02	100.1
C24B	60.5	21.1	3.88	2.61	0.11	3.17	0.09	0.14	8.2	0.15	0.02	99.8
C25A	60.0	20.8	3.82	2.44	0.06	3.63	0.09	0.15	8.6	0.19	0.01	99.6
C25B	58.3	21.6	3.93	2.49	0.05	3.92	0.08	0.15	9.2	0.2	0.02	99.7
C26A	57.2	21.3	4.02	2.44	0.02	4.41	0.09	0.15	10.4	0.19	0.01	100.1
C26B	57.1	21.9	3.95	2.45	0.02	4.57	0.09	0.15	9.8	0.25	0.01	100.1
C31A+B	58.7	18.8	3.75	2.34	1.16	2.14	0.5	0.16	11.8	0.37	0.18	99.5
C32A+B	58.3	18.6	3.67	2.24	0.72	3.02	0.44	0.16	12.4	0.33	0.19	99.8
C33A+34A	53.4	17.1	3.34	2.02	3.63	0.32	0.42	0.14	19.4	0.51	0.18	99.9

Appendix 1. Figures show weight percent of total oxides.

Sample	Al	Ca	Cu	Fe	K	Li	Mg	Mo	Na	P	S	Si	Sr	Ti	Zn	Cl
1,0 M NaCl	0.005	0	0.003	0.009	1.9	0.0	0.01	0.00	23805	0.08	0.03	1.14	0.00	0.00	0.01	35096
0,1 M NaCl	0.000	0	0.002	0.004	1.5	0.0	0.01	0.01	2370	0.00	0.05	0.35	0.00	0.00	0.02	3801
1,0 M NaOH	0.000	0	0.004	0.003	4.1	0.0	0.02	0.00	25155	0.05	0.10	0.10	0.00	0.00	0.01	34594
0,1 M NaOH	0.003	0	0.006	0.003	23.6	0.0	0.03	0.01	1570	0.05	0.12	0.00	0.00	0.00	0.01	3619
CaOH2	0.048	760	0.008	0.005	4.0	0.0	0.03	0.02	0.7	0.00	17.54	0.05	0.46	0.00	0.02	0
S21A	0.023	7.4	0.006	0.001	15.4	0.1	2.04	0.05	SAT	0.02	11.46	3.34	0.39	0.00	0.01	3382
S21B	0.098	0.5	0.014	0.009	9.0	0.0	0.00	0.01	SAT	0.01	0.79	11.78	0.09	0.02	0.02	81
S22A	0.008	16.6	0.006	0.000	16.4	0.0	3.20	0.04	SAT	0.09	3.13	17.41	0.77	0.00	0.00	10380
S22B	0.147	0.0	0.027	0.006	11.9	0.0	0.00	0.01	SAT	0.15	1.43	53.10	0.00	0.15	0.07	111
S23A	0.001	6.8	0.010	0.001	23.7	0.0	0.00	0.04	SAT	0.06	4.24	37.55	0.00	0.00	0.01	33358
S23B	0.338	0.0	0.011	0.008	18.1	0.0	0.00	0.01	SAT	0.19	0.93	117.08	0.00	1.96	0.04	489
S24A	0.071	0.7	0.005	0.010	5.7	0.0	1.59	0.01	SAT	0.03	1.62	4.63	0.02	0.00	0.01	3518
S24B	0.197	0.0	0.015	0.008	4.1	0.0	0.00	0.00	SAT	0.08	0.23	18.60	0.00	0.12	0.01	292
S25A	0.004	1.9	0.005	0.002	8.2	0.0	3.80	0.01	SAT	0.32	0.56	8.70	0.06	0.00	0.00	9987
S25B	0.320	0.0	0.009	0.007	8.1	0.0	0.00	0.00	SAT	0.18	0.30	52.28	0.01	0.26	0.01	134
S26A	0.002	2.3	0.004	0.002	12.6	0.0	0.01	0.01	SAT	0.30	0.74	79.98	0.10	0.00	0.00	36443
S26B	0.567	0.0	0.008	0.007	14.3	0.0	0.00	0.01	SAT	0.21	0.29	119.23	0.00	1.28	0.03	212
31	0.007	30	0.006	0.002	20.0	0.1	9.75	0.01	2450	0.02	3.53	10.41	1.30	0.00	0.02	
32	0.016	75	0.004	0.001	39.0	0.1	13.00	0.02	21750	0.08	4.48	13.23	2.43	0.00	0.00	
33A	0.055	1220	0.011	0.006	30.6	0.1	0.76	0.03	167	0.00	24.61	12.84	3.29	0.00	0.05	

Appendix 2.   Figures show mg/L.

Experimental and numerical study of the T-H-M behaviour of compacted FEBEX bentonite in small-scale tests

M.V. Villar
Centro de Investigaciones Energéticas, Mediambientales y Tecnológicas (CIEMAT), Madrid, Spain

M. Sánchez, A. Lloret, A. Gens & E. Romero
Universitat Politècnica de Catalunya, (UPC), Barcelona, Spain

ABSTRACT: The conditions of the bentonite in an engineered barrier for high-level radioactive waste disposal have been simulated in a series of tests. Cylindrical cells with an inner length of 60 cm and a diameter of 7 cm have been constructed. Inside the cells, six blocks of compacted FEBEX bentonite have been piled-up, giving rise to a total length similar to the thickness of the clay barrier in a repository according to the Spanish concept. The bottom surface of the material was heated at 100°C and the top surface was injected with granitic water. The duration of the tests was 6, 12 and 24 months. The temperatures inside the clay and the water intake were measured during the tests and, at the end, the cells were dismounted and the dry density and water content were measured at different positions. The injection of water generates in the vicinity of the hydration surface an increase of the water content and a decrease of the dry density due to the swelling of the clay, while heating gives rise to an increase of the dry density and a reduction of the water content in the 18 cm closest to the heater, even after 2 years of thermo-hydraulic (TH) treatment. The tests have been modelled using a numerical code especially developed to handle coupled T-H-M (Thermo-Hydro-Mechanical) problems in porous media. The models intend to reproduce the evolution of the main variables of the test (temperature and water intake) and also the results obtained in the postmortem study of the small-scale test.

1 INTRODUCTION

The design of high level radioactive waste (HLW) repositories in deep geological media includes the construction of a barrier around the waste containers constituted by a sealing material (Pusch 1994). Bentonite has been chosen as sealing material in most disposal concepts because of its low permeability, swelling capacity and retention properties, among other features. The behaviour of a HLW repository is determined, to a large extent, by the characteristics of the design and construction of the engineered barriers and especially by the changes that may occur in the mechanical, hydraulic, and geochemical properties as a result of the combined effects of heat generated by the radioactive decay and of the water and solutes supplied by the surrounding rock. Therefore, it is considered of fundamental importance for the evaluation of long-term behaviour that the processes taking place in the near-field be understood and quantified. Much attention has been paid since the 1980's to the performance of tests at different scales, in both the laboratory and the field,

in order to observe the thermo-hydro-mechanical processes taking place in the engineered barriers and the geological medium. The purpose of these experiments has been the direct observation of the phenomena occurring in the barrier and of the behaviour of the system, this providing the information required for the verification and validation of the mathematical models of the coupled processes and their numerical implementation.

However, the performance of large-scale in situ tests is complicated and time-consuming. Another drawback of in situ tests is that the boundary conditions are not always controlled and known. For this reason, laboratory tests of different scales are very useful to identify and quantify processes in a shorter period of time (Melamed *et al.*, 1992, ENRESA 2000).

The work presented here has been performed in the framework of FEBEX, which is a project for the study of the near field for a HLW repository in crystalline rock according to the Spanish concept (ENRESA 1995, 2000). The experimental work of the FEBEX Project consists of three main parts: an in situ test, under natural conditions and at full scale (Grimsel, Switzerland); a mock-up test, at almost full scale (CIEMAT, Madrid); and a series of laboratory tests to complement the information from the two large-scale tests. Among the laboratory tests included in the project are those performed in cells in which the compacted bentonite is subjected simultaneously to heating and hydration, in opposite directions, for different periods of time, in order to simulate the conditions of the clay barrier in the repository.

The second part of the work is oriented to the numerical study of the tests. The *THM* formulation proposed by Olivella *et al.*, 1994 has been adopted as a framework to analyze these experiments. The finite element code CODE_BRIGHT (Olivella *et al.*, 1994) has been used to perform the numerical simulations. CODE_BRIGHT is a useful tool to handled coupled *THM* problems in porous media. The simulations intend to reproduce the evolution of the main variables of the tests (water intake and temperature) and also the experimental data coming form the post-mortem study. Related to these studies, it is important to highlight that the major part of the constitute laws and parameters considered in the analyses are the same adopted for the Operational Base Case (OBC) model of the 'mock-up' test (ENRESA 2000). In that sense the results obtained in this study are a kind of model predictions.

2 MATERIAL

The tests have been performed with a bentonite from the Cortijo de Archidona deposit (Almería, Spain), the FEBEX clay, which is considered by ENRESA, the Spanish agency for radioactive waste management, a suitable material for the backfilling and sealing of HLW repositories. The conditioning of the bentonite in the quarry, and later in the factory, was strictly mechanical (homogenisation, removal of rock fragments, drying to temperatures lower than 60°C and crumbling of clods). Finally, the clay was sieved by a 5-mm mesh to obtain the granulated product to be used in the laboratory. Its main characteristics, summarised beneath, are detailed in ENRESA (1998, 2000) and Villar (2002).

The bentonite has a content of dioctahedric smectite of the montmorillonite type higher than 90 percent. Besides, it contains variable quantities of quartz, plagioclase, cristobalite, potassium feldspar, calcite and trydimite. The cation exchange capacity is of 111 ± 9 meq/100 g, and the exchangeable cations are Ca (38%), Mg (28%), Na (23%) and K (2%). The liquid limit of the bentonite is $102 \pm 4\%$ and the hygroscopic water content of the clay at laboratory conditions (relative humidity $50 \pm 10\%$, temperature $21 \pm 3°C$) is about $13.7 \pm 1.3\%$.

3 EXPERIMENTAL SET-UP FOR THE THERMO-HYDRAULIC TESTS

Cylindrical cells with an inner length of 60 cm and a diameter of 7 cm have been constructed (Fig. 1). They are made of Teflon to prevent as much as possible the lateral heat conduction and externally covered with steel semi-cylindrical pieces to avoid the deformation of the cell by the swelling of the bentonite. Inside the cells, six compacted blocks of bentonite have been piled-up, giving rise to a total length similar to the thickness of the bentonite barrier of the mock-up test of the FEBEX Project. To obtain the blocks, the clay with its hygroscopic water content (around 14 percent) has been uniaxially compacted at a dry density of 1.65 Mg/m³.

The bottom part of the cell is a plane heater whose temperature is fixed to 100°C, which is the expected on the surface of the waste container in the Spanish concept. Over the upper lid of the cell, there is a deposit in which water is circulated at constant temperature (20–30°C). In this way, a constant gradient of 1.1–1.3°C/cm between top and bottom of the sample is imposed.

Granitic water of a salinity 0.02 percent is injected through the upper lid of the cell at a pressure of 1.2 MPa. It simulates the water that saturates the barrier in a repository excavated in granitic rock, and it is the same employed to saturate the mock-up test of the FEBEX Project (ENRESA 2000).

The duration of the tests has been of 6 (test FQ1/2), 12 (test FQ1) and 24 (test FQ2) months. During the tests, the temperatures at different positions inside the clay were measured by thermocouples and recorded, as well as the volume of water intake. The 5 thermo-couples, T1 to T5, were placed at 50, 40, 30, 20 and 10 cm from the heater.

At the end of the thermo-hydraulic treatment, the cell is dismounted and the clay blocks extracted. Special care is taken to avoid any disturbance in the conditions of the clay, particularly its dry density and water content. As the blocks are extracted, they are cut into cylindrical sections of 2.5 cm in thickness in which dry density and water content are determined. The gravimetric water content of each section has been determined at the end of the tests by oven drying at 110°C during 24 hours. Dry density (ρ_d) is defined as the relation between the weight of the dry sample and the volume occupied by it prior to drying. The volume of the specimens has been determined by immersing them in a recipient containing mercury and by weighing the mercury displaced.

Geochemical, mineralogical and fabric studies – described elsewhere (Martín 2002) – have also been performed. The main conclusions of the latter study are: the TH treatment provokes a transport of soluble salts (mainly NaCl and Na_2SO_4) by diffusive and

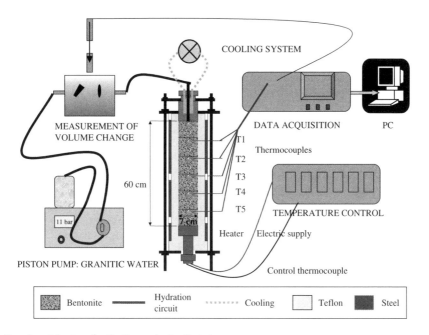

Figure 1. Experimental set-up for the thermo-hydraulic tests.

advective mechanisms; there is an ionic exchange of Na^+ by Ca^{2+} in the smectite (specially near the heater); the hydration with granitic water gives place to the dissolution/precipitation of calcite; and the TH treatment for two years does not originate any mineralogical change in the smectite. Permeability, swelling pressure and swelling under load were also measured in different positions. The results obtained are described in Villar (2001): the swelling capacity and the hydraulic conductivity of the bentonite after the TH treatment are mainly related to the dry density and water content attained during it and no major irreversible modifications of these properties with respect to those of the untreated clay were detected.

4   RESULTS OF THE THERMO-HYDRAULIC TESTS

The evolutions of temperature at different locations inside the clay and of the water intake during the 24-month test are shown in (Fig. 2). The repercussion of the laboratory temperature on the temperatures inside the clay was noticeable.

The average temperatures recorded by each thermocouple for the period between 1000 hours after the beginning of the test and the end of the test are plotted in (Fig. 3) for three tests of different duration. There is a sharp temperature gradient in the vicinity of the heater, the temperature decreasing from 100°C in the heater surface to 50°C at 10 cm

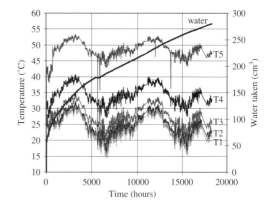

Figure 2. Evolution of temperatures at different locations inside the clay (T1 to T5, 50 to 10 cm from the heater) and of water intake in the 24-month test.

from it inside the clay. On the other hand, a decrease in temperature with saturation (longer tests) is observed, especially in the wetter zones, i.e., within the closest 20 cm to the hydration surface. This is due to the higher heat dissipation of the wet clay.

The final distribution of water content for three tests is shown in Figure 4. The water content reduction by the temperature effect is similar for the three tests, which means that desiccation takes place rather quickly and only affects the closest 18 cm to the heater, where the water contents are lower than the initial ones, which were around 14 percent. In the 6-month test, there is a central zone, between 20 and 40 cm, in which the initial water content is barely modified. The effect of hydration, which is evident only in the closest 20 cm to the hydration surface in the 6-month test, is noticeable in the 40 upper centimetres at 12 months. In the next 12 months, the water content increases in these 40 upper centimetres, but does not change in the closest 18 cm to the heater, probably because in this zone the thermal effect is still predominant.

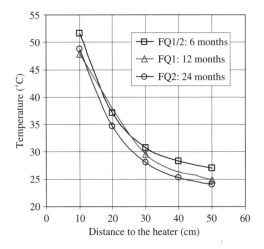

Figure 3. Average steady temperatures at different positions for three tests of different duration.

The final distribution of bentonite dry density is plotted in Figure 5. The dry density decreases from the heater towards the hydration surface with an approximately linear tendency. In the zones affected by hydration, the dry density decreases below the initial value ($1.65\,\text{Mg/m}^3$) due to the expansion caused by saturation. On the contrary, near the heater, the dry density increases, due to the shrinkage caused by desiccation. Despite the dispersion of the data – caused by the initial inhomogeneities and by the determination method – it can be observed that the densities reached near the hydration surface are lower in the longer tests, due to the fact that their saturation and, consequently, their swelling deformation, is higher. Due to this density decrease, the degrees of saturation are lower than what could be expected by the high water content. In fact, a degree of saturation higher than 90 percent is only reached in the 3 cm closest to the hydration surface after 6 months of TH treatment, in 8 cm after 12 months and in 10 cm after 24 months. On the other hand, near the heater (18 cm) the density remains the same from 6 to 24 months, and the degrees of saturation decrease from the initial 55–60 percent to values around 25 percent in the vicinity of the heater.

An overall decrease of dry density has been observed, since the average value measured was $1.57\,\text{Mg/m}^3$ for the 6 and 12-month tests and 1.55 for the 24-month tests, whereas the initial dry density of the blocks was $1.65\,\text{Mg/m}^3$. This is due to the fact that the cells are made of Teflon, a material which slightly deforms because it is not able to withstand the high swelling pressure developed by the clay as it

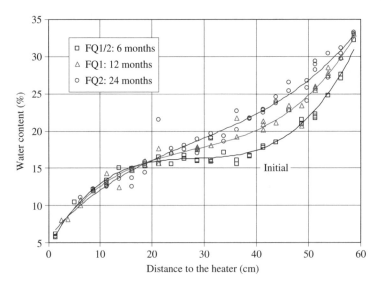

Figure 4. Final distribution of water content for tests of different duration.

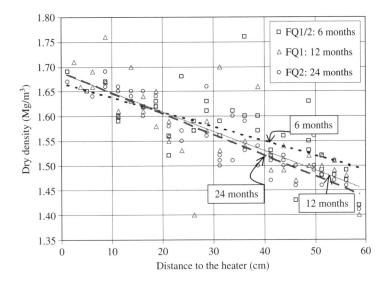

Figure 5. Final distribution of dry density for tests of different duration.

saturates. This overall decrease of dry density is linked to the hydration of the clay, and for this reason is more important for the longer tests that have reached a higher degree of saturation.

Shrinkage cracks near the heater were not observed in any of the tests. The two upper blocks, whose joint was placed at 10 cm from the hydration surface, were sealed in the 1- and 2-year tests, but the next joint, placed at 20 cm from the hydration surface, was not sealed even in the 2-year test.

5 MODELLING

5.1 General

The FEBEX bentonite response under the small-scale test conditions is very complex, due to the physical processes and interactions that take place during the simultaneous heating and hydration of the clay under conditions of practically constant volume. Because of that, a coupled *THM* analysis becomes necessary in order to simulate this problem. The *THM* formulation proposed by (Olivella *et al.*, 1994), has been adopted as general framework for the analyses and the finite element code CODE_BRIGHT, (Olivella *et al.*, 1996), has been used to analyze the small-scale tests as a boundary value problem.

The numerical analyses are focused on the evolution of the main measured variables of the tests (temperature and water intake). The study are also oriented to the critical analysis of the comparisons between the variables (dry density and water content) computed at the end of the simulations

respect the ones obtained experimentally in the post-mortem tests. Especial attention is placed also in the response of other keys variables. This study has an additional interest, and it is related to observe if in the hydration of the cells can be detected similar problems to the ones observed in the 'mock-up' experiment. As it is explained in (Sánchez & Gens 2002) and (Sánchez 2003), there is an apparent decay in the hydration rate of the 'mock-up' test. Since the expansive clay and the THM conditions of the small-scale tests are very similar to the ones of the 'mock-up' test, it is expected that the same problems also appear in these tests.

Previous to introduce the numerical results a brief description of the formulation and the capabilities of the code used in the analyses are presented. Besides, the main constitutive models and parameters adopted in the simulations are introduced.

5.2 Coupled formulation

The theoretical framework is composed of three main parts: balance equations, constitutive equations and equilibrium restrictions. The framework is formulated using a multi-phase, multi-species approach. The subscripts identify the phase ('s' for solid, 'l' for liquid and 'g' for gas). The superscript indicates the species ('h' for mineral, 'w' for water and 'a' for air). The liquid phase may contain water and dissolved air, and the gas phase may be a mixture of dry air and water vapour. Dry air is considered as single specie. Thermal equilibrium between phases is assumed. This means that the three phases are at the same temperature. A detailed derivation is given in

Olivella *et al.*, (1994), an only a brief description is included here.

5.2.1 *Balance equations*

Equations for mass balance were established following the compositional approach, which consists of balancing the species rather than the phases. The mass balance of solid present in the medium is written as:

$$\frac{\partial}{\partial t}\left(\rho_s\left(1-n\right)\right)+\nabla\cdot\left(\mathbf{j}_s\right)=0 \tag{1}$$

where ρ_s is the mass of solid per unit volume of solid, $\mathbf{j}_s$ is the flux of solid, t is time and ∇ is the divergence operator.

Water is present in liquid and gas phases. The total mass balance of water is expressed as:

$$\frac{\partial}{\partial t}\left(\theta_l^w S_l n+\theta_g^w S_g n\right)+\nabla\cdot\left(\mathbf{j}_l^w+\mathbf{j}_g^w\right)=f^w \tag{2}$$

where θ_l^w and θ_g^w are the masses of water per unit volume of liquid an gas respectively. Sl, is the volumetric fraction of pore volume, occupied by the alpha phase ($\alpha=l,g$). $\mathbf{j}_l^w$ and $\mathbf{j}_g^w$ denote the total mass fluxes of water in the liquid and gas phases (water vapour), with respect to a fixed reference system. f^w is an external supply of water.

Dry air is present in liquid and gas phases. The total mass balance of dry air is expressed as:

$$\frac{\partial}{\partial t}\left(\theta_l^a S_l n+\theta_g^a S_g n\right)+\nabla\cdot\left(\mathbf{j}_l^a+\mathbf{j}_g^a\right)=f^a \tag{3}$$

where θ_l^a and θ_g^a are the masses of dry air per unit volume of liquid an gas respectively. S_α, is the volumetric fraction of pore volume, occupied by the alpha phase ($\alpha=l,g$). $\mathbf{j}_l^a$ and $\mathbf{j}_g^a$ denote the total mass fluxes of dry air in the liquid and gas phases, with respect to a fixed reference system. f^a is an external supply of dry air.

Regarding the thermal problem, equilibrium between the phases is assumed. Therefore, the temperature is the same for all the phases. Consequently, only one equation is needed for the energy balance. If either the complexity of the problem or the experimental evidence justifies the necessity of a more detailed treatment of this equation, different temperatures can easily be considered. The total internal energy, per unit volume of porous media, is obtained adding the internal energy of each phase corresponding to each medium. Applying the balance equation to this quantity, the following equation is obtained:

$$\frac{\partial}{\partial t}\left(E_s\rho_s\left(1-\phi\right)+E_l\rho_{lp}S_l\phi+E_g\rho_g S_g\phi\right)$$
$$+\nabla\cdot\left(\mathbf{i}_c+\mathbf{j}_{Es}+\mathbf{j}_{El}+\mathbf{j}_{Eg}\right)=f^E \tag{4}$$

where E_s is the solid specific internal energy; E_l and E_g are specific internal energies corresponding to liquid and gas phase, respectively, ρ_s is the solid density; ρ_l and ρ_g are the liquid and gas phase densities; $\mathbf{i}_c$ is the conductive heat flux; $\mathbf{j}_{Es}$ is the advective energy flux of solid phase respect to a fixed reference system; $\mathbf{j}_{El}$ and $\mathbf{j}_{Eg}$ are the advective energy flux of liquid and gas phases, respectively, with respect to a fixed reference system; f^E are the energies supply per unit volume of medium.

Finally, the balance of momentum for the porous medium reduces to the equilibrium equation in total stresses:

$$\nabla\cdot\sigma+\mathbf{b}=0 \tag{5}$$

where σ is the stress tensor and $\mathbf{b}$ is the vector of body forces.

5.2.2 *Constitutive equations*

The constitutive equations establish the link between the independent variables (or unknowns) and the dependent variables.

Concerning the hydraulic problem it is assumed that the liquid and gas flows follow Darcy's law:

$$\mathbf{q}_\alpha=-\mathbf{K}_\alpha\left(\nabla P_\alpha-\rho_\alpha\mathbf{g}\right);\quad \alpha=l,g \tag{6}$$

where P_l and P_g are liquid and gas pressures respectively, ρ_l is the liquid density, ρ_g is the gas density, $\mathbf{g}$ is the gravity vector and $\mathbf{K}_\alpha$ is the permeability tensor of the alpha phase ($\alpha=l,g$), which is given by:

$$\mathbf{K}_\alpha=\mathbf{k}\,\frac{k_{r\alpha}}{\mu_\alpha};\quad \alpha=l,g \tag{7}$$

The intrinsic permeability tensor ($\mathbf{k}$) depends on the pore structure of the porous medium. $k_{r\alpha}$ is the value of relative permeability that controls variation of permeability in the unsaturated regime and μ_α denotes the dynamic viscosity. α may stand for either l or g depending on whether liquid or gas flow is considered. The variation of intrinsic permeability with porosity is given by:

$$\mathbf{k}=\mathbf{k}_o e^{\left(b\left(\phi-\phi_o\right)\right)} \tag{8}$$

where $\mathbf{k}_o$ is the intrinsic permeability corresponding to ϕ_o (a reference porosity), and b is a material parameter. The relative permeability of liquid (k_{rl}) phases is made dependent on S_e (effective degree of saturation) according to:

$$k_{rl}=S_e^n\;;\;S_e=\frac{S_l-S_{lr}}{S_{ls}-S_{lr}} \tag{9}$$

where S_l is degree of saturation, S_{lr} and S_{ls} are residual and maximum degree of saturation, respectively, and

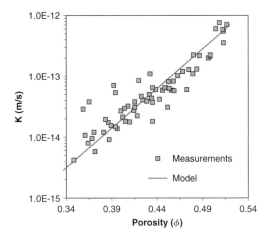

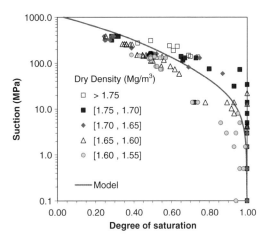

Figure 6. Variation of saturated permeability with porosity. Experimental data and adopted model for the intrinsic permeability law.

Figure 7. Retention curve adopted in the analyses, together with the experimental data of FEBEX bentonite (symbols).

Table 1. Model parameters related to the hydraulic problem.

Retention curve				Intrinsic permeability		
P_o[MPa]	λ_o	P_d [MPa]	λ_d	k_o [m/s]	b	n_o
28	0.18	1100	1.1	$1.9 \, e^{-14}$	30	0.40

n is a material parameter. A value of $n = 3$ has been adopted in the analyses.

The saturated permeability results obtained experimentally in permeability test of the FEBEX bentonite (ENRESA 2000) have been used to fit the exponential model (Eq. 8) adopted for the intrinsic permeability law (Fig. 6). Table 1 collects the adopted model parameters.

Another relevant law related to the hydraulic problem is the retention curve. A modified van Genuchten law has been adopted to model the dependence of the degree of saturation on suction:

$$S_l = \left[1 + \left(\frac{s}{P_o}\right)^{\frac{1}{1-\lambda_o}}\right]^{-\lambda_o} f_d ; \quad f_d = \left(1 - \frac{s}{P_d}\right)^{\lambda_d} \quad (10)$$

where s ($s = P_g - P_l$), is the suction, P_o is a parameter related to the capillary pressure (the air entry value) and λ_0 is a parameter that controls the shape of the curve (van Genuchten 1978). The function f_d is include to obtain more suitable values at high suctions, P_d is related with the suction at 0 degree of saturation and λ_d is a model parameter (when is

null the original model is recovered). In the same figure, the lines correspond to the fit obtained using the purposed model. Table 1 presets the model parameters.

The molecular diffusion of vapour water in gas phase is governed by Fick's law, through:

$$\mathbf{i}_g^w = -\mathbf{D}_g^w \nabla \omega_g^w = - \left(\phi \rho_g S_g \tau D_m^w \mathbf{I} + \rho_g \mathbf{D}_g' \right) \nabla \omega_g^w \quad (11)$$

where $\mathbf{i}_g^w$ is the non advective mass flux of water in gas, $\mathbf{D}_g^{w}$ is the dispersion tensor, ω_g^w is the mass fraction of water in gas, τ is the tortuosity and $\mathbf{D}_g'$ the mechanical dispersion tensor. $\mathbf{D}_m^w$ is the molecular diffusion coefficient of vapour in gas. A value of τ equal to 0.8 has been adopted (ENRESA 2000).

Regarding the thermal problem the Fourier's law has been adopted for the conductive flux of heat. Thermal conductivity depends on the hydration state of the clay and is expressed by a variant of the geometric mean:

$$\mathbf{i}_c = -\lambda \nabla T \quad (12)$$

where:

$$\lambda = \lambda_{sat}^{s_l} \lambda_{dry}^{(1-s_l)} \quad (13)$$

and:

$$\lambda_{dry} = \lambda_{solid}^{(1-\phi)} \lambda_{gas}^{\phi} ; \quad \lambda_{sat} = \lambda_{solid}^{(1-\phi)} \lambda_{liq}^{\phi} \quad (14)$$

where λ is the thermal conductivity. Figure 8 presents the experimental values obtained for the FEBEX bentonite; together with the model result computed using equations 13 and 14. A value of 0.47 is adopted for λ_{dry} and 1.15 for λ_{sat}.

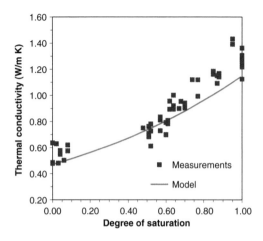

Figure 8. Thermal conductivity: FEBEX bentonite experimental results (symbols) together with the model adjust.

Table 2. Model parameters related to the mechanical problem.

Parameters defining the Barcelona Basic Model (BBM)			
κ	0.04	κ_s	0.25
ν	0.4	α_{is}	−0.003
α_{sp}	−0.147	α_{ss}	0.00
α_o	1.5×10^{-4} (°C^{-1})	α_2	0.00
λ_0	0.15	p_c	0.10 MPa
p_o^*	14 MPa	α	0.395
r	0.75	β	0.05
M	1.5	T_o	20°C
k	0.1	ρ	0.2

Finally, regarding the mechanical problem, the constitutive model known as *BBM* (Barcelona Basic Model) has been adopted. The model parameters have been identified in the preoperational stage of the FEBEX project (FEBEX report 1997) and are indicated in Table 2. The main equations of the model are presented in the Appendix. A more advanced expansive model has been recently proposed for this clay (Lloret *et al.*, 2003), but for the analysis presented here the initial one has been adopted.

5.2.3 Equilibrium restrictions

Other types of relationships that relate dependent variables with unknowns are the equilibrium restrictions. They are obtained assuming chemical equilibrium for dissolution of the different species (dry gas and vapour) in phases (liquid, gas). This assumption is sufficiently adequate because these chemical processes are fast compared to the transport processes that take place in porous media and, for this reason, they are not rate controlling. In this problem the

concentration of water vapour in the gas phase is controlled by the psychometric law; and the solubility of dry gas in water is given by Henry's law (Olivella *et al.*, 1994).

5.3 Numerical code

It is possible to solve the sets of equations presented above in a coupled way, in order to do that one unknown (or state variable) is associated to each balance equation of the formulation (Equations 1 to 5). Then, from the state variables, the dependent variables are calculated using the constitutive equations or the equilibrium restrictions. In this way, the finite element program CODE_BRIGHT (Olivella *et al.*, 1996), has been developed. CODE_BRIGHT is a tool developed to handle coupled *THM* problems in geological media. Finite elements in space and finite differences in time are used for the discretization of the system of equations.

5.4 Simulations

1-D axis-symmetrical models have been used in these analyses. A mesh of one hundred (100) elements has been adopted. A sensitivity analysis has also been carried out to verify that the model results do not depend on the mesh. The initial and boundary conditions of the model have been imposed in order to be the closest possible to the experiments. The initial water content of the bentonite block is close to 14%, from the retention curve adopted an initial value of suction of 110 MPa has been adopted. An initially uniform temperature of 22°C is assumed. And initial hydrostatic stresses of 0.15 MPa have been adopted. Regarding the boundary conditions a temperature of 100°C is imposed in the contact between heater and bentonite (the bottom of the cell). While a constant water pressure of 1.2 MPa is imposed in the other extreme of the cell (upper part). The thermal boundary condition along the sample has been adopted in order to adjust the temperature field, in that sense a temperature of 23°C has been fixed with a radiation coefficient of 1 (one). Regarding the mechanical problem, a small expansion of the cell has been allowed in order to consider the overall decrease of the dry density observed in the experiments (Section 4). Finally, a constant gas pressure (0.1 MPa) has been adopted in the analyses.

The 24-months test has been modelled, two stets of comparisons are presented: comparisons between evolutions of measured and computed variables, and comparisons corresponding to the analysis of the post-mortem study. The Figure 9 presents the evolution of temperature in different points of the cell. The cyclic variations of temperature, due to the changes in the laboratory temperature, are not taken into account

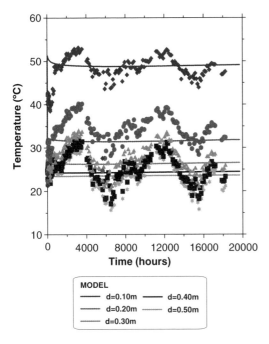

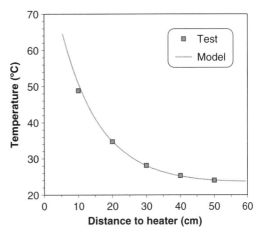

Figure 11. Distribution of temperature along the cell at the end of 24-month test. Registered and computed values.

MODEL

—— d=0.10m	—— d=0.40m
——— d=0.20m	········ d=0.50m
——— d=0.30m	

Figure 9. Evolution of the temperature in the test, computed versus registered values.

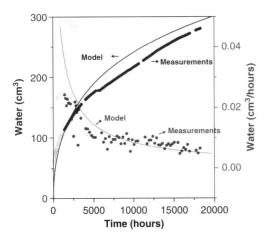

Figure 10. Computed versus registered values of cumulative water intake and rate of water intake.

in the simulations. The model only intends capture the whole thermal evolution of the test. In that sense, the temperature field is quite well reproduced.

The numerical results related to the hydraulic problem are also good, as it can be seen in (Fig. 10), in terms of the cumulative water intake and of the water intake rate, as well. As it was already commented, one of the objectives of these numerical

analyses is to verify if some of the apparent problems detected in the rate of hydration of the 'mock-up test' (Sanchez & Gens 2002) are also present in these small-tests. If the unforeseen behaviour of the 'mock-up' test is ascribed to some phenomena or effects related to the behaviour of the FEBEX bentonite, which were non-contemplated in the original *THM* formulation, these problems may be also present in these tests due to the similar *THM* conditions of both experiments (Section 3). In that sense, is not easy to obtain a strong conclusion, because the global saturation of the cell at the end of the 24 months-test is close to the one corresponding to the 'mock-up' test at dates in which problems begin to appear (Sanchez & Gens 2002).

Concerning the comparisons related to the post-mortem analyses, (Fig. 11) presents the distribution of temperature along the sample at the end of the test. A strong reduction of the temperature in zones close to the heater (twice the imposed temperature at 10 cm) it can be observed. A difference in the thermal problem between the cell tests and to the 'mock-up' experiments is that in the first one the dissipation of heat is greater due to the relatively major amount of external areas.

The comparisons related to the distribution of the water content along the cell (Fig. 12) show that the results are not very good, especially in zones close the heater. It is important to highlight that the simulation does not take into account the cooling phase, at which the samples were submitted during the dismantling of the test (before the determination of the water content values). It is possible that these hot-zones could be the more affected by the omission of this cooling stage. In any case, the simulations capture qualitatively good this variable. As in the

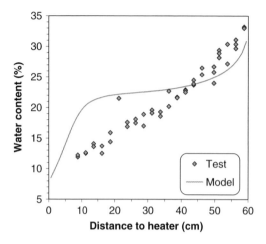

Figure 12. Distribution of water content along the cell at the end of 24-month tests. Registered and computed values.

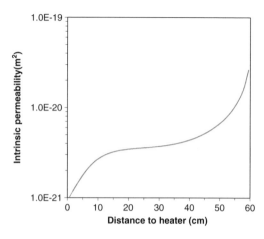

Figure 14. Computed intrinsic permeability field along the cell at the end of 24-month test.

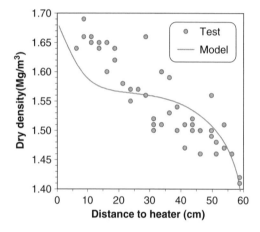

Figure 13. Distribution of dry density along the cell at the end of 24-month test. Registered and computed values.

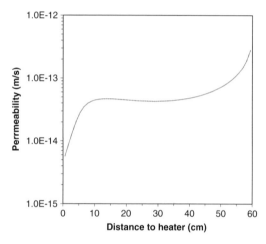

Figure 15. Computed permeability field along the cell at the end of 24-month test.

'mock-up' test, it can be observed the effect of the wetting, in the lower part of the sample (approximately 10 cm), due to the condensation of the water vapour coming form the bottom of the cell (where the heater is located).

Figure 13 shows the measured values of dry clay density together with the model results. It can be observed a clear density reduction in zones close to the hydration front (due to the clay swelling) and higher densities as the distance to the heater diminishes. The model captures qualitatively the global trend of the test; but, as for the case of the water content, the agreements between model results and measurements are not good in zones close to the heater. The omission of the cooling and unload

phases in the analyses could be the main reason for the discrepancies also in this case. On the other hand, it is important to have in mind that the *BBM* mechanical law is not the better one to model the behaviour of expansive clays, perhaps the adoption of a more proper mechanical model for the FEBEX bentonite (*i.e.* Lloret *et al.*, 2003) could help to obtain better results.

Once the main variables of the problem are well captured (Fig. 11 to 13), it is interesting to analyse its influence on other relevant variables, such as the permeability field. Generally, the main thermomechanical couplings considered in the permeability models are related to its dependence on dry density and on temperature (through the liquid viscosity and liquid density). Figure 14 shows the dependence of intrinsic permeability on porosity, in this case through the exponential model proposed (Eq. 8, Fig. 6). The higher intrinsic permeability is computed in zones close to the hydration front, due to the porosity increment in these zones (Fig. 13). In the case of non-saturated flow, the relative permeability law implies also higher values of permeability as the clay saturation progresses (Eq. 9). Figure 15 presents the final distribution of computed permeability, in which the effects of the non-saturation and temperature are included.

6 SUMMARY AND CONCLUSIONS

The conditions of the bentonite in an engineered barrier for HLW disposal have been simulated experimentally in a series of tests. Six compacted blocks of dry density $1.65 \, g/cm^3$ have been piled up in a hermetic cell. The diameter of the blocks was 7 cm and their height 10 cm. Consequently, the total length of clay inside the cell was 60 cm, the same than the thickness of the bentonite barrier of the mock-up test of the FEBEX Project. The bottom surface of the material was heated at 100°C while the top surface was injected with granitic water. The duration of the tests was 6, 12 and 24 months. The temperatures inside the clay and the water intake were measured during the tests and, at the end, the cells were dismounted and the dry density and water were measured in different sections of the blocks. The tests have been modelled as a boundary value problem using a numerical code especially developed to handle coupled T-H-M (Thermo-Hydro-Mechanical) problems in porous media. The following conclusions can be drawn:

- There is a sharp temperature gradient in the vicinity of the heater. The temperatures inside the clay during the TH treatment are a little lower as the duration of the test is longer, what is due to the increase in thermal conductivity of the clay with water content.
- The water content increases within the closest 20 cm from the hydration surface in the 6-month test, and within 40 cm in the 12 and 24-month tests. Despite this increase in water content, a degree of saturation higher than 90 percent is only reached in a few centimetres located closest to the hydration surface (12 cm in the case of the 24-month test). This is because the increase in water content is linked to a reduction in dry density caused by swelling.
- After six months of TH treatment, the closest 18 cm to the heater experience a desiccation that is not recovered after 24 months, hence the water content remains in this zone below the initial value, being close to 6 percent in the vicinity of the heater.
- The numerical model can reproduce the global tendencies observed in the small-scale tests; either the results related to the evolution of the main variables of the test, as the related to the post-mortem study. One important point is that most of constitutive model and parameters assumed in the analysis are the same of the ones adopted for the 'mock-up' test model (OBC, Operation Base Case, ENRESA 2000).
- The study oriented to identify if a kind of locking hydration, similar to the observed in the 'mock-up' test (Sánchez & Gens 2002), is also present in these tests does not give strong conclusions. Mainly due to the relatively short duration of test. In that sense, it is important to mention that the saturation level at the end of the 24-month cell is similar to the one of the 'mock-up' test when these problems begin to appear.
- Tests like the presented in this work, in which the expansive clay and the THMC conditions are very similar to the ones expected in a large-scale test, are very useful to study the response of the bentonite. But, due to relevant processes that take place at high saturation level (Sánchez & Gens 2003, Sánchez 2003) it is necessary to extend the duration of the tests for a very long period of time. It can be concluded that the combinations of these kind of tests with coupled THMC numerical analyse can help to a better understand of the FEBEX bentonite behaviour under actual conditions.

ACKNOWLEDGEMENTS

This research work has been carried out in the context of the FEBEX Project, financed by ENRESA (Spanish National Agency for Waste Management) and the European Commission. R. Campos, J. Aroz and J. Almendrote have performed the laboratory work. Dr. J. Cuevas and P.L. Martín have supported the design of the thermo-hydraulic tests.

APPENDIX

Mechanical constitutive model

The mechanical law adopted is the Barcelona Basic Model (BBM). This model extends the concept of critical state for saturated soils to the unsaturated conditions, including the dependence of yield surface on suction. Two stress variables are considered: the net stresses ($\sigma - P_g \, \boldsymbol{m}$) and capillary suction. Net stress is the excess of total stress over gas pressure. If full saturation is achieved, net mean stress becomes effective stress. For simplicity net stress will also be denoted by σ.

A thermoplastic constitutive law (Alonso et al., 1990; Gens 1995) has been selected based on a generalised yield surface that depends not only on stresses but on temperature and suction as well:

$$f = f\left(\sigma, \varepsilon_v^p, s, \Delta T\right) \tag{A.1}$$

$\Delta T (= T - T_o)$ is the temperature difference with respect to an arbitrary reference temperature, T_o. In terms of invariants:

$$f = f\left(p, J, \theta, \varepsilon_v^p, s, \Delta T\right) \tag{A.2}$$

where:

$$p = \left(\frac{1}{3}\right)\left(\sigma_x + \sigma_y + \sigma_z\right) \tag{A.3}$$

$$J^2 = \tfrac{1}{2}\, \text{trace}\left(s^2\right) \tag{A.4}$$

$$\theta = -\frac{1}{3}\sin^{-1}\left(1.5\sqrt{3}\, det\, s/J^3\right) \tag{A.5}$$

$$s = \sigma - p\,I \tag{A.6}$$

I is the identity tensor. For simplicity, a form of the classical Modified Cam-clay model is taken as the reference isothermal saturated constitutive law:

$$F = \frac{3J^2}{g_y^2} - L_y^2\left(p + P_s\right)\left(P_o - p\right) = 0 \tag{A.7}$$

Volumetric strain is defined as:

$$\varepsilon_v = \varepsilon_x + \varepsilon_y + \varepsilon_z \tag{A.8}$$

and ε_v^p is the plastic volumetric strain.

The basic assumption is that the pre-consolidation pressure, P_o, is made dependent on both suction and temperature:

$$P_o = P^c \left(\frac{P_c}{P^c}\right)^{\frac{\lambda(o)-kio}{\lambda(s)-kio}} \tag{A.9}$$

$$P_c = P_o^* + 2\left(\alpha_1 \Delta T + \alpha_2 \Delta T(T)\right) \tag{A.10}$$

$$Ly = \frac{M}{g_y\left(\theta = -\pi/6\right)} \tag{A.11}$$

$$\lambda(s) = \lambda(o)\left[(1-r)exp\left(-\beta s\right) + r\right] \tag{A.12}$$

In addition:

$$P_s = k\, exp\left(-\rho \Delta T\right)s \tag{A.13}$$

In common with other critical state models, it is assumed that hardening depends on plastic volumetric strain only according to:

$$\dot{P_o} = \frac{1+e}{\lambda(0)-k_{io}}P_o^* \, \dot{\varepsilon}_v^p \tag{A.14}$$

and the plastic potential:

$$G = \frac{3J^2}{g_p^2} - \alpha L_p^2\left(p + P_s\right)\left(P_o - p\right) \tag{A.15}$$

$$L_p = \frac{M}{g_p\left(\theta = -\pi/6\right)} \tag{A.16}$$

If $\alpha = 1$, an associated plastic model results.

The variation of stress-stiffness with suction and, especially, the variation of swelling potential with stress and suction have been carefully considered. The resulting elastic model is as follows:

$$\dot{\varepsilon}_v^e = \frac{k_i}{1+e}\frac{\dot{P}}{P} + \frac{k_s}{1+e}\frac{\dot{s}}{s+0.1} + \left(\alpha_o + 2\alpha_2 \Delta T\right)\dot{T} \tag{A.17}$$

$$k_i = k_{io}\left(1 + \alpha_{is}\, s\right) \tag{A.18}$$

$$k_s = k_{so}\left(1 + \alpha_{sp}\, ln\, p/P_r\right) exp\left(\alpha_{ss}\, s\right) \tag{A.19}$$

$$\dot{E} = \dot{J}/G \quad ; \quad G = E/2(1+v) \tag{A.20}$$

REFERENCES

Alonso, EE, Gens, A & Josa, A (1990) A constitutive model for partially saturated soils. *Géotechnique* 40: 405–430.

ENRESA (1995) Almacenamiento geológico profundo de residuos radiactivos de alta actividad (AGP). Diseños conceptuales genéricos. *Publicación Técnica ENRESA* 11/95. Madrid, 105 pp. (In Spanish).

ENRESA (1998) FEBEX. Bentonite: origin, properties y fabrication of blocks. *Publicación Técnica ENRESA* 05/98. Madrid, 146 pp.

ENRESA (2000) Full-scale engineered barriers experiment for a deep geological repository for high-level radioactive waste in crystalline host rock (FEBEX project). *Nuclear Science and Technology Series, EUR* 19147.

Commission of the European Communities. Luxembourg, 362 pp.

FEBEX 1997. Preoperational Thermo-Hydro-Mechanical (THM) Modelling of the 'MOCK-UP' test. FEBEX Report 70-UPC-M-3-002.

Gens, A (1995) Constitutive laws. In: Gens, A, Jouanna, P, Schrefler, BA (eds) Modern issues in non-saturated soils. *Springer Verlag*. Wien. 129–158.

Lloret, A, Villar, MV, Sánchez, M, Gens, A, Pintado, X & Alonso, EE (2003) Mechanical behaviour of heavily compacted bentonite under high suction changes. *Géotechnique* 53(1): 27–40.

Martín Barca, M (2002) Procesos geoquímicos y modificaciones texturales en bentonita FEBEX compactada sometida a un gradiente termohidráulico. *Ph. D. Thesis*. Universidad Autónoma de Madrid, 293 pp. (In Spanish).

Olivella, S, Carrera, J, Gens, A & Alonso, EE (1994) Nonisothermal Multiphase Flow of Brine and Gas Through Saline Media. *Transport in Porous Media* 15: 271–293.

Olivella, S, Gens, A, Carrera, J & Alonso EE (1996) Numerical formulation for a simulator (CODE-BRIGHT) for the coupled analysis of saline media. *Engineering Computations* 13: 87–112.

Sánchez, M (2003) Thermo-Hydro-Mechanical coupled analysis in low permeability media. *PhD* Thesis, Universitat Politècnica de Catalunya, Spain.

Sánchez, M & Gens, A (2002) Second report on THM modelling results. FEBEX II. UPC Geomechanical Group. ENRESA Report: 70-UPC-L-5-011.

van Genuchten, R (1978) Calculating the unsaturated hydraulic permeability conductivity with a new closed-form analytical mode. *Water Resource Research* 37(11): 21–28.

Villar, MV (2001) Modification of physical, mechanical and hydraulic properties of bentonite by thermo-hydraulic gradients. Proceedings 6th International Workshop Key Issues in Waste Isolation Research. Paris, 28–30 November 2001. pp 3–18.

Villar, MV (2002) Thermo-hydro-mechanical characterisation of a bentonite from Cabo de Gata. A study applied to the use of bentonite as sealing material in high level radioactive waste repositories. *Publicación Técnica ENRESA* 01/2002. Madrid. 258 pp.

Barrier behaviour and THM modelling

Advances in Understanding Engineered Clay Barriers – Alonso & Ledesma (eds)
© 2005 Taylor & Francis Group, London, ISBN 04 1536 544 9

T-H-M modelling of the Prototype Repository Experiment: comparison with current measurements

A. Ledesma & G.J. Chen
Technical University of Catalonia (UPC), Barcelona, Spain

ABSTRACT: The Prototype Repository Project is a full scale test of the KBS-3 Swedish concept for high level radioactive waste, being conducted at Äspö Hard Rock Laboratory and managed by SKB (Swedish Agency for radioactive waste disposal). The project includes the collaboration of different groups from other National Agencies, Companies and Universities, supported by the European Community. The field test involved the excavation of a tunnel and 6 vertical deposition holes in a crystalline rock environment. Each hole includes a canister with heaters surrounded by compacted bentonite blocks. The paper presents the thermo-hydro-mechanical simulation of the experiment performed by one of the groups involved in the project, and includes comparisons with some field measurements. A prediction for the future evolution of some typical variables (temperature, and degree of saturation) is also included. Due to the large amount of variables involved, a sensitivity analysis has been performed, and the effect of some key parameters is described in the text. The agreement between predictions and measurements highlights the ability of the developed models to reproduce the basic coupled processes involved in the experiment.

1 INTRODUCTION

Prototype Repository is the short name for "Full-scale test of the KBS-3 method for deep geological disposal of spent nuclear fuel in crystalline rock", an international EC supported experiment being performed at the Äspö Hard Rock Laboratory (Sweden). The KBS-3 method was initially developed in Sweden and refers to the following main characteristics:

- Copper canister with cast steel insert
- Buffer of bentonite clay surrounding the canisters
- Emplacement in vertical bore holes
- Repository depth of 400–700 m in crystalline rock
- Backfilling of deposition tunnels with bentonite-based materials

The test site is a 65 m long TBM-bored tunnel including six 1.75 m diameter deposition holes of 8 m depth. The outer 25 m long part has two holes and it is separated from the inner 40 m section including 4 holes by means of a tight plug. Details of the work plan may be found in Dahlström (1998), Svemar & Pusch (2000) and Persson & Broman (2000).

The canisters were cylinders of 1.05 m diameter and 4.83 m height, installed in the deposition holes and surrounded by compacted bentonite rings 0.5 m height and 1.65 m of outer diameter. The gap between blocks and the rock was filled up with bentonite

pellets. A small 1 cm gap was left between canister and bentonite blocks as a tolerance for installation purposes. The initial density of the bentonite blocks was $1.78 \, g/cm^3$, reaching $2 \, g/cm^3$ after saturation. The bentonite used was MX80 sodium bentonite from Wyoming.

After the installation of each heater, the bentonite blocks, the instrumentation and backfilling of the surrounding area, the corresponding system was switched on. That is, heaters were turned on according to the installation time schedule. First heater started last September 2001 (Goudarzi & Börgesson, 2003). Since that date, temperature, relative humidity, total pressure and water pressure from the instrumentation have been collected.

This paper presents part of the geo-mechanical modelling work performed by one of the groups involved in the Project. An "in-house" program, CODE_BRIGHT, has been used for that purpose. The main features of the code are described in Olivella et al (1996) and Gens & Olivella (2000), but basically, it is a finite element program that solves the energy balance equation, the water and air balance equations and the equilibrium equation in a porous deformable media. In previous publications, the authors have already presented a predictive modelling of the experiment, as part of the reports corresponding to the Prototype European Project.

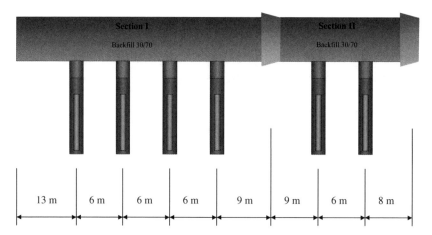

Figure 1. Basic geometry of the Prototype experiment. Hole number 1 is on the left hand side of the figure, in the inner part of section I (Svemar & Pusch, 2000).

Also, some preliminary simulations were presented in Ledesma & Chen (2003). Here, a more elaborated model is presented, due to the fact that as far as field data are available, the parameters and the model itself are improved. An important effort has been devoted to the determination of the parameters of the materials, either from laboratory experiments or from simple back-analysis of the measurements available. Whenever possible, the same set of parameters and initial conditions have been used for all the heaters involved in the experiment.

After a brief description of the main features of the code, the thermo-hydro-mechanical simulation of deposition holes 1 and 3 are presented, including a comparison with the field values available. Additionally, a comprehensive list of the main parameters and physical laws involved in the problem is also presented.

2 DESCRIPTION OF THE CODE

CODE_BRIGHT is a finite element code for the simulation of THM problems in geological media. It was initially developed for the analysis of those problems in saline media (Olivella et al, 1996), but it has been extended to cope with THM behaviour of other materials. In fact that code has been used in recent years to analyse THM problems in the context of radioactive waste disposal (Gens et al, 1995; Gens et al, 1998; Gens & Olivella, 2000). A brief description of the main features of the code is included here for consistency.

A porous medium composed by solid grains, water and gas is considered. Thermal, hydraulic and mechanical aspects are taken into account, including coupling between them in all possible directions. The problem is formulated in a multiphase and multi-species approach. The three phases are solid phase (s), liquid phase (l, water + air dissolved) and gas phase (g, mixture of dry air and water vapour). The three species are solid ($-$), water (w, as liquid or evaporated in the gas phase) and air (a, dry air, as gas or dissolved in the liquid phase).

The following assumptions are considered in the formulation of the problem:

– Dry air is considered a single species and it is the main component of the gaseous phase. Henry's law is used to express equilibrium of dissolved air.
– Thermal equilibrium between phases is assumed. This means that the three phases are at the same temperature.
– Vapour concentration is in equilibrium with the liquid phase, the psychrometric law expresses its concentration.
– State variables (also called unknowns) are: solid displacements, u (three spatial directions); liquid pressure, P_l; gas pressure, P_g; and temperature, T.
– Balance of momentum for the medium as a whole is reduced to the equation of stress equilibrium together with a mechanical constitutive model to relate stresses with strains. Strains are defined in terms of displacements.
– Small strains and small strain rates are assumed for solid deformation. Advective terms due to solid displacement are neglected after the formulation is transformed in terms of material derivatives (in fact, material derivatives are approximated as eulerian time derivatives). In this way, volumetric strain is properly considered.

- Balance of momentum for dissolved species and for fluid phases are reduced to constitutive equations (Fick's law and Darcy's law).
- Physical parameters in constitutive laws are function of pressure and temperature. For example: concentration of vapour under planar surface (in psychrometric law), surface tension (in retention curve), dynamic viscosity (in Darcy's law), are strongly dependent on temperature.

The governing equations that CODE_BRIGHT solves are:

- Mass balance of solid
- Mass balance of water
- Mass balance of air
- Momentum balance for the medium
- Internal energy balance for the medium

The resulting system of Partial Differential Equations is solved numerically. Finite element method is used for the spatial discretization while finite differences are used for the temporal discretization. The discretization in time is linear and the implicit scheme uses two intermediate points, $t^{k+\varepsilon}$ and $t^{k+\theta}$ between the initial t^k and final t^{k+1} times. Finally, since the problems are non-linear, the Newton-Raphson method has been adopted following an iterative scheme.

3 BASIC ASSUMPTIONS OF THE T-H-M ANALYSES

The analyses presented refer to the T-H-M behaviour of the barrier around deposition holes number 1 and number 3. Most of the instruments were installed on these heaters, and therefore, it is reasonable to simulate them first. In previous works (Ledesma & Chen, 2003), six different model geometries were compared, in order to check its effect on some fundamental variables of the problem (i.e. temperature on the canister surface). It was concluded that an axisymmetric model considering only one heater was a compromise between simplicity and quality of the results. In particular, differences in terms of temperature with respect to the full 3D analyses were always less than 5°C. Therefore a "Quasi-3D" geometry (or 2D axisymmetric) was adopted here. However, it should be pointed out that this geometry may underestimate the final temperature, because only one heater is considered. Nevertheless, due to the time delay for switching on the rest of the heaters, the predictions are considered to be realistic, at least for the initial period of the experiment.

Main parameters and boundary and initial conditions of the analyses presented in this paper are summarized in the appendix. Also a brief list of the

physical laws considered is presented. Note that the same parameters have been used for all the analyses with only two exceptions. One refers to the initial conditions, which have been different for hole 1 and hole 3. This is due to the fact that hole number 3 was dry before installation, whereas hole 1 was wet and had an inflow of water due to the intercept of some rock fractures. The second exception refers to the saturated permeability of the bentonite, which is within the measured range in both cases, but that required slight different values in order to match the measured variables properly. Other effects like the influence of the salinity of the water on the bentonite hydraulic conductivity were not taken into account, and could to some extent explain this difference between both deposition holes.

Typical variables recorded used as control variables are temperature and degree of saturation (or suction). Additionally total stresses are also measured, and have been considered in the comparison as well. Only the mid plane of both heaters is considered here, although the analyses provided with results for other sections. However, for those close to the bottom or to the upper part of the deposition hole, the agreement was not as good as for the mid sections. That could be explained by considering different effects: on the one hand, the geometry analysed was not the real one, but an axisymmetric simplification; and on the other hand, there were some changes in geometry in those sections (i.e., in the upper part some small bentonite blocks were installed and adapted to the geometry left by the cables and lids).

One aspect that may become important when dealing with mechanical variables is the distribution of slots in the geometry. Indeed those gaps may change the evolution of the total stress at some particular points, and may create a disagreement between measured and computed quantities. In this case, the 1 cm gap between the heater and the bentonite ring of the barrier has been simulated. For hole 1, this gap has been considered initially wet, in order to reproduce the actual measurements. For hole 3, however, the gap has been assumed totally dry from the beginning.

The code can not deal with contact elements, and a two step analysis has been required. First the gap is included and the deformation of the mesh is controlled to check at what time and where there is full contact. It has been observed that the swelling of the bentonite and the heater dilation gives a full contact almost simultaneously at all nodes between heater and bentonite. In a second step the mesh is changed in such a way that the gap is removed and there is a full contact between both materials. Using this procedure it is expected to reproduce more accurately the evolution of the total stresses in the barrier. Additionally it should be notice that time required for a full contact between heater and bentonite may be different for each hole (depending on the initial water content, etc.). There is a practical difficulty however when comparing the simulation with measurements: some scatter regarding this effect is expected, because the thickness of the gap is assumed constant and equal to 1 cm, whereas in practice that value may be different for each diameter considered (this is in fact a slot left for tolerance when installing the heater after the bentonite rings are placed).

Due to the calculation effort required, computed values of the problem variables do not extend to long times. This is in fact an ongoing work. However, in order to illustrate the long term behaviour of the barrier, some simple analyses for hole 1 and considering 1D conditions have been included. In that case the computational time is reduced and long term analyses become more feasible than in the Quasi-3D models.

4 ANALYSIS OF HOLE 1

Regarding hole number 1, figure 2 presents the time evolution of saturation degree, temperature and total stress for three points of the bentonite: one close to the heater ($r = 0.585$ m), one close to the pellets and the rock ($r = 0.785$ m) and a central one ($r = 0.685$). All of them correspond to the mid plane, as indicated above.

Degree of saturation measurements have been obtained from relative humidity transducers by means of the psychrometic law. It can be observed that the agreement is, in general, good. Temperature records may be eventually influenced by the installation of the rest of heaters. However, the trends and the temperature values seem to be consistent with measurements. For radius = 0.685 m, that is in the central area of the bentonite ring, there are two measurements available that show some scatter.

Saturation degree evolution has been reproduced reasonably. It should be pointed out that near the heater a wetting-drying cycle is observed, which is typical of coupled processes. Indeed, that point first increases its saturation degree, due to the water vaporisation that moves from the heater to the outer bentonite areas, and after a few days it dries because of this effect takes place eventually on that point. Finally water from the rock saturates the bentonite and the degree of saturation increases in a continuous manner. This coupling between temperature and water flow (in liquid or gas form) gives this cyclic response in the bentonite zone closer to the heater. This effect depends partly on the amount of water available for vaporisation. Note that as mentioned in previous section, the slot between the

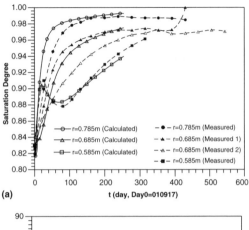

(a)

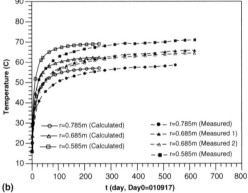

(b)

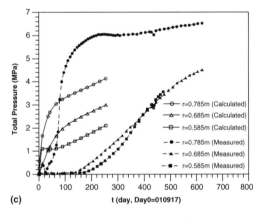

(c)

Figure 2. Computed and measured variables against time for mid section of hole 1 – Quasi3D analyses – (a) Saturation degree, (b) Temperature, (c) Total stress.

heater and the bentonite in hole number 1 has been considered wet, with an initial degree of saturation of 33%. This value was obtained by a trial and error procedure.

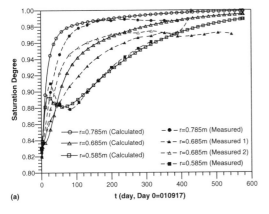

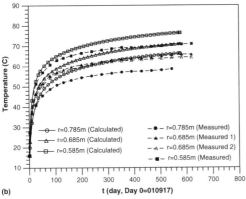

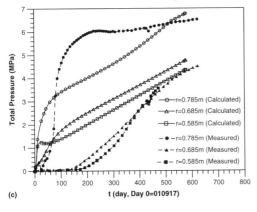

Figure 3. Computed and measured variables against time for mid section of hole 1 – 1D analyses – (a) Saturation degree, (b) Temperature, (c) Total stress.

The total pressure is not reproduced with accuracy (see figure 2c), but this is the case quite often due to different reasons. Sometimes the installation of total pressure cells is not easy and they do not react to the stress instantaneously because of some arching

effects may develop inside the material. Additionally the heterogeneity of gaps and joints between bentonite rings may generate a different pattern of total stress evolution. Despite that, the trends of computed and measured values in figure 2c are quite similar and eventually, in a long term perspective, a new comparison will be required.

For this hole, the gap between heater and bentonite is closed in around 35 days, according to the analyses. This is of course an estimate difficult to check in practice, but it is the value required to reproduce the rest of the measured variables.

Finally, a long term analysis has been performed considering 1D conditions, that is, a line in the bentonite barrier in the mid section of the canister. This analysis was performed in order to obtain results for at least two years of the test, because the computational effort required is much smaller than in the Quasi-3D case. Figure 3 presents the corresponding results for this model.

Note that the trends are similar to those observed in figure 2, and the comparison with current measurements is again reasonable. For the 1D model, the heat power of the heater has been changed in order to take into account the effect of the geometry, as suggested in Ledesma & Chen, 2003.

5 ANALYSIS OF HOLE 3

The analysis of hole 3 is similar from a conceptual point of view than the previous case. The main difference refers to the initial conditions as in this case the deposition hole was dry. This effect has been considered by assuming a very low rock hydraulic conductivity. Figure 4 presents for the mid section of this case, the evolution of degree of saturation, temperature and total stresses in the bentonite barrier. Note that due to this effect, the saturation process after the first year of the experiment is very slow, and may be eventually almost negligible. During the first year, however, the effect of vapour diffusion introduces some changes in the saturation trends. This behaviour is consistent with actual measurements, although a long term perspective is required to complete an accurate picture of this heater.

The temperature in this hole is greater than in hole number 1. This is due partly to the effect of other heaters, but the low thermal conductivity obtained for the drier bentonite has much more influence on that.

The low rate of saturation has also influence on the mechanical response of the bentonite. In fact the simulations indicate that the gap between heater and bentonite is not closed so far. Therefore the total stresses in the buffer are not developed, as indicated in figure 4c.

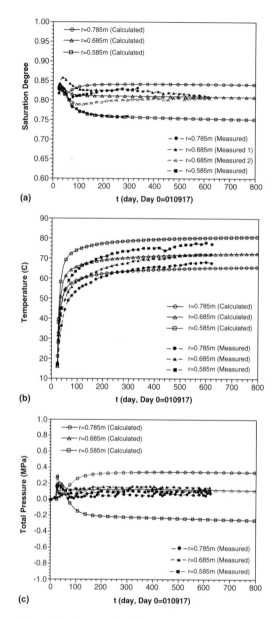

(a)

(b)

(c)

Figure 4. Computed and measured variables against time for mid section of hole 3 – Quasi3D analyses – (a) Saturation degree, (b) Temperature, (c) Total stress.

6 CONCLUSIONS

The paper presents some of the THM simulations carried out in the context of the Prototype Repository Project, a large scale field test being performed at Äspö Hard Rock Laboratory, Sweden. Only a few examples have been presented here. A comparison of the measured and computed evolution of degree of saturation, temperature and total stress for some selected points of the mid plane of hole number 1 and 3 have been presented. The results show that despite the large amount of parameters and the difficulties in characterising those materials, the simulation can reproduce the main aspects of the processes involved, as, for instance, the wetting-drying cycles of the zone close to the heater. This effect has been simulated thanks to the consideration of the vapour flow in the system. This result is also consistent with other THM simulations performed in the context of radioactive waste disposal problems.

Additionally, the influence of gaps in the stress field is considered to be important, and it has been considered in all the analyses. That required performing multi-step calculations, considering first the gap in the mesh, and later assuming full contact between heater and bentonite. In that way, it is easier to reproduce the level of total stresses measured in the experiment.

Finally, it should be pointed out that both deposition holes behave in a different manner. Whereas hole 1 is expected to saturate in 3 years after the initiation of the test, hole number 3 will take much more time, due to the low hydraulic conductivity of the rock surrounding that heater. Those differences between holes may be representative of what could be expected in a real repository under operating conditions. Therefore, the long term evolution of the test will provide valuable information for understanding the behaviour of this type of engineered barriers under different conditions of the host rock.

7 APPENDIX

Parameters, boundary and initial conditions used in the analyses. They have been defined from the data available on the experiment and on the materials involved. In some cases, a personal guess has been used.

7.1 Main parameters

See Table 1.

7.2 Boundary conditions

At $r = 70m$: $T = 16°C, P_l = 4.0MPa, \sigma_r = \sigma_\theta = 26MPa,$ $\sigma_y = 13MPa$

(1) 1D_THM: Heat flux= $280 W/m$;

(2) Quasi3D_THM: Heat flux= $1800 W/_{Heater}$

344

Table 1.

	Bentonite	Rock	Backfill	Heater	Pellets	Slot
Thermal parameters						
Specific heat						
c (J/kgK)	1091	750	1200	460	1091	1091
Thermal conductivity						
λ_{dry} (W/mK)	0.3	2.5	1.5	50.16	0.1	0.1
λ_{sat} (W/mK)	1.3	2.5	1.5	50.16	1.0	0.6
Hydraulical parameters						
Intrinsic permeability						
k_0 (m^2)	0.2×10^{-20} (Hole 1) 0.8×10^{-20} (Hole 3)	3×10^{-19} (Hole 1) 5×10^{-22} (Hole 3)	5×10^{-18}	10^{-30}	10^{-18}	10^{-11}
ϕ_0	0.366	0.003	0.40	0.001	0.706	0.98
Retention curve						
P_0 (MPa)	60.0	4.0	0.12	1000	0.4	0.5
β	0.3	0.56	0.18	0.36	0.4	0.6
Mechanical Parameters						
Thermal elasticity						
b_s (°C^{-1})	10^{-5}	7.8×10^{-6}	10^{-5}	1.2×10^{-5}	10^{-5}	10^{-5}
Elasticity						
E (MPa)		5×10^4	30	2.1×10^5		10^{-3}
ν		0.25	0.3	0.2		0.25
Elasto-plasticity (Barcelona Basic Model)						
Elastic parameters	$\kappa_{i0} = 0.4767$ $\kappa_{s0} = 0.2218$ $K_{min} = 13.33$ MPa $\nu = 0.2$				$\kappa_{i0} = 1.4301$ $\kappa_{s0} = 0$ $K_{min} = 4.43$ MPa $\nu = 0.2$	
Plastic parameters	$\lambda(0) = 1.5$ $r = 0.75$ $\beta = 0.05$ $p^c = 0.1$ MPa $P_0^* = 8$ MPa $M = 1$ $\alpha = 0.395$ $k = 0.1$				$\lambda(0) = 4.5$ $r = 0.75$ $\beta = 0.05$ $p^c = 0.1$ MPa $P_0^* = 8$ MPa $M = 1$ $\alpha = 0.395$ $k = 0.1$	

7.3 *Initial conditions*

See Table 2.

Table 2.

Materials	Temperature (°C)	Stress (MPa)	Porosity	Saturation degree (%)
Bentonite (Hole 1)	16	−0.5/ −0.5/−0.5	0.366	83
Bentonite (Hole 3)				
Rock		−26/ −26/−13	0.003	100
Backfill		−0.5/ −0.5/−0.5	0.4	61
Heater			0.001	0
Pellets (Hole 1) Pellets (Hole 3)			0.706	100 18
Slot (Hole 1) Slot (Hole 3)			0.98	32.83 20

7.4 MAIN LAWS AND EQUATIONS USED:

7.4.1 *Thermal problem*
Thermal conductivity

$$\lambda = \lambda_{dry}^{1-S_r} \cdot \lambda_{sat}^{S_r}$$

7.4.2 *Hydraulic problem*
Retention Curve:

$$S_e = \frac{S_l - S_{rl}}{S_{ls} - S_{rl}} = \left[1 + \left(\frac{P_g - P_l}{P_0} \right)^{\frac{1}{1-\beta}} \right]^{-\beta}$$

Intrinsic Permeability:

$$k = k_0 \frac{\phi^3}{(1-\phi)^2} \frac{(1-\phi_0)^2}{\phi_0^3}$$

Relative Permeability:

$$k_{rl} = S_e^3$$

7.4.3 *Mechanical problem*
Thermal Elasticity:

$$\Delta \varepsilon_v = 3b_s \Delta T$$

Plasto-elasticity (Barcelona Basic Model)

Equations for elastic deformation:

$$d\varepsilon_v^e = \frac{\kappa_i(s)}{1+e} \frac{dp'}{p'} + \frac{\kappa_s(p',s)}{1+e} \frac{ds}{s+0.1}$$

where

$$\kappa_i(s) = \kappa_{i0}(1+\alpha_i s), \kappa_s(p',s)$$
$$= \kappa_{s0}\left(1+\alpha_{sp} \ln(p'/p_{ref})\right)\exp(\alpha_{ss}s)$$

Equations for yielding curve:

$$\left(\frac{p_0}{p^c} \right) = \left(\frac{p_0^*}{p^c} \right)^{[\lambda(0)-\kappa_i]/[\lambda(s)-\kappa_i]}, \ \lambda(s) = \lambda(0)\left[(1-r)\exp(-\beta s)+r\right]$$

where

$$p' = \frac{1}{3}\left(\sigma_x + \sigma_y + \sigma_z\right) - \max\left(p_g, p_l\right)$$

ACKNOWLEDGEMENTS

This work has been performed with the support of the European Union through Research Project FIKW-CT-2000-00055. Moreover, the support from ENRESA (Spanish Agency for radioactive waste disposal) and SKB is also gratefully acknowledged.

REFERENCES

Dahlström, L-O. (1998). *Test Plan for the Prototype Repository*. NCC Teknik, SKB, Sweden.

Börgesson, L. (2001). *Compilation of laboratory data for buffer and backfill materials in the Prototype Repository*. Äspö Hard Rock Laboratory, International Progress Report IPR-01-34, SKB, Sweden.

FEBEX. (2000). FEBEX Project. Full-scale engineered barriers experiment for a deep geological repository for high level radioactive waste in crystalline host rock. Final Report. ENRESA Technical Publication 1/2000. Madrid, Spain.

Gens, A., García-Molina, A.J., Olivella, S., Alonso, E.E. & Huertas, F. (1998). *Analysis of a full scale "in situ" test simulating repository conditions*. Int. Journal Numerical and Analytical methods in Geomechanics, 22, p. 515–548.

Gens, A. & Olivella, S. (2000). *Non-isothermal multiphase flow in deformable porous media. Coupled formulation and application to nuclear waste disposal*. In Developments in Theoretical Geomechanics. Smith & Carter eds., Balkema, Rotterdam, p. 619–640.

Gens, A., Vaunat, J. & Ledesma, A. (1995). *Analysis of hydration of an engineered barrier in a radioactive waste repository scheme using an elastoplastic model*. Proc. 1st Int. Conf. on Unsaturated Soils, Alonso&Délage eds., Balkema, p. 1065–1073.

Goudarzi, R. & Börgesson, L. (2003). *Prototype Repository. Sensors Data Report (Period 010917-030301). Report no 5*. Äspö Hard Rock Laboratory, International Progress Report IPR-03-23, SKB Sweden.

Ledesma, A. & Chen, G. (2003). *T-H-M modelling of the Prototype Experiment at Äspö HRL (Sweden)*. Proc. Int. Conf. GeoProc, Stockholm, October. CD-ROM.

Olivella, S., Gens, A., Carrera, J. & Alonso, E.E. (1996). *Numerical formulation for a simulator (CODE_B-RIGHT) for the coupled analysis of saline media*. Engineering Computations, 13, no. 7, p. 87–112.

Persson, G. & Broman, O. (2000). *Prototype Repository. Project Plan FIKW-CT-2000-00055*. Äspö Hard Rock Laboratory. International Progress Report, IPR-00-31. SKB, Sweden.

Svemar, Ch. & Pusch, R. (2000). *Prototype Repository. Project description FIKW-CT-2000-00055*. Äspö Hard Rock Laboratory. International Progress Report, IPR-00-31. SKB, Sweden.

Advances in Understanding Engineered Clay Barriers – Alonso & Ledesma (eds)
© 2005 Taylor & Francis Group, London, ISBN 04 1536 544 9

Simulation of the Prototype Repository

H.R. Thomas, P.J. Cleall & T.A. Melhuish
Geoenvironmental Research Centre, Cardiff School of Engineering, Cardiff University, UK

ABSTRACT: A fully coupled mechanistic Thermal-Hydraulic-Mechanical model has been applied to simulate the behaviour of the engineered barrier system employed by SKB in the Prototype Repository. The numerical code COMPASS has been used to perform a series of full three-dimensional finite element analyses of the repository. The heterogeneity of the rock mass has been investigated with particular attention being given to the influence of the fractures in the host rock. A range of simulation results is presented detailing the hydraulic, thermal-hydraulic and thermal-hydraulic-mechanical behaviour of the Prototype Repository. Attention has been given to the micro-macro interaction occurring within the MX-80 bentonite buffer as the material begins to saturate and swell. This approach represents the effect of adsorbed water in the micropores reducing the hydraulic conductivity and "choking" the flow of water in the macropores. Attention is also given to the mechanical behaviour of the slot filling material. The simulation results have been compared with the experimentally measured results. It is found that the model is able to capture both the trends and patterns in the thermal field and hydraulic field.

1 INTRODUCTION

The Prototype Repository is an international, EC-supported project with the objective to investigate, on a full-scale, the integrated performance of engineered barriers and near-field rock of a deep repository intended for the disposal of high-level nuclear waste. It has been constructed at the Äspö Hard Rock Laboratory, in Sweden (Dahlström, 1998). The experimental configuration, as the name suggests, is a full-scale prototype repository representing the proposed final SKB concept for disposal. The work program is extremely comprehensive, with system design, characterisation of the rock mass, preparation of emplaced materials, through to handling and deposition of canisters being considered.

The Prototype Repository has been constructed in a 65 m long TBM drift at approximately 450 m below the ground surface in crystalline rock. Section I of the experiment consists of four 1.75 m diameter, 8 m deep full-scale deposition holes. Heater canisters have been emplaced in these holes with a combination of bentonite blocks, bentonite pellets and backfill material used to form a series of engineered barriers in accordance with the KBS-3 concept (Dahlström, 1998). This was installed during 2001 with the activation of the heaters taking place from September 2001. A large range of experimental data pertaining to Section I has been collected and is readily available (Goudarzi and Börgesson, 2003). Section II, consisting of two identical deposition holes, has recently been installed with the heaters being switched on in May 2003. Figure 1 shows a schematic view of the layout of the Prototype Repository and deposition holes.

2 MODELLING APPROACH

The simulation work has been performed using COMPASS (**CO**de for **M**odelling **PA**rtly **S**aturated Soil), a mechanistic model (Thomas and He, 1995, 1998; Thomas and Sansom, 1995; Thomas et al., 1998; Thomas and Cleall, 1999), where the various aspects of soil behaviour under consideration, are included in an additive manner. In this way the approach adopted describes heat transfer, moisture migration, solute transport and air transfer in the material, coupled with stress/strain behaviour.

3 MATERIAL PARAMETERS

The mechanistic model employed in this work requires a number of material parameters to define the behaviour of each individual material present. Due to limitations of space full details of the parameters used and their determination cannot be presented here. A summary of the parameters is given in Table 1.

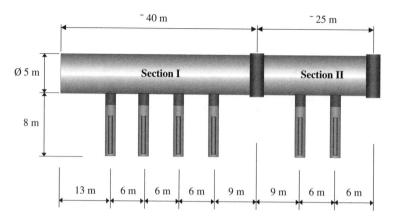

Figure 1. Schematic layout of the Prototype Repository and deposition holes.

Table 1. THM material parameters used in the analyses.

Parameter	MX-80 (Cylinder)	MX-80 (Ring)	MX-80 (Pellets)	Backfill (30/70 mix)	Rock (Äspö diorite)
ρ_d (kg/m³)	1660	1780	1200	1750	2770
e	0.67	0.56	1.55	0.57	0.005
Initial Sr	0.81	0.87	0.23	0.57	1.0
C_{ps} (J/kg.K)	800	800	800	850	750
Retention Curve	$S_r = \left[\dfrac{\theta_{res} + (\theta_{sat} - \theta_{res})}{\left(1 + \left(\dfrac{1000 \ .\alpha\ .S}{\rho\, g}\right)^b\right)^{(1-1/b)}} \right] / n$				See equation (1)
α	4.5×10^{-7}	4.5×10^{-7}	4.5×10^{-7}	0.00014	–
b	1.75	1.75	1.75	1.19	–
θ_{res}	0.0001	0.0001	0.0001	0.000001	–
θ_{sat}	0.4	0.36	0.61	0.363	–
β	–	–	–	–	0.33
Hydraulic Conductivity	$k_{unsat} = k_{sat.}\,(S_r)^3$				See equation (2)
k_{sat} (m/s)	1×10^{-13}	1×10^{-13}	1×10^{-13}	1.5×10^{-10}	10^{-9}–10^{-12}
κ	0.17	0.17	2.0	0.17	–
κ_s	0.094	0.094	1×10^{-7}	0.094	–
G (Pa)	1×10^7	1×10^7	1×10^7	1×10^7	27.6×10^9
ν	0.499	0.499	0.499	0.499	0.25

Five separate materials have been considered. Each of these requires a series of material constants and constitutive relationships to describe their thermal-hydraulic-mechanical behaviour. This information has been obtained from experimental work performed by a number of groups (Patel et al., 1997; Dahlström, 1998; Johannesson et al., 1999; Börgesson and Hernelind, 1999; Börgesson et al., 2001, 2002). In some cases mathematical equations have been fitted to experimental data to cast the material relationships in a form suitable for computer modelling.

4 GEOMETRIC MODEL

A full three-dimensional model of SKB's Prototype Repository incorporating all of the primary features of the tunnel has been developed. The model domain measures 200 m by 100 m by 200 m. This model has

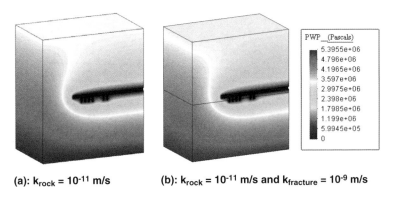

(a): $k_{rock} = 10^{-11}$ m/s   (b): $k_{rock} = 10^{-11}$ m/s and $k_{fracture} = 10^{-9}$ m/s

Figure 2.   (a) $k_{rock} = 10^{-11}$ m/s. (b) $k_{rock} = 10^{-11}$ m/s and $k_{fracture} = 10^{-9}$ m/s.

been discretised using 8 noded hexahedral elements and consists of 158,175 elements and 146,380 nodes. The mesh has been refined in and around the buffer with a coarser mesh discretisation used in the far-field rock. This geometrical model has been used for the majority of the numerical modelling work with smaller three-dimensional and two-dimensional models being implemented to investigate the mechanical response of the buffer and pellets under thermal and hydraulic gradients.

$$S_r = \left[\left(1 + \left(\frac{s}{P_0} \right)^{1/(1-\beta)} \right) \right]^{-\beta} \tag{1}$$

$$k_{unsat} = k_{sat} (S_r^{1/2}(1-(S_r^{1/\beta})^\beta)^2 \tag{2}$$

5   RESULTS

5.1   Hydraulic analysis of the rock mass prior to emplacement

A number of analyses of the rock mass prior to emplacement have been performed with the hydraulic conductivity of the rock ranging from 10^{-11} to 10^{-12} m/s. A set of analyses incorporating a "representative" fracture have also been undertaken. All analyses have been performed using the full 3D tunnel domain. Figures 2(a) and 2(b) show the steady-state pore water pressure contour plots for the full 3D tunnel mesh. Figure 2(a) is taken for a homogeneous rock mass with hydraulic conductivity of 10^{-11} m/s and it was found to reach steady-state pore water pressures after 2 years. Figure 2(b) incorporates a fracture with a hydraulic conductivity of 10^{-9} m/s and it was found that this analysis reached steady-state conditions after 180 days.

5.2   Thermal-hydraulic analysis of Section I

A large number of coupled thermal-hydraulic analysis have been performed to investigate the complex flow patterns that occur in the Prototype Repository following heater activation. In order to establish an understanding of these processes a comprehensive step-wise approach to the modelling has been adopted. A full range of 3D tunnel section and 2D axisymmetric analyses have been undertaken with the results being presented previously (Thomas et al., 2002). However, more recently a full three-dimensional, thermal-hydraulic analysis of Section I has been performed. This analysis captures the thermal and hydraulic behaviour for the first 600 days before Section II was installed. The performance of the full repository after Section II has been installed with the heaters in boreholes 5 and 6 being activated is also being performed.

Recent research has shown that the micro and macro structure of a bentonite buffer material may have a pronounced effect on the saturation rates of the material (Thomas et al., 2003). It was proposed that as water enters the buffer the majority of it becomes adsorbed within the micropores and hence is unavailable for further flow. This adsorption of water results in swelling of the micropore in the buffer. Depending on the degree of mechanical restraint swelling of the micropore will lead to some reduction in the size of the macropores (Pusch, 1998). As the only water available for flow is contained in the macropores, the swelling of the micropore in a restrained material would thus tend to "choke" moisture flow and further reduce the effective hydraulic conductivity of the material. Following the approach presented in Thomas et al., (2003) the hydraulic conductivity relationship used here has been modified accordingly. This modification has assumed that 94% of the water present in the clay would be adsorbed and not available for flow. In

349

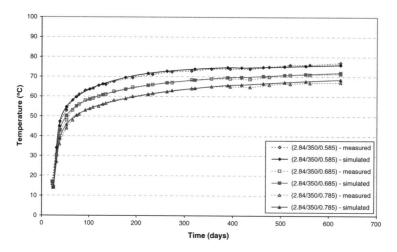

Figure 3. Measured and simulated temperature plots for Borehole 3.

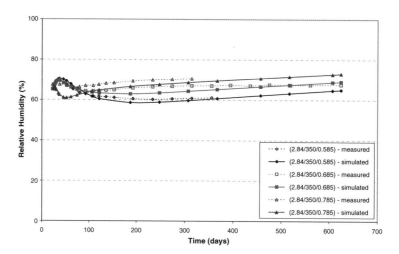

Figure 4. Measured and simulated relative humidity plots for Borehole 3.

the case of the Prototype Repository, compared to the buffer investigated in Thomas et al. (2003), the buffer is relatively free to deform, due to the presence of the pellet region, therefore the effect of microstructural swelling choking of the macro structure has been reduced. This has led to the following relationship being assumed as a first approximation

$$k = k_{unsat} \cdot k_l \cdot \left[S_a + 0.06 S_r \left(\exp(0.01(1 - 0.06) S_r) \right) \right] \quad (3)$$

Figure 3 shows both the simulated and experimentally measured temperature plots after 624 days for 3 different radii positions in Borehole 3. At a radius of 0.585 m the temperature has been simulated to be 76.1°C after 624 days, at a radius of 0.785 m the

temperature is simulated as 68.6°C. The corresponding experimentally measured results are 77.1°C and 67.2°C respectively.

Figure 4 shows simulated and experimentally measured relative humidity's in Borehole 3. It can be seen that initially there is a slight increase in relative humidity for both the simulated and measured results closer to the heater as moisture is drawn out of the buffer further away from the canister. This is followed by a general reduction in the relative humidity's as the temperature begins to increase. At a radius of 0.585 m the relative humidity is simulated to reach a minimum of 58.7% after 190 days. These simulated trends follow the experimentally measured trends where there is a corresponding minimum

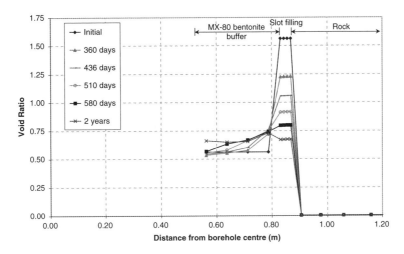

Figure 5. Variation of void ratio through the centre of the buffer and pellets over time.

relative humidity of 60.3% after 230 days. In both cases resaturation take place slowly up to 624 days. The initial over prediction of moisture movement at a radius of 0.785 m is possibly due to an over estimation of suction in the nearby pellet region.

5.3 Thermal-hydraulic-mechanical analysis

The coupled mechanical response of the system due to changes in the hydraulic and temperature fields is of importance particularly in the swelling materials present within the system. These materials although effectively restrained by the rock mass are able to deform locally. As discussed earlier the magnitude of such deformations and development of swelling pressures may have an influence on the flow fields. To investigate these effects a number of analyses have been performed. Results showing the variation of void ratio and stresses in the MX-80 buffer have been presented in previous work (Thomas et al., 2002) and recent work has focused on an investigation into the mechanical properties of the slot filling material (i.e. the pellets plus the air between the pellets).

It is expected that under saturated conditions the buffer and pellets will swell and combine to form a homogenous material with the same void ratio throughout (Dahlström, 1998; Börgesson et al., 2002). In order to capture this behaviour of the slot filling under thermal, hydraulic and stress gradients it was necessary to conduct a series of sensitivity analyses. This was undertaken to ascertain approximate values for the mechanical material parameters for the slot filling material. Figure 5 shows the result of this sensitivity analysis where κ_s, the elastic stiffness parameter for changes in suction of the soil,

was set to the value of 1×10^{-7} and κ, the elastic stiffness parameter for changes in net mean stress of the soil, was set to a value of 2.0. This was achieved using an elastic model based on the model presented by Alonso et al. (1990). It can be seen that over a period of time the buffer swells and its void ratio increases. The consequence of this is that the slot filling is compressed and its void ratio reduces. By the end of the analysis the buffer and pellets have formed a more homogenous material with a void ratio in the range of 0.65 and 0.68.

6 CONCLUSIONS

In conclusion a series of coupled mechanistic thermal-hydraulic-mechanical analysis of the behaviour of the Prototype Repository have been presented. In particular, full three-dimensional analyses of the system have been performed. A full range of simulation results have been presented detailing the hydraulic, thermal-hydraulic and thermal-hydraulic-mechanical behaviour of the Prototype Repository. Particular attention has been given to the micro–macro interaction occurring within the MX-80 bentonite buffer as the material begins to saturate and swell. It was assumed that as the material saturated, 94% of the available water was adsorbed in the micropores hence swelling and reducing the hydraulic conductivity and "choking" the flow of water in the macropores. Comparison of the simulated and experimental results found a good correlation of behaviour in the thermal field and a reasonable correlation in the hydraulic regime. Attention was also given towards the accurate modelling of the mechanical behaviour of the slot filling material.

351

Following a sensitivity analysis if was found that the void ratio in both the buffer and slot filling materials approached similar values and hence formed a more homogenous material as a result of saturation. Hence, the thermal-hydraulic-mechanical analysis was able to simulate the expected mechanical behaviour of the barrier materials. The simulation results have been compared with the experimentally measured results. It was shown that the model was able to capture both the trends and patterns in the thermal field and hydraulic field.

REFERENCES

Alonso, E.E., Gens, A. and Josa, A. (1990) "A constitutive model for partially saturated soils", Geotechnique, 40(3): 405–430.

Börgesson, L. and Hernelind, J. (1999) "Preliminary modelling of the water saturation phase of the buffer and backfill materials" International Progress Report IPR-00-11.

Börgesson, L., Johannesson, L.E. and Sandén, T. (2001) "Compilation of Laboratory Data for Buffer and Backfill Materials in the Prototype Repository", August 2001.

Börgesson, L., Gunnarsson, D., Johannesson, L.E. and Sandén, T. (2002) "Installation of buffer, canisters, backfill and instruments in Section 1" March 2002.

Dahlström, L.O. (1998) "Test Plan for the Prototype Repository", SKB Progress Report HRL-98-24.

Goudarzi, R. and Börgesson, L. (2003) "Sensors data report (Period: 010917-030301) Report No: 5" International Progress Report IPR-03-23.

Johannesson, L.E., Börgesson, L. and Sandén, T. (1999) "Backfill materials based on crushed rock (part 2).

Geotechnical properties determined in laboratory" International Progress Report IPR-99-23.

Patel, S., Dahlström, L.O. and Stenberg, L. (1997) "Characterisation of the rock mass in the Prototype Repository at ÄSPÖ HRL, Stage 1", SKB Progress Report HRL-97-24.

Pusch, R. (1998) "Microstructural evolution of buffer clay" In Proceedings of workshop on microstructural modelling of natural and artificially prepared clay soils with special emphasis on the use of clays for waste isolation, Lund, pp. 31–38.

Thomas, H.R. and He, Y. (1995) "Analysis of coupled heat, moisture and air transfer in a deformable unsaturated soil", Geotechnique, 45(4): 677–689.

Thomas, H.R. and Sansom, M.R. (1995) "Fully coupled analysis of heat, moisture and air transfer in unsaturated soil", ASCE, Journal of Eng. Mech. 121(3): 392–405.

Thomas, H.R. and He, Y. (1998) "Modelling the behaviour of unsaturated soil using an elasto plastic constitutive relationship", Geotechnique, 48(5): 589–603.

Thomas, H.R., He, Y. and Onofrei, C. (1998), "An examination of the validation of a model of the hydro/thermo/mechanical behaviour of engineered clay barriers", International Journal for Numerical and Analytical Methods in Geomechanics, 22: 49–71.

Thomas, H.R. and Cleall, P.J. (1999) "Inclusion of expansive clay behaviour in coupled thermo hydraulic mechanical models", International Journal of Engineering Geology, 54: 93–108.

Thomas, H.R., Cleall, P.J. and Melhuish, T.A. (2002) "Simulation of the Prototype Repository using predictive THMCB modelling" UWC contribution to D34.

Thomas, H.R., Cleall, P.J., Chandler, N., Dixon, D. and Mitchell, H.P. (2003) "Water infiltration into a large-scale in-situ experiment in an underground research laboratory", Geotechnique, 53(2): 207–224.

Advances in Understanding Engineered Clay Barriers – Alonso & Ledesma (eds)
© 2005 Taylor & Francis Group, London, ISBN 04 1536 544 9

Hydraulic bentonite/rock interaction in FEBEX experiment

Lennart Börgesson
Clay Technology AB, Lund, Sweden

Jan Hernelind
FEM-Tech AB, Västerås, Sweden

ABSTRACT: The wetting of the bentonite based buffer material and the hydraulic response of the surrounding rock in a repository for spent nuclear fuel is depending on not only the properties of the rock and the buffer material but also on the interaction between the rock and the buffer. The large field tests in ÄHRL in Sweden (Prototype Repository) and Grimsel in Switzerland (FEBEX) are very well suited for analysing these interactions since a large number of measurements both in the rock and in the buffer are done and considerable modelling is performed before and during the test operation.

The excavation of the test tunnel and the first 1000 days of the FEBEX experiment have been modelled in the framework of the international project DECOVALEX. The modelling has been performed in three stages where the third stage implies THM modelling of both the buffer and the rock and subsequent comparison of the results.

The authors have modelled these processes with the finite element code ABAQUS including the specially made models of water unsaturated buffer materials. The element mesh is divided into three sub meshes with different density of elements. In ABAQUS it is possible to combine and connect two quite different element meshes, which allows building of a model with very large dimensions and yet catch the processes in small parts of the model. The total volume of the model is $600 \times 150 \times 300\,m^3$ while the processes in the near field rock around the test drift in the scale $30 \times 30 \times 70\,m^3$ are studied.

The modelling includes all steps in the FEBEX experiment, i.e. the excavation, buffer and canister placement, heating of the canisters and the subsequent wetting and swelling of the buffer material, temperature increase and mechanical and hydraulic response in the rock. The water pressure, temperature and displacements in the rock have been measured in some points and the results are compared with the measurements.

The conclusions from these comparisons are that most hydraulic modelling results agree rather well with the measurements and thus that the models are relevant. However, there are some inconsistencies that yield the following preliminary conclusions:

− There seems to be a skin zone with reduced hydraulic conductivity in the rock close to the drift, since the measured water pressure around the drift is higher than the predicted.
− The hydraulic interaction between the rock and the buffer is not as good and direct as modelled since the predicted influence from the suction in the buffer is not observed in the measurements.

1 INTRODUCTION

Measurement results from some large-scale field tests have been submitted to the international project DECOVALEX III in order to get an experimental basis for developing and testing coupled models describing THM processes in the near and far field of a repository. DECOVALEX III Task 1 has comprised the full-scale field test of a deposition tunnel in the Grimsel Test Site in Switzerland named FEBEX.

The exercise of Task 1 has been to predict the performance of the buffer material and the near field rock during the first 1000 days. In order to make the calculations stepwise and to understand the rock behaviour the predictions have been divided into the following 3 parts:

Part A. Hydro-mechanical modelling of the rock before installation of the buffer and the canisters
Part B. Thermo-hydro-mechanical analysis of the bentonite behaviour without considering the rock
Part C. Thermo-hydro-mechanical analysis of the rock after installation of the buffer and canisters

The most interesting part is the hydraulic response, since the wetting of the buffer is depending on the

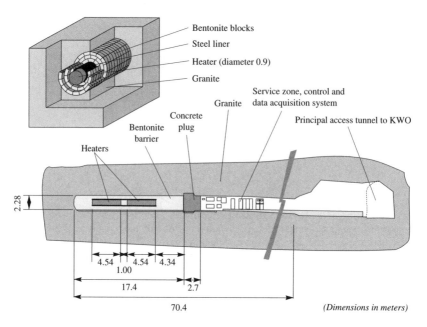

Figure 1. Overview of the FEBEX test.

hydraulic interaction with the near field rock. This article mainly deals with the hydraulic results of tasks A and C but since the THM-processes are coupled some temperature and mechanical results will also be shown. The calculations include thus a fully coupled 3D model of the THM response of the entire test, starting with the excavation of the tunnel and ending 1000 days after installation of the buffer, canisters and plug. Each part has been reported separately (Börgesson & Hernelind, 2000, 2001 and 2003).

2 FEBEX IN-SITU TEST

The Test Site is located at an elevation of 1730 m above sea level, around 450 m beneath the Juchlistock mountain in the granitic rocks of the Aare Massif in central Switzerland. The Grimsel Test Site (GTS tunnel) is located in a tunnel system, which branches off from the main access tunnel (KWO tunnel) to the underground power station of the KWO (Kraftwerke Oberhasli AG). The GTS tunnel (also named the Laboratory tunnel) has a diameter of 3.5 m and a length of almost 1000 m. It was excavated in 1983.

The FEBEX tunnel has the diameter 2.28 m and the length 71.4 m. It was drilled with a tunnel-boring machine in September and October 1995. The FEBEX test is performed in the inner 20 meters of the drift. The test design, which is shown in Fig. 1, consists of two 4.54 m long heaters embedded by a buffer material of high density compacted bentonite

blocks. The canisters are enclosed in a perforated steel liner for installation reasons. Outside the buffer material 17 m from the end of the drift a concrete plug is built. Altogether 632 instruments are placed in the rock, bentonite and heaters. The heaters are heated with a power that is regulated to yield a maximum temperature of 100 degrees in the buffer (FEBEX, Enresa publicacion tecnica num. 01/98).

3 FINITE ELEMENT CODE

The finite element code ABAQUS was used for the calculations. ABAQUS is originally designed for non-linear stress analyses. It contains a capability of modelling a large range of processes in many different materials as well as complicated three-dimensional geometry.

The code includes special material models for rock and soil and ability to model geological formations with infinite boundaries and in situ stresses by e.g. the own weight of the medium. It also includes capability to make substructures with completely different finite element meshes and mesh densities without connecting all nodes. This technique has been used in the presented calculations. Detailed information of the available models, application of the code and the theoretical background is given in the ABAQUS Manuals. A more detailed description of the models is also given in the project reports (Börgesson & Hernelind, 2001 and 2003).

The main components of the code are the following:

- The hydro-mechanical model consists of porous medium and wetting fluid and is based on equilibrium, constitutive equations, energy balance and mass conservation using the effective stress theory.
- The model also includes conservation of energy implied by the first law of thermodynamics, which states that the time rate of change of kinetic energy and internal energy for a fixed body of material is equal to the sum of the rate of work done by the surface and body forces.
- The constitutive equations for the solids and the liquid in the porous medium include also thermal expansion.
- The mass continuity equation for the fluid is combined with the divergence theorem.
- The constitutive behaviour for pore fluid is governed by Darcy's law.
- Vapour flow is modelled as a diffusion process driven by a temperature gradient (coded as UEL user supplied routine with stiffness and flow).
- The heat transfer is uncoupled with the mechanical analysis and the heat conduction is assumed to be governed by the Fourier law.

In ABAQUS the coupling of the thermal and hydro-mechanical solutions is solved through a "staggered solution technique" where the initial thermal analysis is used as input for the initial hydro-mechanical analysis, which in turn affects the thermal properties. In a second thermal analysis the new thermal properties are used and these results used as new input data for the second hydro-mechanical analysis and so on until there is no change in behaviour between two steps.

4 STRUCTURAL MODEL

4.1 Conceptual hydro-geological model

A conceptual model of the rock in the FEBEX area has been generated by Jan-Erik Ludvigson (Ludvigson, 2000). This model is based on different reports (see e.g. Guimerà et al., 1998). Fig. 2 shows the rock structure according to the conceptual model. The area around the FEBEX tunnel is intersected by 5 fractured zones. Two of them are located outside the FEBEX tunnel. Those are the "South shear zone" and the "North shear zone". The other three intersect the FEBEX tunnel. Those are the "Shear and breccia zone" and two "Lamprophyre dykes". Fig. 2 also shows the location of the KWO, GTS, and FEBEX tunnels and four boreholes used to characterise the area (FEBEX and BOUS holes).

Proper limits of the model are the "South and North shear zones", the KWO tunnel and a southwest boundary about 100 m away from the FEBEX tunnel. Since all zones are sub vertical a proper simplification is to model all zones and boundaries vertical.

4.2 Finite element model

With the conceptual model as basis the finite element model was built. The mesh consists of three structures:

- The outer rock (the main structure) with the dimensions $600 \times 150 \times 300 \, m^3$
- The near-field rock (first substructure) with the dimensions $30 \times 30 \times 70 \, m^3$
- The FEBEX tunnel with all installations (second substructure) with the dimensions Ø 2.2 m and length 70 m.

Fig. 3 shows the main structure (entire 3D model) and Fig. 4 shows a horizontal section of the substructures at the test level with the tunnel and installations. A 3D illustration of the fracture zones and the ZEDEX tunnel is shown in Figure 5. The model consists of about 49 790 elements. The elements are 3-dimensional with 8-nodes.

5 MATERIAL MODELS AND BOUNDARY CONDITIONS

5.1 General

The model contains several materials such as the granitic rock matrix, the different fractures or fractured zones, the bentonite buffer, the plug and the canisters. The rock matrix and fracture hydraulic properties were originally derived from the report on FEBEX hydrogeological characterization and modelling (Guimerà et al., 1998) and after calibration calculations only changed for one fracture zone. The thermal and mechanical properties of the rock are taken from the specifications of Task 1A (UPC and ENRESA, 2000) and from own experiences. The bentonite buffer properties were determined in Task 1B and identical properties are used for Task 1C (Börgesson & Hernelind, 2001 and 2003).

5.2 All materials except the buffer

5.2.1 Material properties

All materials except the buffer are modelled as linear elastic materials. In addition the rock and fractures are modelled as porous materials with a hydraulic model governed by Darcy's law. The properties of all

FEBEX TUNNEL - Fracture zones

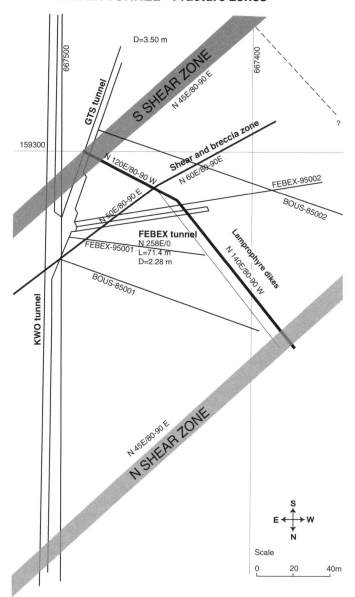

Figure 2. Conceptual model of the rock structure (Ludvigsson, 2000).

materials except the buffer are shown in Table 1, where

K_1 = *hydraulic conductivity in the horizontal direction parallel to the KWO tunnel*
K_2 = hydraulic conductivity in the horizontal direction parallel to the KWO tunnel

K_3 = hydraulic conductivity in the axial direction
E = *Young's modulus*
ν = *Poisson's ratio*
λ = *thermal conductivity*
ρ = *bulk density*
c = *specific heat*
α = *coefficient of thermal expansion*

356

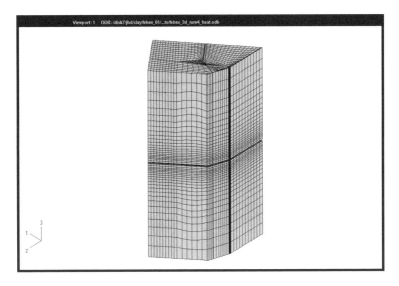

Figure 3.   Element mesh. The KWO and GTS tunnels are the black horizontal band in the centre of the model.

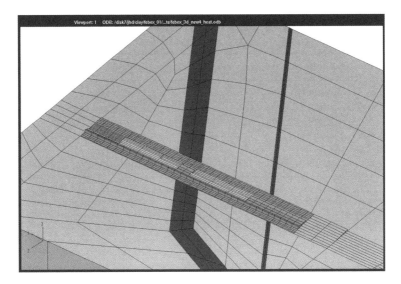

Figure 4.   Horizontal section of the element mesh including the buffer, plug and canisters and the surrounding near field rock. The difference in mesh density between the rock and the second substructure (the buffer etc.) is obvious.

Only the rock and fractures are modelled as porous medias with hydraulic conductivity. The water in the porous model has also elastic properties with an E-modulus and a thermal expansion coefficient.

Five fracture zones with different hydraulic properties are modelled. Table 2 shows the hydraulic properties used in the predictions.

5.2.2   Boundary conditions
The following *hydraulic boundary* conditions were considered suitable (Börgesson & Hernelind, 2000)

and used for the rock:

– Upper horizontal boundary: Constant water pressure of −1350 kPa.
– Lower horizontal boundary: No flow
– Boundary along the "South shear zone": No flow
– Boundary along the "North shear zone": No flow
– Boundary along the KWO and GTS tunnels: No flow
– Southwest boundary: No flow
– Inner boundaries of the open tunnels: Constant water pressure of 0 kPa.

357

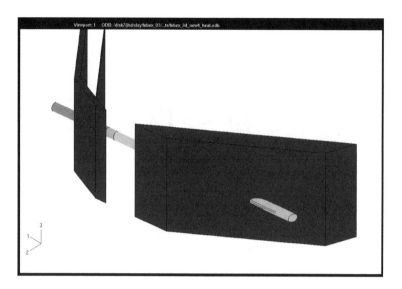

Figure 5. 3D illustration of the fracture zones and the ZEDEX tunnel inside the first substructure.

Table 1. Properties of all materials except the bentonite.

Material	$K_1 \cdot 10^{-12}$ (m/s)	$K_2 \cdot 10^{-12}$ (m/s)	$K_{>3} \cdot 10^{-12}$ (m/s)	E (kPa)	ν	λ (W/m,K)	ρ (kg/m^3)	c (Ws/kg,K)	$\alpha \cdot 10^{-6}$
Rock	4.6	9.2	6.9	$2.5 \cdot 10^7$	0.25	3.6	2400	920	8.3
Fractures	(1)	(1)	(1)	$2.5 \cdot 10^7$	0.25	2.7	2400	920	8.3
Water	–	–	–	$2.1 \cdot 10^6$ (2)	–	–	1000	–	300 (3)
Plug	–	–	–	$2.0 \cdot 10^6$	0.3	1.7	2400	900	8.3
Canisters	–	–	–	$2.0 \cdot 10^6$	0.3	450	7850	460	11.5

(1) See Table 2 (2) bulk modulus; (3) volumetric.

Table 2. Values of width and hydraulic conductivity of the five fracture zones.

Fracture	Width (m)	Hydraulic conductivity (m/s)	Transmissivity (m^2/s)
Shear and breccia zone (F1)	5	$1.38 \cdot 10^{-9}$	$6.9 \cdot 10^{-9}$
Lamprophyre dyke (F2)	1.5	$2.3 \cdot 10^{-10}$	$3.45 \cdot 10^{-10}$
Lamprophyre dyke (F3)	0.25	$2.3 \cdot 10^{-10}$	$5.75 \cdot 10^{-9}$
South shear zone (F4)	5	$2.2 \cdot 10^{-9}$	$1.1 \cdot 10^{-8}$
North shear zone (F5)	5	$2.2 \cdot 10^{-9}$	$1.1 \cdot 10^{-8}$

The *mechanical boundary* conditions were fixed displacements for the top and bottom surface. For the vertical surfaces the in-plane displacements are fixed. All other boundaries inside the structure were either free (inside the drift during excavation) or coupled to the neighbouring materials.

The *thermal* boundary conditions were constant temperature 12°C at all inner and outer boundaries.

5.2.3 Initial conditions

The calculations were made in several steps and the excavation of the ZEDEX tunnel was simulated before the heating and the wetting of the buffer.

The *hydraulic* boundary conditions were applied at the start of the simulated excavation. The excavation was then started with the steady state hydraulic situation of the rock as initial conditions, while the initial conditions of the buffer and the coupling of the bentonite mesh to the rock mesh (simulating the installation) were applied later according to the calculation sequence.

For the *mechanical* part the following stresses were used as initial conditions:

$\sigma_v = 10$ MPa (vertical stress)
$\sigma_H = 40$ MPa (maximum horizontal stress)
$\sigma_h = 20$ MPa (minimum horizontal stress)

The direction of σ_H is perpendicular to the North and South shear zones and the direction of σ_h is thus parallel to those zones.

The initial *temperature* was 12°C in all parts.

5.3 *Bentonite buffer*

5.3.1 *Introduction*

Completely coupled THM-modelling of water-unsaturated bentonite buffer material has been performed in the calculations. The properties of the bentonite have been derived from laboratory tests and calibration and validation calculations that were preformed in Task 1B (Börgesson & Hernelind, 2001; Pintado, Ledesma & Lloret; Villar, 1999).

5.3.2 *Reference material*

The existence of slots between the bentonite blocks makes the average density of the buffer lower than the average density of the blocks.

The basic mean properties of the blocks are the following:

– dry density: $\rho_d = 1.69 \text{ g/cm}^3$ and
– water ratio: $w = 0.144$

which yield (using the density of solids $\rho_s = 2.7 \text{ g/cm}^3$ and the density of water $\rho_w = 1.00 \text{ g/cm}^3$)

– void ratio: $e = 0.60$ and
– degree of saturation: $S_r = 0.65$.

5.53% of the total volume were gaps. About 2.6% of the total volume came from the 30 mm gap in the roof. The rest of the volume is evenly distributed in the buffer. If all gaps are included the average dry density will be $\rho_d = 1.60 \text{ g/cm}^3$. If the gap at the roof is excluded the average dry density will be $\rho_d = 1.64 \text{ g/cm}^3$. The average of those values was chosen yielding the following initial conditions used for the calculations:

– dry density: $\rho_d = 1.62 \text{ g/cm}^3$ and
– water ratio: $w = 0.144$

which yield

– void ratio: $e = 0.67$ and
– degree of saturation: $S_r = 0.58$.

The water ratio at water saturation is for this void ratio $w = 0.248$.

The latter values of the void ratio and degree of saturation have been used both for deriving parameter values for the material model and as initial conditions in the calculations.

5.3.3 *Thermal properties*

The thermal conductivity of FEBEX bentonite has been measured and the following expression suggested /2/.

Table 3. Hydraulic conductivity K as a function of void ratio e.

T (°C)	e	K (m/s)
20	0.47	$0.048 \cdot 10^{-13}$
20	0.57	$0.11 \cdot 10^{-13}$
20	0.69	$0.25 \cdot 10^{-13}$
20	0.82	$0.55 \cdot 10^{-13}$

$$\lambda = 1.28 - 0.71/(1+\exp((S_r - 0.65)/0.1)) \qquad (1)$$

The *specific heat* has been calculated as the weight average of the specific heat of water and particles according to Eqn 2.

$$c = 800/(1+w) + 4200w/(1+w) \qquad (2)$$

5.3.4 *Hydraulic properties*

Hydraulic conductivity

The water flux in the liquid phase is modelled to be governed by Darcy's law. The hydraulic conductivity of FEBEX bentonite has been measured for a large range of densities. Table 3 shows the values at 20°C used in the calculations. The dependency on temperature is adapted to the change in viscosity of water.

The influence of the degree of saturation S_r is governed by the parameter δ in Eqn 3.

$$K_p = (S_r)^\delta K \qquad (3)$$

where

$K_p =$ hydraulic conductivity of partly saturated soil (m/s)
$K =$ hydraulic conductivity of completely saturated soil (m/s)
$\delta = 3$

Thermal vapour flow diffusivity

The water vapour flux is modelled as a diffusion process driven by the temperature gradient and the water vapour pressure gradient (at isothermal conditions) according to Eqn 4:

$$q_v = -D_{Tv} \nabla T - D_{pv} \nabla p_v \qquad (4)$$

where

$q_v =$ vapour flow
$D_{Tv} =$ thermal vapour flow diffusivity
$T =$ temperature
$D_{pv} =$ isothermal vapour flow diffusivity
$p_v =$ vapour pressure

The isothermal vapour flow is neglected and thus $D_{pv} = 0$.

The thermal water vapour diffusivity D_{Tv} can be evaluated from moisture redistribution tests by calibration calculations. The following relations were found to yield acceptable results (Börgesson & Johannesson, 1995):

$$D_{Tv} = D_{Tvb} \qquad 0.3 \le S_r \le 0.7 \qquad (5)$$

$$D_{Tv} = D_{Tvb} \cdot \cos^a\left(\frac{S_r - 0.7}{0.3} \cdot \frac{\pi}{2}\right) \qquad S_r \ge 0.7 \qquad (6)$$

$$D_{Tv} = D_{Tvb} \cdot \sin^b\left(\frac{S_r}{0.3} \cdot \frac{\pi}{2}\right) \qquad S_r \le 0.3 \qquad (7)$$

a and b are factors that regulates the decreased vapour flux at high and low degree of saturation. The thermal vapour flow diffusivity D_{Tvb} and the parameters a and b according to Eqns 4 to 7 have been determined with calibration calculations (Börgesson & Hernelind, 2001).

$D_{Tvb} = 0.4 \cdot 10^{-11}$ m^2/s,K

$a = 6$

$b = 10$

Water retention curve
The water retention properties have been determined both on unconfined and confined samples /2/. One relation was measured on confined samples with the dry density between 1.60 and 1.65 g/cm^3 and fitted to the modified van Genuchten expression in Eqn 8.

$$S_r = S_{r_0} + \left(S_{r_{max}} - S_{r_0}\right)\left[1 + \left(s/P_0\right)^{1/(1-\lambda)}\right]^{-\lambda}\left[1 - s/P_s\right]^{\lambda_s} \quad (8)$$

where S_{r0} and S_{rmax} are the residual and maximum degree of saturation and P_0 (MPa), P_s (MPa), λ and λ_s are material parameters with the following values:

$S_{r0} = 0.01$

$S_{rmax} = 1.00$

$P_0 = 3.5 \cdot 10^4$ (kPa)

$P_s = 4.0 \cdot 10^6$ (kPa)

$\lambda = 0.30$

$\lambda_s = 1.5$

Since the water transport in ABAQUS is governed by the pore water pressure (u_w) but the measurements and requested results are in relative humidity (R_f) a conversion from calculated negative pore water pressure to relative humidity has to be done. The conversion according to Eqn 9, which is derived from thermodynamic considerations, has been used (see e.g. Fredlund and Rahardjo, 1993).

$$R_f = \exp(u_w/135\ 022) \qquad (9)$$

5.3.5 Mechanical properties
The following data has been used for the **Porous Elastic** model (logarithmic relation between the void ratio e and the average effective stress p) (Börgesson, Johannesson, Sandén & Hernelind, 1995):

$$\Delta e = \kappa \Delta \ln p \qquad (10)$$

$\kappa = 0.165$

$v = 0.4$

The value $\kappa = 0.165$ was recalculated from the compression index $C_c = 0.38$ derived from oedometer tests.

The following standard values has been used for the **properties of water and solid phases**:

$B_w = 2.1 \cdot 10^6$ kPa (bulk modulus of water)

$B_s = 2.1 \cdot 10^8$ kPa (bulk modulus of solids)

$\alpha_w = 3.0 \cdot 10^{-4}$ (coefficient of thermal volumetric expansion of water)

$\alpha_s = 0$ (coefficient of thermal expansion of solids)

$\rho_w = 1000$ kg/m^3 (density of water)

$\rho_s = 2700$ kg/m^3 (density of solids)

The **effective stress concept** according to Bishop is used for modelling the mechanical behaviour of the water-unsaturated buffer material, defining the relation between the effective stress p and total stress p_{tot} according to Eqn 11:

$$p = (p_{tot} - u_a) + \chi(u_a - u_w) \qquad (11)$$

Eqn 11 is simplified in the following way:
$u_a = 0$ (no account is taken to the pressure of enclosed air)

$\chi = S_r$

The shortcomings of the effective stress theory can be compensated in ABAQUS by a correction called **moisture swelling procedure**. This procedure changes the volumetric strain ε_v by adding a strain that can be made a function of the degree of saturation S_r:

$$\Delta\varepsilon_V = f(S_r) = \ln(p_0/p) \cdot \kappa/(1+e_0) \qquad (12)$$

$$p = p_{tot} - u_w \cdot S_r \qquad (13)$$

where

ε_V = volumetric strain

p_0 = initial effective stress taken from the initial conditions

p = actual effective stress

κ = porous bulk modulus (from Eqn 10)

e_0 = initial void ratio

p_{tot} = actual total stress

u_w = pore water pressure

S_r = degree of water saturation

The moisture swelling relation (*M.S.*) that is needed as input is the logarithmic volumetric strain according to Eqn 14 where $\Delta\varepsilon_v$ is taken from Eqn 12 (Börgesson & Hernelind, 2001).

$$M.S. = \ln(1 + \Delta\varepsilon_v) \qquad (14)$$

The data for the moisture swelling procedure is derived from the following assumption:

The relation between total stress and suction of a confined sample (constant volume) is linear when suction is decreased to 0. During a decrease in suction from the initial value 99.46 MPa to 0 MPa the total pressure increases from 0 to 7 MPa, which yields to the relation given by Eqn 15.

$$p_{tot} = 7000 + 0.07038 \cdot u_w \qquad (15)$$

The moisture swelling procedure (*M.S.*) is then derived from Eqns 12–14.

5.3.6 Initial conditions

The following initial bentonite conditions were applied:

$e = 0.67$ (void ratio)
$S_r = 0.58$ (degree of saturation)
$u = -99460$ kPa (pore pressure)
$p = 57687$ kPa (average effective stress)

6 CALCULATION SEQUENCE

The calculations were done in the following sequence:

1. Application of initial HM-conditions before excavation of the FEBEX drift.
2. Transient HM calculation of the excavation of the FEBEX drift and the subsequent period of 335 days (before the installation of the buffer).
3. Validation and calibration of the hydraulic model by comparing inflow and water pressure measurements with calculated results.
4. Installation of the buffer, canisters and plug.
5. Thermal calculation covering the first 1000 days of the test from start heating.
6. Application of initial HM-conditions for the buffer.
7. Transient HM calculation including the applied temperatures received from the thermal calculation. Calculation run for 1000 days.

The planned staggered repetition of the thermal and HM-calculations were omitted since it was concluded that change of the temperature in the rock and the subsequent change in HM-response would be insignificant. If the purpose had been to study the buffer in more detail these steps would have been required.

7 RESULTS AND COMPARISONS

7.1 Hydraulic behaviour before installation of the buffer

Two element models were used for the calculations. The validation and calibration calculations of the hydraulic behaviour of the rock were initially made with a rock model that was similar to the one shown in Fig. 3 but without the substructure with the refined mesh of the near field rock. The modelled pore water pressure distribution before excavation and 100 days after start excavation corresponding to 63 days after finished excavation are shown in Fig. 5. The water pressure at the inner part of the intended drift is about 800 kPa all the way from the drift to the north and south shear zones before start excavation while the zone influenced by the excavation with decreased water pressure reaches 40–80 m away from the drift.

The inflow to different parts of the site has been measured and is compared to the calculated inflow in Table 4, where also the measured water pressure in the near field rock is compared to the calculations. The results validated the model in large and the only calibration or changes done after the first prediction (where all values were taken from the literature) were that the transmissivity of the "Shear breccia zone" was reduced with a factor 10, the water table was lowered 120 m from the top of the model and the south-west boundary was moved 100 m further away from the test site.

The calculated water pressure before and after excavation of the FEBEX drift is shown in Fig. 6. In Table 5 the measured water pressure in the borehole that runs parallel to the FEBEX drift at a distance of only 3.8 m from the centre of the FEBEX drift is compared to the calculated pressure. The agreement is good between 50 and 100 m from the entrance but rather poor close to the entrance and far away. It is

Table 4. Comparison between measured and calculated inflow into the tunnels after calibration (after excavation but before installation of the buffer).

Location of measured inflow	Measured inflow (ml/min.)	Calculated inflow (ml/min.)
Inflow from N shear zone into KWO tunnel	62.5	53.4
Inflow from S shear zone into GTS tunnel	23	50.2
Inflow from "Shear breccia zone" into FEBEX after excavation	33.3	38.9
Total inflow from into GTS and KWO between the shear zones after excavation	27.8	72.9

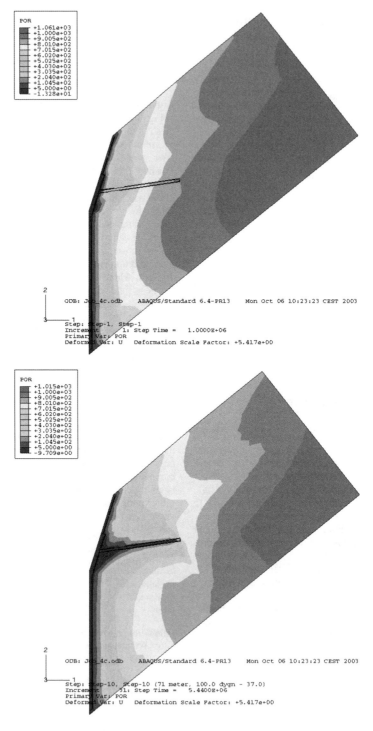

Figure 6.   Modelled water pressure distribution (kPa) in a horizontal section of the model before (upper) and after (lower) excavation of the FEBEX drift.

362

though important that the pressures agree in the location of the FEBEX installation.

After completed calibration the water pressure drop in sections 3 and 4 during excavation and the total inflow into the test section of the FEBEX drift (54–71 m) were predicted. The water pressure was measured during the mining operation in those sections. The results of the measurements and the predictions are shown in Table 6.

The results show that the predicted inflow is well within the range of the measured but also that the predicted water pressure decrease is larger than the measured. This difference is probably caused by a skin zone that seems to appear after excavation. Such a zone is not included in the model.

7.2 Thermo-hydro-mechanical behaviour after installation of the buffer

7.2.1 General

The refined element mesh was used for the calculations predicting the effects of the installation of the buffer and heating of the canisters (step 4–7), but before then the excavation was simulated (steps 1–3) and the results checked. This check showed that the results from the two models were very similar.

Two types of results of the calculation of the installation of the buffer are shown. At first some general results not related to measurements are shown. These results will be shown as contour plots of the water pressure at some interesting times and

locations. Then time-history plots of water pressure at some points were the calculations can be compared to the measurements will be shown. Although the main item of this article is the hydraulic results, examples of temperature and mechanical results will also be presented since the processes are coupled and both temperature and strain may affect the hydraulic results.

7.2.2 Some results of general interest

Some results of the temperature calculations are shown in Fig. 7. The temperature 15 degrees have reached the border of the near field rock (first rock substructure) after 1000 days.

Some contour plots of the calculated water pressure results are shown in Figs. 8 and 9. It is interesting to note that after 100 days there is a negative water pressure at the rock/buffer contact on the rock surface, which derives from the suction of the buffer, while after 300 days the negative water pressure has changed to positive in large parts of the rock surface at the buffer/rock contact due to the inflow of water that cannot be absorbed by the buffer. After 1000 days the minimum water pressure at the rock buffer contact is +250 kPa.

The wetting of the buffer is included in the calculation, since all models are completely coupled and the hydraulic interaction between the buffer and the rock is important for the evolution of the water pressure in the rock. Fig. 10 shows a contour plot of the calculated degree of water saturation at the end of the period (after 1000 days). The figure shows both the result of the coupled calculation with the large rock model described in this article and results from detailed calculations of only the buffer made in task 1B (Börgesson & Hernelind, 2001). The latter calculations were done without considering the interaction with the rock, since water was assumed to be freely available at the rock surface at the water pressure 0 kPa. These results were checked, compared with measurements and found to well represent the real measured behaviour. The results seem to agree rather well although the boundary conditions are completely different and a very large element mesh is used in the 3D calculation.

Table 5. Comparison between measured and calculated water pressure after calibration in borehole FEBEX 95.002 (before excavation).

Measuring section	Measuring interval (meters from entrance)	Measured pressure (kPa)	Calculated average pressure (kPa)
5	23–49	167	620
4	50–61	611	790
3	62–74	759	840
2	75–105	902	880
1	106–132	1568	940

Table 6. Comparison between measured and calculated water pressure in borehole FEBEX 95.002 during the mine by operation and water inflow into the test section of the FEBEX drift after excavation.

Measuring section	Measuring interval (meters from entrance)	Measured pressure/inflow		Predicted pressure/inflow	
		Before excavation	After excavation	Before excavation	After excavation
4	50–61	680 kPa	480 kPa	790 kPa	300 kPa
3	62–74	790 kPa	670 kPa	840 kPa	300 kPa
FEBEX	54–71	4.5–8.5 ml/min		7.0 ml/min	

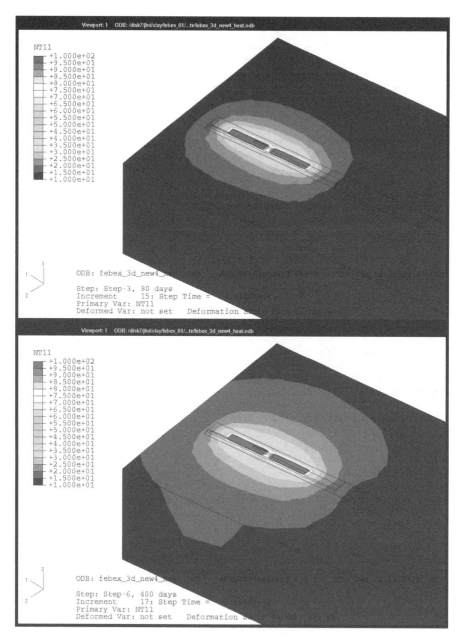

Figure 7. Temperature distribution (°C) in the near field rock and FEBEX tunnel (substructures 1 and 2) after 180 days (upper) and 1000 days.

7.2.3 *Some results that can be compared with measurements*

The evolution of **water pressure** with time in boreholes SF21 and SF23 in different points within 1000 days from start heating is shown in Fig. 11. The calculated values are generally 300–500 kPa lower than the measured ones. The reason is mainly caused by a similar difference in start values. Another difference is that the calculation predicts a decrease in pressure in the beginning (cf. Figs. 8 and 9).

The distribution of water pressure along boreholes SF21 and SF23 at 3 different times is shown in Fig. 12

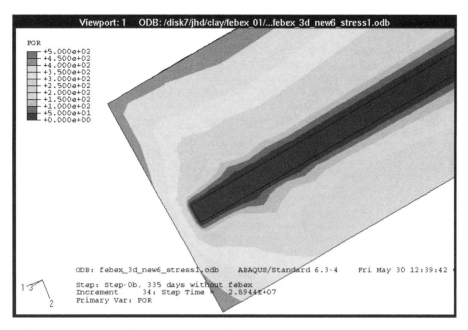

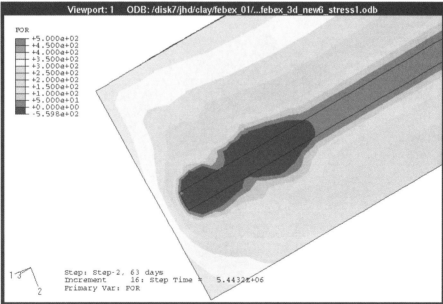

Figure 8. Pore water pressure distribution (kPa) in the near field rock and in the rock surface of the FEBEX tunnel just before installation of the buffer (upper) and 100 days after.

together with measured results. The difference is rather general and mainly caused by a difference in start pressure, while the measured change in pressure is better captured by the calculations.

The calculated evolution of *temperature* with time in boreholes SF21 and SF22 in points 1–4 within

1000 days from start heating is shown in Fig. 13 together with measured temperatures. The agreement between measured and calculated values is very good.

The evolution of the changes of the three normal components of *stress* in borehole SG2 and the

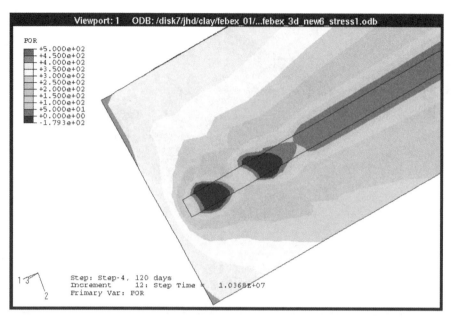

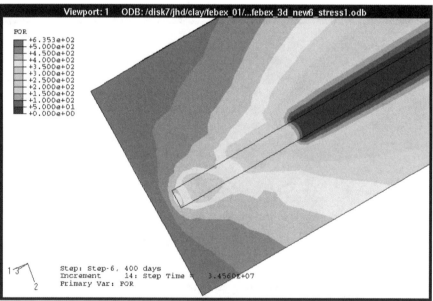

Figure 9. Pore water pressure distribution (kPa) in the near field rock and in the rock surface of the FEBEX tunnel 300 days (upper) and 1000 days after installation of the buffer.

evolution of radial *displacements* in borehole SI1 within 1000 days from start heating is shown in Fig. 14 together with measured results. The range of stress change agrees well but there is not a very good correlation for separate measuring points and the peak stress that occurs within 200 days in some points is not captured by the calculations. The calculated displacements in borehole SI1 are generally a factor 2 larger than the measured ones except for point 1, which agrees very well. These types of stress and displacement measurements are however difficult to perform and the results may be questioned.

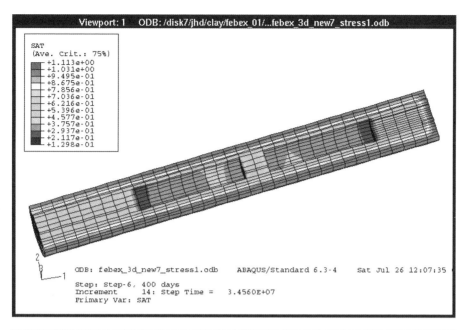

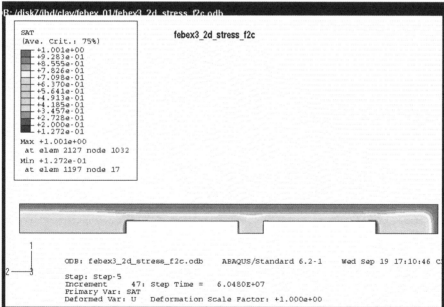

Figure 10. Degree of water saturation of the buffer material after 1000 days according to the calculations with the large-scale 3D model of the entire rock and according to another calculations with a 2D axial symmetric model of only the buffer.

8 COMMENTS AND CONCLUSIONS

The issue has been to model the thermal, hydraulic and mechanical response of the rock to the excavation of the FEBEX tunnel and the installation of the buffer and heating of the canisters. The calculations of the interaction between the near field rock and the buffer material in the FEBEX tunnel were done by sub-structuring of a very large model of the rock. Although the model has a dimension of 600 m the

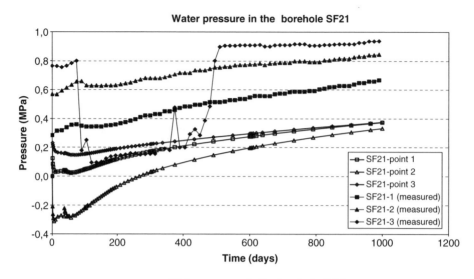

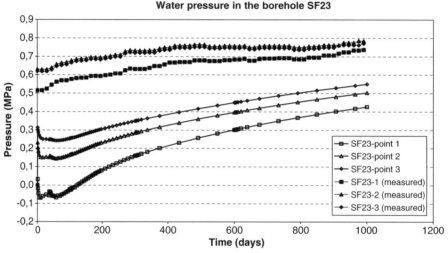

Figure 11. Calculated and measured water pressure in boreholes SF21 and SF23. Open symbols represent calculated results and filled represent measured results.

buffer and near field rock processes could be captured, thanks to the ability of the code ABAQUS to connect structures with different element mesh and element density. A large amount of instruments made it possible to compare the calculations with measurements. The comparison showed that:

– the modelled water pressure in the rock in the FEBEX area before excavation of the FEBEX tunnel and the water inflow into the tunnels agree well with measurements after changing the properties of one fractured zone and reducing the level of the ground water table

– the predicted decrease in water pressure in the near field rock at excavation of the FEBEX tunnel was larger than the measured decrease while the predicted and measured inflow into the FEBEX tunnel agreed well

– the predicted change in water pressure in the near field rock after installation of the buffer is in good agreement with the measurements but the pressure level is lower due to that the initial water pressure before installation of the buffer is only about half the measured ones

– the predicted early decrease in water pressure in the rock caused by the suction of the bentonite is not observed by the measurements

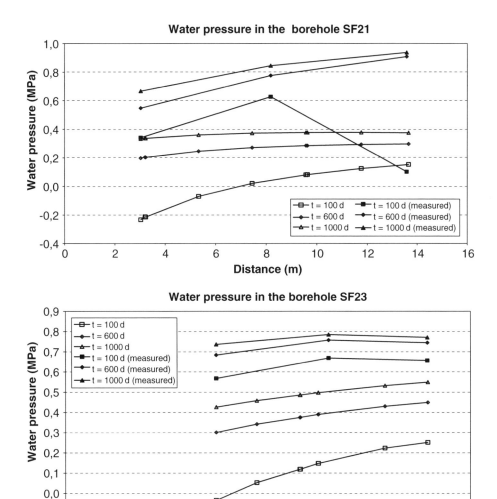

Figure 12.   Calculated and measured water pressure in borehole SF21 as a function of the z-coordinate and in borehole SF23 as a function of the y-coordinate. Open symbols represent calculated results and filled represent measured results.

– the temperature predictions agree well with the measurements
– the range in predicted displacements and change of total stress in the near field rock is in good agreement with the range of measured displacement and stress but the correlation between predictions and measurements in the specific points are poor.

Some of the *hydraulic* predictions after installation of the buffer were not so good although the change in water pressure in the near field rock was well predicted. The main reason for the discrepancy is that the initial water pressure derived after excavation of the

FEBEX tunnel was too low. The conclusion is that there is a skin effect with different properties in the rock boundary around the drift that is not included in the model. That such a skin effect exists has been found in other large-scale tests as the Buffer Mass Test in Stripa (Börgesson et al., 1992) and the Backfill and Plug Test in Äspö HRL (Börgesson & Hernelind, 1999).

Another discrepancy is that the predicted reduction in water pressure in the boreholes of between 50 and 300 kPa during the first months was not observed in the measurements. The reason for the latter discrepancy can be e.g.

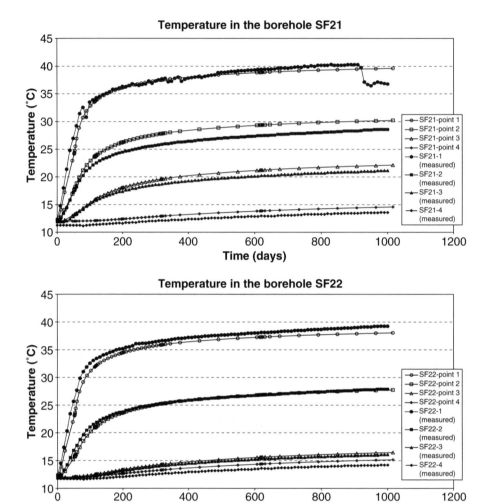

Figure 13.   Calculated and measured temperatures in boreholes SF21 and SF22. Open symbols represent calculated results and filled represent measured results.

- a difference in behaviour of the modelled and real contact between the buffer and the rock. The negative pore pressure in the buffer might not be distributed to the rock if the water flow in the rock mainly takes place in fractures instead of in the rock matrix as modelled
- a delayed rock/buffer interaction caused by the slot between the rock and the buffer
- measurement problems that may be induced by the large water volume and possible air pockets between the packers in the boreholes.

The *temperature* predictions were as expected good. It is generally known that temperature is rather easy to predict.

The results of the *mechanical* predictions are ambiguous. The range in predicted change of total stress in the near field rock is in good agreement with the range of measured stress but the correlation between prediction and measurement in the specific points are poor. There are two possible main reasons for the discrepancies and both are probably part of the explanation:

- The rock model is not accurate enough, since the rock structure is more complicated than the model.
- Measuring stresses and strains in rock is difficult and the results may be unreliable.

370

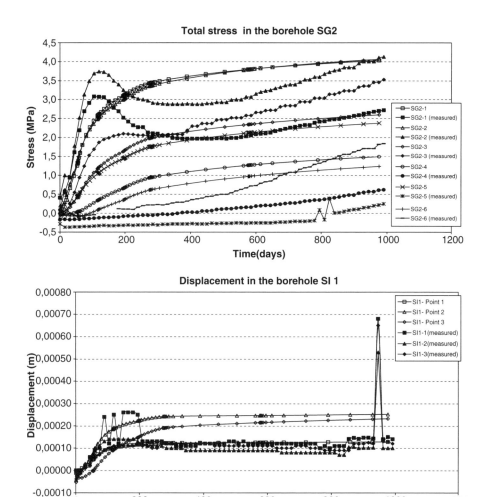

Figure 14.   Calculated and measured change in total stress in borehole SG2 (upper) and displacements in boreholes SI1.
Open symbols represent calculated results.

REFERENCES

ABAQUS manuals. ABAQUS Inc.
Börgesson L., Johannesson L.-E, Sandén T. and Hernelind
    J., 1995. Modelling of the physical behaviour of water
    saturated clay barriers. Laboratory tests, material models
    and finite element application. SKB Technical Report TR
    95-20, SKB, Stockholm.
Börgesson L. and Johannesson L.-E. 1995. Thermo-Hydro-
    Mechanical modelling of water unsaturated buffer
    materials. Status 1995. SKB Arbetsrapport AR 95-32.
Börgesson L. and Hernelind J. 1999. Preparatory modelling
    for the backfill and plug test. Scoping calculations of
    H-M processes. SKB ÄHRL IPR-99-11.

Börgesson L. and Hernelind J. 2000. DECOVALEX III,
    TASK 1. Modelling of FEBEX in-situ test. PART A:
    HYDRO-MECHANICAL   MODELLING   OF   THE
    ROCK.
Börgesson L. and Hernelind J. 2001. DECOVALEX III,
    TASK 1B. FEBEX in-situ test. THERMO-HYDRO-
    MECHANICAL MODELLING OF THE BUFFER
    MATERIAL.
Börgesson L. and Hernelind J. 2003. DECOVALEX III,
    TASK 1. FEBEX in-situ test. PART C: THERMO-
    HYDRO-MECHANICAL ANALYSES OF THE ROCK.
Börgesson L., Pusch R., Fredriksson A., Hökmark H.,
    Karnland O. and Sandén T. 1992. Final Report of the
    Rock Sealing Project – Identification of Zones Disturbed

by Blasting and Stress Release. SKB Stripa Project TR 92-21.

DECOVALEX III, Task 1: Modelling of FEBEX in-situ test. PART B: THERMO-HYDRO-MECHANICAL ANALYSIS OF THE BENTONITE BEHAVIOUR.

FEBEX. Full scale engineered barriers experiment in crystalline host rock. Pre-operational stage. Summary report. ENRESA publicacion tecnica num. 01/98.

Fredlund D.G. and Rahardjo H. 1993. Soil Mechanics for unsaturated soils. John Wiley & Sons Inc., New York.

Guimerà J., Carrera J., MartÚnez L., Vázquez E., Fierz T., Buhler C., Vives L., Meier P., Medina A., Saaltink M., Ruiz B. and Pardillo J. 1998: FEBEX Hydrogeological characterization and modelling. UPC, 70-UPC-M-0-1001, Jan. 1998.

Ludvigson J.-E. 2000. Conceptualisation of geological features at the FEBEX site for groundwater flow modelling

Pintado X. Ledesma A. and Lloret A. Backanalysis of thermohydraulic bentonite properties from laboratory tests. UPC.

UPC and ENRESA, 2000. DECOVALEX III, Task 1: Modelling of FEBEX in-situ test. PART A: HYDRO-MECHANICAL MODELLING OF THE ROCK (Rev. 3).

UPC and ENRESA, 2003. DECOVALEX III, Task 1: Modelling of FEBEX in-situ test. PART C: THERMO-HYDRO-MECHANICAL ANAYSIS OF THE ROCK (Rev. 2).

Villar M.V. 1999. Ensayos THM para el proyecto FEBEX. CIEMAT report 70-IMA-L-0-66.

Advances in Understanding Engineered Clay Barriers – Alonso & Ledesma (eds)
© *2005 Taylor & Francis Group, London, ISBN 04 1536 544 9*

Temperature Buffer Test – comparison of modelling results/experimental findings: causes of differences

H. Hökmark

Clay Technology AB, Ideon Research Centre, Lund, Sweden

ABSTRACT: The French program for research on disposal of Nuclear Waste includes the option of disposal of exothermic waste in different rock-types. The TBT experiment which recently has been set up and initialized in a vertical 8 m deep disposal pit at 420 m depth in the Äspö HRL, southern Sweden, is a part of the granitic rock research. The TBT project is owned by ANDRA and carried out with SKB as organizing partner, taking advantage of SKB's experience with field experiments in hard rocks and of the existing infrastructure in the Äspö HRL. The main and overall test objective is to improve the understanding of the Thermo-Hydro-Mechanical properties and behaviour of bentonite-based buffer materials, in particular in the high temperature range, above 100°C. This paper gives an overview of the test, the aims of the test and of the considerations that have led to the present test design.

A number of modelling teams have made blind predictions of the Thermo-Hydro-Mechanical evolution of the barriers. Here comparisons are made between a few different predictions, selected from a large number of results. In particular predictions of the scope and extent of drying in the innermost, hottest parts of the buffer appear to differ considerably between the teams. Comparisons are also made between modelling results and experimental results that have been recorded after the test start in late March 2003. Differences are highlighted and attempts are made to identify the causes of these differences and to evaluate their significance.

1 INTRODUCTION

1.1 *General*

For ANDRA's program for geological disposal of nuclear waste, there is an interest in extending the experimental work to include field-scale testing of bentonite-based buffer materials under high temperature conditions. ANDRA has recently started an experiment in the Äspö Hard Rock Laboratory to meet that interest (Thorsager, 2002). The experiment, the Temperature Buffer Test (TBT), is being conducted with SKB as organizing partner with the intention to make as good use as possible of the existing infrastructure in the Äspö HRL and of the experience that SKB has accumulated over the years in the area of large-scale field tests at large depths.

1.2 *Test objectives*

The overall objective of the TBT experiment is to improve the understanding of the Thermo-Hydro-Mechanical properties and behaviour of bentonite-based buffer materials, in particular in the high temperature range above 100°C. A modelling program has been initiated for this purpose. Knowledge to be acquired from TBT could help in designing

solutions for vitrified waste packages, which generate intense heat but for a relatively short time, giving a short intense thermal pulse with high temperatures.

Effects of such high temperatures on the buffer confinement properties will be analyzed by sampling and laboratory testing after completion of the test. During the test, physical quantities like temperature, saturation etc., need to be monitored in order to establish the THM histories of the sampled specimens. Instrumentation schemes (Sandén et al., 2002), data acquisition schemes (Karlsson, 2002) and data management schemes (Börgesson, 2002) have been elaborated for this purpose.

A secondary objective is to test the option of a sand shield between canister and bentonite buffer. A sand shield will increase the temperature at the canister surface and protect the surrounding bentonite in decreasing its maximum temperature. Sand will also facilitate canister retrieval.

2 TEST DESCRIPTION

2.1 *Location and background conditions*

An existing KBS3-type deposition hole was selected for the experiment. The hole has a diameter of about

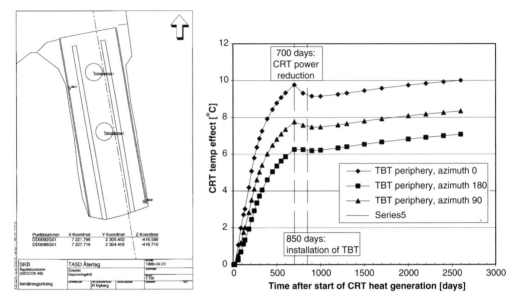

Figure 1. Left: location of TBT pit (DD0086G01) about 6 m from the CRT pit. Right: estimate of CRT effects at the periphery of the TBT pit as function of time elapsed after the CRT start.

1750 mm and depth of a little more than 8 m, and is located in the floor of an experimental room called the CRT tunnel at 420 m depth below ground surface. Figure 1, left part, shows the CRT tunnel and the test pit. It is at a distance of about 6 m from the Canister Retrieval Test (CRT) pit, meaning that the heat generated in the CRT one influences the TBT experiment. The right part shows a simple estimate of that thermal influence. At the time of the TBT installation, the CRT experiment will have increased the temperature at the periphery of the TBT hole by about 8 degrees. During the following years, the changes caused by the CRT experiment will be minor. Temperature measurements made in the rock surrounding the TBT hole confirm that there is a general increase of about 10°C above the general background temperature.

2.2 Preliminary design considerations

It was early decided as a primary design guideline to accommodate two parallel experiments in the available pit and to keep the experimental conditions as similar as possible for both sub-experiments. One should address the pure bentonite buffer option and the other experiment the sand-shield option. The two sub-experiments should be installed on top of each other and be powered and instrumented together as one experiment. The heater diameter should be about 0.6 m. To extend the portion at experiment mid-height where the thermal flow is approximately radial

(the thermal plateau), an auxiliary heater rod was planned to be installed below the existing hole.

To achieve the experimental goals it was postulated that the temperature should reach about 100°C at a radial distance of 0.2 m from the surface of the heaters.

Preliminary design modelling based on these guidelines showed, in short, the following (Hökmark, 2001; Börgesson and Hernelind, 2002; Hökmark, 2002):

- The power should be between 400 and 500 W per meter of heater length.
- If the two sub-experiments are powered by one common, continuous heat source, there will be an axial heat transport from the poorly cooled sand-shield section to the better-cooled section surrounded by pure bentonite.
- High radial temperature gradients may promote the formation of a saturation front, in particular around the lower heater, such that the innermost parts of the buffer may dehydrate and remain dry for very long periods of time.

2.3 Design

Figure 2 shows the design of the experiment as it is in fact being run at present. The heater rod extension has been omitted such that the experiment is now completely confined in the existing 8 m deep hole. The two heaters are separated by a 0.5 m high bentonite block in order to limit the axial heat transfer between the two sections. The two heaters are

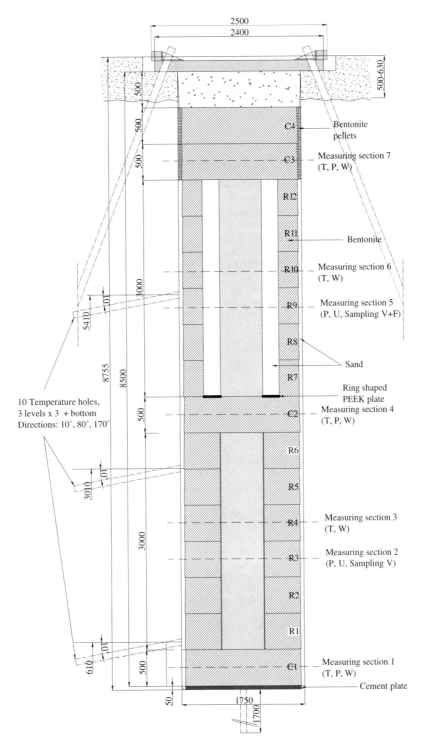

Figure 2. Experiment layout.

powered individually but, at present (October 2003), with the same power (1500 W each). The upper section is the composite barrier part, i.e. with a sand-shield between heater and bentonite buffer.

There is a 0.05 m sand filter in the annular space between the stack of bentonite blocks and the walls of the deposition hole, except for in the uppermost parts, where there is a bentonite seal, applied in the form of a pellet filling. The function of the sand filter is to provide a controlled and uniform hydraulic boundary, which could be an atmospheric groundwater pressure or, if necessary to accelerate the hydration, an elevated ground water pressure. The function of the bentonite seal is to confine the water contained in the filter and make it possible to increase that pressure.

The bentonite has been installed as pre-compacted 0.5 m high blocks with an initial moisture content of 17.5%. The blocks are ring-shaped (R1 through R12) or cylinder-shaped (C1 through C4). There is an air-filled, about 10 mm wide, gap between blocks and heater in the lower section.

To provide mechanical confinement there is a concrete plug and a steel lid on top of the package. The steel lid is anchored to the rock by nine cables distributed around the periphery of the hole (Bäck, 2002).

2.4 Instrumentation

The instrumentation of the experiment is shown in Fig. 2. A detailed description is given by Sandén and others in a paper submitted to the Sitges workshop on large-scale field experiments (Sandén et al., 2003). In particular the sections at mid-height of the two heaters, i.e. Section 3 and Section 6, are densely instrumented.

2.5 TBT running conditions

At present (October 2003), both heaters are operating at their nominal working power (1500 W each). The outer sand filter has been filled up with water and the saturation process is in progress. Sensor readings have been collected for about 200 days and are being regularly reported in sensor data reports (Goudarzi and Börgesson, 2003).

3 MODELLING

3.1 General

A first set of calculations were conducted in order to explore the general feasibility of the test and to help in the design process. The design modelling phase included three studies:

– Study 1. Analytical 1-D thermal analyses aimed at finding relations between power and temperature

requirements. Separate analyses were made for the two barrier options (pure bentonite and sand-shield).
– Study 2. Numerical 2-D axis-symmetric thermal analyses aimed at examining the effects of running the two sub-experiments in one common experiment pit.
– Study 3. Numerical 1-D hydrothermal analyses aimed at getting a first estimate of the hydration time-scale. Separate analyses were made for the two barrier options (pure bentonite and sand shield).

These calculations have been described in separate reports (Hökmark, 2001; Börgesson and Hernelind, 2002; Hökmark, 2000) and will not be detailed or referenced further in this paper other than the general statements made above in the preliminary design consideration section (Section 2.2).

A predictive modelling phase was initiated in late 2002 after finalizing the design process, and was completed by late May 2003. The outcome of blind predictions made within the predictive modelling phase by a small number of modelling groups is the main issue of this paper.

After some time of heat generation, when systematic trends of the measured data can be discerned, a phase of evaluation modelling is intended to follow. The objective of the evaluation modelling phase will not merely be to recalibrate the model parameters to achieve a better fit with the experimental findings, but more to explore the validity of the conceptual models used in the predictive modelling phase.

3.2 Predictive modelling program

For the predictive modelling phase, a program was issued in November 2002. The program specified the geological environment, the geometry of the experiment, material properties and the thermal load conditions (Hökmark and Fälth, 2003).

Predictions of a number of variables (temperature, saturation, pressure, RH etc.) were asked at a number of points and for a number of vertical and horizontal scan-lines, as shown in Fig. 3.

It was left to the teams that were invited to participate to decide upon the level effort. The teams could leave out, for instance, mechanical aspects or some of the specified scan-lines at their own discretion. However, a least common denominator should be prediction of temperatures and saturation on the horizontal scan-lines A1–A4 and B1–B4. The thermo-hydraulic evolution on these two scan-lines is an issue of paramount interest here and captures the essence of the problems being addressed in the TBT experiment.

Predictions were asked for a number of times after the test start: 10, 30, 60, 90, 120, 270, 500, 750, 1000 and 1200 days.

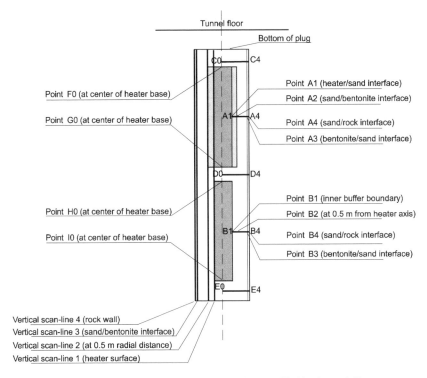

Figure 3.   Schematics of experiment layout with points and scan-lines specified in the modelling program.

3.3   *Modelling activities*

A number of modelling groups gave predictions of the THM evolution at the points and scan-lines described in the predictive modelling program. These were:

– EDF team, organized by ANDRA, using code ASTER
– EuroGeomat team, organized by ANDRA, using code Cleo
– UPC/DM Iberia team, organized by ENRESA, using Code_Bright
– Clay Technology team, organized by SKB, using code ABAQUS and Code_Bright.

This means that there are now five sets of results altogether. All simulations were performed with quasi 3-D axi-symmetric model geometries, which included the deposition hole as well as a substantial portion of the surrounding rock, except for the ClayTech Code_Bright simulations. These were performed with 1-D radial models, each for one of the horizontal mid-sections of the two heaters. However, the quasi 3-D aspects of the problem were indirectly accounted for by using ABAQUS results to define time-dependent thermal boundary conditions.

All analyses except the Code_Bright analyses (ENRESA and Clay Technology) addressed all issues

(T, H and M). The Code_Bright models addressed T-H issues only.

The format of the specifications given in the modelling program regarding material data, geological setting etc. is not that of clear manual for a benchmark exercise. Nor does it warrant that the information is complete, sufficient, or fully relevant for all aspects of the problem. It was a question for the teams to make all necessary idealizations and simplifications.

4   RESULTS AND RESULTS COMPARISONS

In this section, a number of selected results are compiled. As far as possible, the results are presented such that reasonably fair comparisons can be made. (All teams did not provide predictions exactly for the times and for the scan-lines specified in the program). The full sets of results will be found in the predictive modelling report.

The focus here is on the T-H behaviour, in particular that of the densely instrumented regions around the two heater mid-sections, i.e. at the scan-lines A1–A4 and B1–B4 (c.f. Fig. 3).

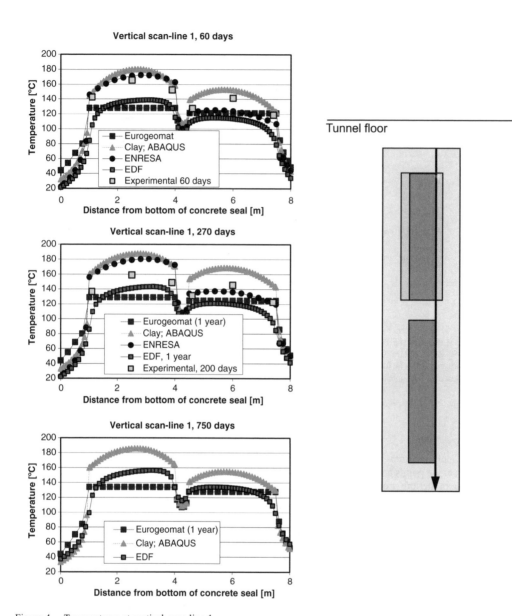

Figure 4. Temperatures at vertical scan-line 1.

4.1 Temperatures

4.1.1 Large scale – vertical scan-lines

Figure 4 shows temperatures at the vertical scan-line along the heater surfaces. The experimental values are those provided by the thermocouples built in to the surface of the heaters. The ENRESA model was not analyzed for more than 1 year.

The effect of the poor cooling provided by the sand-shield is evident. At the beginning (after 60 days), the surface of the upper heater is more than 20 degrees hotter than that of the lower heater according to the experimental results and to most modelling results.

It is interesting to note that the difference in temperature between the two sections is reduced with time (according to the measurements from about 23°C after 60 days to about 14°C after 200 days) and that this is reproduced qualitatively by ENRE-SAS's model and the ClayTech ABAQUS model. The reason is probably that these models predict some

378

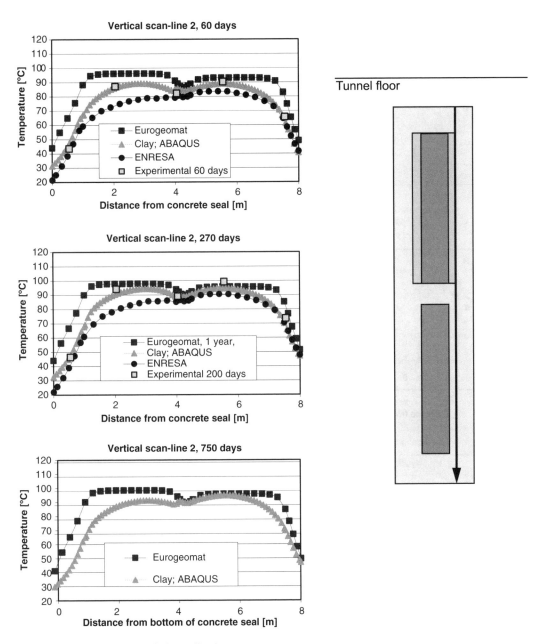

Figure 5. Temperatures at vertical scan-line 2.

desaturation around the lower heater, which reduces the heat flow from the upper to the lower heater. In the ABAQUS model, that difference has increased again after 750 days, which is consistent with the resaturation of the lower section predicted by that model.

For models that did not predict any desaturation (EDF, Eurogeomat) the difference between the two sections is approximately constant over time.

Quantitatively there are considerable differences. Most of these differences are due to the way the interaction with the tunnel floor was represented in the axi-symmetric models, and the way the heat generated by the CRT experiment was accounted for.

Figure 5 shows the temperature at about 0.2 m radial distance from the heater surfaces. The measured temperatures here are a few degrees

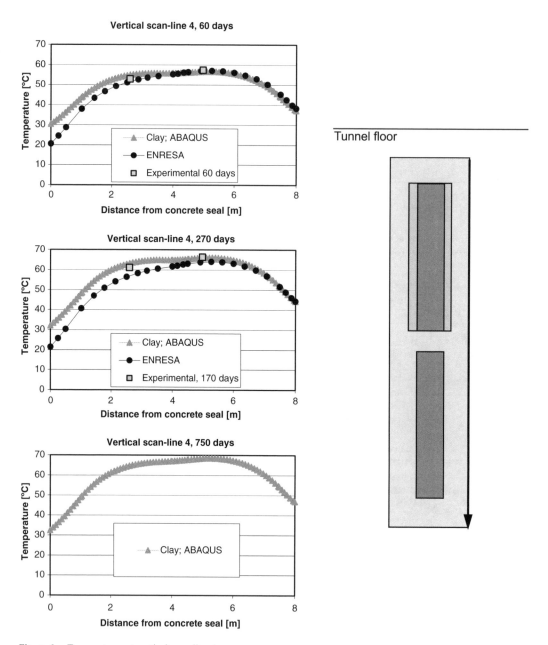

Figure 6. Temperatures at vertical scan-line 4.

below 100°C. This may mean that the power of upper heater will have to be increased later. Otherwise, after completion of the experiment, this part of the bentonite will not have been exposed to as high temperatures as intended.

Figure 6 shows the temperature at the rock wall. As opposed to the conditions close to the heater surfaces

(Fig. 4), here the lower section is hotter. There are two reasons for this:

– Because of the poor cooling of the upper heater, there is heat flux from the upper heater to the lower one. This means that the radial heat flux across the space between canister surface and rock wall is

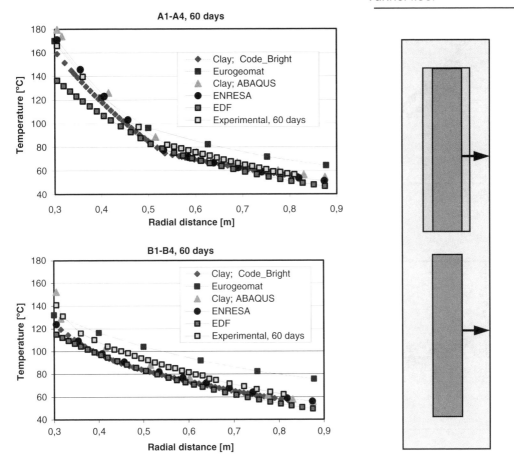

Figure 7. Temperatures after 60 days at horizontal scan-lines A1–A4 and B1–B4.

higher in the lower section. This gives higher rock wall temperatures.
– The rock surrounding the upper heater is closer to the cooled tunnel floor.

4.1.2 *Deposition hole scale – horizontal scan lines at heater mid-height*

Figure 7 shows the temperature at the horizontal scan-lines at the mid-section of the two heaters after 60 days. For the upper heater, the difference in thermal conductivity between the bentonite and the sand-shield gives a clear difference in slope. This is clear from the experimental results and is nicely reproduced by all models. The dense arrays of thermocouples in the two sections do in fact allow for qualitative estimates of the scope and extent of bentonite dehydration, for instance close to the surface of the lower heater (Sandén et al., 2003).

Figure 8 shows the temperature at the horizontal scan-lines at the mid-section of the two heaters after 270 days. The results do not differ much from those shown in the previous figure (Fig. 7).

Note that the ABAQUS model gives a distinct slope excess close to the lower heater. This is because the gap between heater and bentonite was explicitly accounted for in the ABAQUS model.

In the EDF model, the sand-shield slope is less steep than in the other models (see also Fig. 7 and Fig. 9). This is because the sand-shield material was given higher heat conductivity than suggested in the modelling program. (New EDF calculations, recently performed with the same sand-shield heat conductivity, show a good agreement with the thermal results of the other teams).

Figure 9 shows the temperature at the horizontal scan-lines at the mid-section of the two heaters

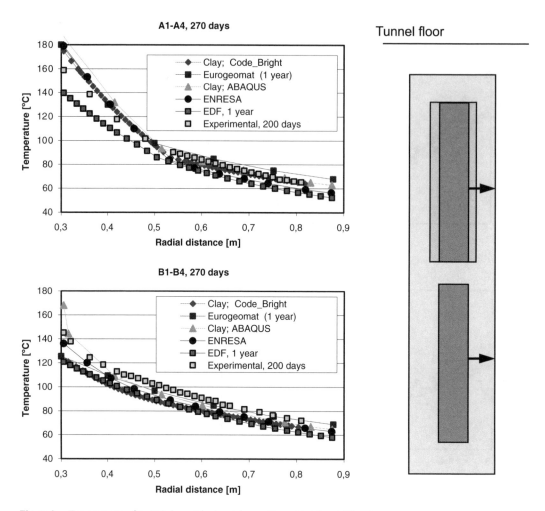

A1-A4, 270 days

Temperature [°C] vs. Radial distance [m]

- Clay; Code_Bright
- Eurogeomat (1 year)
- Clay; ABAQUS
- ENRESA
- EDF, 1 year
- Experimental, 200 days

B1-B4, 270 days

Temperature [°C] vs. Radial distance [m]

- Clay; Code_Bright
- Eurogeomat (1 year)
- Clay; ABAQUS
- ENRESA
- EDF, 1 year
- Experimental, 200 days

Tunnel floor

Figure 8. Temperatures after 270 days at horizontal scan-lines A1–A4 and B1–B4.

after 750 days. The results do not differ much from those shown in the previous figures (Fig. 7 and Fig. 8).

Around the lower heater, the ClayTech Code_Bright 1-D model shows a slightly steeper slope some 0.15 m from the heater surface. (See also Figs 7 and 8). This is an effect of the dehydration predicted to take place here in this model. The ABAQUS model shows the same effect, however more pronounced after 270 days (Fig. 8) than after 750 days, when some resaturation has taken place.

4.2 Relative humidity

Figure 10 shows the relative humidity in the two horizontal heater mid-height sections after 60 days.

All RH sensors were then still in working order. The two Code_Bright models seem to overpredict the humidity in the upper section. For the 1-D ClayTech model, part of the explanation is that the sand shield was in fact drier than assumed (and specified in the modelling program). In the lower section, the ABAQUS model and both Code_Bright models reproduce the experimentally found RH reduction close to the heater reasonably well, at least qualitatively.

Figure 11 shows the relative humidity after 270 days. Unfortunately, most signals from the lower section have disappeared. In the upper section, the ABAQUS model and the 1-D Code_Bright model seem to match the experimental results. For the Code_Bright model, this may be more of a lucky

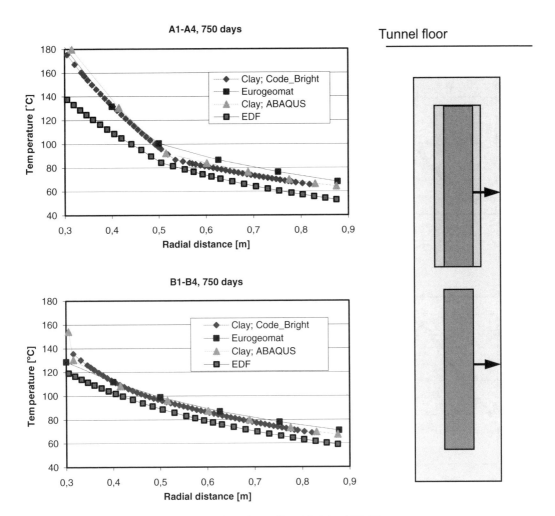

Figure 9.   Temperatures after 750 days at horizontal scan-lines A1–A4 and B1–B4.

coincidence, since part of the effect (i.e. the increasing RH) is due to an overestimation of the amount of water that evaporates in the sand–shield and diffuses into the bentonite. (In reality, the sand–shield is drier than assumed and specified in the modelling program.)

4.3   Degree of saturation

Figure 12 shows the degree of saturation at the two horizontal heater mid-height sections after 60 days. The ABAQUS model and the ENRESA Code_Bright model did not account explicitly for the hydraulic behaviour of the sand-shield. The ClayTech Code_Bright model overestimates the saturation of the bentonite in the upper section, because of the water assumed to be contained initially in that shield (c.f.

RH plots in the previous section). In the lower section, there is a significant difference between the models. The ABAQUS model and both Code_Bright models predict vapour movement from the hottest parts to colder parts, while the other models do not.

Figure 13 shows the degree of saturation at the two horizontal heater mid-height sections after 270 days. The general picture is very much the same as in the previous figure (Fig. 11). The ABACUS model and the ENRESA Code_Bright model predict a modest desaturation close to the sand shield. In the lower section, the ABAQUS model and both Code_Bright models predict a very clear dehydrated region, while the EDF and Eurogeomat models show a gentle slope with a slow increase of the saturation at all radial

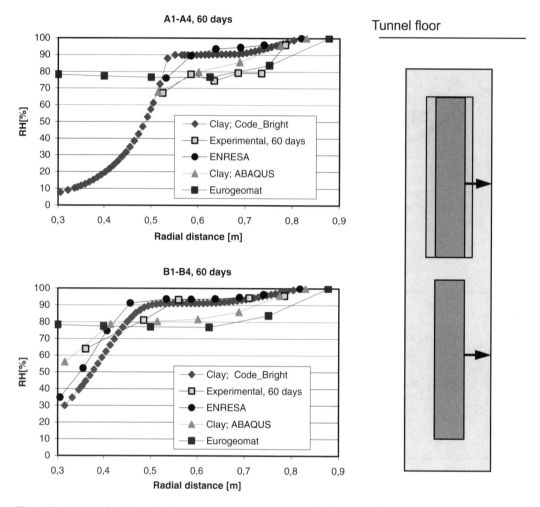

Figure 10. Relative humidity after 60 days at horizontal scan-lines A1–A4 and B1–B4.

distances. In the EDF model, desaturation did not take place because the pore gas was assumed to be confined. Had the gas been allowed to escape, then the EDF model would, as confirmed by recent simulations, have predicted desaturation similar to the way it was predicted here by, for instance, the Code_Bright models.

Figure 14 shows the degree of saturation at the two horizontal heater mid-height sections after 750 days. For the upper section, the ClayTech models predict the saturation to be completed, while the EDF and Eurogeomat models show a slower process with gentle saturation gradients.

For the lower section, the ClayTech Code_Bright model predicts very little changes to have taken place over the last year (c.f. Fig. 13), while the ABAQUS

model predicts a slight resaturation during that time also in the hottest parts. This means that the Code_Bright model predicts the development of a quasi-stationary state with a stable saturation front.

4.4 Deformations

The focus here is on the saturation process. Mechanical processes will influence the saturation. They were included by the ClayTech ABAQUS team, the EDF team and the Eurogeomat team. None of the teams using Code_Bright included mechanical aspects. A full account of the mechanical aspects will be given in the predictive modelling report. Comparisons and a discussion will have to be presented in

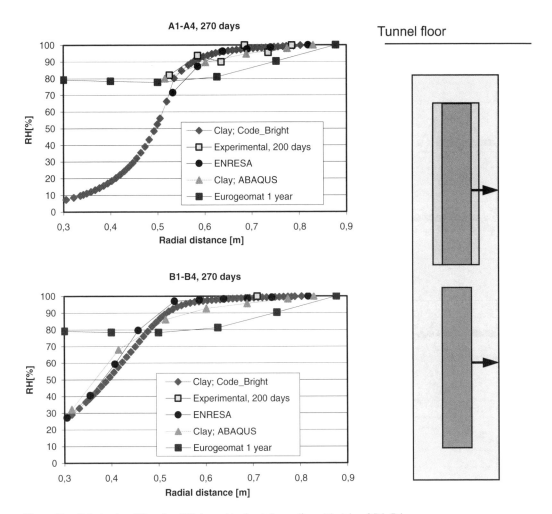

Figure 11.   Relative humidity after 270 days at horizontal scan-lines A1–A4 and B1–B4.

following papers. However, as an example of mechanical results, the radial displacements along the horizontal scan-line B1–B4 through the lower section are shown in Fig. 15.

In all results we see that the deformations are negative, i.e. directed towards the heater. The deformations are caused by swelling that begins at the interface between sand filter and bentonite. There is no account of any compression of the sand filter.

In the ABAQUS model, a contact element was used to represent the gap that will exist between heater and bentonite blocks initially. After 60 days, that gap has closed almost completely, meaning that the ABAQUS model predicts a net increase in porosity. At the hot side, the porosity decreases but

not sufficiently to balance the effects of swelling in the wet and colder parts.

In the EDF and Eurogeomat models, the bentonite-canister interface was fixed without gaps such that all deformations found in the bentonite correspond to a redistribution of the initial porosity.

The comparison shows that the gap effect, i.e. the volume expansion, makes out a substantial part of the radial bentonite deformations. Therefore, it would be of interest to see if there are any signs of gap closure in the experimental data. The thermocouple readings for the lower section (Fig. 7 and Fig. 8) show that a slight decrease of the temperature offset between heater surface and bentonite may have taken place between 60 and 200 days. This may indicate that the

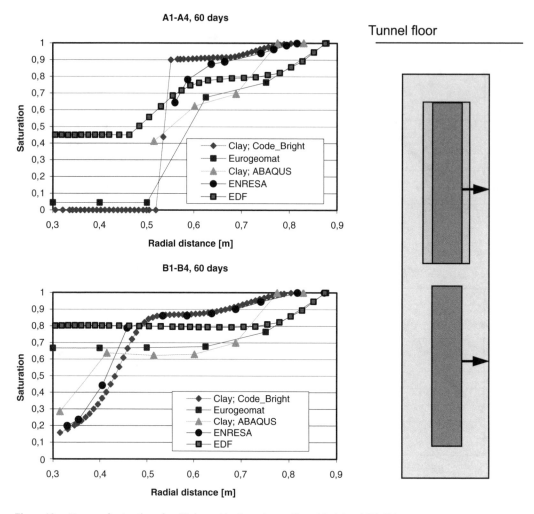

Figure 12. Degree of saturation after 60 days at horizontal scan-lines A1–A4 and B1–B4.

gap is closing, however not as fast as predicted by the ABAQUS model.

The measured radial stresses on the B1–B4 scan-line amount to about 3 or 4 MPa after 200 days. This level of stress is also predicted by the ABAQUS model and the Eurogeomat model, while the EDF model predicts a little more, about 10 MPa.

5 DISCUSSION

At present (October 2003), we have experimental results for only about 200 days after test start. The experimental results seem, however, to support the following preliminary conclusions (Sandén et al., 2003):

– Soon after the test start a dehydrated, about 0.15 m wide, zone has formed around the lower heater.
– After formation of the dehydrated zone, all following processes are slow. According to RH measurements, there may be signs of slow saturation of the bentonite surrounding the sand shield, but this is uncertain. According to interpretations of the temperature gradients, changes in the hydration status are almost zero.

The possibility must be considered that the system (or parts of the system, i.e. the lower bentonite

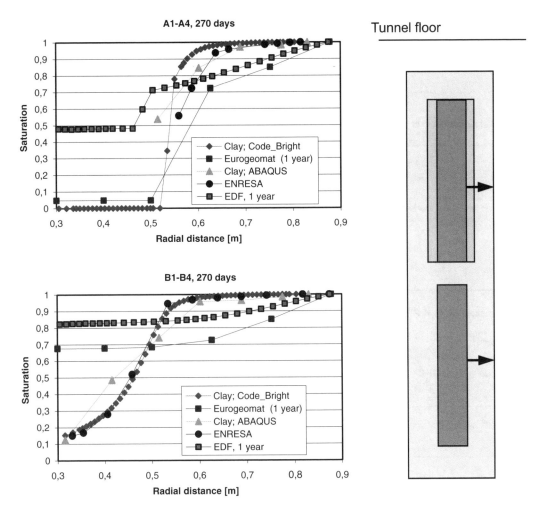

Figure 13. Degree of saturation after 270 days at horizontal scan-lines A1–A4 and B1–B4.

barrier) has reached a quasi-stationary state. This is consistent with the following conceptual description:

– Water is being evaporated in the hot parts and moves by diffusion towards colder parts.
– The vapour transport is efficient enough to balance the flow of liquid water in the opposite direction.
– The distance to the groundwater pressure boundary is sufficiently large that the supply of liquid water to the front region is too slow to prevent the front from forming.

The above will promote formation of a stable saturation front.

Some of the models did predict such a saturation front, at least qualitatively. This was the case for both Code_Bright models, where no signs of resaturation

of the hottest parts were found within the simulated time range (c.f. Fig. 13 and Fig. 14). It was to some extent the case also for the ABAQUS model. The ABAQUS model did however predict some resaturation after 750 days.

The Cleo model and the Aster model did not predict any desaturation at all.

If the experimental results are representative of the evolution of the system, and not caused by equipment errors and artefacts, then it seems to be clear that there is indeed evaporation and vapour transport. It may yet be too early to draw conclusions regarding the general relevance and validity of the models used to represent vapour transport in the different codes:

– For instance, if the 1-D Code_Bright model of the B1–B4 scan-line had been defined such that the

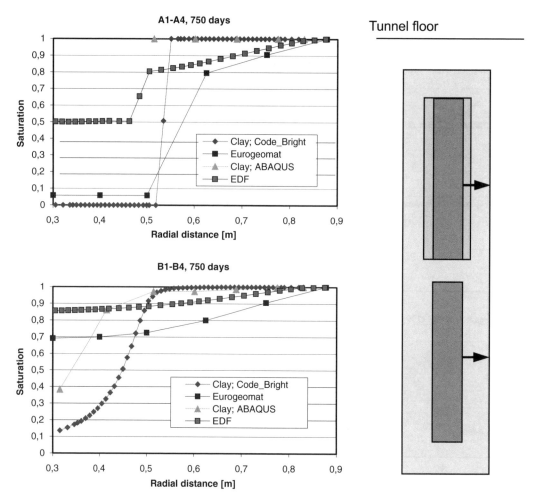

Figure 14.   Degree of saturation after 750 days at horizontal scan-lines A1–A4 and B1–B4.

pore gas (vapour and dry air) were completely confined, then the gas pressure would have choked the vapour transport and inhibited most of the predicted drying. This would have given results qualitatively similar to those obtained here by code ASTER and code Cleo.

– The ABAQUS code does not represent the vapour phase explicitly at all. The agreement between prediction and experiment is largely a question of experienced use of a pragmatic (and successful) approach of representing effects of temperature gradients on the moisture distribution.

For the vapour transport in Code_Bright, the transport coefficient depends explicitly on temperature and saturation. The driving potential is the vapour mass fraction. Development of high pore gas pressures will tend to even out vapour mass fraction gradients. In the EDF model, for instance, the transport coefficient is a constant Fick's diffusivity, but the driving potential is the same as in Code_Bright, i.e. the vapour mass fraction. In the EDF model gas pressures were high (giving modest mass fraction gradients), whereas they were much lower in the Code_Bright models. The example points to the possibility that different assumptions regarding gas escape or gas confinement may be part of the explanation for the differences found between the results here. The gas escape issue is consequently an area that needs more attention in the future, in particular for systems where high temperatures must be expected.

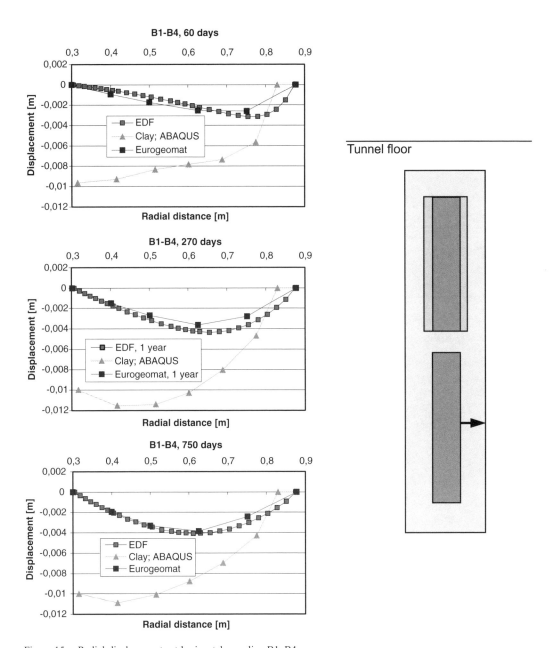

Figure 15. Radial displacements at horizontal scan-line B1–B4.

REFERENCES

Bäck R. (2002). Temperature Buffer Test. Report for retaining plug and anchoring. SKB International Progress Report IPR-02-64

Börgesson L. (2002). Temperature Buffer Test. Data management. SKB International Progress Report IPR-03-11

Börgesson L., Hernelind J. (2002). Test of vitrified wastes. Study 2 – Temperature field. Clay Technology AB, Lund, Sweden

Goudarzi R., Börgesson L. (2003). TBT sensors data report. Clay Technology AB, Lund, Sweden

Hökmark H., Fälth B. (2002). Predictive modelling program rev3. Clay Technology AB, Lund, Sweden

Hökmark H. (2001). Test of vitrified wastes. Study 1 – Thermal. Clay Technology AB, Lund, Sweden

Hökmark H. (2002). Test of vitrified wastes. Study 3 – Hydration. Clay Technology AB, Lund, Sweden

Karlsson T. (2002). Temperature Buffer Test. Report for data acquisition equipment. SKB International Progress Report IPR-03-12

Sandén T., De Combarieu M., Hökmark H. (2003). Description of the instrumentation installed in Temperature Buffer Test. Clay Technology AB, Lund, Sweden

Sandén T., Börgesson L. (2002). Temperature Buffer Test. Report on instruments and their positions for THM measurements in buffer and rock and preparation of bentonite blocks for instruments and cables. SKB International Technical Document ITD-02-05

Thorsager P. (2002). Temperature Buffer Test. Detailed design materials. Foundation and artificial saturation. SKB international Progress Report IPR-02-62

Advances in Understanding Engineered Clay Barriers – Alonso & Ledesma (eds)
© 2005 Taylor & Francis Group, London, ISBN 04 1536 544 9

Effects of over-heating on the performance of the engineering clayed barriers of the mock-up test

P.L. Martín & J.M. Barcala

Centro de Investigaciones Energéticas, Medioambientales y Tecnológicas, Madrid (Spain)

ABSTRACT: The mock-up test is an experiment that simulates the ENRESA's AGP Granite concept at almost full scale and under controlled boundary conditions. It was expected to reach full steady state conditions, both thermal and hydraulic, but due to a wrong function the total heating power was supplied during more than 36 hours after 1400 days of experiment. Some regions of the bentonite, the closest to the heaters, became dried because the temperature achieved values higher than 200°C. This thermal pulse (over-heating) and the coupled processes provide an exceptional opportunity to compose an exercise to calibrate the numerical THM codes.

1 INTRODUCTION

FEBEX is a demonstration and research project, which is being carried out by an international consortium led by the Spanish agency ENRESA and simulates components of the engineering barrier system in accordance with the ENRESA's AGP (Deep Geological Disposal, ENRESA, 1994) Granite reference disposal concept. The project includes tests on three scales: an "in-situ" test at full scale in natural conditions; a "mock-up" test at almost full scale in controlled conditions; and a series of laboratory test to complement the information from the two large-scale test.

2 EXPERIMENTAL SETUP

The main components of the mock-up experiment are described as follows (Figure 1): a confining structure simulates the gallery, through which hydration takes place; a Programmable Logic Control (PLC) manages the heater control system (HCS), composed of two electric heaters (0.17 m in diameter) concentric to the confining structure, that simulate the heat generation of the waste canisters; a hydration system supplies the water to hydrate the bentonite mass at a constant controlled N_2-pressure from two tanks working alternately; a 0.62 m-thick engineered barrier composed of two rings of compacted smectite

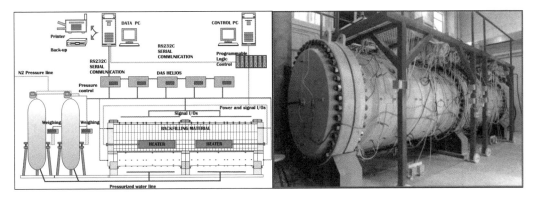

Figure 1. General layout and external view of the mock-up experiment at CIEMAT.

Table 1. Sensors installed and associated parameters measured in the mock-up test.

Measured parameter	Sensor type	Installation		
		Bentonite	Structure	External
Temperature	RTD Pt100	328	20	
Room temperature	RTD Pt100			1
Injection pressure				
Manometer	DIGIBAR II			1
Water pressure	DRUCK 1400PTX		2	
Total pressure	KULITE BG0234			
Radial		14		
Tangential		14		
Axial		22		
Fluid pressure	KULITE KHM375	20		
RH + temperature	VAISALA HMP233	40		
Extensiometric gauges	HBM		19	
Values from the PLC				
Temperature	RTD Pt100			18
Average temperature	Calculation			2
Power supplied	Calculation			2
Voltage CC to the sensors				2
Total number		438	41	26

blocks surrounding the heaters; instrumentation monitors the boundary conditions and the system behaviour by sensors installed within and outside the buffer material; and a monitoring and control system records the data by a Data Acquisition System (DAS). Detailed description of these components and the aims of the project are shown in ENRESA (2000).

More than 500 sensors have been installed within the experiment to measure the most important parameters such as temperature, total pressure, pore pressure, water injection pressure, relative humidity (water content) and strains within the structure (Table 1). They have been chosen to withstand mechanical stress, high temperature, humidity and salinity coming from harsh experimental conditions within the barrier.

The test design must support long-time operation with the instrumentation placed in harsh conditions, and most of the manufacturers cannot supply components with the required performances. Also, the experiment scale affects the equipment that supports them and the complexity of the associated infrastructure, both in the component types as in their operative connectivity (ENRESA, 2000).

Chemical tracers, as well as corrosion probes including different types of material, have also been placed within the bentonite.

3 OPERATION

The installation phase of the mock-up has been carried out between October '96 and January '97. The operational stage – simultaneous hydration and heating – started in February 4th 1997 at 12:45 P.M (ENRESA, 2000), "day 0" in the time scale. It was initially planned to last for three years, but it has been decided to extend the operational phase to get as close as possible complete saturation of the buffer, and continues to date.

Three days before the beginning of the operation phase, a volume of 634 l of water injected at high flow (3.5 l/min.) flooded completely the voids, and made possible the closure of the gaps between blocks by the swelling of the bentonite.

3.1 Hydration

The pressure of the hydration system was initially set to 0.55 MPa. The water injection pressure has been between 0.5–0.6 MPa since that date. The rate of injection follows an asymptotic curve, as expected, with lower rates as the time goes by. The hydration system has not been affected by major changes. Injection pressure-controller has been installed to establish the nominal values.

3.2 Heating

The power supply to the heater has been made in three phases: 250 W/canister during the first 6 days of operation (to 65°C measured in the heater-bentonite interface), 500 W/canister during the next 4 days (to 95°C measured), and automatically controlled from there (to fit the target temperature of 100°C at this interface). The average power consumption was about 550 W/canister. As shown in Figure 8, the temperature steady state in the clay barrier is achieved in a short time (15 days) and generates temperature gradients with maximum values around 0.03°C/m, close to the heater surface.

3.3 *Overheating*

Due to a wrong function, towards the end of November 2000 (day 1391), the control program supplied the total heating power (around 2700 W/canister) during more than 36 hours, until the temperature at the sensors in the heaters reached the resistance safety value (300°C).

The over-heating was automatically stopped but temperature within the bentonite had achieved values higher than 200°C (Figure 9). This dried the bentonite around the heaters and generates a thermal pulse (Martín & Barcala, 2001a).

4 TEST EVOLUTION: BOUNDARY CONDITIONS AND THM BEHAVIOUR

The following data, from last six and a half years, show good homogeneity throughout the test, between the two halves of the experiment, zones A and B, and between points located at similar radial distances, which is the main requirement to be worth to validate and verify the models. Then, the dismantling of the mock-up and the study of the clay barrier within will supply more information to validate the conceptual THM and THG models and to verify and optimise the fitting between model predictions and the actual geochemical evolution at large scale.

The data are grouped in (Martín & Barcala, 2003):

– Thermal boundary conditions: power supply and heater's temperature, outside structure temperature, and room temperature.
– Hydraulic boundary conditions: injection pressure and volume.
– THM parameters within the bentonite: temperature, total pressure, fluid pressure and RHE (relative humidity of equilibrium).

The radial symmetry is preserved. Some graphics have a broken Y-axis to show more clearly the effects of the over-heating.

4.1 *Thermal boundary conditions*

The power supply automatically was increasing slightly during the test time (Figure 2), exception made of over-heating, from the initial 425 W/heater to 575 W/heater. Then, due to problems with the isolation of the electric resistances, heater operation passed to constant power supply (700 W/heater).

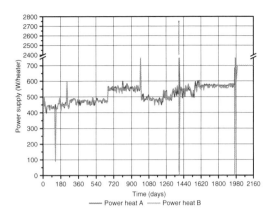

Figure 2. Boundary conditions: power supply.

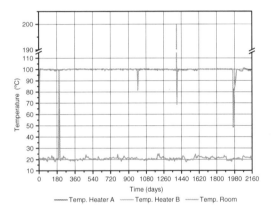

Figure 3. Boundary conditions: control and room temperatures.

Control temperature is close to the target, 100°C (Figure 3). The maximum 200°C value in the graphic is not the actual value.

Heater temperature sensors indicate values close to 90°C at the ends, with 10°C difference respect to the control (Figure 4). This difference increases slightly with time, indicating the variation of thermal conductivity close to the heaters (related to change in water content or structure).

Temperatures outside the structure depend on the external climate variations, but remain in the ±2°C band, around the average values of each distance (Figure 5).

4.2 *Hydraulic boundary conditions*

The injected water volume is calculated from the masses of the water tanks, corrected by the nitrogen injection pressure (Figure 6).

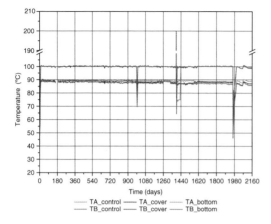

Figure 4. Thermal boundary conditions: heater temperature.

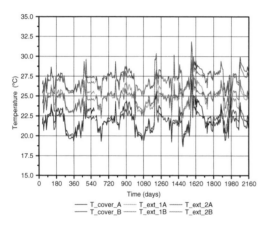

Figure 5. Thermal boundary conditions: temperature on the structure.

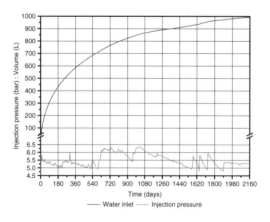

Figure 6. Hydraulic boundary conditions: Injection pressure and injected volume.

4.3 THM behaviour

Continuous recording of temperatures, volume of injected water, relative humidity, fluid pressure and total pressure (Figures 7 to 15) allows to ponder the processes, fit the numerical codes and address the specific laboratory experiments.

After the analysis of the time series from the over-heating, in Cañamón et al. (2002), with the mathematical tools described in Cañamón et al. (2001), it is established that the high-frequency perturbation in the temperature signals becomes negligible in a short time (from 25 to 75 hours).

The same type of analysis for the sensors of temperature, total pressure and relative humidity, before and after the over-heating, show that the frequency components of the signals are not modified in a remarked way, in spite of the very different periods they need to recover the previous state (temperature sensors need several days, total pressure sensors almost a year).

So, we can establish the values coming from sensors in the experiment are reliable in the whole time range. This is the first requirement to begin the analysis of the experiment THM behaviour.

The deviations of temperature values are neglected related to the average values for each radial distance; so, the average values of the temperature sensors in operation are presented. In the case of RHE sensors, the values are referred to the average initial value of the bentonite block within they are installed; so, it is possible to work with the increments of the values. Total and fluid pressures are presented as raw data.

4.3.1 Temperature

After a transient state of two weeks, the test was conducted in a quasi-stationary temperature state. After overheating, quasi-stationary temperature state was completely recovered in less than 18 days, so there is no indication on the curves, with exception of the peaks, which values are not coincident with the maxima (Figure 7). The thermal gradient within the bentonite around the heater is 3°C/cm.

The slight temperature waves observed are caused by external variations due to the difficulty of conditioning the test room. These fluctuations are evident in the sensors placed in the outer radii.

The data, in general, show good homogeneity throughout the test. The initial variations observed among sensors located at selected radial distances in the same section (0.20, 0.39, 0.58, and 0.77 m, respectively) are less than 2°C.

The properties that control the clay behaviour under a thermal load are the thermal conductivity, the specific heat and the thermal expansion coefficient. In addition to these factors, the final temperatures in the barrier are strongly related to the degree of saturation of the bentonite.

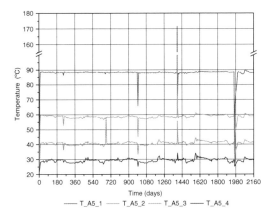

Figure 7. Temperatures in section 5: module A.

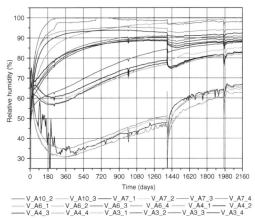

Figure 8. RHE sensors (except AB plane): A module.

4.3.2 Relative humidity (RH)

The RH sensors allow the monitoring of the hydration drying-wetting process. Figure 10 shows the evolution of the relative humidity (RH) at several distances from the heater in a "hot" section.

Depending on the location of the RH measurement (distance to the hydration surface and thermal gradient imposed on the sensor) different behaviours can be observed. Values measured at 0.05 and 0.20 m from the heater surface show clearly the effects of evaporation (drying) and condensation (wetting) in different zones of the barrier. Values measured at 10 cm from the hydration surface of the structure suggest that bentonite in this external zone is fully saturated.

In sensors located close to the heater the following phases can be distinguished: (i) a sharp increasing RH value appears related to the vapour phase generated by the initial heating; (ii). continuous heat transfer from heaters produces the drying of clay that causes decreasing RH; (iii) after some time, hydration reaches the dried zone and overcomes the drying process, increasing the RH values.

This transient indicates the redistribution of water, process with slower dynamics than the thermal transfer. So, it needs 360 days to reach the minimum value of RHE and 1080 days to recover the initial values in the level closer to the heater; however, level 2 achieve a relative minimum to 270 days but without retrieving the original value, indicating in this way the adsorption of water through the vapour phase.

During overheating, sharp peak values of RH were observed again in the inner rings (day 1400). This fact supports the idea that a new drying of bentonite was produced and generated enough vapour to flow through the location of the sensor. This vapour outflow from the heaters seems to be radial and redistributes water within the bentonite.

The sensor placed in zones without heater ("cold" sections) show higher RH values, reaching almost full saturation according to retention curves.

The RH evolution curves in all sections show tendencies similar to those described above. The only differences are related to the relative radial and longitudinal positions with respect to the heaters (hot and cold zones).

The hydraulic conductivity and the retention curve, both critical parameters to evaluate the full-saturation time, regulate the hydration of the bentonite.

The synchronous evolution of the different RH sensors indicates that their behaviour is a real process in the RHE (relative humidity of equilibrium) of the material. It is not an artefact. This evolution is shown as a decrease of magnitude in the more saturated zones of the external ring in the barrier (Figure 8). From above, the water inflow is radial and homogeneous, only modified by the implicit THM behaviour of the barrier. This was a specification of the design.

4.3.3 Total pressure

Under hydration, bentonite develops a mechanical pressure by swelling recorded by the total pressure sensors located at selected radial distances (35.0 cm and 66.5 cm), and orientated in the three main directions.

Axial pressures range from 4.5 to 9.0 MPa, radial pressures from 5.0 to 8.0 MPa and tangential pressures from 5.0 to 10.0 MPa. The pressures are in accordance with swelling pressure values of the bentonite measured in the laboratory: 10 and 6 MPa for dry densities of 1700 and 1600 Kg/m^3, respectively.

The higher-pressure values are located within the outer ring due to the evolution of the saturation front (Figure 9). In fact, the average-pressure values in this zone and their slow increase indicate full saturation of the buffer.

On the other hand, the smaller pressure values in the inner rings could indicate two different processes: the redistribution of stresses coming from the outer rings or the extension of saturation into this zone. Small variations of the pressure evolution appear to be linked to external temperature variations.

Overheating produced small pressure value peaks and redistributed total stresses and water content within the bentonite by the heating-cooling cycle generated. Recovery of the values previous to overheating is not so quickly as in the temperature or RHE sensors, but it is more pronounced in the sensors placed in the outer ring. In general, it is needed around one year to recover values in the saturated part of the outer ring (Figure 9), both cold or hot zones, 220 days in the no-saturated part of the hot zone, and 180 days in the no-saturated part of the cold zones. Sensors measuring axial and tangential stresses seem to be more affected.

The more saturated materials suffer a lower pressure peak than the less saturated ones, but the absolute decrease respect to the previous values is higher. It seems that pressure loss is more intense for the higher degrees of saturation, and longer for the hotter materials. This could indicate the competition between two transfer processes, thermal and hydraulic, and their location in the experiment (the thermal in the outer saturated zone and the hydraulic, in vapour phase, in the inner no-saturated zone).

Some local extremely high-pressure values (up to 10.5 MPa) can be explained by considering the locally combined effects: XYZ co-ordinates, dry density of the block (higher than 1800 Kg/m^3 in the core blocks), sensor installation (place carved in the

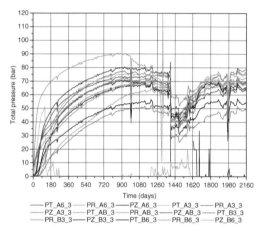

Figure 9.   Total pressure sensors in the hot zone: outer ring.

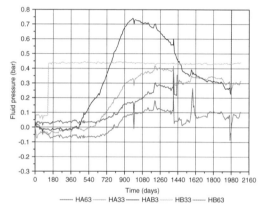

Figure 11.   Fluid pressure sensors in the hot zone: outer ring.

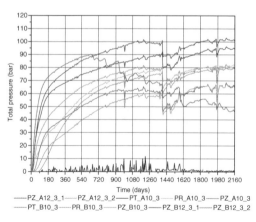

Figure 10.   Total pressure sensors in the cold zone: outer ring.

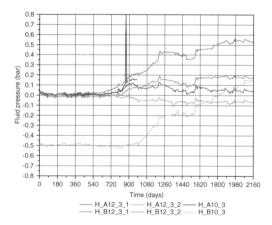

Figure 12.   Fluid pressure sensors in the cold zone: outer ring.

block), local swelling pressure (depending of hydration and temperature), and/or development and concentration of local stresses due to readjustment between blocks.

4.3.4 Fluid pressure

The measured values cannot be assigned to water or gas pressure without uncertainty, mainly in the inner ring. Only some variations seem to be related to the arrival of water, always for sensors located in the outer rings. In all cases, the fluctuations seem to be related to temperature variations.

It is necessary almost 700 days to observe some increase in the fluid pressure within the outer rings, both in the hot and in the cold zones. This time achieves 900 days in the inner rings of the cold sections, showing a dependence of the fluid temperature (through water viscosity).

During overheating, sharp peak values of fluid pressure, generated in the inner rings, support the idea introduced above related to the drying of bentonite. Values are recovered in the outer rings after overheating; but values in the inner ring continue to vary slowly or remain almost negligible.

5 OVER-HEATING DETAIL

Herewith, detailed data from over-heating are presented with some comments, only for the more interesting values. Data series include the period from 25/11/2000, before fail, to 14/12/2000, including a cooling phase of more 100 hours (Martín & Barcala, 2001). The time series present 30-minutes data. Reference time (0 time) in graphics is 13:00 on 26/11/2000.

5.1 Thermal boundary conditions

As a consequence of supplying the total power during almost 40 hours (Figure 13), heaters have produced an over-heating. Registered values indicate temperatures close to 200°C at the ends of the heaters, and temperature at the control section has been estimated in 240°C (Figure 14). This temperature has not been recorded by the system, but obtained by the cross fitting of the heating and cooling curves. So, the differential temperature between the ends and the control zone of the heaters was increased from usual 10°C to 40°C.

One control sensor of heater A was damaged (CPT_8_A) and eliminated from the control algorithm.

5.2 THM behaviour

Presentation of data follows the same rules that the general case above.

5.2.1 Temperature

Thermal waves generated by over-heating cross the barrier, from the heaters to the structure, within all the instrumented sections. Peaks are sharper in the hottest sections and the differences, related to the quasi-steady state of each level, are maximal in the sections A5 y B5, central zone of the heaters (closed to 110°C, Figure 15), and are decreasing when we move away from them (from 90 to 95°C in the bottom of the heaters, sections A2, A8, B2 y B8, and from 30 to 35°C in sections A1, A9, B1 and B9), to the covers of the structure (less than 4°C in sections A11 y B11).

The previous temperatures are completely recovered in 18 days, and most of the sections in 11 days. Test behaviour is similar in the two halves.

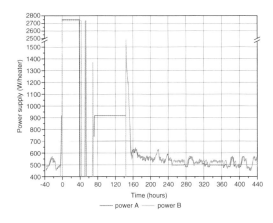

Figure 13. Over-heating: Thermal boundary conditions: power supply.

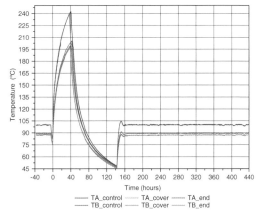

Figure 14. Over-heating: Thermal boundary conditions: heater temperature.

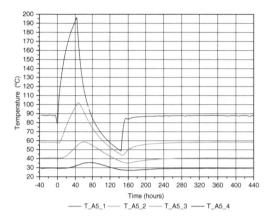

Figure 15. Over-heating: temperatures in section 5: A module.

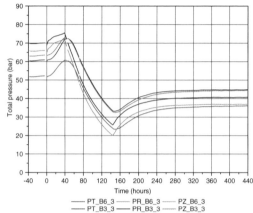

Figure 17. Over-heating: total pressure in the hot zone: outer ring: B module.

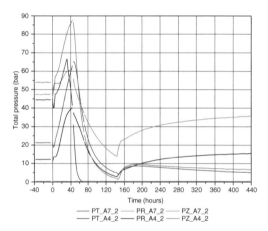

Figure 16. Over-heating: total pressure in the hot zone: inner ring: A module.

5.2.2 *Total pressure*

The overall effect has been the drying of the bentonite close to the heaters and the redistribution of the stresses within it. We can observe three phases:

The temperature increase has generated total pressure peaks, caused by the thermal expansion of the phases in the mineral, while water in the material evaporates and it becomes drier.

The cooling phase, after the breakdown of the power supply, produce the retraction of the mineral phases that remain in place, that in addition to the decrease of suction, reduce the total pressure to lower values than previously recorded.

The temperature recovering does not produce the total recovering of the pressures, by the thermal expansion of the material, as a result the pressures remain lower. Consequently, the loss of pressure is a function of the degree of saturation and the maximal temperature in the material.

The less saturated materials, in the inner rings of the hot zone (Figure 16), support a sharper peak than those materials more saturated, in the outer rings of the hot zone (Figure 17) or in the cold zone, because the temperature is lower here. However, the pressure peaks in every level are not coincident with the local temperature maxima, but with the heater temperature maxima, indicating a direct transmission of the pressures within the solid phase (good contact between different elements).

PZ and PT sensors are more affected, close to the heaters where water evaporates (Figure 16). The curve shape is similar in all the zones with higher degree of saturation and lower temperatures (without evaporation): discontinuity in the tendency curve (simultaneous with the heaters switch off) in the values of axial PZ, that becomes less sharp in the radial pressure PR, and it is smoothed for the tangential pressure PT (Figure 17). However, the differential pressure (between peak and valley) for the saturated zones in every section are dependent of the achieved maximal local temperature, by the combination of the thermal expansion and the generated increase of suction. The more important variations are located in the inner ring (Figure 16), at the contact with the heaters, supporting the drying of the bentonite. The recovery rate is higher and simultaneous with the heaters switch on for the radial pressures.

The over-heating has damaged the PZ_A6_3 sensor.

5.2.3 *Fluid pressure*

Most of sensors in the outer ring only show small variations in these records, but sensors in the inner

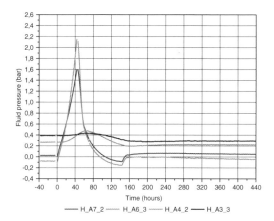

Figure 18. Over-heating: fluid pressure in the hot zone: inner and outer rings: A module.

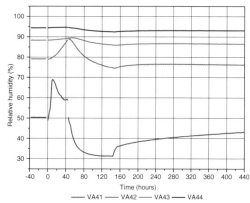

Figure 19. Over-heating: RHE in the hot zone: section 4, A module.

rings show extremely sharp peaks of fluid pressure (Figure 18).

Again, this conducts us to the drying of the bentonite close to the heater. The water evaporates and supplies enough vapour to increase the fluid pressure that drives a vapour flow through the barrier, crossing the position of the sensors. Other hand, the material in the outer ring is saturated in a great extend and becomes impervious to the vapour flow.

So, the fluid pressures are a confirmation of the saturation of the barrier material, at least with values higher than 0.80 within the outer ring. In the inner ring, the highest values corresponding to the values of saturated vapour pressure at the temperature in the position of the sensor, produced by evaporation of the pore water near the sensors.

After cooling, the values decrease according to a redistribution of water by transport in vapour phase o by suction increase in the barrier core. The driven forces that control these processes are different: the gradient of vapour pressure in the first case, and the suction of the bentonite in the second one. This is true because we are observing a transient state of the system.

5.2.4 *Relative humidity*
The differences observed between sensors placed at the same distance from the heater are negligible, in the two parts of the experiment. The vapour flows are radial.

In agreement with the hypothesis related to the fluid pressure, the higher RHE values, measured in the two sensors installed in the outer rings (sensors 3 and 4), scarcely show variations that indicate vapour flow. Meanwhile, the barrier in the hottest zone, between heater and sensors, has been dried, generating vapour that goes through the sensors 1 and

2 and produces peaks in the signal of the sensors (Figure 19).

These peaks are sharper and two-folded in the sensors closer to the heater (level 1). These sensors do not recover their previous RHE values after recover the temperatures; so, confirm the bentonite drying. The rest of the levels recover the values in a week.

We consider that the sensors installed in the outer ring are surrounded by completely saturated material, and the effect of variation of temperature is the cause of the observed variations in the sensors.

5.3 *THM processes in the barrier*

The THM phenomena occurring in the barrier present a complex analysis due to the coupling between processes of diverse origin and the complicated response of the bentonite, related to its micro-structure.

After Gens & Alonso (1992) and Alonso (1998), the mechanical behaviour of the bentonite is the result of the interactions between the change of volume of aggregates of highly-expansive clay-mineral particles (micro-structural level) and the spatial re-organization of the skeleton composed by these aggregates, considered as granulates (macro-structural level).

5.3.1 *THM evolution of the experiment*
The total pressure (swelling pressure) also shows this decreasing behaviour in the sensors of the external ring for the long term (Figure 9).

There is a differential behaviour between the bentonite sections that includes the heaters (hot sections: A3, A4, A6, A7, B3, B4, B6 y B7), and those without heaters (cold sections: A12, A10, B12 y B10). The decreasing tendency is only clearly noted in the external sensors of the hot sections, but it seems that similar processes of smaller magnitude have

399

begun both in the internal sensors of the hot sections (Figure 9) and in the external sensors of the cold sections (Figure 10). This evolution have been unfortunately masked by the over-heating.

This kind of different behaviour is also seen in the fluid pressure sensors, separating the two zones as a function of the temperature field. The external rings show gradual increment of pressure, that achieve higher and more stable values before that the corresponding ones in the cold zones (Figure 11 and Figure 12); meanwhile, the internal rings show a similar behaviour in both zones, hot and cold, (Figures 11 and 12).

All this conducts to the idea that the physico-chemical processes driving the evolution of the measured parameters suffer deviations related to the temperature field generated in the barrier. As thermal gradient-driven processes (thermo-hydraulic coupled phenomena), or as temperature increased-rate processes (chemical ones or Arrhenius type).

Anyway, it is noticed the great interest of the time-series analysis tools to extract the information and to interpret the results in experiments with complex coupling of processes or long duration.

5.3.2 Analysis of thermally-induced THM transient

The over-heating has induced a THM transient in the experiment that has affected to the measured parameters with different intensity and duration.

The behaviour of the RHE sensors can be divided in hot zones (coincident with the cross-section of the heaters), cold zones (external to the heaters) and intermediate zone (between heaters):

– The increase of RHE, associated to the temperature increase, becomes smaller when we move away from the heat source. In the inner ring, a two-folded transient appears followed by a loss of water by evaporation, and a redistribution driven by the generated gradient of pressure. In the outer ring, the thermal disequilibrium produces a small initial increase that becomes a final loss of water.
– The behaviours in the cold and intermediate zones are similar. First, a small increase of RHE with the temperature wave, then a slow decrease that indicates redistribution of water. This behaviour is smoothed in the outer ring, with lower temperatures.

The same rational can be applied to the fluid pressure, dividing the test in hot, cold and inter-mediate zones.

– There are pressure increases associated to local increments of temperature, which decrease in magnitude when we move away from the thermal source. So, the increments in the inner ring are from

1.4–1.6 bars in sections #7 to 2.2–2.3 bars in sections #4 (with higher temperature at the same radius). Meanwhile, the outer ring shows scarcely increase, from 0.05 bars in sections #3, to 0.2 bars in sections #6. The inner ring show final values slightly higher that come back slowly to the previous ones, while the outer ring does not recover the previous values, indicating water migration from the periphery to the core of the barrier.
– In the cold zones, the inner ring increases the pressure in a similar way for the extend of the experiment, from 0.7 to 0.8 bars in sections #10 and #12, while the values in the outer ring are practically constant (less than 0.05 bars). The later behaviour is similar to the hot zones, the internal values are higher, and the outer ones are lower.
– The whole data set indicates a redistribution of water within the barrier material by two different ways: (1) a process that takes advantage of the vapour generated around the heater to distribute it along the longitudinal axis of the experiment (unsaturated inner core), driven by the vapour pressure and increasing the vapour pressure values in the inner ring of the cold zones, and, so, increasing the saturation there; and (2) other process that takes advantage of the fluid pressure increase in the saturated zone of the outer ring to inject water by the combined effect of the pressure and the increased suction in the inner ring.

The fluid pressure presents behaviour with a strong coupling with temperature distribution, while the RHE coupling is smoothed by the intrinsic behaviour of the material (hydro-mechanical transfers, chemical effects on the water activity, etc.)

5.3.3 Drying of bentonite

The two-folded peaks in the RHE values during the over-heating (Figure 20) seem to correspond to the

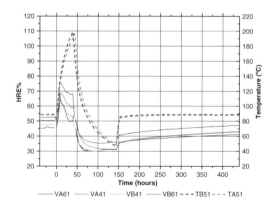

Figure 20. Evolution temperature/RHE during over-heating: heater/sensor distance 3/5 cm.

two-folded shape observed in the differential thermal analysis and in the thermo-gravimetric studies of the montmorillonites.

The curve shape indicates the presence of divalent cations, particularly Ca and Mg (Caillère *et al.*, 1982). It is produced by the lost of the hydration water of the cations that represents 4 to 5 weight percentage. The temperatures of occurrence indicate the energy of the second and first hydration layers of the cations. These temperatures are 160 and 210°C for Ca, and 190 and 220°C for Mg (MacKenzie, 1957).

The two-folded peak in level 1 is also observed in level 2, but only for the RHE sensors placed below the heaters (Figure 21, sections 3 and 6). This behaviour, that could indicate a difference in the degree of saturation of the material or a better thermal contact in this zone, cannot be compared with the fluid pressure values, because there are not this type of sensors in the analogous positions. The observed fluctuations in the signal of the level-1 sensors can be attributed to local phenomena of evaporation-condensation around them.

5.3.4 *Saturated vapour pressure vs. temperature*

By fitting the values of the saturated water vapour pressure, as function of temperature, and applying it to the values of estimated temperature in the position of the sensors, of the internal ring in the hot zone (sections 4 and 7, to 0.35 m from heaters), is observed a very good agreement between the values of saturated water vapour pressure (P vs c), and the fluid pressures during the over-heating around the sensor (P vs m), corrected from the initial (Figure 22). This fact indicates that the saturation state prevents the vapour to migrate. Small discrepancies can be attributed to the losses from adsorption in the bentonite and the chemical composition of the pore water. The estimation of temperatures has weighted the actual values in the closest points to each pressure sensor (sections A8, A5 and A2), by the distance.

5.4 *TH equilibrium conditions*

The adsorption of water is an exothermic reaction. During adsorption, the material is in a temporary equilibrium with the vapour pressure, at the temperature generated from the adsorption heat. Later, during the cooling of the material, water intake increases the hydration of the material. Here, water intake could be considered as a cooling process.

If material is at constant temperature (or in a quasi-constant temperature distribution, as in our experiment in normal operation conditions), the adsorption heat does not produce an additional temperature

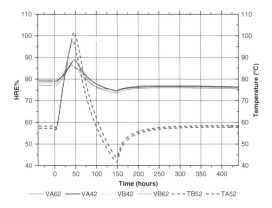

Figure 21.   Evolution temperature/RHE during over-heating: heater/sensor distance 21/20 cm.

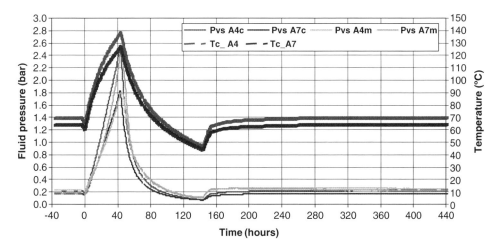

Figure 22.   Water vapour pressure in the hot zone: calculated values (c) vs. measured values (m).

increase and the water intake is directly related to the external vapour pressure at the given temperature (as in the measurement of adsorption isotherms). If the material, in thermal and saturation equilibria, is exposed to sudden environmental changes, can return to equilibrium by two different ways (Crank, 1975):

– Increasing its water content to the value of equilibrium with the RH of surroundings.
– Increasing its temperature (all the material phases: bentonite, pore water and air) to the value that corresponds to the RH in equilibrium with the surrounding RH. So, the water content remains unchanged.

From both possibilities, the first one, which means an important increase in the water content of the material, can only be achieve after a long equilibrium time that permit the redistribution of water. The second way, which supposes a temperature increase, is easily achieved due to the adsorption heat and the distribution of temperature imposed by the heating system. A relatively small increase in water content could be enough to obtain the needed temperature.

Hence, the first equilibrium is the thermal one with water content practically constant. It is pseudo-equilibrium, because the final temperature and the final water content are imposed. Therefore, the essential characteristic of the propagation of changes of temperature and humidity in porous material is the co-existence of two equilibrium states – one temporary, quickly established, without water intake, and other permanent, slowly established, with changes in the water content.

5.4.1 Propagation of perturbations

When hydration progresses, a front that separates the initial conditions and the pseudo-equilibrium moves through the material with a certain velocity (water o vapour flow velocity if the heat capacity is not taken into account). So, a fast perturbation, with changes of temperature and HR, without water intake into the material, will be followed by other slower perturbation with water intake (and stabilization of the RHE). Both perturbations are observed in the over-heating episode. On one hand, the fast perturbation associated to the described pseudo-equilibrium is shown as a vapour front driven by its pressure gradient; on the other hand, the slow perturbation associated to drying of material and the water intake in the zones far away from the thermal source.

It is usual to consider in models that bentonite achieve hydraulic equilibrium with the surrounding RH almost instantaneously, though there are evidences of its slow structural evolution during hydration. Most of the liquid-water intake is instantaneous, but this is not true with the vapour-water phase, where equilibrate the material with the imposed RH could

take more than 40 days (in samples 0.025 m thickness). This difference in behaviour is taken into account in some THM models, as the double structure or porosity models, as a water transfer between the porous structures defined into the material, from the macro-porosity (extra-aggregates) to the micro-porosity (intra-aggregates), and vice versa. However, the changes in RHE continue being considered instantaneous in the time range of the THM processes

6 INTERACTION MICRO-STRUCTURE/ MACRO-STRUCTURE

The behaviour micro-structural of the bentonite is not affected by the macro-structure, but the inverse one is not true: the behaviour macro-structural can be affected by micro-structural strains, after in a irreversible way (Gens & Alonso, 1992). Usually, it is assumed that the irreversible macro-structural strains are proportional to the micro-structural ones. It is expected that the interaction between macro and micro-structure will not be instantaneous, but it presents certain kinetics in the transfer to the suction equilibrium.

6.1 Measured THM-parameter distributions

Distributions of temperature and RH have been estimated at several times; also distribution of suction has been calculated at these times. The estimations are made by construction of a regular grid by a linear kriging method with an anisotropy factor.

The thermodynamic relation between suction and partial vapour pressure of the pore water is (Fredlund & Rahardjo, 1988):

$$\psi = -\frac{RT}{v_{w0}\,\omega_v} \ln\!\left(\frac{\overline{u_v}}{u_{v0}}\right)$$

where:

ψ Total suction (kPa).
R Gas universal constant [8.31432 (J/mol · K)].
T Absolute temperature [T = 273.16 + t⁰(K)].
t^0 Temperature (°C).
v_{w0} Specific volume of water [1/ρ_w(m³/kg)].
ρ_w Water density (998 kg/m³ a 20°C).
w_v Molecular weight of water (18.016 kg/kmol).
u_v Partial vapour pressure of pore water (kPa).
u_{v0} Saturated vapour pressure of pore water over a plane surface at the given temperature.

In our calculation, the value inside the ln() operator is replaced by the estimation obtained from the measured RHs, the absolute temperature values are estimated from the measured distribution of temperatures, and water density is taken as a constant.

The suction is calculated in an element-to-element scheme and converted to MPa.

In the pictures of thermal evolution during over-heating, it is shown that the points closed to the heaters, the bentonite reach values higher than 200°C (Figure 23, 42 h). After the heater switch-off, temperatures go down quickly to values around 45°C (Figure 24, 142 h). The variations in temperature far away from the heater are not so intense. Finally, the operational thermal state is established: 100°C constant temperature in the contact bentonite-heater (Figure 24, 440 h and following).

In the analogue figures of the HR evolution, it is shown the vapour front that increases temporary the values (Figure 25, 11 h). Later, the increased temperature dries the material and reduces the RHE (Figure 25, 80 h). The previous values are recovered after 1000 h. Figures show the 40% and 80% RH lines.

From calculated suction closed to heaters, it could be deduced that the vapour front reduces suction values from 110 MPa to 55 MPa (Figure 27, 11 h). However, from above, it does not seem possible to interpret directly these values as a hydration of the material. Because the front wave last less than 24 h, it is considered a pseudo-equilibrium transient.

Later, due to the intense drying, with temperature >200°C and time enough to transfer vapour, suction increases to 165 MPa (Figure 27, 80 h). Finally the previous values are recovered (Figure 28, 280 h). Figures show the 50-MPa and 100-MPa lines.

As it was noted above, the evolution of the stresses is interesting, mainly the radial ones. The over-heating generates a sudden and moderate increment, followed by an important decrease (around 2 MPa). The radial stresses recover partially after establishing the RH values, but they remain far lower of the previous values. There is a net decrease of the radial stresses, important and irreversible.

The synthetic generalised stress path near the heaters corresponds to four phases (five stages, modified from Sánchez et al., 2001): the thermally-induced expansion of the material, the intense drying associated to the reached maximal temperature and the duration of the over-heating, the cooling of the barrier to the observed minimal temperature, and, finally, the recovery of the previous values of temperature and RHE.

– In the first step, the mechanisms related to the increase of temperature, associated to the thermal expansion, tend to increase the stresses.
– Second, the increase of temperature goes on, but related to the drying of the material, and increases the suction. The drying controls the retraction of the material reducing the stresses. This mechanism produces irreversible macro-structural changes, mainly an increment of density in the macro-structure.

– The third phase is characterized by the decrease of temperatures and the hydration of the material that recover the lost water. This produces the expansion of the material.
– Finally, the recovery of the previous conditions allow hydration to continue and temperatures to increase, both factors produce the expansion.

The zones adjacent to the heater have lost water that move away towards the structure, both radial and longitudinal directions. At the same time, the water is coming from the structure (liquid hydration front). The combined effects of these processes have modified the quasi-steady state of water potential within the barrier: modification of the temperature and the suction gradients, possible chemical effects on the pore water potential, irreversible changes in material structure, etc.

In the cold and hot zones of the experiment, the pore pressure increase could explain the water intake. The thermal expansion of the pore water cannot be dissipated if the permeability is very low, and the pore pressures increase. In the mock-up, this "non drainage" condition is true in the saturated periphery of the barrier, but the core barrier remains unsaturated and constitutes the only possibility of water migration driven by the gradient of water pressure. This "pumping" mechanism could work in three phases:

– First, the pore pressure goes up in the periphery during the over-heating and this has injected water in the core barrier.
– Second, the periphery of the barrier takes water from the hydration system during the cooling of the barrier, by means of the increased suction.
– Third, the pore pressure goes up again in the periphery during the reestablishment of the previous thermal conditions, and water is newly injected to the core.

These processes will be enhanced by the thermally induced reduction of the water viscosity.

Besides, in the hot zones, the migration of vapour towards colder zones where it cans condensate can form almost saturated zones (at least in the sense to be impervious to vapour flow) at intermediate positions within the barrier. So, the effect is the reduction of the geometry and the local suction gradient, just in the zone where the core barrier extract water from periphery. This process could also occur during the normal operation of the test, doing the water-driving potentials lower and lower as the test goes on.

The global effect seems to have increased the velocity of the point of pseudo-equilibrium of the THM coupled processes (in the sense described by Carnaham, 1987) during same time, till a new pseudo-equilibrium has been established, deeper inside the core barrier.

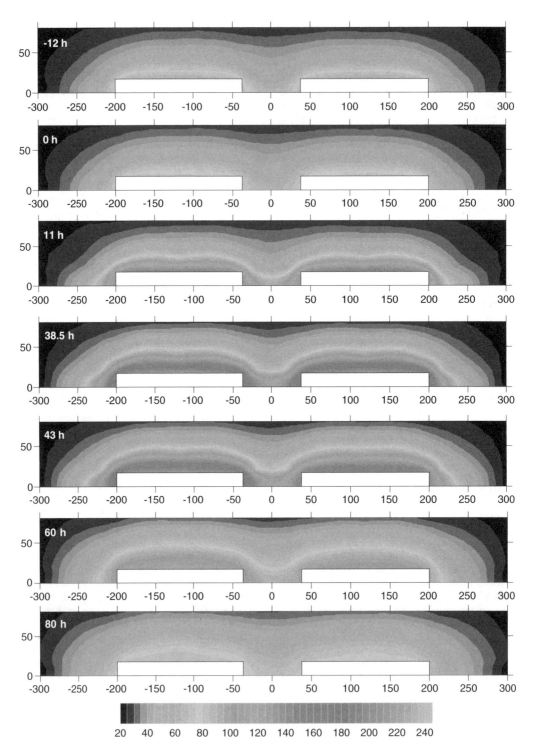

Figure 23. Temperature evolution (°C) during the over-heating (38.5 h) and cooling.

404

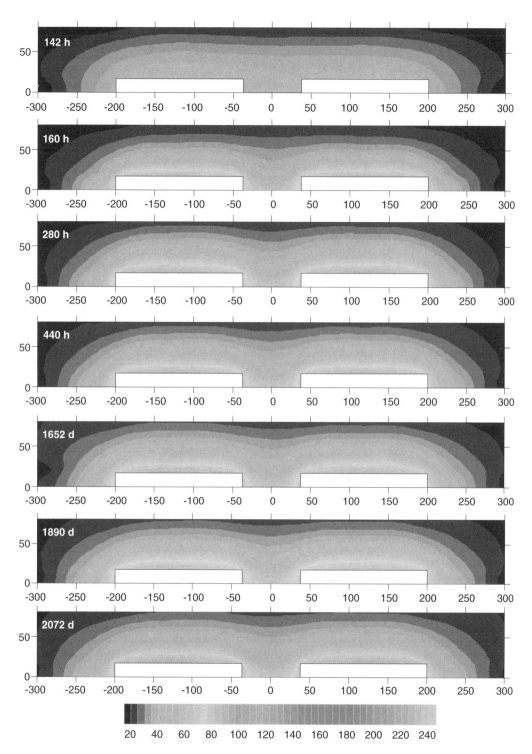

Figure 24.   Temperature evolution (°C) during the later cooling.

405

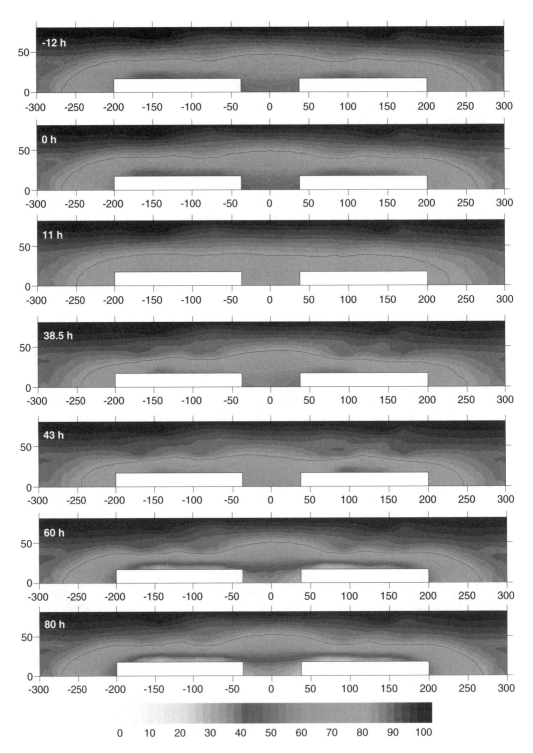

Figure 25.   RH evolution (%) during the over-heating (38.5 h) and cooling.

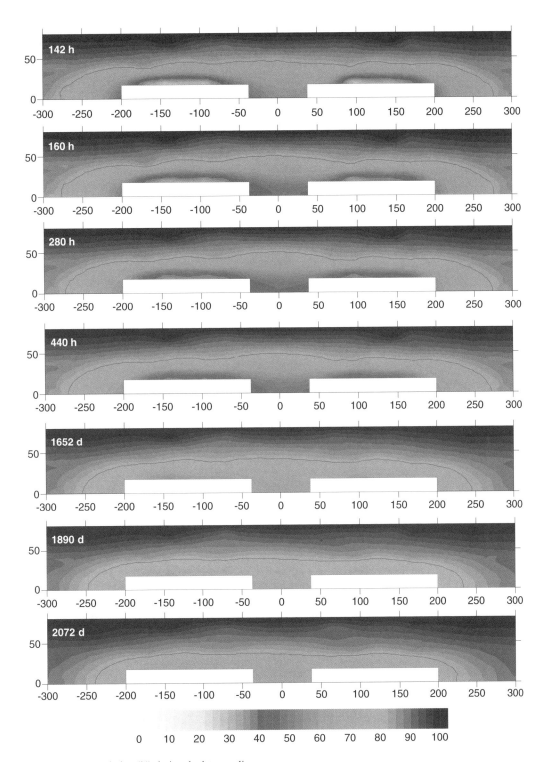

Figure 26.   RH evolution (%) during the later cooling.

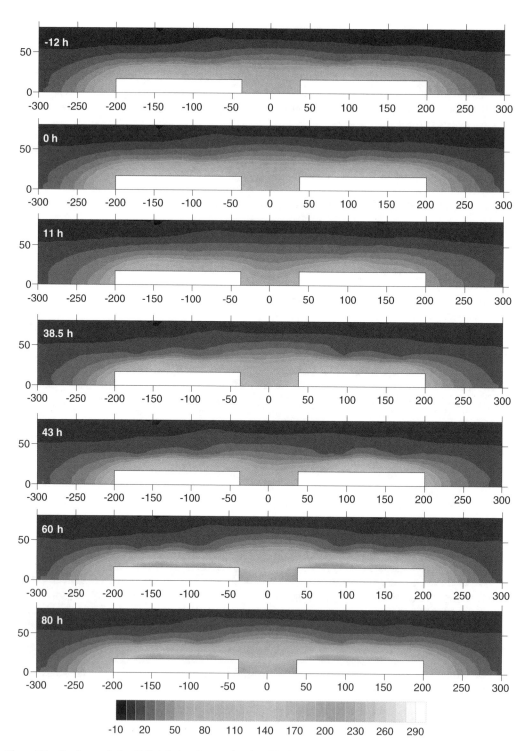

Figure 27. Suction evolution (MPa) during the over-heating (38.5 h) and cooling.

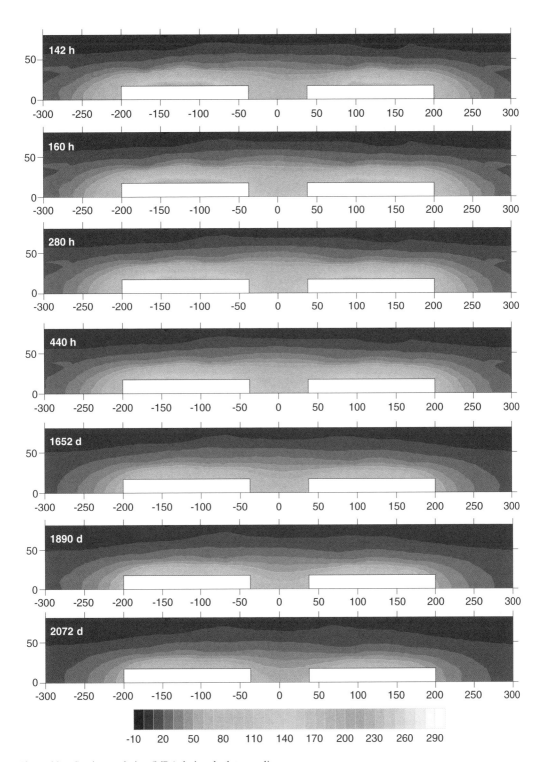

Figure 28. Suction evolution (MPa) during the later cooling.

This experiment has supplied important data to evaluate the THM models in a qualitative way, mainly due to the over-heating episode.

7 DISCUSSION AND CONCLUSIONS

The Mock-up test was initially planned to last 3 years, but the reliability of all the test components (instruments, heaters and control systems) and high amount and quality of the data being generated persuaded us to prolong the operational phase to get as close as possible to the full saturation of the buffer.

During these phase an incident, an over-heating, has unleashed a series of transient processes in the THM behaviour of the test. These coupled phenomena compose an exceptional exercise to calibrate the numerical codes.

Six and a half years after the start of the operational phase of this test, it can be concluded that it is a valuable instrument for establishing the viability of the reference concept and making progress in the understanding and evaluation of behaviour in the near field, especially the clay barrier.

7.1 *Operational behaviour of the experiment components*

The mock-up test overpasses its design operative time, 3 years, functioning correctly and with most of the sensors still in operation.

– The reliability of the instruments is fairly good. More than 85% of the sensors remain operative, in spite of over-pass the operation expectancy. However, it is not possible to guarantee their future behaviour.
– The reliability and performance of the heaters are sufficiently high. The change in the functioning of the heaters (one resistance with filtered power supply) must extend their life expectancy and prevent them from any further damage.
– The data acquisition and control systems are working satisfactorily.

7.2 *Conclusions on the measurement of THM parameters*

The definitive evaluation of the test will be only made after the dismantling. However, some qualitative conclusions may be established, mainly the differences in the behaviour between "hot" and "cold" zones of the buffer, indicating major implications of the thermal aspects in the transport processes. Also some THM couplings have been identified at large scale.

7.2.1 *Temperature*
– The thermal regime is homogeneous and symmetric; both with respect to the central section and the longitudinal axis. Fluctuations of temperature are observed near the structure and are related to external variations.
– The temperature distribution is controlled by thermal conduction, due to the slow rate transport processes involved in the saturation of the buffer material.
– Temperature calculation based on thermal conductivity, tuned by saturation of the material, produces good fittings to data (ENRESA, 2000). More complex models are not needed.
– A limited number of sensors are enough to display the temperature distribution and evolution.

7.2.2 *Relative humidity*

– The RH sensors work well within the RH and temperature ranges of the test. This allows the monitoring of the hydration process during the time of this experiment.
– Both, water inflow and water vapour outflow, behave mainly in a radial direction.
– The saturation of the bentonite barrier is apparently controlled by the hydraulic properties of the bentonite and the thermal gradient imposed by heating.

7.2.3 *Total pressure*

– The total pressure measurement depends on the XYZ co-ordinates, the dry density of the block, the sensor installation, and/or the development and concentration of local stresses (due to the reactions between blocks).
– The pressure increases seem to be associated with the arrival of the hydration front. Measured values become stable as the buffer material saturates and the material reaction changes its structure.
– Values for the different directions (PR, PZ and PT) are converging in average values.

7.2.4 *Fluid pressure*

– The values registered are so close to zero that they are greatly affected by the fluctuations of the external temperatures and the thermal waste due to natural convection on the structure.
– The fluid pressure measurements seem to be due to liquid in the external rings and to gas in the internal rings. It would be necessary to get enough free water (full saturation) within the inner rings of the buffer material to consider this pressure value as water pressure.
– Gas pressure depends on location and water pressure gives an average value.

7.2.5 *Interaction with models*

– In a first moment, hydration progress through the whole barrier, with higher rates in the outer ring, apparently controlled by the hydraulic properties of the bentonite affected by the local temperatures. The regime was regular and symmetric, and there was agreement with the THM model.

– Then, with the evolution of the system, hydration show clear differences between "hot" and "cold" zones of the buffer, verified with measurements of the injected flow. The regime is regular and symmetric, except the different magnitude. Modifications of the model are needed to take into account this behaviour.

– The critical revision of the experiment has not detected artefacts that could modify the response of the sensor in the experiment (CIEMAT, 2001-a and 2001-b). Statistical tools have demonstrated that the observed signals are representative of the THM behaviour of the system.

7.3 Over-heating consequences

The thermal increase and the coupled processes generated by overheating compose an exceptional exercise to calibrate the numerical codes, because:

– Differences in the behaviour between "hot" and "cold" zones have been confirmed.
– It has generated a vapour flow through saturated and non-saturated material.
– A homogeneous THM behaviour of the system has been recorded with agreement between the RH and fluid pressure values.
– A redistribution of water in vapour phase, originated in the thermal source, seems to have occurred within the bentonite. The core barrier acted like a "pipe" to distribute vapour to the cold zones at the ends of the structure.
– Changes in suction are recovered.
– Total pressures have become more homogeneous, but the losses of magnitude indicate irreversible structural changes. Couture (1985) indicated that steam presence reduces the swelling capacity of bentonite, even without mineralogical changes.
– The heating-cooling cycle seem to work as a "pump" in the semi-permeable system composed by the structure and the barrier.
– The system recovering gives a good chance to compare actual values with models.
– No important damages can be attribute to the over-heating incident: the experiment seems to have worked like a thermal pulse test.

The test has continued to work correctly to date. No variation has been observed in the sensor functioning.

7.4 Implications for the Performance Assessment

The FEBEX mock-up test is being extremely useful to fit the main parameters of the clay behaviour and the calibration of the numerical THM code, CODE-BRIGHT developed by UPC (ENRESA, 1998c), and the thermo-hydro-geochemical code, CORE-LE developed by UDC. The major implications of the thermal aspects in the transport processes, including irreversible structural changes are a main issue.

So, this Mock-up test is a source of valuable data to improve the knowledge of the THM processes in the EBS, and to apply them to obtain reliable numerical codes to accomplish the Performance Assessment of a Deep Geological Disposal.

ACKNOWLEDGEMENTS

This research work has been carried out in the context of the FEBEX and FEBEX II Project, financed by ENRESA (Spanish National Agency for Waste Management) and the European Commission under contracts FI4W-CT-95-0006 and FIKW-CT-2000-0016.

REFERENCES

Alonso, E (1998) Modelling expansive soil behaviour. Keynote Lecture . *Proc. II Int. Conference on Unsaturated Soils.* International Academic Publishers.

Alonso, E, Gens, A & Lloret, A (1991) Double structure model for the prediction of long-term movements in expansive materials. Computer methods and advances in Geomechanics. Beer G., Booker J.R. & Carter, J.P. edts. *Balkema*. Rotterdam, 541–548.

Caillère, S, Hénin, S & Rautureau, M (1982) Minéralogie des argiles: 2. classification et nomenclature. INRA. *Masson*. Paris.

Cañamón, I, Calle, O, Elorza, FJ, Rodríguez, R & Borregón, JL (2001) Aplicación del análisis espectral a la modelización acoplada Termo-Hidro-Mecánica. XVII Congreso de Ecuaciones Diferenciales y Aplicaciones y VII Congreso de Matemática Aplicada. Universidad de Salamanca, 24–28 de Septiembre de 2001. Pp. 687. *L. Ferragut y A. Santos.* Salamanca.

Cañamón, I, Elorza, FJ, Rodríguez, R, Valeriano, L & López Iraza, V (2002) Técnicas de análisis de datos en experimentos relacionados con el almacenamiento de residuos radiactivos de alta actividad.

CIEMAT-70-IMA-D-0-17 (2001-a) Funcionamiento del ensayo en maqueta y recomendaciones sobre el programa de desmantelamiento. CIEMAT.

CIEMAT-70-IMA-D-0-18 (2001-b) Programa de estudios hasta el desmantelamiento del ensayo in situ. CIEMAT.

Couture, RA (1985) Steam rapidly reduces the swelling capacity of bentonite. *Nature*, Vol. 318, No.6041: 50–52.

Crank, J (1975) The mathematics of diffusion. 2nd edition. *Oxford University Press*, Oxford.

ENRESA (1994) Almacenamiento geológico profundo de residuos radiactivos de alta actividad (AGP). Conceptos preliminares de referencia. *Publicación Técnica ENRESA* 07/94. Madrid.

ENRESA (2000) FEBEX project. Final Report. *Publicación Técnica ENRESA* 1/2000. Madrid.

411

Gens, A & Alonso, EE (1992) A framework for the behaviour of unsaturated expansive clays. *Can. Geotech. J.* 29: 1013–1032.

MacKenzie, RC (1957) The montmorillonite DTA curve (1). *Bull. Gr. Franc.des Argiles* 9: 7–15.

Martín, PL & Barcala, JM (2001) Report on over-heating failure. Mock-up test. *Informe técnico*. CIEMAT/DIAE/54520/1/2003.

Martín, PL & Barcala, JM (2003) 23rd Data report. Mock-up test. *Informe técnico*. CIEMAT/DIAE/54520/3/2001.

Sánchez, M, Gens, A, Guimaraes, L do N, & Olivella, S (2001) Generalised plasticity model for THM simulations involving expansive clays. In: *6th International Workshop on Key Issues on Waste Isolation Research*. KIWIR. Paris, November 2001.

Advances in Understanding Engineered Clay Barriers – Alonso & Ledesma (eds)
© 2005 Taylor & Francis Group, London, ISBN 04 1536 544 9

Impact of canister corrosion on barrier behaviour

S. Olivella, J. Alcoverro & E.E. Alonso
Universitat Politènica de Catalunya, Barcelona (Spain)

ABSTRACT: As the steel canister corrodes it expands and the bentonite barrier may be subjected to compression on its inner boundary. At mid term, hydration of bentonite and swelling pressure development take place. For long term conditions, a deviatoric stress increment is induced in the bentonite barrier by the corrosion induced expansion of the canister.

1 INTRODUCTION

In the context of THM analyses simulating repository conditions the program CODE_BRIGHT (Olivella et al 1995) is used to perform the numerical modelling of the corrosion process of a steel canister containing a radioactive waste and surrounded by a bentonite buffer. As a result of the corrosion process of the canister, steel will be transformed into magnetite. The ratio of the volume of the produced magnetite divided by the volume of the steel that has been corroded will be assumed to be given by a constant that is greater than 1.

A rigorous modelling of the corrosion of the canister would require a suitable mathematical model of the relevant physical phenomena involved in the corrosion of the steel. However, the available data provides estimations of the velocity of corrosion obtained from a series of corrosion tests. Therefore, the approach followed in order to implement the corrosion of the canister in CODE_BRIGHT will be essentially kinematic. In the present case, two approaches have been proposed: (1) prescribed displacements on the bentonite and (2) prescribed swelling in the canister. These approaches will be described in the following section. The second one has been implemented in CODE_BRIGHT to perform a series of analyses to evaluate the impact of the *corrosion induced expansion* on the behaviour of the bentonite barrier.

2 MODELLING CORROSION INDUCED EXPANSION

The modelling of corrosion induced expansion in a simplified way requires to study in detail the assumptions that can be performed and find a simple way to implement the solution in CODE_BRIGHT. After that, a coupled THM analysis will be possible

and useful to investigate mechanical effects of canister corrosion on bentonite under different conditions.

Alcoverro et al (2003, this issue) have investigated the distribution of stresses and strains in a canister induced by corrosion. Their results may serve to improve the way to treat the canister subjected to corrosion. Here, a simplified way is considered and it is described below.

2.1 Controlled displacement approach versus controlled deformation

Two possible ways to treat the corrosion induced expansion are discussed in this section. The reasons for choosing one of them are given.

In the first approach, it is assumed that (1) the steel and the magnetite are much stiffer than the bentonite and (2) the deformations due to thermal dilation are negligible. Therefore, the field of displacements in the canister will be due only to corrosion. In these conditions, the action of the canister against the bentonite buffer is equivalent to prescribe displacements at the contact between the canister and the bentonite.

The law of prescribed displacements is determined by requiring them to be equal to the displacements of the external surface of the (partly corroded) canister due to the corrosion of the steel. At a certain time t, the relationship between the volume of corroded steel and the volume of produced magnetite reads:

$$\pi\left(r_e^2 - r_b^2\right) = k\pi\left(r_{e0}^2 - r_{b0}^2\right) = k\pi\left(r_{e0}^2 - r_b^2\right) \tag{1}$$

where r_{e0} and r_{b0} are, respectively, the initial (before the corrosion) radial coordinates of the interior surface of the canister and radial coordinate of the current boundary between steel and magnetite and r_e and r_b are the corresponding values after the corrosion. To get the second equality, it has been taken into account that due to the assumed high stiffness of the steel, $r_b = r_{b0}$. Differentiation of this expression yields:

$$dr_e = -(k-1)\frac{r_b}{r_e}dr_b = -(k-1)\frac{r_b}{\sqrt{kr_{e0}^2 - (k-1)r_b^2}}dr_b \quad (2)$$

Therefore, the radial component of the velocity of the outer surface of the canister v_e and the radial component of the velocity of the boundary between steel and magnetite v_b (the rate of corrosion) are related by:

$$v_e = -(k-1)\frac{r_b}{r_e}v_b = -(k-1)\frac{r_b}{\sqrt{kr_{e0}^2 - (k-1)r_b^2}}v_b \quad (3)$$

Since v_b depends at least on temperature, the motion of the external surface of the canister will be the result of a computation in which the boundary conditions (in this case, prescribed displacements) depend on the displacement itself.

The second way to treat the canister corrosion from the mechanical point of view consists in assuming that a volumetric strain can be imposed in the steel elements. In this case, it will be assumed that (1) the corrosion process does not induce internal stresses in the steel nor in the magnetite and (2) the magnetite has the same mechanical properties (elastic parameters) as the steel. Under these conditions, the (partly corroded) canister may be replaced by a homogeneous material that has the same mechanical properties as the steel and that swells homogeneously according to certain law. However, it is clear that the swelling will be only in the direction normal to the surface of the canister.

The swelling law will be determined by requiring that the displacements of the external surface of the (partly corroded) canister due to the corrosion of the steel must be equal with the displacements of the homogeneous material due to the aforementioned swelling. Since the swelling due to the corrosion occurs only in the radial direction, the radial deformation due to corrosion ε_r^c will be equal to the volumetric deformation due to corrosion ε_v^c. Therefore, the radial deformation due to the corrosion can be calculated according to the following formula:

$$\varepsilon_r^c = \varepsilon_v^c = \frac{\pi(r_e^2 - r_i^2) - \pi(r_{e0}^2 - r_{i0}^2)}{\pi(r_{e0}^2 - r_{i0}^2)} = \frac{r_e^2 - r_i^2}{r_{e0}^2 - r_{i0}^2} - 1 \quad (4)$$

where r_{i0} and r_{e0} are the initial radial coordinates of the internal and external surfaces of the canister and r_i and r_e are the corresponding values after the swelling. To get the last equality, it has been taken into account

that the inner surface of the canister does not move due to the swelling process, that is $r_i = r_{i0}$. Differentiation of this expression yields

$$d\varepsilon_r^c = \frac{2r_e dr_e}{(r_{e0} + r_{i0})(r_{e0} - r_{i0})} = -(k-1)$$

$$\frac{r_b dr_b}{\left(\frac{r_{e0} + r_{i0}}{2}\right)(r_{e0} - r_{i0})} = -(k-1)\frac{dr_b}{r_{e0} - r_{i0}} \quad (5)$$

where r_b is the radial coordinate of the current boundary between steel and magnetite. To get the second equality, is has been taken into account that, as shown before, $r_e dr_e = -(k-1)r_b dr_b$. To get the last equality, it has been assumed that the thickness of the canister is small in comparison with the mean radius, so that $r_b \Box (r_{e0} + r_{i0})/2$.

2.2 Approach used and implementation in CODE_BRIGHT

The first approach, prescribed displacements on the bentonite, disregards completely the deformations of the steel canister and the effect of bentonite swelling pressure on the canister. In contrast, the second approach, prescribed swelling in the canister, allows taking into account the deformations due to temperature and pressure in the steel. This means that the swelling of the canister is coupled to the swelling of the bentonite and the movement of the outer boundary of the canister is influenced by both swelling processes. Therefore, it has been this last approach the one that has been implemented in CODE_BRIGHT.

In order to implement the second approach in CODE_BRIGHT, a radial strain induced by the change in volume of steel due to corrosion is incorporated. The increment of radial strain can be written in the following form:

$$d\varepsilon_r^c = -\frac{(k-1)}{r_{e0} - r_{i0}}dr_b = -\frac{(k-1)}{e}v_b dt \quad (6)$$

where e is the thickness of the steel element, k is the coefficient of volume change due to corrosion (for instance $k = 2.1$ when magnetite is produced from steel) and v_b is the measured rate of corrosion. As mentioned before, this formula assumes that the radius of the steel element is large in comparison with its thickness.

Finally, the rate of corrosion can be expressed as a function of time and temperature as:

$$v_b = -(a + b\exp(-ct))\exp\left(-\frac{d}{T + 273.15}\right) \quad (7)$$

414

where t is time and T is temperature. The parameters given by ENRESA are:

$$a = 4.28 \times 10^{-11}\,\text{m/s}$$
$$b = 2.85 \times 10^{-10}\,\text{m/s}$$
$$c = 0.57 \times 10^{-6}\,\text{s}^{-1}$$
$$d = 1500\,\text{K}$$

and have been written in the units that CODE_BRIGHT uses internally. It can be observed that this function depends on temperature, so the rate of corrosion will be affected by this variable.

3 ANALYSIS OF BENTONITE BEHAVIOUR AS CORROSION INDUCED EXPANSION TAKES PLACE

This is a typical 1-D THM analysis of a scheme for waste disposal in which an unsaturated bentonite barrier is subjected to heating induced by the waste. The case is based on the schematic representation of a repository depicted in Figure 1.

In order to investigate the processes taking place through scoping calculations, a 1-D mesh is defined. The finite element mesh is described in Table 1. It assumes axisymetry and the outer boundary of the domain is situated at 9 m from the center of the canister. A bentonite barrier of 0.7 m is considered. The power of the canister is defined by a decaying

function of time in the following way:

$$Q = Q_o \exp(-\alpha t) = 250\,\text{J/s/m} \ \exp(-5 \times 10^{-10}\,\text{s}^{-1} t) \quad (8)$$

The properties of the FEBEX bentonite are described elsewhere (swelling pressure of the

Table 1. Position of nodes in the 1-D finite element mesh.

Node	Radial coordinate (m)	Comments
1	0.35	Inner node of the *canister + liner*, where the power is applied
2	0.48	Outer node of *canister + liner*
3	0.502	First node in the *bentonite* near the liner
15	0.768	Node in the center of *bentonite*
28	1.057	Node in the *bentonite* near the host rock
34	1.19	Node between bentonite and host rock
40	1.44	Node in the *host rock* near the bentonite
49	2.90	Node in the *host rock*
55	6.7	Node in the *host rock* far from the bentonite
58	9.0	End node where constant temperature (28°C), constant liquid pressure (0.1 MPa) and constant stress (12.5 MPa) are prescribed

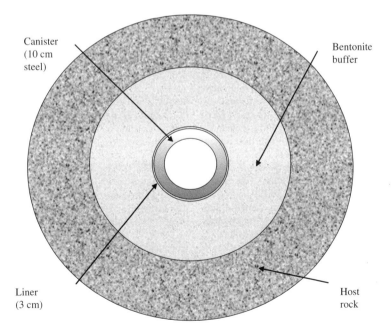

Figure 1. Schematic representation of the repository in which the liner and the canister are made of steel that may undergo corrosion and increase its volume.

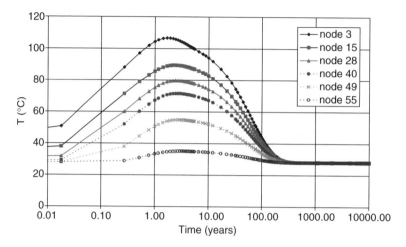

Figure 2.   Time evolution of temperature.

order of 5 MPa) and the host rock is granite ($E = 25000$ MPa and $k = 10^{-18}$ m^2).

Several analyses were performed and included in a report by Olivella et al (2003). For space reasons, only a representative case can be included in this paper. It consists in a case in which it has been considered that the canister and the liner undergo corrosion.

Corrosion implies expansion of the canister (10 cm thick) and liner (3 cm thick). This *corrosion induced expansion* can develop movements of the canister in several ways. The swelling stress developed in the bentonite together with its stiffness may imply that the canister + liner *corrosion induced expansion* generates movements both in the outer boundary and in the inner boundary. The movements depend on the stiffness of the bentonite and the stiffness of the canister + liner system. It is assumed that the canister and liner maintain the elastic properties when corrosion takes place. Here, however, the canister has the inner boundary fixed. This is based on the assumption that the materials inside the canister are very dense and cannot compress.

During corrosion of the liner (and canister) the corrosion rate is increased by a factor of 3 in order to take into account the existence of several surfaces of corrosion: inner liner, outer liner and outer canister. It is assumed that the liner permits circulation or water and hence the canister is also corroded. Later, when the liner has been fully corroded, the rate is reset to the nominal value, because there is only one surface available for corrosion.

Figures 2 and 3 show temperature and degree of saturation. The thermal and hydraulic behaviour are practically not influenced by the *corrosion induced*

expansion because these results are practically identical to the ones obtained in the case of no corrosion.

Figure 4 shows the displacement for different points. Significant displacements are developed compared to the less than 0.01 m values usually calculated in the absence of corrosion. These displacements are induced by the canister + liner movement. Figure 5 shows that the displacement of the inner boundary is zero while the outer boundary moves significantly up to 0.14 m (14 cm). If corrosion was not taken into account, the displacement of the canister + liner element, due to thermal and mechanical effects would have been of the order of 0.001 m (1 mm), i.e. negligible compared to the movement taking place induced by corrosion.

The rate of corrosion is 3 times faster during 2300 years and then decreases to nominal value when the liner is fully corroded. This reduction of the corrosion rate explains the change in the displacement evolution observed in Figure 5. Since the total steel available is 13 cm, the maximum movement will be $(2.1 - 1) \times 13$ cm $= 14.3$ cm. According to Figure 5, full corrosion of the liner and canister is achieved at 10000 years, approximately.

Figure 6 shows that porosity near the canister decreases as the bentonite is compressed. This compression gradually reduces as the radial distance increases which is due to axisymmetry effects.

Figures 7 to 9 show radial stress, circumferential stress and mean-deviatoric stress. Normally, these stresses approach a value of 6 MPa because this is the swelling pressure of the bentonite. Here, however, radial stress increases up to 15 MPa in the center of the bentonite which is much higher than the swell-

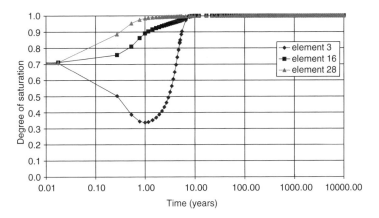

Figure 3. Time evolution of liquid saturation in the bentonite.

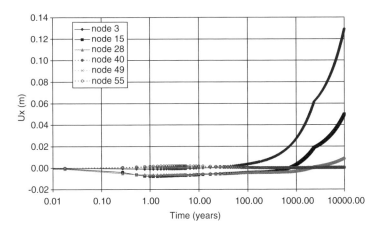

Figure 4. Time evolution of displacements in the bentonite. The movement of Node 3, near the liner + canister, is similar to the movement of the liner + canister outer boundary (see Figure 5).

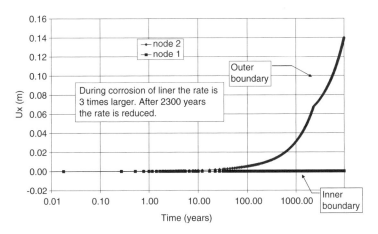

Figure 5. Time evolution of displacements in the inner and outer boundary of the steel element that simulates the canister + liner (the maximum movement achievable by corrosion is 0.143 m).

417

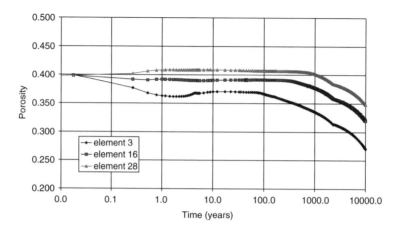

Figure 6. Time evolution of porosity in the bentonite. The decrease that takes place from 100 years is due to expansion of the steel as corrosion takes place.

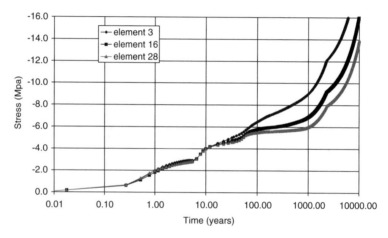

Figure 7. Time evolution of radial stress in the bentonite.

ing pressure indicating that an extra compression of the barrier has taken place. Radial compression of the bentonite, however, implies a reduction of the circumferential stress. The circumferential stress achieves 5 MPa at 100 years due to hydration, then it decreases because the radial displacement induces extension in the circumferential direction. Afterwards, as compression continues, circumferential stress increases again. In this situation, radial stress is much higher than circumferential stress. Figure 9 shows that both deviatoric and mean stresses increase. At 10 years mean stress has increased up to 4 MPa due to hydration and the deviatoric stress is practically negligible. This situation is roughly maintained up to 80 years approximately because the effect of corrosion is small during this period. But

afterwards, the deviatoric stress starts to increase and mean stress decreases somewhat. After 1000 years, both increase due to corrosion induced compression.

The effect of corrosion on bentonite is quite significant because, as indicated before, if no corrosion is considered then deviatoric stress remains very small and mean stress approaches the swelling pressure of 6 MPa.

These variations of mean stress and the deviatoric stress can be better understood if the stress path is represented. This has been done in Figure 10. It can be seen that yielding of the bentonite is approached due to radial compression of the barrier. As the crititcal state line is approached the means stress increases, and this is due to the confinement provided by the host rock.

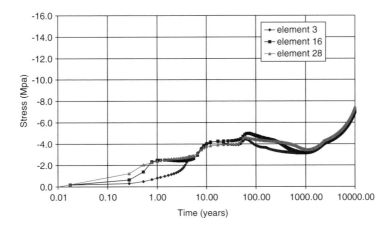

Figure 8.   Time evolution of radial stress in the bentonite.

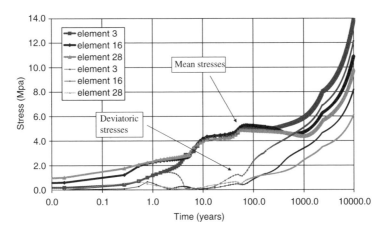

Figure 9.   Time evolution of mean and deviatoric stresses in the bentonite.

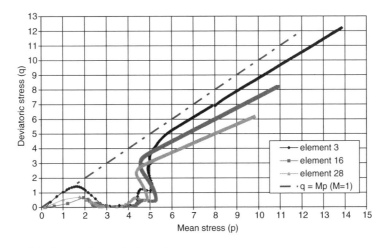

Figure 10.   Stress path in the q-p plane in the bentonite.

4 CONCLUSIONS

Corrosion of canisters is expected in the long term conditions. Corrosion of canisters is associated with several processes which include interactions with the bentonite barrier. This work investigates the effect of *corrosion induced expansion* on the bentonite barrier.

Specific changes have been introduced in CODE_BRIGHT to analyse the *corrosion induced expansion*. The basic idea is that, as corrosion progresses, the volume of the canister increases. The variation of volume is calculated from the rate of corrosion which in turn is a function of temperature and time. This rate has been measured and calibrated parameters are available.

Once the *corrosion induced expansion* has been implemented in CODE_BRIGHT a series of analyses have been conducted and a representative case is presented in this paper. This analysis consists in heating an unsaturated clay barrier with a 3 cm liner plus 10 cm canister which undergoes drying and hydration. For long term conditions, the bentonite buffer develops a swelling pressure due to full saturation.

The case presented in the paper considered fixed inner boundary and a faster corrosion rate has been assumed to take place during corrosion of the liner because three corrosion fronts may be simultaneously active. When the liner is fully corroded, the corrosion rate comes back to the rate equivalent to a single active corrosion front. Expansion of the steel as it transforms into magnetite induces a movement of the liner outer boundary up to 14 cm. This implies a significant compression of the bentonite and, consequently, high mean stress and deviatoric stresses are developed, well above of the swelling pressure which would prevail in case without taking into account the effects of corrosion.

ACKNOWLEDGEMENTS

This work has been performed in the framework of the project BENIPA funded by ENRESA.

REFERENCES

Alcoverro, J., S. Olivella and E.E. Alonso, (2003) Stresses and strains in the canister induced by corrosion. This issue. Large Scale Field Tests In Granite.

Olivella, S., A. Gens, J. Carrera, and E.E. Alonso, (1995). Numerical formulation for a simulator (CODE_B-RIGHT) for the coupled analysis of saline media. In Engineering Computations, n. 13, pp. 87–112.

Olivella, S., J. Alcoverro and E.E. Alonso, (2003) ENRESA Technical Report 70-UPC-L-1-001, Project BENIPA.

Advances in Understanding Engineered Clay Barriers – Alonso & Ledesma (eds)
© 2005 Taylor & Francis Group, London, ISBN 04 1536 544 9

Stresses and strains in a canister induced by corrosion

J. Alcoverro, S. Olivella & E.E. Alonso
Universitat Politènica de Catalunya, Barcelona (Spain)

ABSTRACT: A simplified model for the stresses and strains induced by the corrosion process of a steel canister is presented. Using this model, the influence of some parameters on the strain and stress fields in the canister is analysed. The analysis shows the development of stresses in the canister, which depend on the state of corrosion. The magnitude of these induced stresses depends on the properties of both steel and magnetite, and may lead to failure at the boundary between the uncorroded and corroded zones of the canister.

1 INTRODUCTION

After the emplacement of a steel canister containing radioactive waste into a bentonite barrier, a number of phenomena take place as a consequence of the interactions of the canister with its environment. One of these processes is canister corrosion, whereby steel (Fe) is transformed into magnetite (Fe_3O_4) with an increase of volume. Canister corrosion initiates on the exterior surface of the canister and proceeds inwards, eventually leading to the complete corrosion of the canister. This is a comparatively slow process, with a characteristic time larger than the characteristic time of hydration of the bentonite barrier. In a companion paper (Olivella et al., 2003), the (mechanical) effect of the canister corrosion on the bentonite barrier has been analysed. The present paper focuses on the (mechanical) effect that the canister corrosion induces on the canister itself, whereby a simplified analysis has been conducted that allows to obtain in a closed form the stress and strain distributions both in the uncorroded (steel) and in the corroded (magnetite) zones of the steel canister.

2 MECHANICAL MODEL OF THE CANISTER CORROSION

2.1 *Simplifying assumptions*

The model of the canister corrosion that will be used is based on the following simplifying assumptions:

– Plain strain and axisymmetry about the (cylindrical) canister axis.
– Infinitesimal strains.
– Static conditions.
– Constant temperature.

– Due to corrosion, steel is transformed into magnetite.
– Steel is an isotropic material. Its strain tensor is related to the stress tensor by Hooke's law.
– Magnetite is an isotropic material. Its strain tensor is the sum of an elastic strain (related to the stress tensor by Hooke's law) and a constant spherical strain (due to corrosion).
– There is no gap between the uncorroded (steel) and the corroded (magnetite) zones.

This set of assumptions, that seems reasonable as a first approximation to the problem at hand, makes the problem one-dimensional and solvable in closed form.

2.2 *Field equations*

When dealing with plain strain and axisymmetric conditions, it is convenient to introduce a system of cylindrical coordinates (r, θ, z), with its origin located on the axisymmetry axis and its z-axis parallel to it. Here, r is the radial coordinate (distance to the axisymmetry axis), θ is the angular coordinate (angle that the projection of the position vector makes with a reference direction perpendicular to the axisymmetry axis) and z is the axial coordinate (distance to the plane perpendicular to the axisymmetry axis passing through the origin).

As a consequence of the plane strain, axisymmetry and static conditions, all fields depend only on the radial coordinate r. Moreover, the (local) coordinate directions (r, θ, z) are principal directions of the (infinitesimal) strain tensor ε and of the stress tensor σ. In what follows, we will use the subindices r, θ and z to denote the corresponding (physical) principal components of these tensors and the (physical) components of the displacement vector $\boldsymbol{u}$.

In these conditions, the (physical) principal components of the (infinitesimal) strain tensor are related to the radial displacement u_r by the following relations:

$$\varepsilon_r = \frac{du_r}{dr}, \quad \varepsilon_\theta = \frac{u_r}{r} \quad \text{and} \quad \varepsilon_z = 0. \tag{1}$$

On the other hand, the radial component of the balance of momentum (the angular and axial components are identically satisfied in the assumed conditions) reads:

$$\frac{d\sigma_r}{dr} + \frac{\sigma_r - \sigma_\theta}{r} = 0. \tag{2}$$

According to the assumptions made, the (infinitesimal) strain tensor is the sum of: (1) an elastic strain given by Hooke's law; and (2) a constant spherical strain due to corrosion, that is zero for the steel. If the reference configuration is stress free, then the constitutive relations of the steel and of the magnetite are of the form:

$$\varepsilon_r = \frac{1}{E}\left[\sigma_r - v(\sigma_\theta + \sigma_z)\right] + \alpha, \tag{3a}$$

$$\varepsilon_\theta = \frac{1}{E}\left[\sigma_\theta - v(\sigma_r + \sigma_z)\right] + \alpha, \tag{3b}$$

$$\varepsilon_z = \frac{1}{E}\left[\sigma_z - v(\sigma_r + \sigma_\theta)\right] + \alpha, \tag{3c}$$

where E is Young's modulus, v is Poisson's modulus and α is the constant spherical strain due to corrosion, that is zero for the steel. From the constitutive relations (3) and the plane strain condition (1c), it follows that the axial stress is:

$$\sigma_z = v(\sigma_r + \sigma_\theta) - \alpha E. \tag{4}$$

Substituting (4) in (3a–b), solving for σ_r and σ_θ as a function of ε_r and ε_θ, substituting them in (1) and taking into account (1a–b) yields an ordinary differential equation of second order:

$$r^2 \frac{d^2 u_r}{dr^2} + r \frac{du_r}{dr} - u_r = 0, \tag{5}$$

whose general solution is of the form:

$$u_r = Ar + \frac{B}{r}, \tag{6}$$

where A and B are constants to be determined by use of suitable boundary conditions. Substitution of

(6) into (1) yields

$$\varepsilon_r = A - \frac{B}{r^2}, \quad \varepsilon_\theta = A + \frac{B}{r} \quad \text{and} \quad \varepsilon_z = 0. \tag{7}$$

Finally, substitution of (7) into (3) and solution of the resulting system, yields:

$$\sigma_r = \frac{E}{(1+v)(1-2v)}\left[A - (1-2v)\frac{B}{r^2} - (1+v)\alpha\right], \tag{8a}$$

$$\sigma_\theta = \frac{E}{(1+v)(1-2v)}\left[A + (1-2v)\frac{B}{r^2} - (1+v)\alpha\right], \tag{8b}$$

$$\sigma_z = \frac{E}{(1+v)(1-2v)}\left[2Av - (1+v)\alpha\right]. \tag{8c}$$

2.3 Solution of the problem

Both in the uncorroded (steel) and in the corroded (magnetite) zones, the form of the radial displacement field is given by (6). However, the constants A and B will not be the same in both zones. In what follows, we will use the superindices s and m to indicate, respectively, pertinence to the uncorroded (steel) or to the corroded (magnetite) zones. In order to solve the problem at hand, we must supply boundary conditions for both the uncorroded and the corroded zones.

Rather generally, we will assume that the canister is subjected to both an internal and an external pressure. The internal pressure is due to the action of the radioactive waste on the canister, whereas the external pressure is due to the action of the bentonite barrier on the canister. The values of these pressures will vary with time as a result of the mechanical interactions among those elements. In the present analysis, however, they will be taken as parameters.

Therefore, we will consider the following boundary conditions: at the interior boundary of the canister, a prescribed internal pressure p_i; at the boundary between the uncorroded and corroded zones, continuity of the radial displacement field and continuity of the radial stress field; and at the exterior boundary of the canister, a prescribed external pressure p_e. More precisely, we will use the following boundary conditions:

$$\sigma_r^s(r_i) = -p_i, \tag{9a}$$

$$\sigma_r^s(r_b) = \sigma_r^m(r_b) \quad \text{and} \quad u_r^s(r_b) = u_r^m(r_b) \tag{9b,c}$$

$$\sigma_r^m(r_e) = -p_e, \tag{9d}$$

where r_i is the radial coordinate of the interior boundary of the canister, r_b is the radial coordinate of the boundary between the uncorroded and the corroded zones and r_e is the radial coordinate of the exterior boundary of the canister. Note that conditions (9b-c) at the boundary between the uncorroded and the corroded zones result from the assumption made that there is no gap at that boundary.

Using expressions (6) and (8a), these boundary conditions may be written in the following form:

$$\frac{E^s}{(1+v^s)(1-2v^s)}\left[A^s - (1-2v^s)\frac{B^s}{r_i^2}\right] = -p_i, \quad (10a)$$

$$\frac{E^s}{(1+v^s)(1-2v^s)}\left[A^s - (1-2v^s)\frac{B^s}{r_b^2}\right] = \frac{E^m}{(1+v^m)(1-2v^m)}$$

$$\left[A^m - (1-2v^m)\frac{B^m}{r_b^2} - (1+v^m)\alpha^m\right] \quad (10b)$$

$$A^s r_b + \frac{B^s}{r_b} = A^m r_b + \frac{B^m}{r_b} \quad (10c)$$

$$\frac{E^m}{(1+v^m)(1-2v^m)}\left[A^m - (1-2v^m)\frac{B^m}{r_e^2}\right.$$

$$\left. - (1+v^m)\alpha^m\right] = -p_e, \quad (10d)$$

where it has been taken into account that the constant spherical strain due to corrosion of the steel α^s is zero. By solving the linear system of equations (10), we obtain the values of A^s, B^s, A^m and B^m that, when replaced in (6), (7) and (8), give the radial displacement, the strain and the stress fields in the uncorroded (steel) and in the corroded (magnetite) zones of the steel canister.

3 RESULTS PREDICTED BY THE MODEL

Using the model developed, we have analysed the mechanical effects that the corrosion induces in the steel canister by assigning values to the various model parameters. The values considered in these analyses are summarised in Table 1. In selecting them, we have considered the following:

– Pressures (p_i, p_e). Since the main concern here is the effect that corrosion induces in the canister, we have taken both the internal and the external pressures equal to zero.
– Canister geometry (r_i, r_e). Similar to the geometry of the canisters considered by ENRESA.
– State of corrosion (r_b). Corrosion of the outer 25% of the steel canister.
– Steel properties (E^s, v^s). Values taken from the standard ranges for steel.
– Magnetite properties (E^m, v^m). As we did not have reliable values, we selected "reference" values for them, defined by $E^m/E^s = 0.10$ and $v^m/v^s = 0.10$, and we carried out two series of analyses. In the first, the reference value of E^m was reduced by factors of 2 and 10. In the second, the reference value of v^m was increased by factors of 5 and 10.
– Strain due to corrosion (α^m). A value provided by ENRESA has been used.

In order to illustrate the results of these analyses, we have selected plots showing the distributions of displacements and stresses in the canister. More precisely, the selected plots are:

– Figures 1a and 1b. Radial displacement u_r.
– Figures 2a and 2b. Radial stress σ_r.
– Figures 3a and 3b. Circumferencial stress σ_θ.
– Figures 4a and 4b. Axial stress σ_z.
– Figures 5a and 5b. Mean deviatoric stress $\sqrt{3J_2}$.

Type "a" figures refer to the sensitivity analysis with respect to E^m and type "b" figures refer to the sensitivity analysis with respect to v^m. Concerning

Table 1. Values of the parameters used in the analyses.

	Reference	0.5 E^m	0.1 E^m	5 v^m	10 v^m	Units
p_i	0.000	0.000	0.000	0.000	0.000	MPa
p_e	0.000	0.000	0.000	0.000	0.000	MPa
r_i	0.350	0.350	0.350	0.350	0.350	m
r_b	0.425	0.425	0.425	0.425	0.425	m
r_e	0.450	0.450	0.450	0.450	0.450	m
E^s	$183 \cdot 10^3$	$183 \cdot 10^3$	$183 \cdot 10^3$	$183 \cdot 10^3$	$183 \cdot 10^3$	MPa
v^s	0.300	0.300	0.300	0.300	0.300	–
E^m	$183 \cdot 10^2$	$91.5 \cdot 10^2$	$18.3 \cdot 10^2$	$183 \cdot 10^2$	$183 \cdot 10^2$	MPa
v^m	0.030	0.030	0.030	0.150	0.300	–
α^m	0.367	0.367	0.367	0.367	0.367	–

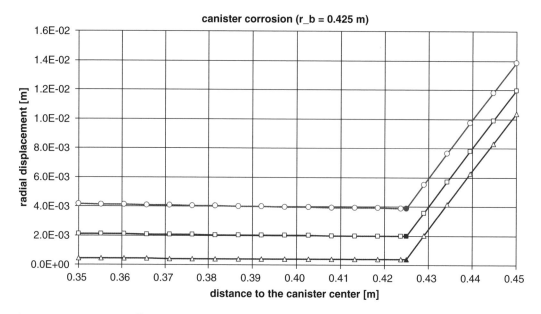

Figure 1a. Sensitivity to E^m of the radial displacement u_r distribution. $E^m/E^s = 0.10$ (circles), 0.05 (squares), 0.01 (triangles) and $v^m/v^s = 0.10$.

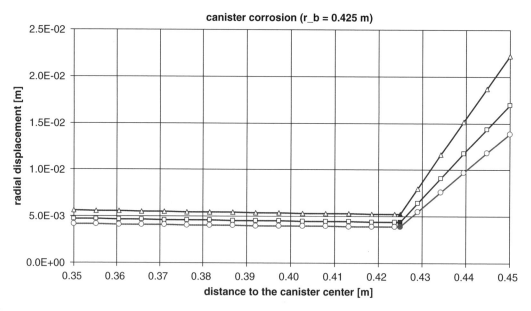

Figure 1b. Sensitivity to v^m of the radial displacement u_r distribution. $E^m/E^s = 0.10$ and $v^m/v^s = 0.10$ (circles), 0.50 (squares), 1.00 (triangles).

the symbols, we use: (1) circles for the "reference" case; (2) squares for the parameter variation (E^m or v^m) closer to the "reference" case; and (3) triangles for the parameter variation (E^m or v^m) farther to the "reference" case.

Looking at the results, we may summarise the observed trends as follows:

– (Fig. 1a) smaller E^m lead to smaller u_r, and (Fig. 1b) larger v^m lead to larger u_r (mainly in the corroded zone).

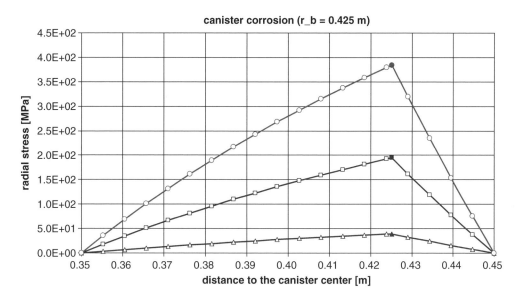

Figure 2a. Sensitivity to E^m of the radial stress σ_r distribution. $E^m/E^s = 0.10$ (circles), 0.05 (squares), 0.01 (triangles) and $v^m/v^s = 0.10$.

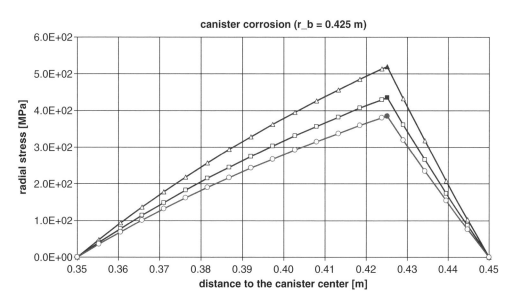

Figure 2b. Sensitivity to v^m of the radial stress σ_r distribution. $E^m/E^s = 0.10$ and $v^m/v^s = 0.10$ (circles), 0.50 (squares), 1.00 (triangles).

– (Fig. 2a) smaller E^m lead to smaller (in absolute value) σ_r, whereas (Fig. 2b) larger v^m lead to larger σ_r (in absolute value).
– (Fig. 3a) smaller E^m lead to smaller (in absolute value) σ_θ (mainly in the corroded zone), whereas

(Fig. 3b) larger v^m lead to larger (in absolute value) σ_θ (mainly in the corroded zone).
– (Fig. 4a) smaller E^m lead to smaller (in absolute value) σ_z (mainly in the corroded zone), whereas (Fig. 4b) larger v^m lead to larger (in absolute value) σ_z (mainly in the corroded zone).

425

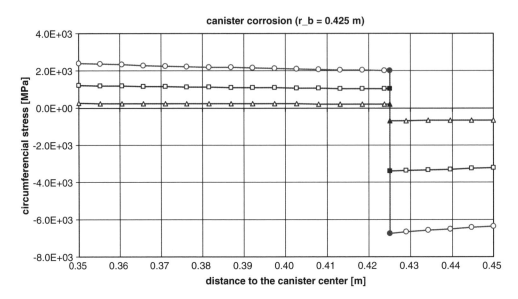

Figure 3a. Sensitivity to E^m of the circumferencial stress σ_θ distribution. $E^m/E^s = 0.10$ (circles), 0.05 (squares), 0.01 (triangles) and $v^m/v^s = 0.10$.

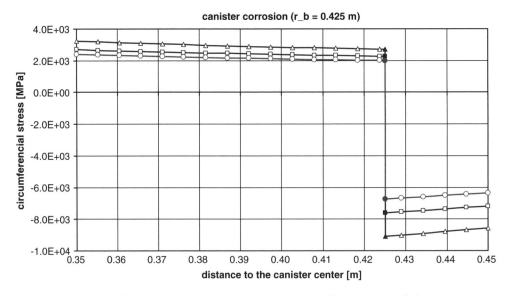

Figure 3b. Sensitivity to v^m of the circumferencial stress σ_θ distribution. $E^m/E^s = 0.10$ and $v^m/v^s = 0.10$ (circles), 0.50 (squares), 1.00 (triangles).

– (Fig. 5a) smaller E^m lead to smaller $\sqrt{3J_2}$ (mainly in the corroded zone), whereas (Fig. 5b) larger v^m lead to larger $\sqrt{3J_2}$ (mainly in the corroded zone).

Although the present model is very simple, it points out some aspects that might be relevant for a more accurate modelling of canister corrosion in the context of canister disposal into a bentonite barrier, namely:

– The values of E^m and v^m have a large influence on the displacement field in the canister (see Fig. 1a and Fig. 1b) and on the stiffness of the canister.

426

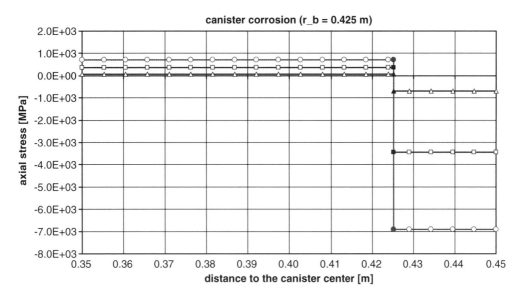

Figure 4a. Sensitivity to E^m of the axial stress σ_z distribution. $E^m/E^s = 0.10$ (circles), 0.05 (squares), 0.01 (triangles) and $v^m/v^s = 0.10$.

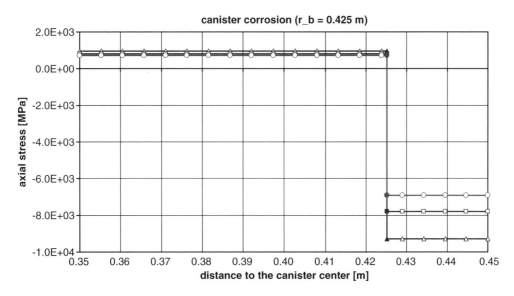

Figure 4b. Sensitivity to v^m of the axial stress σ_z distribution. $E^m/E^s = 0.10$ and $v^m/v^s = 0.10$ (circles), 0.50 (squares), 1.00 (triangles).

– If magnetite can be appropriately modelled as a von Mises material, then the boundary between the uncorroded and corroded zones is closer to failure (see Fig. 5a and Fig. 5b).

Finally, it should be mentioned that the results obtained are also dependent on the state of corrosion of the canister. In this regard, the state of corrosion of 25% that has been chosen seems reasonable as a first approximation to the problem.

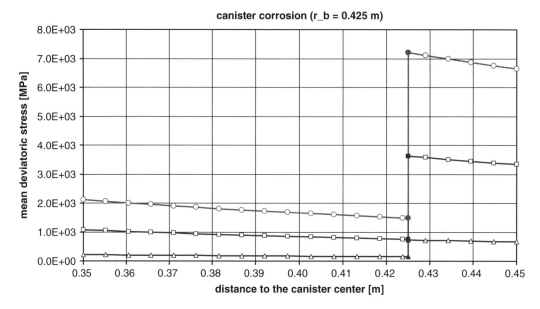

Figure 5a. Sensitivity to E^m of the mean deviatoric stress $\sqrt{3J_2}$ distribution. $E^m/E^s = 0.10$ (circles), 0.05 (squares), 0.01 (triangles) and $v^m/v^s = 0.10$.

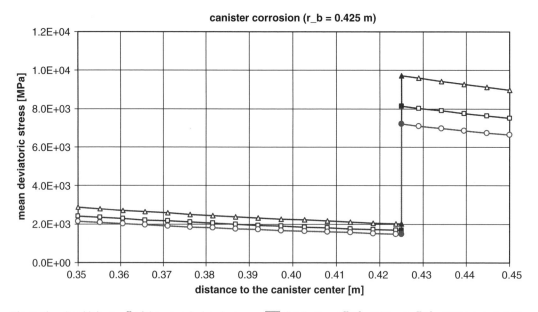

Figure 5b. Sensitivity to v^m of the mean deviatoric stress $\sqrt{3J_2}$ distribution. $E^m/E^s = 0.10$ and $v^m/v^s = 0.10$ (circles), 0.50 (squares), 1.00 (triangles).

4 CONCLUSIONS

A simplified model for the corrosion of a steel canister that has a closed form solution has been developed. In this model, both the uncorroded (steel) and the corroded (magnetite) zones of the steel canister have been modelled as linear elastic materials, whereby associated to corrosion there is a constant spherical strain. The model has been used with realistic values for its parameters with the exception of the elastic properties of the magnetite, that have been varied in order to analyse their influence. Although the analysis is only a first approximation to the problem, it is useful in that it indicates some aspects that may be relevant in a more accurate modelling of a steel canister placed into a bentonite barrier, such as the large influence of the elastic properties of the magnetite on the stiffness of the canister and the proximity to failure of the boundary between the uncorroded and the corroded zones of the steel canister.

ACKNOWLEDGEMENTS

This work has been performed in the framework of the project BENIPA funded by ENRESA.

REFERENCES

Olivella S, Alcoverro, J and Alonso, EE (2003) Impact of Canister Corrosion on Barrier Behaviour. Proceedings of the Symposium *Large Scale Field Tests in Granite*, November, 12th–14th, Sitges (this issue).

Advances in Understanding Engineered Clay Barriers – Alonso & Ledesma (eds)
© 2005 Taylor & Francis Group, London, ISBN 04 1536 544 9

Thermo-hydro-mechanical simulation of the concrete bulkhead during the heating phase of the tunnel sealing experiment

R. Guo, B. Kjartanson & J. Martino

Geotechnical Science and Engineering Branch, Atomic Energy of Canada Limited, Pinawa, Canada

ABSTRACT: A numerical analysis for the concrete bulkhead and surrounding rock of the Tunnel Sealing Experiment (TSX) was conducted using MOTIF. MOTIF is a fully coupled thermo-hydro-mechanical analysis program developed by AECL. The fully coupled response of the concrete bulkhead and adjacent rock to heating obtained from this modelling is compared with measured data. Specifically, the modelled and measured spatial and temporal responses include the temperature, hydraulic head and displacement evolution.

1 INTRODUCTION

The waste emplacement configurations proposed for a deep geologic repository for the permanent isolation of nuclear fuel waste in Canada have a number of features in common. The repository configuration consists of a horizontal array of waste emplacement rooms at a nominal depth of 500 to 1000 m in plutonic rock of the Canadian Shield. Each waste emplacement room contains used fuel within corrosion-resistant containers placed either vertically within boreholes drilled into the floors of the rooms (in-floor borehole emplacement method) or horizontally within the confines of each room (in-room emplacement method) (Russell and Simmons, 2003). The containers are isolated from the rock by repository sealing materials (e.g., clay-based buffer and back-fills). The long-term safety of an underground repository for heat-generating used fuel relies on the ability of a series of barriers to limit the rates of container corrosion, used-fuel dissolution and radionuclide release. Both natural and engineered barriers are to be used to achieve this purpose.

The Tunnel Sealing Experiment (TSX) (Chandler et al., 2002b) was constructed in Atomic Energy of Canada Limited's (AECL's) Underground Research Laboratory (URL) to evaluate the ability of two full-sized bulkhead seals to minimize water flow in the axial direction of a tunnel under ambient and elevated temperature conditions.

In this paper the results of analysis conducted for the concrete bulkhead for the heating phase of AECL's TSX are presented and are compared with data collected during the first 180 days of heating (when the paper was written, the TSX was only heated for 180 days).

2 THE TUNNEL SEALING EXPERIMENT

The TSX (Chandler et al., 2002a) is located on the 420 Level of AECL's Underground Research Laboratory (URL). It is an international project conducted jointly by JNC of Japan, ANDRA of France and AECL to investigate technologies for construction of repository seals. The configuration of the TSX is presented in Figure 1. Two types of bulkheads have been constructed, one composed of highly compacted sand-bentonite blocks and the other constructed using low-heat high-performance (LHHP) concrete (Fig. 1). The LHHP concrete developed by AECL has the properties of high strength, low heat of hydration, low shrinkage and low pH, all of which are desirable for use in nuclear

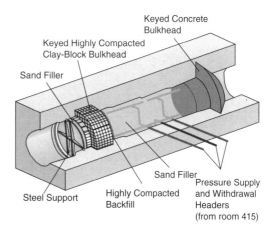

Figure 1. Configuration of the tunnel sealing experiment.

waste disposal. The 11.45-m-long region between the two bulkheads was filled with sand and water. The water was pressurized to 4 MPa, which is similar to the ambient pore pressure in the far-field rock. The objectives of the TSX are described by Chandler et al. (2002b). One of the objectives is to evaluate the influence of elevated temperature on the performance of the bulkheads and adjacent rock. In order to raise the temperature of the TSX a circulating water system was installed as shown in Figure 1. The heated water is pumped from heaters located in near-by room through the inlet pipe to the inlet headers located at the bottom of the TSX chamber, and flows through the chamber sand to the outlet headers located at the top of the TSX chamber and then through the outlet pipes back to the heaters.

The tunnel dimensions are representative of a full-scale excavation in a repository. The elliptical shaped tunnel with a width-to-height ratio of 1.25 (4.375-m width to 3.5-m height) was selected to limit the formation of an Excavation Damaged Zone (EDZ) under the known *in situ* rock stresses. The concrete and clay bulkheads were keyed into the rock to cut off flow through the EDZ. A cement-based grout was injected into the interface between the concrete bulkhead and the rock to reduce flow along the interface.

Numerical modelling shows that the vertical faces on a key would minimize damage in the rock surrounding the key (Chandler et al., 2002b). Modelling also shows a rigid conical-shaped concrete bulkhead (Fig. 2) would better transfer some of the load from the pressure chamber onto the rock. It was decided to construct a vertical key for the clay

bulkhead and a conical key for the concrete bulkhead. A different shape was selected for the clay bulkhead as it was less rigid than the concrete bulkhead and was easier to construct.

Some instrument locations on a vertical cross section through the axis of the TSX are shown in Figure 2. Five thermistors (VSM02T, VSM08T, VSM19T, VSM24T, and VSM27T) were installed to measure the temperature in the concrete bulkhead and a thermo-couple TPC28T was installed in the interface between the concrete bulkhead and the concrete backfill. HGT1 (Fig. 3) includes five packer isolated zones, in which the pore pressure is measured

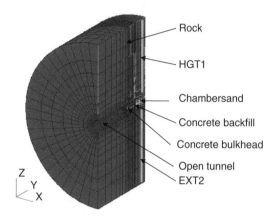

Figure 3. Finite element discretization of the concrete model.

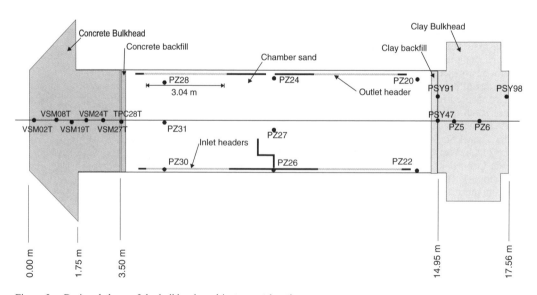

Figure 2. Designed shape of the bulkheads and instrument locations.

by pressure transducers. EXT2 (Fig. 3) has six linear-variable-displacement-transformers (LVDT) measuring displacement between fixed anchors and five thermistors measuring temperature.

3 NUMERICAL MODELLING FOR THE TSX

3.1 General assumptions and parameters used in the modelling

One of the objectives of the thermo-hydro-mechanical (T-H-M) modelling of the TSX was to simulate heating response in the bulkheads and surrounding rock. Heat transfer by thermal convection, which is the primary mechanism for heat transfer in the TSX chamber, is not accommodated by many commercially available analysis codes. MOTIF (Model Of Transport In Fracture/porous media) was developed by AECL to investigate T-H-M processes in fractured rock (Guvanasen and Chan, 1992, 2000) but was also used successfully to simulate thermal convection in the sand chamber of the TSX (Guo and Chandler, 2002, Guo et al., 2002, Guo et al., 2003).

Owing to the large computational memory requirements for three-dimensional T-H-M modelling, the TSX simulation had to be divided into two parts. One focuses on the clay bulkhead and is referred to as the TSX clay model, while the other focuses on the concrete bulkhead and is referred to as the TSX concrete model. Only the T-H-M simulation for the concrete bulkhead during heating is described.

In the concrete bulkhead modelling, the rock mass surrounding the chamber of the TSX, the concrete bulkhead and the sand emplaced in the chamber were all assumed to be homogeneous, isotropic and linearly elastic. Water viscosity and density are nonlinear functions of temperature included in MOTIF (Chan et al., 1999). The other thermal, hydraulic, and mechanical properties of the various components in the concrete model (Table 1) were assumed to be temperature-independent. The compressibility of water was set to $4.5 \times 10^{-10}\,\mathrm{Pa}^{-1}$.

3.2 Model geometries, boundary conditions and initial conditions

The three-dimensional concrete model geometry is illustrated in Figure 3. The axial dimension is 24.115 m and the radial dimension is 25 m. The tunnel is assumed to have a circular cross-section with an average radius of 2 m. The chamber length is 5.3 m. There are 8908 eight-noded elements and 10148 nodes in the concrete model.

Guo et al. (2003) describe modelling of the pressurization of the water in the TSX chamber under ambient temperature conditions. The initial hydraulic-head distribution for the model was based on the results from the hydro-mechanical modelling for chamber pressurization. The initial temperature for the model was set to 14.5°C. The initial mechanical condition for the simulation was a stress of 0 MPa for the whole model.

Four hydraulic-boundary conditions are defined for the simulations.

– The hydraulic head on the outside curved surface of the model is set to 400 m.
– No hydraulic flow occurs across the vertical symmetric surface through the axis of the tunnel and the two end surfaces perpendicular to the axis of tunnel.
– The hydraulic heads at the nodes on the surface of the open-portion of the tunnel and the downstream surface of the bulkheads are constrained to 0 m.
– The hydraulic heads at the inlet-header nodes are constrained to 410 m and at the outlet-header nodes are constrained to 400 m. Positive fluid flows are set at the inlet-header nodes. During heating the total flow rate was set at 7 L/min during the first 40 days. The flow rate was adjusted to 8 L/min after 40 days and again to 8.3 L/min after 90 days.

Table 1. Thermal and mechanical properties of components in the concrete model.

	Concrete	Concrete backfill	Rock	Chamber sand
Hydraulic conductivity (m/s)	3×10^{-14}	1×10^{-5}	1×10^{-14}	6.25×10^{-5}
Thermal conductivity (W/(m°C))	1.8	1.0	3.5	2
Specific heat capacity (J/(kg°C))	900	900	1015	820
Young modulus (MPa)	3.6×10^4	1×10^3	5.759×10^4	100
Poisson's ratio	0.155	0.155	0.207	0.3
Thermal expansivity (1/°C)	1.0×10^{-5}	1.0×10^{-5}	1×10^{-6}	0
Density (kg/m³)	2430	2430	2650	1850 (dry)
Porosity	0.1	0.1	0.005	0.287
Biot's coefficient	0.2	1.0	0.2	1.0

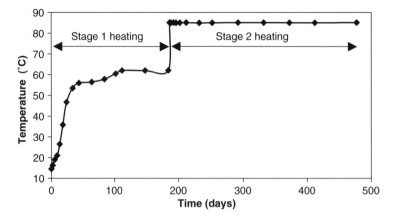

Figure 4. Temperature of the water in the inlet header.

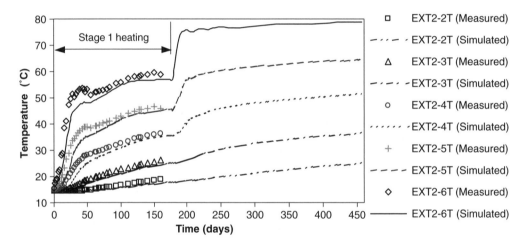

Figure 5. Comparison of measured and simulated temperatures at the extensometer 2 thermistor locations.

Two thermal-boundary conditions are imposed in the simulations.

- No heat loss occurs across the two end surfaces, the outside curved surface, the vertical symmetric plane, the downstream surface of the concrete bulkhead and the inner surface of the open-portion of the tunnel.
- The temperature of the inlet-header nodes follows the temperature of the water before going into the inlet headers, as shown in Figure 4.

Four mechanical-boundary conditions were imposed in the simulation.

- The nodes on the vertical symmetric surface through the axis of the tunnel are fixed in the X-direction. (refer to Fig. 3 for the X-, Y- and Z-directions).
- The nodes on the end surface near the empty tunnel are fixed in the Y-direction.

- The nodes on the end surface crossing the chamber are fixed in the Y-direction.
- The nodes on the outside curved surface are fixed in the X- and Z-directions.

In addition, the concrete is bonded to the rock for these analyses.

4 SIMULATED THERMAL RESULTS AND COMPARISONS WITH MEASURED DATA

4.1 Simulated thermal results and comparisons with measured data

Figure 5 compares the measured and simulated rock temperatures with time at different locations along extensometer string EXT2 (refer to Fig. 3 for the

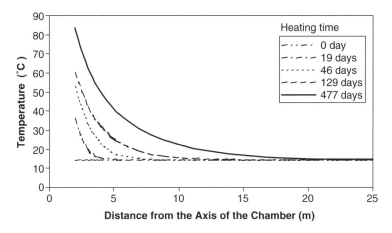

Figure 6. Simulated temperature distributions along EXT2 under the inlet header in the rock.

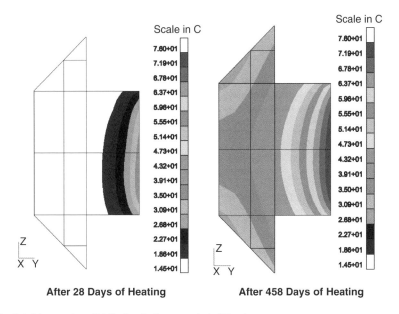

Figure 7. Simulated temperature distribution in the concrete bulkhead.

locations) below the floor of the pressure chamber. The simulated results match the measured results very well. The simulated results show that the temperature in the rock is predicted to continue increasing after 458 days of heating.

Figure 6 shows the simulated-temperature distributions along EXT2 under the inlet header in the rock at six different times. The rock temperature increased very quickly within 2 m of the surface of the chamber. The thermal front did not reach the model's outer boundary, meaning that the outer

radius of 25 m is sufficiently large to conduct the thermal simulation.

Figure 7 shows the simulated temperature distributions on the vertical longitudinal section through the axis of the concrete bulkhead after 28 days and 458 days of heating. The temperature on the upstream surface of the concrete bulkhead was 37°C after 28 days of heating and 77°C and 37°C on the upstream and downstream surfaces, respectively, after 458 days of heating. Note that an adiabatic boundary condition was assigned to the downstream surface of

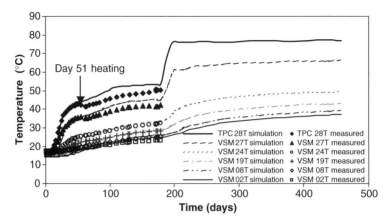

Figure 8. Comparison of measured and simulated temperature at different thermistor locations during the first 180 days of heating.

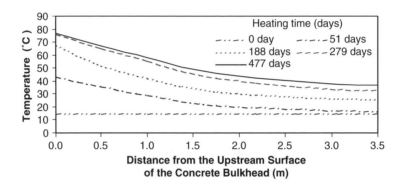

Figure 9. Simulated temperature distributions along the central axial of the concrete bulkhead remaining key uncertainties in simulation of thermally-induced hydraulic response.

the concrete bulkhead. An insulated wall was constructed close to the face of the concrete bulkhead, providing justification for using the adiabatic boundary condition.

Figure 8 shows the temperature increase with time at different thermistor locations along the axis of the concrete bulkhead and provides a comparison between the simulated temperatures and measured data. For thermistors TPC28T and VSM27T (Fig. 2), the simulated temperatures matched the measured temperature very well during the first 50 days of heating. After day 51, the simulated temperatures were 3 to 5°C higher than those measured. This may have been caused by a temporary shutdown of the heated water circulation system and by a loss of heated water through a leak at day 51 of heating. Generally, the trend of the simulated results matched the trend of the measured data. The sudden loss of hydraulic head in the chamber may have caused strain

reversals in the materials, affecting their thermal and hydraulic conductivities.

Figure 9 shows the simulated temperature distributions along the axis of the concrete bulkhead at six different times. After 458 days of heating the temperature of the concrete bulkhead is predicted to reach 77°C at the upstream side and 37°C at the downstream side.

4.2 *Simulated hydraulic results and comparisons with measured data*

Figures 10 and 11 show the measured and simulated hydraulic head changes, respectively, at locations corresponding to HGT1-1 to HGT1-5 during heating. The simulated hydraulic head changes for HGT1-1 to HGT1-5 have the same pattern as the measured data, but the simulated values are smaller than the measured data. The selection of 0.2 for Biot's coefficient is

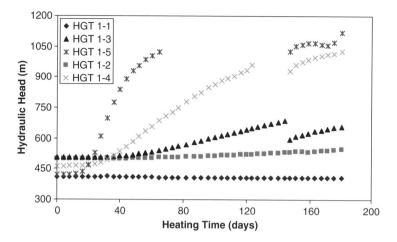

Figure 10.   Measured hydraulic heads at packer locations HGT1-1 to HGT1-5 during heating.

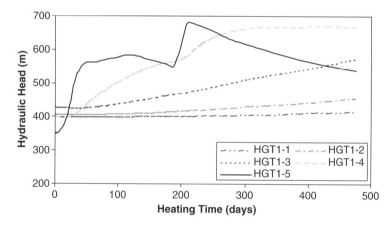

Figure 11.   Simulated hydraulic head changes at packer locations HGT1-1 to HGT1-5 during heating using a Biot's coefficient of 0.2.

somewhat uncertain as the value has been shown to be greatly dependent upon the confining stress, varying from 0.7 in unconfined samples to 0.2 at high confinement (Chandler, 2001). However, using a higher value for Biot's coefficient resulted in a lower increase in hydraulic head; therefore, other factors that may affect the head increase need to be investigated.

There was a notable difference between the measured hydraulic head and the simulated thermally induced hydraulic head in the rock surrounding the TSX. Issues that should be examined by physical testing include the following.

1. Does the rock temperature increase affect the permeability of the surrounding rock?
2. Do the packers and vibrating wire piezometers used to measure thermally induced hydraulic head reliably record the actual hydraulic-head increase?

3. What is the influence of the selected value for porosity of the rock on the computed thermally induced hydraulic head? Does a thermally induced porosity change influence the actual hydraulic-head change?

Figure 12 shows the simulated hydraulic-head distributions on the vertical longitudinal section through the axis of the concrete bulkhead after 28 days and 458 days of heating. Figures 13 and 14 show the simulated hydraulic-head distribution along the axis of the concrete bulkhead at different times and simulated hydraulic head with time at different locations, respectively. A Biot's coefficient of 0.2 is used in these simulations.

The largest simulated hydraulic head in the concrete bulkhead during the first stage of heating (chamber temperature at 50°C) occurred after 28 days

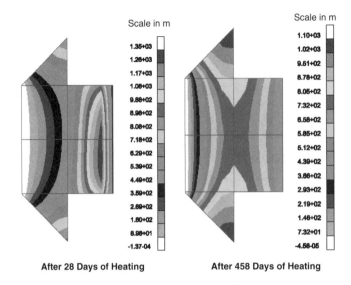

After 28 Days of Heating   After 458 Days of Heating

Figure 12. Simulated hydraulic head after 28 days and 458 days of heating.

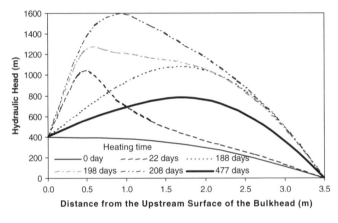

Figure 13. Simulated hydraulic head distributions along the axis of the concrete bulkhead at different times during heating.

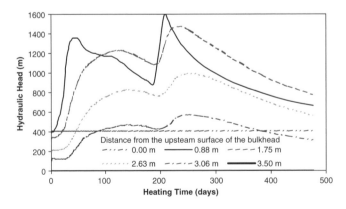

Figure 14. Simulated hydraulic head with time at different locations along the axis of the concrete bulkhead during heating.

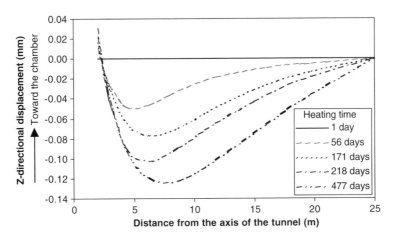

Figure 15. Simulated thermally-induced Z-directional displacement distributions along EXT2 at different times.

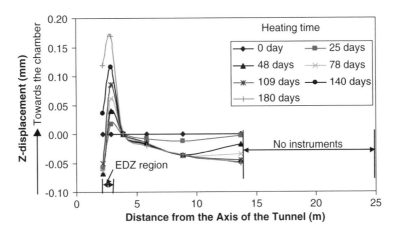

Figure 16. Measured Z-displacements along EXT2 in the rock during heating.

and was about 1350 m at 0.5 m from the upstream surface of the concrete bulkhead. During the second stage of heating (80°C chamber temperature) the predicted hydraulic head in the concrete bulkhead reached a maximum of about 1600 m at 0.85 m from the upstream side. As the thermal front advanced into the concrete with time, the pore pressure further away from the upstream face also increased. At the same time, the pore pressure at 0.5 m from the upstream surface dissipated slowly with time. The greatest hydraulic head after 111 days of heating was 1250 m at the central part of the concrete bulkhead (Fig. 14).

Finally, after 477 days of heating, most of the excess pore pressure had dissipated. By then, the highest hydraulic head along the axis of the concrete bulkhead had decreased to about 700 m (Fig. 14).

4.3 Simulated mechanical results and its comparisons with measured data

Figure 15 shows simulated vertical-displacement (Z-direction) distributions in the rock at different times along EXT2. Due to the axisymetric geometric conditions in the rock around the TSX, the thermally induced displacements along EXT2 moved away from the chamber (i.e., negative displacement) except near the chamber surface.

Figure 16 shows the measured vertical displacements at different times along EXT2. Like the simulated displacement, the measured displacement is also towards the chamber near the rock surface, except in the region of the EDZ (The thickness of EDZ is about 0.5 to 1.0 m depending on locations around the tunnel) (Martino and Chandler, 2004). The measured displacement direction is away from the tunnel at greater radial depths, as was observed for the simulated displacements. The measured pattern and magnitude of the displacements are somewhat different from the simulated displacements in the EDZ.

Scale in Pa

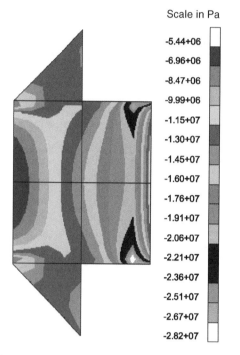

-5.44+06	
-6.96+06	
-8.47+06	
-9.99+06	
-1.15+07	
-1.30+07	
-1.45+07	
-1.60+07	
-1.76+07	
-1.91+07	
-2.06+07	
-2.21+07	
-2.36+07	
-2.51+07	
-2.67+07	
-2.82+07	

Figure 17. Simulated thermally-induced maximum principal stress distributions on the vertical longitudinal section through the axis of the concrete bulkhead after 477 days of heating.

Figure 17 shows simulation of the thermally induced effective maximum-principal-stress distributions on the vertical longitudinal section through the axis of the concrete bulkhead after 477 days of heating. All stresses in the bulkhead are compressive. The largest thermally induced simulated stress in the concrete bulkhead is 28 MPa, occurring near the bottom of the tunnel about 0.5 m from the upstream surface of the concrete bulkhead. Compared with the concrete compressive strength, the thermally induced stresses are small. No thermally induced hydraulic fracturing should be possible.

Figure 18 shows the simulated thermally induced axial displacement distributions along the axis of the concrete bulkhead at different times following the start of heating. The part of the concrete bulkhead within 0.4 m from the upstream surface moved toward the chamber as a result of thermal expansion. At the same time, the rest of the concrete bulkhead moved toward the downstream side due to thermal expansion. After 477 days of heating, the axial displacement on the centre of the downstream surface moved about 1.5 mm toward the downstream side.

5 SUMMARY

MOTIF was successfully used to conduct the fully coupled T-H-M simulation with thermal advection. There is a very good match between the simulated and measured thermal response in the concrete bulkhead and surrounding rock.

There is a reasonable match between the simulated and measured mechanical results. The apparent influence of the EDZ on the rock stress and displacement requires further study.

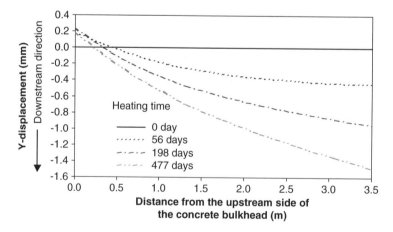

Figure 18. Simulated thermally- induced axial (Y-direction) displacement distributions along the axis of the concrete bulkhead at different times.

Comparisons of measured and computed hydraulic data show that the pattern of the simulated hydraulic response computed using MOTIF matched the measured results reasonably well, although there is still a difference in the magnitude of hydraulic response. This may be caused by the choice of parameter values, suggesting that additional back-analyses are necessary if the in-situ material properties in the experiment are to be better defined.

ACKNOWLEDGEMENTS

Ontario Power Generation funded the numerical modelling of the TSX and preparation of this paper.

REFERENCES

Chan, T, Scheier, NW & Guvanasen, V (1999) MOTIF code version 3.2 theory manual. Ontario Hydro, Nuclear Waste Management Division Report 06819-REP-01200-0091-R00.

Chandler, NA (2001) The incorporation of rock pore pressure in repository design and excavation stability analysis. Ontario Power Generation Inc., Nuclear Waste Management Division Report 06819-REP-01200-10068-R00.

Chandler, N, Martino, J & Dixon, D (2002a) The Tunnel Sealing Experiment. In Proc. of 6th International Workshop on Design and Construction of Final Repositories. Session 4, Number 11. Brussels. ONDRAF-NIRAS.

Chandler, N, Cournut, A, Dixon, D, Fairhurst, C, Hansen, F, Gray, M, Hara, K, Ishijima, Y, Kozak, E, Martino, J, Masumoto, K, McCrank, G, Sugita, Y, Thompson, P, Tillerson, J & Vignal, B (2002b) The five year report of the Tunnel Sealing Experiment: An international project of AECL, JNC, ANDRA and WIPP. Atomic Energy of Canada Limited Report AECL-12727.

Guo, R & Chandler, N (2002) Thermal-hydraulic numerical modelling of the flow of heated water through a sand-filled tunnel in granite. In Proc. 55th Canadian Geotechnical Conference, Niagra Falls, Ontario, pp.497–504. Canadian Geotechnical Society.

Guo, R, Chandler, NA & Dixon, DA (2002) Modelling the thermally induced hydraulic and mechanical response for the heated phase of the Tunnel Sealing Experiment. Ontario Power Generation Inc., Nuclear Waste Management Division Report 06819-REP-01200-10095-R00.

Guo, R, Chandler, N, Martino, J & Dixon, D (2003) Thermo-hydro-mechanical numerical modelling of the TSX with comparisons to measurements during stage 1 heating. Ontario Power Generation Inc., Nuclear Waste Management Division Report 06819-REP-01300-10095-R00.

Guvanasen, V & Chan, T (1992) A Three-dimensional finite-element solution for heat and energy transport in deformable rock masses with discrete fractures. In Proc. of the 7th International Conference on Advances in Computational Geomechanics, Cairns, Australia, 6–10 May, 1991. Balkema, Rotterdam, 1992.

Guvanasen, V & Chan, T (2000) A three-dimensional numerical model for thermohydromechanical deformation with hysteresis in a fractured rock mass. International Journal of Rock Mechanics and Mining Sciences, 37, 89–106.

Martino, JB & Chandler, NA (2004) Excavation induced-damage studies at the Underground Research Laboratory. Special Publications of IJRM.

Russell, SB & Simmons, GR (2003) Engineered barrier system for a deep geologic Repository. In Proc. 2003 International High-Level Radioactive Waste Management Conference, Las Vegas, USA, 2003 March 30–April 02.

Gas generation and transport measurements in the FEBEX project

Norbert Jockwer & Klaus Wieczorek
Gesellschaft für Anlagen- und Reaktorsicherheit mbH (GRS), Braunschweig, Germany

ABSTRACT: At the Grimsel Test Site (Switzerland) the disposal of high level radioactive waste in granite formations is investigated and demonstrated. For sealing the residual volume between the waste canisters and the surrounding host rock it is backfilled with a buffer of highly compacted bentonite blocks. One aspect of the investigation programme is to determine the gas generation and release from the host rock, the buffer, and the waste containers at ambient and elevated temperature. Gases are of importance, as in sealed areas the inner pore pressure may increase above the frac pressure of the host rock and of the sealing, an ignitable atmosphere may be created, and gases may be transport medium for released radionuclides.

The investigations indicated that in a repository sealed with bentonite significant amounts of carbon dioxide by decomposition of organic material and hydrogen by corrosion of the waste containers will be generated. In the non-water-saturated stage of the buffer it is high permeable and released gases can migrate into open areas of the repository. In these open areas an ignitable atmosphere generated by the released hydrogen and oxygen from the air has to be avoided. In the water-saturated stage of the buffer it is almost impermeable to gas and the inner pore pressure will increase as a result of further gas generation. Inner pore pressures which affect the integrity of the seal or the host rock have to be avoided.

1   INTRODUCTION

At the Grimsel Test Site, ENRESA performs a test for disposing high-level radioactive waste in granite formations in co-operation with NAGRA. The test field was installed in 1996 and heating started in February 1997. In February 2002, heater 1 was switched off, while heater 2 still continues running. The concrete plug, part of the buffer, and heater 1 were removed for laboratory investigations. In 2003, further equipment will be installed at heater 2 for determining the long term behaviour of the buffer. The test field will be sealed again by a concrete plug.

One aspect of that test is the gas generation and the gas migration in the buffer material which is investigated by GRS in the test field and an additional laboratory programme.

Figure 1 shows a principle drawing of the test field phase I with the two heaters, the buffer, and the installation for determining the gas pressure and collecting gas samples out of the pore volume of the buffer for analysis.

2   INVESTIGATIONS IN THE TEST FIELD

2.1   *Instrumentation*

In order to determine the pore pressure and the gases generated and released from the buffer material by thermal and microbial degradation or by corrosion of the metallic components, six ceramic draining pipes having an outer diameter of 60 mm and a length of 3 m were installed surrounding the electrical heater 1. Three of these were close to the surface of the heater and three mounted to the gallery wall (see figure 1). From both ends of these draining pipes, PFA tubes run through the buffer and the concrete plug to a valve panel in the open gallery, where the gas pressure is recorded by a data collection system and gas samples can be taken for qualitative and quantitative analyses. Additionally, permeability measurements by gas injection tests via the valve panel and the draining pipes were performed in the non-flooded and in the flooded buffer and by pressure recovery tests in the flooded buffer.

2.2   *Investigation and results*

2.2.1   *Pore pressure*

The pore pressure was measured in the six filter pipes for gas sampling (see figure 1). In more than five years of measurement, a pressure increase was detected only in those three of the six filter pipes which are located at the gallery wall (GF-SL01, GF-SL02, GF-SL03), while the pressure in the pipes GF-SL04, GF-SL05, and GF-SL06 close to the heater remained at atmospheric pressure.

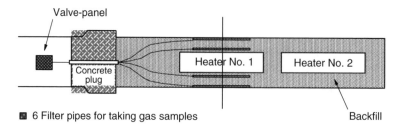

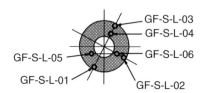

Cross section at Heater No. 1

Figure 1.   Principle drawing of the FEBEX test gallery with the draining pipes for gas sampling (Phase I).

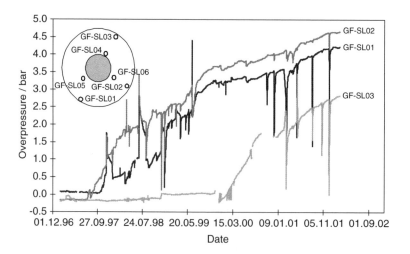

Figure 2.   Gas pressure development in the gas sampling pipes GF-SL01, GF-SL02, and GF-SL03.

The pressure development in the three filter pipes GF-SL01, GF-SL02, and GF-SL03 is shown in Figure 2. Until July, 1997, no significant pressure increase was detected. From July the pressure started increasing in the pipe GF-SL02; the same was detected in GF-SL01 in October 1997 and in GF-SL03 in February 2000. The pressure increase coincided with the detection of formation water in the pipes from the surrounding host rock. The different pressures of 0.45 MPa and 0.25 MPa represent the different

formation water pressures in the test field and they indicate furthermore that the water saturated buffer close to the gallery wall as well as the interface between the buffer and the granite became watertight as a result of the bentonite swelling. Otherwise the different draining pipes would became flooded at the same time and its water pressure would be almost the same.

The draining pipes close to the heater and those at the gallery wall before they became flooded showed

444

atmospheric in the range between 80,4 and 84,2 kPa and a fluctuation of 0.5 to 3.0 kPa within one week. This indicates that neither the concrete plug nor the unsaturated bentonite buffer is gastight. A variation of the atmospheric pressure which is also detected in the pipes of 0.8 kPa extracts and replaces about 1% of the gas in the pore volume of the buffer. The gas is thus rarefied continuously and is exchange at least once a year.

The peaks and drops of the pressure in the flooded draining pipes shown in Figure 2 are caused by gas injections for determining the gas permeability and by pressure recovery tests for determining the water permeability (see further down).

2.2.2 Gas generation and release

Gas samples were taken from the six draining pipes after the start of heating at first daily, later on monthly, and then four times a year. The gas samples were transferred by a manual pump into gas sampling bags which were sent to the laboratories for analyses. The draining pipes mounted to the gallery wall became flooded in 1997 and 2000 (see above), therefore gas sampling was not possible any more. The draining pipes which are located close to the heater with a distance to the gallery wall of 0.5 m did not become flooded within more than five years. They were still filled with gas at atmospheric pressure and gas sampling was possible over the whole time. The analyses indicated that the components carbon monoxide, carbon dioxide, hydrogen, methane, ethane, and propane were released by oxidation and by thermal or microbial decomposition of the organic material within the bentonite buffer as well as by corrosion of the metallic components (heater and installation). Figure 3 shows the concentration versus time of these gas components in draining pipe GF-SL06 (mid level close to the heater). The gas concentrations and their development with time in the other draining pipes are similar.

The investigation of gas generation and release in the test field can be summarized:

- Carbon dioxide with a concentration of about 700 vpm (content in the mine air) was already present prior to switching on the heaters and it has increased to more than 100 000 vpm (10 vol%) as a result of desorption and oxidation of the hydro-carbons.
- Carbon monoxide is increasing parallel with the increase of carbon dioxide and represents almost the equilibrium between these components.
- Hydrogen with a concentration of more than 1000 vpm (0.1 vol%) was already present prior to switching on the heaters as a result of corrosion of the metallic components (heaters, installation) in the test field. During operation of the test field it

has decreased to a level less than 10 vpm. This means that corrosion has decreased and the already present hydrogen might have escaped through the not-gastight concrete plug. Beginning of 1997, when the outer draining pipes became flooded, the release of hydrogen increased again. This is obviously due to a higher corrosion rate as a result of a higher humidity in the pore volume of the buffer. The decrease in 2001 is a result of higher ventilation rate in the entrance gallery preceding the dismantling of heater 1.

- Oxygen has decreased from 20 vol% (air concentration) to 3 vol% as a result of consumption by oxidation of organic and metallic components.
- Methane with a concentration of about 10 vpm was already present prior to switching on the heaters and it has increased to 450 vpm (0.045 vol%) as a result of thermal decomposition of long-chained hydro-carbons.
- Ethane has always been found in concentrations of 1 vpm prior to heating and it increased up to 40 vpm when the interface between the host rock and the buffer became flooded, afterwards it decreased again.
- Propane with a concentration of about 100 vpm was already present prior to switching on the heaters and it has decreased to 5 vpm as a result of thermal decomposition and oxidation to carbon dioxide. When the interface between the host rock and the buffer became flooded it increased to 140 vpm, afterwards it decreased again.

2.2.3 Gas permeability

Gas injection tests for determining the buffer's effective permeability to gas were performed twice every year. In the dry buffer, the permeability remained at estimated values above 10^{-13} m^2 (the permeability was too high for exact measurement). In the water-saturated buffer around the pipes GF-SL01, GF-SL02, and GF-SL03, the effective permeability to gas decreased to extremely low values. While a measurement at beginning of the saturation at GF-SL03 resulted in 10^{-15} m^2, it decreased to values below 10^{-21} m^2 when the interface between host rock and the buffer was saturated.

2.2.4 Water permeability

In the water-filled pipes near the gallery wall, pressure recovery tests were performed in order to determine the permeability of the saturated buffer to water. Like the gas injection tests, the pressure recovery tests were evaluated using the computer code WELTEST (Schlumberger-Geoquest, 1997).

First evaluation indicated integral values in the range of 10^{-17} m^2, with a high skin factor representing a flow obstruction near the pipes. This can be

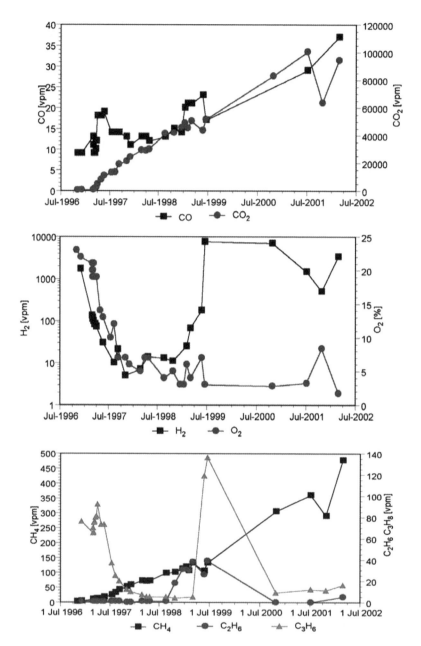

Figure 3. Concentration of the gas components carbon monoxide, carbon dioxide, hydrogen, oxygen, methane, ethane, and propane, in the draining pipe GF-SL06.

interpreted as an influence of the unsaturated buffer farther away from the pipes, while the completely saturated bentonite near the pipes has a much lower permeability. Later, water permeability decreased to values below $10^{-19}\,\mathrm{m}^2$ with no skin factor, which may be a hint to larger portions of the buffer being saturated. A typical measurement curve with the corresponding modelled pressure recovery curve is shown in Figure 4 for a measurement performed at GF-SL02 in August 2001.

GF-SL02:Linear plot of Build up 1 :0.00055200001 to 66.514442 hr

■ ■ P   -v- Delta t
———— Reg23_P -v- Delta t

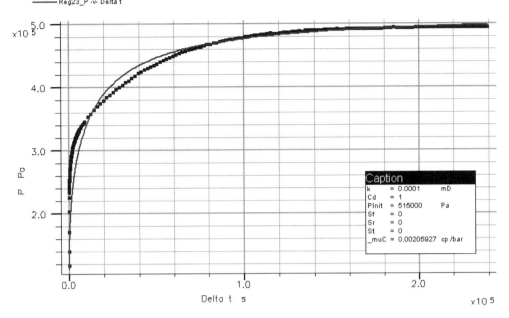

Figure 4.   Pressure recovery test at GF-SL02, August 2001.

3   INVESTIGATIONS IN THE LABORATORY

3.1   *Methods*

In addition to the investigations in the test field, laboratory investigations were performed regarding the gas generation from the bentonite at defined physico-chemical conditions. For these investigations glass ampoules with a volume of 500 ml, as shown in Figure 5, were used. Via the injection tube at the top 10 or 100 grams of the ground highly compacted bentonite blocks were filled into the ampoules. The residual volumes were filled with laboratory air or with nitrogen. To one half of the ampoules 10 ml of water was added while the other remained in the natural dry stage. Afterwards the injection tube was welded gastight.

For gas generation and release, the gastight sealed ampoules were placed in an oven. Within one investigation programme the temperatures were of 20, 50, or 100°C and the exposure time was 100 days. Within another investigation programme the temperature was constant 100°C and the exposure periods were 1 to 1000 days. After the exposure time at the envisaged temperature the ampoules were withdrawn from the oven for analyses of the gases generated and released into the residual volume. The gas was

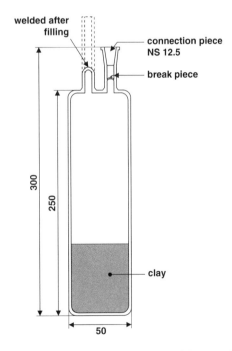

Figure 5.   Ampoule for the investigation of the generation and release of gases from the clay as a result of elevated temperature.

extracted by a pump via the opened break seal and injected into a gas chromatograph.

3.2 Results

The measurement results indicate that carbon dioxide is the only gas component of importance released from the buffer material in the temperature range between 20 and 100°C at aerobic or anaerobic conditions.

Figure 6 shows the amount of carbon dioxide generated in the bentonite for the exposure temperatures of 100°C and disposure periods up to 300 days. At the wet stage up to $2.0 m^3$ carbon dioxide per 1000 kg bentonite were released within 300 days. For longer exposure periods to investigation is still running.

In the natural dry stage of the bentonite the amount of carbon dioxide released into the residual volume is by a factor of 20 less than in the wet stage, when adding 10 ml of water to 10 g of bentonite. The reason might be that in the dry stage, carbon dioxide is partly adsorbed to the internal surfaces of the clay. In the wet stage, water is adsorbed, and carbon dioxide is desorbed and released into the residual volume.

Whether air or nitrogen was in the residual volume of the ampoule (aerobic or anaerobic conditions, respectively) did not influence the generation of carbon dioxide. The reason may be that the buffer material has adsorbed oxygen on the internal surfaces of the clay or that air is trapped in the pore volume during fabrication and storage of the highly compacted bentonite blocks. This oxygen is then consumed for oxidation of the organic material in the bentonite, resulting in the generation of carbon dioxide. This means that for the oxidation at 100°C no external oxygen is necessary. Furthermore, carbon dioxide may also be generated by thermal decomposition of carbonates in the bentonite. The variation of the temperature indicated that with decreasing temperature, the velocity of gas generation decreases significantly.

The amount of the hydro-carbons released is less than 10 ml per 1000 kg bentonite. The storage conditions (temperature, gas in the residual volume of the ampoules and additional water) do not influence the released amount significantly. It might be possible that the hydro-carbons generated by thermal decomposition of the organic material oxidises instantaneously and carbon dioxide is generated.

Hydrogen was not determined which indicates that it is generated only by corrosion of metallic components installed in the test field.

4 CONCLUSIONS

The investigation on gas generation and gas migration indicated that in the test field carbon dioxide and hydro-carbons are released from the heated bentonite buffer. As a result of corrosion of the metallic components in the test field (heater, installation) hydrogen is generated. The determination of the gas pressure inside the sealed test field in the non-saturated stage of the buffer indicated always atmospheric pressure in the range between 80.4 and 84.2 kPa with a fluctuation of 0.5 to 3 kPa within one week. This indicate that neither the bentonite buffer nor the concrete plug are gastight and the released gases could migrate into the open gallery.

A variation of 0.8 kPa atmospheric pressure which could also be determined in the test field extracts and

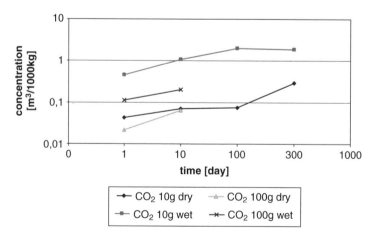

Figure 6. Release of carbon dioxide from the bentonite versus exposure time when tempering at 100°C in the dry and wet stage with air in the residual volume of the ampoules.

replaces about 1% of the gas in the pore volume of the buffer via the non-gastight concrete plug. The gas in the pore volume is thus rarefied continuously and is exchanged at least once within one year.

Permeability measurements in the water-saturated contact region between the buffer and the host rock showed values to water below $10^{-19}\,m^2$ and to gas below $10^{-21}\,m^2$. These values indicate that the disposal field will become gastight when the whole buffer becomes water-saturated. If gas generation continues, it will create increased inner pore gas pressure which may be a safety issue for the sealed parts of the repository. In the meantime, no critical gas pressure will be generated in the non-saturated stage.

In a real repository with disposal areas much bigger than the FEBEX test field, the generation of hydrogen and hydro-carbons and their release into the open galleries with much higher oxygen concentration may lead to an ignitable atmosphere which has to be avoided.

The results obtained in the test field and the laboratory indicate that further experimental investigations and model calculations on gas generation and gas migration in repositories for radioactive waste are essential.

ACKNOWLEDGEMENTS

This work was co-funded by the German Bundesministerium für Wirtschaft und Technologie (BMWi) under the contract No. 02 E 9390 and by the Commission of the European Communities (CEC) under the contracts No. FI4W-CT95-0006 and FIKS-CT-2000-00016. The authors would like to thank for this support.

REFERENCES

Huertas, F., Fuentes-Cantillana, J.L., Jullien, F., Rivas, P., Linares, J., Farina, P., Ghoreychi, M., Jockwer, N., Kickmaier, W., Martinez, M.A., Samper, J., Alonso, E., Elorza, F.J. (2000) Full-Scale engineered barriers experiment for a deep geological repositories for high-level radioactive waste in crystalline host rock (FEBEX project), Final Report, European Commission, Contract No. FI4W-CT95-0006, EUR 19147 EN.

Schlumberger-Geoquest, (1997) Weltest 200 Technical Description, Logined BV.

Villar Galicia, M.V. (2002) Thermo-hydro-mechanical characterisation of a bentonite from Cabo de Gatta. ENRESA, Madrid, Spain.

Advances in Understanding Engineered Clay Barriers – Alonso & Ledesma (eds)
© *2005 Taylor & Francis Group, London, ISBN 04 1536 544 9*

Coupled thermo-hydro-mechanical analysis of the Prototype Repository Project using numerical code THAMES

M. Chijimatsu

Hazama Corporation, Tokyo (Japan)

H. Kurikami & Y. Sugita

Japan Nuclear Cycle Development Institute (JNC), Tokai (Japan)

ABSTRACT: JNC has joined the Prototype Repository Project (PRP). JNC has performed a predictive analysis of the coupled thermo-hydro-mechanical (THM) behaviors in the PRP using a numerical code THAMES. The PRP is the full-scale engineered barrier system (EBS) experiment including back fill of the experimental tunnel above test pits. Authors performed two kinds of predictive analyses. One was for the discussion about the influence of the pellet properties installed in the gap between the buffer block and the rock mass. The other one was for the discussion about the influence of the boundary conditions.

1 INTRODUCTION

The PRP is the full-scale and in-situ experiment to validate the engineered barrier system (EBS) on geological disposal of high-level radioactive waste, and is one of the R&D programs of European Commission projects (Svemar and Pusch, 2000). The PRP has studied the vertical emplacement option of the EBS. The PRP has six test holes and the test tunnel over the test holes is backfilled to simulate the expecting repository environment. In the PRP, some numerical codes are applied to assess the barrier performances of the EBS. JNC has joined the PRP and has performed the simulation of the EBS using a fully coupled THM numerical code. However, the numerical code is unable to consider the real experimental conditions in the PRP, especially the clearance, which is filled with bentonite pellet, between buffer block and rock mass. Therefore, certain assumptions on the experimental conditions are required. Such assumptions in the numerical simulation have to be validated. Then, the properties of bentonite pellet, which is not considered in the simulation and is filled in the clearance between buffer block and rock mass, were examined by laboratory tests.

2 DESCRIPTION OF THE NUMERICAL CODE THAMES

Analysis of the coupled thermal, hydraulic and mechanical processes is carried out with the computer code named THAMES (Ohnishi, et al, 1985). THAMES is

a finite element code for the analyses of coupled thermal, hydraulic and mechanical behaviors of a saturated-unsaturated medium. THAMES has been extended to take into account of the behaviors in the buffer materials such as the water movement due to thermal gradient and the swelling phenomena. The unknown variables are total pressure, displacement vector and temperature. The mathematical formulation for the model utilizes Biot's theory, with the Duhamel-Neuman's form of Hooke's law, and energy balance equation. The governing equations are derived with the fully coupled thermal, hydraulic and mechanical relationships. This code has been validated with the data of the laboratory tests (Chijimatsu, et al, 1998), the engineered scale tests (Chijimatsu, et al, 2000a) and the in-situ experiments (Chijimatsu, et al, 2000b). To treat the water/vapor movement and heat induced water movement, the continuity equation used in the extended THAMES code is as follows;

$$\left\{ \frac{\rho_l^2 g k_p K}{\mu_l} h_{,i} \right\} + \left\{ \rho_l D_T T_{,i} \right\}_{,i}$$

$$- \rho_{lo} n S_r \rho_l g \beta_P \frac{\partial h}{\partial t} - \rho_l \frac{\partial \theta}{\partial \psi} \frac{\partial h}{\partial t}$$

$$- \rho_l S_r \frac{\partial u_{i,i}}{\partial t} + \rho_{lo} n S_r \beta_T \frac{\partial T}{\partial t} = 0 \qquad (1)$$

where K is the intrinsic permeability, k_p is the relative permeability, μ_l is the viscosity of water, ρ_l is the density of water, g is the gravitational acceleration.

D_T is the thermal water diffusivity, n is the porosity, S_r is the degree of saturation θ is the volumetric water content, ψ is the water potential head, β_P is the compressibility of water, β_T the thermal expansion coefficient of water and z is the elevation head u_i is the displacement vector, T is temperature, h is the total head and t is time. The subscript 0 means the reference state. This equation means that the water flow in the unsaturated zone is expressed by the diffusion equation and in the saturated zone by the Darcy's law.

The energy conservation equation has to treat the energy change by evaporation. The equation is given as

$$
\begin{aligned}
&(\rho C_v)_m \frac{\partial T}{\partial t} + nS_r \rho_l C_{vl} V_{li} T_{,i} - K_{Tm} T_{,ii} \\
&+ nS_r T \frac{\beta_T}{\beta_P} \left\{ \frac{\rho_l g k_p K}{\mu_l} h_{,i} + D_T T_{,i} \right\}_{,i} \\
&+ \frac{1}{2}(1-n)\beta T \frac{\partial}{\partial t}(u_{i,j} + u_{j,i})\delta_{ij} = 0
\end{aligned} \tag{2}
$$

where $(\rho C_v)_m$ is the specific heat of the material consisting of water and the soil particles, C_{vl} is the specific heat of water, V_{li} is the velocity vector of water, K_{Tm} is the thermal conductivity of consisting of water and the solid particles.

The equilibrium equation has to take the swelling behavior into account.

$$
\begin{aligned}
&\left[\frac{1}{2} C_{ijkl}(u_{k,l} + u_{l,k}) - F\pi\delta_{ij} - \beta\delta_{ij}(T - T_o) + \chi\delta_{ij}\rho_l g(h - z) \right] \\
&+ \rho b_i = 0
\end{aligned} \tag{3}
$$

where C_{ijkl} is the elastic matrix, ρ is the density of the medium and b_i is the body force. χ is the parameter for the effective stress, $\chi = 0$ at the unsaturated zone, $\chi = 1$ at the saturated zone. The symbol F is the coefficient relating to the swelling pressure process and $\beta = (3\lambda + 2\mu)\alpha_s$, where λ and μ are Lame's constants and α_s is the thermal expansion coefficient.

The swelling pressure π can be assumed to be the function of water potential head (ψ) as follows;

$$
\pi(\theta_1) = \rho_l g(\Delta\psi) = \rho_l g(\psi(\theta_1) - \psi(\theta_0)) = \rho_l g \int_{\theta_0}^{\theta_1} \frac{\partial \psi}{\partial \theta} d\theta \tag{4}
$$

where θ_0 is the volumetric water content at the initial state. This is based on the theory that swelling pressure is equivalent to the water potential.

3 PARAMETERS OF BENTONITE MX-80 FOR SIMULATION

For the simulation of Prototype Repository, SKB carried out many laboratory experiments about the bentonite MX-80. The initial conditions of bentonite

MX-80 at the laboratory experiments are as follows.

Dry density: $\rho_d = 1.67\,\mathrm{g/cm^3}$,
Water content: $\omega = 0.17$,
Void ratio: $e = 0.77$,
Degree of saturation: $S_r = 0.61$

In this chapter, we show the parameters of bentonite MX-80 for simulation by THAMES. Almost parameter except for the hydraulic conductivity, thermal vapor flow diffusivity and swelling pressure parameter are the same with those used for the simulation conducted by SKB (Börgesson and Herneilind, 1999).

3.1 Thermal property

The thermal conductivity K_{Tm} of MX-80 is obtained as a function of degree of saturation S_r. For the simulation, we used the equation (5).

$$
\begin{aligned}
K_{Tm} = &0.300 - 0.221S_r - 1.28S_r^2 + 16.4S_r^3 - 23.1S_r^4 \\
&+ 8.44S_r^5 + 0.793S_r^6
\end{aligned} \tag{5}
$$

Specific heat c (kJ/kgK) is a function of water content ω as show in equation (6).

$$
c = \frac{80.0 + 4.2\omega}{100 + \omega} \tag{6}
$$

3.2 Hydraulic property

The hydraulic conductivities (k) of MX-80 are measured with different void ratio (e) under the different temperature (T) conditions. Fig. 1 shows the relationship between the void ratio and the hydraulic conductivity obtained by the experiments. From these experiment results, SKB used the tabulated data for the simulation where, hydraulic conductivity is a function of void ratio and temperature. From this table, the relationship between the hydraulic conductivity and temperature at each void ratio is calculated as shown in Fig. 2. From this figure, it is

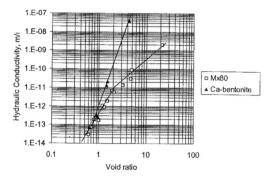

Figure 1. Measured hydraulic conductivity.

known that hydraulic conductivity increase with temperature increasing. Therefore, we estimated the intrinsic permeability $K(\mathrm{m}^2)$ from the hydraulic conductivity $k(\mathrm{m/s})$ by using the equation (7).

$$K = \frac{k\mu}{\rho g} \tag{7}$$

where, $\mu(\mathrm{Pa\,s})$ is the viscosity of water, $\rho(\mathrm{kg/m}^3)$ is the density of water and $g(\mathrm{m/s}^2)$ is the gravitational acceleration. Fig. 3 shows the relationship between the intrinsic permeability and the temperature at each void ratio. From this figure, it is known that the intrinsic permeability is a function of void ratio. Fig. 4 shows relationship between the intrinsic permeability and the void ratio. This relationship is expressed as following equation.

For the simulation by THAMES, equation (8) is used as the intrinsic permeability.

$$K = 1.81 \times 10^{-20} \left(e\right)^{4.30} \tag{8}$$

The unsaturated permeability is a function of degree of saturation as shown in equation (9).

$$k_p = \left(S_r\right)^{\delta} k \tag{9}$$

where, k_p is the hydraulic conductivity of partly saturated soil, k is the hydraulic conductivity of completely saturated soil and δ is the parameter. Parameter δ is determined by back analysis of infiltration test carried out by SKB and determined value is 2.2. Comparison of simulation and measured results of infiltration test is shown in Fig. 5.

As water retention curve, we use the van Genuchten model. Parameters of van Genuchten model for simulation are shown as follows and comparison of water retention curve between SKB and van Genuchten model is shown in Fig. 6.

$$S_r = \left\{1 + \left|\alpha\psi\right|^n\right\}^{\left(-(1-1/n)\right)}, \alpha = 5.2 \times 10^{-4}\left(1/m\right), n = 1.70 \tag{10}$$

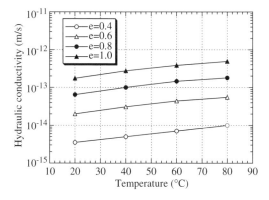

Figure 2. Temperature dependency of hydraulic conductivity.

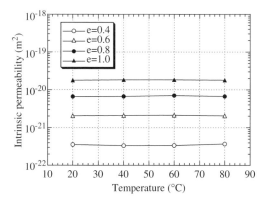

Figure 3. Temperature dependency of intrinsic permeability.

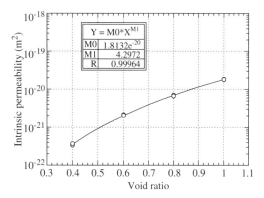

Figure 4. Relationship between intrinsic permeability and void ratio.

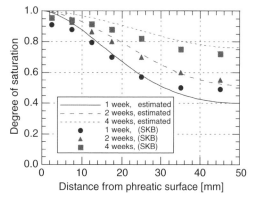

Figure 5. Comparison of infiltration test.

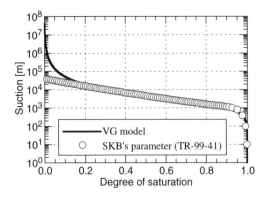

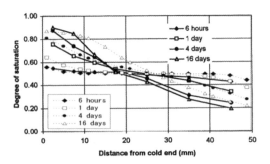

Figure 6. Water retention curve of MX-80.

Figure 8. Comparison of saturation distribution in the specimen between measured and simulated results of temperature gradient test(straight line; measurement results, dotted line; simulation results).

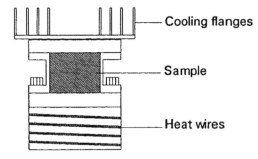

Figure 7. Test apparatus for temperature gradient test.

The thermal vapor flow diffusivity is determined by temperature gradient test conducted by SKB. Test apparatus is shown in Fig. 7. The size of specimen is 50 mm in diameter and 50 mm in height. The temperature of top and bottom side of specimen was controlled at fixed different temperature. After the several time, specimen was picked up and measured the water content distribution in the specimen.

SKB used the equation (11), (12) and (13) as the thermal vapor flow diffusivity. In these equations, the thermal vapor flow diffusivity D_{Tv} is a function of the degree of saturation.

$$D_{Tv} = D_{Tvb} \qquad (0.3 \le S_r \le 0.7) \qquad (11)$$

$$D_{Tv} = D_{Tvb} \cdot \cos^a\left(\frac{S_r - 0.7}{0.3} \cdot \frac{\pi}{2}\right)(0.7 \le S_r) \qquad (12)$$

$$D_{Tv} = D_{Tvb} \cdot \sin^b\left(\frac{S_r}{0.3} \cdot \frac{\pi}{2}\right) \ (S_r \le 0.3) \qquad (13)$$

The parameters D_{Tvb}, a and b were determined by back analysis of temperature gradient test. We re-determined the D_{Tvb} by our analysis code THAMES. Fig. 8 shows the comparison of saturation distribution

in the specimen between measurement results and simulation results of temperature gradient test. Initial degree of saturation of the specimen is 50%. The determined values are shown in following equations. Here, the parameter D_{Tvb} is only determined and other parameters a and b are the same with SKB values.

$$D_{Tv} = D_{Tvb} = 2.0 \times 10^{-13} m^2 / sK, a = 6, b = 6 \qquad (14)$$

3.3 Mechanical property

Young's modulus is 13 MPa and Poison's ratio is 0.4.

Maximum swelling pressure of MX-80 is obtained by a theoretical model for swelling characteristics by Komine et al. (2003) By this model, maximum swelling pressure of MX-80 is a function of dry density and temperature. The parameter for swelling pressure (F in equation (3)) is a function of degree of saturation as shown in equation (15).

$$F = 2\sigma_{sw\,max}\left(1 - S_r\right)\frac{\partial S_r}{\partial \psi} \qquad (15)$$

4 INFLUENCES BY THE DIFFERENCE IN THE MODELING

In this chapter, we show the influences by the difference in the modeling. Analyses are carried out by two-dimensional model. Fig. 9 shows the model geometry. Fig. 9 (a) is the whole geometry and (b) is the geometry of engineered barrier. Analysis region is 11 m in width and 74 m in height. Initial conditions and boundary conditions for the simulation are shown in Fig. 10.

Table 1 shows the analysis case. Case0-1 is the base case. The initial void ratio of buffer material is 0.77 and it is not considered the gap between the

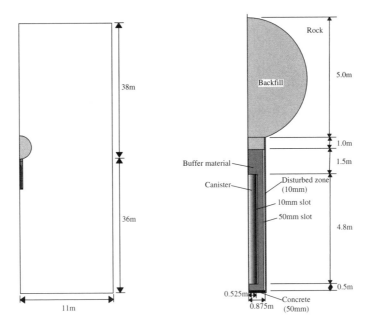

(a) Whole geometry

(b) Geometry of engineering barrier

Figure 9.   Model geometry.

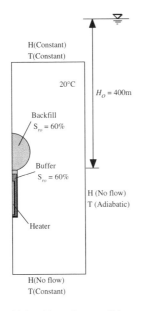

Figure 10.   Initial and boundary conditions.

heater and the buffer and between the rock and the buffer. This void ratio value of buffer material corresponds to that after saturation. The hydraulic conductivity of the rock mass is 10^{-10} m/s. Case1-1,

Case1-2 and Case1-3 are the cases to research the difference between the permeability of the rock mass. The hydraulic conductivities of the rock mass are 10^{-10} m/s, 10^{-12} m/s, 10^{-14} m/s, respectively. The initial void ratio of the buffer material is 0.64. This value corresponds to the bentonite block before the installation to the test pit. In these cases, it is not considered the gap between the heater and the buffer and between the rock and the buffer. Since the bentonite of Case1-1 corresponds to that before the installation and the bentonite of Case0-1 corresponds to that after saturation (after swelling), real phenomena will be occurred between these two cases. Case2-1 is a case with the gap between the heater and the buffer and that between the rock and the buffer. The thermal property of the gap is the same as water.

Fig. 11 (a) shows the time history of the degree of saturation in the buffer with different permeability of the rock mass. Closed legend is the results of the outer side of the buffer (near rock mass) and open one is the results of the inner side of the buffer (near the heater). Fig. 11 (b) shows the time history of temperature at the same points as Fig. 11 (a). It is also indicated from these figures that the effect of the permeability of the rock mass upon the distribution of saturation and temperature in the buffer is small.

Fig. 12 (a) shows the time history of degree of saturation in the buffer with different model for the

455

gap between the buffer and the rock mass. At the inner part of buffer that will take the most time to reach the saturation, the re-saturation time of Case2-1 is located between the results of Case0-0 and Case1-1. Fig. 12 (b) shows the time history of temperature with different model for the gap between the buffer and the rock mass. Regarding temperature distribution, there is little difference among the models for the gaps between the rock mass and the buffer.

5 INFLUENCES BY BOUNDARY CONDITION

Analyses were carried out by three-dimensional model to examine the effects of the boundary conditions. Figs. 13 and 14 show the model geometries. Fig. 13 is a one-heater model and Fig. 14 is two-heater model. The analysis region of the one heater model is 3 m in y direction (direction along the test

Table 1. Analysis case.

Case	Initial void ratio of buffer material	Hydraulic conductivity of rock mass (m/s)	Consideration of gap
Case0-1	0.77	10^{-10}	No
Case1-1	0.64	10^{-10}	No
Case1-2	0.64	10^{-12}	No
Case1-3	0.64	10^{-14}	No
Case2-1	0.64	10^{-10}	Yes

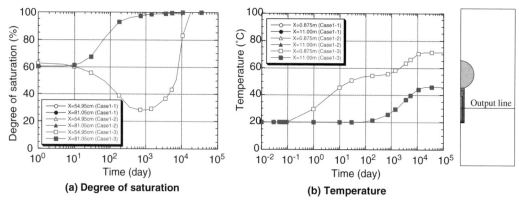

(a) Degree of saturation   (b) Temperature

Figure 11. Comparison with different permeability of rock mass (Case1-X).

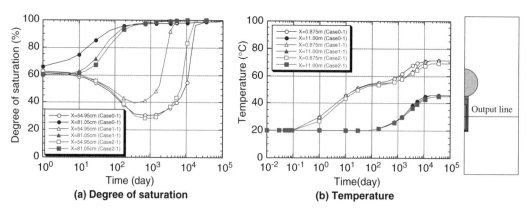

(a) Degree of saturation   (b) Temperature

Figure 12. Comparison with different permeability model for gap (CaseX-1).

drift), 5 m in x direction and 60 m in height. The analysis region of the two-heater model is 12 m in y direction, 5 m in x direction and 60 m in height. The initial conditions are the same as those in chapter 4. The boundary conditions for the simulation are summarized in Fig. 15.

Fig. 16 shows the simulation results by the one-heater model. Fig. 16 (a) shows the time history of temperature at the output points in the result of one-heater model simulation. The output points are along the line y = 0 m and z = −3.1 m as shown in Fig. 13. Temperature increases after 1000 days because the output power of the heater is constant. Fig. 16 (b) shows the time history of degree of saturation in the buffer and the rock mass in the result of one-heater simulation. Degree of saturation near the heater decreased until 200 days after the start of heating and the minimum value is about 0.4. After that, it increases and becomes almost full saturation after 1000 days from the start of heating. Degree of saturation in the buffer close to the rock mass increased from the start of heating due to the infiltration of groundwater from the rock mass. Fig. 16 (c) shows the time history of stress in the buffer close to the heater and the rock mass. Both stresses increase with time because of swelling of the buffer. Stress in the buffer close to the rock mass is larger than that close to the heater and it becomes about 6.5 MPa after 1000 days from the start of heating.

Fig. 17 shows the time history of temperature at the output points in the result of two-heater model simulation. Fig. 17 (a) is the result along the line y = 3 m, z = −3.1 m and (b) is the result along the line y = 9 m, z = −3.1 m. Temperature is lower comparing with the one-heater model, because temperature at the boundary x = 5 m is constant. Furthermore, it degreases gradually after it shows the peak value. Temperature distribution around the each experimental hole is different, because the temperature at the boundary y = 12 m is constant. Fig. 18 shows the time history of degree of saturation in the each experimental hole. Difference between the two holes is due to the difference of temperature distribution, but this difference is very small.

Fig. 19 shows the comparison of temperature evolution and saturation evolution in the buffer between different models. Temperature in the one-heater model is higher than that of two-heater model. This is effect of the boundary conditions. In the one-heater

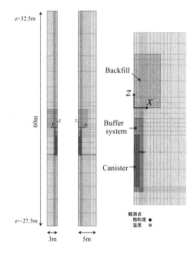

Figure 13.   Model geometry (One-heater model).

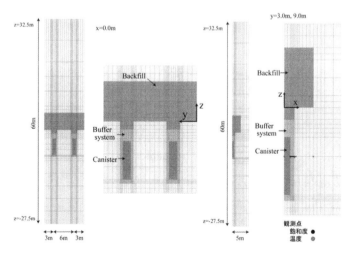

Figure 14.   Model geometry (Two-heaters model).

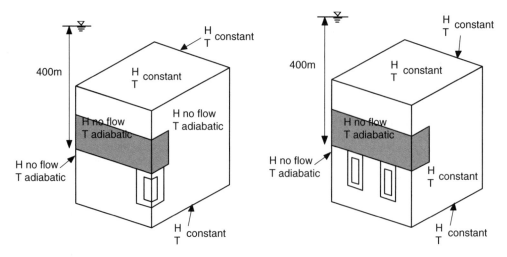

Figure 15. Boundary conditions.

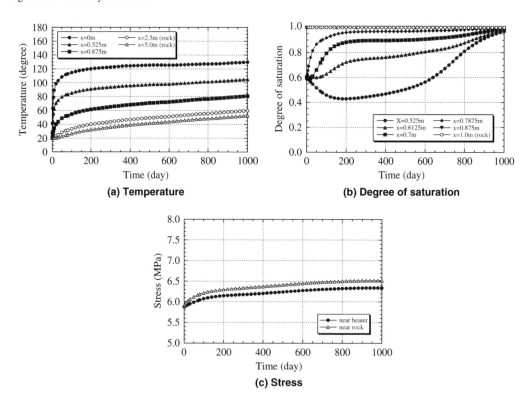

(a) Temperature

(b) Degree of saturation

(c) Stress

Figure 16. Simulation results by one-heater model (Cont.).

model, all side boundaries are adiabatic but two-heater model has temperature constant boundary. In the case of two-heater model, temperature around the heater at the y = 9 m (Left heater in Fig. 14) is lower than that around the heater at the y = 3 m (Right heater in Fig. 14). It is the reason that the left side of the model is temperature constant boundary. Re-saturation time in one-heater model is earlier than the one in two-heaters model. It is the reason that water diffusivity of bentonite is higher due to high

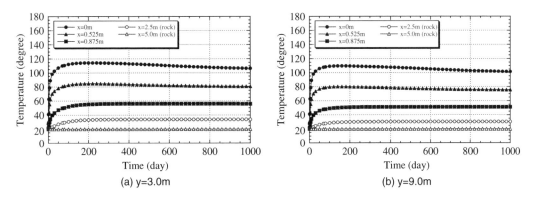

Figure 17. Time history of temperature in buffer and rock mass (Two heaters model).

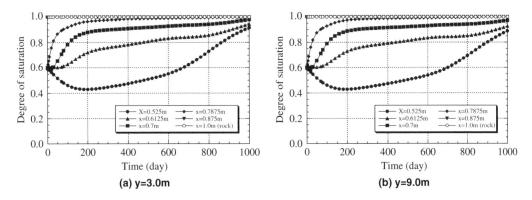

Figure 18. Time history of saturation in buffer and rock mass (Two heaters model).

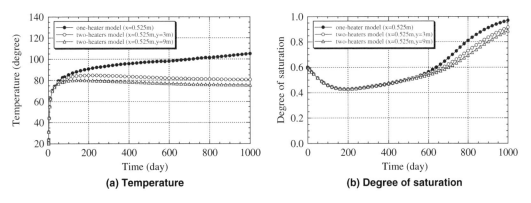

Figure 19. Comparison between different models.

temperature in the one-heater model than the one in the two-heater model. Re-saturation time of the buffer in the left hole in Fig. 14 (y = 9 m) is earlier than that in the right hole (y = 3 m). It is also dependent on temperature.

6 CONCLUSIONS

From this study, following results were obtained.

– Re-saturation phenomena in the buffer are not dependent on the permeability of rock mass if the

459

hydraulic conductivity of the rock mass is in the 10^{-10} to 10^{-14} m/s range.

- The re-saturation time of the buffer depends on the initial void ratio of the buffer.
- There is not so large difference about the temperature distribution between the some cases for modeling of the gap compared with the distribution of degree of saturation in the buffer.
- It is concluded that when the gap is not considered in the model it is better to use the property at the high density of bentonite before installing into the disposal pit. However, these are the results when we did not consider the structure change inside the bentonite due to the swelling in detail. In order to achieve the more detail evaluation, it needs to consider the structure change inside the bentonite and the parameter change during the swelling.
- Temperature distribution around the experimental hole is highly dependent on the boundary conditions. Furthermore, the re-saturation time is dependent on temperature distribution. Therefore, it is important to use the suitable boundary condition for the accurate simulation. Seepage in the buffer is dependent on permeability of the rock mass and pore pressure distribution in the rock mass. It is important to understand these characteristics in order to evaluate the saturation phenomena in the buffer.

REFERENCES

Börgesson, L. and Hernelind, J. (1999): Preliminary modeling of the water-saturation phase of the buffer and back fill material, SKB IPR-00-11.

Chijimatsu, M., Fujita, T., Kobayashi, A. and Nakano, M. (1998): Calibration and Validation of Thermal, Hydraulic and Mechanical Model for Buffer Material, JNC Technical report JNC TW8400 98-017.

Chijimatsu, M., Fujita, T., Kobayashi, A. and Nakano, M. (2000a): Experiment and validation of numerical simulation of coupled thermal, hydraulic and mechanical behavior in the engineered buffer materials, Int. J. for Numer. Anal. Meth. Geomech, 24, pp. 403–424.

Chijimatsu, M., Fujita, T., Sugita, Y., Amemiya, K. and Kobayashi, A. (2000b): Field experiment, results and THM behavior in the Kamaishi mine experiment, Int. J. of Rock Mec. & Min. Sci., 38, pp. 67–78.

Komine, H., Kurikami, H., Chijimatsu, M., Kobayashi, A., Sugita, Y. and Ohnishi, Y. (2003): Coupled thermal, hydraulic and mechanical simulation with a theoretical model for swelling characteristics, GeoProc 2003, International Conference on Coupled T-H-M-C Processes in Geo-systems; Fundamentals, Modeling, Experiments & Applications, pp. 550–555.

Ohnishi, Y., Shibata, H. and Kobayashi, A. (1985): Development of finite element code for the analysis of coupled thermo-hydro-mechanical behaviors of a saturated-unsaturated medium, Proc. of Int. Symp. on Coupled Process Affecting the Performance of a Nuclear Waste Repository, Berkeley, pp. 263–268.

Svemar, C. and Pusch, R. (2000): Project description FIKW-CT-2000-00055, SKB International Progress Report IPR-00-30.

Advances in Understanding Engineered Clay Barriers – Alonso & Ledesma (eds)
© *2005 Taylor & Francis Group, London, ISBN 04 1536 544 9*

THM predictive modelling of the Temperature Buffer Test – Clay Technology's contribution

B. Fälth, L. Börgesson & H. Hökmark
Clay Technology AB, Lund, Sweden

J. Hernelind
FemTech AB, Västerås, Sweden

ABSTRACT: ANDRA is performing a Temperature Buffer Test (TBT) in granitic rock at Äspö Hard Rock Laboratory, Sweden. The overall objective of the experiment is to investigate how well the bentonite buffer can endure high temperatures. As a part of the TBT project, predictive modelling work has been done.

This paper describes Clay Technology's contribution to that predictive modelling work, which has been done by adoption of a two-step approach. One large axially symmetrical THM model was analysed with ABAQUS and in addition, two 1D axially symmetrical TH-models were analysed with Code_Bright. The Code_Bright models correspond to the mid-height sections of the two canisters. These models used thermal boundary conditions provided by the ABAQUS model. In this way effects of axial heat transport could be accounted for also in these 1D models.

One fundamental observation is that both codes predict a considerable desaturation of the innermost hottest part of the buffer surrounding the lower heater. This is in particular the case for the Code_Bright model.

1 INTRODUCTION

1.1 General

ANDRA is performing a Temperature Buffer Test in granitic rock at Äspö Hard Rock Laboratory, Sweden. The overall objective of the experiment is to investigate how well the bentonite buffer can endure high temperatures. A drawing of the experiment setup is provided (Fig. 1). A sand shield around the canister will reduce the temperature to which the bentonite will be exposed. Without sand shield, the bentonite will be exposed to temperatures well over 100°C. Both options are tried in the TBT experiment.

1.2 Predictive modelling objectives

As a part of the TBT project, predictive modelling work has been performed. The predictive modelling was intended to be a test of conceptual models suggested to describe the THM behaviour of the bentonite buffer during hydration. In the predictive modelling phase, models that are based on the final design decisions were to be analyzed. Material models and model parameter values that represent state-of-the art knowledge were to be used.

All THM modelling predictive work performed within the TBT project is based upon a Predictive Modelling Program (Hökmark & Fälth, 2003), issued soon after finalization of the test design in November 2003. In that program a number of scan-lines and history points were defined (Fig. 2). Predictions of degree of saturation, relative humidity, temperature, porosity, stresses, displacements and gas pressure were asked for at these lines and points, or for selections of these.

1.3 Modelling approach

The modelling work presented here is Clay Technology's contribution to the TBT predictive modelling work and has been done by adoption of a two-step approach.

One axially symmetrical THM model analysed with ABAQUS (all scan-lines and history points).

Two 1D axially symmetrical TH-models were analysed with the finite element code Code_Bright. The models correspond to the mid-height sections of the two canisters (scan-lines A1–A4, B1–B4). These models used thermal boundary conditions provided by the ABAQUS model. In this way effects of axial heat transport could be accounted for also in these models, at least indirectly.

The ABAQUS model and the Code_Bright models are presented in chapter 2 and chapter 3, respectively. In chapter 4, the modelling results are presented.

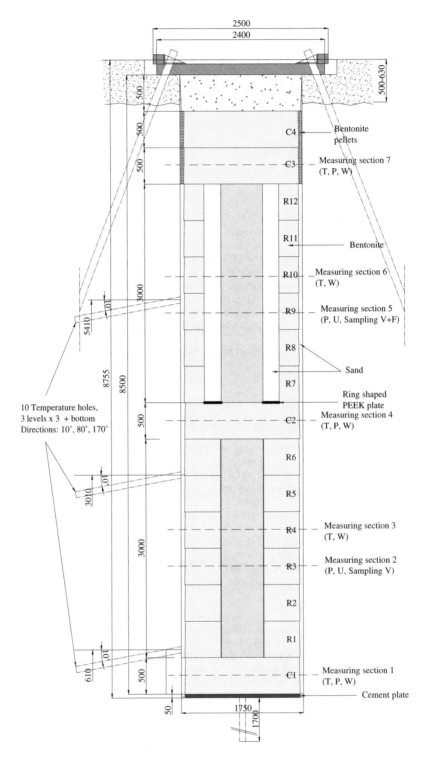

Figure 1. TBT experiment setup.

462

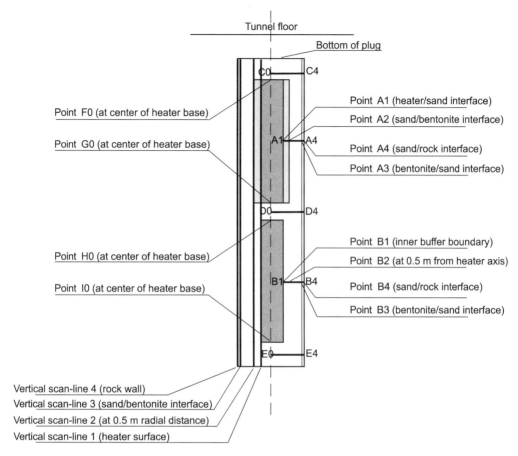

Tunnel floor

Bottom of plug

C0　C4

Point F0 (at center of heater base)

Point A1 (heater/sand interface)
Point A2 (sand/bentonite interface)

A1　A4

Point G0 (at center of heater base)

Point A4 (sand/rock interface)

Point A3 (bentonite/sand interface)

D0　D4

Point H0 (at center of heater base)

Point B1 (inner buffer boundary)
Point B2 (at 0.5 m from heater axis)

B1　B4

Point I0 (at center of heater base)

Point B4 (sand/rock interface)

Point B3 (bentonite/sand interface)

E0　E4

Vertical scan-line 4 (rock wall)
Vertical scan-line 3 (sand/bentonite interface)
Vertical scan-line 2 (at 0.5 m radial distance)
Vertical scan-line 1 (heater surface)

Figure 2.　Proposed scan-lines.

2 THERMO-HYDROMECHANIC AXIALLY SYMMETRICAL ABAQUS MODEL

2.1 General

An axially symmetrical 2D THM-model was created and analysed with the finite element code ABAQUS. Here, a brief description of the conceptual model used is given. A more detailed description of the material model is given in other works (Börgesson & Hernelind, 1999; Börgesson & Hernelind, 2003). Detailed information on the models available in the code, application of the code and the theoretical background is given in the ABAQUS Manuals (Hibbitt, Karlsson & Sorensen).

2.2 Geometry and materials

The model was modelling the experiment as a whole (Figs 3, 4).

The materials used in this model are denoted as

- *Bentonite 1: Buffer blocks.*
- *Bentonite 2: Pellet filling.*
- *Concrete: Concrete plug.*
- *Cement: Concrete in bottom of deposition hole.*
- *Heater: Steel in heaters.*
- *Sand shield: Sand surrounding the upper heater.*
- *Outer sand: Sand in outer sand slot.*
- *Air: Air in air slot initially surrounding lower heater.*
- *Rock: CRT tunnel host rock.*

Between the lower heater and the bentonite buffer, the initial 10 mm air slot was modelled. This slot was allowed to close as the bentonite was swelling.

2.3 Initial- and boundary conditions

2.3.1 Initial conditions
Overall temperature 24°C

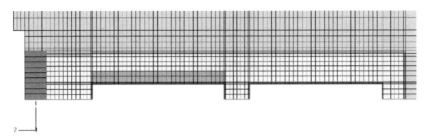

Figure 3. The central part of the ABAQUS model.

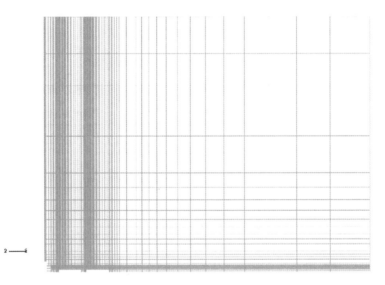

Figure 4. The overall mesh of the ABAQUS model.

Void ratio in *Bentonite 1* and *Bentonite 2*: 0.77
Water saturation in *Bentonite 1*: 0.61
Water saturation in *Bentonite 2*: 1

2.3.2 *Boundary conditions*

At all boundaries except along the symmetry axis, convection boundary conditions were applied. The convection coefficient used was $10 \, \text{W}/(\text{m}^2, °\text{C})$ and the surrounding temperature was set to $24°\text{C}$. Constant volume heat loads corresponding to 1500 W/ heater were uniformly applied in the heater material.

The pore pressure in the *Outer sand*:

0.1 MPa day 1–365
1 MPa day 365–730
2 MPa day 730–1096
0.1 MPa day 1096–1200

Mechanical boundary conditions:

The *Outer sand* was mechanically locked to the rock wall

The *Sand shield* had no movement restrictions
The heaters were permitted to move in the axial direction
The air slot surrounding the lower heater was permitted to close as the bentonite swelled. (Bentonite and heater surface had contact capabilities)
The outer boundaries of the rock were locked.

2.4 *Material models and material data*

2.4.1 *Retention*

The pore pressure u_w of the unsaturated buffer material is a function of the degree of saturation S_r independent of the void ratio:

$$u_w = f(S_r) \tag{2-1}$$

The pore air pressure was not modelled.

2.4.2 *Liquid flow*

The constitutive behaviour for pore fluid is governed by Darcy's law, which is generally applicable to low fluid velocities. Darcy's law states that, under uniform conditions, the volumetric flow rate of the wetting liquid through a unit area of the medium, $\mathbf{q_w}$, is

$$\mathbf{q_w} = -\mathbf{K}\frac{1}{g\rho_w}\left(\frac{\partial u_w}{\partial \mathbf{x}} - \rho_w \mathbf{g}\right), \qquad (2\text{-}2)$$

where

$\mathbf{K}$ = hydraulic conductivity of the medium
g = gravity acceleration
ρ_w = water density
u_w = pore pressure
$\mathbf{x}$ = spatial position

The hydraulic conductivity for a partly saturated medium $\mathbf{K_p}$ is

$$\mathbf{K_p} = (S_r)^\delta \mathbf{K}, \qquad (2\text{-}3)$$

where δ is a parameter (usually between 3 and 10).

2.4.3 Vapour flow
Vapour flow is modelled as a diffusion process driven by a temperature gradient:

$$\mathbf{q_v} = -D_{Tv}\frac{\partial T}{\partial \mathbf{x}} \qquad (2\text{-}4)$$

$$D_{Tv} = D_{Tvb}\cdot \sin^b\left(\frac{S_r}{0.3}\cdot\frac{\pi}{2}\right) \qquad S_r \leq 0.3 \qquad (2\text{-}5)$$

$$D_{Tv} = D_{Tvb} \qquad 0.3 \leq S_r \leq 0.7 \qquad (2\text{-}6)$$

$$D_{Tv} = D_{Tvb}\cdot\cos^a\left(\frac{S_r - 0.7}{0.3}\cdot\frac{\pi}{2}\right) \qquad S_r \geq 0.7 \qquad (2\text{-}7)$$

where

$\mathbf{q_v}$ = vapour flux
T = temperature
$D_{Tvb} = 0.7\cdot 10^{-11}$ m^2/s, the basic thermal vapour diffusivity
$a = b = 6$

2.4.4 Heat conduction
The only thermal flux that is modelled is thermal conduction, which is assumed to be governed by the Fourier law.

$$\mathbf{f} = -\mathbf{k}\frac{\partial T}{\partial \mathbf{x}} \qquad (2\text{-}8)$$

where

$\mathbf{f}$ = heat conductive flux
$\mathbf{k} = \mathbf{k}(T)$ = heat conductivity matrix

The conductivity can be fully anisotropic, orthotropic, or isotropic.

2.4.5 Mechanical behaviour of the structure
The mechanical behaviour has been modelled with a non-linear Porous Elastic Model and Drucker-Prager Plasticity model.

The Porous Elastic Model implies a logarithmic relation between the void ratio e and the average effective stress p according to

$$\Delta e = \kappa \Delta \ln p \qquad (2\text{-}9)$$

where κ = porous bulk modulus = 0.2. Poisson's ratio v is also required and was set to 0.4. The Drucker-Prager Plasticity model contains the following parameters:

friction angle in the p-q plane $\beta = 16°$
cohesion in the p-q plane d = 100 kPa
dilation angle $\Psi = 2°$
yield function $q = f(\varepsilon_{pl}^d)$

The yield function is the relation between Mises' stress q and the plastic deviatoric strain ε_{pl}^d at a specified stress path. The dilation angle determines the volume change during shear.

The volume change caused by the thermal expansion of water and particles can be modelled with the parameters

α_s = coefficient of thermal expansion of solids
α_w = coefficient of thermal expansion of water

Only the expansion of the separate phases is taken into account. The possible change in volume of the structure by thermal expansion (not caused by expansion of the separate phases) is not modelled. However, a thermal expansion in water volume will change the degree of saturation, which in turn will change the volume of the structure.

The water and the particles are mechanically modelled as separate phases with linear elastic behaviour. The pore air is not mechanically modelled.

The effective stress concept according to Bishop is used for modelling the mechanical behaviour of the water-unsaturated buffer material:

$$s_e = (s - u_a) + \chi(u_a - u_w) \qquad (2\text{-}10)$$

where

u_a = pressure of enclosed air = 0
$\chi = S_r$

The shortcomings of the effective stress theory can be partly compensated in ABAQUS by a correction called "moisture swelling". This procedure changes

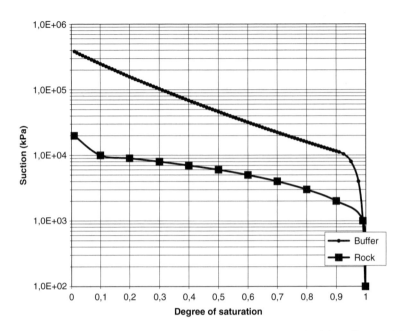

Figure 5. Relation between suction and degree of saturation used in the material models of the buffer material and the rock.

Table 1. Densities, thermal and mechanical properties.

Material	ρ (kg/m^3)	λ (W/m,K)	c (J/kg,K)	E ($\cdot$ 106 kPa)	W ($\cdot$ 10−5 °C−1)	A ($\cdot$ 10−5 °C−1)
Concrete	2400	1.7	900	3	0.3	1.2
Cement	2400	1.7	900	3	0.3	1.2
Heater	7850	50	460			1.15
Rock	2600	2.7	920	1.85	0.3	0.83
Sand shield	1850	0.4	800	stiff	–	–
Outer sand	2150	1.7	1600	stiff	–	–
Air	1.3	0.048	1000	–	–	–

the volumetric strain ε_v by adding a strain that can be made a function of the degree of saturation S_r:

$$\Delta\varepsilon_v = f(S_r) \qquad (2\text{-}11)$$

2.4.6 *Material data*
In this section, parameter values for the material models are presented. The retention properties used for the buffer material (*Bentonite 1*) and for *Rock* are presented as curves (Fig. 5). The densities, thermal and mechanical properties for all materials except the buffer material were regarded as constants (Table 1). The hydraulic conductivity K of the buffer was made temperature- and void ratio dependent (Table 2). The thermal conductivity of the buffer is a function of the water saturation and its heat capacity is a function of the water content (Table 3). The density of this material was set to 2000 kg/m^3. In Table 4, the yield

function values are presented and in Table 5 data for the moisture swelling procedure are given.

3 THERMO-HYDRAULIC AXI-SYMMETRIC 1D CODE_BRIGHT MODELS

3.1 *General*

Two different thermo-hydraulic axially symmetrical models were made for studying the saturation process at the heaters mid-height sections (scan-lines A1–A4, B1–B4, cf. Fig. 2). The assumption, which these models were based on, is that the water- and heat transports in these sections are almost purely radial. The models were analysed with Code_Bright, version 2.1CT (CIMNE, 2002). In this chapter, the geometry, initial conditions, boundary conditions and constitutive models are described.

Table 2. Hydraulic conductivity K as function of void ratio e and temperature T for *Bentonite 1* and *Bentonite 2*.

T (°C)	e	K (m/s)	T (°C)	e	K (m/s)
20	0.4	$0.035 \cdot 10^{-13}$	60	0.4	$0.07 \cdot 10^{-13}$
20	0.6	$0.2 \cdot 10^{-13}$	60	0.6	$0.44 \cdot 10^{-13}$
20	0.8	$0.65 \cdot 10^{-13}$	60	0.8	$1.45 \cdot 10^{-13}$
20	1.0	$1.75 \cdot 10^{-13}$	60	1.0	$3.85 \cdot 10^{-13}$
40	0.4	$0.05 \cdot 10^{-13}$	80	0.4	$0.1 \cdot 10^{-13}$
40	0.6	$0.31 \cdot 10^{-13}$	80	0.6	$0.55 \cdot 10^{-13}$
40	0.8	$1.0 \cdot 10^{-13}$	80	0.8	$1.8 \cdot 10^{-13}$
40	1.0	$2.75 \cdot 10^{-13}$	80	1.0	$4.9 \cdot 10^{-13}$

Table 3. To the left: thermal conductivity λ of *Bentonite 1* and *Bentonite 2* as a function of the degree of saturation S_r. To the right: heat capacity as function of water content for *Bentonite 1* and *Bentonite 2*.

S_r	λ(W/m,K)	w	c (J/kg,K)
0	0.3	0	800
0.2	0.3	0.1	1109
0.3	0.4	0.2	1367
0.4	0.55	0.3	1585
0.5	0.75	1.0	2500
0.6	0.95		
0.7	1.1		
0.8	1.2		
0.9	1.25		
1.0	1.3		

Table 5. Change in volumetric strain ε_v as a function of the degree of saturation S_r.

S_r	$\Delta\varepsilon_v$
0	−0.2
0.1	−0.01
0.2	0.02
0.3	0.03
0.4	0.02
0.5	0.01
0.6	0
0.7	−0.02
0.8	−0.03
0.88	−0.04
0.94	−0.06
0.97	−0.11
0.99	−0.24
1	−0.81

Table 4. Yield function.

q (kPa)	ε_{pl}
1	0
50	0.005
100	0.02
150	0.04
200	0.1

3.2 *Geometry*

The dimensions of the models were in accordance with the experiment setup (Fig. 6, Table 6) (Hökmark & Fälth, 2003).

3.3 *Initial- and boundary conditions*

3.3.1 *Initial conditions*

The void ratios and degrees of saturation values of the bentonite were the values of the blocks as they were installed in the experiment (Table 7). In both models, the initial gas pressure was set to 0.1 MPa, and the initial temperature was set to 24°C.

3.3.2 *Thermal boundary conditions*

To account for the heat flux redistribution between the two heaters, the boundary conditions were derived from the ABAQUS 2D THM-analysis results. The canister surface radial heat fluxes q(t) and rock wall temperature histories T(t) at each heater's mid height were used and applied at the Code_Bright model's boundaries (Figs 7, 8, 9).

3.3.3 *Hydraulic boundary conditions*

The water pressure in the outer sand slot was varied according to the scheme given in the TBT-predictive modelling program (Hökmark & Fälth, 2003). As mentioned above, the pressure was set to 0.1 MPa at test start. After 365 days, it was increased to one MPa and then increased to two MPa after 730 days. After 1096 days, the pressure was reduced to 0.1 MPa. The gas pressure was kept at a constant level of 0.1 MPa.

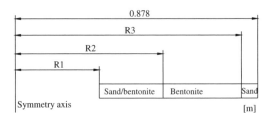

Figure 6.   Models basic geometry.

Table 6.   Model dimensions.

Model	R1 (m)	R2 (m)	R3 (m)	Comment
Upper	0.305	0.535	0.819	Scan-line A1–A4
Lower	0.315	–	0.820	Scan-line B1–B4

Table 7.   Initial conditions.

Model	Sand shield e	S_r	Bentonite e	S_r	Outer sand e	S_r
Upper	0.429	0.17	0.582	0.83	0.563	1
Lower	–	–	0.638	0.76	0.563	1

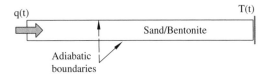

Figure 7.   Thermal boundary conditions. T(t) was applied at the rock wall.

3.4   Material models and material data

3.4.1   Hydraulic properties

The retention curve model that was used is of van Genuchten-type. The van Genuchten relation is given by

$$S_r = \left\{ 1 + \left(\frac{P_g - P}{P_0 \cdot \left(\frac{\sigma}{\sigma_0} \right)} \right)^{\frac{1}{1-\lambda}} \right\}^{-\lambda}, \qquad (3\text{-}1)$$

where

P = pore pressure
P_g = gas pressure
P_0 = reference pressure

λ = shape parameter
σ/σ_0 = surface tension ratio

P_0 and λ were set by fitting Eqn. 3-1 to experimental values (Table 8). To get appropriate parameter values, the experimental retention curve for e = 0.77 given in the TBT-predictive modelling program was used. For this particular void ratio, P_0 and λ were determined by curve fitting. The approximate theory here used for other void ratios is that, for a given material, the suction is determined by the water content. Thus, for a given suction level (or water content), the relation between void ratio e and saturation S_r is

$$S_{r2} = S_{r1} \times \frac{e_1}{e_2} \qquad (3\text{-}2)$$

Using Eqn. 3-2, the experimental suction curve was translated into curves, which correspond to the actual void ratios. To avoid numerical problems, the retention parameters for the outer sand slot were set to be the same as for the bentonite. This could be done without influencing the results because the sand slot was by definition water saturated all the time.

For water transport, there are two mechanisms:

Moisture moves from the rock wall towards the heater driven by pore pressure gradients, obeying the Darcy's law.

468

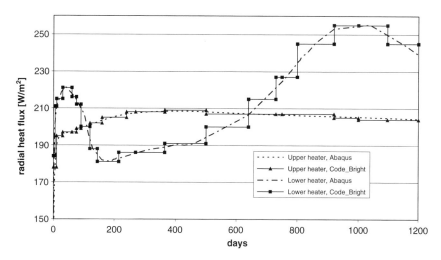

Figure 8. Heat flux histories as derived from the ABAQUS results and the curve fitting used as boundary conditions in the Code_Bright models.

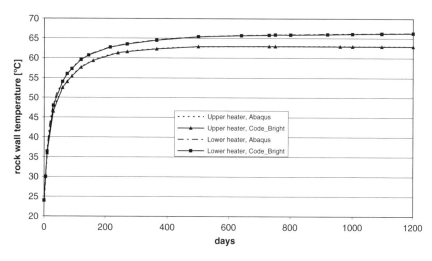

Figure 9. Temperature T(t) as derived from ABAQUS results and the curve fitting used as boundary conditions in the Code_Bright models.

Table 8. Retention parameters and intrinsic permeabilities.

Model	Sand shield			Bentonite			Outer sand		
	P_0 (MPa)	λ	k (m^2)	P_0 (MPa)	λ	k (m^2)	P_0 (MPa)	λ	k (m^2)
Upper	0.005	0.95	10^{-13}	40	0.5	$0.16 \cdot 10^{-20}$	40	0.5	10^{-13}
Lower	–	–		30	0.45	$0.29 \cdot 10^{-20}$	30	0.45	10^{-13}

Table 9. Thermal conductivities.

Model	Sand shield		Bentonite		Outer sand	
	λ_{dry} (W/m,K)	λ_{sat} (W/m,K)	λ_{dry} (W/m,K)	λ_{sat} (W/m,K)	λ_{dry} (W/m,K)	λ_{sat} (W/m,K)
Upper	0.4	1.7	0.3	1.28	0.4	1.7
Lower	–	–	0.3	1.28	0.4	1.7

Moisture moves in the opposite direction in vapour form, driven by vapour concentration gradients, obeying laws of molecular diffusion.

The Darcy flux q is given by

$$q = -\frac{kk_r}{\mu}\nabla P \,, \qquad (3-3)$$

where

k = is the intrinsic permeability
k_r = the relative permeability
μ = dynamic viscosity of water

The pore pressures (and pore pressure gradients ∇P) are calculated from the retention relation. Values of k (Table 8) for the bentonite were determined by fitting to given values (Hökmark & Fälth, 2003).

The relative permeability k_r was assumed to depend on the water saturation S_r according to the generalized power law

$$k_r = S_r^3 \qquad (3-4)$$

The viscosity of water depends on the temperature according to

$$\mu = 2.1 \cdot 10^{-12} \exp\left(\frac{1808.5}{273.15 + T}\right), \qquad (3-5)$$

where T = temperature.

Non-advective flux i of vapour is driven by vapour mass fraction gradients according to

$$i = -(\phi \cdot (1 - S_r) \cdot \rho \cdot D_v) \cdot \nabla \omega \qquad (3-6)$$

where

ϕ = porosity
ρ = gas phase density
ω = mass fraction of vapour in the gas phase
D_v = the diffusion coefficient, which is given by

$$D_v = \tau \cdot 5.9 \cdot 10^{-6} \cdot \left(\frac{(273.15 + T)^{2.3}}{P_g}\right), \qquad (3-7)$$

where τ = tortuosity factor set to 1.

3.4.2 Heat transfer

Code_Bright considers both conductive heat flux and advective heat flux associated with mass motions

(solid, liquid and vapour). For the thermal conductivity λ the relation

$$\lambda = \lambda_{sat} \cdot \sqrt{S_r} + \lambda_{dry} \cdot \left(1 - \sqrt{S_r}\right) \qquad (3-8)$$

was used, where
λ_{sat} = heat conductivity at full saturation
λ_{dry} = heat conductivity in the dry state

The values of λ_{sat} and λ_{dry} (Table 9) were obtained by fitting to given values (Hökmark & Fälth, 2003).

4 RESULTS

4.1 General

In this chapter, the results from the ABAQUS model and from the Code_Bright models are presented. The emphasis in the presentation of the ABAQUS results will be on the results regarding the two canisters mid height sections. The mechanical part of the ABAQUS model is in a preliminary state, but will be refined in the further TBT modelling work. Thus, the mechanic results presented here should be regarded as preliminary.

4.2 ABAQUS results

The buffer around the upper heater experience desaturation during the first time period but gets fully saturated after approximately 500 days (Fig. 10). The temperature increases to reach its maximum values within almost 60 days. The maximum temperature at the heater surface reaches almost 190°C (Fig. 11).

The buffer closest the lower heater is highly desaturated during the first time period, but gets almost fully saturated after 1200 days (Fig. 16).

4.3 Code_Bright results

The buffer closest to the sand shield around the upper heater gets a higher saturation than the initial value during the first tenth of days. This is due that the initial pore water in the sand is absorbed by the bentonite. After this, there is some desaturation, but the bentonite is fully saturated after 750 days. After

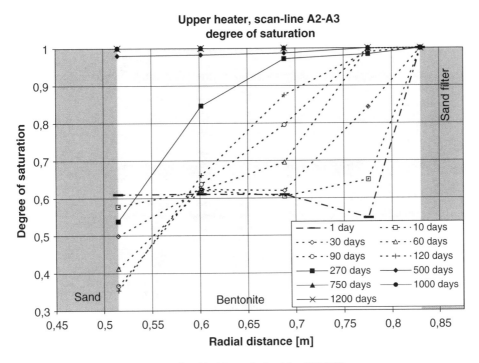

Figure 10.   Degree of saturation along scan-line A2–A3 as calculated by ABAQUS.

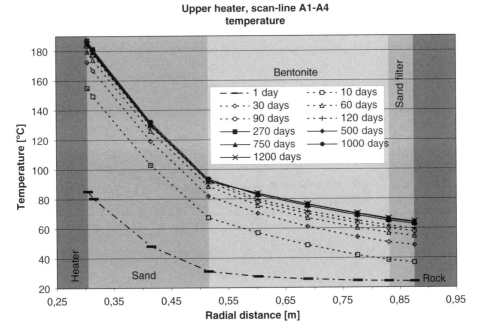

Figure 11.   Temperature along scan-line A1–A4 as calculated by ABAQUS.

471

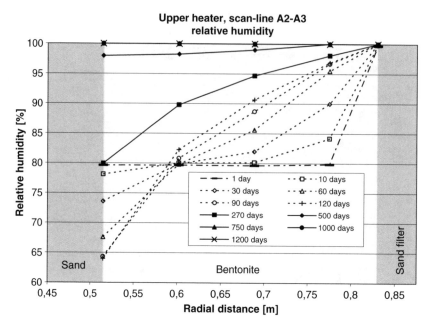

Figure 12. Relative humidity along scan-line A2–A3 as calculated by ABAQUS.

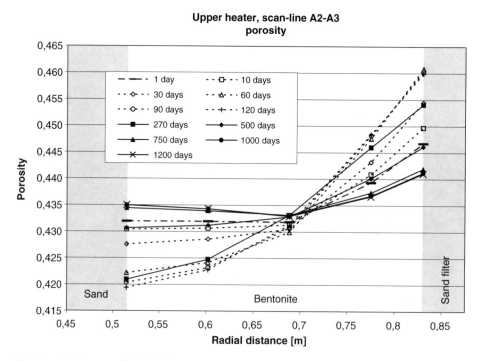

Figure 13. Porosity along scan-line A2–A3 as calculated by ABAQUS.

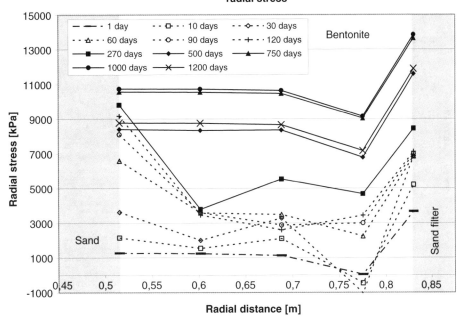

Figure 14. Radial stress along scan-line A2–A3 as calculated by ABAQUS.

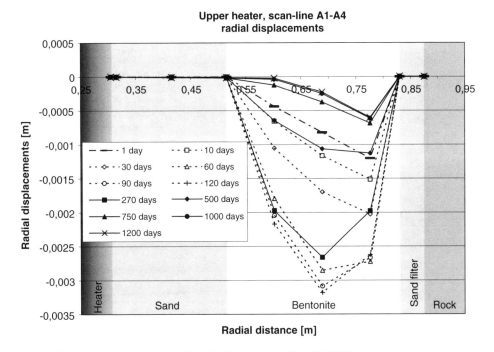

Figure 15. Radial displacements along scan-line A1–A4 as calculated by ABAQUS.

473

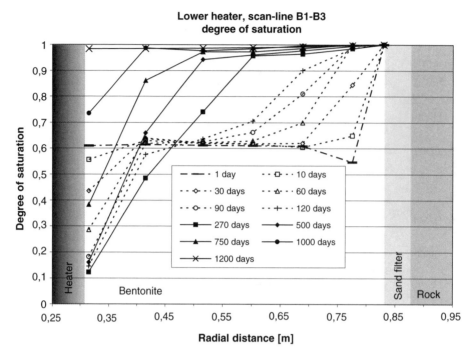

Figure 16. Degree of saturation along scan-line B1–B3 as calculated by ABAQUS.

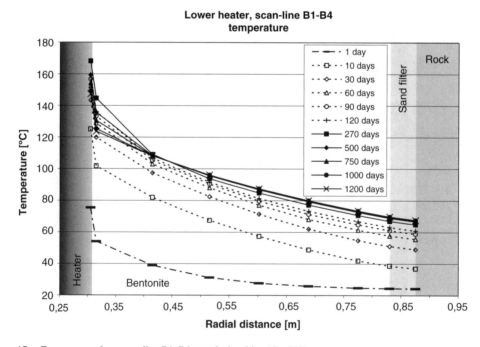

Figure 17. Temperature along scan-line B1–B4 as calculated by ABAQUS.

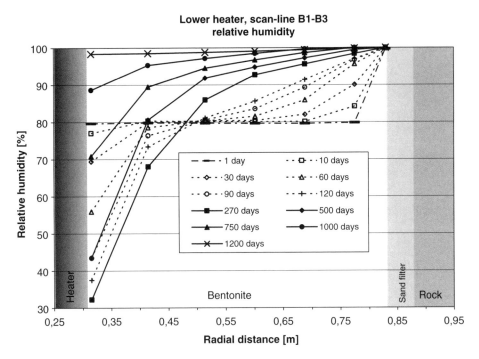

Figure 18.   Relative humidity along scan-line B1–B3 as calculated by ABAQUS.

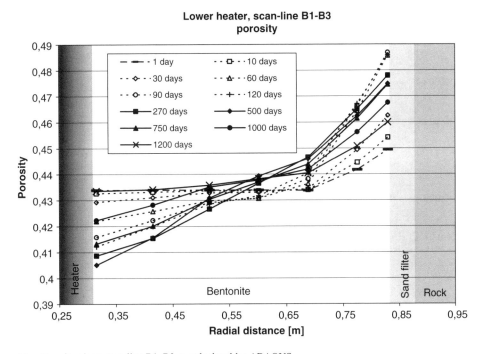

Figure 19.   Porosity along scan-line B1–B3 as calculated by ABAQUS.

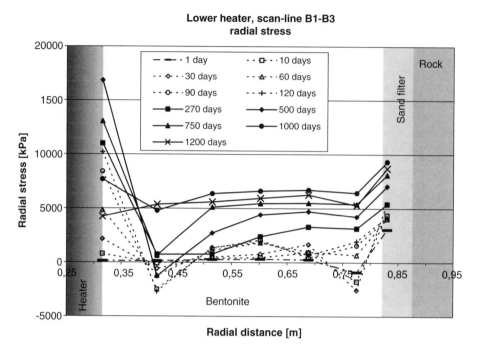

Figure 20. Radial stress along scan-line B1–B3 as calculated by ABAQUS.

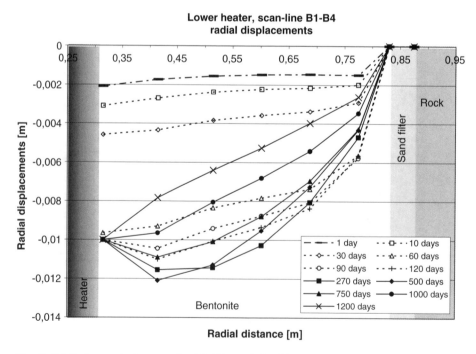

Figure 21. Radial displacements along scan-line B1–B4 as calculated by ABAQUS.

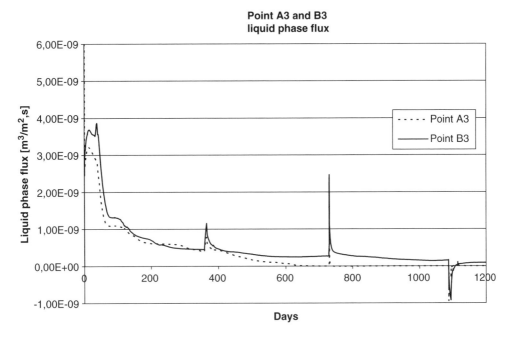

Figure 22.   Liquid phase flux at point A3 and B3 as calculated by ABAQUS.

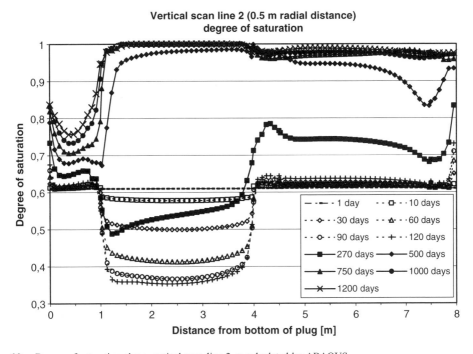

Figure 23.   Degree of saturation along vertical scan-line 2 as calculated by ABAQUS.

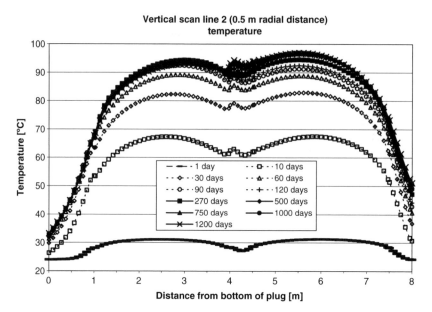

Figure 24.   Temperature along vertical scan-line 2 as calculated by ABAQUS.

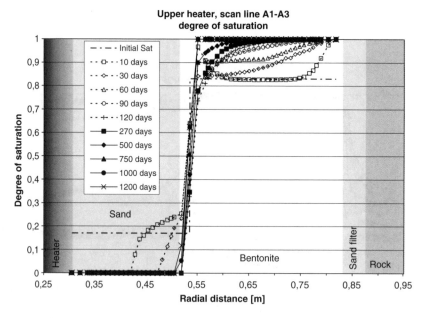

Figure 25.   Degree of saturation along scan-line A1–A3 as calculated by Code_Bright.

1200 days, one can observe a small resaturation of the sand shield (Fig. 25). The temperature around the upper heater is calculated to be lower than in the corresponding ABAQUS results (Fig. 26).

There is a difference between the ABAQUS- and the Code_Bright results when it comes to the water saturation around the lower heater. In the ABAQUS model, the buffer is almost saturated after 1200 days,

478

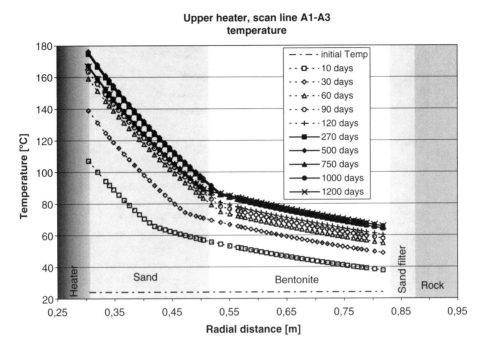

Figure 26.   Temperature along scan-line A1–A3 as calculated by Code_Bright.

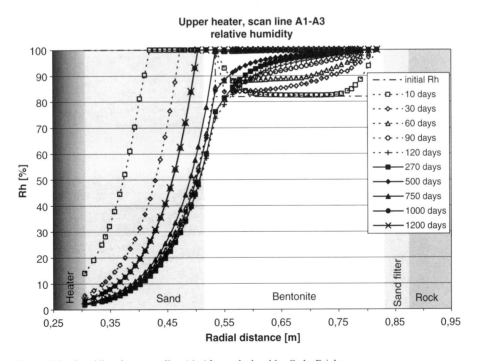

Figure 27.   Relative humidity along scan-line A1–A3 as calculated by Code_Bright.

479

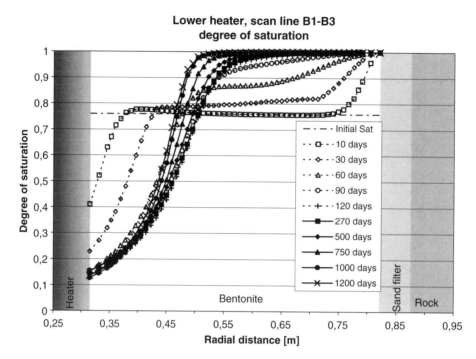

Figure 28. Degree of saturation along scan-line B1–B3 as calculated by Code_Bright.

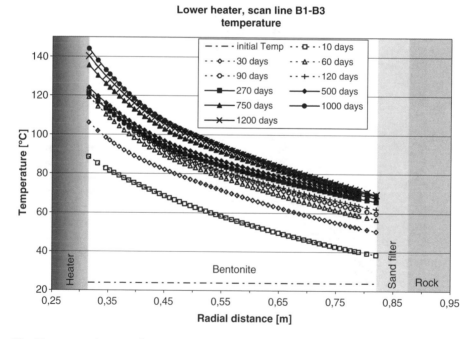

Figure 29. Temperature along scan-line B1–B3 as calculated by Code_Bright.

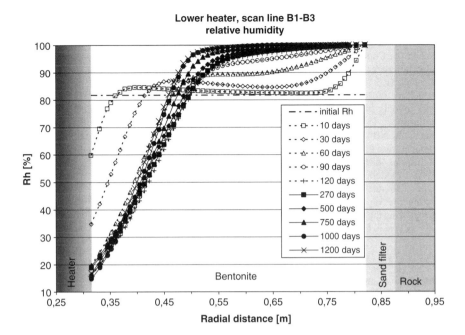

Figure 30.   Relative humidity along scan-line B1–B3 as calculated by Code_Bright.

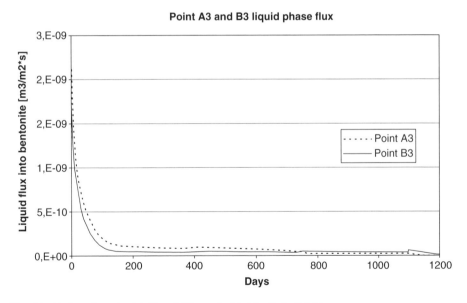

Figure 31.   Liquid phase flux at point A3 and B3 as calculated by Code_Bright.

but this is not the case with the Code_Bright model. In the Code_Bright model, the innermost buffer material remains unsaturated. There is a "saturation front", which first moves towards the heater, but then stops at 0.5 m radial distance (Fig. 28).

5   DISCUSSION

Modelling results from two different codes have been presented in this paper. The most significant difference between the codes is how they represent

481

the waters saturation of the buffer. For both codes, a significant desaturation of the bentonite closest to the lower heater can be observed. This desaturated area is not resaturated in the Code_Bright model within the 1200 days modelled here. The high vapour transport rate keeps the hottest part of the buffer dry. In the ABAQUS model, the buffer becomes fully saturated.

Code_Bright models the vapour phase explicitly and vapour transport is driven by vapour mass fraction gradients. In ABAQUS, the vapour transport is driven by temperature gradients. Both codes have a non-constant vapour diffusivity D, but its state dependence is not modelled in the same way.

Code_Bright also uses a tortuosity factor τ when calculating the vapour diffusivity. There are uncertainties regarding the tortuosity. Laboratory tests could possibly be used to calibrate this parameter further.

The pore air pressure is not modelled the models. The air pressure is considered to be constant at 0.1 MPa. If the air pressure would be allowed to vary in the Code_Bright models, the vapour transport would be influenced. A higher air pressure gives decreased vapour transport, which in turn gives decreased desaturation. In the further modelling work with Code_Bright, the air pressure should be taken into account.

In addition to the model's conceptual uncertainties, which have been discussed here, there are also some uncertainties regarding the initial conditions in the experiment. It is shown in the Code_Bright results that the initial water content in the sand shield has an impact on the saturation process of the buffer surrounding the upper heater. The water content value accounted for here turned out to be higher than in the real sand shield. Thus, the calculated saturation time was probably somewhat shorter than if the real sand water content value would be accounted for.

REFERENCES

CIMNE (2002) Code_Bright user's guide
Börgesson, L, & Hernelind, J, (1999) Coupled thermo-hydro-mechanical calculations of the water saturation phase of a KBS3 deposition hole. Influence of hydraulic rock properties on the water saturation phase, *SKB Technical Report TR-99-41, SKB, Stockholm*
Börgesson, L, & Hernelind, J, (2003) Hydraulic bentonite/rock interaction in FEBEX experiment, *Large Scale Field Tests in Granite, Sitges Symposium*
Hibbitt, Karlsson & Sorensen, ABAQUS manuals, HKS Inc.
Hökmark, H, & Fälth, B, (2003) TBT-predictive modelling program

Advances in Understanding Engineered Clay Barriers – Alonso & Ledesma (eds)
© 2005 Taylor & Francis Group, London, ISBN 04 1536 544 9

Mock-up for studying THM behaviour of swelling clay MX80 at temperature >100°C

Claude Gatabin
CEA Saclay

Jean Claude Robinet
Eurogéomat Consulting, Orléans

1 INTRODUCTION

In 2001, French National Agency of Nuclear Waste, ANDRA, has contracted to CEA's laboratory, LECBA, a study concerning the THM characterisation and behaviour of bentonite MX80. The project was mainly based on the knowledge of the effects of high temperatures on high compacted bentonite samples in non-saturated and saturated conditions.

In order to bring a global response to this problem, a pool of laboratories was created in which the members are CEA (French Atomic Energy Agency) for the experimental part, Eurogeomat Consulting for the physical modelling part and EDF (French Electricity Company) for the numerical modelling part.

Therefore the decision to build up a mock-up with small size but strongly instrumented in which boundary conditions are well known has been decided.

The first results are today available and commented on this paper. However this preliminary refections can be interpreted differently after more time analysis.

2 OBJECTIVES

The major objective of the mock-up is to understand the phenomenological behaviour of compacted MX80 in engineered barrier position but in higher limit conditions to amplify phenomena.

It will be necessary to approach the physic state of water in a such weakly porous media, the hydrous transfers, the thermical transport with high temperature and thermal gradient, the hydraulical transport and his coupling with mechanics.

Since we need acquisition of high accuracy datas for verifying or develop THM models.

The second objective consists in a laboratory support for TBT in situ (in Äspöe HRL) experiment.

2.1 Principle

The experiment includes two stages. The first stage named **THM 1G**, for Thermo-hydro-mechanical One gradient is a one-dimensional experiment in high compacted cylindrical core where a thermical gradient

THM 1G: 2 water contents, thermical gradient only

THM 2G: 2 gradients, thermical & hydraulical gradients

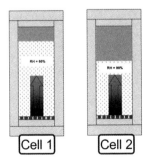 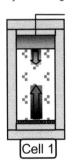 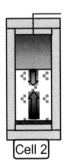

Figure 1. Principle of THM 1G & 2G.

is applied on the lower face of the cylinder, whereas a cold constant temperature is held on the other face.

The second stage named **THM 2G**, for Thermo-Hydro-Mechanical two gradients, consists, when the maximal thermical gradient of THM1G reached an equilibrium, in injection of pure water on the cold face of the clay sample (Figure 1).

In any case the experiment is conducted at constant volume.

Two completely similar mock-up allow to study in the same time two water content samples. The first reason is to determine if the water transfer or transport is a function of the initial saturation degree.

3 DESIGN

For this purpose, it involves a large adaptability and an easy use device. The cell is composed of a 316 stainless steel cylinder lined with a PTFE cylinder for thermal insalution, a 316 SS fixed base including the heater and a moving piston including the hydration and the frozen circuits. The bentonite sample is a ortho-cylinder of 200 mm diameter and height. These sizes were chosen to avoid boundary effects on the measures and to allow the set up of a maximum of sensors.

Figure 2 shows a simple draft of the design.

3.1 Instrumentation and sensors

The sensors are disposed in the bentonite by drilling through the casing as far as a 6 cm diameter middle axis cylinder, perpendicular to the vertical axis (to the gradient). Each cell includes 14 temperature RTD sensors (20 temperature measurements), 7 Relative Humidity capacitive sensors, 8 radial total pressure sensors, and 1 axial stress sensor between the piston and the clamping cover.

The heater is a 500 watt heating cable spirally wound in a PTFE shealted copper disk. The temperature is controlled with a high accuracy RTD switched into the device. Positions (Y) on vertical axis are given in the Table 1. The origin is the bottom of the sample.

3.2 Experimental protocol

Samples have been compacted by isostatic pressing and carefully machined so that there were no gaps or voids between the sample and the casing.

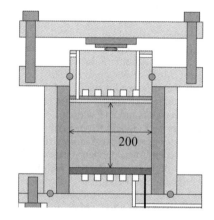

Figure 2. Draft of a cell.

Table 1. Sensors positions in the mock-up.

Temperature		Relative humidity		Radial pressure	
n° sensor	Y (mm)	n° sensor	Y (mm)	n° sensor	Y (mm)
T0	0	HR1/HRT1	22.5	PT1	15.0
T1	2.5	HR2HRT2	37.5	PT2	39.0
T2	18.75	HR3/HRT3	52.5	PT3	63.0
T3	35.0	HR4/HRT4	72.5	PT4	87.0
T4	51.25	HR5/HRT5	92.5	PT5	101.0
T5	67.5	HR6/HRT6	112.5	PT6	125.0
T6	83.75	HR7/HRT7	132.5	PT7	149.0
T7	100			PT8	173.0
T8	116.25				
T9	132.5				
T10	148.75				
		Pore pressure			
T11	165	PI1	20.0		
T12	181.25	PI2	52.0		
T13	197.5	PI3	84.0		
T14	206*	PI4	116.0		

Samples specifications are given in Table 2.

The whole experiment, with the two cells completely assembled, wired and blanketed is shown on Figure 3.

In order to have a good monitoring, we have decided to respect the following protocol for THM 1G:

- Temperature held as constant on the top of the sample $= 20°C$
- Increasing temperature step per step with a slope of 10°C per day under 100°C, 5°C per day above 100°C (*max 150°C*)
- Waiting stabilization between two steps: minimum when Relative Humidity measures $\approx$ constant, maximum time no more longer than five days.

THM1G started the 26th of may, 2003 and stopped the 16th of September, 2003.

4 RESULTS AND COMMENTS

All the results are given on the following graphs, comparing cell 1 (initial HR 60%) with cell 2 (initial HR 75%).

Note that, in cell 2, RH sensor n°2 failed at 140°C and provides a little vapor leak through its cable.

The same behaviour is observed in the two cells with a global displacement to the high humidities for cell 2.

RH sensors close to the heater show a desaturation below 60°C (heater) corresponding to a 1.5°C/cm

Table 2. Samples specifications.

	Cell 1: 1858iA	Cell 2: 1857iA
Conditioning powder		
Relative Humidity	60%	90%
Compaction pressure	33 MPa	33 MPa
Sample weight	13332 g	13395 g
Water content	13.66%	17.86%
Diameter	202.7 mm	202.7 mm
Length	203.0 mm	203.0 mm
Bulk density	2.035 g/cm^3	2.045 g/cm^3
Dry density	1.791 g/cm^3	1.735 g/cm^3
Porosity	0.3242	0.3453
Void ratio	0.48	0.527
Saturation degree	0.755	0.897
Swelling pressure at saturation	24.5 MPa	18.2 MPa

Figure 3. Picture of THM 1G experiment.

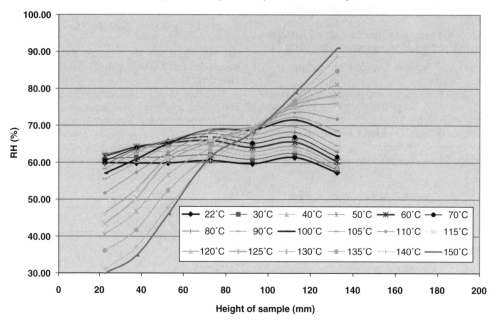

Figure 4.   THM 1G, Cell 1, Relative Humidity versus height of the sample at different temperatures of the heater.

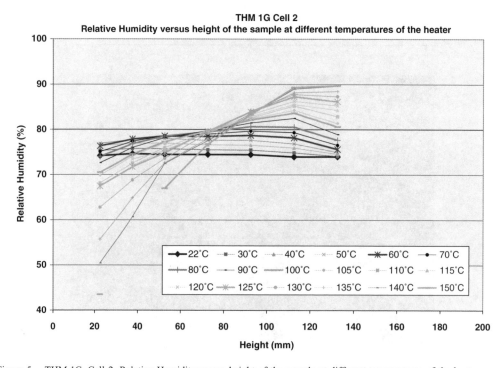

Figure 5.   THM 1G, Cell 2, Relative Humidity versus height of the sample at different temperatures of the heater.

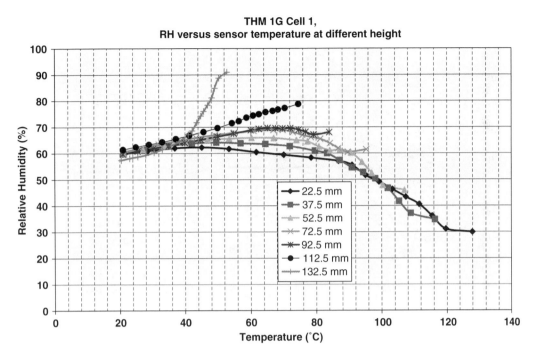

Figure 6. THM 1G, Cell 1, Relative Humidity versus sensor temperature at different heights of the sample.

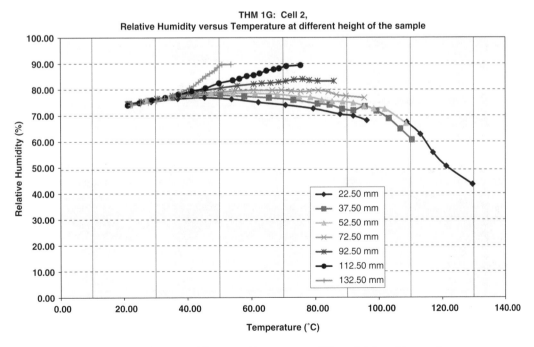

Figure 7. THM 1G, Cell 2, Relative Humidity versus Temperature at different heights of the sample.

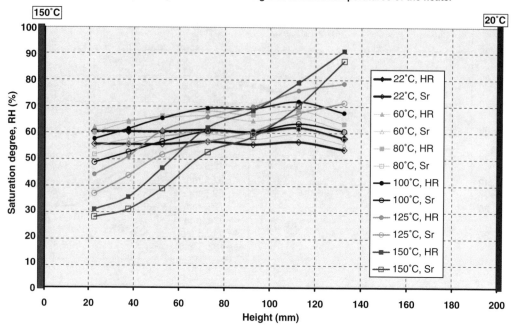

Figure 8. THM 1G, Cell 1, Saturation degree versus height at different temperatures of the heater.

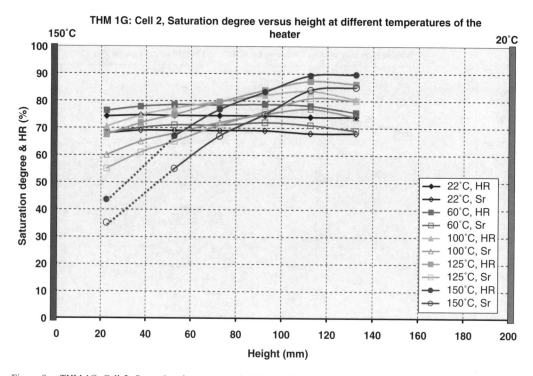

Figure 9. THM 1G, Cell 2, Saturation degree versus height at different temperatures of the heater.

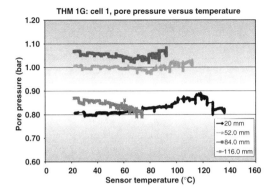

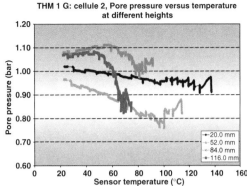

Figure 10 & 11. THM 1G, cells 1 & 2: pore pressure versus temperature at different heights of the sample.

gradient while upper sensors record resaturation (Figures 4, 5, 6 & 7).

Figures 8 & 9 show RH measures expressed in saturation degree (Sr), using Kelvin's law and *Eurogeomat* retention curve fitted with a Vachaud-Vauclin expression at different temperatures of the heater.

$$Sr = a/a + [Pc/\rho w.g]^b$$
$$a = 0.007 \ \& \ b = 1.04$$
$$\rho w.g \approx 10^4$$

When the temperature gradient reaches the maximum of 6.5°C/cm, the first 8 cm are desaturated while the last 12 cm are resaturated.

The pore pressure (Figure 10 & 11) stayed constant during all the test, with a small decrease in cell 2.

5 CONCLUSION

The results of THM1G are not yet completely discussed with partners but there are no uncertainties about the presence of a desaturation close to the heater, in accordance with TBT experiment.

THM1G shows clearly a water transfer from the lower part to the upper part of the sample. This transfer is induced and amplified by the thermal gradient, but it also depends both of the geometry (i.e. the thickness) of the material and its initial saturation degree. The transfer of water seems to be done in liquid phase, in any case at constant atmospheric pressure. The existence or not of a temperature threshold for initiating the desaturation should be confirmed.

Chemical effects. HC and THMC modelling

Direct and inverse modelling of multicomponent reactive transport in single and dual porosity media

J. Samper, L. Zheng & L. Montenegro
ETS Ingenieros de Caminos. Universidad de La Coruña, La Coruña

A.M. Fernández & P. Rivas
CIEMAT. Avenida Complutense, Madrid

Z. Dai
Department of Geological Sciences. Wright State University, USA

ABSTRACT: A methodology for solving the inverse problem of coupled water flow and multicomponent reactive solute transport in porous media implemented in INVERSE-CORE^{2D} is applied here to the interpretation of a long-term permeability test performed on a sample of FEBEX compacted bentonite. FEBEX (Full-scale Engineered Barrier EXperiment) is a demonstration and research project dealing with the bentonite engineered barrier designed for sealing and containment of waste in a high level radioactive waste repository. Hydrogeochemical modelling of porewaters indicate that the main geochemical processes controlling the chemistry of the bentonite are: acid-base reactions, aqueous complexation, cation exchange, sorption protonation/deprotonation dissolution/ex-solution of CO_2 and dissolution/precipitation of highly soluble minerals such as calcite, dolomite, chalcedony and gypsum/anhydrite. All these processes are assumed to take place under equilibrium conditions. The initial saline porewater of the bentonite is flushed with a fresh water with a chemical composition typical of granite formation. Water flux and chemical data of effluent waters were measured during the experiment. Direct and inverse modelling of this experiment has been carried out. Models of chloride breakthrough curves indicate that bentonite exhibits a double porosity behavior. The model reproduces the trends of most measured data except for bicarbonate which is affected by uncertainties in the evolution of $CO_{2(g)}$ pressures.

1 INTRODUCTION

Compacted bentonites are being used in many countries as backfill and sealing material in high-level radioactive waste disposal (HLW) concepts due its low permeability, high swelling capacity and high plasticity. Knowing the pore water composition in the clay barrier is essential for performance assessment purposes because the pore water chemical composition controls the processes involved in the release and transport of the radionuclides: corrosion of the canisters, oxidation-dissolution of the waste matrix and sorption on mineral surfaces.

FEBEX (Full-scale Engineered Barrier EXperiment) is a demonstration and research project dealing with the bentonite engineered barrier designed for sealing and containment of waste in a high level radioactive waste repository (ENRESA, 1998). Hydrogeochemical modelling of porewaters indicate that the main geochemical processes controlling the

chemistry of the bentonite are (Fernández et al., 1999; ENRESA, 2000): acid-base reactions, aqueous complexation, cation exchange, dissolution/ex-solution of CO_2 and dissolution/precipitation of highly soluble minerals such as calcite, dolomite, chalcedony and gypsum/anhydrite.

Obtaining reliable data for pore water chemistry of compacted bentonite (dry density of $1.65 \, g/cm^3$) at initial and saturated conditions of the clay barrier with water contents as low as 14% and 23.8%, respectively, is very difficult. The different laboratory techniques employed so far to obtain pore water alter the system in several ways and introduce sampling artefacts into the resulting data. Namely, squeezing at high pressures may produce the oxidation and dissolution of the accessory minerals present in the bentonite, the outgassing of $CO_{2(g)}$, and some type of chemical fractionation. Furthermore, squeezing does not allow to extract pore water from bentonites with water contents below 20%. The aqueous extract method provides a method to quantify the total content of soluble salts because it works with a low solid-to-liquid ratio. During the extraction, mineral dissolution of various soluble minerals as well as cation exchange processes take place which affect the concentrations of dissolved species. For this reason, indirect methods based on hydrogeochemical modelling must be used to infer the chemical composition of the pore water (Fernández et al., 2004).

FEBEX porewater chemistry data have been analyzed in great detail (Fernández et al., 2001). A permeability test was performed by CIEMAT (Fernández et al., 2002) in order to obtain directly the pore water chemistry of the FEBEX bentonite. The scope of this test is to push out the original solution of a compacted and saturated bentonite sample by means of an incoming granitic water. The test was perfomed during 5 years. Water fluxes, electrical conductivity, pH, Cl^-, SO_4^{2-}, NO_3^-, HCO_3^- Na^+, K^+, Ca^{2+}, Mg^{2+}, Sr^{2+}, Al^{3+} and Zn^{2+} were monitored during this time.

Here we present the interpretation of the permeability test using the inverse methodology of Dai and Samper (2000; 2004) for solving the inverse problem of coupled water flow and multicomponent reactive solute transport in porous media, which was implemented in INVERSE-CORE^{2D} (Dai, 2000). Two conceptual models are considered: single and double porosity media. The experiment has been modelled with the CORE^{2D} V4. (Samper et al., 2000; 2003d) and INVERSE-CORE^{2D} V1 (Dai, 2000). CORE^{2D} solves for groundwater flow, heat transport and multi-component reactive solute transport while INVERSE-CORE^{2D} V1 estimates flow, transport and geochemical parameters based on minimizing a generalized least-squares criterion by means of a Gauss-Newton-Levenberg-Marquardt method.

After a forward model was set up, the sensitivity of model results to model parameters is analyzed. Some parameters which cannot be obtained in the experiment are estimated. Different boundary conditions are tested. Three geochemical conceptual models which were used for modeling other experiments of FEBEX are also tested. Finally, a double porosity model is proposed. Besides describing the permeability test, the main experimental results are shown. After that, several possible conceptual geochemical models are presented. In the next section the numerical model and the computer codes, CORE^{2D}

and INVERSE-CORE^{2D}, are described. Finally, numerical results of the permeability test are presented for both single and double porosity models.

2 DESCRIPTION OF THE PERMEABILITY TEST

The experimental column used for the permeability test consists on a stainless steel cell in which a sample of compacted clay is subject to water flow (Figure 1). The internal diameter of the cell is 50 mm and the length is 25 mm. A HPLC pump inyects a solution at a pressure of 40 bars through a porous stainless steel filter providing a flow rate that after stabilization reached a nearly-constant value of approximately 2 mL/month. Outflowing water comes out through another stainless steel filter and is recovered at the end of the compacted clay and sampled inside a syringe (Fernández et al., 2002). The main characteristics of this test are listed in Table 1 (Fernández et al., 2002). Porewaters of the compacted clay began to be displaced on September 29th 1998 by means of the injection of a granitic water.

The chemical composition of the inflow granitic water is listed in Table 2 (Fernández et al., 2002). The

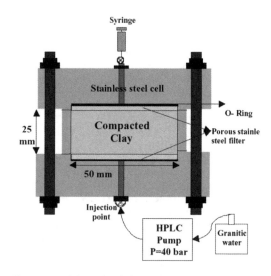

Figure 1. Schematic design of the permeability test (Fernández et al., 2002).

Table 1. Main characteristics of the permeability test.

Initial time	Initial ρ_d (g/cm^3)	Initial gravimetric water content (%)	Total clay mass (g)	Pore volume (cm^3)	Injection pressure (cm H$_2$O)
29/9/98	1.65	23.1	99.7	19.09	$4.08 \cdot 10^4$

initial pore water chemical composition of the FEBEX compacted bentonite at a dry density of $1.65\,g/cm^3$ (to which it corresponds a water content of 23.8%) depends of the conceptual geochemical model used on the THG modeling (see Section 4).

3 EXPERIMENTAL RESULTS

Hydrodynamic and chemical data were monitored during the duration of the experiment. Nearly 2 mL were needed for a complete chemical analysis; therefore samples could only be taken once the cell delivered enough amounts of water. The first sample of 2 mL was obtained after 69 days, the second after 107 days and the rest of the aliquots were obtained, regularly, each month with an approximate flow rate of 2 mL/month. Possible reasons for the fact that the first two samples took longer times to be collected include: (1) The compacted bentonite was not fully saturated, therefore some time was needed to saturate the sample, and (2) The permeability of the sample increases with time.

The clay sample has a dry density of 1.65 kg/L, a porosity of 0.39. The pore water volume is 19.09 mL and a total volume of 91.13 mL was collected after 1177 days. The quasi steady-state hydraulic conductivity of the clays is $2.7 \cdot 10^{-14}\,m/s$. The electrical

conductivity of the effluent water decreased from 15536 in the first alliquot to $700\,\mu S/cm$ in the last one. Most chemical species show clear dilution trends which are shown later. Chloride for instance decreases from 3500 to 64 mg/L.

4 CONCEPTUAL MODELS

Since the flow model for the permeability test is rather simple, efforts are concentrated on establishing the conceptual geochemical model (CGM). Such CGM requires the identification of relevant geochemical processes (aqueous complexation, acid-base, mineral dissolution/precipitation, cation exchange, and gas dissolution/exsolution) and the chemical composition of pore waters. Four possible conceptual geochemical models have been proposed, including models CGM-0, CGM-1 and CGM-2 of Samper et al. (2003a) and that of Fernández et al. (1999, 2001) which will be denoted as CGM-CIEMAT. Table 3 to 5 show the main characteristics of CGM-0, CGM-1 and CGM-2 models.

The volume of water accessible for geochemical reactions is considered to be equal to the total water content (23.8%). Table 6 shows the estimated chemical composition of the FEBEX bentonite pore water at a water content of 23.8% for models CGM-0,

Table 2. Chemical composition (in mg/L) of the inflow granitic water.

Cations	mg/L	Anions	mg/L
Na^+	11.5 ± 0.7	F^-	0.20
K^+	1 ± 0.0	Cl^-	13.5 ± 0.7
Ca^{2+}	40 ± 7.1	Br^-	<0.10
Mg^2	8.9 ± 0.7	NO_3^-	4.3 ± 0.7
Sr^+	0.09 ± 0.01	HCO_3^-	144.5 ± 0.7
Ba^+	0.04 ± 0.01	SO_4^{2-}	14 ± 0.0
Cs^+	$5.6 \cdot 10^{-3}$	SiO_2	22.1 ± 0.1
Al^{3+}	$80 \cdot 10^{-3}$	I^-	$10.7 \cdot 10^{-3} \pm 3 \cdot 10^{-3}$
B^{3+}	$31 \cdot 10^{-3} \pm 41 \cdot 10^{-3}$	pH	8.15 ± 0.21
Mn^{2+}	<0.03	Eh	262 mV
Th	$0.18 \cdot 10^{-3}$	Fe	<0.03
U	$8.7 \cdot 10^{-3} \pm 0.7 \cdot 10^{-3}$	Elec. conductivity	$291\,\mu S/cm$

Table 3. Chemical species considered in model CGM-0. Numbers of species are given within parentheses (Samper et al., 2003a).

Primary species (10)	H_2O, H^+, Ca^{+2}, Mg^{+2}, Na^+, K^+, Cl^-, SO_4^{2-}, HCO_3^-, $SiO_{2(aq)}$
Aqueous complexes (17)	OH^-, $CaSO_{4(aq)}$, $CaCl^+$, $MgCl^+$, $NaCl_{(aq)}$, $MgHCO_3^+$, $NaHCO_{3(aq)}$, $CaHCO_3^+$, $MgCO_{3(aq)}$, $CaCO_{3(aq)}$, $CO_{2(aq)}$, CO_3^{-2}, KSO_4^-, $MgSO_{4(aq)}$, $NaSO_4^-$, $H_2SiO_4^{2-}$, $HSiO_3^-$
Minerals (4)	Calcite, chalcedony (chemical equilibrium), anhydrite (equilibrium, only precipitation)
Exchangeable cations (5)	H^+, Ca^{+2}, Mg^{+2}, Na^+, K^+ (selectivity constants)
Protonation	$XOH_2^+ = XOH + H^+$ and $XO^- + H^+ = XOH$

Table 4. Chemical species considered in model CGM-1 (Samper et al., 2003a).

Primary species (10)	H_2O, H^+, Ca^{+2}, Mg^{+2}, Na^+, K^+, Cl^-, SO_4^{2-}, HCO_3^-, $SiO_{2(aq)}$
Aqueous complexes (17)	OH^-, $CaSO_{4(aq)}$, $CaCl^+$, $MgCl^+$, $NaCl_{(aq)}$, $MgHCO_3^+$, $NaHCO_{3(aq)}$, $CaHCO_3^+$, $MgCO_{3(aq)}$, $CaCO_{3(aq)}$, $CO_{2(aq)}$, CO_3^{-2}, KSO_4^-, $MgSO_{4(aq)}$, $NaSO_4^-$, $H_2SiO_4^2$, $HSiO_3^-$
Minerals (4)	Calcite, dolomite (kinetics), chalcedony and gypsum (at chemical equilibrium)
Exchangeable cations (5)	H^+, Ca^{+2}, Mg^{+2}, Na^+, K^+ (selectivity constant)
Gases (1)	Fixed pressure of $CO_{2(g)}$
Protonation	$XOH_2^+ = XOH + H^+$ and $XO^- + H^+ = XOH$

Table 5. Chemical species considered in model CGM-2 (Samper et al., 2003a).

Primary species (10)	H_2O, H^+, Ca^{+2}, Mg^{+2}, Na^+, K^+, Cl^-, SO_4^{2-}, HCO_3^-, $SiO_{2(aq)}$
Aqueous complexes (17)	OH^-, $CaSO_{4(aq)}$, $CaCl^+$, $MgCl^+$, $NaCl_{(aq)}$, $MgHCO_3^+$, $NaHCO_{3(aq)}$, $CaHCO_3^+$, $MgCO_{3(aq)}$, $CaCO_{3(aq)}$, $CO_{2(aq)}$, CO_3^{-2}, KSO_4^-, $MgSO_{4(aq)}$, $NaSO_4^-$, $H_2SiO_4^{2-}$, $HSiO_3^-$
Minerals (4)	Calcite, dolomite (kinetics), chalcedony (chemical equilibrium) and gypsum (only precipitation)
Exchangeable cations (5)	H^+, Ca^{+2}, Mg^{+2}, Na^+, K^+ (selectivity constant)
Gases (1)	Fixed pressure of CO_2
Protonation	$XOH_2^+ = XOH + H^+$ and $XO^- + H^+ = XOH$

Table 6. FEBEX bentonite pore water composition (in mol/L) at water a content of 23.8% estimated with different CGM models.

Species	CGM-0	CGM-1	CGM-2	CGM-CIEMAT
Cl^-	$1.12 \cdot 10^{-1}$	$1.10 \cdot 10^{-1}$	$1.39 \cdot 10^{-1}$	$1.13 \cdot 10^{-1}$
SO_4^{2-}	$1.27 \cdot 10^{-2}$	$4.37 \cdot 10^{-2}$	$1.65 \cdot 10^{-2}$	$4.30 \cdot 10^{-2}$
HCO_3^-	$6.45 \cdot 10^{-4}$	$4.15 \cdot 10^{-4}$	$4.55 \cdot 10^{-4}$	$1.54 \cdot 10^{-3}$
Ca^{2+}	$6.34 \cdot 10^{-3}$	$1.92 \cdot 10^{-2}$	$1.66 \cdot 10^{-2}$	$1.54 \cdot 10^{-2}$
Mg^2	$8.82 \cdot 10^{-3}$	$2.14 \cdot 10^{-2}$	$2.16 \cdot 10^{-2}$	$1.94 \cdot 10^{-2}$
Na^+	$1.14 \cdot 10^{-1}$	$1.20 \cdot 10^{-1}$	$1.06 \cdot 10^{-1}$	$1.29 \cdot 10^{-1}$
K^+	$1.04 \cdot 10^{-3}$	$2.15 \cdot 10^{-3}$	$1.64 \cdot 10^{-3}$	$1.24 \cdot 10^{-3}$
pH	7.94	7.65	7.69	7.43

CGM-1 and CGM-2 (Samper et al., 2003a) and CGM-CIEMAT (Fernández et al., 1999, 2001, 2002).

5 NUMERICAL MODEL

The permeability test is modelled using CORE^{2D} (Samper et al., 2000, 2003d) and INVERSE-CORE^{2D} V1 (Dai, 2000). CORE^{2D} is a finite element code which solves for groundwater flow, heat transport and multi-component reactive solute transport. It accounts for the following chemical equilibrium reactions: acid-base, redox, aqueous complexation, surface adsorption via surface complexation and K_d-approach, cation exchange, mineral dissolution-precipitation and gas dissolution-exsolution. It also considers chemical kinetics for dissolution/

precipitation of minerals as well as radioactive decay. INVERSE-CORE^{2D} V1 uses the Gauss-Newton-Levenberg-Marquardt algorithm to minimize the objective function following an iterative process.

6 NUMERICAL RESULTS

6.1 Single porosity model

In the single porosity model the system is divided in to two material zones, one for the bentonite and another for the stainless steel filter. Parameters used for these two material zones are summarized in Table 7. Saturated hydraulic conductivity is calculated from experimental results (Fernández et al., 2002). Total transport porosity and density were measured directly (Fernández et al., 2002). Diffusion

Table 7.   Summary of parameters for FEBEX compacted bentonite and stainless steel filter.

Property		Compacted bentonite	Steel filter
Saturated hydraulic conductivity	K	$2.19 \cdot 10^{-9}$ m/day	$2.19 \cdot 10^{-3}$ m/day
Total transport porosity	Φ	0.39	0.80
Molecular diffusion coefficient	D_0	$1.27 \cdot 10^{-5}$ m^2/day	$1.27 \cdot 10^{-5}$ m^2/day
Longitudinal dispersivity	DS_L	10^{-4} m	10^{-4} m
Transverse dispersivity	DS_T	10^{-4} m	10^{-4} m
Dry density	ρ	$1.65 \cdot 10^3$ kg/m^3	$1.20 \cdot 10^3$ kg/m^3
Tortuosity	τ	$\theta^{7/3}/\Phi^2$	$\theta^{7/3}/\Phi^2$

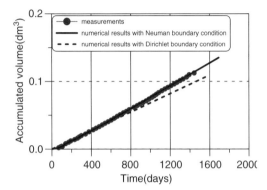

Figure 2.   Modeling results of accumulated water outflow volume with Dirichlet flow boundary condition and Neuman flow boundary condition.

coefficients were taken from those used for the coupled thermo-hydro-geochemical models of the 'mock-up' and 'in situ' FEBEX experiments (ENRESA, 2000; Samper et al., 2003b,c). Longitudinal and transverse dispersivities were estimated with INVERSE-core by fitting model results to measured chloride data. Two types of flow boundary conditions have been tested: (1) Dirichlet and (2) Neuman. There are small differences in cumulative water inflows computed with these two conditions (Figure 2). Although a Dirichlet condition seems more appropriate to represent a constant injection pressure, the fact that the flow rate decreased during the experiment for reasons yet to be explained, we believe that the model represents better the experiment by using Neuman boundary condition with prescribed fluxes qual to measured values.

For solute transport a Neuman transport boundary condition is used according to which solute flux through the boundary is equal to the product of solute concentration times water flux. Anion exclusion is assumed for chloride. An accesible porosity of 0.19 is estimated with INVERSE-CORE^{2D} for chloride. It should be recalled that the total porosity is 0.39.

Breakthrough curves of chloride computed with the CGM-CIEMAT model and using Dirichlet and Neuman flow boundary conditions are compared to measurements in Figure 3. Although the overall fit of model results to measured data is good, a semi-log plot of concentrations reveals that this model does not reproduce well the long tail of the measured values. The model tends to underestimate the concentrations at late times.

Based on the three conceptual geochemical models, CGM-0, CGM-1 and CGM-2, computed results are compared to measurements for the main chemical species. Figure 4 shows the results for sulfate and chloride. It can be seen that model results for sulfate show a wide spread. Measured data fall within the band of computed values. Model results for chloride are similar for the three models and for the most part reproduce the measured data.

Models CGM-1 and CGM-2 fit better measured calcium data (Figure 5). All models are unable to fit sodium data during intermediate times (Figure 5). The results for magnesium and potassium are shown in Figure 6. The models fail to reproduce the trends of bicarbonate possibly due to an inadequate definition of CO_2 pressures (see Figure 7).

All models reproduce the trends of measured pH (Figure 7) when proton sorption/desorption is considered. As illustrated by Samper et al. (2003d) models which do not account for surface sorption protonation fail to reproduce measured pH.

6.2   Double porosity model

Generally the term 'double porosity' is used to represent a conceptual model, in which, based on the microscopic heterogeneities, the medium is divided into two or more domains which are coupled by an interaction term for modeling flow and solute transport. This kind of conceptual model can be called 'dual-domain', 'dual-region' or 'multi-region' and the sub-regions can be two or more mobile regions (Gwo et al., 1995) or one mobile and one immobile region.

Dual or double porosity models, initially introduced to simulate single-phase flow in fissured

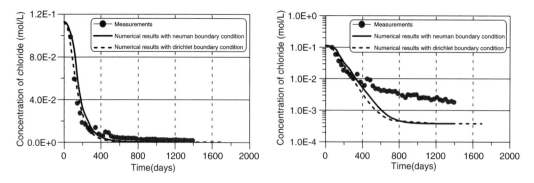

Figure 3. Measured and computed breakthrough curve of chloride: arithmetic (left) and logarithmic scale (right).

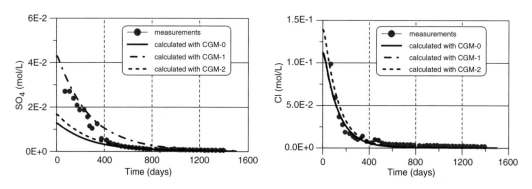

Figure 4. Model results for sulfate (left) and chloride (right) using models CGM-0, -1 and -2.

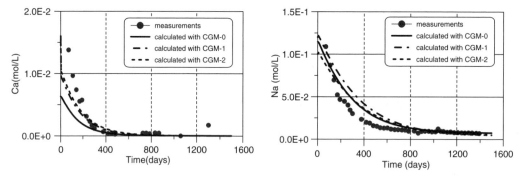

Figure 5. Model results for calcium (left) and sodium (right) using models CGM-0, -1 and -2.

groundwater reservoirs (e.g. Barenblatt et al., 1960), assume that a porous medium consists of two separate but connected continua. Of these, one continuum is associated with a system or network of fractures, fissures, macropores, or intraggregate pores, while the other continuum involves the porous matrix blocks or soil aggregates. Hence, dual porosity

models usually involve two flow/transport equations which are coupled by means of a sink/source term to account for water/solute transfer between the pore systems.

The dual-porosity concept has been commonly used to describe the preferential movement of water and solutes at the macroscopic scale, a phenomenon that is

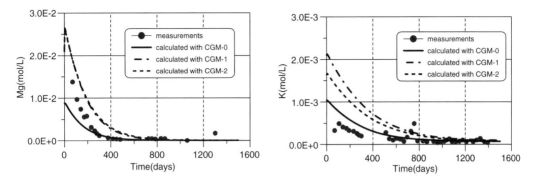

Figure 6. Modeling results of magnesium (left) and potassium (right) using models CGM-0, -1 and -2.

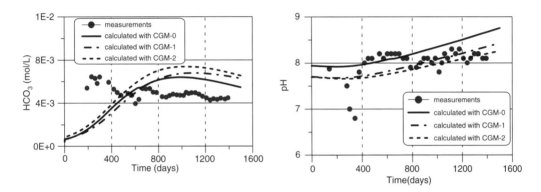

Figure 7. Model results for bicarbonate (left) and pH (right) using models CGM-0, -1 and -2.

widely believed to occur in most natural (undisturbed) media (e.g. Schwartz et al., 2000). The double porosity model has been used to predict water flow in fractured media and solute transport of both steady and transient flow. Recently, attempts have been made to extend the concept to variably saturated fractured rocks and structured soils (Gerke and van Genuchten, 1993a, 1993b, 1996; Ray et al., 1997; Larson and Jarvis, 1999; Saxena et al., 1994). Nowadays, some soil scientists used this conceptual model to predict the leaching of pollutants into soils and groundwaters (Hantush et al., 2000, 2002) and the transport of solutes through clay soils (Jaivis et al., 1991a; 1991b).

Compacted clay also exhibits a double-porosity behaviour (Alonso et al., 1995; Kim et al., 1997). As shown in Figure 8, double porosity model includes two parts: macro-pores through which water flows and therefore is considered the mobile part and micro-pores through which there is no flow and can be considered as immobile part. An increased level of complexity can be built into the single porosity model by assuming that the equilibrium is not reached between the mobile part (macro-pores) and the immobile part (micro-pores which may not be well

connected). Furthermore, it is assumed that water flows through the mobile part but a mass transfer (by molecular diffusion) will exist between mobile and immobile parts. In order to account for the two regions, three material zones are defined: material 1 which corresponds to the mobile part; material 2 for the immobile part while material 3 is used to model the two stainless steel filters located upstream and downstream the clay sample (Figure 8).

Let f_m and f_{im} be the fractions of the total volume of the sample occupied by the mobile and immobile domains, respectively. By trial and error we estimated that $f_m = 0.3$ and $f_{im} = 0.7$. In other words, the mobile zone occupies 30% of the total volume of the sample, while the immobile zone of the clay sample is 70%.

It should be noticed that in our formulation of double porosity, model porosities for mobile and immobile zones (materials 1 and 2) are not global porosities, but relative porosities. Let Φ_m and Φ_{im} be the relative porosities of the mobile and immobile zones which are defined as the ratio of void volume to total volume of each domain. According to this approach, the total porosity of the sample is given by $\Phi = \Phi_m f_m + \Phi_{im} f_{im}$. Parameters Φ_m and Φ_{im} have a

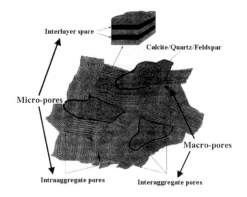 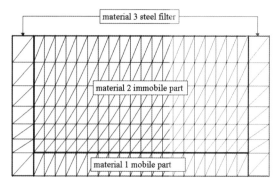

Figure 8. Schematic representation of 'macro-pores' and 'micro-pores' in a double porosity model (left) and finite element mesh used in the double porosity model.

Table 8. Material zone parameters for double porosity models. K_1 is hydraulic conductivity in the direction of the flow while K_2 is the conductivity in the transverse direction. DS_L and DS_T are longitudinal and transverse dispersivities.

Material	K_1 (m/day)	K_2 (m/day)	Φ, Φ_m, Φ_{im}	DS_L(m)	DS_T(m)
1	$7.27 \cdot 10^{-9}$	$7.27 \cdot 10^{-9}$	0.7	$2.50 \cdot 10^{-4}$	$5.00 \cdot 10^{-5}$
2	$2.18 \cdot 10^{-12}$	$2.18 \cdot 10^{-12}$	0.257	$2.50 \cdot 10^{-4}$	$5.00 \cdot 10^{-6}$
3	$2.19 \cdot 10^{-3}$	0.219	0.8	$1.00 \cdot 10^{-4}$	$1.00 \cdot 10^{-3}$
Physical value	$2.18 \cdot 10^{-9}$	$2.20 \cdot 10^{-10}$	0.39	$2.50 \cdot 10^{-4}$	$1.85 \cdot 10^{-5}$

clear physical meaning, although they are difficult to measure. It should be noticed that the overall porosity of the mobile region can be computed as $\Phi_m f_m$. Similarly, $\Phi_{im} f_{im}$ represents the overall porosity of the immobile region. According to our experience, this way of parameterizing double porosity is not affected by the geometry adopted in the model. Therefore, the values $\Phi_m f_m$ and $\Phi_{im} f_{im}$ are independent of model geometry. INVERSE-CORE is used to estimate total and accessible porosities, diffusion coefficients, and dispersivities for each material zone. Parameters of each material zone are shown in Tables 8, 9 and 10.

Since transport in the compacted bentonite is a diffusion-dominated process, to make the model numerically double porosity, the diffusion coefficients in the immobile part must be much smaller than those in the mobile part which is advection-dominated. In the double porosity model, different diffusion coefficients are assumed for each species (see Table 9). Anion exclusion is taken into account in both mobile and immobile parts. Accessible porosities for each domain are listed in Table 10. The double porosity model leads to better results for chloride, bicarbonate and sodium (see Figure 9 where model results for single and double porosity models are compared).

Table 9. List of effective diffusion coefficients for the double porosity model.

Species	Effective diffusion coefficients (cm^2/day)		
	Material 1	Material 2	Material 3
H^+	$1.5 \cdot 10^{-1}$	$3.79 \cdot 10^{-8}$	$5.97 \cdot 10^{-2}$
Ca^{+2}	$2.75 \cdot 10^{-4}$	$3.79 \cdot 10^{-11}$	$5.09 \cdot 10^{-3}$
Mg^{+2}	$2.75 \cdot 10^{-4}$	$3.79 \cdot 10^{-11}$	$4.52 \cdot 10^{-3}$
Na^+	$2.75 \cdot 10^{-4}$	$3.79 \cdot 10^{-11}$	$8.54 \cdot 10^{-3}$
K^+	$2.75 \cdot 10^{-4}$	$3.79 \cdot 10^{-11}$	$1.26 \cdot 10^{-2}$
Cl^-	$1.51 \cdot 10^{-6}$	$6.30 \cdot 10^{-9}$	$1.30 \cdot 10^{-2}$
SO_4^{2-}	$2.75 \cdot 10^{-4}$	$3.79 \cdot 10^{-11}$	$6.86 \cdot 10^{-3}$
HCO_3^-	$2.81 \cdot 10^{-4}$	$3.79 \cdot 10^{-11}$	$1.30 \cdot 10^{-2}$

7 DISCUSSION OF RESULTS

There are uncertainties on the appropriate flow boundary conditions for the permeability test. Although a Dirichlet condition seems more appropriate to represent a constant injection pressure, the fact that the flow rate decreased during the experiment, we believe that a Neuman condition represents better the flow through the bentonite samples (see Figure 2). However, the Neuman condition leads to computed

Table 10. Total and accessible porosities in mobile and immobile parts.

Material 1 (mobile part)		Material 2 (immobile part)	
Total porosity $\Phi_m f_m$	Accessible porosity	Total porosity $\Phi_{im} f_{im}$	Accessible porosity
0.21	0.0735	0.1799	0.105

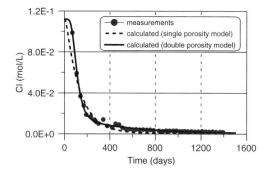
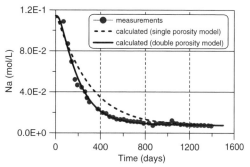

Figure 9. Modeling results of chloride (left) and sodium (right) with using single and double porosity models and model CGM-0 (Dirichlet condition is used for water flow).

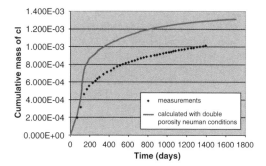

Figure 10. Cumulative Cl mass outflow: computed values (solid line) versus measured values (symbols) with a double porosity model using a Neuman condition for water flow.

Figure 11. Modeling results of chloride using a double porosity model, model CGM-0 and with Dirichlet or Neuman conditions. Notice that the Dirichlet conditions fit better measured data.

solute fluxes which exceed measured fluxes (see Figure 10). Figure 11 shows a comparison of model results for chloride using a double porosity model (model CGM-0) with Dirichlet and Neuman conditions. It should be noticed that the model with a Dirichlet condition fits measured data better than the model with a Neuman condition. The Dirichlet condition, however, overestimates also solute mass fluxes (See Figure 12). Such overestimation occurs mainly during the first month of the test during which probably the sample was not yet fully water saturated. Figure 12 illustrates clearly that computed cumulative mass after the first 44 days agrees for the most with measured mass fluxes. This discrepancy could be

attributed to the fact that the bentonite sample was not fully saturated at the beginning of the test. A numerical model with unsaturated flow should be constructed in order to test this hypothesis.

Anion exclusion is a relevant process for chloride transport. Models without exclusion fail to reproduce the breakthrough curve of chloride. The accessible porosity for chloride is only 48% of total porosity in a single porosity model. In the double porosity model anion exclusion is taken into account in both mobile and immobile parts. According to our results, chloride can access 35% of the mobile-part pores and 58% of the immobile pores. These accessible porosities are an order of magnitude larger than those obtained by

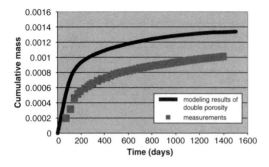 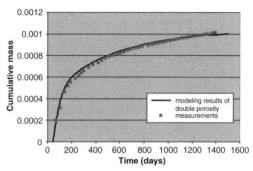

Figure 12. Cumulative Cl mass outflow: computed values (solid line) versus measured values (symbols) with a double porosity model using a Neuman condition for water flow. The figure on the right corresponds to the cumulative mass computed after the first 44 days.

CIEMAT (García-Gutiérrez et al., 2002) using isotopic chloride. Reasons for such a discrepancy remain to be found.

Measured data fall within the bands of computed breakthrough curves with conceptual models CGM-0, 1 and 2 and using a single porosity model, except and bicarbonate. Changes in the pressure of $CO_{2(g)}$ may affect bicarbonate. It should be pointed out that our model considers a fixed pressure of $CO_{2(g)}$ during the test and, therefore, cannot reproduce measured bicarbonate data. Changes in $CO_{2(g)}$ pressure during the test should be taken into account in order to explain the trends of bicarbonate. This is an ongoing task.

8 CONCLUSIONS

The interpretation of a long-term permeability test has been performed numerically using single and double porosity models and three conceptual geochemical models. Model results allow us to draw the following conclusions:

1. There are uncertainties on the appropriate flow boundary conditions. Although a Dirichlet condition seems more appropriate to represent a constant injection pressure, the fact that the flow rate decreased during the experiment leads us to believe that a Neuman condition represents better the flow through the bentonite sample. However, the Neuman condition leads to solute fluxes which exceed measured fluxes. A numerical model with unsaturated flow should be constructed in order to test this hypothesis.
2. Anion exclusion is a relevant process for chloride transport. Models without exclusion fail to reproduce the breakthrough curve of chloride.

Chloride can access 35% of the mobile-part pores and 58% of the immobile pores. These accessible porosities are an order of magnitude larger than those obtained by CIEMAT (García-Gutiérrez et al., 2002). Reasons for such a discrepancy remain to be found.

3. Measured data fall within the bands of computed breakthrough curves with conceptual models CGM-0, 1 and 2 and using a single porosity model.
4. Major discrepancies between model results and measured data are found for bicarbonate. As pointed out by Samper et al. (2003a), they might be related to changes in the pressure of $CO_{2(g)}$ which are not considered in the model.
5. Chloride and sodium data tend to support the double porosity model. The relevance of double porosity for other species remains to be ascertained.

Considering the complexity of the porosity of compacted bentonite, a double porosity model would be more attractive than a single porosity model, but the increase in model complexity introduces additional uncertainties. The following issues need to be addressed in future studies: (1) The ratio between mobile and immobile parts; (2) Other possible ways to describe the interactions between mobile and immobile parts; and (3) Finding ways to relate the properties of mobile and immobile zones to the overall properties of the clay sample.

ACKNOWLEDGMENTS

The work presented here has been funded by ENRESA. The project as a whole was supported by the EC (Contracts FI4W-CT95-0006 and FIKW-CT-2000-0016 of the Nuclear Fission Safety Programme).

REFERENCES

Alonso, E.E., Lloret, A., Gens, A. & Yang, D.Q. (1995) Experimental behaviour of highly expansive double-structure clay. Unsaturated Soils, 11–16.

Barenblatt, G.I., Zheltov, I. & Kochina, I. (1960) Basic concepts in the theory of seepage of homogeneous liquids in fissured rocks. Prikl.Mat. Mekh., 24, 852–864.

Dai, Z. (2000) Inverse Problem of Water Flow and Reactive Solute Transport in Variably Saturated Porous Media, Ph.D. Dissertation, *University of La Coruña*, La Coruña, Spain, 334 pp.

Dai, Z. & Samper, J. (2004) Inverse problem of coupled water flow and multicomponent reactive transport in porous media: Formulation and Application, *Water Resour. Res.*, 2004 (7), Art. No. W07407.

ENRESA (1998) FEBEX Full-scale engineered barriers experiment in crystalline rock: Pre-operational Stage Summary Report, ENRESA Tech. Pub. 1/98.

ENRESA (2000) Full-scale engineered barriers experiment for a deep geological repository in crystalline host rock (FEBEX Project), European Commission, EUR 19147 EN. 403 pp.

Fernández, A.Mª., Baeyens, B., Bradbury, M. & Rivas, P. (2004). Analysis of the pore water chemical composition of a Spanish compacted bentonite used in an engineered barrier. Physics and Chemistry of the Earth. 29/1: 105–118.

Fernández A.M., Cuevas, J. & Rivas, P. (1999) Estudio del agua intersticial de la arcilla FEBEX. Informe Técnico Proyecto FEBEX 70-IMA-L-0-44. CIEMAT.

Fernández, A.M, Cuevas, J. & Rivas, P. (2001) Pore water chemistry of the FEBEX bentonite. Mat. Res. Soc. Symp. Pro. 663, 573–588.

Fernández, A.M., Montenegro, L., Rivas, P. & Samper, J. (2002) Pore water chemical composition obtained in a permeability cell: infiltration flow test, Techn. Report FEBEX. CIEMAT.

García-Gutiérrez, M., Missana, T. & Cormenzana, J.L. (2002) Diffusion coefficients and accessible porosity for HTO and ^{36}Cl in compacted bentonite. International Meeting on Clay in natural and engineered barriers for radioactive waste confinement, Reims, France, pp. 331–332.

Gerke, H.H. & Van Genuchten, M.Th. (1993a) A dual porosity model for simulating the preferential movement of water and solutes in structured porous media. *Water Resour. Res.* 29, 305–319.

Gerke, H.H. & Van Genuchten, M.Th. (1993b) Evaluation of a first order water transfer term for variably saturated dual porosity flow models. *Water Resour. Res.* 29, 1225–1238.

Gerke, H.H. & Van Genuchten, M.Th. (1996) Macroscopic representation of structural geometry of simulating water and solute movement in dual-porosity media. Advances in Water Resources. 19, 343–357.

Gwo, J.P., Jardine, P.M., Wilson, G.V. & Yeh, G.T. (1995) A multiple-pore-region concept to modeling mass transfer in subsurface media. *J. Hydrol.*, 164, 217–237.

Hantush, M.M., Mariño, M.A. & Islam, M.R. (2000) Models for leaching of pesticides in soils and groundwater. *J. Hydrol.*, 227, 66–83.

Hantush, M.M., Govindaraju, R.S., Mariño, M.A. & Zhonglong, Z. (2002) Screening mode for volatile pollutants in dual porosity soils. *J. Hydrol.*, 260, 58–74.

Kim, J.Y., Edil, T.B. & Park, J.K. (1997) Effective porosity and seepage velocity in column tests on compacted clay. *J. Geotechnical and Geoenvironmental Engineering,* vol. 123, No. 12, 1135–1141.

Jaivis, N.J., Jansson, P.E., Dik, P.E. & Messing, I. (1991a) Modelling water and solute transport in macroporous soil. Model description and sensitivity analysis. *J. Soil Science*, 42, 59–70.

Jaivis, N.J., Bergstrom, L. & Dik, P.E. (1991b) Modelling water and solute transport in macroporous soil. Chloride breakthrough under non-steady flow. *J. Soil Science*, 42, 71–81.

Larsson M.H. & Jaivis, N.J. (1999) Evaluation of a dual-porosity model to predict field scale solute transport in a macroporous soil. *J. Hydrol.*, 215, 153–171.

Ray, C., Ellsworth, T.R., Valocchi, A.J. & Boast, C.W. (1997) An improved dual porosity model for chemical transport in macroporous soils. *J. Hydrol.*, 193: 270–292.

Samper, J., Juncosa, R., Delgado, J. & Montenegro, L. (2000) CORE^{2D}: A code for non-isothermal water flow and reactive solute transport, Users manual version 2, Universidad de La Coruña, Tech. Publ. 6/2000. ENRESA, 131 pp.

Samper, J., Vázquez, A. & Montenegro, L. (2003a) Modelización hidrogeoquímica de extractos acuosos para la obtención de la composición química del agua intersticial. Jornadas sobre Zona No saturada, ZNS 2003, Valladolid, Spain, 251–260.

Samper, J., Fernández, A.M., Zheng, L., Montenegro, L., Rivas, P. & Vázquez, A. (2003b) Testing and validation of coupled thermo-hydro-geochemical models with geochemical data from FEBEX in situ test, in this workshop.

Samper, J., Zheng, L., Molinero, J., Fernández, A.M., Montenegro, L. & Rivas, P. (2003c) Reactive solute transport mechanisms in nonisothermal unsaturated compacted clays, in this workshop.

Samper, J., Yang, Ch.B. & Montenegro, L. (2003d) Users Manual of CORE^{2D} V4. A code for non-isothermal water flow and reactive solute transport, Informe Técnico para ENRESA. Universidad de La Coruña, 131 pp.

Samper, J., Fernández, A.M., Zheng, L., Montenegro, L., Rivas, P. & Dai Z. (2003d) Modelización directa e inversa del transporte reactivo multicomponente en medios de doble porosidad, In: *VI Jornadas de Zona No Saturada, Valladolid. Temas de Investigación en Zona no Saturada, Valladolid*, pp. 261–268, 2003.

Saxena, R.K., Jaivis, N.J. & Bgergstrom, L. (1994) Interpreting non-steady state tracer breakthrough experiments in sand and clay soils sing a dual-porosity model. *J. Hydrol.*, 162, 279–298.

Schwartz, R.C., Juo, A.S.R. & McInnes, K.J. (2000) Estimating parameters for a dual-porosity model to describe non-equilibrium, reactive transport in a fine-textured soil. *J. Hydrol.*, 229, 149–167.

Advances in Understanding Engineered Clay Barriers – Alonso & Ledesma (eds)
© *2005 Taylor & Francis Group, London, ISBN 04 1536 544 9*

Pore water chemistry of saturated FEBEX bentonite compacted at different densities

Ana María Fernández & Pedro Rivas
CIEMAT, Madrid (Spain)

ABSTRACT: The understanding of the pore water chemical compositions in the bentonite barrier is one of the main issues for performance assessment of nuclear repositories, since the pore water composition strongly influences the release and transport of radionuclides.

The pore water chemistry of compacted FEBEX bentonite at different dry densities has been modelled using a thermodynamic model. The modelling is based on Bradbury & Baeyens's procedure but, in this paper, the accessible porosities to chloride are not used as free water. The external water is used to determine the solid to liquid ratio in the compacted bentonite-water system, which depends on the total porosity and the amount of water in interlayer or internal positions for each dry density.

Considering the chemical composition of the pore water, some properties of the bentonite watersystem have been calculated: ionic strength, surface potential, DDL thickness, water activity, osmotic potential and swelling pressure.

1 INTRODUCTION

Compacted bentonites are being considered in many countries as a backfill material in high-level radio-active waste disposal (HLW) concepts because of their low permeability, high swelling capacity, high plasticity and high sorption.

The understanding of the pore water chemistry in the clay barrier is essential since the pore water composition influences the release and transport of the radionuclides. However, obtaining reliable data on the pore water chemistry of compacted bentonite (dry density of $1650 \, kg/m^3$) under initial and saturated conditions of a repository where the water contents (w.c.) are so low (14 wt.% and 23.8 wt.% of the dry mass, respectively), is very difficult. Many of the different laboratory techniques employed so far to obtain pore water compositions tend to perturb the system and introduce sampling artefacts into the measured data.

Squeezing at high pressures in saturated bentonites with high expansible smectite contents produces the dilution of water samples because some part of the interlayer water is extracted during the tests, even with pressures of 10–60 MPa (Fernández et al., in press-b). The leaching experiments or aqueous extraction experiments are generally carried out at unrealistically low solid to liquid (S:L) ratios and the unconstrained dissolution of highly soluble salts and sparingly soluble minerals, together with cation exchange reactions on the montmorillonite, leads to

water compositions and cation occupancies which are very dependent on the experimental conditions (Bradbury and Baeyens, 1998; Fernández et al., 2001). Furthermore, other experimental conditions, apart from the S:L ratios, differ somewhat from those expected in the repository: the oxidation potential and the partial pressure of CO_2 are not usually controlled in the different experiments.

For this reason, it is necessary to apply indirect methods based on geochemical modelling to deduce the chemical composition of the pore water. There are numerous papers concerning the modelling of the pore water in compacted bentonite and argillaceous rocks (Wersin, 2003; Bradbury & Baeyens, 1998, 2002, 2003; Fernández et al., 2001; Wanner et al., 1992; Muurinen & Lehikoinen, 1999; Pusch et al., 1999; Bruno et al., 1999; Wieland et al., 1994; Curti, 1993). However, the modelling task is also difficult due to the physico-chemical complexity of the clay minerals. Most of the models does not take into account all the specific properties of these minerals.

The pore water chemistry in bentonites is the result of different interactions occurring in the clay/water system: interactions between water, solutes and clay. For this reason, in order to perform a more realistic modelling, it is necessary to know the mineralogical and chemical components of the clay system; its physico-chemical characteristics, the hydration mechanisms, as well as the types of waters, porosity, microstructure, and the ion diffusion pathways of compacted bentonites.

A bentonite-water system must be defined, among other parameters, by a water/rock ratio, since the concentrations in the aqueous phase strongly depend on this parameter. Thus, an important issue in calculating the pore water chemistry in high clay mineral content media is to distinguish between the types of water and their distribution in the clay/water system (absorbed water, interlayer water and pore water), and the respective volumes accessible to cations and anions. The water volume available for chemical reactions or the amount of free water is required to calculate the porewater composition of compacted saturated bentonite. Usually, this value is taken to be the chloride accessible porosity obtained from Cl^- *through-diffusion* tests (Pearson, 1998; Bradbury & Baeyens, 2002, 2003). However, the water/rock ratio can be also described in terms of the external porosity, when the types of water and their distribution in the clay/water system is known (Fernández & Rivas, this volume). Then, the concentration of chloride to be used in the modelling calculations can be fixed. When the bentonite-water system reaches equilibrium, the chemical composition of the pore water will be controlled by dissolution of trace minerals in the bentonite, by the influence of the high effective S:L ratio and by the buffering effects of exchangeable cations and $\equiv$SOH sites (Bradbury and Baeyens, 2003; Fernández et al., in press-a).

Furthermore, when the thermodynamic equilibrium is approached, the main physico-chemical parameters and the chemical potential of the chemical components controlling the main water-rock interaction processes are fixed. These processes affect the solubility, redox state, complexation and sorption reaction of the radionuclides.

In this paper, the pore water composition of the FEBEX bentonite is presented, as a function of the dry density, which corresponds to the conditions expected in a HLW repository.

2 MATERIALS

The Spanish concept for the geological disposal of radioactive wastes considers granite formations as a suitable host rock for underground repositories. The canisters containing the spent fuel are placed horizontally in drifts and surrounded by a clay barrier constructed with highly-compacted bentonite blocks (ENRESA, 1994). The clay material used is a bentonite from Cortijo de Archidona deposit (Almería, SE Spain) which has been selected by ENRESA (Empresa Nacional de Residuos Radiactivos) as a suitable material for the sealing and backfilling of HLRW repositories.

The conditioning of the bentonite in the quarry, and later in the factory, was strictly mechanical

(homogeneization, rock fragment removal, crushing, drying at 50–60°C and sieving) to obtain a granulated material with the specified characteristics: grain-size distribution ($\leq$5 mm) and 14 wt.% water content. Some 300 tons of this bentonite were acquired for material studies and manufacturing of bentonite blocks compacted at 1.65 g/cm^3 of dry density used as an engineered barrier within the FEBEX project during the *in situ* (Grimsel, Switzerland) and *mock-up* (Madrid, Spain) tests (Huertas et al., 2000).

The major mineral phase (92%) of the FEBEX bentonite is montmorillonite (dioctaedric aluminic smectite with a 2:1 structure, which consists of two silica tetrahedral sheets with a central alumina/magnesium octahedral sheet). This bentonite was formed by alteration of pyroclastic volcanic rocks, such as poligenitic tuffs and porphyritic agglomerates of rhyolitic-rhyodacitic character (Fernández-Soler, 1992).

The main physico-chemical characteristics are summarised as follows. Other characteristics can be found in Huertas et al. (2000), Villar (2002) and Fernández et al. (in press a,b). The specific weight is 2.70 ± 0.04 g/cm^3, the liquid limit is $102 \pm 4\%$, the plastic limit is $53 \pm 3\%$, the saturated permeability at 1.65 g/cm^3 is $3 \cdot 10^{-14}$ m/s, the equilibrium gravimetric water content of the clay under laboratory conditions (relative humidity $50 \pm 10\%$) is about $13.7 \pm 1.3\%$.

In spite of the high smectite content, the bentonite contains numerous accessory minerals. Some of them are neoformed minerals (calcite: 0.6 ± 0.1 wt.%, cristobalite: 2 ± 1 wt.%), and others are remains of the original volcanic rock minerals (quartz: 2 ± 1 wt.%, plagioclase: 3 ± 1 wt.%, K-felspars: traces, tridimite: traces) which appear nearly unaltered. Other accessory minerals are mica (biotite, sericite, muscovite), chlorite, non-differentiated Al, K, Fe, Mg and Mn silicates, Augite-diopside, hypersthene, hornblende, oxides (ilmenite, rutile, magnetite, Fe-oxides), phosphates (apatite, xenotime, monacite) and non-differentiated titanium minerals and rare earths, which have been determined by weight from dense concentrates and SEM identification, their total content being 0.8 wt.%. Some minerals, such as carbonates (calcite, dolomite), chlorides (0.13 ± 0.02 wt.%) and sulphates (gypsum (0.14 ± 0.01 wt.%), barite, celestite (0.019 ± 0.004 wt.%)), have been determined by a normative calculation and SEM identification. Attention is brought to the content of these last minerals present at trace levels in the bentonite because of their influence on the chemistry of the pore water (Fernández et al., 1999, 2001).

Based on XRD and the positions (CuK$_\alpha$) of the 001/002 and 002/003 reflections (Moore & Reynolds, 1989), the smectitic phases of the FEBEX bentonite are actually made up of a smectiteillite mixed layer with $\sim$11% of illite layers ($\Delta 2\theta = 5.502$). In general, the clay mineral in this deposit has an average interstratification value of 7–15% (Cuadros and Linares, 1996). The thickness of the FEBEX smectite quasicrystals is around 102 ± 5 Å, calculated by means of the Scherrer equation, and quasicrystals of saturated FEBEX smectite consist on 6 lamellae or layers stacked along the crystallographic c-axis. The Biscaye index of smectite crystallinity is 0.966.

Based on chemical analyses, the structural formula or unit-cell formula of the Ca conditioned FEBEX smectite is (Fernández et al., in press a,b):

$(Si_{7.78} Al_{0.22})^{IV} (Al_{2.77} Fe^{3+}_{0.33} Fe^{2+}_{0.02} Mg_{0.81} Ti_{0.02})^{VI}$
$O_{20} (OH)_4 (Ca_{0.50} Na_{0.08} K_{0.11})_{1.19}$

Tetrahedral charge: -0.22, Octahedral charge: -0.97, interlayer charge: +1.19

18% of the charge is due to tetrahedral substitution, and 82% to octahedral replacement. According to the chemical analysis, the theoretical exchange capacity is 1.05 eq/kg and, therefore, the total charge per unit-cell is 0.79. The 0.11 K ions per formula unit are not exchangeable and belong to the 11 wt.% illitic layers of the smectite-illite mixed layers of the FEBEX bentonite clay. The *surface charge density* (σ), i.e., the excess charge per unit surface area of this smectite is 0.136 C/m^2.

The total specific surface area was determined by two methods. The first one is based on the unit cell parameters ($a = 5.156$ Å and $b = 9.0$ Å), obtained from the XRD analyses, and the unit weight (752.46 g per 44 atoms of oxygen) (Horseman et al., 1996; Newman, 1987). The calculated value was 746 ± 4 m^2/g. The second method is based on Keeling's hygroscopic method from an adsorption isotherm of water in a constant relative humidity atmosphere (75%) of saturated NaCl. In this method, the weight variation of a dried clay material is analysed over a period of several weeks. The total specific surface area of the FEBEX bentonite obtained by this method was $S_{TOTAL} = 725 \pm 47$ m^2/g. The external specific surface, which was calculated by means of the nitrogen adsorption/desorption isotherms, is 61.53 m^2/g.

3 MODEL DEVELOPMENT

3.1 *Model principles*

3.1.1 *Thermodynamic principles*
Chemical thermodynamic equilibrium is a valid model for the generalization of many processes governing chemical reactivity. The free energy concept of chemical thermodynamics is a particularly powerful tool to predict the nature and extent of reactions as they approach equilibrium in the systems. The partial molar free energy or chemical potential, μ_i, describes the free

energy contribution of the ith component driving the system toward equilibrium (Stumm & Morgan, 1981):

$$\mu_i(T,P) = \bar{G}_i = \left(\frac{\partial G_i}{\partial n_i}\right)_{T,P,n_j} \tag{1}$$

where, T is the absolute temperature, P is the pressure and n, is the composition expressed as mole number.

The change in Gibbs free energy for a system of i components and α phases is obtained by summation of all components and all phases:

$$dG = -\sum_\alpha S^\alpha dT + \sum_\alpha V^\alpha dP + \sum_\alpha \sum_i \mu_i dn_i \tag{2}$$

where, S is the entropy and V the volume.

In a closed system of i components and α phases at constant T and P, the thermodynamic equilibrium state is defined by the following relationship:

$$dG = \sum_\alpha \sum_i \mu_i dn_i = 0 \tag{3}$$

Thus, for an i component in equilibrium with a vapour phase, the equilibrium criterion requires that the two phases be in equilibrium:

$$\mu_i(l) = \mu_i(g) \tag{4}$$

$$\mu_i^o(g) + RT \ln P_i = \mu_i^o(l) + RT \ln x_i \tag{5}$$

The chemical composition of the pore water (mobile water), when expressed in ion activity units, is a measure of the electrochemical potential of ions in the clay/water system. If the pore water approaches a steady state with all the phases (solids, gases, ...) of the system, the chemical composition of the pore water can be used to predict the solid phase components, or other phases, controlling the chemical equilibrium in the bentonite/water system.

The potential energy of water is commonly described in terms of water potential, which is the difference between chemical and mechanical potentials between bentonite-water system and pure water at the same temperature. The water potential is the chemical potential of water expressed on a volume basis and has the contribution of the matric potential, the solute or osmotic potential and the pressure potential.

3.1.2 *Basis for the calculations*
In a previous work (Fernández et al., in press-a), Bradbury & Baeyens's procedure (Bradbury & Baeyens, 2002) was used in order to model the pore water chemical composition under two conditions: i) initial bentonite powder with a water content of 14% wt and, ii) saturated compacted bentonite with a dry density of 1650 kg/m^3.

This procedure is based on physico-chemical characterisation and geochemical modelling. It

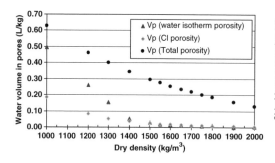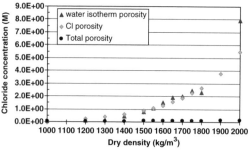

Figure 1.   Water volume in interparticle pores and the associated chloride concentration as a function of the dry density.

takes into consideration factors, such as the industrial preparation of the "as received" bentonite powder, montmorillonite swelling, semi-permeable membrane effects, very low free water volumes and the highly effective buffering characteristics of the exchangeable cations and the amphoteric edge sites in the compacted material.

As a summary, the main ideas are described here: besides knowing the physico-chemical characterisation of the bentonite material, two quantities need to be fixed in order to calculate a unique aqueous chemistry and to fully define the bentonite-water system: the pH (or pCO_2) and the chloride concentration. The initial pH is fixed in the compacted material by the high buffering capacity given by the amphoteric edge sites, $\equiv$SOH sites, of the montmorillonite. The pH of the porewater directly depends on the speciation of these sites, (i.e., the proportions of sites present as $\equiv$SOH, $\equiv$SOH$_2^+$, $\equiv$SO$^-$), and this is fixed in the powdered source material through equilibration with air (compaction does not alter the state of the $\equiv$SOH sites). The chloride concentration is specified by the free water volume. In these calculations, the free water volume, i.e., the porewater volume in compacted bentonite, is supposed to be equivalent to the accessible porosity to chloride in *through-diffusion* experiments.

In this work, the same procedure has been followed. However, the external porosity calculated in section 3.2 has been considered as the accessible porosity or free water volume. In order to calculate the initial pH of the compacted material, the free water volume in the bentonite powder was supposed to be the external water at 50% of relative humidity ($\sim$5 wt.%) (Fernández & Rivas, this volume).

3.2   Water in the compacted bentonite-water system

A bentonite-water system must be defined, among other parameters, by a water/rock ratio, since the concentrations in the aqueous phase strongly depend on this parameter. This water/rock ratio can be conveniently described by a quantity analogous to the porosity. In compacted bentonite, the total porosity (physical porosity) can be divided in: i) internal porosity (interlayer water) and, ii) external porosity (external water) which corresponds to free water plus DDL water. The external porosity is called effective porosity (Pusch et al., 1990) or geochemical porosity (Pearson, 1998). The amount of water involved in the geochemical processes in the engineering barrier corresponds to the free water or inter-particle water, i.e., the external water, taking into account that the amount of this water depends on the modifications of the dry density and the saline concentration in the system.

A distribution of the external and internal porosity as a function of dry density was obtained in a previous paper (Fernández & Rivas, this volume). These values are calculated based on the types of water and their distribution in the FEBEX smectite and on the statistical distribution of the water in the compacted material (external/internal water) as a function of the total porosity. The external porosities calculated have been found to be similar to the accessible porosities determined in through-diffusion tests (García-Gutiérrez et al., 2002). Most of the water is situated in interlayer space, but there is always a small fraction of external water in the compacted saturated bentonitesystem.

The volume of water in the inter-particle pores (external water) and the chloride concentration as a function of the dry density are shown in Fig. 1 (called water isotherm porosity). The values are compared with those obtained from total porosity and chloride accessible porosity. As it can be observed, the chloride accessible porosity and the calculated external porosity are quite similar at dry densities between 1400 and 1800 kg/m^3. At densities lower than 1400 kg/m^3, the bentonite-water system is better described by the calculated external porosity, because the through-diffusion experiments are dependent of the salinity of the injected water (see Muurinen et al., 1987). At low densities (in this case < 1400 kg/m^3), where the efficiency of the bentonite functioning as a semi-permeable membrane is much lower than 100%, the external fluid composition may have some influence above a certain concentration. However, at high densities and membrane efficiencies approaching 100%, the re-saturating fluid composition plays little or no role (Bradbury & Baeyens, 2002; Dixon, 2000). At densities lower than $1400 \, kg/m^3$, the ratio external/internal water depends on the saline concentration. The amount of external water is a function of the osmotic equilibrium between interlayer composition and the external salinity. As much saline is the interparticle water (pore water), the interlayer tends to expel water up to reach the osmotic equilibrium, increasing the amount of external water. If the salinity decreases, the smectite tends to acquire the maximum amount of water that it can, decreasing the amount of external water.

The ionic strengths are greater than 3 M at dry densities higher than $1800 \, kg/m^3$. This poses a conceptual problem in order to modelling the pore water system and to understand the chemical equilibrium of the bentonite-water system.

3.3 Input data

The main input data used for modelling purposes are described as follow:

3.3.1 Chloride inventory and cation exchange population

The chloride and sulphate inventories, as well as the cation exchange capacities of the FEBEX bentonite are shown in Table 1 and Table 2.

3.3.2 Selectivity coefficients and protolysis constants

The selective coefficients, protolysis sites and protolysis constants are shown in Table 3 and Table 4.

3.4 Model implementation

The calculations were carried out using the geochemical code PHREEQC2 (Parkhurst & Appelo, 1999). The thermodynamic database of the code was modified in order to take into account the high salinities that exist in the FEBEX bentonite-water system. The activity coefficients were calculated according to Truesdell & Jones equation (Truesdell & Jones, 1974) (valid for solutions with I < 2.5 M), using the a and b parameters determined by Parkhurst (1990) and Kielland (1937) (Langmuir, 1997). The Setchenow equation was used for non-charged molecular species, introducing K_i parameters found in Langmuir (1997).

The main geochemical processes of the FEBEX bentonite-water system to be considered in the modeling come from experimental results obtained

Table 1. Cl^- and and SO_4^{2-} inventories of the FEBEX bentonite.

Cl^- (mmol/kg)	SO_4^{2-} (mmol/kg)
21.85 ± 3.96	10.26 ± 0.68

Table 2. Cation exchange capacities on FEBEX bentonite, in meq/kg (PSI lab).

NaX	269.5
CaX$_2$	331.0
MgX$_2$	331.5
KX	22.9
SrX$_2$	4.3
CEC	959.2

Table 3. Log K_C for exchange reactions for FEBEX bentonite (PSI data).

Na-mont. + K ⇔ K-mont + Na	1.0253
2Na-mont. + Ca ⇔ Ca-mont + 2Na	1.1072
2Na-mont. + Mg ⇔ Mg-mont + 2Na	1.0294
2Na-mont. + Sr ⇔ Sr-mont + 2Na	1.1072

Table 4. Site types, site capacities and protolysis constants of the FEBEX bentonite taken from Bradbury & Baeyen's data (Bradbury & Baeyens, 1997).

Site types	Site capacities (*)
≡SSOH	$2.34 \cdot 10^{-3} \, mol/kg$
≡SW1OH	$4.68 \cdot 10^{-2} \, mol/kg$
≡SW2OH	$4.68 \cdot 10^{-2} \, mol/kg$
Surface complexation reaction	**log K(int)**
$\equiv S^S OH + H^+ \Leftrightarrow \equiv S^S OH_2^+$	4.5
$\equiv S^S OH \Leftrightarrow \equiv S^S O^- + H^+$	−7.9
$\equiv S^W OH + H^+ \Leftrightarrow \equiv S^W OH_2^+$	4.5
$\equiv S^W OH \Leftrightarrow \equiv S^W O^- + H^+$	−7.9
$\equiv S^{W2} OH + H^+ \Leftrightarrow \equiv S^{W2} OH_2^+$	6.0
$\equiv S^{W2} OH \Leftrightarrow \equiv S^{W2} O^- + H^+$	−10.5

(*) Site capacities values for the FEBEX bentonite have been scaled over their CEC value from the Wyoming CEC value.

in experiments performed at high (squeezing tests) and low (aqueous extracts solutions) solid to liquid ratios (Fernández et al., 2001):

– Equilibrium with respect to calcite
– Equilibrium with respect to gypsum and celestite
– Equilibrium with respect to chalcedony
– Dilution of chlorides
– Ionic exchange reactions

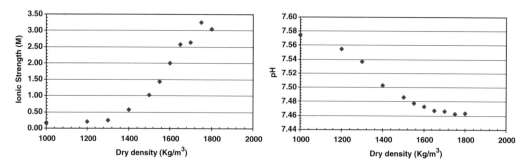

Figure 2. Ionic strengths and pH of the pore water as a function of the dry density.

- Surface complexation reactions
- Thermodynamic stability of the clay fraction (smectite) at the expected pH conditions (7 < pH < 11).

4 RESULTS AND DISCUSSION

4.1 Pore water chemical composition as a function of the dry density

The results obtained in the modelling are shown in Fig. 2, Fig. 3, Fig. 4 and Fig. 5. The pore waters are Na-Cl type with ionic strengths ranging from 0.17 to 3.3 M. The pH values varies from 7.57 to 7.46 as a function of the dry density. The variation of the sulphate concentration as a function of the free water used in the modelling (external porosity or total porosity) is worthy to note. As it was expected, the exchangeable sodium concentration decreases and exchangeable calcium concentration increases when the pore water volume increases in the bentonite-water system.

4.2 Properties of the compacted bentonite-water system: surface potential, osmotic potential and swelling pressure

4.2.1 Surface potential and double layer thickness

The surface potential of the FEBEX smectite and the double layer thickness in the bentonite-water system have been approximately calculated from the pore water ionic strength and the surface charge density of the FEBEX smectite by applying the Diffuse Double Layer Theory. The results are shown in Fig. 6, Fig. 7 and Fig. 8.

As it was expected, the DDL thickness decreases when the ionic strength increases. In compacted bentonite systems at dry densities between 1400 and 1700 kg/m³, the DDL thickness is lower than the Stern layer. For this reason, either the application of the DDL models is not valid in these systems, or

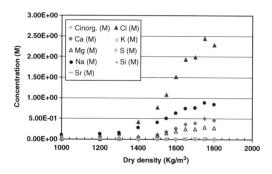

Figure 3. Chemical composition of the pore water as a function of the dry density.

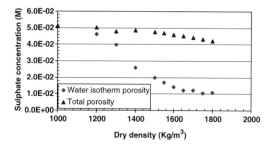

Figure 4. Sulphate concentration in the pore water as a function of the dry density.

the DDL water type and free water cannot be differentiated.

4.2.2 Swelling pressure

According to the principles of thermodynamic equilibrium, the swelling pressure, P_s, and relative partial free energy of water in the clay/water system are related by the relationship (Low & Anderson, 1958; Kahr et al., 1990, Karnland, 1997):

$$\bar{g}_w - \bar{g}_w^o = -\bar{v}_w \times P_s \qquad \text{(J/mol)} \qquad (6)$$

510

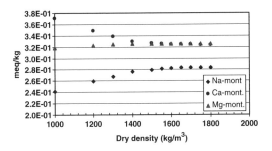

Figure 5. Sodium, calcium and magnesium in exchange positions as a function of the dry density.

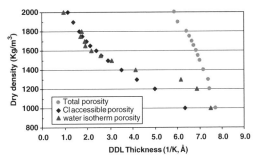

Figure 8. Dry density versus DDL thickness in the FEBEX compacted bentonite.

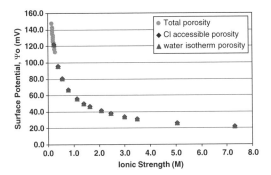

Figure 6. Smectite particles surface potential versus ionic strength of the pore water.

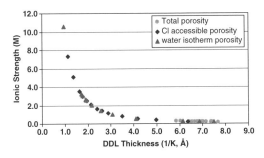

Figure 7. Ionic strength versus DDL thickness.

The term $\bar{v}_w = (\delta V / \delta m_w)_{T,P}$ represents the partial specific volume of water. This quantity can differ substantially from the value for pure water at near room temperature ($v_w = 10^{-3}$ m³/kg), in particular in bentonite systems with low contents of water (Oliphant & Low, 1982).

The difference in free energy is also related to P, the partial pressure of water vapour in the system and

the vapour pressure, P_o, of pure water at temperature T of the experiment:

$$\overline{G}_w - \overline{G}_w^o = M_w\left(\overline{g}_w - \overline{g}_w^o\right) = RT \ln \frac{P}{P_o} \quad \text{(J/mol)} \quad (7)$$

where, R is the gas constant (8.31 J/mol·K) and M_w is the molecular weight of water ($1.8 \cdot 10^{-2}$ kg/mol). Thus, the changes in free energy of water in bentonites and the swelling pressure can be determined from the water vapour isotherm. The swelling pressure as a function of the water content is:

$$P_s = -\frac{RT}{M_w v_w} \ln \frac{P}{P_o} \quad (8)$$

For low water contents, Oliphant & Low (1982) found an empirical correction for the value of the partial specific volume of water in bentonite-water systems:

$$\bar{v}_w = 1.002 \exp\left(0.036 \frac{m_w}{m_s}\right) \quad (9)$$

Applying the above equations and knowing the relationship m_w/m_s as a function of the dry density (see Fernández & Rivas, this volume), ν_w (cm³/g) for the FEBEX bentonite-water system can be easily calculated. Then, if the water activity at equilibrium in the compacted bentonite-water system is known, the swelling pressure can be determined.

On the other hand, the water potential or suction in the bentonite-water system can be calculated by the following relationship:

$$s = -1 \cdot 10^{-6} \cdot \frac{RT}{v_w} \ln \frac{P}{P_o} \quad \text{(MPa)} \quad (10)$$

where, R = 8.31441 J/mol·K; T is the absolute temperature and $v_w = 1.80 \cdot 10^{-5}$ m³/mol).

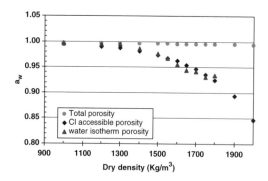

Figure 9. Water activity as a function of the dry density.

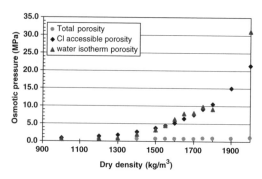

Figure 10. Osmotic pressure as a function of the dry density.

According to Thermodynamics, the chemical composition of the pore water (mobile water), when expressed in ion activity units, is a measure of the electrochemical potential of ions in the clay/water system. At equilibrium, the chemical potential of all the phases is the same. As the pore water composition was determined taking into account the whole bentonite-water system (smectite, solutes, solid phases, internal water, external water, etc.), the water activity determined in the pore water modelling for each dry density could represent the water activity of the bentonite-water system. The pore water activity as a function of the dry density is shown in Fig. 9. Also, the osmotic potential can be calculated from the chemical composition of the pore water Fig.10. The comparison between osmotic potential and water potential is shown in Fig. 11. The swelling pressure of the FEBEX-bentonite system has been calculated taking into account the water activity of the pore water. The results are shown in Fig. 12 and compared with experimental data obtained in conventional oedometer tests (Villar, 2002). As can be observed, the osmotic pressure in the bentonite-water system explains most of the swelling pressure of the compacted and saturated FEBEX bentonite with dry densities below 1650 kg/m³. The osmotic pressure is related to the concentration of the solvated exchangeable cations in osmotic equilibrium with the salinities of the external pore water. At densities higher than 1700 kg/m³, another mechanism seems to occcur, which is probably related to the hydration energy of the exchangeable cations. At these high densities, some layers in the quasicrystals are in a state in which only monolayers of water are present. Thus, the exchangeable cations cannot dehydrate to maintain the osmotic equilibrium. Their behavior could be to uptake water to acquire the second solvation shell, because the hydration energy of the bivalent cations is very high ($\sim$500 $-$ 400 kcal/g). For this reason, the energy to be applied to the system so that the bentonite is not capable to adsorb water is very high.

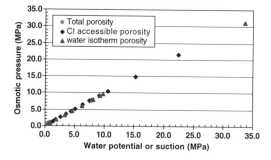

Figure 11. Osmotic pressure versus water potential or suction.

5 CONCLUSIONS

The pore water chemistry of compacted FEBEX bentonite at different dry densities was calculated. Free water was assimilated to the water obtained as external water in a system where the total porosity is described in terms of internal or interlayer porosity and external porosity. Most of the water in compacted bentonite is internal water (interlayer porosity). The external water constitutes the pathway through which the ions move.

The thermodynamic formulation is a powerful tool to calculate the properties of the bentonite-water system and all the possible interactions among water, clay and solutes. The pore water calculated describes the main physico-chemical parameters of the bentonite-water system, because this pore water is at thermodynamic equilibrium with the whole system, and the chemical potential of all the phases must be equal.

Macroscopic properties, such as accessible porosities to chloride and swelling pressure, can be determined by improving the knowledge of the chemical composition of the pore water, which

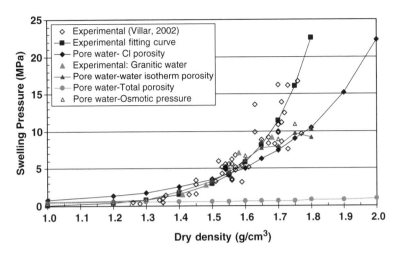

Figure 12. Swelling pressure as a function of the dry density.

reflects the thermodynamic equilibrium of the microscopic system. The swelling pressures can be explained by osmotic pressures in the bentonite-water system. Because the DDL thickness at dry densities higher than $1400\,kg/m^3$ is lower than $3\,\text{Å}$, the application of the DDL models could be not valid in these systems.

ACKNOWLEDGEMENTS

This work has been supported by CIEMAT-ENRESA and the European Commission 5th framework programme within the FEBEX project. The contribution of Dr. B. Bart and Dr. M. Bradbury from PSI are gratefully acknowledged.

REFERENCES

Bradbury, M.H., Baeyens, B., 2003. Porewater chemistry in compacted resaturated MX-80 bentonite. J. Contaminant Hydrology, Vol. 61, pp. 329–338.

Bradbury, M.H., Baeyens, B., 2002. Porewater chemistry in compacted re-saturated MX-80 bentonite: Physico-chemical characterisation and geochemical modelling. PSI Bericht Nr. 02-10 Paul Scherrer Institut, Villigen, Switzerland and Nagra Technical Report NTB 01-08, Nagra, Wettingen, Switzerland.

Bradbury, M.H., Baeyens, B., 1998. A physico-chemical characterisation and geochemical modelling approach for determining porewater chemistries in argillaceous rocks. Geochim. Cosmochim. Acta 62, pp. 783–795.

Bradbury, M.H., Baeyens, B., 1997. A mechanistic description of Ni and Zn sorption on Namontmorillonite. Part II: Modelling. J. Contaminant Hydrology 27, pp. 223–248.

Bruno, J., Arcos, D., Duro, L., 1999. Processes and features affecting the near field hydrochemistry. Groundwater-bentonite interaction. SKB Tech. Rep. 99-29.

Cuadros, J., Linares, J., 1996. Experimental kinetic study of the smectite to illite transformation. Geochimica et Cosmochimica Acta, 60, No. 3, pp. 439–453.

Curti E., 1993. Modelling bentonite porewaters for the Swiss High-level Radioactive Waste Repository. Nagra Technical Report NTB 93-45. 85 pp.

Dixon, D.A., 2000. Porewater salinity and the development of swelling pressure in bentonite based buffer and backfill materials. POSIVA Report 2000–04, Posiva Oy, Helsink, Finland.

ENRESA, 1994. Almacenamiento geológico profundo de residuos radiactivos de alta actividad (AGP). Conceptos preliminares de referencia. ENRESA. Publicación Técnica 07-94.

Fernández, A.M^a., Rivas, P., Cuevas, J., 1997. Estudio de las aguas intersticiales de bentonitas españolas de referencia. CIEMAT Interim Report CIEMAT/IMA/54A40/1/97, 40, pp.

Fernández, A.M^a, Cuevas, J., Rivas, P., 2001. Pore water chemistry of the FEBEX bentonite. Mat. Res. Soc. Symp. Pro. 663, pp. 573–588.

Fernández, A.M^a., Baeyens, B., Bradbury, M., Rivas, P., in press-a. Analysis of the pore water chemical composition of a Spanish compacted bentonite used in an engineered barrier. Submitted to Applied Clay Science.

Fernández A.M^a, in press-b. Caracterización y modelización del agua intersticial en materiales arcillosos: Estudio de la bentonita de Cortijo de Archidona. Ph. D. Thesis.

Fernández, A.M^a, Rivas, P., this volume. Analysis and distribution of waters in the compacted FEBEX bentonite: Pore water chemistry and adsorbed water properties.

Fernández-Soler, J.M., 1992. El volcanismo calco-alcalino de Cabo de Gata (Almería). Ph. D. Thesis. Universidad de Granada. 243, pp.

García-Gutiérrez, M., Missana, T., Cormenzana, J.L., 2002. Diffusion coefficients and accessible porosity for HTO and ^{36}Cl in compacted bentonite. International Meeting at REIMS Clay in natural and engineered barriers for radioactive waste confinement, pp. 331–332.

Horseman, S.T., Higgo, J.J., Alexander, J., Harrington, J.F., 1996. Water, gas and solute movement through argillaceous media. Report CC-96/1. Nuclear Energy Agency, OECD, Paris, France.

Huertas, F., Fuentes-Cantillana, J.L., Jullien, F., Rivas, P., Linares, J., Fariña, P., Ghoreychi, M., Jockwer, N., Kickmaier, W., Martínez, M.A., Samper, J., Alonso, E., Elorza, F.J., 2000. Full-scale engineered barriers experiment for a deep geological repository for high-level radioactive waste in crystalline host rock (FEBEX project). European Commission Report EUR 19147 EN.

Kahr, G., Kraehenbuehl, F., Stoeckli, H.F., Müllervon Moos, 1990. Study of the water-bentonite system by vapor adsorption, immersion calorimetry and X-ray techniques. II. Heats of immersion, swelling pressures and thermodynamic properties. Clay Minerals 25, pp. 499–506.

Karnland, O., 1997. Bentonite swelling pressure in strong NaCl solutions. Correlation between model calculations and experimentally determined data. SKB technical report 97-31, 30, pp.

Kielland, A.G., 1937. Individual activity coefficients of ions in aqueous solutions. J. Am. Chem. Soc. 59, pp. 1675–1735.

Langmuir, D., 1997. Aqueous Environmental Geochemistry. 600 pp. Prentice Hall. New Jersey.

Low, P.F., Anderson, D.M., 1958. Osmotic pressure equation for determining thermodynamic properties of soil water. Soil Sci. 86, pp. 251.

Muurinen, A., Penttila-Hiltunen, P., Rantanen, J., Uusheimo, K., 1987. Diffusion of uranium and chloride in compacted Na-bentonite. Report YJT-87-14. Nuclear waste Commission of Finnish Power Companies, Helsinki, Finland.

Muurinen, A., Lehikoinen, J., 1999. Pore water chemistry in compacted bentonite. Engineering Geology, 54, pp. 207–214.

Newman, A.C.D., 1987. Chemistry of Clays and Clay Minerals. Mineral Society. Monograph N° 6. Longman Scientific & Technical, 480, pp.

Oliphant, J.L., Low, P.F., 1982. The relative partial specific enthalpy of water in montmorillonite-water systems. J. Colloid and Interface Sci. 89, pp. 366.

Parkhurst D.L., Appelo C.A.J., 1999. PHREEQC2 – A computer program for speciation reaction-path, 1D-Transport and inverse geochemical calculations. U.S. Geological Survey.

Parkhurst, D.L., 1990. Ion-Association models and mean activity coefficients of various salts. In: Chemical modeling of aqueous systems II, ed. D.C. Melchior and R.L. Bassett, Am. Chem. Soc. Symp. Ser. 416, pp. 30–43. Washington DC.

Pearson, F.J., 1998. Geochemical and other porosity types in clay-rich rocks. Water-Rock interaction, Arehart & Hulston (eds.), pp. 259–262. Balkema. Rotterdam.

Pusch, R., Muurinen, A., Lehikoinen, J., Bors, J., Eriksen, T. 1999. Microstructural and chemical parameters of bentonite as determinants of waste isolation efficiency. European Commission. Nuclear Science and Technology. Project Report EUR 18950 EN.

Stumm, W., Morgan, J.J., 1981. Aquatic chemistry. 2nd. Ed. John Wiley & Sons. New York.

Truesdell, A.H., Jones, B.F., 1974. WATEQ, A computer program for calculating chemical equilibria of natural waters: Journal of Research, U.S. Geological Survey, v. 2, pp. 233–274.

Villar, M.V., 2002. Thermo-hydro-mechanical characterisation of a bentonite from Cabo de Gata. ENRESA Publicación Técnica 04/2002.

Wanner, H., Wersin, P., Sierro, N., 1992. Thermodynamic modelling of bentonite-groundwater interaction and implications for near field chemistry in a repository for spent fuel. SKB 92-37, 28 pp.

Wersin, P., 2003. Geochemical modelling of bentonite porewater in high-level waste repositories. Journal of Contaminant Hydrology 61, pp. 405–422.

Wieland, E., Wanner, H., Albinsson, Y., Wersin, P., Karland, O., 1994. A surface chemical model of the bentonite-water interface and its implications for modelling the near field chemistry in a repository for spent fuel. SKB Technical Report 94-26. 64, pp.

Advances in Understanding Engineered Clay Barriers – Alonso & Ledesma (eds)
© *2005 Taylor & Francis Group, London, ISBN 04 1536 544 9*

Diffusion coefficients and accessible porosity for HTO and chloride in compacted FEBEX bentonite

M. García-Gutiérrez, T. Missana & M. Mingarro
CIEMAT, Madrid, Spain

J.L. Cormenzana
Empresarios Agrupados, Madrid, Spain

ABSTRACT: The transport properties of a Spanish bentonite (FEBEX bentonite) were studied. The behaviour of the conservative anionic species Cl^- is compared with HTO (tritiated water, neutral and conservative tracer). Apparent and effective diffusion coefficients, as well as the accessible porosity, were determined in a wide range of clay dry densities. Effective diffusion coefficients for HTO and Cl^- were measured in through-diffusion experiments. Apparent diffusion coefficients were measured for HTO with in-diffusion experiments.

A very important parameter for describing the transport of radionuclides is the accessible porosity, which is often difficult to determine for the anions, especially at high clay densities. Therefore, it is very important to use different methods to compare the results obtained in order to understand the limitations of each technique and finally provide reliable results. Furthermore, the use of different techniques and the independent determination of each transport parameter can be used to validate the results.

Apparent and effective diffusion coefficients for both HTO and Cl^- showed an exponential decrease when the dry density increased, with the decrease being more pronounced in the case of Cl^-. The accessible porosity for HTO is equal to the total porosity. The accessible porosity for chloride is only a small fraction of total porosity (2–3% at a dry density of $1.65 \, g/cm^3$, when the total porosity is about 40%).

1 INTRODUCTION

Compacted bentonite is used as an engineered barrier in high-level radioactive waste (HLRW) repositories because it is a material of very low permeability and high sorption capability for many solutes. The transport of radionuclides through the compacted bentonite is a diffusion-controlled process retarded by sorption. For the performance assessment calculations of a repository, apparent and effective diffusion coefficients of relevant radionuclides in the bentonite are necessary. In addition, performance assessment calculations of spent fuel repositories have shown that doses are mainly controlled by non-sorbing anionic species ($^{129}I^-$ and $^{36}Cl^-$). As a consequence, it is necessary to have an in-depth knowledge of the behaviour of anionic species in the compacted bentonite.

Transport of anionic species in compacted clays is affected by electrostatic forces between the negatively charged clay surfaces and the anions. This results in a smaller volume of pore water available for anion transport, a phenomenon called anionic exclusion. In general, the diffusion behaviour of radionuclides in porous media is a complex process affected by many parameters, such as the properties of the diffusing species, the properties of the solid itself (pore structure, degree of compaction, adsorption properties, dry density, (Kozaki *et al.*, 2001)), the geochemistry of the system (pore water chemistry), and the temperature. The effective diffusion coefficient (D_e) of nuclides in bentonite depends on the diffusing species, particularly on its charge. The apparent diffusion coefficient (D_a) also depends on the sorption characteristics of the bentonite clay. By means of through-diffusion and in-diffusion experiments both diffusion coefficients can be obtained.

Another important parameter to be determined is the diffusion-accessible porosity (ϕ), in particular for anionic species that suffer anion exclusion. The term "accessible porosity" is used to denote the proportion of the total volume of a porous material that is available for diffusion. For anions the accessible porosity is generally lower than the total porosity (ε), while for cations and neutral species all porosity is usually assumed to be accessible (Muurinen & Lehikoinen, 1995). For conservative tracers, the ratio between D_e and D_a corresponds to the accessible porosity species (Yu & Neretnieks, 1997).

In principle, different methods can be used for the determination of the accessible porosity but, the determination of the accessible porosity for anionic species at high clay density is particularly complicated, from an experimental point of view. Through-diffusion experiments at steady-state conditions, using the time-lag method (Bourke et al., 1993), have been shown to be very useful for the determination of D_e but they are less precise in the determination of the accessible porosity. Saturation techniques could provide very good results in the case of neutral species like HTO, but they are extremely time-consuming for anionic species like chloride. Therefore, it is very important to use different methods to compare the results obtained in order to understand the limitations of each technique and finally provide reliable results.

In this work, accessible porosities for HTO (for all clay densities) and Cl⁻ (for clay densities up to 1.2 g/cm³) were directly obtained by means of saturation experiments. In these experiments the water-saturated bentonite samples were introduced into a reservoir with the tracer until equilibrium concentration was reached. The measurement of the tracer activity in the clay pore water gives a direct measurement of the accessible porosity. Alternative estimations of accessible porosity were done in through-diffusion experiments with constant concentration and concentration profile experiments, which provided redundant, although less precise, methods for its estimation. Similar values of accessible porosities were obtained with the three methods.

2 MATERIALS AND METHODS

2.1 Materials

FEBEX bentonite is the clay selected as Spanish reference buffer material and was used in all the experiments. This clay has a smectite content greater than 90% ($93 \pm 2\%$), with quartz ($2 \pm 1\%$), plagioclase ($3 \pm 1\%$), cristobalite ($2 \pm 1\%$), potassic feldspar, calcite, and trydimite as accessory minerals. The specific weight of the FEBEX bentonite is 2.7 g/cm³. An exhaustive description of this clay can be found elsewhere (Huertas et al., 2000). The clay was compacted at different dry densities (from 1.0 to 1.7 g/cm³) with special emphasis on the density of 1.65 g/cm³, because this is the density considered in the Spanish reference concept for radioactive waste repositories in granite and clay.

Through-diffusion experiments, with variable (TDV) and constant (TDC) concentration in the inlet reservoir, were performed with 50-mm diameter samples and different thicknesses (5.3, 8.3, 10 and 20 mm). In-diffusion (ID) experiments with constant concentration were carried out with 38-mm diameter samples and 50-mm thickness. Saturation experiments were performed with 50-mm diameter samples having a thickness of either 20 mm (HTO) or 5.3 or 8.3 mm (Cl⁻).

In all the experiments, the bentonite samples were saturated with deionised water, requiring approximately 2.5 months. Complete water saturation of the samples was verified by a final, stable weight at the end of the experiment.

Due to the fact that deionised water is used for saturating the sample, the final ionic strength of the bentonite pore water is expected to be lower than the initial. The measured conductivity in the reservoir ranged between 1 and 3 mS/cm. The details on the preparation of the samples can be found elsewhere (García-Gutiérrez et al., 2001).

HTO and ³⁶Cl⁻ were used as tracers, and both were measured by liquid scintillation counting using a TR-2700 Packard apparatus. When it was necessary to extract the tracer from the clay, the plug was sliced and the solid of each slice was re-suspended in water and the tracer activity measured after centrifugation of the supernatant.

2.2 Theoretical description

In a porous medium, such as a compacted clay, the diffusion process is different from the diffusion in free water, since it is affected by the length of the diffusion path, tortuosity (τ), and the form of the pores, constrictivity (δ). Then, the pore diffusion coefficient, D_p, is related to the diffusion coefficient in free water, D_w, by:

$$D_p = \frac{\delta}{\tau^2} D_w \qquad (1)$$

In a diffusion experiment, only the pores that are connected and contribute to the transport of the dissolved species, from one side to the other of the clay sample, are important. Therefore, for each ion an "accessible porosity" has to be considered. The effective diffusion coefficient, D_e, defined by:

$$D_e = \phi \cdot D_p \qquad (2)$$

(where ϕ is the diffusion-accessible porosity) can be experimentally obtained when the steady state is reached.

In the case of sorbing elements, reaching the steady state can be very time consuming, so an apparent diffusion coefficient, D_a, can be calculated from the diffusion profile into the sample. Thus, D_a takes into account implicitly the retardation of the solute due to the interactions with the porous material, and is defined by:

$$D_a = \frac{D_p}{R_f} \tag{3}$$

where R_f is the retardation factor. If lineal sorption is considered, the sorption isotherm can be described by the simple relation $S = K_d \cdot C$, where S is the tracer concentration on the solid phase; C, the tracer concentration in the liquid phase; and K_d, the distribution coefficient. In this case, R_f is:

$$R_f = 1 + \frac{\rho_d}{\varepsilon} K_d \tag{4}$$

where ρ_d is the dry density of the material. According with these definitions, the following equation can be finally written:

$$D_a = \frac{D_p}{R_f} = \frac{D_p}{1 + \frac{\rho_d \cdot K_d}{\phi}} = \frac{\phi \cdot D_p}{\phi + \rho_d \cdot K_d} = \frac{D_e}{\alpha} \tag{5}$$

where $\alpha = \phi + \rho_d \cdot K_d$ is a dimensionless parameter called the capacity factor. For non-sorbing solutes, such as HTO and Cl$^-$, $K_d = 0$, so that the capacity factor is equal to the accessible porosity, ϕ. Equation (5) shows the relation between the different diffusion coefficients (D_a and D_e) to be determined.

The total porosity of the sample, ε, is the ratio of the pore volume to the total volume of a representative sample of the medium. For a medium with a dry density, ρ_d, and solid density, ρ_s, the total porosity can be obtained by:

$$\varepsilon = 1 - \frac{\rho_d}{\rho_s} \tag{6}$$

The accessible porosity, ϕ, represents the fraction of the total porosity used for solute transport.

The one-dimensional diffusion equation for a porous medium, according to Fick's second law, is expressed by:

$$\frac{\partial C}{\partial t} = D_a \frac{\partial^2 C}{\partial x^2} \tag{7}$$

where C is the concentration of the tracer in the porewater, t is the diffusing time, D_a is the apparent diffusion coefficient, and x is the distance from the source.

As mentioned before, equation (5) shows the relationship between D_e, D_a, and ϕ. If two parameters are measured, the third one can be calculated, as has been done in this article for Cl$^-$ (D_e and ϕ were measured and D_a was calculated). When all three parameters can be measured and they satisfy equation (5), the consistency of the values obtained is proved, as was done in this article for HTO.

2.3 Experimental methods

2.3.1 Through-diffusion experiments, TD

In TD experiments, a stainless-steel diffusion cell is connected to two reservoirs (inlet and outlet) where the solution is continuously stirred. The cell contains a ring with the compacted bentonite placed between two sinters. After the saturation of the clay sample with water, the water in the inlet reservoir is spiked with the tracer. Two different configurations are used: the first one is the variable concentration (TDV) case, using inlet and outlet reservoirs of 100 mL. In this configuration, the concentrations in both reservoirs change with time and they are periodically measured. The second one is the constant concentration gradient (TDC) case, in which a large (1 L) inlet reservoir and a very small (20 mL) outlet reservoir are used. In the inlet reservoir, the large, the concentration is practically constant. The outlet reservoir is periodically changed in order to keep the concentration close to zero. In order to measure diffusion parameters, the TDC method requires that steady-state conditions are reached.

In the TDV configuration, the effective diffusion coefficient, D_e, is estimated from the theoretical time evolution of the concentrations in the reservoirs for different values of D_e and α using a computer code. The code provides the concentration in inlet reservoir $G_1(D_e, \alpha, t)$ and outlet reservoir $G_2(D_e, \alpha, t)$ for each sampling instant t. The difference between experimental and theoretical concentrations is quantified using the function Φ, defined as:

$$\Phi(D_e, \alpha) = \sum_{i=1}^{n} \left(C_{1i} - G_1(D_e, \alpha, t_i) \right)^2$$
$$+ \sum_{i=1}^{n} \left(C_{2i} - G_2(D_e, \alpha, t_i) \right)^2 \tag{8}$$

where C_{1i} and C_{2i} are, respectively, the experimental values of the concentrations in inlet and outlet reservoirs at instant t_i. The couple of values (D_e, α) that minimise function Φ are taken as D_e and α. For non-sorbing species, such as HTO and Cl$^-$, the values of α are quite small, and for small values of α the function $\Phi(D_e, \alpha)$ is insensitive to changes in α, although very sensitive to changes in D_e. As a consequence, the values of α estimated with this method are not considered and only the D_e value will be taken from the TDV experiments.

TDV experiments were carried out for HTO in 1- and 2-cm thick clay plugs compacted at the dry densities of 1.1, 1.3, 1.5, 1.65, and 1.7 g/cm^3. In the case of ^{36}Cl$^-$ the plugs were 0.53 and 0.83 cm thick and compacted at the densities of 1.0, 1.2, 1.4, 1.6, and 1.65 g/cm^3.

In TDC experiments, the expression of the cumulative amount of solute that has passed to outlet reservoir (Q) as a function of the time is (Crank, 1975;

Bourke *et al.*, 1993):

$$Q = A \cdot L \cdot C_0 \left[\frac{D_e}{L^2} t - \frac{\alpha}{6} - \frac{2\alpha}{\pi^2} \sum_{1}^{\infty} \frac{(-1)^n}{n^2} \right.$$

$$\left. \exp\left(\frac{-D_e n^2 \pi^2 t}{L^2 \alpha} \right) \right] \tag{9}$$

where A is the cross-sectional area, L is the thickness of the sample, and C_0 is the tracer concentration in the inlet reservoir. The other parameters in equation (9) were already defined: D_e is the effective diffusion coefficient, and α is the capacity factor, which is equivalent to the accessible porosity for non-sorbing tracers. For long times, the steady-state condition is reached, the exponential tends to zero, and the equation (9) becomes:

$$Q = A \cdot L \cdot C_0 \left[\frac{D_e}{L^2} t - \frac{\alpha}{6} \right] \tag{10}$$

The effective diffusion coefficient can be calculated from the slope of the straight-line fitting the long-term behaviour of Q, equation (10) *vs.* time. The porosity accessible to the tracer can be calculated from the intercept of the fitting straight line to the time-axis, time-lag t_e (Bourke *et al.*, 1993). These experiments were performed only with $^{36}Cl^-$ in a clay plug of 1.65 g/cm^3 and for cells of 1 cm thickness.

2.3.2 In-diffusion experiments, ID

In ID experiments, one or both sides of the water saturated bentonite cell was contacted with a large volume of tracer solution. After a given time, the diffusion cell was disassembled, the bentonite plug cut into slices approximately 1 mm thick, and the activity in each slice measured to obtain a concentration profile in the bentonite plug.

Since the concentration in the reservoir remains practically constant during the experiment and the bentonite plug is long enough, the concentration profile within the plug can be fit by the following analytical solution (valid for a constant concentration in the boundary and semi-infinite medium) (Crank 1975; Eriksen & Jacobsson 1984; Idemitsu *et al.*, 2000):

$$\frac{C}{C_0} = erfc\left(\frac{x}{2\sqrt{D_a t}} \right) \tag{11}$$

where C is the tracer concentration in the bentonite plug pore water, C_0 the constant concentration in the reservoir, D_a the apparent diffusion coefficient, x the distance, and t the diffusion time.

For several values of D_a, the theoretical concentration profiles in the plug are obtained using equation

(11). The comparison of the experimental profile and the theoretical profiles, for different D_a values, allows identifying the value of D_a for which the fitting is the best. ID experiments provide the D_a values. These experiments were done only for HTO, using 5-cm thick bentonite plugs of different dry densities.

2.3.3 Saturation experiments, SAT

SAT experiments are performed introducing the diffusion cell into a bath; in these experiments a constant concentration profile in the clay plug has to be obtained. In order to ensure that the concentration profile into the plug is effectively constant, the concentration in the saturation solution is periodically monitored until a constant concentration is measured. After the experiment, the plug is sliced and each slice is weighed to determine its volume. Then the clay is transferred to a centrifuge tube and a volume of water added. The tubes are kept in continuous stirring during three days and, after this period, the samples are centrifuged (14000 rpm, 30 min) and the activity in the supernatant is measured. When the experiment finishes, the tracer concentration in the saturation solution is equal to the tracer concentration in the pore water of the bentonite plug, and the following equation can be used:

$$A_F = V \cdot C_R \cdot \phi \tag{12}$$

where A_F is the activity in the bentonite slice; V, the volume of the slice; C_R, the concentration in the reservoir; and ϕ, the accessible porosity. Each slice provides a value of the accessible porosity, which allows calculating a mean value and its error from a single experiment. SAT experiments have been performed for HTO, at all the clay densities considered, and for $^{36}Cl^-$ at the clay density of 1.0 and 1.2 g/cm^3.

2.3.4 Concentration profile experiments, PRO

In TDV experiments with chloride and high clay densities, the final concentrations in the inlet and outlet reservoirs are usually quite different. However, since tracer concentration in the outlet reservoir is significant, it can be ensured that the concentration profile in the clay plug pore water is linear. Pore water concentration at any point in the sample can be calculated interpolating between concentrations at the extremes (inlet and outlet reservoirs concentrations). As a consequence, for each slice the average tracer concentration in the pore water can be predicted, and using equation (12) the accessible porosity can be calculated. Each slice provides a value of the accessible porosity, which allows calculating a mean value and its error from a single experiment. In these samples, 5 or 6 slices (approximately 1 mm thick) could be obtained. PRO experiments have been performed for $^{36}Cl^-$ at a clay dry density of 1.4, 1.6 and 1.65 g/cm^3.

3 RESULTS AND DISCUSSION

3.1 Through-diffusion experiments

The methodology explained in the previous section, equation (10), was applied at all the through-diffusion experiments with variable concentration (TDV), and the D_e estimations results are included in (Table 1) for HTO and (Table 2) for Cl$^-$. An example of the TDV experimental results, are shown in (Fig. 1). (Fig. 1a) shows the results with HTO and (Fig. 1b) shows the results with $^{36}Cl^-$. The figures show the experimental results of only one cell, for each clay density, but the results obtained in different cells with the same dry density were always very similar. The upper graphs correspond to the concentration evolution in the inlet reservoir, and the lower graphs correspond to concentration evolution in the outlet reservoir.

Figure 2 shows an example of the experimental results of TDC experiments with Cl$^-$ (clay density 1.65 g/cm^3) where the lineal fit in the steady-state region and the time-lag value are indicated. Accessible porosity and effective diffusion coefficient can be obtained from this fitting line. Using equation (5), the apparent diffusion coefficient can be calculated. The results of this experiment are also included in (Table 2).

Table 1. D_e for HTO obtained by TDV experiments, at different dry density.

Dry density (g/cm^3)	Sample 1 (m^2/s)	Sample 2 (m^2/s)
1.10	1.92E-10	2.01E-10
1.30	1.36E-10	1.56E-10
1.50	8.80E-11	8.93E-11
1.65	5.8 ± 0.2E-11	5.8 ± 0.2E-11
1.70	5.76E-11	4.57E-11

Table 2. D_e for Cl$^-$ at different dry density. The experimental method is indicated.

Dry density (g/cm^3)	Sample 1 (m^2/s)	Sample 2 (m^2/s)	Method
1.00	2.72E-11	3.68E-11	TDV
1.20	1.08E-11	2.52E-11	TDV
1.40	2.90E-12	1.70E-12	TDV
1.60	8.70E-13	–	TDV
1.65	1.10E-12	7.70E-13	TDC

3.2 In-diffusion experiments

The summary of the D_a values for HTO, obtained with in-diffusion experiments, are shown in (Table 3). (Fig. 3) shows the experimental results of ID experiments with HTO and the fit obtained for different clay dry densities from 1.1 to 1.7 g/cm^3. (Fig. 3a) shows the experimental concentration profile in the sample for a dry density of 1.65 g/cm^3 and the theoretical profiles for three values of D_a. The experimental values fit very well to the theoretical profile for a mean $D_a = 1.7 \cdot 10^{-10}$ m^2/s, and all the experimental points can be described with a small variation in $D_a(\pm 0.1 \cdot 10^{-10})$. (Fig. 3b) shows the experimental results and the best fit to the theoretical curves for the other clay densities. The agreement between experimental and theoretical profiles is very good. No ID experiments were performed for $^{36}Cl^-$, and all the $D_a(Cl^-)$ values shown in this article have been calculated from D_e and ϕ values using equation (5).

3.3 Saturation and concentration profile experiments

The results of the accessible porosity obtained for HTO and Cl$^-$ by means of SAT and PRO experiments are summarised in (Table 4) and (Table 5). The experimental results show that for HTO the accessible porosity is always approximately the same as the total porosity. In the case of Cl$^-$ the values are significantly smaller.

3.4 General results

No ID experiments were performed for $^{36}Cl^-$. D_a values for Cl$^-$ were calculated according with the equation (5) and using the D_e and ϕ values presented in (Table 2) and (Table 5) respectively. (Table 6) includes these calculated D_a values.

The summary of all the results obtained in the different experiments for HTO and ^{36}Cl are show in (Figs 4 5, and 6). (Fig. 4) shows the effective diffusion coefficients, (Fig. 5) shows the accessible porosity and (Fig. 6) shows the apparent diffusion coefficients, all the previous mentioned parameters plotted as a function of the clay dry density.

As can be seen in the figures, the effective/apparent diffusion coefficients and the accessible porosity show an exponential decrease when the clay density increases. The decrease is significantly more pronounced in the case of Cl$^-$. Therefore for each parameter, the experimental values can be adjusted using exponential functions of the form $D_{a,e} = A \cdot e^{-B\rho}$, where A and B are constants and ρ is the bentonite dry density. Accessible porosity, apparent and effective diffusion coefficients in FEBEX bentonite at any dry density can be therefore easily estimated by interpolation.

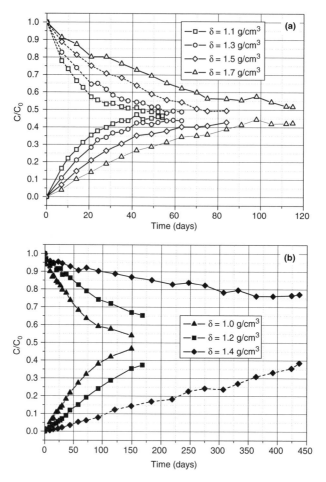

Figure 1. Concentration evolution for inlet and outlet reservoirs obtained in TDV experiments in clay plugs at different densities. a) HTO, b) $^{36}Cl^-$.

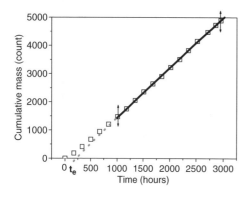

Figure 2. TDC experiments with $^{36}Cl^-$. (Clay dry density 1.65 g/cm³). The initial fit of the steady-state region and the *time-lag* are indicated.

Table 3. D_a for HTO obtained by ID experiments.

Dry density (g/cm³)	Sample 1 (m²/s)	Sample 2 (m²/s)
1.10	3.3E-10	–
1.30	3.1E-10	–
1.50	2.6E-10	–
1.65	1.7E-10	1.6E-10
1.70	2.6E-10	–

Results in (Fig. 5) clearly show that the measured accessible porosity for HTO is roughly equal to total porosity in FEBEX bentonite at different dry densities. Since total porosity is easy to calculate according to the equation (6), if one of the diffusion coefficients (effective or apparent) is measured the

520

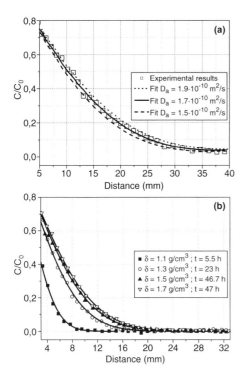

Table 4. ϕ for HTO obtained by SAT experiments.

Dry density (g/cm³)	Sample 1 (%)	Sample 2 (%)
1.0	65.4	64.5
1.2	56.6	58.3
1.4	51.9	51.0
1.6	42.8	43.3

Table 5. ϕ for Cl⁻. The experimental method used is indicated.

Dry density (g/cm³)	Sample 1 (%)	Sample 2 (%)	Method
1.00	18.0	15.3	SAT
1.20	10.5	15.2	SAT
1.40	5.9 ± 1.8	3.4 ± 1.4	PRO
1.60	2.4 ± 1.1	–	PRO
1.65	2.2	2.3	TDC
1.65	2.8 ± 0.4	2.5 ± 0.3	PRO

Table 6. Calculated values of D_a for ^{36}Cl⁻.

Dry density (g/cm³)	Sample 1 (m²/s)	Sample 2 (m²/s)
1.00	1.51E-10	2.41E-10
1.20	1.03E-10	1.66E-10
1.40	4.90E-11	5.00E-11
1.60	3.60E-11	–
1.65	5.30E-11	3.30E-11

Figure 3. ID experiments for HTO. (a) The method used for the evaluation of D_a and its error are shown (Clay dry density 1.65 g/cm³). Experimental time t = 190 h.

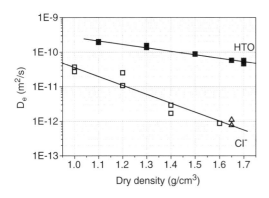

Figure 4. Effective diffusion coefficients obtained with different methods for HTO and ^{36}Cl⁻ as a function of clay density. The squares correspond to values obtained by TDV experiments. The triangles correspond to values obtained by TDC experiments. The exponential fits are included as continuous lines.

other coefficient can be calculated using equation (5) and giving to ϕ the value of the total porosity (ε).

Chloride ions have a very different behaviour than HTO in FEBEX bentonite. In fact, even at low densities only a fraction of total porosity is accessible for chloride. If dry density increases, the accessible porosity for chloride decreases rapidly, and at the density of 1.65 g/cm³ its value is between 2% and 3%, while the total porosity is close to 40%. These results show that a significant anionic exclusion occurs and that its effects are more pronounced when the dry density increases. As a consequence, only a small fraction of the total porosity is accessible for chloride.

It is usually accepted that anionic exclusion decreases when the pore water ionic strength increases. The electrostatic effect may be higher at lower clay density because the overlapping of the double layers of the pore surfaces does not occur, but at higher clay density the ionic strength effect is expected to be less significant. However, the effect of pore water composition on anionic exclusion will be studied in next future works.

Figure 6 shows the values of D_a(HTO) measured in in-diffusion experiments, and the exponential function that best fits these experimental results for HTO. Using equation (5) with the exponential functions

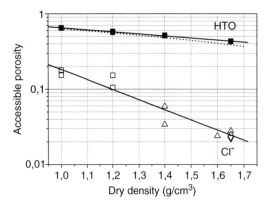

Figure 5. Accessible porosities obtained with different methods for HTO and Cl⁻ as a function of clay density. The squares correspond to values obtained by SAT experiments; up triangles, to values obtained by PRO experiments; and down triangles, to values obtained by the *time-lag* technique. The continuous lines correspond to the exponential fits of the experimental values and the dotted line corresponds to the total porosity of the clay, equation (6).

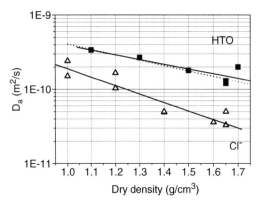

Figure 6. Apparent diffusion coefficients obtained with different methods for HTO and Cl⁻ as a function of clay density. The squares correspond to values obtained by in-diffusion experiments with HTO; and triangles, to values calculated using equation (5) from the D_e and ϕ for Cl⁻. The continuous lines correspond to the exponential fits of the experimental values and the dotted line is obtained by the ratio of D_e and ϕ fits obtained in the previous figures.

shown in (Fig. 4) and (Fig. 5) for ϕ (HTO) and D_e(HTO), an alternative exponential function for D_a(HTO) can be created. (Fig. 6) shows that both exponential functions are very similar, giving confidence that the parameter values measured for HTO are correct and that a simple diffusion model in pore water fully explains experimental results for HTO.

4 CONCLUSIONS

In FEBEX bentonite the accessible porosity for HTO agrees very well with total porosity, which implies that all the pores in compacted bentonite are available for diffusion of neutral species. Different results were obtained in the case of Cl⁻, for which the accessible porosity is significantly smaller than total porosity, even at the lower densities. Results for chloride clearly show that FEBEX bentonite displays a significant anionic exclusion. The effect of pore water composition on anionic exclusion will be studied in the future. For HTO the values of D_a, D_e and ϕ measured independently satisfy the relation between effective and apparent diffusion coefficients for conservative tracers ($D_e = D_a \cdot \alpha$). The consistency of the transport parameters measured using different methods confirms that these methods are appropriate and the values obtained are correct.

ACKNOWLEDGEMENTS

This work was carried out within the CIEMAT-ENRESA association, and partially supported by the FEBEX II project co-funded by the European Commission and performed as part of the fifth EURATOM framework programme, key action Nuclear Fission (1998–2002).

REFERENCES

Bourke, P.J., Jefferies, N.L., Lever, D.A., Lineham, T.R. (1993) "Mass transfer mechanisms in compacted clays" in Geochemistry of Clay-Pore Fluid Interactions, Eds. by D.A.C. Manning, P.L. Hall and C.R. Hughes, Chapman & Hall, London.

Crank, J. (1975) "Mathematics of diffusion" 2nd edition, Clarendon Press, Oxford.

Eriksen, T.E., Jacobsson, A. (1984) "Diffusion in clay, experimental techniques and theoretical models" SKB Technical Report 84-05.

García-Gutiérrez, M., Missana, T., Mingarro, M., Samper, J., Dai, Z., Molinero, J. (2001) "Solute transport properties of compacted Ca-bentonite used in FEBEX project" J. of Contaminant Hydrology 47, p.127–137.

Huertas, F., Fuentes-Cantillana, J.L., Jullien, F., Rivas, P., Linares, J., Fariña, P., Ghoreychi, M., Jockwer, N., Kickmaier, W., Martinez, M.A., Samper, J., Alonso, E., Elorza, F.J. (2000) "Full-scale engineered barriers experiment for a deep geological repository for high-level radioactive waste in crystalline host rock (FEBEX project)" Nuclear Science and Technology, EUR 19147.

Idemitsu, K., Xia, X., Ichishima, T., Furuya, H., Inagaki, Y., Arima, T. (2000) "Diffusion of plutonium in compacted bentonites in the reducing condition with corrosion products of iron" Material Research Society, vol. 608, Scientific Basis for Nuclear Waste Management XXIII, p. 261–266.

Kozaki, T., Inada, K., Sato, S., Ohashi, H. (2001) "Diffusion mechanism of chloride ions in sodium montmorillonite" J. of Contaminant Hydrology 47, p.159–170.

Muurinen, A., Lehikoinen, J. (1995) "Evaluation of phenomena affecting diffusion of cations in compacted bentonite" Nuclear Waste Commission of Finnish Power Companies, Report YJT-95-05.

Yu, J-W., Neretnieks, I. (1997) "Diffusion and sorption properties of radionuclides in compacted bentonite" SKB, Technical Report 97-12.

Advances in Understanding Engineered Clay Barriers – Alonso & Ledesma (eds)
© 2005 Taylor & Francis Group, London, ISBN 04 1536 544 9

Reactive solute transport mechanisms in nonisothermal unsaturated compacted clays

J. Samper, L. Zheng, J. Molinero & L. Montenegro
Civil Engineering School, University of La Coruña, Campus de Elviña s/n, La Coruña, Spain

A.Mª. Fernández & P. Rivas
CIEMAT, Avenida Complutense, Madrid, Spain

ABSTRACT: A review of relevant reactive solute transport mechanisms through compacted clays is presented. Issues which have risen during the FEBEX project include: (1) Porosity structure of compacted bentonites with consideration of complex porosity structures such as dual porosity; (2) Anion exclusion in partly-saturated clays; (3) Constitutive laws for solute diffusion (tortuosities) in unsaturated clays; (4) Cross diffusion which refers to the transport of a chemical species in response to concentration gradients of other species; and (5) Chemical and hydrodynamic couplings such as the dependence of permeability on pore water chemical composition. The relevance of these issues for FEBEX compacted bentonite is analyzed. Numerical analyses indicate that cross-diffusion is relevant near the heaters where large concentrations are attained for a relatively large period of time.

1 INTRODUCTION

FEBEX (Full-scale Engineered Barrier EXperiment) is a demonstration and research project dealing with the bentonite engineered barrier designed for sealing and containment of waste in a high-level radioactive waste repository (Huertas et al., 2000). It includes two main experiments: an in situ test performed at Grimsel (Switzerland) and a mock-up test operating since February 1997 at CIEMAT facilities in Madrid (Spain). One of the objectives of FEBEX project is the development and testing of conceptual and numerical models for the thermal, hydrodynamic, and geochemical (THG) processes taking place in engineered clay barriers.

Here we present a review of relevant reactive solute transport mechanisms through compacted clays which have risen during the FEBEX project. They include: (1) Porosity structure of compacted bentonites with consideration of complex porosity structures such as dual porosity versus a single porosity model; (2) Anion exclusion in unsaturated clays; (3) Constitutive laws for solute diffusion in partly-saturated media; (4) Cross diffusion; and (5) Chemical and hydrodynamic couplings such as the dependence of permeability on pore water chemical composition. First we present the mathematical formulation of water flow and reactive solute transport through unsaturated compacted clays. Then we address the issues of porosity structure, anion exclusion, cross diffusion and chemical and hydrodynamic couplings. The relevance of these issues for FEBEX compacted bentonite is analyzed.

2 MATHEMATICAL FORMULATION OF WATER FLOW AND SOLUTE TRANSPORT

2.1 Water flow

Water may flow through compacted bentonite in the liquid phase in response to hydraulic gradients according to Darcy's Law. It can also flow as water vapor due to moisture gradients (Fick's Law) and air convection. Fluxes of liquid water and vapor are related through evaporation and condensation. The energy transfer due to liquid water/vapor transition is important due to the significant evaporation/condensation enthalpy (585 cal/g, at 20°C) increment. Most of the energy transported through the medium must be transferred through both processes. Our formulation takes into account the following processes: a) the flow of liquid water (advection), b) the flow of water vapor (advection and diffusion), c) the flow of gaseous species different from steam (i.e. 'dry' air) (advection and diffusion), d) the flow of air dissolved in the water (advection), e) the transport of heat through the solid (conduction), f) the transport of heat through the liquid phase (convection) and g) the transport of heat in the gaseous phase (convection). The general mass balance equation for water, air and heat is given by (Juncosa et al., 2002a)

$$\frac{\partial m^i}{\partial t} = -\nabla \cdot \mathbf{q}_{tot}^i + r^i$$

where the mass of water m and the flux $\mathbf{q}$ are given by

$$m^w = \phi \, S_l \rho_l X_l^w + \phi \, S_g \rho_g X_g^v$$

$$\mathbf{q}_{tot}^w = \rho_l X_l^w \mathbf{q}_l + \rho_g X_g^v \mathbf{q}_g + \mathbf{j}_g^v$$

For air one has

$$m^a = \phi \, S_l \rho_l X_l^a + \phi \, S_g \rho_g X_g^a$$

$$\mathbf{q}_{tot}^a = \rho_l X_l^a \mathbf{q}_l + \rho_g X_g^a \mathbf{q}_g + \mathbf{j}_g^a$$

where fluxes are given by

$$\mathbf{q}_k = -\frac{\mathbf{K}_{ik} k_{rk}}{\mu_k}\left(\nabla P_k + \rho_k g \nabla z\right) \quad k = l, g \text{ (Darcy's law)}$$

$$\mathbf{j}_g^i = -\mathbf{D}_a^v \nabla X_g^i \qquad i = a, v \text{ (vapor)}$$

Similarly, mass and flux for heat are given by:

$$m^h = \rho_r (1-\phi) h_r + \sum_{k=l,g} \left(\phi \, S_k \rho_k h_k\right)$$

$$\mathbf{q}_{tot}^h = \Lambda \nabla T + \sum_{k=l,g} h_k \rho_k \mathbf{q}_k + \sum_{i=w,a} \mathbf{j}_g^i h_i$$

where **D** is the diffusivity tensor, ϕ is porosity, g is gravitational acceleration, μ is viscosity, h is enthalpy, ρ is density, j is diffusive flux, Λ is thermal conductivity, $\mathbf{K}_i$ is the tensor of intrinsic permeability, k_r is the relative permeability, P is pressure, $\mathbf{q}$ is Darcy velocity, $\mathbf{q}_{tot}$ is the total mass flux, r is the source/sink, S is saturation, T is temperature, X is mass fraction and z is the vertical coordinate. Subscripts a refers to air, g to gas, h to heat, I to gaseous species (a or v), k to phase (l or g), l to liquid phase, r to solid phase, v to vapor and w to water.

2.2 Solute transport

The solute transport model includes the following processes: (a) advection, (b) molecular diffusion, and, (c) mechanical dispersion. Each of these processes produces a solute flux per unit surface and unit time. There are as many transport equations as chemical components (primary species) in the system. Mass balance equation for the j-th component is given by

$$\rho_l X_l^w \theta_l \frac{\partial C_j}{\partial t} + \frac{\partial\left(\rho_l X_l^w \theta_l P_j\right)}{\partial t} + \frac{\partial\left(\rho_l X_l^w \theta_l W_j\right)}{\partial t}$$

$$+ \frac{\partial\left(\rho_l X_l^w \theta_l Y_j\right)}{\partial t} = L^*\!\left(C_j\right) + r_i\!\left(C_j^0 - C_j\right)$$

where

$$L^*(\cdot) = \nabla \cdot \left(\rho_l X_l^w \theta_l \mathbf{D}^j \nabla(\cdot)\right) - \rho_l X_l^w \mathbf{q}_l \nabla(\cdot)$$

$$+ \left(r_e - r_c\right)(\cdot)$$

C_j is the total dissolved concentration of the j-th species, r_e is the evaporation rate, r_i is the sink term, C_j^0 is the dissolved concentration of j-th species in the sink term r_i. W_j is the total exchanged concentration of j-th component, D^j is the dispersion coefficient, Y_j is the total sorbed concentration of j-th component, N_c is the number of solutes, P_j is the precipitated concentration of the j-th component, θ is water content, and r_c is the condensation rate.

A chemical species is a chemical entity that is able to be distinguished from the rest, due to both its chemical composition and the phase it is in. The chemical system can be defined through a knowledge of the concentration of its components (or primary species). Once the total concentration of each system's components are known, it is possible to compute the concentration of any aqueous chemical species (or secondary species) through appropriate mass balance and mass action equations. Similarly, the concentration of mineral phases, exchange complex and adsorbed species can be computed using similar equations. Secondary aqueous species are obtained using chemical mass-action equations. The source/sink chemical term for the transport equation takes into account the amount of mass transferred from/to solution due to dissolution/precipitation of minerals, ion exchange and adsorption.

The relevant equations linking primary to secondary species as well as those for adsorbed or exchanged mineral phases are shown in Table 1. In this table a is activity, C is the total dissolved concentration of the primary species, v_{ij}^x stoichiometric coefficient of the primary species j in the dissociation reaction of the secondary aqueous species i, K_{eq} is the equilibrium constant, N_p is the number of minerals involved in the dissolution/precipitation reactions, v_{ij}^p is the stoichiometric coefficient of the primary species j in the dissolution reaction of the precipitated species I, N_y is the total number of adsorption species, N_x is the umber of secondary aqueous species, p_i is molality of the i-th

Table 1. Expressions of total dissolved, precipitated, exchanged and adsorbed concentrations of primary species (Xu et al. 1999).

Law of mass action	$K_{eq} = \prod_{j=1}^{N_c} (a_j)^{v_j}$
Total analytic concentration	$T_j = C_j + Y_j + W_j + P_j$
Total dissolved concentration	$C_j = c_j + \sum_{i=1}^{N_x} v_{ij}^x x_i$
Total precipitation	$P_j = \sum_{i=1}^{N_p} v_{ij}^p p_i$
Total exchanged concentration	$W_j = w_j$
Total adsorbed concentration	$Y_j = \sum_{i=1}^{N_y} v_{ij}^y y_i$

mineral, v_{ij}^y is stoichiometric coefficient of the primary species j in the desorption reaction of the surface complex I, T is total concentration, w_j is concentration of the j-th exchange cation, x_i is the concentration of the i-th secondary aqueous species, y_i is molal concentration of the i-th adsorbed species and subscript j refers to the j-th primary species (see Xu et al., 1999).

The previous formulation of water flow and reactive solute transport has been implemented in FADES-CORE© (Juncosa et al., 2002a,b).

3 DOUBLE POROSITY

The term 'double porosity' refers to a conceptual model in which the medium is divided into two or more domains which are coupled by an interaction term for modeling flow and solute transport. This kind of conceptual model can be called 'dual-domain', 'dual region' or 'multi-region' and the sub-regions can be two or more mobile regions (Gwo et al., 1995) or one mobile and one immobile region.

Dual or double porosity models, initially introduced to simulate single-phase flow in fissured groundwater reservoirs (e.g. Barenblatt et al., 1960), assume that a porous medium consists of two separate but connected continua. Of these, one continuum is associated with a system or network of fractures, fissures, macropores, or intraggregate pores, while the other continuum involves the porous matrix blocks or soil aggregates. Hence, dual porosity models usually involve two flow/transport equations which are coupled by means of a sink/source term to account for water/solute transfer between the pore systems.

The dual-porosity concept has been commonly used to describe the preferential movement of water and solutes at the macroscopic scale, a phenomenon that is widely believed to occur in most natural (undisturbed) media (e.g. Schwartz et al., 2000). The double porosity model has been used to predict water flow in fractured media and solute transport of during both steady and transient flow. Recently, attempts have been made to extend the concept to variably

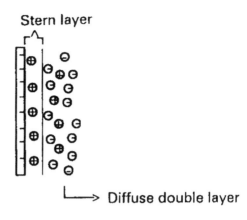

Stern layer

Diffuse double layer

Figure 1. Distribution of ions in a pore solution (after Corapcioglu & Lingam, 1994).

saturated fractured rocks and structured soils (Gerke and van Genuchten 1993a, 1993b, 1996; Saxena et al., 1994; Ray. et al., 1997; Larson and Jarvis 1999). Nowadays, some soil scientists use this conceptual model to predict the leaching of pollutants into soils and groundwaters (Hantush et al., 2000, 2002) and the transport of solutes through clay soils (Jaivis et al., 1991a, 1991b).

Compacted clay also exhibits a double-porosity behaviour (Alonso et al., 1995; Kim et al., 1997). As shown in Figure 1, double porosity model includes two parts: macro-pores through which water flows and therefore is considered the mobile part and micro-pores through which there is no flow and can be considered as immobile part. An increased level of complexity can be built into the double porosity model by assuming that the equilibrium is not reached between the mobile part (macro-pores) and the immobile part (micro-pores which may not be well connected). Furthermore, it is assumed that almost all of the water will flow through the mobile part but a mass transfer (by molecular diffusion) will exist between mobile and immobile parts. Samper et al. (2004a, in this volume) present a double-porosity model using CORE^{2D} (Samper et al., 2000) which accounts for the two regions by using three material zones: material 1 which corresponds to the mobile part; material 2 for the immobile part.

Let f_m and f_{im} be the fractions of the total volume of the sample occupied by the mobile and immobile domains, respectively. By trial and error Samper et al. (2004a, in this volume) estimated for FEBEX bentonite that $f_m = 0.3$ and $f_{im} = 0.7$. In other words, the mobile zone occupies 30% of the total volume of the sample, while the immobile zone of the clay sample is 70%.

It should be noticed that in this formulation of double porosity, model porosities for mobile and immobile zones are not global porosities, but relative porosities. Let Φ_m and Φ_{im} be the relative porosities of the mobile and immobile zones which are defined as the ratio of void volume to total volume of each domain. According to this approach, the total porosity of the sample is given by $\Phi = \Phi_m f_m + \Phi_{im} f_{im}$. Parameters Φ_m and Φ_{im} have a clear physical meaning, although they are difficult to measure. It should be noticed that the overall porosity of the mobile region can be computed as $\Phi_m f_m$. Similarly, $\Phi_{im} f_{im}$ represents the overall porosity of the immobile region. Samper et al. (2004a, in this volume) present the numerical interpretation of a long-term permeability test using single and double porosity models and several conceptual geochemical models. Their study concludes that chloride and sodium data cannot be adequately represented unless a double porosity model is used. The relevance of double porosity for other species remains to be ascertained.

4 ANION EXCLUSION

Anion exclusion is a phenomenon which causes a deficit of anions in the close vicinity of negatively charged clay surfaces. Electrostatic interactions between ions in the pore fluid and clay particles lead to the deficiency of anions close to a negatively charged surface. Soils physicists also refer to this phenomenon as negative adsorption, anion repulsion or anion expulsion, but the term anion exclusion seems to be the most used.

Anion exclusion is observed to be pronounced in soils with high cation exchange capacity (CEC) and is dependent on the soil-solute, solute-solvent and solvent-soil interactions. The soil-solute interactions are the most important and are included of two different phenomena. One is the repulsion between anions and negative charges on the clay surfaces. The other is bonding of the anions to sites located on the broken edges of clay minerals. However, the planar surface area of clay usually grossly exceeds the edge surface area and therefore the negative adsorption often masks the presence of any positive adsorption of the anions. Figure 1 shows a typical distribution of ions in a pore fluid, as presented by the Diffuse Double Layer Theory of Stern (1924). The anions are repelled by negatively charged surface and pushed towards the center of the pore. Through the anions in the bulk solution repel each other, the repulsion between the negatively charged surface and the anions is much greater than the repulsion between the anions. On the other hand, the velocity of the water in a single pore is maximum at the center and minimum on the particle surface (Figure 2). Hence, the anions which are pushed towards the center possess a greater

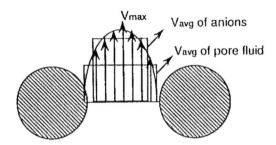

Figure 2. Velocity profile in a soil pore (after Corapcioglu & Lingam, 1994).

velocity of advection than the average pore water velocity. Therefore, negatively charged pollutants can migrate faster than predicted on the basis of average pore water velocity when advection is the dominating transport mechanisms. On the contrary, if diffusive transport dominates in comparison with advection, then anions will move slower than predicted on the basis of totally-accesible porosity.

Anion exclusion occurs both in saturated and unsaturated zones. Hence, the behaviour of the Diffuse Double Layer (DDL) at high and low water contents is important. In the case of high liquid content a full development of DDL can be expected reaching a given distance in the pore space which is not influenced by the surface charge. The width of DDL is usually taken as the zone where the ionic concentration differs significantly with respect to the equilibrium pore water. In unsaturated conditions when the thickness of the liquid layer is lower than the theoretical width of the DDL, a truncated DDL is formed by the readjustment of the ionic concentration. The truncated DDL has a tendency to absorb water to form a DDL of full extent (De Haan & Bolt, 1963).

Incorporating the effect of the anion exclusion in solute transport models requires evaluating the thickness of the DDL in order to estimate the amount of the pore space which is not available for a given dissolved anion. Based on DDL theory De Haan (1965) and Bolt & De Haan (1979) derived a theoretical model to estimate the equivalent distance of exclusion of a given anion. Knowing the distance of exclusion, accessible porosity values can be estimated and used for solute transport modeling.

Samper et al. (2004a, in this volume) have analyzed the relevance of anion exclusion for chloride in a long-term permeability test. They find that anion exclusion is a relevant process for chloride transport. Models without exclusion fail to reproduce the breakthrough curve of chloride. Chloride can access 35% of the mobile pores and 58% of the immobile pores.

Table 2. Cross diffusion coefficient ($m^2 \cdot s^{-1}$) calculated for the THG model of mock-up test. Shaded columns identify the largest cross-diffusion coefficients.

Species	H^+	Ca^{+2}	Mg^{+2}	Na^+	K^+	Cl^-	SO_4^{2-}	HCO_3^-
H^+	$9.31 \cdot 10^{-09}$	$-2.08 \cdot 10^{-10}$	$-2.37 \cdot 10^{-10}$	$-2.91 \cdot 10^{-09}$	$-3.83 \cdot 10^{-11}$	$4.42 \cdot 10^{-9}$	$5.28 \cdot 10^{-10}$	$7.96 \cdot 10^{-12}$
Ca^{+2}	$-2.06 \cdot 10^{-16}$	$6.15 \cdot 10^{-10}$	$-3.27 \cdot 10^{-11}$	$-4.02 \cdot 10^{-10}$	$-5.29 \cdot 10^{-12}$	$6.10 \cdot 10^{-10}$	$7.30 \cdot 10^{-11}$	$1.10 \cdot 10^{-12}$
Mg^{+2}	$-1.85 \cdot 10^{-16}$	$-2.58 \cdot 10^{-11}$	$5.48 \cdot 10^{-10}$	$-3.60 \cdot 10^{-10}$	$-4.75 \cdot 10^{-12}$	$5.47 \cdot 10^{-10}$	$6.55 \cdot 10^{-11}$	$9.86 \cdot 10^{-13}$
Na^+	$-1.73 \cdot 10^{-16}$	$-2.42 \cdot 10^{-11}$	$-2.75 \cdot 10^{-11}$	$7.44 \cdot 10^{-10}$	$-4.45 \cdot 10^{-12}$	$5.13 \cdot 10^{-10}$	$6.13 \cdot 10^{-11}$	$9.24 \cdot 10^{-13}$
K^+	$-3.14 \cdot 10^{-16}$	$-4.37 \cdot 10^{-11}$	$-4.97 \cdot 10^{-11}$	$-6.11 \cdot 10^{-10}$	$1.95 \cdot 10^{-09}$	$9.27 \cdot 10^{-10}$	$1.11 \cdot 10^{-10}$	$1.67 \cdot 10^{-12}$
Cl^-	$2.58 \cdot 10^{-16}$	$3.59 \cdot 10^{-11}$	$4.09 \cdot 10^{-11}$	$5.02 \cdot 10^{-10}$	$6.61 \cdot 10^{-12}$	$8.46 \cdot 10^{-10}$	$-9.12 \cdot 10^{-11}$	$-1.38 \cdot 10^{-12}$
SO_4^{2-}	$1.61 \cdot 10^{-16}$	$2.24 \cdot 10^{-11}$	$2.54 \cdot 10^{-11}$	$3.13 \cdot 10^{-10}$	$4.12 \cdot 10^{-12}$	$-4.75 \cdot 10^{-10}$	$4.44 \cdot 10^{-10}$	$-8.56 \cdot 10^{-13}$
HCO_3^-	$1.89 \cdot 10^{-16}$	$2.63 \cdot 10^{-11}$	$3.00 \cdot 10^{-11}$	$3.68 \cdot 10^{-10}$	$4.85 \cdot 10^{-12}$	$-5.59 \cdot 10^{-10}$	$-6.69 \cdot 10^{-11}$	$1.18 \cdot 10^{-09}$

5 CROSS-DIFFUSION

5.1 Cross-diffusion theory

The effective diffusion coefficient of a species or an chemical component in pure solution can be computed by considering the forces driving diffusional flux in a multicomponent solution. The forces driving the diffusion of an aqueous species within a solvent stem from both chemical potential and electronic potential gradients. According to Lichtner et al. (1996) the diffusive flux J_i of the ith aqueous species can be computed from

$$J_i = D_i^0 (1 + \frac{\partial \ln \gamma_i}{\partial \ln c_i}) \nabla c_i$$

$$- \left\{ \frac{z_i D_i^0 c_i}{\sum\limits_{k=1}^{N_s} z_k^2 D_k^0 c_k} \sum\limits_{k=1}^{n_s} z_k D_k^0 (1 + \frac{\partial \ln \gamma_k}{\partial \ln c_k}) \nabla c_k \right\}$$

where D_i^0 refers to the tracer diffusion coefficient of the subscripted species, γ_i refers to the activity coefficient of the subscripted species, z_i refers to the charge of the i-th species, n_s designates the number of aqueous species in the solution, and c_i is the concentration of i-th species. The diffusive flux J_i can be combined with Fick's law by

$$J_i = -\sum\limits_{j=1}^{N_s} D_{ij} \nabla c_j$$

where D_{ij} designates the diffusion coefficient of the i-th species in response to the j-th concentration gradient in a pure electrolyte solutions. Cross-diffusion coefficients it can be expressed as

$$D_{ij} = \delta_{ij} D_i^0 (1 + \frac{\partial \ln \gamma_i}{\partial \ln c_i})$$

$$- \left\{ \frac{z_i D_i^0 c_i}{\sum\limits_{k=1}^{N_s} z_k^2 D_k^0 c_k} z_j D_j^0 (1 + \frac{\partial \ln \gamma_j}{\partial \ln c_j}) \right\}$$

By assuming that $\frac{\partial \ln \gamma_j}{\partial \ln c_j} = 0$, then the latter equation reduces to

$$D_{ij} = \delta_{ij} D_i^0 - \frac{z_i D_i^0 c_i}{\sum\limits_{k=1}^{N_s} z_k^2 D_k^0 c_k} z_j D_j^0$$

5.2 Relevance of cross diffusion for the THG model of the mock-up test

Cross diffusion has been implemented in FADES-CORE© and its relevance has been tested in the thermo-hydro-geochemical (THG) model of FEBEX mock-up test. Table 2 contains the list of cross-diffusion coefficients of i-th species in response to j-th concentration gradients, D_{ij}, for the THG conditions of the mock-up test. Since the FEBEX bentonite pore waters have concentrations which are largest for chloride and sodium, cross-diffusion coefficients D_{ij} are largest for $i =$ chloride and $j =$ sodium (see shaded columns in Table 2).

When cross diffusion is taken into account, the computed charge balance of solutions improves. Modeling results with cross diffusion are markedly different than those of obtained without cross diffusion (see Figure 3). Chloride profiles with cross diffusion are steeper than those obtained without cross diffusion. Similar results are found when cross diffusion is analyzed in a heating and hydration experiment in a small cell (CT23).

Although the initial motivation for taking into consideration cross diffusion was the preservation of charge balance in TGH models, our numerical analyses of the THG model of mock-up test and

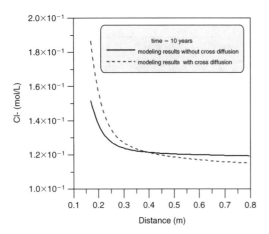

Figure 3. Comparison of chloride modeling results with and without cross diffusion coefficient for the mock up test after 10 years of heating and hydration.

cell CT23, indicate that model results may change significantly due to cross diffusion.

As expected there is an improvement in charge balance when cross diffusion is considered. However, there are other factors affecting charge balance which overlap with the effect of cross diffusion. These factors include:

1. Anion exclusion. When an anion is excluded from some parts of pore space, charge neutrality is not preserved.
2. Transport of aqueous complexes which may have different diffusion coefficients.
3. The formulation of cross diffusion in the diffusive flux J_i of the i-th aqueous species is valid only for infinitely dilute solutions. For pore waters having large ion concentrations, this simplification may cause some problems.

6 CHEMICAL-HYDRODYNAMIC COUPLINGS

6.1 Theory

According to the Gouy-Chapman theory, the thickness of the diffuse double layer T depends on different factors (Thomas & Egloffstein, 2001) according to

$$T \propto \left[D / n_0 v^2 \right]^{1/2}$$

where D is the dielectric constant of the pore solution, n_0 is the electrolytic concentration in the pore solution and v is the valence of the cations in the pore solution.

The thickness of the diffuse double layer influences the permeability by reducing or increasing the extension of the flow paths between two clay particles. Thomas & Egloffstein (2001) proposed a formulation to show that

$$K \propto \left[n_0 v^2 / D \right]^{1/2}$$

In a pore solution with several cations, one can use the ionic strength I to represent $n_0 v^2$. Charles et al. (2000) show the results of flexible-wall permeameter tests which show that the hydraulic conductivity increases with the increase of the concentration of the permeated liquids. Based on the theory of Thomas & Egloffstein (2001) and the experimental results of Charles et al. (2000) two formulations of variable permeability depending ionic strength are proposed here.

6.2 Formulation

Let a be a proportionality factor. Based on the last equation, one can get the following formulation

$$K = a \left[n_0 v^2 / D \right]^{1/2}$$

which can be reformulated as

$$K = a \left(\frac{2}{D} \right)^{1/2} \left[\frac{1}{2} n_0 v^2 \right]^{1/2}$$

Let $A = a(2/D)^{1/2}$ and taking into account all the chemical species using the ionic strength I instead of the concentration of one chemical, we get

$$K = A(I)^{1/2}$$

This formulation can be written in dimensionless forms by introducing a reference or initial permeability K_0, a reference ionic strength I_0, and a reference proportionality factor A_0

$$K_0 = A_0 (I_0)^{1/2}$$

Dividing the last two equations, we get

$$\frac{K}{K_0} = \frac{A}{A_0} \left(\frac{I}{I_0} \right)^{1/2}$$

A/A_0 is the ratio of the two proportionality factors that can be denoted by $a = A/A_0$. Therefore, we get the final expression of our formulation of variable permeability

$$\frac{K}{K_0} = a \left(\frac{I}{I_0} \right)^{1/2}$$

Published data can be fitted using expression similar to the last equation for variable permeability but

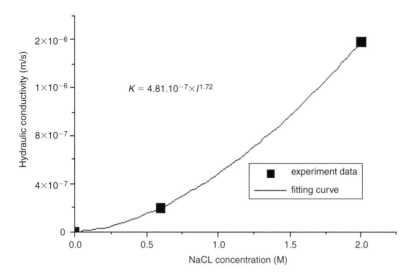

Figure 4. Formulation of variable permeability depending on ionic strength obtained by fitting experimental data of NaCl of Charles et al. (2000).

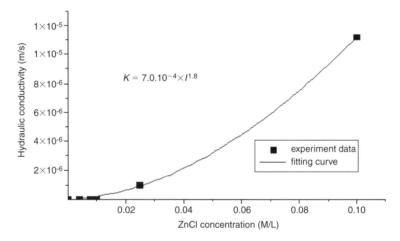

Figure 5. Formulation of variable permeability depending on ionic strength obtained by fitting experiment data of ZnCl of Charles et al. (2000).

allowing the exponent to be different than 1/2. Figures 4 and 5 show the results of such fittings which lead to the following expressions: $K = 4.81 \cdot 10^{-7} \times I^{1.72}$ (see Figure 4) and $K = 7.0 \cdot 10^{-4} \times I^{1.8}$ (see Figure 5).

6.3 Modeling results

Constitutive equations of variable permeability $K/K_0 = a(I/I_0)^{1/2}$ and $K/K_0 = a(I/I_0)^{1.72}$ coupled with flow equation, have been implemented in FADES-CORE©. The THG model of mock-up test with permeability depending on ionic strength leads to

an improvement in modeling results of cumulative water inflow and relative humidity (Figures 6 and 7). Figure 6 shows the comparison of modeling results of cumulative water inflow obtained with models of constant and variable permeability. Figure 7 shows the comparison of modeling results of relative humidity. Although the model of variable permeability improves the results of the constant permeability model, there are still discrepancies between modeling results and measurements. Double porosity behaviour (not yet accounted for in this model) and swelling phenomena could be at the core of such discrepancies.

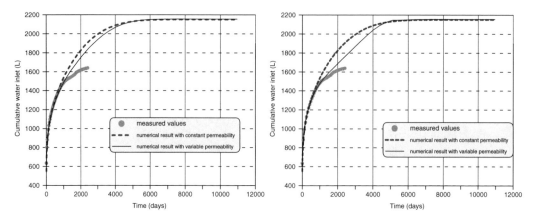

Figure 6. Comparison of modeling results of cumulative water inflow for the mock-up test obtained with models of constant and variable permeability $(K/K_0) = a(I/I_0)^{1/2}$ (left) and $(K/K_0) = a(I/I_0)^{1.72}$ (right).

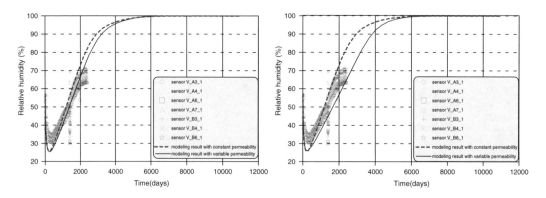

Figure 7. Comparison of modeling results of relative humidity for the mock-up test obtained with models of constant and variable permeability $K/K_0 = a(I/I_0)^{1/2}$ (left) and $K/K_0 = a(I/I_0)^{1.72}$ (right).

7 CONCLUSIONS

Water flow and reactive solute transport mechanisms through compacted clays have been reviewed using data from the FEBEX project. The results of such review indicate that double porosity models are appropriate at least for chloride and sodium. The relevance of double porosity for other species remains to be ascertained.

Anion exclusion is a relevant process for chloride transport. Models without exclusion fail to reproduce the breakthrough curve of chloride. Chloride can access 35% of the mobile-part pores and 58% of the immobile pores.

The relevance of cross diffusion has been addressed for the purpose of preservation of charge balance in THG models. Our numerical analyses of the THG model of mock-up test and cell CT23 indicate that model results change significantly due to cross diffusion. Cross diffusion coefficients are specially relevant for chloride and sodium.

There is a large evidence on the dependence of permeability on pore water chemistry. Reported published data have been fitted to theoretical expressions relating permeability to ionic strength of pore water. Such formulations have been implemented in FADES-CORE and tested with the THG model of the mock up test. The THG model of mock-up test with variable permeability leads to an improvement in modeling results of cumulative water inflow and relative humidity. However, there are still discrepancies between modeling results and measurements which could be caused by a double porosity behaviour and swelling phenomena (not accounted for in this model)

ACKNOWLEDGMENTS

The work presented here has been funded by
ENRESA. The project as a whole was supported by
the EC (Contracts FI4W-CT95-0006 and FIKW-CT-
2000-0016 of the Nuclear Fission Safety Programme).

REFERENCES

Alonso, E.E., Lloret, A., Gens, A. & Yang, D.Q. (1995).
Experimental behaviour of highly expansive double-
structure clay. Unsaturated Soils, 11–16.

Barenblatt, G.I., Zheltov, I. & Kochina, I. (1960). Basic
concepts in the theory of seepage of homogeneous
liquids in fissured rocks. Prikl. Mat. Mekh., 24, 852–864.

Bradbury, M.H. & Baeyens, B. (2003) Porewater chemistry
in compacted re-saturated MX-80 bentonite. *Journal of
Contaminant Hydrology*, 61, 329–338.

Bolt, G.H. & De Haan, F.A.M. (1979). Soil Chemistry. B.
Physico-Chemical Models. G.H. Bolt, eds. Elsevier.
233–257.

Charles, D., Shackelford, C., Benson, R., Katsumi, T. Edil,
T.B. & Lin, L. (2000). Evaluating the hydraulic
conductivity of GCLs permeated with non-standard
liquids. Geotextiles and Geomembranes, 18, 133–161.

Corapcioglu, M.Y. & Lingam, R. (1994). Advances in
porous media. Vol. 2. M.Y. Corapcioglu, eds. Elsevier.
151–168.

De Haan, F.A.M. (1965). Determination of the specific
surface area of soils on the basis of anion exclusion
measurements. Soil Science, 99, no 6, 379–386.

De Haan, F.A.M & Bolt, G.H. (1963). Determination of
anion adsorption by clays. Soil Science Society Proceedings 1963, 636–640.

García-Gutiérrez et al., 2002.

Gerke, H.H. & van Genuchten, M.Th. (1993a). A dual
porosity model for simulating the preferential movement
of water and solutes in structured porous media. Water
Resources Research, 29, 305–319.

Gerke, H.H. & van Genuchten, M.Th. (1993b). Evaluation
of a first order water transfer term for variably saturated
dual porosity flow models. Water Resources. Research,
29, 1225–1238.

Gerke, H.H. & van Genuchten, M.Th. (1996). Macroscopic
representation of structural geometry of simulating water
and solute movement in dual-porosity media. Advances
in Water Resources, 19, 343–357.

Gwo, J.P., Jardine, P.M., Wilson, G.V. & Yeh, G.T. (1995).
A multiple-pore-region concept to modeling mass
transfer in subsurface media. J. Hydrol., 164, 217–237.

Larsson M.H. & Jaivis, N.J. (1999). Evaluation of a dual-
porosity model to predict field scale solute transport in a
macroporous soil. J. Hydrol., 215, 153–171.

Hantush, M.M., Mariño, M.A. & Islam, M.R. (2000).
Models for leaching of pesticides in soils and groundwater. J. Hydrol., 227, 66–83.

Hantush, M.M., Govindaraju, R.S., Mariño, M.A. &
Zhonglong, Z. (2002). Screening mode for volatile
pollutants in dual porosity soils. J. Hydrol., 260, 58–74.

Huertas et al. (2000). Full-scale engineered barriers
experiment for a deep geological repository in crystalline
host rock (FEBEX Project). European Commission. EUR
19147 EN.

Jaivis, N.J., Jansson, P.E., Dik, P.E. & Messing, I. (1991a).
Modelling water and solute transport in macroporous
soil. I. model description and sensitivity analysis. Journal
of Soil Science, 42, 59–70.

Jaivis, N.J., Bergstrom, L. & Dik, P.E. (1991b). Modelling
water and solute transport in macroporous soil. II,
Chloride breakthrough under non-steady flow. Journal
of Soil Science, 42, 71–81.

Juncosa, R., Samper, J., Navarro, V. & Delgado, J. (2002a).
Modelos de flujo multifásico no isotermo y transporte
reactivo multicomponente en medios porosos: 1. Formulación físico-matemática. Ingeniería del Agua, 9, no 4,
423–436.

Juncosa, R., Samper, J., Navarro, V. & Delgado, J. (2002b).
Formulación numérica en elementos finitos de problemas
de flujo multifásico no isotermo y transporte de solutos
reactivos en medios porosos. Revista Internacional de
Métodos Numéricos para Cálculo y Diseño en Ingeniería,
18, no 3, 415–432.

Kim, J.Y., Edil, T.B. & Park, J.K. (1997). Effective porosity
and seepage velocity in column tests on compacted clay.
J. Geotechnical and Geoenvironmental Engineering, 123,
no 12, 1135–1141.

Lichtner, P.C., Steefel, C.I. & Oelkers, E.H. (1996). Reactive
transport in porous media. Mineralogical Society of
America. Reviews in Mineralogy, 34, 148–158.

Ray, C., Ellsworth, T.R., Valocchi, A.J., & Boast, C:W.
(1997). An improved dual porosity model for chemical
transport in macroporous soils. Journal of Hydrology,
193, 270–292.

Samper, J., Juncosa, R., Delgado, J. & Montenegro, L.
(2000). CORE^{2D}: A code for non-isothermal water flow
and reactive transport. Users manual version 2. ENRESA
Technical Report 6/2000.

Samper, J., Fernández, A.Mª., Zheng, L., Montenegro, L.,
Rivas, P. & Dai, Z. (2004a). Direct and inverse modelling
of multicomponent reactive transport in single and dual
porosity media. In this volume.

Samper, J., Vázquez, A. & Montenegro, L. (2004b). Inverse
hydrochemical modelling of aqueous extract experiments
for the estimation of FEBEX bentonite pore water
chemistry. In this volume.

Saxena, R.K., Jaivis, N.J. & Bgergstrom, L. (1994).
Interpreting non-steady state tracer breakthrough experi-
ments in sand and clay soils sing a dual-porosity model.
J. Hydrol., 162, 279–298.

Schwartz, R.C., Juo, A.S.R. & McInnes, K.J. (2000).
Estimating parameters for a dual-porosity model to
describe non-equilibrium reactive transport in a fine-
textured soil. J. Hydrol., 229, 149–167.

Stern, O. (1924). Zur theory der electrolytischen dop-
pelschicht. Electrochem, 30. 508–516.

Thomas, A. & Egloffstein (2001). Natural bentonites: influence of the ion exchange and partial desiccation on
permeability and self-healing capacity of bentonites
used in GCLs. Geotextiles and Geomembranes, 19,
427–444.

Xu, T., J. Samper, C. Ayora, M. Manzano and E. Custodio,
Modeling of non-isothermal multicomponent reactive
transport in field scale porous media flow systems, *J. of
Hydrol.*, Vol. 214, 144–164, 1999.

Advances in Understanding Engineered Clay Barriers – Alonso & Ledesma (eds)
© 2005 Taylor & Francis Group, London, ISBN 04 1536 544 9

Predictive model of geochemical changes in porewater, buffer and backfill in an engineered barrier system

A. Luukkonen
VTT Building and Transport, Espoo (Finland)

ABSTRACT: The modelling attempt is batch reaction oriented. Calculations consider changes that occur during the wetting of canister buffer and tunnel backfill of a repository engineered barrier system (EBS). The modelling deals also with time-dependent changes at the EBS boundaries. All calculations follow, however, the equilibrium thermodynamic assumption, and are not tied to any strict time-span.

The approach concentrates on the major element compositions of porewaters, and the changes in solid phases of the repository. The initial properties of canister buffer resemble the estimations for compacted Wyoming MX-80 sodium bentonite. The tunnel backfill is assumed to consist of sodium bentonite (30%), and crushed diorite rock (70%) components. The geochemistry of backfill has been estimated in accordance with mineral quantities present in the components of the backfill mixture. The initial groundwater sucked into the EBS at the repository boundaries is Na-Ca-(HCO_3^-)-SO_4^{2-}-Cl-water having a reference to brackish seawater origin.

The geochemical reactions considered are cation exchange and surface complexation in montmorillonite, and dissolution/precipitation of certain minerals (calcite, gypsum, pyrite, quartz, and goethite). As an initial condition, the undersaturated pore volumes of buffer and backfill contain entrapped air (O_2 content 20%). In the buffer modelling cases, the porewater speciation and dissolution/precipitation equilibria are considered at $65°C$, and in the backfill modelling cases the temperature is assumed to be $40°C$.

The results indicate changes in porewater and solid phase compositions in the repository EBS. During the first wetting of the EBS, entrapped oxygen is consumed in the pyrite dissolution reaction. Most dissolved iron is precipitated as goethite. At the repository boundaries, gypsum runs out after couple reaction cycles causing changes in the sulphate and cation contents of porewater. Other minerals do not run out during the 40 batch reaction cycles considered. However, the EBS boundary calculations indicate that porewater compositions created, and sucked deeper into EBS, evolve moderately with time.

1 INTRODUCTION

In the Prototype repository, the electrically heated canisters cause a heat flow from canisters to the engineered barrier system (EBS). According to Pusch & Börgesson (2001), the MX-80 bentonite buffer surrounding the heated canisters experiences a maximum temperature of $90°C$. The temperature in the tunnel backfill is expected to stay below $40–50°C$ during the execution of experiment (King et al., 2001; Pusch & Börgesson, 2001).

A simplified cross-section of the Prototype is presented in Figure 1. The first wetted volumes of buffer and backfill are created at the boundaries of the Prototype EBS. These porewaters are sucked deeper into the undersaturated volumes of buffer and backfill and represent the first wetting of EBS. At the same time, new porewater compositions are repeatedly generated at the EBS boundary. These waters, as well, are sucked deeper into EBS until the suction power vanishes to the EBS water saturation.

The equilibrium-modelling attempt is batch reaction oriented. It illustrates how the geochemistry evolves in the EBS interior and boundaries as a function of batch reaction cycles. The calculations model major elemental compositions for porewaters and related compositional changes in the EBS solid phases. The modelling code utilised is PHREEQC-2 version 2.5 with associated thermodynamic database Wateq4f (Parkhurst & Appelo, 1999).

2 WYOMING MX-80 BENTONITE

2.1 Material properties

The MX-80 sodium bentonite is used as buffer in the canister deposition holes. Material properties that can be associated to the bentonite blocks after compaction are presented in Table 1 (Börgesson & Hernelind, 1999).

For modelling purposes, all properties of bentonite are defined for $1 dm^3$ of pore volume. Initially a

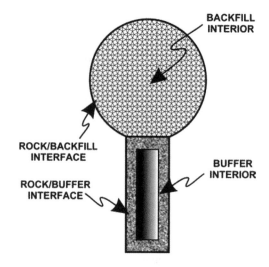

BACKFILL INTERIOR

ROCK/BACKFILL INTERFACE

ROCK/BUFFER INTERFACE

BUFFER INTERIOR

Figure 1. Schematic cross-section of the Prototype repository with indications of four special interest areas of current geochemical modelling.

Table 1. Material properties for initial compacted bentonite (Börgesson & Hernelind, 1999). Water content (m_{water}/m_{solid}), porosity (V_{void}/V_{tot}), and degree of saturation (V_{water}/V_{void}).

Dry density g/cm^3	Water content	Porosity	Degree of saturation
1.57	0.17	0.435	0.61

material block contains 0.61 dm^3 water and 0.39 dm^3 air (Table 1). During the first reaction cycle external water completely replaces the earlier pore volume and the block saturates with water (the gas phase reacts with water in a closed system). In compacted bentonite, the initial 0.39-dm^3 volume of air corresponds a 5.89-dm^3 volume of water and bentonite, that weights 9.25 kg. The weight of bentonite is 8.64 kg.

2.2 Geochemical properties

The composition of MX-80 bentonite varies as presented in Table 2. Based on this variation an average mineralogical composition has been estimated for current purposes. The cation exchange capacity (CEC) for bentonite is around 787 meq/kg (Bradbury & Baeyens, 2003). The amount potentially reactive species in 8.64 kg of bentonite are presented in the last column of Table 2. The specific surface area in MX-80 bentonite available for surface complexation is around 31.5 m^{2}g (Wieland et al., 1994; Bradbury & Baeyens, 2002; Wersin, 2003).

3 TUNNEL BACKFILL

3.1 Material properties

The tunnel backfill consist of 30% bentonite mixed with 70% crushed Tunnel Boring Machine (TBM) muck. The material properties presented in Table 3 are after Börgesson & Hernelind (1999).

At start of the wetting process, a material block contains 0.58 dm^3 of water and 0.42 dm^3 of air in the

Table 2. The composition of the Volclay MX-80 bentonite.

	Unit wt g/mol	Bentonite[a] ca. Wt%	Current estim. Wt%	Bentonite moles/kg	Buffer moles/(1 dm^3 pore water)
Montmorillonite	744	75.0	75.0	1.01	8.71
Illite	778	0.0–4.0	2.0	0.03	0.22
Kaolinite	258	1.0–7.0	4.0	0.15	1.34
Albite	262	5.0–9.0	7.0	0.27	2.31
Quartz	60.1	10.0	10.0	1.66	14.4
Gypsum	136	0.3	0.3	0.02	0.22
Pyrite	120	0.3	0.3	0.03	0.22
Calcite	100	1.4	1.4	0.14	1.21
CH$_2$O[b]	30.0		0.4	0.13	1.15
Cation occupancies in the exchange sites[c]					
Ca^{2+}				0.13	1.14
Mg^{2+}				0.08	0.69
Na$^+$				0.67	5.77
K$^+$				0.01	0.11
Surface site capacities[d]					
≡S^{s}OH				0.002	0.017
≡S^{w1}OH				0.04	0.36
≡S^{w2}OH				0.04	0.35

According to [a]Bruno et al. (1999), [b]Pirhonen (1986), [c]Bradbury & Baeyens (2003), and [d]Bradbury & Baeyens (1997).

pore volume. When the block saturates with external groundwater, the gas phase essentially disappears. The initial 0.42-dm^3 volume of air in a compacted block results a 6.56-dm^3 volume, and a weight of 11.48 kg, of water and backfill. The weight of backfill is 10.90 kg.

3.2 Geochemical properties

The TBM-muck is mostly diorite and its modal composition can be found from Patel et al. (1997). Äspö diorite contains frequently narrow fracture and broader hydrothermal alteration zones. Currently it is estimated that 2% of rock is completely altered. This increase the amount of secondary minerals (chlorite, goethite, calcite, epidote, fluorite, and quartz) compared to the fresh rock modal compositions. A judged partial composition for the TBM-muck is

presented in Table 4. The CEC of the muck is at the most 5–10% of the CEC of bentonite. Therefore, in a 70/30 mixture, bentonite is still mostly responsible (more than 80%) of the exchange properties of the backfill. In this study, the cation exchange properties and surface geochemistry of backfill are based on the bentonite fraction only. The amounts of potentially reactive species in 10.90 kg of backfill are presented in Table 4.

4 GEOCHEMICAL REACTIONS

The fractures of the near-field bedrock feed external water into EBS. The external water used in the calculations is a salinity median of the samplings March '98–June '99 (Andersson & Säfvestad, 2000) from the near field of the Prototype repository drift. The concentrations in water are shown in Table 5.

The first intrusion of external water into a batch volume differs from all subsequent cycles because initially batches contain entrapped air. An under-saturated bentonite cell contains 2.8 mmol/L O_2 and 0.004 mmol/L CO_2 (0.39 dm^3 air at 1 atm pressure) while an undersaturated backfill cell contains 3.3 mmol/L O_2 and 0.006 mmol/L CO_2. Oxygen is consumed during the first batch cycle (dissolution of pyrite) and CO_2 is dissolved for further reactions.

Table 3. Material properties for initial compacted tunnel backfill (Börgesson & Hernelind, 1999). Water content (m_{water}/m_{solid}), porosity (V_{void}/V_{tot}), and degree of saturation (V_{water}/V_{void}).

Dry density g/cm^3	Water content	Porosity	Degree of saturation
1.75	0.19	0.363	0.58

Table 4. Potentially reactive solids of the tunnel backfill of the Prototype experiment.

	Unit wt g/mol	TBM-muck mol/kg	Bentonite mol/kg	70/30 Backfill mol/kg	70/30 Backfill mol/(1 dm^3 water)
Montmorillonite	744	–	1.01	0.30	3.30
Illite	778	–	0.03	0.01	0.08
Kaolinite	258	–	0.15	0.05	0.51
Biotite	986	0.16	–	0.11	1.22
K-feldspar	278	0.49	–	0.34	3.72
Albite	262	1.58	0.27	1.18	12.9
Quartz	60.1	2.19	1.66	2.03	22.2
Epidote	469	0.16	–	0.11	1.21
Amphibole	877	0.01	–	0.01	0.11
Pyrite	120	0.03	0.03	0.03	0.30
Goethite	88.9	0.07	–	0.05	0.57
Calcite	100	0.04	0.14	0.07	0.77
Gypsum	136	–	0.02	0.01	0.08
Fluorite	78.1	0.01	–	0.01	0.10
CH$_2$O	30.0	0.13	–	0.09	1.02
Cation occupancies in the exchange sites					
Ca^{2+}			0.13	0.04	0.43
Mg^{2+}			0.08	0.02	0.26
Na$^+$			0.67	0.20	2.18
K$^+$			0.01	0.00	0.04
Surface site capacities					
$\equiv$S^{s}OH			0.002	0.0006	0.007
$\equiv$S^{w1}OH			0.04	0.01	0.14
$\equiv$S^{w2}OH			0.04	0.01	0.13

Table 5. The main composition of sample KA3542G02/2587 (Andersson & Säfvestad, 2000). The pe value is estimated. All concentrations are in mg/l except pH and pe values.

pH	7.4	pe	−3.0
Na^+	1620	HCO_3^-	188
K^+	9.5	Cl^-	3890
Ca^{2+}	624	SO_4^{2-}	296
Mg^{2+}	79.4	HS^-	0.0
Fe^{2+}	0.23		
Si^{4+}	7.8		

Table 6. Exchange and surface parameters for the thermodynamic models considered.

Parameter	Unit/reaction	Value
Cation exchange[a]		
log K	$Ca^{2+} + 2NaX \Leftrightarrow CaX_2 + 2Na^+$	0.41
log K	$Mg^{2+} + 2NaX \Leftrightarrow MgX_2 + 2Na^+$	0.34
log K	$K^+ + NaX \Leftrightarrow KX + Na^+$	0.60
Surface complexation[b]		
log K	$\equiv S^SOH + H^+ \Leftrightarrow \equiv S^SOH_2^+$	4.5
log K	$\equiv S^SOH \Leftrightarrow \equiv S^SO^- + H^+$	−7.9
log K	$\equiv S^{W1}OH + H^+ \Leftrightarrow \equiv S^{W1}OH_2^+$	4.5
log K	$\equiv S^{W1}OH \Leftrightarrow \equiv S^{W1}O^- + H^+$	−7.9
log K	$\equiv S^{W2}OH + H^+ \Leftrightarrow \equiv S^{W2}OH_2^+$	6.0
log K	$\equiv S^{W2}OH \Leftrightarrow S^{W2}O^- + H^+$	−10.5

[a] According to Wersin (2003), Gaines-Thomas convention.
[b] According to Bradbury & Baeyens (1997).

The modelling tries to take into account both cation exchange and surface complexation. The constants for equilibria are presented in Table 6. The surface complexation follows a mechanistic approach presented by Bradbury & Baeyens (1997). The pH dependent surface processes affect at small extent to charge balancing of interacting solution. In real, outer sphere complexes compensate charged surfaces (e.g. $\equiv SOH_2^+$, $\equiv SO^-$; Davis & Kent, 1990), but these counter-ion balances have not been included in the modelling. In repeated batch reactions this omission will cause propagating charge balance errors in the modelled water compositions. The issue is dealt in more detail in the modelling results. According to Bradbury & Baeyens (2002), proton exchange is not significant among the cation exchange processes. Therefore, proton exchange has been omitted from the cation exchange processes (Table 6).

The calculations equilibrate the porewater compositions to following solids: calcite, gypsum, quartz, pyrite and goethite (cf. Muurinen & Lehikoinen, 1999; Pusch et al., 1999; Bruno et al., 1999). Concentrations of these solids in the initial buffer and backfill cells are presented in Tables 2 and 4.

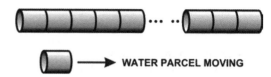

WATER PARCEL MOVING

Figure 2. Illustration of the modelling method for the EBS interior. A porewater parcel is repeatedly transferred to a new undersaturated material cell.

Calculations make a partial attempt to take into account the effect of temperature. In the buffer calculations the assigned temperature is 65°C while in the backfill calculations the temperature is 40°C. With the current knowledge, temperature dependencies are available for gas-phase and aqueous speciation calculations, and for dissolution/precipitation reactions of calcite, gypsum, quartz and pyrite (Parkhurst & Appelo, 1999). However, the cation exchange and surface geochemical reactions (Table 6), and goethite precipitation are estimated with equilibrium constants defined at 25°C only.

5 MODELLING APPROACH

5.1 *EBS interior*

In both buffer and backfill cases porewater parcels are transported through successive material cells. An illustration of the modelling is presented in Figure 2. The calculation starts from the EBS boundary. The system is equilibrated and then porewater is transported into a next undersaturated material cell deeper in EBS. In repeated process porewater equilibrates with successive volumes.

In this first wetting simulation, entrapped air is consumed away from the cells. In real, however, formation of gas bubbles within and around the repository is possible. After the first wetting, the repeated external recharge at the EBS boundary defines what kinds of water compositions are sent deeper in EBS. The later cycles of EBS interior transport, however, are not modelled here.

5.2 *EBS boundary*

The calculations consider porewater and solid phase evolution in one cell at the EBS boundary as new parcels of external water repeatedly fill the pore spaces of an edge-cell volume. An illustration of the modelling method is shown in Figure 3. At start of the modelling, the first parcel of external water fills the undersaturated cell. After the first cycle, next refills occur into the saturated cell. The porewaters created in successive cycles are sent deeper into the EBS.

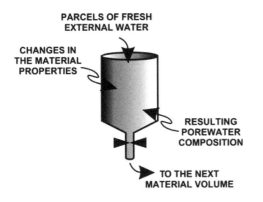

PARCELS OF FRESH EXTERNAL WATER

CHANGES IN THE MATERIAL PROPERTIES

RESULTING POREWATER COMPOSITION

TO THE NEXT MATERIAL VOLUME

Figure 3. Illustration of the modelling method at the EBS boundary. The studied cell volume is repeatedly filled with external water.

6 MODELLING RESULTS

6.1 *EBS interior*

The evolution occurring in the buffer and backfill cells (cf. Fig. 2) are illustrated in Figures 4 and 5. Presented diagrams illustrate how porewater composition and solid materials evolve as the wetting front advances as a function of cell number. The initial external water composition and the initial material concentrations are presented on the left border of each diagram.

As external water enters the first cell distinct changes occur (Figs 4 and 5). The undersaturated cell volume contains oxygen that is consumed with pyrite dissolution. Since the amount of oxygen is constant in every subsequent cell, the amount of pyrite dissolution stabilises to a constant level causing as well stabilisation of pe and pH. Dissolution of pyrite is a constant source of Fe. Significant part of the Fe input is removed from porewater with goethite precipitation.

In both buffer and backfill cases (Figs 4 and 5) the first cells indicate distinct dissolution of gypsum. Dissolution increases sulphate concentrations in porewater to the level of gypsum saturation. Later, cells precipitate small amounts of gypsum indicating that a part of sulphur produced in the dissolution of pyrite is removed from porewater.

Constant input of CO_2 during every reaction cycle causes continuous increase in alkalinity of porewater. In the buffer case (Fig. 4) the equilibrated pH is a little higher than in the backfill case (Fig. 5). Only the buffer case precipitates a little calcite in the first cell. Otherwise, all cells dissolve small amounts of calcite as a response for CO_2 input.

There is a general decreasing trend among major cations, and increasing trend among certain anions of porewater (Figs 4 and 5). It is evident that there occur

cumulative negative charge balance errors in the resulting porewaters. After 40 calculation cycles the imbalance is about -20%. At pH around 6.9, the surface complexation model (cf. Bradbury & Baeyens, 2002) constantly adsorbs small amounts of protons leading to positively charged surfaces. The modelling does not counter-balance these surfaces. It was simply believed that counter-balancing occurs in the system predominantly with Cl^-. This is simplification, however. In the equilibrated porewaters two major anions, SO_4^{2-} and Cl^-, occur practically at equal concentrations. Therefore, also SO_4^{2-} is likely adsorbed onto the charged surfaces, and consequently the growing trend of dissolved SO_4^{2-} in Figures 4 and 5 possibly conflicts with this deduction.

Considering the dissolution of quartz, it can be concluded that it dissolves to solubility limit in the first cell, and then practically nothing happens in the subsequent cells. This feature can be generalised to other silicates as well. Once dissolved silicon concentration in porewater attains levels high enough, it begins to inhibit dissolution of silicates.

6.2 *EBS boundary*

The simulated geochemical evolutions at the EBS boundaries (buffer and backfill) are illustrated in Figures 6 and 7. The studied boundary cell (Fig. 3) is refilled 40 times with external water. The first batch reaction cycle is similar to the first reaction cycle of the wetting front studies (Figs 4 and 5). The initial undersaturated cell contains entrapped air that is consumed away. Subsequent reaction cycles occur in the anoxic saturated conditions.

During the first reaction cycle pyrite dissolution is evident. The consumption of O_2 causes an initial drop in pH and a part of produced protons are adsorbed in the surface complexation sites. Without surface complexation buffering the pH drop would be somewhat larger. The dissolution of pyrite produces Fe in solution that is removed with goethite precipitation.

During the first reaction cycles gypsum is dissolved to its solubility limit leading to significant production of dissolved sulphate and Ca in water. The role of cation exchange is significant during high Ca production. Cation exchange is a partial sink for produced Ca and it releases other major cations into solution. As soon as gypsum resources run out, all major cation and sulphate concentrations drop to a lower level in porewater. Similarly, as sulphate runs out from porewater, the pe values have a chance to find a lower level. The amount of dissolved Fe is a simple function of pe.

The different CEC between the buffer and backfill is evident while comparing the major cation concentrations in water and cation occupation in the exchange sites. In the backfill case (Fig. 7) the fewer exchange sites are filled faster with Ca than in the

539

BUFFER WETTING FRONT

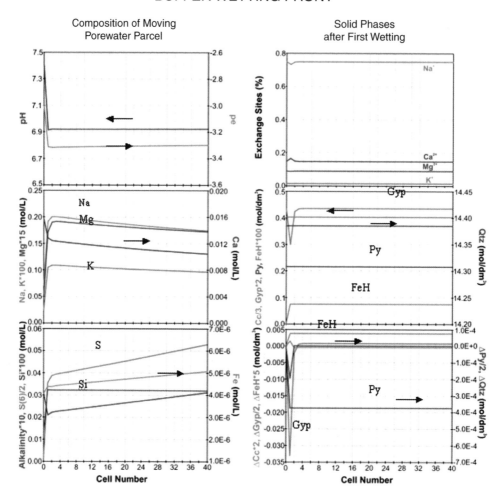

Figure 4. Geochemical evolution of moving porewater parcel and compositions of material volumes after first wetting of buffer. The porewater parcel is transferred 40 times to a new undersaturated buffer cell. The equilibrium temperature assumption is 65°C. Cc = calcite, Gyp = gypsum, Py = pyrite, FeH = goethite.

buffer case (Fig. 6) In the backfill case, the exchange sites have smaller capability to regulate cation concentrations in water. Therefore, Ca concentrations grow faster towards initial external water values than in the buffer case.

The buffer and backfill simulations indicate contradictory behaviour in respect of calcite during the first batch reaction cycles. In the backfill case (Fig. 7) the first cycles dissolve calcite that cause sharp increase in alkalinity and an increase in pH. Due to surface complexation, pH values exhibit a hysteresis effect to the system changes. In the buffer case (Fig.

6) as long as dissolving gypsum keeps the dissolved Ca at a high level the system precipitates calcite. As the Ca levels drop in the solution, the dissolution of calcite becomes possible for a while. In the long run, both buffer and backfill boundaries precipitate small amounts of calcite. This is possible because of relatively high alkalinity and Ca stemming from the external water. Due to temperature rise to 65°C (buffer) and 40°C (backfill) calcite becomes supersaturated in respect to the external water composition (Table 5). This causes the calcite precipitation tendency at the EBS boundaries and concordant pH drop.

540

BACKFILL WETTING FRONT

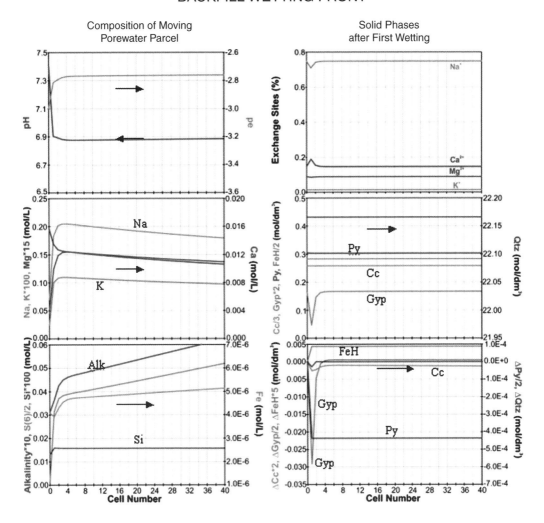

Figure 5. Geochemical evolution of moving porewater parcel and compositions of material volumes after first wetting of backfill. The porewater parcel is transferred 40 times to a new undersaturated backfill cell. The equilibrium temperature assumption is 40°C. Cc = calcite, Gyp = gypsum, Py = pyrite, FeH = goethite.

Considering the overall behaviour of equilibrium minerals, it can be concluded that the 40 batch cycles, in most cases, have negligible effect on the total mineral amounts available in the cells. Notable exception is gypsum that is dissolved away during the simulations. Like gypsum, quartz dissolves to its solubility limit during each batch reaction cycle but the reserve is large. Based on Tables 2 and 4, it can be concluded that the amounts of most silicates are so high that dramatic changes in the silicate reserves are not probable in the conditions studied. However,

silicate dissolution reactions might contribute to resulting porewater compositions.

7 DISCUSSION AND CONCLUSIONS

It seems unlikely that organic matter would be the principal reducer of the entrapped oxygen in the studied systems. As Tables 2 and 4 indicate there is carbon (CH_2O) available. However, if oxygen is consumed in aerobic respiration instead of dissolution

BUFFER BOUNDARY EVOLUTION

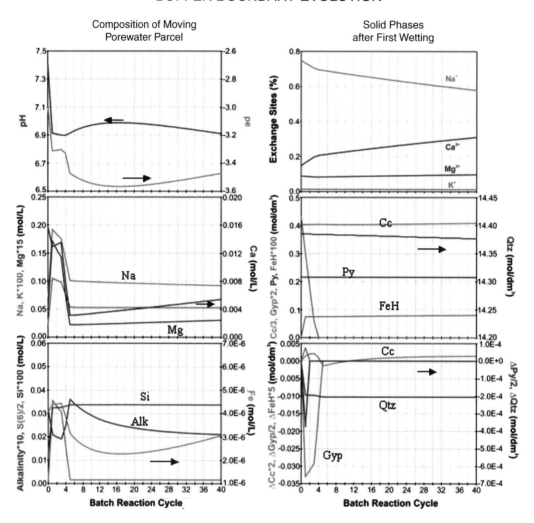

Figure 6. Geochemical evolution in material properties and resulting porewater compositions in an EBS cell volume at the rock-buffer boundary. The cell volume is refilled 40 times with water presented in Table 5. The equilibrium temperature assumption is 65°C. Cc = calcite, Gyp = gypsum, Py = pyrite, FeH = goethite.

of pyrite, alkalinities in the modelled porewaters grow to unreasonable levels. In order to compensate this, at least partially, a strong precipitation of calcite is needed. The source of Ca for strong calcite precipitation appears to be gypsum that is dissolved instantaneously. As a conclusion, the sulphide oxidation has been assumed to be the dominant process consuming oxygen in the calculations. This is in accordance with the results of Pedersen (2000). The amount of viable microbes decreases rapidly during swelling of bentonite and leads to eradication of life in water saturation.

In addition to surface protonation/deprotonation reactions also other cations adsorb on the hydrous oxide surfaces (e.g. Bradbury & Baeyens, 1997). However, considering the current calculations more difficult problem arises from the charged surfaces. Based on the modelling results it seems probable that at least Cl^- and SO_4^{2-} participate to counter-balancing of positively charged surfaces. Adsorption of SO_4^{2-} on the surfaces may increase solubility gypsum and this in turn means higher production of Ca available for cation exchange and calcite reactions. Therefore, the

542

BACKLFILL BOUNDARY EVOLUTION

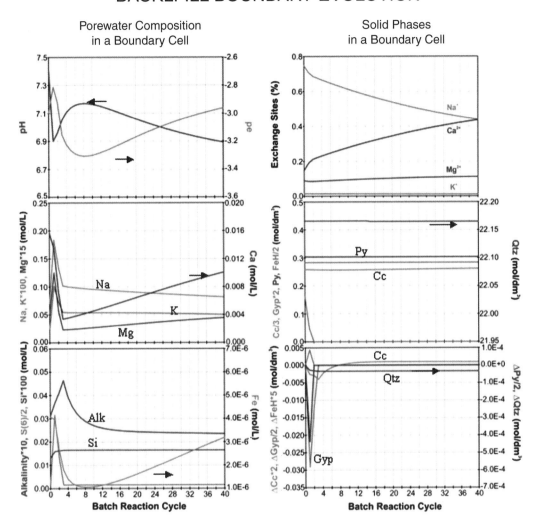

Figure 7. Geochemical evolution in material properties and resulting porewater compositions in an EBS cell volume at the rock-backfill boundary. The cell volume is refilled 40 times with water presented in Table 5. The equilibrium temperature assumption is 40°C. Cc = calcite, Gyp = gypsum, Py = pyrite, FeH = goethite.

trends for dissolved cations and anions presented in Figures 4 and 5 are speculative.

The backfill contains 70% crushed TBM-muck. It was assumed that all cation and surface exchange properties of backfill originate from bentonite. This judgement has its justification but it is undeniable that crushed rock contributes as well at small extent to cation and surface exchange properties of backfill.

The effect of temperature was taken partially into account in the calculations. But for example, no temperature dependency is available for goethite

dissolution/precipitation. However, elevated temperatures may have unexpected effects to the systems because batch reaction times are potentially extensive. Considering the silicates available in the buffer and backfill (Tables 2 and 4) there are several phases susceptible for partial alteration or gradual decomposition (e.g. montmorillonite, illite, amphibole, plagioclase, and K-feldspar).

The evolution of repository as a function of time is handled qualitatively only. At the beginning of the repository operation, batch cycles follow each other more quickly than at the later stages of operation. External water infiltration causes high porewater pressures at the repository boundaries creating strong gradients in the saturated part of EBS, and drives water deeper into EBS (Börgesson & Hernelind, 1999). At the same time, there are high negative porewater pressures in the undersaturated EBS. Therefore, the suction gradients created are possibly strong. Later the gradients begin to diminish and finally vanish. This modifies the later stages of the systems and probably thermal/geochemical gradients will be then the significant driving forces. In any case, near the saturation level porewater moves only by diffusion meaning very slow transport.

Finally, initial water composition utilised in calculations has potentially large significance to simulation results. If the water composition presented in Table 5 changes with time it has effect on the geochemical evolution. In the view of long-term stability considerations these aspects have an important role in safety assessment (e.g. Guimerà et al., 1998; Puigdomenech et al., 2001). Similarly, concrete, cast or injected into walls of repository tunnels and deposition holes, likely alters the infiltrating external water composition (e.g. Huertas et al., 2000). However, these problems are not topical in the view of Prototype experiment.

REFERENCES

Andersson, C & Säfvestad, A (2000) Äspö Hard Rock Laboratory. Compilation of groundwater chemistry data from the Prototype repository. March 1998–June 1999 (unpubl. report). *Svensk Kärnbränslehantering AB (SKB), Stockholm, Sweden. International Technical Document 00-04*: 6 p.

Bradbury, MH & Baeyens, B (1997) A mechanistic description of Ni and Zn sorption on Namontmorillonite. Part II: modelling. *Journal of Contaminant Hydrology 27*: 223–248.

Bradbury, MH & Baeyens, B (2002) Porewater chemistry in compacted re-saturated MX-80 bentonite: Physicochemical characterisation and geochemical modelling. *Paul Scherrer Institute, Villingen, Swizerland, PSI Bericht 02-10*: 41 p.

Bradbury, MH & Baeyens, B (2003) Porewater chemistry in compacted re-saturated MX-80 bentonite. *Journal of Contaminant Hydrology 61*: 329–338.

Börgesson, L & Hernelind, J (1999) Äspö Hard Rock Laboratory. Prototype Repository. Preliminary modelling of the water-saturation phase of the buffer and backfill materials. *Svensk Kärnbränslehantering AB (SKB), Stockholm, Sweden. International Progress Report 00-11*: 99 p.

Bruno, J, Arcos, D & Duro, L (1999) Processes and features affecting the near field hydrochemistry. Groundwater-bentonite interaction. *Svensk Kärnbränslehantering AB (SKB), Stockholm, Sweden. Technical Report 99-29*: 56 p.

Davis, JA & Kent, DB (1990) Surface complexation modeling in aqueous geochemistry. In: (ed. Ribbe, P.H.) Mineral-water interface geochemistry. *Mineralogical Society of America, Washington D.C., U.S.A. Reviews in Mineralogy 23*: 177–260.

Guimerà, J, Duro, L, Jordana, S & Bruno, J (1998) Effects of ice melting and redox front migration in low permeability media. In: Characterization and evaluation of sites for deep geological disposal of radioactive waste in fractured rocks. Proceedings from the 3rd Äspö International Seminar, Oskarshamn, June 10–12, 1998. *Svensk Kärnbränslehantering AB (SKB), Stockholm, Sweden. Technical Report 98-10*: 253–262.

Huertas, F, Farias, J, Griffault, L, Leguey, S, Cuevas, J, Ramírez, S, Vigil de la Villa, R, Cobeña, J, Andrade, C, Alonso, MC, Hidalgo, A, Parneix, JC, Rassineux, F, Bouchet, A, Meunier, A, Decarreau, A, Petit, S & Viellard, P (2000) Effects of cement on clay barrier performance. Ecoclay project. *European Commission, Luxembourg. Nuclear science and technology series EUR 19609*: 140 p.

King, F, Ahonen, L, Taxén, C, Vuorinen, U & Werme, L (2001) Copper corrosion under expected conditions in a deep geologic repository. *Svensk Kärnbränslehantering AB (SKB), Stockholm, Sweden. Technical Report 01-23*: 176 p.

Muurinen, A & Lehikoinen, J (1999) Porewater chemistry in compacted bentonite. Posiva Oy, Eurajoki, Finland. *Posiva Report POSIVA 99-20*: 46 p.

Parkhurst, DL & Appelo, CAJ (1999) User's guide to PHREEQC (Version 2) – A computer program for speciation, batch-reaction, one-dimensional transport, and inverse geochemical calculations. *U.S. Geological Survey, Denver, Colorado. Water-Resources Investigations Report 99-4259*: 312 p.

Patel, S, Dahlström, L-O & Stenberg, L (1997) Äspö Hard Rock Laboratory. Characterisation of the rock mass in the Prototype Repository at Äspö HRL, Stage 1 (unpubl. report). *Svensk Kärnbränslehantering AB (SKB), Stockholm, Sweden. Progress Report HRL-97-24*: 62 p.

Pedersen, K (2000) Microbial processes in radioactive waste disposal. *Svensk Kärnbränslehantering AB (SKB), Stockholm, Sweden. Technical Report 00-04*: 97 p.

Pirhonen, V (1986) Bentoniittitäyteaineen ominaisuudet. Puristetun bentoniitin koostumus ja fysikaaliset ominaisuudet – koe-erä R11–85 (unpubl. report in Finnish). *Teollisuuden Voima Oy, Eurajoki, Finland. TVO/KPA – Turvallisuus ja tekniikka, Work Report 86-03*: 6 p.

Puigdomenech, I (ed.), Gurban, I, Laaksoharju, M, Luukkonen, A, Löfman, J, Pitkänen, P, Rhén, I, Ruotsalainen, P, Smellie, J, Snellman, M, Svensson, U, Tullborg, E-L, Wallin, B, Vuorinen, U & Wikberg, P (2001) Hydrochemical stability of groundwaters surrounding a spent nuclear fuel repository in a 100,000 year perspective. *Posiva Oy, Eurajoki, Finland. Posiva Report POSIVA 2001-20*: 46 p.

Pusch, R & Börgesson, L (2001) Äspö Hard Rock Laboratory. Prototype Repository. Instrumentation of buffer and backfill in Section I. *Svensk Kärnbränslehantering AB (SKB), Stockholm, Sweden. International Progress Report 01-60*: 28 p.

Pusch, R, Muurinen, A, Lehikoinen, J, Bors, J & Eriksen, T (1999) Microstructural and chemical parameters of bentonite as determinants of waste isolation efficiency. *European Commission, Luxembourg. Nuclear science and technology series EUR 18950*: 121 p.

Wersin, P (2003) Geochemical modelling of bentonite porewater in high-level waste repositories. *Journal of Contaminant Hydrology 61*: 405–422.

Wieland, E, Wanner, H, Albinsson, Y, Wersin, P & Karnland, O (1994) A surface chemical model of the bentonite-water interface and its implications for modelling the near field chemistry in repository for spent fuel. *Svensk Kärnbränslehantering AB (SKB), Stockholm, Sweden. Technical Report 94-26*: 64 p.

Advances in Understanding Engineered Clay Barriers – Alonso & Ledesma (eds)
© *2005 Taylor & Francis Group, London, ISBN 04 1536 544 9*

Thermo-hydraulic-chemical-mechanical behaviour of unsaturated bentonite

H.R. Thomas, P.J. Cleall & S.C. Seetharam
Geoenvironmental Research Centre, Cardiff School of Engineering, Cardiff University, UK

ABSTRACT: This paper explores the Thermo-Hydraulic-Chemical-Mechanical behaviour of unsaturated buffer via numerical modelling. Coupled flow relationships accommodate a number of mechanisms: (i) heat transfer; (ii) moisture transfer (iii) dry air flow, and (iv) chemical transfer by both advection and hydrodynamic dispersion, which includes thermal diffusion. In addition chemical species are also assumed to be reactive and subjected to geochemical reactions in the soil-water system. As regards mechanical behaviour, an elasto-plastic stress/strain relationship is implemented. The coupled thermo/hydro/chemical/mechanical model presented is applied to simulate a laboratory-based experiment in order to investigate the transport behaviour of major ions in bentonite pore water. For the simulation of the thermal field it is claimed that the modelling approach is able to represent the mechanisms occurring within the soil. With respect to the flow of moisture it is claimed that the models can represent the important mechanisms occurring in the soil. Considering the behaviour of major ions, the model is able to qualitatively simulate the experimental behaviour. This suggests that the influence of moisture and heat flow on chemical transport has been captured. Therefore, confidence in the coupling of the heat, moisture and chemical variables has been gained. Moreover, since the ions are also undergoing simultaneous geochemical reactions, with the main processes being ion exchange and precipitation/dissolution reactions, confidence in the coupling between the geochemical and the chemical transport model has also been gained.

1 INTRODUCTION

Bentonitic clay has been a popular choice as an engineered barrier in high level nuclear waste repositories in Europe, North America and Japan. This is due to its favourable thermal, hydraulic, mechanical and geochemical properties. Since engineered barriers are expected to provide long term waste isolation, understanding their behaviour in a typical repository condition is of paramount importance in order to ensure a sound repository design. Typically, in the repository conditions, a thermo-hydraulic gradient exists due to the heat generated by the waste canisters and progressive wetting of the initially unsaturated bentonitic clay by the water table in the host formation. Since bentonitic clays are highly expansive, a thermo-hydraulic gradient can cause swelling/shrinkage. Furthermore, the presence of various ions in bentonite pore water, coupled with the thermo-hydraulic gradient will lead to redistribution of ions, which may influence the geochemical properties of bentonitic clays. Therefore, in order to assess the performance of bentonitic clays as engineered barriers, the complex coupling that exists between the thermal, hydraulic, chemical and mechanical behaviour should be considered.

This paper presents an application of the numerical model, COMPASS, to simulate the coupled thermo/hydro/chemical/mechanical behaviour of unsaturated bentonitic clay. The numerical model incorporates a coupled thermo/hydro/chemical/mechanical formulation. Heat transfer formulation includes flow due to conduction, convection, and latent heat of vaporisation (Thomas, 1985). Moisture movement in both liquid and vapour form, caused by pressure gradients is governed by Darcy's law, and vapour transfer due to diffusion is represented by a modified Philip and de Vries approach, (Philip and de Vries, 1957; Ewen and Thomas, 1989). Chemical transport is assumed to be driven by advective, diffusive and dispersive fluxes. It is also assumed that the chemicals are reactive and interact with the soil surface. The reactions are assumed to be instantaneous and hence an equilibrium-based approach is adopted to solve mass action equations. MINTEQA2, a geochemical model is utilised to solve the chemical equilibrium equations (Allison et al., 1991). Finally, the stress/strain behaviour of the soil is represented with an elasto-plastic constitutive relationship and is governed by stress equilibrium equation (Thomas and He, 1995).

The problem considered is a thermo-hydraulic experiment carried out on a highly compacted bentonite. This is one of a series of laboratory based experiments carried out by University of Madrid and CIEMAT as part of the FEBEX project (Cuevas et al., 2002), in order to gain a better understanding of the thermo/hydro/mechanical and geochemical behaviour of the compacted bentonite. The specific aim of the modelling exercise is to simulate the transport behaviour of major soluble ions under the influence of thermo-hydraulic gradient in a deformable soil. The nature of the experiment is such that an accurate prediction of the transport behaviour is only possible by considering a coupled thermo/hydro/chemical/mechanical analysis. This experiment therefore acts as an appropriate benchmark problem to examine the numerical model.

In the following sections, the experimental work and experimental observations are introduced first. The assumptions in the modelling work, material parameters and initial/boundary conditions are then presented. Finally, a comparison of the simulated results against the experimental results is presented for moisture, two major ions (chloride and sodium) and exchange complex. Results for other ions (sulphate, bicarbonate, magnesium and calcium) and deformation behaviour have not been presented due to space constraint.

2 SMALL CELL BENTONITE EXPERIMENT

The experiments carried out by University of Madrid and CIEMAT (Cuevas et al., 2002) have been performed in hermetic cells in which a compacted block of bentonite is hydrated on the top while a thermal gradient is applied from the bottom. A schematic diagram of a typical hermetic cell was shown in Figure 1 (Cuevas et al., 2002). The specimen is 2.5 cm long and 5.0 cm diameter. A temperature gradient of 12°C/cm is applied across the bentonite cell. The temperature at the hydration end was kept constant at 30°C by the circulation water, whilst a constant temperature of 65°C was applied by a heater at the opposite end. Granitic water was supplied at a pressure of 1 MPa.

The influence of the chemistry of the hydration water on the processes achieved by verified by using granitic water, which simulate the conditions of the outer part of the barrier (Cuevas et al., 2002). A series of tests were performed for different durations related to the saturation time – $t_{1/4}$, $t_{1/2}$, t_1, t_2, and t_5, which corresponds to a duration of 4 days, 8 days, 16 days,

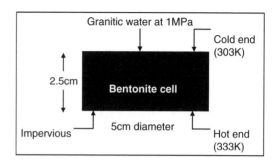

Figure 1. Conceptual diagram of the thermo-hydraulic experiment (Cuevas et al. 2002).

32 days and 80 days respectively. At the end of each experiment the bentonite cells have been sliced horizontally into 5 sections of 0.5 cm each. In these samples, the concentration of major ions and pH in the pore water were determined by 1:4 bentonite-water aqueous extract.

According to Cuevas et al. (2002), the bentonite cell reaches saturation in 16 days. As regards, the behaviour of ions under a thermo-hydraulic gradient the following major processes were observed (Cuevas et al. 2002); (i) advective movement of ions from the cold end towards the hot end during initial stages, (ii) consequent build up ion concentrations at the hot end, (iii) backward diffusion of ions from hot end towards cold end as the bentonite cell reaches saturation, and (iv) equilibrium distribution of ions after saturation. Although these are the typical processes that all ions undergo, ions such as sulphate, sodium, calcium and potassium are also affected by dissolution-precipitation and ion exchange reactions. As stated earlier, the simulated behaviour of only chloride and sodium are discussed in this paper.

3 SIMULATION OF A SMALL CELL HEATING AND HYDRATION EXPERIMENT

The experimentally observed behaviour suggests that there are four variables that interact in a complex manner. These are moisture, heat, chemicals and displacement (due to swelling/compression). Therefore, a simulation of the coupled thermo/hydro/chemical/mechanical (*THCM*) behaviour of the bentonite clay has been undertaken. Several assumptions have been made in the analysis and these are discussed below. As regards the numerical strategy for coupling the transport with the geochemical reactions, a sequential non-iterative approach has been adopted. Although this method is found to yield less accurate results when heterogeneous reactions are involved, the

Table 1. Initial composition of dissolved and precipitated ions in the initial pore water and the exchange complex in the bentonite solid.

Ions/exchange complex (mmol/100 g)	Cl^-	SO_4^{2-}	CO_3^{2-}	Na^+	K^+	Mg^{2+}	Ca^{2+}	CaX_2	MgX_2	NaX	KX
Dissolved	2.23	0.63	0.008	2.66	0.036	0.18	0.21	18.86	15.97	30.76	1.89
Precipitated	0.00	0.403	6.00	0.00	0.00	0.85	5.56	–	–	–	–

efficiency of the computational time has been the motivating factor in choosing this method.

Specific attention is drawn to the initial conditions for the chemicals in the bentonite water, and the initial conditions for ion exchange complexes in the bentonite solid. The initial conditions of the chemicals in the pore water and the exchange complex were determined using MINTEQA2 (Allison et al., 1991). This was necessary, as it is extremely difficult to determine experimentally the pore water composition at very low water contents (Fernandez et al., 2000). Due to limitations of space, the geochemical modelling using MINTEQA2 is not presented in this paper.

3.1 Assumptions

A number of assumptions have been made as regards the chemical processes and deformation behaviour. These are: (i) dissolution-precipitation of only halite, calcite, dolomite and anhydrite is considered, (ii) ion exchange reactions involve only calcium, magnesium, potassium and sodium ions, (iii) ion exchange of hydrogen ions and protonation/deprotonation reactions are ignored, (iv) thermodynamic stability of the clay fraction (at $7 < pH < 11$) is considered, (v) the geochemical reactions are modelled on the basis of an instantaneous equilibrium assumption, (vi) carbon dioxide transport in the gaseous phase has been ignored, and (vii) an elastic model for the deformation behaviour has been assumed, since this is a first attempt at modelling coupled deformation and reactive multicomponent chemical transport.

3.2 Material parameters

There are a series of sets of material parameters required for the purpose of simulation, covering flow, geochemical and deformation analysis. Five of these concern flow analysis and are; (i) moisture retention relationship, (ii) hydraulic conductivity relationship, (iii) thermal conductivity relationship, (iv) diffusion coefficient, and (v) material constants. The sixth set concerns geochemical parameters, whilst the seventh concerns deformation parameters. The information pertaining to the first four sets of material data are based on Huertas et al. (2000) and Martin et al. (2000). The material constants have been extracted from Mayhew and Rogers (1976) and Huertas et al. (2000).

The information regarding the geochemical parameters and the deformation parameters have been taken from Fernandez et al. (2000) and Mitchell (2002) respectively.

3.3 Initial and boundary conditions

The bentonite cell is initially set to be in equilibrium with the laboratory conditions (14% gravimetric water content) (Cuevas et al., 2002). It is assumed that the temperature is in equilibrium with the laboratory conditions and is equal to 298 K. The initial conditions for the chemical components are presented in Table 1. These are based on the geochemical modelling performed using MINTEQA2. For the deformation analysis, an initial uniform isotropic pressure of 200 kPa has been assumed. As regards boundary conditions, the pore water pressure is fixed at the top face (Dirichlet), whilst for temperature both the top and bottom face are fixed (Dirichlet). For the ions, a Cauchy's boundary condition is applied, as the influx of chemicals into the bentonite cell depends on the influx of hydration water. Displacements are restrained in the z direction on the top and bottom edges of the domain (Figure 1). Whilst, on the left and the right edges, displacement in the x direction is restrained.

3.4 Simulation results

The simulation results for the temperature, moisture, chloride and sodium ions, and the exchange capacity are discussed in this section. As stated earlier, (section 2), once the experiment has been completed, the bentonite plug is sliced into 5 sections of 5 mm each and the sample is subjected to an 1:4 aqueous extract analysis in order to determine the concentration of ions. However, due to this method, the final bentonite sample will come into contact with an additional quantity of water, which could affect the equilibrium composition of ions due to further dissolution of trace minerals. This could also trigger ion exchange reactions. Therefore, in order to compare the model results with the experimental ones (1:4 aqueous extract method), it is necessary to re-equilibrate the model results with the additional amount of water and then compare the re-equilibrated

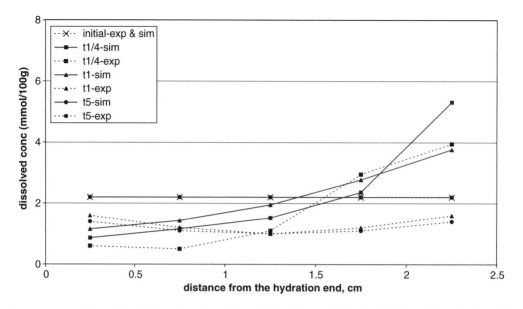

Figure 2. Comparison of simulated and experimental results for Cl⁻ions at 4 days ($t_{1/4}$), 16 days (t_1) and 80 days (t_5) – 1:4 aqueous extract results.

results with the experimental results. Hence in this paper re-equilibrated results are presented for comparison purpose.

3.4.1 *Temperature and moisture distribution*
The simulation results predicted the temperature distribution to nearly reach an equilibrium distribution during the first 1-hour and a final equilibrium distribution after 4 days. This is due to the fact that at the hot end the thermal conductivity drops due to drying. However, as resaturation occurs, the temperature distribution tends towards equilibrium. As regards the moisture field, the model predicted 85% saturation in 8 days and 100% saturation in 16 days. This is in agreement with the experimental observation (Cuevas et al., 2002).

3.4.2 *Chloride (Cl⁻)*
Figure 2 presents a comparison of re-equilibrated model results against the experimental results for the dissolved chloride ion concentration at three different time intervals – $t_{1/4}$, t_1 and t_5. The results presented are the average concentration for each slice (5 mm). From Figure 2, it can be seen that there is a good correlation between the experimental and model results at $t_{1/4}$. However, the model slightly overpredicts the dissolved concentration both at the hot end and the cold end compared to experimental results. Mass balance calculations suggest that the average experimental chloride concentration is less than the initial chloride concentration. This suggests

that some chloride ions are lost from the system. Cuevas et al. (2002) suggested that such a decrease in the mass of the chloride ions in the bentonite cell may be attributed to ion exclusion. Therefore, this leads to an overprediction by the model. At t_1 the model results show an overprediction at the hot end. However, the experimental results show that the chloride concentration has nearly achieved a steady state value. Once again the key feature of the comparison is the discrepancy in the mass balance between the model and the experimental results. At t_5 it is obvious that the chloride concentration has halved as per the experiment. This means half of the mass of chloride ions has been lost from the system due to anion exclusion. This feature has not been captured by the model as the effect of anion exclusion has been ignored.

3.4.3 *Sodium (Na⁺)*
Figure 3 presents a comparison of re-equilibrated model results against the experimental results for the dissolved sodium ion concentration at three different time intervals – $t_{1/4}$, t_1 and t_5. It can be seen that there is a good qualitative agreement between the model and the experimental results at $t_{1/4}$. However, quantitatively the model underpredicts the dissolved sodium ions in the first 2 cm, but there is also a corresponding overprediction in the last 0.5 cm. It is believed that the underprediction at the cold end is due to the instantaneous equilibrium approach implemented, which will result in an instantaneous

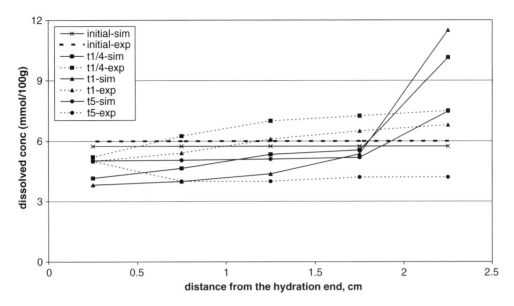

Figure 3. Comparison of simulated and experimental results for Na^+ ions at 4 days ($t_{1/4}$), 16 days (t_1) and 32 days (t_5) – 1:4 aqueous extract results.

dissolution of trace minerals such as calcite, dolomite and anhydrite, which in turn will trigger release of sodium ions from the bentonite solid due to ion exchange equilibrium with other ions. This will lead to an overestimation of the sodium ions in the dissolved phase. As a result these overpredicted sodium ions are carried towards the hot end by the fast rate of hydration. Nevertheless, the model is still able to show a good qualitative trend. At t_1, it can be seen that the model continues to predict a greater accumulation of sodium concentration at the hot end compared to the experimental results, which represents full saturation. The model also shows a corresponding underprediction in the first 2 cm of the bentonite cell. This therefore suggests that the experimental diffusion coefficient used in the simulation may not be high enough to accelerate the backward diffusion of sodium ions so as to reach nearly steady state condition at saturation (t_1).

At t_5, it can be seen that there is a reasonable agreement between the experimental and model results at t_5. However, the model is unable to capture the slight higher concentration at the hydration end as per the experimental behaviour. Looking at the experimental results at the hydration end, it is difficult to explain as to why there is a slightly higher concentration of sodium. Cuevas et al. (2002) suggests some sort of a contamination from the porous sinter material or a specific reaction involving sulphur species in the clay which may be causing this phenomenon. Furthermore, the model overpredicts

the sodium ion concentration by approximately 1 mmol/100 g compared to the experimental results at t_5. This suggests that the model overpredicts the replacement of sodium ions by other cations in the pore water, which could be triggered by the excessive dissolution (due to the instantaneous equilibrium approach) of trace minerals (calcite, dolomite and anhydrite). Nevertheless, it is possible to conclude that the model shows a reasonable correlation with respect to experimental results at t_5.

3.4.4 Cation exchange capacity

Table 2 presents a comparison of simulated (re-equilibrated) initial and final concentration of the cation exchange complex – NaX, KX, MgX$_2$ and CaX$_2$. It can be seen that there are no significant differences between the initial and final concentrations of the exchange complex in the bentonite cell. This suggests that any change in the average concentration of major cations (Na^+, K^+, Mg^{2+} and Ca^{2+}) in the pore water can be attributed to precipitation/dissolution of the trace minerals and the incoming granitic water into the bentonite cell. As regards to experimental data for the final concentration, Cuevas et al. (2002) have presented the results in terms of 1 M ammonium acetate extraction method. However, the exchange complex used in the modelling is based on Na/Mg exchange method. Since, cation exchange capacity is an operationally defined value, the model and the experimental results cannot be compared in this instance. Nevertheless, Cuevas

Table 2. Simulated initial and final concentration of exchange complex in the bentonite cell (corresponding to 1:4 aqueous extract analysis).

Exchange complex	Model results (meq/100 g)	
	Initial concentration	Final concentration
NaX	24.43	25.47
KX	2.18	1.68
MgX_2	16.4	17.26
CaX_2	20.99	19.33

et al. (2002) have confirmed that there is no significant change in the experimental final concentration compared to the initial conditions.

4 CONCLUSIONS

This paper has demonstrated the capability of the numerical model, COMPASS, via the simulation the small cell heating and hydration experiment carried out as part of the FEBEX project. The nature of the experiment demanded that a thermo/hydro/chemical/ mechanical analysis be carried out in order to successfully predict the behaviour of various components in the bentonite experiment. Specifically, the behaviour of three important primary variables were presented, they were; the temperature, moisture and chemicals (Cl^- and Na^+).

Both temperature and moisture results were predicted correctly. As regards to chemical transport, the behaviour of an anion and a cation were presented. The Cl^- and Na^+ ions showed a good qualitative and a reasonable quantitative trend with respect to experimental results. However, one of the key observations was an overprediction of ion concentration at the hot end and an underprediction at the hydration end for all the ions. This was mainly attributed to the initial fast rate of hydration and also due to the instantaneous equilibrium approach implemented in the numerical model, which might have influenced dissolution/precipitation to a large extent. Nevertheless even with this implementation of instantaneous equilibrium approach, the model still showed good qualitative trends in all the cases.

The model also predicted the final concentration of the exchange complex to be close to that of the initial values, suggesting that any change in the average concentration of major cations (Na^+, K^+, Mg^{2+} and Ca^{2+}) in the pore water can be attributed to precipitation/dissolution of the trace minerals and the incoming granitic water into the bentonite cell.

This was in agreement with the experimental results where there was no significant change in the final concentration of ion exchange complexes compared to the initial conditions. Finally, it is acknowledged that several simplifications have been made in the simulation work, and hence it is expected that improved solutions can be achieved by including more complexities into the model. Nevertheless, the model results have been satisfactory and thus have provided improved confidence in the model's ability to simulate the thermo/hydro/chemical/mechanical behaviour of bentonite type soils.

REFERENCES

Allison, J.D., Brown, D.S. and Kevin J. Novo-Gradac (1991). *MINTEQA2 user manual version 3.0*, Environmental Research Laboratory, US EPA.

Cuevas, J., Villar, M.V., Martín, M., Cobeña, J.C. and Leguey, S. (2002). Thermo-hydraulic gradients on bentonite: distribution of soluble salts, microstructure and modification of the hydraulic and mechanical behaviour. Applied Clay Science, vol. 22(1–2), November 2002, pp. 25–38.

Ewen, J. and Thomas, H.R. (1989). *Heating unsaturated medium sand*, Geotechnique, 39, No. 3, 455–470.

Fernandez, A.M., Cuevas, J. and Rivas, P. (2000). *Pore water chemistry of the Febex bentonite*. Scientific Basis for Nuclear Waste Management XXIV, (www.mrs.org/ members/proceedings/nucl_waste/7/7_10.pdf, accessed on 10th January, 2001).

Huertas, F., Fuentes-Cantillana, J.L., Jullien, F., Rivas, P., Linares, J., Farina, P., Ghorechi, M., Jockwer, N., Kickmaier, W., Martinez, M.A., Samper, J., Alonso, E. and Elorza, F.J. (2000). *Full- Scale engineered barriers experiment for a deep geological repository for high-level radioactive waste in crystalline host rock (FEBEX project)*. Euratom.

Langmuir, D. (1997). *Aqueous environmental geochemistry*. Prentice-Hall, Inc., pp. 600.

Mayhew, Y.R. and Rogers, G.F.C. (1976). *Thermodynamic and transport properties of fluids*. 2nd edn, Oxford: Blackwell.

Martin, M., Cuevas, J. and Leguey, S. (2000). Diffusion of soluble salts under a temperature gradient after the hydration of compacted bentonite. Applied clay science, vol.17, pp. 55–70.

Philip J.R. and de Vries D.A. (1957). *Moisture movements in porous materials under temperature gradients*. Transactions, American Geophysical Union, 38, No 2, 222–232.

Thomas, H.R. (1985). *Modelling two-dimensional heat and moisture transfer in unsaturated soils, including gravity effects*. International Journal of Analytical Methods in Geomechanics, vol.9, 573–588.

Thomas, H.R. and He, Y. (1995). Analysis of coupled heat, moisture and air transfer in a deformable unsaturated soil. Geotechnique, vol. 45(4), pp. 677–689.

Advances in Understanding Engineered Clay Barriers – Alonso & Ledesma (eds)
© *2005 Taylor & Francis Group, London, ISBN 04 1536 544 9*

Inverse hydrochemical modelling of aqueous extract experiments for the estimation of FEBEX bentonite pore water chemistry

J. Samper, A. Vázquez & L. Montenegro
Civil Engineering School, University of La Coruña, Spain

ABSTRACT: A numerical method is presented for the interpretation of aqueous extract experiments. The numerical model which is solved with CORE^{2D} accounts for: 1) dilution, 2) aqueous complexation; 3) acid-base reactions 4) kinetic mineral dissolution and precipitation processes; 5) equilibrium mineral dissolution and precipitation processes; 6) cation exchange and 7) gas dissolution/ex-solution. The methodology has been successfully applied to model aqueous extract experiments performed on FEBEX bentonite samples with different durations (from 1 to 30 days) and S/L ratios ranging from 1:1 to 1:16.

1 INTRODUCTION

Compacted bentonite is foreseen as backfill and sealing material in the reference concept of a Deep Geological Disposal (DGD) of High-Level Radioactive Waste (HLRW). This material constitutes one of the engineered barriers in the DGD concept. ENRESA launched in 1995 with the collaboration of other European organizations the FEBEX project (Full-Scale Engineered Barriers Experiment; Huertas et al., 2000) with the aims of testing the feasibility of this reference concept and developing the knowledge and conceptual and numerical models for the thermal, hydrodynamic, mechanical and geochemical (THMG) processes expected to take place in engineered clay barriers. FEBEX consists of an "in situ" test performed under natural conditions and at full scale in a gallery excavated in a granitic massif at Grimsel (Switzerland), an almost full-scale "mock-up" test operated at the Ciemat facilities in Madrid (Spain), a series of laboratory tests to complement the information from the two large-scale tests and development of thermo-hydro-mechanical and thermo-hydro-geochemical models (Samper et al., 2004, in this volume). One of the objectives of the project is the characterisation and study of the evolution of the bentonite thermal, hydrodynamic, mechanical and geochemical properties under real conditions.

There are numerous experimental studies of water-clay interactions (Fritz & Kam, 1985; Grambow et al., 1985; Wanner, 1986; Wanner et al., 1992; Beaucaire et al., 1995; Sasaki et al., 1995; Cuevas et al., 1997; Snellman et al., 1987; Cranga et al., 1998; Muurinen & Lehikoinen, 1998; Rivas et al., 1998). Geochemical modelling of pore water in clays is also an active

field of work (Curti, 1993; Wieland et al., 1994; Bradbury & Baeyens, 1998, 2002, 2003; Pearson et al., 1998, Fernández et al., 1999, 2001, 2004, Muurinen & Lehikoinen, 1999, Pusch et al., 1999, Vázquez et al., 2001, Fernández and Rivas, 2003). These studies have concluded that the evolution of clay pore water chemistry is controlled by cation exchange and the dissolution/precipitation of soluble accessory minerals of the clay and depends greatly on the solid to liquid ratio, S/L (Fernández et al., 1999, 2001; Samper et al., 2003a).

Geochemical characterization of clays can be performed in situ (field techniques) and ex situ by means of laboratory techniques. Obtaining reliable data for clay pore water chemistry is a difficult task. Squeezing at high pressures and aqueous extracts have been used to obtain pore water from compacted FEBEX bentonite. These two ex situ techniques may alter the water-clay system in several ways and introduce sampling artefacts in measured data. Squeezing at high pressures may induce oxidation and dissolution of accessory minerals of the bentonite, outgassing of $CO_{2(g)}$, and chemical fractionation (Sachi and Michelot, 2000, Pearson et al., 2003). Furthermore, squeezing does not allow to extract pore water from bentonites with water contents below 20% Cuevas et al., 1997, Fernández et al., 1999, 2001). Therefore, the squeezing method cannot be used for characterizing bentonite at ambient conditions with a water content of 14%, for which it is necessary to resort to the aqueous extract technique. The aqueous extract technique provides a method to quantify the total content of soluble salts (Parker, 1921) because it works with a low solid to liquid ratio, S/L. An 1:R aqueous extract test consists on adding to a mass M of clay sample a mass of distilled water equal to R times M. The mixture of solid and water is allowed to react during a period of time (usually 2 days). After that, liquid and solid phases are separated by centrifugation. Chemical analyses are performed on the supernatant solution. It should be pointed out that the ratio S/L is equal to the aqueous extract ratio 1:R when the clay sample is fully dry. If the clay sample has an initial gravimetric water content, w_0, then the ratio S/L is given by:

$$\frac{S}{L} = \frac{1}{w_0 + R(1 + w_0)} \tag{1}$$

During the extraction, mineral dissolution of various soluble minerals (halite, carbonates and gypsum) as well as cation exchange processes take place which affect the concentrations of dissolved species (Bradbury and Baeyens, 1998, Fernández et al., 2001). These processes complicate the interpretation of aqueous extract data.

Ciemat carried out aqueous extract tests on intact FEBEX bentonite samples at the following 1:R ratios: 1:1, 1:2, 1:4, 1:8, 1:10 and 1:16 with distilled and granitic water (Fernández et al., 1999, 2001). Tests were also performed with different reaction times ranging from 2 to 30 days for a 1:4 ratio in order to evaluate the kinetics of mineral dissolution/precipitation.

Since laboratory techniques used to obtain bentonite pore water may alter the geochemical system and introduce sampling artefacts, indirect methods based on hydrogeochemical modelling are needed in order to infer the chemical composition of bentonite pore water. The University of La Coruña has developed a methodology for the interpretation of aqueous extract tests. Bentonite initial pore water is obtained by inverse hydrogeochemical modelling using a reactive transport code, CORE^{2D} (Samper et al., 2000; Samper et al., 2003b). Initial compositions can be estimated automatically by solving the inverse problem (Montenegro et al., 2003) or by trial and error. Here we present a methodology to interpret the different aqueous extract tests performed by Ciemat. Results obtained from these tests are compared to values derived from squeezing tests.

2 METHODOLOGY

Bentonite initial pore water is obtained by trial and error hydrogeochemical modelling using the reactive transport code, CORE^{2D} (Samper et al., 2000). Starting with the initial chemical composition of bentonite pore water at a water content of 14%, distilled water is added until the appropriate S/L ratio of the test is attained.

The numerical model considers a sample of unsaturated bentonite of $0.1 \times 0.1 \times 0.1$ dm^3. The mesh has 5 nodes and 4 elements. Distilled water is added at all nodes.

2.1 Conceptual geochemical model

Our methodology requires the definition of a conceptual geochemical model (CGM). The CGM is defined in terms of: 1) relevant geochemical processes, and, 2) chemical composition of the pore water at a given water content (usually at dry conditions of 14%).

Different conceptual geochemical models can be proposed depending on the type of data used to construct them: squeezing and/or aqueous extract data. Samper et al. (2001) relied on squeezing data for THG modelling of FEBEX experiments. Fernández et al. (1999, 2001) used aqueous extract data to derive another geochemical model of FEBEX bentonite. They tested the model with squeezing data. The model showed significant discrepancies when compared to sulfate and sodium concentrations of squeezed waters. The conceptual model of Samper et al. (2001) will be denoted as CGM-0 while that of Fernández et al. (1999, 2001) will be referred to as CGM-Ciemat.

Here we evaluate the possibility of developing a single conceptual model that can reproduce simultaneously squeezing and aqueous extract data. First we start by developing a conceptual model by using only aqueous extract data. This is denoted as model CGM-1. Modifications are introduced in this model in order to fit not only aqueous extract but also squeezing data. This modified model is termed CGM-2.

Therefore, up to four conceptual geochemical models can be proposed for the FEBEX bentonite which include CGM-Ciemat of Fernández et al. (1999, 2001) and models CGM-0 (Samper et al., 2001), CGM-1 and CGM-2.

Model CGM-1 considers the following geochemical processes: aqueous complexation, acid-base, mineral dissolution/precipitation with chemical equilibrium or kinetics, cation exchange, and gas dissolution/ex-solution. Table 1 shows chemical species considered in model CGM-1 used to simulate these geochemical processes.

Bidistilled water added to bentonite samples has a pH of 7 and the chemical concentration of all species is 10^{-20} mol/l. CO$_{2(g)}$ is fixed at atmospheric pressure ($10^{-3.5}$ bar). Table 2 contains the kinetic laws and specific surface used for calcite (Plummer et al., 1979) and disordered dolomite (Ayora et al., 1994). Kinetic rate constants are listed in Table 3. Gypsum and chalcedony are assumed at chemical equilibrium.

Table 1. Chemical species considered in model CGM-1. Number of species are given within parentheses.

Primary species (10)	H_2O, H^+, Ca^{2+}, Mg^{2+}, Na^+, K^+, Cl^-, SO_4^{2-}, HCO_3^-, $SiO_{2(aq)}$
Aqueous complexes (17)	OH^-, $CaSO_{4(aq)}$, $CaCl^+$, $MgCl^+$, $NaCl_{(aq)}$, $MgHCO_3^+$, $NaHCO_{3(aq)}$, $CaHCO_3^+$, $MgCO_{3(aq)}$, $CaCO_{3(aq)}$, $CO_{2(aq)}$, CO_3^{-2}, KSO_4^-, $MgSO_{4(aq)}$, $NaSO_4^-$, $H_2SiO_4^{2-}$, $HSiO_3^-$
Minerals (4)	Calcite and Dolomite (kinetics) Chalcedony and Gypsum/Anhydrite (equilibrium)
Exchangeable cations (5)	H^+, Ca^{2+}, Mg^{2+}, Na^+, K^+ (selectivity coefficients)
Gas (1)	$CO_{2(g)}$

Table 2. Kinetic laws used for calcite (Plummer et al., 1979) and disorder dolomite (Ayora et al., 1994) Here k is the kinetic constant (see Table 3) and a is activity of the chemical species.

Mineral	Kinetic law	Specific surface (m^2/m^3)
Calcite	$R_{dis} = k_1 \cdot a_{H+} + k_2 \cdot a_{H2CO_3^-} + k_3 \cdot a_{H2O} - k_4 \cdot a_{Ca2+} \cdot a_{HCO3^-}$	0.1
Disordered dolomite	$R_{dis} = k$	0.5

Table 3. Kinetic constants for calcite and dolomite.

	Calcite ($mol/m^2 s$)	Disordered dolomite ($mol/m^2 s$)
k_1	$8.91 \cdot 10^{-5}$	$2.2 \cdot 10^{-6}$
k_2	$5.01 \cdot 10^{-8}$	–
k_3	$6.46 \cdot 10^{-11}$	–
k_4	$1.86 \cdot 10^{-02}$	–

Table 4. Selectivity coefficients and cation exchange capacity used in the model.

	Selectivity coefficients	Exchangeable cations ($meq/100\,g$)
NaX	1	26.95
KX	$1.43 \cdot 10^{-1}$	2.29
Mg_2X	$4.40 \cdot 10^{-1}$	33.15
Ca_2X	$5.63 \cdot 10^{-1}$	33.10
HX	$8.40 \cdot 10^{-6}$	1

Cation exchange has been modelled using selectivity coefficients calculated according the mass action law. The Gaines-Thomas convention is used for cation exchange. Experimental data for cation exchange capacity (CEC) were taken from Fernández et al. (2003). Table 4 shows the selectivity coefficients and CEC values.

Table 5 shows the initial chemical composition of FEBEX bentonite pore water at 14% water content inferred with model CGM-1. This chemical composition was derived from the hydrogeochemical

Table 5. Chemical composition of the FEBEX bentonite pore water at 14% water content inferred by hydrogeochemical modelling.

Species	Concentrations (mg/l)
Cl^-	5502
SO_4^{2-}	2976
HCO_3^-	17
SiO_2	16.8
Ca^{2+}	840
Mg^{2+}	535
Na^+	2944
K^+	89.7
pH	7.62

Table 6. List of experiments carried out by Ciemat (Fernández et al., 1999, 2001).

Experiments	S/L ratio	Final water content (%)	Time (days)
Squeezing	3.3:1	30.0	24
	3.8:1	26.5	34
	4.2:1	23.8	21
Aqueous extract at different 1:R ratios	1:1.28	128	2
	1:2.42	242	2
	1:4.70	470	2
	1:9.25	925	2
	1:11.53	1153	2
	1:1.18.4	1840	2
Aqueous extract at 1:4 and different durations	1:4.7	470	2, 4, 6, 10, 30

modelling of different aqueous extracts tests the characteristics of which are listed in Table 6 (Fernández et al., 1999, 2001).

3 NUMERICAL RESULTS AND DISCUSSION

3.1 Aqueous extracts of different durations

Figures 1 to 4 show the results of aqueous extracts 1:4.7 ratio of different durations. Lines represent numerical results while symbols are used for

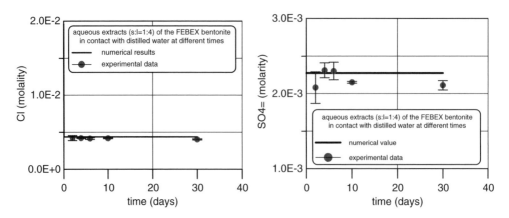

Figure 1.   Computed and measured chloride (left) and sulphate (right) data in aqueous extract tests of FEBEX bentonite in contact with distilled water at different times.

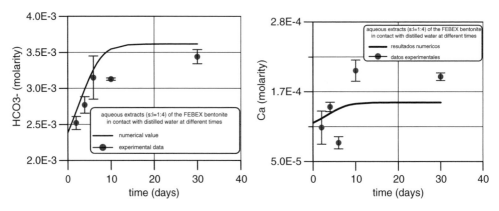

Figure 2.   Computed and measured of bicarbonate (left) and calcium (right) data in aqueous extract tests of FEBEX bentonite in contact with distilled water at different times.

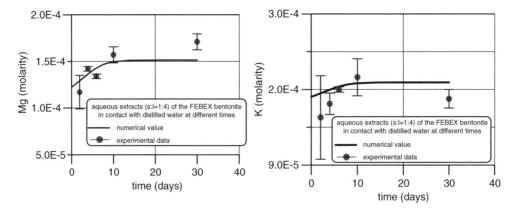

Figure 3.   Computed and measured of magnesium (left) and potassium (right) data in aqueous extract tests of FEBEX bentonite in contact with distilled water at different times.

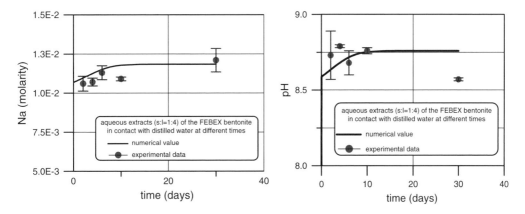

Figure 4. Computed and measured of sodium (left) and pH (right) data in aqueous extract tests of FEBEX bentonite in contact with distilled water at different times.

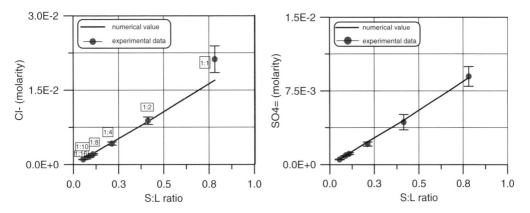

Figure 5. Computed and measured aqueous extract data at different S/L ratios for Cl^- (left) and SO_4^{-2} (right). Values of 1:R are given within the squares.

experimental data. Error bars indicate the experimental measurement error. The model assumes that distilled water is added in the first 20 minutes during which a strong dilution is produced for all chemical species. Such dilution is shown for chloride in Figure 1. This figure indicates clearly that chloride and sulphate are not affected by kinetic processes. Numerical results reproduce chloride data and pass through the scattered sulphate data. Computed bicarbonate concentrations (Figure 2) show an increase within the first 6 days due to the dissolution of calcite and disordered dolomite. Numerical results agree with experimental data. Ca^{2+} data also show an increase with time. In this case, the model under-predicts the steady values. Mg^{2+}, Na^+ and K^+ show also increasing trends associated to mineral dissolution and cation exchange. Figure 4 shows that computed pH fit experimental data except for the value at 30 days.

3.2 Aqueous extracts at different S/L ratios

Figures 5 to 7 show computed results of aqueous extracts tests for different S/L ratios with distilled water. Computed values of chloride in Figure 5 follow a straight line, indicating that this species is conservative. The model reproduces all measured data except that of 1:1. Sulphate also shows a conservative behaviour (Figure 5). Ca^{+2}, Mg^{+2}, Na^+ and K^+ are affected by mineral dissolution/precipitation and cation exchange. Therefore, their trends deviate from straight lines of conservative species. Computed values for these cations match measured data (Figures 6 and 7).

Figure 8 illustrates that computed bicarbonate concentrations have a trend opposite to that of measured data, possibly due to sample $CO_{2(g)}$ degassing during aqueous extract tests. Degassing causes calcite precipitation and a decrease of

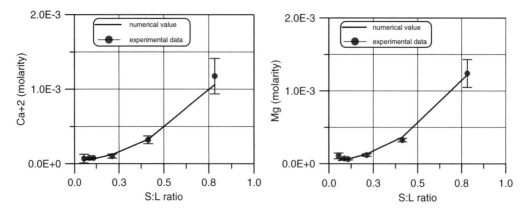

Figure 6.   Computed and measured aqueous extract data at different S/L ratios for Ca^{+2} (left) and Mg^{+2} (right).

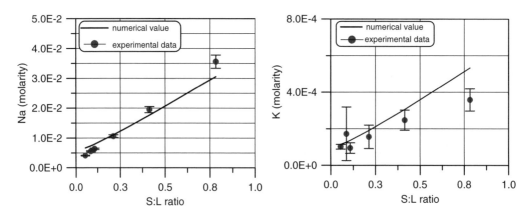

Figure 7.   Computed and measured aqueous extract data at different S/L ratios for Na^+ (left) and K^+ (right).

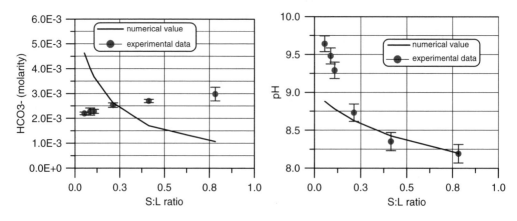

Figure 8.   Computed and measured aqueous extract data at different S/L ratios for HCO_3^- (left) and pH (right).

bicarbonate concentration. It should be pointed out that the model considers a fixed pressure of $CO_{2(g)}$ during the test and, therefore, cannot reproduce measured bicarbonate data. Figure 8 also shows that calculated pH agrees well with measured data at high S/L ratios, however, at low S/L ratios computed values deviate from experimental data, possibly also due to $CO_{2(g)}$ degassing.

Figure 9 shows the sensitivity of computed bicarbonate to changes in $CO_{2(g)}$ pressure. Measured data are between the curves corresponding to $CO_{2(g)}$ pressures of $10^{-4.4}$ and 10^{-3} bar. These results tend to indicate that measured bicarbonate at different S/L ratios could be reproduced by using different gas pressures for each S/L ratio.

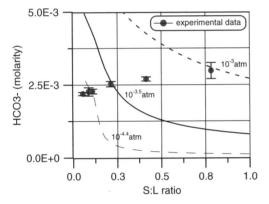

Figure 9. Sensitivity of computed HCO_3^- concentrations to changes in specified $CO_{2(g)}$ pressures.

4 COMPARISON WITH OTHER MODELS

When model CGM-1, which is based on aqueous extract data, is used to model squeezing data at water content of 23.5% and 30% it is found that computed sulphate concentrations largely overpredict measured squeezing data (see Figure 10) due possibly to an excessive dissolution of gypsum.

In order to overcome this discrepancy, a modified version of model CGM-1 is postulated in which it is assumed that no initial gypsum is present in the clay sample at water content of 14% but its precipitation is allowed. The rest of the chemical species considered in the model are the same than in model CGM-1 (see Table 1). In this modified model which is denoted CGM-2 the concentration of sulphate at 14% water content is assumed to be smaller than that of model CGM-1. Optimum concentrations of other species at this water content were derived by trial and error hydrogeochemical modelling using $CORE^{2D}$ (Samper et al., 2000). Chloride, calcium and magnesium concentrations at 14% water content for model CGM-2 are greater than those of CGM-1.

Figures 10 to 13 show the comparison of concentrations computed with models CGM-1 and CGM-2 at different S/L ratios and their fit to experimental data. Figure 10 shows the results for sulphate. While model CGM-1, which considers the presence of gypsum in the system, is able to fit aqueous extract data it fails to match measured squeezing data. On the other hand, computed concentrations of sulphate with model CGM-2, which does not consider gypsum in the system (although allows for its precipitation) match measured squeezing data but largely underpredict measured aqueous extract data. For chloride both models give similar

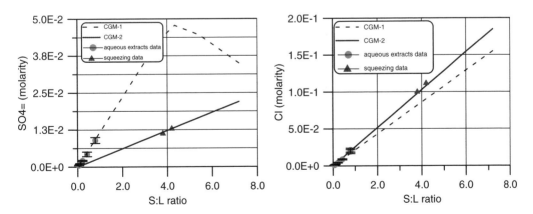

Figure 10. Computed (lines) and measured (symbols) squeezing and aqueous extract concentrations of sulphate (left) and Cl^- (right) at different S/L ratios for models CGM-1 and CGM-2.

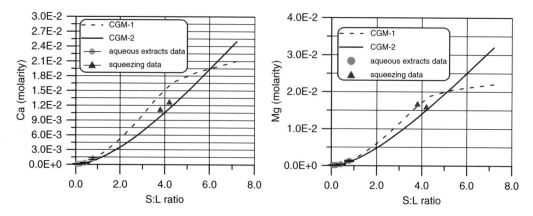

Figure 11. Computed (lines) and measured (symbols) squeezing and aqueous extract concentrations of Ca^{+2} (left) and Mg^{+2} (right) at different S/L ratios for models CGM-1 and CGM-2.

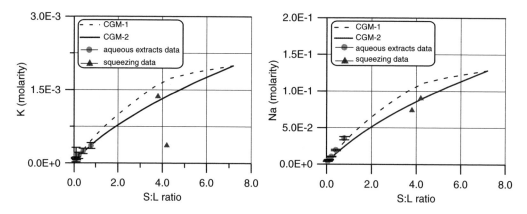

Figure 12. Computed (lines) and measured (symbols) squeezing and aqueous extract concentrations of K^+ (left) and Na^+ (right) at different S/L ratios for models CGM-1 and CGM-2.

results, although model CGM-1 underpredicts squeezing data (Figure 10).

In general, model CGM-2 fits better aqueous extract and squeezing calcium and sodium data than model CGM-1 (Figures 11 and 12). The curves of computed calcium concentrations for models CGM-1 and CGM-2 cross each other at a ratio of S/L = 6 (Figure 11) which corresponds to 16.6% water content. Magnesium shows a similar pattern. Figure 12 illustrates clearly that even though the sodium initial concentration (at 14% water content) is similar in both models, the evolutions of sodium concentration with decreasing S/L ratios predicted by the two models are different.

Figure 13 illustrates that both models predict a decrease of bicarbonate concentration with increasing S/L ratio. Such a decrease is coherent with both squeezing and aqueous extract data.

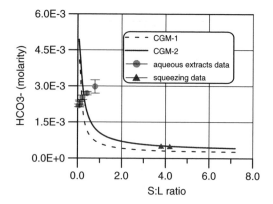

Figure 13. Computed (lines) and measured (symbols) squeezing and aqueous extract concentrations of pH at different S/L ratios for models CGM-1 and CGM-2.

560

Table 7. Chemical species considered in model CGM-0. Number of species in parentheses.

Primary species (10)	H_2O, H^+, Ca^{2+}, Mg^{2+}, Na^+, K^+, Cl^-, SO_4^{2-}, HCO_3^-, $SiO_{2(aq)}$
Aqueous complexes (17)	OH^-, $CaSO_{4(aq)}$, $CaCl^+$, $MgCl^+$, $NaCl_{(aq)}$, $MgHCO_3^+$, $NaHCO_{3(aq)}$, $CaHCO_3^+$, $MgCO_{3(aq)}$, $CaCO_{3(aq)}$, $CO_{2(aq)}$, CO_3^{-2}, KSO_4^-, $MgSO_{4(aq)}$, $NaSO_4^-$, $H_2SiO_4^{2-}$, $HSiO_3^-$
Minerals (3)	Calcite and Chalcedony (equilibrium), Gypsum (only precipitation)
Exchangeable cations (5)	H^+, Ca^{2+}, Mg^{2+}, Na^+, K^+ (selectivity coefficients)

Table 8. Inferred chemical composition of FEBEX bentonite pore water at 14% water content obtained with different conceptual geochemical models. (values in mol/l, except pH).

	CGM-0	CGM-1	CGM-2	CGM-Ciemat
Cl^-	$1.8 \cdot 10^{-1}$	$1.55 \cdot 10^{-1}$	$1.85 \cdot 10^{-1}$	$1.6 \cdot 10^{-1}$
SO_4^{2-}	$2.1 \cdot 10^{-2}$	$3.1 \cdot 10^{-2}$	$2.2 \cdot 10^{-2}$	$3.2 \cdot 10^{-2}$
HCO_3^-	$5.7 \cdot 10^{-4}$	$3.9 \cdot 10^{-4}$	$4.1 \cdot 10^{-4}$	$4.1 \cdot 10^{-4}$
Ca^{2+}	$1.1 \cdot 10^{-2}$	$2.1 \cdot 10^{-2}$	$2.5 \cdot 10^{-2}$	$2.2 \cdot 10^{-2}$
Mg^2	$1.4 \cdot 10^{-2}$	$2.2 \cdot 10^{-2}$	$3.2 \cdot 10^{-2}$	$2.3 \cdot 10^{-2}$
Na^+	$1.8 \cdot 10^{-1}$	$1.28 \cdot 10^{-1}$	$1.28 \cdot 10^{-1}$	$1.3 \cdot 10^{-1}$
K^+	$1.6 \cdot 10^{-3}$	$2.3 \cdot 10^{-3}$	$2.3 \cdot 10^{-3}$	$1.7 \cdot 10^{-3}$
pH	7.86	7.62	7.62	7.72

Table 9. Inferred chemical composition of FEBEX bentonite pore water at $w = 23.8\%$ obtained with different conceptual geochemical models. (values in mol/l, except pH).

	CGM-0	CGM-1	CGM-2	CGM-Ciemat
Cl^-	$1.12 \cdot 10^{-1}$	$1.10 \cdot 10^{-1}$	$1.39 \cdot 10^{-1}$	$1.13 \cdot 10^{-1}$
SO_4^{2-}	$1.27 \cdot 10^{-2}$	$4.37 \cdot 10^{-2}$	$1.65 \cdot 10^{-2}$	$4.30 \cdot 10^{-2}$
HCO_3^-	$6.45 \cdot 10^{-4}$	$4.15 \cdot 10^{-4}$	$4.55 \cdot 10^{-4}$	$1.54 \cdot 10^{-3}$
Ca^{2+}	$6.34 \cdot 10^{-3}$	$1.92 \cdot 10^{-2}$	$1.66 \cdot 10^{-2}$	$1.54 \cdot 10^{-2}$
Mg^2	$8.82 \cdot 10^{-3}$	$2.14 \cdot 10^{-2}$	$2.16 \cdot 10^{-2}$	$1.94 \cdot 10^{-2}$
Na^+	$1.14 \cdot 10^{-1}$	$1.20 \cdot 10^{-1}$	$1.06 \cdot 10^{-1}$	$1.29 \cdot 10^{-1}$
K^+	$1.04 \cdot 10^{-3}$	$2.15 \cdot 10^{-3}$	$1.64 \cdot 10^{-3}$	$1.24 \cdot 10^{-3}$
pH	7.94	7.65	7.69	7.43

Previous to the development of models CGM-1 and CGM-2, Samper et al. (2001) used a conceptual geochemical model CGM-0 based on squeezing data. Starting from a composition corresponding to saturated bentonite (water content of 26–30%), they modelled the decrease of water content with EQ3/6 (Wolery, 1992). Model CGM-0 does not consider dolomite and assumes chemical equilibrium for all mineral phases in the system (calcite, chalcedony and gypsum). Geochemical processes consider in this conceptual model are the same as those of models CGM-1 and CGM-2, except for gas dissolution/ex-solution (see Table 7).

Tables 8 and 9 show the chemical compositions of FEBEX bentonite pore water at ambient (14% water content) and saturated conditions (23.8% water content) corresponding to models CGM-0, CGM-1 and CGM-2. These tables contain also the values calculated by Fernández et al. (1999, 2001) with model CGM-Ciemat. In general, inferred concentrations at 14% water content with different models are similar. At saturation, 23.8% water content, there are discrepancies which are significant for sulphate, bicarbonate and calcium.

5 CONCLUSIONS

Inasmuch as laboratory techniques used to obtain the chemical composition of FEBEX bentonite pore water may alter the geochemical system and introduce sampling artefacts, indirect methods based on hydrogeochemical modelling have been developed to infer the chemical composition of the bentonite pore water at ambient conditions (14% water content). A methodology has been presented for the interpretation of aqueous extract tests which is based on solving the inverse hydrogeochemical model and accounting for the geochemical processes suffered by the sample during the test by means of a reactive transport code, $CORE^{2D}$ (Samper et al., 2000). A conceptual geochemical model, CGM-1, has been derived from aqueous extract data at various durations and 1:R ratios. This model accounts for kinetic mineral dissolution and assumes a constant and fixed $CO_{2(g)}$ pressure.

Model CGM-1 derived from aqueous extract data reproduces adequately measured data of tests performed at different durations (up to 30 days). Most chemical species reach steady equilibrium before 10 days. Therefore, calcite and dolomite kinetics are rather fast. For the most part, this model CGM-1 also fits measured data at different S/L ratios. However, it fails to reproduce the trend of measured bicarbonate and pH data, possibly due to changes in $CO_{2(g)}$ pressure which are not considered in the model. Computed results with model CGM-1 show slight deviations for the 1:1 test for chloride, sodium and potassium. In general this model cannot fit squeezing data.

Starting from model CGM-1, a modified model CGM-2 has been derived in order to fit simultaneously squeezing and aqueous extract data. Model CGM-2 does not consider initial gypsum in the system but its precipitation is allowed. This model CGM-2 fits simultaneously both squeezing and aqueous extract measured concentrations for most chemical species, although it presents some clear deviations for sulphate.

Measured data tend to reject the hypothesis of a constant pressure of $CO_{2(g)}$ during the tests. This hypothesis will have be revised in future works.

Models presented here assume that all water content (14% as initial) is available for geochemical reactions. However, such water content may include some internal water which is not amenable for chemical processes (Bradbury and Baeyens, 2002, 2003, Fernández and Rivas, 2003, Fernández et al., 2004). Work is in progress in collaboration with CIEMAT to explore conceptual geochemical models in which a distinction is made between internal (interlayer) and external (double layer water and pore water) water for the purpose of hydrogeochemical modelling. Other potential improvements for the hydrogeochemical model include the consideration of proton sorption/desorption on clay edges which has been found relevant for the MX-80 bentonite by Bradbury & Baeyens (2003).

ACKNOWLEDGMENTS

This work is part of FEBEX project and has been funded by ENRESA and the European Union through the contracts FI4W-CT95-0006 and FIKW-CT-2000-0016 of the Nuclear Fission Safety programme.

REFERENCES

Ayora, C., Taberner, C. & Samper, J. (1994). Modelización de transporte reactivo: Aplicación a la dedolomitización. Estudios Geológicos, 50, 397–410.

Beaucaire, C., Toulhoat, P. & Pitsch, H. (1995). Chemical characterization and geochemical modelling approach for determining porewater chemistries in argillaceous rocks. Water-Rock Interaction, Balkema, 779–782.

Bradbury, M. & Baeyens, B. (1998). A physicochemical characterization and geochemical modelling approach for determining porewater chemistries in argillaceous rocks. Geochimica et Cosmochimica Acta, 62 (5), 783–795.

Bradbury, M. & Baeyens, B. (2002). Porewater chemistry in compacted re-saturated MX-80 bentonite: Physico-chemical characterisation and geochemical modelling. PSI Report 02–10.

Bradbury, M. & Baeyens, B. (2003). Porewater chemistry in compacted re-saturated MX-80 bentonite. Journal of Contaminant Hydrology, 61, 329–338.

Cranga, M., Trotignon, L., Martial, C. & Castelier, E. (1998). Simulation of the evolution of a Clay Engineered Barrier by interations with granitic groundwater: Dynamics and Characteristic Timescales. Mat. Res. Soc. Symp. Proc., Vol. 506, 629–636.

Cuevas, J., Villar, M.V., Fernández, A.M., Gómez, P. & Martín, P.L. (1997). Pore waters extracted from compacted bentonite subjected to simultaneous heating and hydration. Applied Geochemistry, 12, 473–481.

Curti, E. (1993). Modelling bentonite porewaters for the Swiss High-level Radioactive Waste Repository. Nagra Technical Report 93–45.

Huertas, et al. (2000). Full-scale engineered barriers experiment for a deep geological repository in crystalline host rock (FEBEX Project). European Commission. EUR 19147 EN.

Fernández, A.Mª. & Rivas, P. (2003). Barrera de arcilla. Propiedades y estabilidad geoquímica. ENRESA-2003. GTI-5: Evolución Geoquímica. Versión 1. CIEMAT/DIAE/54460/4/03.

Fernández, A.Mª., Rivas, P. & Cuevas, J. (1999). Estudio del agua intersticial de la arcilla FEBEX. Ciemat Technical Report 70-IMA-L-0-44.

Fernández, A.Mª., Cuevas, J. & Rivas, P. (2001). Pore water chemistry of the FEBEX bentonite. Mat. Res. Soc. Symp. Proc., Vol. 603, 573–588.

Fernández, A.Mª., Baeyens, B., Bradbury, M. & Rivas, P. (2004). Analysis of the porewater chemical composition of a Spanish compacted bentonite used in an engineered barrier. Physics and Chemistry of the Earth, 29, 105–118.

Fritz, B. & Kam, M. (1985). Chemical interactions between the bentonite and the natural solutions from the granite near a repository for spent nuclear fuel. SKB Technical Report 85–10.

Grambow, B., Hermansson, H.P., Björner, I.K. & Werme, L. (1985). Glass/water reaction with and without bentonite present-experiment and model. Mat. Res. Soc. Symp. Proc., Vol. 50, 187–194.

Montenegro, L., Samper, J., Zheng, L., Vázquez, A. & Yang, Ch. (2003). Modelización hidrogeoquímica de la arcilla española de referencia. University of La Coruña Technical Report. University of La Coruña, Spain.

Muurinen, A. & Lehikoinen, J. (1998). Evolution of the porewater chemistry in compacted bentonite. Mat. Res. Soc. Symp. Proc., Vol. 506, 415–422.

Muurinen, A. & Lehikoinen, J. (1999). Pore water chemistry in compacted bentonite. Engineering Geology, 54, 207–214.

Parker, F.W. (1921). Methods of studying the concentration and composition of soil solution. Soil Science, 12, 209.

Pearson, F.J., Waber, H.N. & Scholtis, A. (1998). Modelling the chemical evolution of porewater in the Palfris Marl, Wellenber, Central Switzerland. Mat. Res. Soc. Symp. Proc., Vol. 506, 789–796.

Pearson, F.J., Arcos, D., Bath, A., Boisson, J.Y., Fernández, A.M., Gäbler, H.E., Gaucher, E., Gautschi, A., Griffault, L., Hernán, P. & Waber, H.N. (2003). Geochemistry of water in the Opalinus clay formation at the Mont Terri rock laboratory. Mont Terri Project-Technical Report 2003-03. Swiss Federal Office for Water and Geology. Geology Series, no 5.

Plummer, N.L., Parkhurst, D.L. & Wigley, T.M.L. (1979). Critical review of the kinetics of calcite dissolution and precipitation. Chemical Modelling in Aqueous Systems. Ed. E.A. Jenne. Am. Chem. Soc. Symp. 93, 537–573.

Pusch, R., Muurinen, A., Lehikoinen, J., Bors, J. & Eriksen, T. (1999). Microstructural and chemical parameters of bentonite as determinants of waste isolation efficiency. European Commission. Nuclear Science and Technology. Project Report EUR 18950 EN.

Rivas, P., García, M., Fernández, A.M., Villar, M.V., Leguey, S., Cuevas, J., Martín, M. & Cobeña, J.C. (1998). Thermo-hydro-geochemical tests for the FEBEX project. 2nd Annual FEBEX meeting. September 1998.

Sacchi, E. & Michelot, J.L. (2000). Porewater extraction from argillaceous rocks for geochemical characterization. Methods and interpretations. NEA-OECD.

Samper, J. & Ayora, C. (1993). Acoplamiento de modelos de transporte de solutos y de modelos de reacciones químicas. Estudios Geológicos, 49, 233–251.

Samper, J., Juncosa, R., Delgado, J. & Montenegro, L. (2000). CORE^{2D}: A code for non-isothermal water flow and reative transport. Users manual version 2. ENRESA Technical Report 6/2000.

Samper, J., Juncosa, R., Montenegro, L. & Vázquez, A. (2001). Update of the thermo-hydro-geochemical model of the mock-up test. FEBEX Project Technical Report 70-ULC-L-5-022. University of La Coruña, Spain.

Samper, J., Vázquez, A. & Montenegro, L. (2003a). Modelización hidrogeoquímica de extractos acuosos para la obtención de la composición química del agua intersticial. Jornadas sobre Zona No saturada, ZNS 2003, Valladolid, Spain, 251–260.

Samper, J., Yang, Ch.B. & Montenegro, L. (2003b). Users Manual of CORE^{2D} V4. A code for non-isothermal water flow and reactive solute transport. University of La Coruña Technical Report. University of La Coruña, Spain.

Samper, J., Fernández, A.M^a., Zheng, L., Montenegro, L., Rivas, P. & Vázquez, A. (2004). Testing coupled thermo-hydro-geochemical models with geochemical data from FEBEX in situ test. In this volume.

Sasaki, Y., Shibata, M., Yui, M. & Ishikawa, H. (1995). Experimental studies on the interaction of groundwater with bentonite. Mat. Res. Soc. Symp. Proc., Vol. 353, 337–344.

Snellman, M., Uotila, H. & Rantanen, J. (1987). Laboratory and modelling studies of sodium bentonite groundwater interaction. Mat. Res. Soc. Symp. Proc., Vol. 84, 781–790.

Vázquez, A., Fernández, A.M^a., Samper, J., Montenegro, L., Rivas, P. & Dai, Z. (2001). Modelización hidrogeoquímica de la arcilla FEBEX. Las Caras del Agua Subterránea, Barcelona, Spain, 183–189.

Wanner, H. (1986). Modelling interaction of deep groundwaters with bentonite and radionuclide speciation. Nagra Technical Report 86–21.

Wanner, H., Wersin, P. & Sierro, N. (1992). Thermodynamic modelling of bentonite groundwater interaction and implications for near field chemistry in a repository for spent fuel. SKB Technical Report 92–37.

Wieland, E., Wanner, H., Albinsson, Y., Wersin, P. & Karnland, O. (1994). A surface chemical model of the bentonite-water interface and its implications for modelling the near field chemistry in a repository for spent fuel. SKB Technical Report 94–26.

Wolery, T.J. (1992). EQ3/6. A software Package for Geochemical modelling of Aqueous Systems: Package Overview and Installation Guide (Version 7.0). UCRL-MA-110662-PT-I, Lawrence Livermore National Laboratory, Livermore, California.

Testing coupled thermo-hydro-geochemical models with geochemical data from FEBEX in situ test

J. Samper, L. Zheng, L. Montenegro & A. Vázquez
Civil Engineering School, University of La Coruña, Campus de Elviña s/n, 15192-La Coruña, Spain

A.Mª. Fernández & P. Rivas
CIEMAT, Avenida Complutense 22, 28040-Madrid, Spain.

ABSTRACT: FEBEX (Full-scale Engineered Barrier EXperiment) is a demonstration and research project dealing with the bentonite engineered barrier designed for sealing and containment of waste in a high level radioactive waste repository. It includes two main experiments: an "in situ" full-scale test performed at Grimsel, Switzerland and a "mock-up" test operating since February 1997 at CIEMAT facilities in Madrid, Spain. One of the objectives of FEBEX is the development and testing of conceptual and numerical models for the thermal, hydrodynamic, and geochemical (THG) processes expected to take place in engineered barrier systems. Numerical THG models of heating and hydration experiments performed on small-scale lab cells provide excellent results for temperatures, water inflow and final water content in the cells. These models have been used to simulate the evolution of the "in situ" and "mock-up" tests. Once calibrated the TH aspects of the model, predictions of the THG evolution of both tests have been performed. These predictions are compared to measured geochemical and deuterium data collected after dismantling heater 1 of the "in situ" test in the summer of 2002. Most chemical data have been obtained with aqueous extract methods which require the use of geochemical models for their interpretation in order to account for the geochemical processes suffered by bentonite samples during aqueous extraction. Re-interpretated data are compared to THG model predictions of the "in situ" test performed with 3 possible conceptual geochemical models. A preliminary comparison of predicted and measured values indicates that predictions capture the trends of most chemical species.

1 INTRODUCTION

FEBEX (Full-scale Engineered Barrier EXperiment) is a demonstration and research project dealing with the bentonite engineered barrier designed for sealing and containment of waste in a high-level radioactive waste repository (Huertas et al., 2000). FEBEX is based on the Spanish reference concept for disposal of radioactive waste in crystalline rock. Canisters are emplaced in horizontal drifts and surrounded by a highly-compacted bentonite clay barrier (ENRESA, 1994). Bentonite has the multiple purpose of providing the canister mechanical stability by absorbing stress and deformations, sealing discontinuities in the adjacent rock, retarding the arrival of groundwater at the canister and retarding the migration of radionuclides released after failure of the canister and lixiviation of spent fuel.

FEBEX includes two main experiments: an "in situ" full-scale test performed at Grimsel, Switzerland and a "mock-up" test operating since February 1997 at CIEMAT facilities in Madrid, Spain (Huertas et al., 2000) . One of the objectives of FEBEX is the development and testing of conceptual and numerical models for the thermal, hydrodynamic, and geochemical (THG) processes taking place in engineered clay barriers. The ability of these models to reproduce the observed THG patterns in a wide range of THG conditions should enhance the confidence in their predictive capabilities. Numerical THG models of heating and hydration experiments performed on small-scale lab cells provide excellent results for temperatures, water inflow and final water content in the cells. These models have been used to simulate the evolution of the "in situ" and "mock-up" tests. Once calibrated the TH aspects of the model, predictions of the THG evolution of "in situ" test have been performed. These predictions are compared to measured geochemical and deuterium data collected after dismantling in the summer of 2002 heater 1 of the in situ test.

We first describe how data were collected after dismantling. Then, the technique of aqueous extract to obtain chemical data of FEBEX bentonite pore water is described. Since aqueous extract data cannot be used directly these tests must be interpreted using hydrogeochemical modelling as described by Samper et al. (2004a,b,c in this volume) using a reactive transport code, CORE^{2D} (Samper et al., 2000). Re-interpretated data are compared to THG model predictions of the "in situ" test performed with 3 possible conceptual geochemical models. Finally, we present models for the transport of deuterium, an artificial tracer added in section 29 of the "in situ" experiment. Model results are compared to measured data. Preliminary conclusions on model validity are presented.

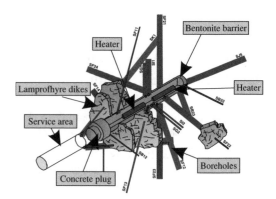

Figure 1.   Scheme of the "in situ" test.

2   DISMANTLING DATA

The FEBEX "in situ" test is being performed in a gallery excavated in the northern zone of the Grimsel underground laboratory operated by NAGRA in Switzerland (Figure 1). The physical components of the tests consists of five basic units: the drift, the heating system, the clay barrier (bentonite), the instrumentation and the monitoring and control system. The drift is 70.4 m long and 2.28 m in diameter. The test zone is located at the last 17.4 m of the drift where heaters, bentonite and instrumentation were installed. The test zone was sealed with a concrete plug. The main elements of the heating system are two heaters, separated horizontally by 1 m, which simulate full-sized canisters. Heaters were placed inside a cylindrical steel liner. Each heater is made of carbon steel, measures 4.54 m in length, 0.9 m in diameter and has a wall thickness of 0.1 m. The objetive of the heaters is to maintain a maximum temperature of 100°C at the steel liner/bentonite interface. The clay barrier includes 5331 blocks of highly compacted bentonite. The weighted average values of dry density and water content of bentonite blocks are 1.7 g/cm^3 and 14.4%, respectively (Huertas et al., 2000).

The "in situ" test began in February 27th, 1997. Heater 1 was switched-off in February 2002. Dismantling of section 1 of the "in situ" test was performed from May to September 2002. A comprehensive post-mortem bentonite sampling and analysis program was designed. Such program envisages the characterization of solid and liquid phases in order to confirm predictions and validate THM and THG models (Villar et al., 2004). Clay samples were taken from different parts of the clay buffer in order to: (a) Analyze the physical and chemical changes induced by the combined effect of heat and hydration; (b) Study the migration patterns of different chemical and isotopic tracers that were emplaced in the clay buffer and at the clay/rock contact during the

installation phase, and; (c) Calibrate sensors installed in the barrier.

Bentonite sampling was designed in order to obtain a good characterization of hydration and physical and chemical patterns and alterations. Samples were taken in vertical sections normal to the axis of the tunnel. In each section samples were taken along different radii. One of the sampling sections is section 29 (S29) where 5 glass ampoules (GA) containing deuterium (G29-1 to G29-5) were placed in 5 bentonite blocks (BB29-5 to BB29-9).

Bentonite blocks of section 29 collected after dismantling were labeled as BB29-5 to BB29-13 (Figure 2). These blocks include: 5 bentonite blocks containing the glass ampoules (BB29-5 to BB29-9) and 4 blocks above them completing the two central radii (BB29-10 to BB29-13). A total of 9 bentonite blocks were sampled (Figure 2).

2.1   Deuterium data

Sampling of selected bentonite blocks (BB29-5 to BB29-13) was carried out in CIEMAT. It should be pointed out that glass ampoules were found only in blocks BB29-7, BB29-8 and BB29-9. Most of the selected bentonite blocks were very damaged, probably during the extraction in Grimsel (Figure 3). Glass ampoules, if found, were broken. Only for BB29-7 all the pieces of the glass ampoule were found (Figure 3).

A total of 124 samples for deuterium analyses were collected from bentonite blocks BB29-7, BB29-8, BB29-9, BB29-10, BB29-11, BB29-12 and BB29-13. Most of the samples correspond to BB29-7, BB29-12 and BB29-11. Samples selected for deuterium isotopic determination include; **BB29-7**: 2.5, 2.6, 2.7, 2.8, 1.9, 1.10, 1.11, 1.12; **BB29-12**: 2.1, 2.2, 2.3, 2.4; **BB29-11**: 2.1, 2.2, 2.3, 2.4 and 2 samples near the glass ampoule of **BB29-7**.

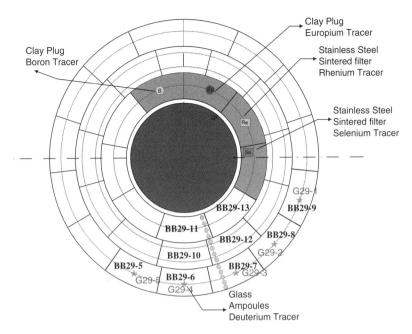

Figure 2.   Location of deuterium ampoules and deuterium bentonite block samples.

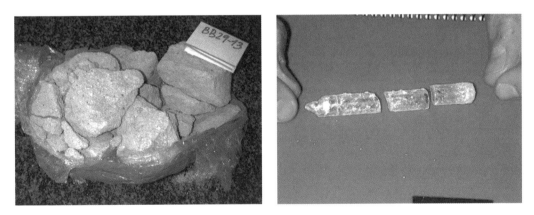

Figure 3.   Bentonite block BB29-13 (left) and deuterium glass ampoule found in bentonite block BB29-7 (right).

Bentonite pore water for deuterium isotopic determination was obtained by adding 40 mL of water to 10 g of bentonite (Fernández et al., 1999). The supernatant solutions were sent for isotopic and chemical analysis. Two additional samples were sent for deuterium analysis: a sample of another section, **BB19-14** without deuterium: 5.3 and a sample of the **bidistilled water** used for aqueous extracts. Deuterium isotopic analyses (δD) were performed by Isotope Ratio Mass Spectrometry (IRMS). Table 1 and Figure 4 show the isotopic data of aqueous extract samples

(1:4 ratio) at different radial distances from the heater. All samples have values between −51.1 and −53.2. Figure 4 shows also error bands for each sample.

2.2 Geochemical data

Supernatant solutions of aqueous extract tests were sent for chemical analysis. Table 2 shows the water content and the chemical composition of bentonite aqueous extract data at various radial distances performed by CIEMAT.

Table 1. Deuterium data of aqueous extract samples (1:4 ratio). Precision of data are 0.8‰.

Identification	Radial distance (cm)	δD_{SMOW}(‰)	Identification	Radial distance (cm)	δD_{SMOW}(‰)
BB29-7/2-5	106	−51.6	BB29-12/2-3	77.5	−52.6
BB29-7/2-6	106	−51.6	BB29-12/2-4	72.5	−52.1
BB29-7/2-7	101.5	−52.0	BB29-11/2-1	66	−53.2
BB29-7/2-8	101.5	−51.6	BB29-11/2-2	61	−52.3
BB29-7/1-9	97.5	−52.2	BB29-11/2-3	56	−51.1
BB29-7/1-10	97.5	−51.5	BB29-11/2-4	51	−52.6
BB29-7/1-11	93.5	−52.6	Pieces near the glass	103.5	−51.5
BB29-7/1-12	93.5	−52.4	Pieces near the glass	103.5	−52.1
BB29-12/2-1	87.5	−52.2	BB19-14/5.3	77.25	−52.3
BB29-12/2-2	82.5	−51.6	Bidistilled water	–	−52.4

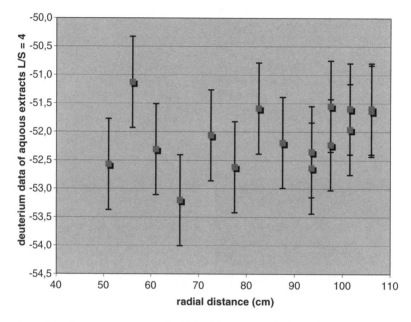

Figure 4. Deuterium data of aqueous extracts samples (1:4 ratio) versus radial distance from heater axis in bentonite blocks BB29-7, BB29-11 and BB29-12.

3 CONCEPTUAL GEOCHEMICAL MODEL

Our methodology requires the definition of a conceptual geochemical model (CGM) in terms of: (1) relevant geochemical processes, (2) chemical composition of the pore water at a at given water content (usually at dry conditions of 14%).

Different conceptual geochemical models can be proposed depending on the type of data used to construct them: squeezing and/or aqueous extract data. Samper et al. (2001) relied on squeezing data for THG modelling of FEBEX experiments. Fernández et al. (1999, 2001) used aqueous extract data to derive another geochemical model of FEBEX bentonite. They tested the model with squeezing data. The model

showed significant discrepancies when compared to sulphate and sodium concentrations of squeezed waters. The conceptual model of Samper et al. (2001) will be denoted as CGM-0 while that of Fernández et al. (1999, 2001) will be referred to as CGM-Ciemat.

Samper et al. (2004a, in this volume) present a conceptual geochemical model developed by using only aqueous extract data (model CGM-1). A modified version of this model intended to fit not only aqueous extract but also squeezing data gives rise to a model CGM-2. More details on conceptual geochemical models can be found in Samper et al., (2004a,b,c in this volume). Table 3 shows the main characteristics of models CGM-0, CGM-1 and

568

Table 2. Chemical composition of bentonite aqueous extract samples (1:4 ratio). Concentrations in mg/l.

Identification	R. dist.(cm)	w (%)	pH	Cl^-	SO_4^{2-}	HCO_3^-	SiO_2	Na^+	K^+	Ca^{2+}	Mg^2
Pieces near GA	103.5	21.2	7.2	2.7	2.3	36.9	–	16	3.6	<1	1.4
BB29-7/2-5	106	26.8	8.4	22	77	207	24.0	155	31	42	144
BB29-7/2-6	106	27.4	8.9	21	70	216	56.4	168	44	65	35
BB29-7/2-7	101.5	26.6	8.3	26	132	176	31.4	143	8.1	10	49
BB29-7/2-8	101.5	25.5	8.4	27	118	184	37.6	148	16	21	70
BB29-7/1-9	97.5	26.5	8.1	24	259	136	23.2	185	2.9	<1	1.1
BB29-7/1-10	97.5	26.5	8.2	29	156	162	28.2	148	3.2	1.1	2
BB29-7/1-11	93.5	25.8	8.0	38	260	141	26.1	185	2.9	<1	1.1
BB29-7/1-12	93.5	26	8.1	33	264	136	25.7	185	3.0	<1	1.1
BB29-12/2-1	87.5	21	8.1	59	270	143	26.3	215	3.1	1.1	1.6
BB29-12/2-2	82.5	22	8.1	76	293	139	27.6	219	3.6	1.9	2.1
BB29-12/2-3	77.5	20.3	8.0	164	255	121	27.1	250	4.1	2.8	2.6
BB29-12/2-4	72.5	19.9	8.0	280	207	108	26.8	280	5.4	6.2	4.3
BB29-11/2-1	66	16.3	8.1	348	137	119	31.5	300	6.4	7.9	5.1
BB29-11/2-2	61	15.2	7.9	370	171	111	32.4	310	6.9	9.1	5.6
BB29-11/2-3	56	14.2	8.1	348	192	123	34.6	310	7.0	8.1	5.4
BB29-11/2-4	51	14.2	8.2	315	209	153	35.7	280	6.7	5.1	3.4

Table 3. Chemical species considered in models CGM-0, CGM-1 and CGM-2. Numbers of species are given within parentheses.

	CGM-0	CGM-1	CGM-2
Primary species (10)	H_2O, H^+, Ca^{2+}, Mg^{2+}, Na^+, K^+, Cl^-, SO_4^{2-}, HCO_3^-, $SiO_2(aq)$,		
Aqueous complexes (17)	OH^-, $CaSO_4(aq)$, $CaCl^+$, $MgCl^+$, $NaCl_{(aq)}$, $MgHCO_3^-$, $NaHCO_{3(aq)}$, $CaHCO_3^+$, $MgCO_{3(aq)}$, $CaCO_{3(aq)}$, $CO_{2(aq)}$, CO_3^{-2}, KSO_4^-, $MgSO_{4(aq)}$, $NaSO_4^-$, $H_2SiO_4^{2-}$, $HSiO_3^-$		
Minerals (3–4)	Calcite, Chalcedony (equilibrium) Gypsum/Anhydrite (equilibrium, only precipitation)	Calcite, Dolomite (kinetic) Chalcedony (equilibrium) Gypsum/Anhydrite (equilibrium)	Calcite, Dolomite (kinetic) Chalcedony (equilibrium) Gypsum/Anhydrite (equilibrium, only precipit.)
Exchangable cations (5)	H^+, Ca^{2+}, Mg^{2+}, Na^+, K^+ (selectivity constants)		
Gas (1)	No condition on $CO_{2(g)}$	Fixed pressure of $CO_{2(g)}$	

CGM-2. Models presented here consider the following geochemical processes: aqueous complexation, acid-base, mineral disolution/precipitation with chemical equilibrium or kinetic, cation exchange, and gas dissolution/ex-solution (except for CGM-0). Proton sorption/desorption on clay edges has been found to be relevant in controlling the evolution of pH and bicarbonate by Samper et al (2004c) in this volume.

4 INFERRED DEUTERIUM AND CHEMICAL COMPOSITION OF FEBEX BENTONITE PORE WATER

4.1 Inferred deuterium data

Deuterium concentration (delta values) of aqueous extract samples at 1:4 ratio were transformed into delta values of bentonite pore water by means of:

$$c_b = c_{ae} \cdot \frac{5 \cdot w + 4}{w} - c_{bd} \cdot \frac{4 + 4 \cdot w}{w} \quad (1)$$

where c_b is bentonite pore water concentration, c_{ae} is aqueous extract concentration, c_{bd} is bidistilled water concentration and w is the water content of the clay sample. Values of c_b are listed in Table 4 and plotted in Figure 5 with their corresponding error bands which are computed from:

$$\varepsilon_b = \varepsilon_{ae} \sqrt{\frac{1}{f^2} + \frac{1}{(1-f)^2}} \quad (2)$$

where ε_b is the analytical error for c_b and f is the mixing fraction of bidistilled water and bentonite

Table 4. Deuterium data of bentonite pore water (ratio 1:4) and analytical error for c_b, ε_b.

Samples	Identification	Radial dist. (cm)	C_b δD_{SMOW}(‰)	ε_b
1	BB29-7/2-5	106	−37.15	15.96
2	BB29-7/2-6	106	−36.66	15.70
3	BB29-7/2-7	101.5	−43.56	16.05
4	BB29-7/2-8	101.5	−35.78	16.57
5	BB29-7/1-9	97.5	−48.96	16.10
6	BB29-7/1-10	97.5	−35.31	16.10
7	BB29-7/1-11	93.5	−57.29	16.42
8	BB29-7/1-12	93.5	−51.43	16.33
9	BB29-12/2-1	87.5	−46.26	19.26
10	BB29-12/2-2	82.5	−26.94	18.56
11	BB29-12/2-3	77.5	−59.69	19.78
12	BB29-12/2-4	72.5	−41.18	20.10
13	BB29-11/2-1	66	−71.73	23.65
14	BB29-11/2-2	61	−50.26	25.07
15	BB29-11/2-3	56	−20.99	26.55
16	BB29-11/2-4	51	−56.72	26.55

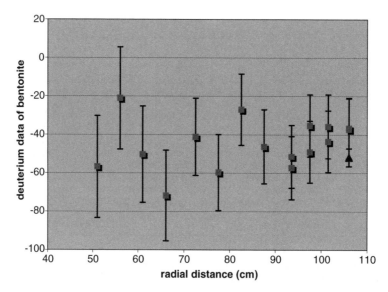

Figure 5. Delta values of bentonite pore waters versus radial distance ($r = 0$ corresponds to heater axis). Error bands have been computed by multiplying the lab precision of 0.8‰ by the multiplicative error factor considering Equation 2. The error band is much smaller for 1:1 ratio (triangle at a distance of 106 cm).

pore water which is given by:

$$f = \frac{w}{5 \cdot w + 4} \quad (3)$$

The term within the square root in Equation 2 is a multiplicative factor which propagates the errors in the aqueous extract to errors in the inferred deuterium composition. One can see in Figure 5 that the error bands for a 1:4 ratio are too large. Error bands could be smaller if samples had been prepared with a 1:1 ratio as is illustrated in Figure 5.

4.2 *Inferred chemical composition*

The chemical composition of bentonite pore water at different water contents was estimated with the inverse methodology described by Samper et al. (2004a, in this volume) which uses the reactive transport code CORE[2D] (Samper et al., 2000). Tables 5 and 6 show the chemical composition of two bentonite samples BB29-7/1-12 and BB29-11/2-3 which have water contents of 26% (almost saturated) and 14.2% (similar to ambient conditions), respectively, together with calculated aqueous extract data.

Table 5. Inferred (26% water content) and calculated aqueous extract resulting data (1:4 ratio) chemical composition of bentonite sample BB29-7/1-12. Concentrations in mol/l.

	CGM-0		CGM-1		CGM-2		
	Inferred	Calculated	Inferred	Calculated	Inferred	Calculated	Measured
Ca^{2+}	$8.3 \cdot 10^{-3}$	$2.5 \cdot 10^{-5}$	$7.8 \cdot 10^{-3}$	$2.7 \cdot 10^{-5}$	$6.3 \cdot 10^{-3}$	$2.4 \cdot 10^{-5}$	$2.5 \cdot 10^{-5}$
Cl^{-}	$1.9 \cdot 10^{-2}$	$9.2 \cdot 10^{-4}$	$2.0 \cdot 10^{-2}$	$9.7 \cdot 10^{-4}$	$1.9 \cdot 10^{-2}$	$9.2 \cdot 10^{-4}$	$9.3 \cdot 10^{-4}$
HCO_3^{-}	$4.6 \cdot 10^{-2}$	$2.2 \cdot 10^{-3}$	$1.1 \cdot 10^{-4}$	$7.2 \cdot 10^{-4}$	$1.7 \cdot 10^{-4}$	$1.2 \cdot 10^{-3}$	$2.2 \cdot 10^{-3}$
K^{+}	$1.1 \cdot 10^{-3}$	$7.6 \cdot 10^{-5}$	$1.1 \cdot 10^{-3}$	$8.0 \cdot 10^{-5}$	$1.0 \cdot 10^{-3}$	$7.6 \cdot 10^{-5}$	$7.6 \cdot 10^{-5}$
Mg^{2+}	$1.5 \cdot 10^{-2}$	$4.4 \cdot 10^{-5}$	$1.5 \cdot 10^{-2}$	$4.9 \cdot 10^{-5}$	$1.3 \cdot 10^{-2}$	$4.6 \cdot 10^{-5}$	$4.5 \cdot 10^{-5}$
Na^{+}	$1.2 \cdot 10^{-1}$	$8.0 \cdot 10^{-3}$	$1.2 \cdot 10^{-1}$	$8.4 \cdot 10^{-3}$	$1.0 \cdot 10^{-1}$	$8.1 \cdot 10^{-3}$	$8.0 \cdot 10^{-3}$
SO_4^{2-}	$5.7 \cdot 10^{-2}$	$2.7 \cdot 10^{-3}$	$6.0 \cdot 10^{-2}$	$2.9 \cdot 10^{-3}$	$5.7 \cdot 10^{-2}$	$2.7 \cdot 10^{-3}$	$2.7 \cdot 10^{-3}$

Table 6. Inferred (14.2% water content) and calculated aqueous extract resulting data (1:4 ratio) chemical composition of bentonite sample BB29-11/2-3. Concentrations in mol/l.

	CGM-0		CGM-1		CGM-2		
	Inferred	Calculated	Inferred	Calculated	Inferred	Calculated	Measured
Ca^{2+}	$5.1 \cdot 10^{-2}$	$1.9 \cdot 10^{-4}$	$4.3 \cdot 10^{-2}$	$2.0 \cdot 10^{-4}$	$5.0 \cdot 10^{-2}$	$2.1 \cdot 10^{-4}$	$2.0 \cdot 10^{-4}$
Cl^{-}	$3.2 \cdot 10^{-1}$	$9.8 \cdot 10^{-3}$	$3.2 \cdot 10^{-1}$	$9.8 \cdot 10^{-3}$	$3.2 \cdot 10^{-1}$	$9.8 \cdot 10^{-3}$	$9.8 \cdot 10^{-3}$
HCO_3^{-}	$6.7 \cdot 10^{-2}$	$2.1 \cdot 10^{-4}$	$1.4 \cdot 10^{-4}$	$8.2 \cdot 10^{-4}$	$1.4 \cdot 10^{-4}$	$9.1 \cdot 10^{-4}$	$2.0 \cdot 10^{-4}$
K^{+}	$2.5 \cdot 10^{-3}$	$1.8 \cdot 10^{-4}$	$2.4 \cdot 10^{-3}$	$1.8 \cdot 10^{-4}$	$2.5 \cdot 10^{-3}$	$1.8 \cdot 10^{-4}$	$1.8 \cdot 10^{-4}$
Mg^{2+}	$5.1 \cdot 10^{-2}$	$2.1 \cdot 10^{-4}$	$4.0 \cdot 10^{-2}$	$2.2 \cdot 10^{-4}$	$4.6 \cdot 10^{-2}$	$2.1 \cdot 10^{-4}$	$2.2 \cdot 10^{-4}$
Na^{+}	$1.7 \cdot 10^{-1}$	$1.3 \cdot 10^{-2}$	$1.6 \cdot 10^{-1}$	$1.3 \cdot 10^{-2}$	$1.7 \cdot 10^{-1}$	$1.3 \cdot 10^{-2}$	$1.3 \cdot 10^{-2}$
SO_4^{2-}	$3.5 \cdot 10^{-2}$	$1.9 \cdot 10^{-3}$	$2.0 \cdot 10^{-2}$	$2.0 \cdot 10^{-3}$	$3.3 \cdot 10^{-2}$	$1.9 \cdot 10^{-3}$	$2.0 \cdot 10^{-3}$

All models are able to reproduce the measured data (compare calculated and measured columns in Tables 5 and 6), except for model CGM-1 which cannot reproduce accurately the bicarbonate calculated concentration. This model shows some limitations when used for the interpretation of "in situ" test data. It should be noticed that all models lead to similar inferred results for all chemical species, except for bicarbonate and sulphate. Models CGM-0 and CGM-2 both of which were derived from squeezing data (Samper et al., 2004a, in this volume) provide very similar results, except for bicarbonate. For this specie inferred concentrations obtained with model CGM-0 are almost two orders of magnitude greater than those of models CGM-1 and CGM-2.

5 TESTING PREDICTIONS WITH INFERRED DEUTERIUM AND GEOCHEMICAL DATA

THG predictions are compared to inferred deuterium and geochemical data from collected data during the dismantling of section 1 of the "in situ" test. Such comparison provides a unique opportunity to test and validate current THG models of the engineered barrier system.

In this paper only a preliminary comparison is presented which is based on data collected mostly at section 29 (Fernández and Rivas, 2003) located in the middle of heater 1. Since squeezing data on pore water chemistry are not available at this section, we resorted to data from section 19 where data from both techniques, aqueous extract and squeezing, are available. First we present the comparison of predictions for deuterium and then that of chemical data.

THG predictions of the "in situ" test have been performed with FADES-CORE (Juncosa, 2001), a code for non-isothermal water flow and multicomponent reactive solute transport which is solved with an efficient finite element method (Juncosa et al., 2002).

5.1 Predictions of deuterium

Glass ampoules containing deuterium were located at the center of five blocks. They were found broken in bentonite blocks BB29-7, BB29-8 and BB29-9. It is not possible to distinguish whether the ampoules broke during dismantling or during the hydration phase. Due to this uncertainty in sampling, predictions of deuterium transport were made under two hypotheses. In the first one it is assumed that the ampoules broke during the early stages of the "in situ" test while the second hypothesis assumed they broke during dismantling. Our predictions assume that deuterium is conservative and suffers no isotopic fractionation.

5.1.1 Predictions assuming ampoules broke during dismantling

Prior estimates of δD (deuterium ratio referred to SMOW standard) for the bentonite pore water is $-31‰$. This delta value corresponds to a D/H ratio of $1.52133 \cdot 10^{-4}$. Similarly, the granitic pore waters have a δD of $-90‰$ which amounts to a correspond to a D/H ratio of $1.4287 \cdot 10^{-4}$. Deuterium predictions were obtained using the TH model calibrated with data from the "mock-up" test and using a longitudinal dispersivity of 0.03 m for the bentonite and the granite and a molecular diffusion coefficient (in water) D_0, of $1.5910^{-10} \, m^2/s$. Figure 6 shows modeling results of δD at different times as well as the comparison of computed δD values to measured δD after 1930 days of heating.

5.1.2 Predictions assuming ampoules broke at an early time

Predictions were also made by assuming that 10 ml of deuterium having a concentration of $2.355 \cdot 10^{-4} \, mol/l$ are added initially at the center of the external block. Adding this mass is equivalent to specifying an initial concentration of $3.66 \cdot 10^{-4} \, mol/l$. Figure 7 shows the predictions at different times. It can be seen that a deuterium pulse develops in addition to the "isotopic dilution" caused by the flow of granitic water into the bentonite. Predictions of deuterium derived with both hypotheses, assuming ampoules broke during dismantling and at an early time, are compared to measurements in Figure 8.

5.2 THG predictions of geochemical data

Thermal and hydrodynamic parameters of the THG model were calibrated from TH data from the "mock-up" test and "in situ" tests. These parameters are summarized in Table 7. A longitudinal dispersivity of 0.03 m was used for the bentonite and the granite while the molecular diffusion coefficient (in water) D_0, was taken equal to $2 \cdot 10^{-10} \, m^2/s$ for all chemical species.

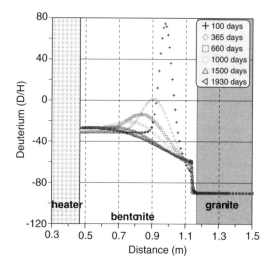

Figure 7. Profiles of computed δD at different times for the case in which the glass ampoule breaks initially.

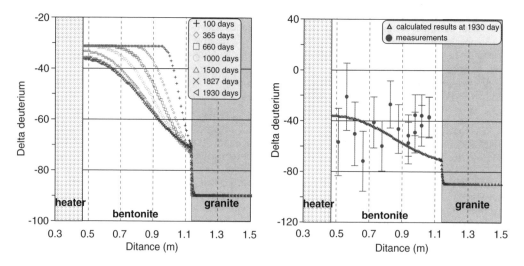

Figure 6. Profiles of computed δD at different times (left) and comparison of computed to measured data after 1930 days (June 2002) of heating (right).

THG predictions were performed with the 3 conceptual geochemical models CGM-0, CGM-1 and CGM-2 described in Section 3. Figure 9 shows a comparison of modeling results of water content with measured data. Measured data show an increase in water content above the value corresponding to 100% saturation (23%) due to bentonite swelling. Our THG model does not account for swelling and therefore cannot reproduce such large water content

values near the granite-bentonite interphase. The model tends to overpredict the water content near the heater. This is due to the fact that our model is 1-D axisymmetric and has a fixed temperature at the heater which is equal to the average value of measured temperatures at the heater surface which range from 75 to 100°C. Should one use a higher temperature of 100°C at the heater-bentonite interphase, the model should provide a better fit to measured water contents near the heater.

Figures 10 to 15 show THG modeling predictions after 1930 days of heating (which corresponds to the dismantling date). These predictions are compared to data inferred from aqueous extracts at section 29. Also shown are the values obtained at section 19 at which both squeezing and aqueous extract data are available. Table 8 shows a comparison between measured chemical composition using squeezing and inferred aqueous extract data. By comparing samples BB19-1 (squeezing) and BB19-13/2.1 (inferred aqueous extracts), which have similar water contents, one can see that the squeezing method leads to much smaller concentrations for sulphate, magnesium, sodium, and potassium.

The comparison of THG predictions and measured data indicates that for the most part the THG model reproduces the trends of measured data. Chloride concentrations are larger near the heater and decrease with radial distance (Figure 10). Although the model tends to underpredict chloride concentrations near the granite interphase at section 29, available data at section 19 fall closer to the band of computed values with models CGM-0, CGM-1 and CGM-2.

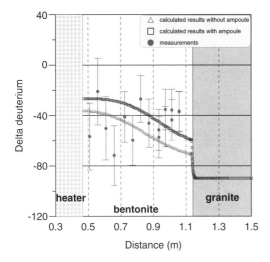

Figure 8. Predicted δD assuming ampoules broke during dismantling and at an early time and comparison to measured data after 1930 days of heating (June 2002).

Table 7. Thermal and hydrodynamic parameters for the THG model of the "in situ" test.

Intrinsic permeability of liquid	$\frac{K}{K_0} = a(\frac{I}{I_0})^{1/2}$ with $a = 1.097$, $I_0 = 0.2674\,\mathrm{mol/L}$, $K_0 = 2.75 \cdot 10^{-21}\mathrm{m}^2$
Relative permeability of liquid	$k_{rl} = S_l^3$
Retention curve	$S_l = ((1 - \Psi/1100000)^{1.1})/((1 + (5 \cdot 10^{-5}\psi)^{1/0.82})^{0.18}$
Liquid viscosity (kg/m·s)	$661.2 \cdot 10^{-3}\,(T - 229)^{-1.562}$
Liquid density (kg/m^3)	$998.2 \cdot \mathrm{e}^{(5 \cdot 10^{-7}(P_1 - 100) - 2.1 \cdot 10^{-4}(T - \mathrm{Tref}))}$
Intrinsic permeability of gas	$5 \cdot 10^{-10}\mathrm{m}^2$
Relative permeability of gas	$k_{rl} = (1 - S_1)^3$
Vapor tortuosity	0.5
Gas viscosity (kg/m·s)	$1.76 \cdot 10^{-5}$
Solid density (kg/m^3)	$2780 \cdot \mathrm{e}^{(-2 \cdot 10^{-6}(T - \mathrm{Tref}))}$
Porosity	0.41
Thermal conductivity of liquid (W/m·°C)	1.5
Specific heat of liquid (J/kg·°C)	4202
Reference temperature (°C)	12
Thermal conductivity of air (W/m·°C)	$2.6 \cdot 10^{-2}$
Air specific heat (J/kg·°C)	1000
Thermal conductivity of vapor (W/m·°C)	$4.2 \cdot 10^{-2}$
Vapor specific heat (J/kg·°C)	1620
Vaporization enthalpy (J/kg)	$2.454 \cdot 10^6$
Thermal conductivity of solid (W/m·°C)	0.8
Specific heat of solid (J/kg·°C)	835.5

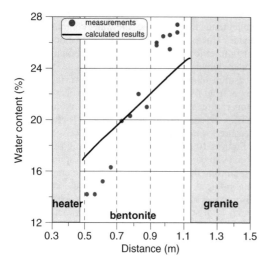

Figure 9. Comparison of modeling results of water content with measured data.

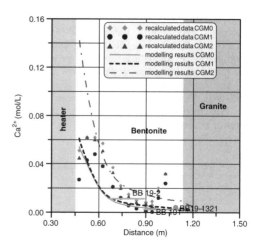

Figure 11. Comparison of predictions of Ca^{2+} with inferred aqueous extract data at section 29 (symbols). Also shown are the values obtained at section 19 with squeezing (horizontal lines) and aqueous extract (symbols).

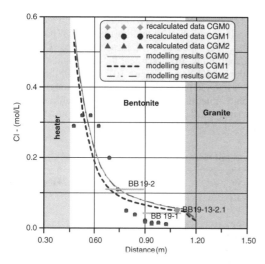

Figure 10. Comparison of predictions of Cl^- with inferred aqueous extract data at section 29 (symbols). Also shown are the values obtained at section 19 with squeezing (horizontal lines) and aqueous extract (symbols).

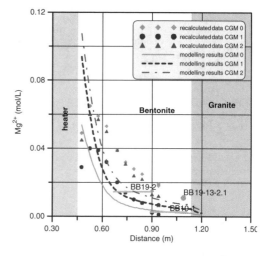

Figure 12. Comparison of predictions of Mg^{2+} with inferred aqueous extract data at section 29 (symbols). Also shown are the values obtained at section 19 with squeezing (horizontal lines) and aqueous extract (symbols).

Calcium (Figure 11) shows similar trends to those of chloride. Aqueous extract data from section 29 show an increase of calcium concentrations near the granite-bentonite interphase. Such an increase, however, is not observed at section 19. Models CGM-0 and CGM-1 seem to fit calcium measured data better than model CGM-2.

Magnesium (Figure 12), sodium (Figure 13) and potassium (Figure 14) also shows a general pattern of

higher concentrations near the heater and lower concentration in the interphase bentonite-granite. Computed results with models CGM-0, CGM-1 and CGM-2 reproduce the general trends of measured data. Available data at section 19 fall closer to the band of computed values with the three models. Only squeezing values for potassium at section 19 are much smaller than expected for reasons yet to be ascertained.

Sulphate aqueous extract data (Figure 15) show different trends. Model predictions are more coherent

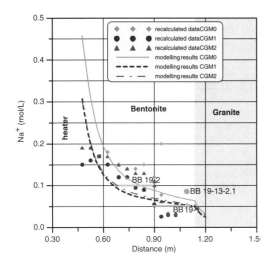

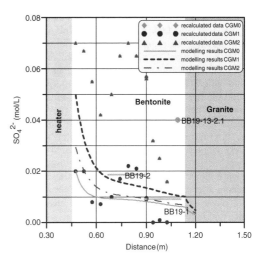

Figure 13. Comparison of predictions of Na$^+$ with inferred aqueous extract data at section 29 (symbols). Also shown are the values obtained at section 19 with squeezing (horizontal lines) and aqueous extract (symbols).

Figure 15. Comparison of predictions of SO$_4^{2-}$ with inferred aqueous extract data at section 29 (symbols). Also shown are the values obtained at section 19 with squeezing (horizontal lines) and aqueous extract (symbols).

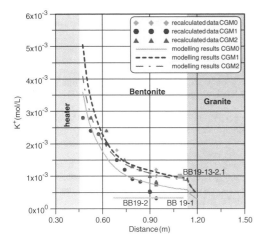

Figure 14. Comparison of predictions of K$^+$ with inferred aqueous extract data at section 29 (symbols). Also shown are the values obtained at section 19 with squeezing (horizontal lines) and aqueous extract (symbols).

Table 8. Comparison of chemical composition at section 19 obtained by squeezing and inferred aqueous extract.

	BB19-1 (Squeezing)	BB19-2 (Squeezing)	BB19-13/2.1 (Aqueous extract)
W (%)	28.7	23.5	26.9
pH	7.5	7.7	7.65
Na$^+$ (mol/l)	$5.21 \cdot 10^{-2}$	$1.00 \cdot 10^{-1}$	$8.67 \cdot 10^{-2}$
K$^+$ (mol/l)	$3.32 \cdot 10^{-4}$	$3.32 \cdot 10^{-4}$	$9.40 \cdot 10^{-4}$
Mg^{2+} (mol/l)	$2.75 \cdot 10^{-3}$	$1.45 \cdot 10^{-2}$	$1.14 \cdot 10^{-2}$
Ca^{2+} (mol/l)	$2.65 \cdot 10^{-3}$	$1.13 \cdot 10^{-2}$	$4.26 \cdot 10^{-3}$
Cl$^-$ (mol/l)	$4.22 \cdot 10^{-2}$	$1.10 \cdot 10^{-1}$	$5.26 \cdot 10^{-2}$
SO$_4^{2-}$ (mol/l)	$9.20 \cdot 10^{-3}$	$1.87 \cdot 10^{-2}$	$4.00 \cdot 10^{-2}$
HCO$_3^-$ (mol/l)	$2.57 \cdot 10^{-3}$	$2.64 \cdot 10^{-3}$	$3.63 \cdot 10^{-4}$

with squeezing data from section 19 than to aqueous extract data. Here we recall the uncertainties of our conceptual geochemical models for sulphate (Samper et al., 2004a, in this volume).

6 CONCLUSIONS

The analysis of available deuterium data from section 29 does not allow to draw clear and firm conclusions

due to deuterium data precision and uncertainties on glass ampoules breaking. Although precision of deuterium data of aqueous extract samples, c_{ae}, at 1:4 ratio is 0.8‰, this analytical error in measuring c_{ae} translates into large errors for bentonite pore water values. Such error decreases when a 1:1 ratio is used. Only 3 glass ampoules were found broken. Most likely they broke during the experiment because if they had broken during dismantling clay samples near glass pieces should still show large deuterium ratios. However, available deuterium data near the glass are similar to deuterium data at other locations. Deuterium added in glass ampoules most likely diffused so much through the bentonite that no signs of the initial pulse remain after five years and a half of diffusion. Model predictions of deuterium transport with and without ampoule breaking pass through available data. Re-interpreted deuterium data show a large scatter and lack a clear trend. This does not allow drawing clear conclusions on model testing.

We should mention that our model for deuterium may lack some relevant features such as isotopic fractionation. Bentonite pore water deuterium concentration has been derived from aqueous extract samples. There are uncertainties on the isotopic transfer (fractionation) between internal (interlayer) and external water (double layer water and free water). Deuterium analyses were performed in liquid phase but it should be explored the possibility to measure deuterium directly in bentonite samples. Vacuum techniques for measuring deuterium from solid phases should provide additional insight into deuterium behaviour.

THG predictions of the "in situ" test have been compared to measured geochemical data collected after dismantling of heater 1 of the "in situ" test. Most chemical data have been obtained with aqueous extract methods which require the use of geochemical models for their interpretation in order to account for the geochemical processes suffered by bentonite samples during aqueous extraction. During re-interpretation of aqueous extract data it is found that model CGM-1 cannot reproduce accurately bicarbonate concentrations.

Re-interpretated data were compared to THG model predictions of the "in situ" test performed with 3 possible conceptual geochemical models. A preliminary comparison of predicted and measured values using data from sections 19 and 29 indicates that:

a) THG predictions capture the trends of most chemical species.
b) Concentrations of most chemical species are larger near the heater and decrease with radial distance
c) Although the model tends to underpredict chloride concentrations near the granite interphase at section 29, available data at section 19 fall closer to the band of computed values with models CGM0, 1 and 2.
d) Models CGM1 and 2 fit Ca, K and Mg measured data better than model CGM-0.
e) The three models capture the general trend of magnesium, sodium and potassium measured data.
f) Aqueous extract sulphate data show a trend which is different than that of squeezing data. Model predictions are more coherent with squeezing data from section 19 than to aqueous extract data.

ACKNOWLEDGMENTS

The work presented here has been funded by ENRESA. The project as a whole was supported by the EC (Contracts FI4W-CT95-0006 and FIKW-CT-2000-0016 of the Nuclear Fission Safety Programme).

REFERENCES

ENRESA, (1994). Almacenamiento geológico profundo de residuos radiactivos de alta actividad (AGP). Conceptos preliminares de referencia. ENRESA. Publicación Técnica 07–94.
Huertas et al. (2000). Full-scale engineered barriers experiment for a deep geological repository in crystalline host rock (FEBEX Project). European Commission. EUR 19147 EN.
Fernández, A.Mª., Rivas, P. & Cuevas, J. (1999). Estudio del agua intersticial de la arcilla FEBEX. Ciemat Technical Report 70-IMA-L-0-44.
Fernández, A.Mª., Cuevas, J. & Rivas, P. (2001). Pore water chemistry of the Febex bentonite. Mat. Res. Soc. Symp. Proc., 603, 573–588.
Fernández, A.Mª. & Rivas, P. (2003).Task 141: Postmortem bentonite análisis. Geochemical behaviour. CIEMAT Internal Note 70-IMA-L-0-107 v0.
Juncosa R. (2001). Modelos de flujo multifásico no isotermo y de transporte reactivo multicomponente en medios porosos. Ph. D. Thesis (in spanish). Universidad Politécnica de Madrid. ENRESA Publicación Técnica 01/2001, 287 pp.
Juncosa, R., Samper, J., Navarro, V. & Delgado, J. (2002). Formulación numérica en elementos finitos de problemas de flujo multifásico no isotermo y transporte de solutos reactivos en medios porosos. Revista Internacional de Métodos Numéricos para Cálculo y Diseño en Ingeniería, 18, no 3, 415–432.
Samper, J., Juncosa, R., Delgado, J. & Montenegro, L. (2000). CORE2D: A code for non-isothermal water flow and reative transport. Users manual version 2. ENRESA Technical Report 6/2000.
Samper, J., Juncosa, R., Montenegro, L. & Vázquez, A. (2001). Update of the thermo-hydro-geochemical model of the mock-up test. FEBEX Project Technical Report 70-ULC-L-5-022. University of La Coruña, Spain.
Samper, J., Vázquez, A. & Montenegro, L. (2004a) Inverse hydrochemical modelling of aqueous extract experiments for the estimation of FEBEX bentonite pore water chemistry. In this volume.
Samper, J., Zheng, L., Molinero, J., Fernández, A.Mª., Montenegro, L. & Rivas, P. (2004b). Reactive solute transport mechanisms in nonisothermal unsaturated compacted clays. In this volume.
Samper, J., Fernández, A.Mª., Zheng, L., Montenegro, L., Rivas, P. & Dai, Z. (2004c). Direct and inverse modelling of multicomponent reactive transport in single and dual porosity media. In this volume.
Villar, M.V., Fernández, A.Mª., Rivas, P., Lloret, A., Daucausse, D., Montarges-Pelletier, E., Devineau, K., Villieras, F., Hynková, E., Cechova, Z., Montenegro, L., Samper, J., Zheng, L., Robinet, J.-C., Muurinen, A., Weber, H.P., Börgesson, L. & Sandén, T. (2004). Task 141: Postmortem bentonite analysis. Version 0. Technical Report 70-IMA-M-6-7 v0.

Closing conference

Advances in Understanding Engineered Clay Barriers – Alonso & Ledesma (eds)
© 2005 Taylor & Francis Group, London, ISBN 04 1536 544 9

Nuclear repositories in granite. Research and development needs in the near future

C. Svemar
SKB, Figeholm, Sweden

F. Huertas
ENRESA, Madrid, Spain

R. Pusch
Geodevelopment AB, Lund, Sweden

ABSTRACT: The two full scale tests – FEBEX and Prototype Repository – are in focus in this paper, but the result they produce is very much linked to general results and conclusions drawn from other experiments conducted earlier and in other places.

They both have addressed the feasibility of disposal technologies – horizontal as well as vertical emplacement modes – and been able to verify the technical feasibility of both.

They have both addressed the capability of numerical modelling of thermo-hydro-mechanical-chemical (THMC) processes taking place in the rock and the buffer and backfill during saturation. So far the results indicate a good enough conceptual understanding of the processes for supporting the performance assessment, which is made for the long-term safety case. The results, however, also indicate a lack of complete conceptual understanding in the evolution of the engineered barriers with respect to the M and C components, when focus is on prediction of a day-to-day behaviour of the buffer.

In order to make full use of the two in situ tests it would be most fruitful if they were run to full saturation before being decommissioned.

1 INTRODUCTION

The two main demonstration projects on repository technology for crystalline rock – FEBEX and Prototype Repository Project – have both addressed practical installation issues, quality assurance procedures, instrumentation and component performance and component interaction as well as THMC modelling. The objectives are in detail:

FEBEX I
– Verify manufacturing and emplacement feasibility
– Develop capability of thermo-hydro-mechanical (THM) and thermo-hydro-chemical (THC) modelling.

FEBEX II
– Study of links between THM and C
– Study of changes of engineered barrier system (EBS) parameters
– Gas generation and transport processes in the EBS

– Change in rock's H, especially in excavation disturbed zone (EDZ)
– Metal corrosion
– Evaluation of instrument system performance
– Address retrievability and indicate potential problems.

Prototype Repository
– Simulate a part of a KBS-3 repository
– Verify construction and emplacement standards
– Simulate interaction of repository components
– Verify modelling capability.

A substantial wealth of data has been gathered as well as a lot of experience in practical disposal technology. In many cases have the *in situ* tests resulted in good enough understanding with minor issues remaining to clarify. But, also new issues for deepened studies have been detected in addition to the ones not fully addressed but on the list for further investigations.

2 WHAT HAS BEEN DONE AND WHO IS ASKING FOR MORE INFORMATION?

2.1 *Disposal technology*

Both projects have the objective of demonstrating the feasibility in handling, construction and retrieval, and in both cases has it been validated that full scale canisters can be emplaced inside the bentonite buffer with the tolerances predicted, and in case of the KBS-3 concept, that tunnels can be satisfactorily backfilled. Plugs can be cast with concrete and later decommissioned for retrieval of canisters in the in-tunnel emplacement mode. Demonstration of canister retrieval has been done in FEBEX, and will in the Prototype Repository case be done later. A specific project, the Canister Retrieval Test, will also demonstrate the retrieval method for in-hole emplacement mode.

2.2 *Test set-ups and their representation of a future repository*

The aim of making field tests underground is to simulate the real conditions – environmental, geometrical, chemical, rock mechanical etc. – in a future repository. All major parameters can be simulated but one: the laboratories are not allowed to use real high-level waste. The heat emission can be simulated by electrical heaters, but the radioactivity cannot.

Another complication in laboratory work is that instruments needed are equipped with cables, so that connections are opened up between the measuring points and the registration units outside the test areas. Those cables are excellent pathways for water, which may introduce recording difficulties that are not present in a real repository.

2.3 *Capability of numerical modelling*

The other major objective in each of the two projects was to verify/validate coupled or uncoupled THMC codes and constitutive relations.

This objective has been focused on in the planning and execution of the work and several different expert teams have been applying codes, which in the first place were used for predictive modelling and later for adjustment, when needed, to fit the readings made once the installations were in place. FEBEX had an advantage in this context as this project was commissioned 3 years earlier than the first Prototype Repository heater. On the other hand, the Prototype Repository Project and associated tests at AEspoe gained from the comprehensive personal experience provided by the international Stripa Project.

The wealth of data has provided good opportunities for the modellers to verify/validate there codes both for porous Grimsel rock (porosity approx 1%) and tight AEspoe rock (porosity approx 0.2%).

2.4 *Who is asking for more information?*

There are several actors, who are engaged in the issue and who have a major impact on the development process from now on and up to the closing of the future repository. Some major actors are:

– Performance assessors
– Repository designers and operators
– Regulators
– General public and opinion-makers
– Landowners and public directly/indirectly affected by a repository project.

Primarily the two first mentioned groups have been engaged in the two projects. The regulators are informed but have not been confronted with the task to file official comments on the work and its results. The last two groups are heavily involved in the Swedish programme, and do take a thorough and genuinely interested part in issues of even detailed science or engineering. The mentioned actors have different focus and their asking is therefore different and may develop to even more special issues in the future.

Performance assessors are looking into the long-term behaviour of the repository system and are thereby focusing on the processes occurring once saturation has been achieved and the "steady-state" condition reached. The principle issues from PA are about discrepancies caused during construction and saturation, which may alter the "starting" point for PA. Another issue is the chemical alteration of components in the system, which may have an important development already during the saturation phase (oxidising conditions).

Repository planners are focused on a short time period and look for the practical information in well established ways. Relevant data have been collected in the projects and are easily accessible.

Regulators would be expected to focus on the same issues as the performance assessors. The interested areas concern the long-term performance rather than the short term saturation processes. However, at least one additional issue is addressed: the radiation dose to man during operation, but this was not part of the so far conducted work in the two projects.

The public in Sweden is very much asking about the consequences the initial activities, like construction and operation/transportation may cause, and the citizens and organisations in the municipalities, where site investigations are going on, specially focus on the consequences for them,. But there is also a great concern about possibilities to make adjustments and decisions in the future. This introduces two issues: monitoring of buffer and emplaced canisters and possibility to retrieve canisters in case this would be the best solution in the future.

3 WHERE DO WE STAND?

The projects have demonstrated the capacity in verifying/validating techniques in full scale to such an extent that it is shown possible to handle the canisters, the buffer and the backfill. The methods and equipment used are not the ones that are going to be used in real repositories, but the message is clear that the remaining work is engineering adjustments and optimisation. The methods and principles of technology have been tested and found to meet the expectations.

The capability to model has been focused on the saturation phase with numerical tools capable of dealing with fully coupled processes. The following general conclusions can be drawn from the application of the THMCB models:

- The temperature evolution is similar for all the models but the ultimate temperature and the time to reach it vary.
- The redistribution of pore water in the early water saturation phase of the buffer is similar for all the models.
- The rate of water saturation in later phases is different for different models. The discrepancies are probably due to different assumptions with respect to the performance of the EDZ.
- The stress/strain processes associated with the redistribution of pore water are slightly differently treated by the modellers. Some use the Cam Clay concept while the more complex way of combining a non-linear Porous Elastic Model and the Drucker/Prager Plasticity Model is preferred by others. Considering the presently poorly understood hydration mechanisms the complexity of the THM models appear to make the sophisticated calculations out of proportion.
- The impact of chemical processes has been largely neglected in the modelling work. Thus, neither the altered physical behaviour of the buffer caused by cation exchange or dissolution/precipitation of quartz and other mineral constituents has been sufficiently well considered. They may be controlling the performance of the buffers and backfills already in a few decades and should be investigated in detail.
- Gas and RN migration tests in FEBEX have been difficult to evaluate due to the complicated nature of all affecting parameters. Other projects and test set-ups with focus on each of these issues provide more accurate interpretations of measured data.

4 REMAINING RESEARCH AND TECHNICAL DEVELOPMENT (RTD) ISSUES

4.1 Disposal technology

The results clearly suggest the technical feasibility of both the horizontal and the vertical emplacement methods. The further development of design and manufacturing may be based on standard engineering procedures. As the technical development progresses with time it is necessary to carry out the work as late as possible. One issue of concern, though, is the impact of water seepage into deposition hole or tunnel. Where are the limits with respect to amount and concentration?

4.2 Capability of numerical modelling

Only the initial transient process during saturation has so far been studied in the two projects. But combining the information from the FEBEX Mock-up and the AEspoe Canister Retrieval Test and Long Term Test of Buffer Materials (LOT) some information is gained also concerning physico/chemical processes that take place in the later part of the saturation process and thereafter.

Both the Mock-up and the Canister Retrieval Test have artificial saturation, i.e. known boundary condition for the supply of water, and the THM codes applied for saturation of both systems show fair agreement with reality with respect to T and H while M is less accurate but still acceptable. Suction is deemed to be the driving mechanism for water saturation. In the FEBEX in situ test the supply of water from the rock seems to have been sufficient for allowing saturation of the bentonite with suction in the bentonite as the controlling mechanism, while in the Prototype Repository the supply has been limited and the saturation process largely retained by de-hydration of the near-field rock. A balance is assumed to take place between suction in bentonite and the near-field rock resulting in de-hydrated of the rock mass including EDZ. For the latter type of rock the understanding of the behaviour of the rock in general and the EDZ in particular needs further RTD.

The LOT project with only 100 mm thick buffer has been saturated and it is expected that possible physico/chemical changes can be seen when some of the 4 parcels are dismantled. The expectation from preceding analysis of test-parcels is that the saturated buffer shows no change in properties of any importance compared to untreated bentonite buffer.

An identified problem is the rather poor agreement between predicted and measured mechanical evolution in the buffer, indicating a lack of some fundamental understanding of the H and M interaction.

Such work is presently done in the laboratory for comparison with data from the *in situ* tests.

As two types of bentonite have been used and the differences observed are within expectations, the judgement is that relevant parameters for prediction of THM performance of other types of bentonite buffers can be based on laboratory scale testing.

The geochemistry has been studied at depth in many *in situ* tests and efforts have been made in the FEBEX project to analyse and model chemical alterations and physical performance under increased temperature. Three main chemical issues remain to be investigated:

- Impact on corrosion of canister (steel as well as copper) by the buffer
- Degradation of the bentonite *per se* and induced by the canisters
- Migration of radio nuclides in the buffer and backfill and in the near-field rock.

The results in FEBEX are useful in analysing all three issues, but the work has also indicated the difficult route of coupling the THM evolution with the evolution of C. Work on understanding and modelling different aspects on C, which addresses the issues, are ongoing. Efforts are also put into combined modelling of THMC. However, it may be too early to gain full capability to include C in a combined model before the capability to model THM is better developed.

The difficulties in modelling THM are connected to the modelling of the saturation process. In the long run the temperature pulse has gone and the system adapts to a HM system with small variations in M. In order to get as much data as possible from the FEBEX and Prototype Repository *in situ* tests it is essential to run the tests to saturation and somewhat beyond that.

4.3 *Consensus in international cooperation*

Maybe the most important issue in the future is to use the data, experience and model results as basis for wide and detailed analysis in forum were research teams can meet, compare and discuss their results, and thereby lay the ground for consensus in respecting modelling capability. This has been and is being done for many topics. Today, however, a forum is lacking for THM as well as THMC modelling of bentonite buffer.

5 CONCLUSIONS

The full scale tests have indicated accurately enough that both horizontal and vertical disposal modes are technically feasible, and consequently that engineer-ing measures are judged to be available for adaptation of both modes to different host rock conditions. A remaining issue, however, is to verify any impact from radioactive packages that must be handled in repositories.

Numerical modelling of THM and THMC is totally focused on processes taking place during the transient saturation phase, which includes many more complex interactions than after saturation. The knowledge of the main mechanisms and processes that drive the saturation of bentonite-based buffer and backfill are verified in the two *in situ* tests, and the understanding of possible impacts of the saturation processes is thereby good enough for assessment of the long-term safety of a repository.

But, the understanding of the processes is not accurate enough for predicting the detailed evolution on a day-to-day basis during saturation, which is the condition the future repository will be in for a long time, during which observation and checking of the performance may be made by the owner, the regulating authorities and the public. As the mechanical evolution in the buffer is not conceptually well understood today, there is a doubt about the present capability to model coupled THM and THMC during saturation. More detailed understanding is judged to be required, especially in relation to the issues of surveying and monitoring during and after completed disposal.

The *in situ* tests will enhance the basis for understanding the THM processes, if the tests are run to complete saturation and somewhat beyond that. For FEBEX is it judged feasible to predict the time to saturation with fair accuracy, and also by natural or artificial means of supplying water experience this accuracy. For the Prototype Repository estimation of the required time for complete buffer saturation depends on the hydraulic interaction between the buffer and the rock, which is not known with great accuracy. Some of the buffers are estimated to reach saturation before the AEspoe HRL will be closed, while some are expected to be on the edge of reaching a high degree of saturation.

FEBEX and Prototype Repository fulfil the need for large scale verification of buffer's THM properties at saturation as well as of the THM development during saturation, although more supporting laboratory investigations are required for more accurate capability of modelling. The two tests are, however, not totally well equipped for in-depth verification of chemical processes taking place or gas migration through the buffer. Large scale *in situ* tests for these complex sets of mechanisms and processes remain to be organised.

Author index